J. vom Scheidt
Stochastic Equations of Mathematical Physics

Mathematische Lehrbücher und Monographien
Herausgegeben von der Akademie der Wissenschaften der DDR
Karl-Weierstraß-Institut für Mathematik

II. Abteilung
Mathematische Monographien
Band 76

Stochastic Equations of Mathematical Physics

von J. vom Scheidt

Stochastic Equations of Mathematical Physics

by Jürgen vom Scheidt

with 147 Figures and 33 Tables

Akademie-Verlag Berlin

Verfasser:

Prof. Dr. sc. nat. Jürgen vom Scheidt

Technische Hochschule Zwickau
Sektion Mathematik/Naturwissenschaften/Informatik

ISBN 3-05-500609-7
ISSN 0076-5430

Erschienen im Akademie-Verlag Berlin, Leipziger Straße 3—4, Berlin, DDR-1086

Lizenznummer: 202 · 100/90
Printed in the German Democratic Republic
Gesamtherstellung: VEB Druckhaus „Maxim Gorki", Altenburg, DDR-7400
Lektor: Dr. Reinhard Höppner
LSV 1075
Bestellnummer: 763 918 6 (9164)

Preface

At present, the theory of differential equations with random parameters is a very active area of mathematical research where the concrete problems arise from physics and technological sciences. In recent years, the development in stochastic analysis proves the importance of the stochastic nature of many physical and technological problems.

In the present book the theory of weakly correlated random functions is applied to problems in connection with random differential equations. This class of random functions can be characterized as functions without "distant effect" and used for modelling of many interesting random functions of physics and engineering. Simulation results are compared with theoretically obtained statistical characteristics with respect to the correlation length, the order of the theoretical expansions, and the number of realizations, as well. This theory of weakly correlated random functions is applied to systems of ordinary differential equations with random input functions (in connection with vibrations of vehicles), to boundary-initial-value problems of hyperbolic differential equations (in connection with vibrations of strings, bars, membranes, and plates) and parabolic equations (in connection with brakes and couplings of vehicles), and to boundary-value problems of elliptic differential equations (e.g. in connection with stationary transverse vibrations).

The book is the result of my research work in the field of random differential equations which was inspired, in general, by physical and technological tasks. It is based on some papers and lectures which appeared in the past few years.

Finally, I would like to express my gratitute to all those who have contributed to the writing and publication of this book. In particular, I offer my sincere thanks to the Technische Hochschule Zwickau and to the Akademie-Verlag Berlin for generously supporting this project.

Zwickau, November 1987 J. vom Scheidt

Contents

Introduction

In recent years the application of mathematical methods in the natural and technological sciences is characterized among others by the fact that the mathematical models based on simplified assumptions are being replaced by models based on more realistic assumptions which, in turn, are more complicated. These considerations also result in the stochastic nature of many quantities being contained in interesting problems. The development in stochastic analysis during recent years proves the importance of the stochastic nature of many physical and technological problems. Some examples are: random vibrations (e.g. see FABIAN [1]; MITSCHKE [1]; PISZCZEK, NIZIOL [1]), random wave propagation (e.g. see SOBCZYK [3]) or random heat equations (e.g. see PARKUS [1, 2]).

Basic notations and methods of the functional analysis and the theory of equations of mathematical physics can be often transferred very simply to the stochastic case. However, characteristic difficulties concerning the essential stochastic questions exist, e.g. the determination of the distribution law of the solution with respect to a stochastic equation involves essential problems in many cases (cf. BHARUCHA-REID [3]). Attemps to overcome these difficulties have been only successful in certain very special cases. Hence we are obliged to pursue simpler goals, for instance the calculation of the first two moments of solutions, or to assume special properties of stochastic quantities involved in the problem. The present book selects the way of special random functions which determine the problem investigated, for example, as coefficients or inhomogeneous terms of a differential equation or as random boundary conditions. By means of these special random functions the distribution law of the solution is obliged to be determined. These special random functions are demanded to possess mathematical properties which are not too restricting and to have physical and technological applications. For this purpose weakly correlated random functions are defined. The essential property of such a weakly correlated random function can be illustrated as follows: In physics and technological sciences random functions are often found to have the property that the influence of the random function does not reach far, i.e. the values of the random function at two points do not correlate when the distance of these points exceeds a certain quantity $\varepsilon > 0$. This number ε is called the correlation length of the random function. In applications ε is always assumed to be sufficiently small. Hence, weakly correlated random functions can also be characterized as functions without "distant effect" or as functions of "noise-natured character". Examples for such random functions are the force which acts on a Brownian particle, the fluctuation of the force which is produced by a turbulent medium (for example, on a membrane, plate or a beam), the fluctuation of a technological surface, the fluctuation of parameters within technological materials, and many others.

The classical paper of UHLENBECK, ORNSTEIN [1] which deals with Brownian motion must be mentioned first. Uhlenbeck and Ornstein investigated the differential equation

$$m \frac{\mathrm{d}u}{\mathrm{d}t} + fu = F(t) \tag{0.1}$$

for the velocity of a Brownian particle where $F(t)$ denotes the random force which is caused by the impacts of molecules of the surrounding medium. This force $F(t)$ is assumed to be weakly correlated (in the present terminology) since Uhlenbeck and Ornstein write: "There will be correlation between the values of $F(t)$ at different times t_1 and t_2 only when $|t_1 - t_2|$ is very small." According to that, it is supposed that the correlation function $\varphi(t)$ "is a function with a very small maximum at $t = 0$". For the higher moments of $F(t)$ they assume a decomposition into groups of lower moments according to the model of the "cluster condition" in modern physics. This heuristical concept of a weakly correlated function was applied by BOYCE [2] to a non-homogeneous Sturm-Liouville problem and the exact definition is contained in PURKERT, VOM SCHEIDT [1] or also in VOM SCHEIDT, PURKERT [3].

The main result of Uhlenbeck and Ornstein consists in the fact that the solution $u(t)$ of (0.1) for each fixed t has an asymptotic normal distribution. Another paper which deals with similar problems was written by VAN LEAR, UHLENBECK [1]. With reference to Uhlenbeck and Ornstein, BOYCE [2] has shown that the solution $u(x)$ of a boundary value problem for a deterministic Sturm-Liouville operator for each fixed x converges to a normally distributed random variable as $\varepsilon \downarrow 0$ if the inhomogeneous term of the differential equation is assumed to be a weakly correlated process with correlation length ε.

Assuming the weak correlation of the random functions which are contained in the physical or technological problems this book concerns with the determination of the distribution law as to the solution of the given problem as an expansion of the distribution density with respect to the correlation length ε.

The random influences on a technological system can have essentially two origins: first, random external elements of the considered technological system (e.g. external forces effecting on the system); secondly, random internal elements (e.g. densities, surfaces, structures). From a mathematical point of view these problems demand the investigation of operator equations (e.g. differential equations, integral equations) where in the first case random inhomogeneous terms have to be dealt with and in the second case random operators. By means of the represented mathematical method it is possible to solve problems of both types.

All random variables are considered on a fixed probability space $\{\Omega, \mathfrak{A}, \mathsf{P}\}$, and characterized by the notation ω. For example, a random function is denoted by $f(x, \omega)$, $x \in \mathbb{R}^d$. The expectation is written by peaked brackets, that is, for example, $\langle X(\omega) \rangle = \langle X(.) \rangle = \langle X \rangle$ denotes the mean of the random variable $X(\omega)$. The norm of vectors and matrices is denoted by bold-faced vertical lines $|.|$, and the norm on function spaces by the usual notation.

The applications of the results as to the weakly correlated random functions are limited to differential equations with random parameters. A random function f is said to be a solution of the differential equation if a.s. sample functions of f satisfy the differential equation (see also BUNKE [2], or SOONG [1]). For this, the derivatives and integrals of a random function are defined by a.s. sample functions.

Chapter 1 plays the central role in the book and contains the theory of the weakly correlated functions. There is presented the exact definition and some basic properties of a weakly correlated function, a weakly correlated connected vector function, and a partially weakly correlated function. Furthermore, linear functionals are investigated having the form

$$r_{ij\varepsilon}(\omega) \doteq \int_{\mathcal{D}_i} F_{ij}(x)\, f_{j\varepsilon}(x,\omega)\,\mathrm{d}x, \quad i = 1, 2, \ldots, n; \quad j = 1, 2, \ldots, l; \quad \mathcal{D}_i \subset \mathbb{R}^m, \tag{0.2}$$

where $\big(f_{1\varepsilon}(x,\omega), f_{2\varepsilon}(x,\omega), \ldots, f_{l\varepsilon}(x,\omega)\big)$ denotes a weakly correlated connected random vector with correlation length ε and $F_{ij}(x)$ deterministic functions. This chapter shows the expansion of moments

$$\left\langle \prod_{p=1}^{k} r_{i_p j_p \varepsilon} \right\rangle, \qquad i_p \in \{1, 2, \ldots, n\}; \quad j_p \in \{1, 2, \ldots, l\}, \tag{0.3}$$

with respect to the correlation length ε which can be written as functions of statistical characteristics ${}^{p}A_q$ depending on the weakly correlated connected random vector and the deterministic functions F_{ij}. These expansions are used for the expansion of the distribution density of the random vector

$$\big(r_{1\varepsilon}(\omega), r_{2\varepsilon}(\omega), \ldots, r_{n\varepsilon}(\omega)\big) \tag{0.4}$$

where $r_{i\varepsilon}(\omega)$ is defined by

$$r_{i\varepsilon}(\omega) \doteq \sum_{j=1}^{l} r_{ij\varepsilon}(\omega). \tag{0.5}$$

The main result is contained in (1.76). In particular, random vectors (0.4) with a normally distributed, weakly correlated function $f_\varepsilon(x,\omega)$ are dealt with. Theorem 1.6 contains the determination of the statistical characteristic ${}^2A_1$ under weak assumptions. The lowest order of the expansions of the moments (0.3) and of the distribution density as to (0.4) is characterized by ${}^2A_1$. The convergence in distribution of

$$\frac{1}{\sqrt{\varepsilon^m}} \big(r_{1\varepsilon}(\omega), r_{2\varepsilon}(\omega), \ldots, r_{n\varepsilon}(\omega)\big)$$

to a Gaussian vector as $\varepsilon \downarrow 0$ can be verified (see Theorem 1.8). The terms of higher order as to the expansions mentioned above are determined by ${}^2A_2$, ${}^3A_2$, and ${}^4A_3$. ${}^2A_2$ is given by Theorem 1.9 for $m = 1$ and by Theorem 1.10 for $m \geqq 2$. ${}^3A_2$ and ${}^4A_3$ are investigated in Section 1.6.6. Some examples are added.

The following considerations in Chapter 1 are devoted to functions of functionals (0.5),

$$d_k(\omega) \doteq d_k(r_{1\varepsilon}, r_{2\varepsilon}, \ldots, r_{n\varepsilon}), \qquad k = 1, 2, \ldots, s.$$

The first two terms of an expansion as to ε of the moment

$$\left\langle \prod_{i=1}^{s} d_i^{k_i} \right\rangle$$

are written in (1.164). Some special considerations for $s = 1, 2$ are deduced with concrete formulas for first and second moments of the random vector

$$\big(d_1(\omega), d_2(\omega), \ldots, d_s(\omega)\big). \tag{0.6}$$

Finally, based on the expansion of the moments the expansion of the density function of (0.6) is given up to terms of second order as to ε. The cases $s = 1, 2$ are considered in a detailed manner.

Such expansions of first order are contained in UHLENBECK, ORNSTEIN [1]; BOYCE [2]; MEUSEL [1]; MEUSEL, VOM SCHEIDT [2]; PURKERT, VOM SCHEIDT [1, 3, 4]; VOM SCHEIDT [2, 3, 5]; VOM SCHEIDT, PURKERT [1, 2, 3]; VOM SCHEIDT, WÖHRL [5]; WÖHRL, VOM SCHEIDT [5].

Some inspirations as to the expansions of a higher order than first one are given by BOYCE, NING-MAO XIA [1]. The papers VOM SCHEIDT [8, 10, 11]; VOM SCHEIDT, FELLENBERG, WÖHRL [1]; VOM SCHEIDT, FELLENBERG [1, 3, 4]; VOM SCHEIDT, WÖHRL [6] deal with expansions of a higher order than first.

The paper LIESE, VOM SCHEIDT [1] is concerned with sequences of random functions $X_n(x)$, $x \in \mathbb{R}^1$, having the property that for each $\varepsilon > 0$ the random variables $X_n(x)$ and $X_n(y)$, $|x - y| > \varepsilon$, are approximately independent for large n. In order to give a more precise formulation of "approximately independent" mixing coefficients are used. The main result refers to the study of the asymptotical behaviour of the sequence of distribution laws of the processes

$$W_n(x, \omega) = \int_0^x X_n(s, \omega) \, ds .$$

Since, roughly speaking, $W_n(x, \omega)$ is the sum of approximately independent random variables for large n it can be shown, under some technical assumptions, that the distribution law $L(W_n)$ converges weakly to the distribution law of a continuous Gaussian process with independent increments. These results include the expansion of linear functionals for weakly correlated processes up to terms of the first order. Furthermore, a limit theorem for stochastic processes of the form

$$Y_n(x, \omega) = \int_0^1 F(x, y) \, X_n(y, \omega) \, dy$$

can be derived. The finite dimensional distributions of $Y_n(x, \omega)$ converge to the corresponding distributions of a Gaussian process. Under some additional assumptions this result can be improved and the weak convergence of the distribution laws can be deduced. These results are generalized to the multidimensional case $X_n(x), x \in \mathbb{R}^d$, by LIESE [1]. The investigations of this kind are not included in this book since the expansions of higher order in connection with functions of linear functionals play a central role. On the other hand, the convergence of the finite-dimensional distributions is sufficient for many physical and technological applications.

Chapter 2 is concerned with simulation results as to weakly correlated processes. Weakly correlated processes of the form

$$f_\varepsilon(x, \omega) \doteq \bar{b}_i(x) \, \zeta_i(\omega) + b_i(x) \, \zeta_{i+1}(\omega) \quad \text{for} \quad x \in [a_i, a_{i+1}], \quad i = 0, 1, \ldots, k - 1,$$

on (α, β) are used where

$$a_i = \alpha + \frac{i}{k} (\beta - \alpha), \qquad i = 0, 1, \ldots, k,$$

$$b_i(x) \doteq \frac{x - a_i}{a_{i+1} - a_i}, \qquad \bar{b}_i(x) \doteq 1 - b_i(x), \qquad i = 0, 1, \ldots, k - 1,$$

and ζ_i, $i = 0, 1, \ldots, k$, denote independent random variables for which the moments $\langle \zeta_i^p \rangle$, $p = 1, 2, \ldots$, exist. The above mentioned statistical characteristics ${}^pA_q$ for the linear functionals

$$r_{i\varepsilon}(\omega) = \int_\alpha^\beta F_i(x) \, f_\varepsilon(x, \omega) \, \mathrm{d}x \quad \text{for} \quad i = 1, 2, 3, 4 \tag{0.7}$$

are computed. Three cases are investigated with respect to the distribution of ζ_p:

(i) normal distribution,
(ii) uniform distribution,
(iii) logarithmic normal distribution.

For concrete simulation results the functions

$$F_1(x) = 1; \quad F_2(x) = x; \quad F_3(x) = \sin(2\pi x); \quad F_4(x) = x^2 + 2x - 1$$

are applied using $(\alpha, \beta) = (0, 1)$. In the case of (i) exact results as to moments of linear functionals (0.7) and their distribution densities can be compared with the approximative determination of moments and densities from the theory of weakly correlated processes. A good coincidence is observed. Furthermore, in this chapter moments of (0.7) from expansions using Chapter 1 are compared with simulation results of these moments concerning all three cases (i), (ii), and (iii). The same comparison is carried out for the distribution densities of $r_{i\varepsilon}(\omega)$, $i = 1, 2, 3, 4$, and of $\big(r_{s\varepsilon}(\omega), r_{s+1,\varepsilon}(\omega)\big)$, $s = 1, 3$, and the corresponding probabilities. In particular, the dependence of the comparing results on the number of realizations m, on the correlation length ε, and on the order of the approximation is investigated by a series of numerical results. For this comparison the χ^2-values play an important role.

In a further section an application of the results as to functions of linear functionals of weakly correlated processes to a simple eigenvalue problem is described. For example, using the method of Ritz it is possible to obtain an eigenvalue problem of the form

$$\big(A_0 + B(\omega)\big) \, U = \Lambda U \tag{0.8}$$

with the eigenvalues Λ and eigenfunctions U. This eigenvalue problem (0.8) with

$$A_0 = \begin{pmatrix} 2 & -2 \\ -2 & -1 \end{pmatrix}, \qquad B(\omega) = \begin{pmatrix} r_{s\varepsilon}(\omega) & 0 \\ 0 & r_{s+1\varepsilon}(\omega) \end{pmatrix} \qquad \text{for} \quad s = 1, 3$$

is considered where $r_{i\varepsilon}(\omega)$, $i = 1, 2, 3, 4$, are given by (0.7). The eigenvalues $\Lambda_{1/2}(\omega)$ of (0.8) are functions of $r_{i\varepsilon}(\omega)$ and it is

$$\Lambda_{1/2}(\omega) = \frac{1}{2} \left(r_{s\varepsilon}(\omega) + r_{s+1\varepsilon}(\omega) \mp \sqrt{(3 + r_{s\varepsilon} - r_{s+1\varepsilon})^2 + 16}\right).$$

A similar comparison between the approximative quantities as to $\big(\Lambda_1(\omega), \Lambda_2(\omega)\big)$ from the theory of weakly correlated processes and the estimations from simulation can be produced as in the above case of linear functionals. Essential distinctions between the comparing results of the linear functionals and of the functions of linear functionals cannot be found.

Simulation results of this kind are contained in vom Scheidt [6, 10, 11]; vom Scheidt, Fellenberg [1, 2]; vom Scheidt, Purkert [3].

Chapter 3 deals with random vibrations of continua in a first part and of discrete mechanical systems in a second part. In the first part, partial differential equations of the form

$$\frac{\partial^2 u}{\partial t^2} + 2\beta \frac{\partial u}{\partial t} + Lu = p(t, x, \omega); \qquad (t, x) \in [0, T] \times \mathcal{D}, \qquad \mathcal{D} \subset \mathbb{R}^n, \qquad (0.9)$$

are treated assuming initial conditions and boundary conditions are given. The operator L is supposed to be elliptic and $p(t, x, \omega)$ denotes a random function on $[0, T] \times \mathbb{R}^n$ given by the external loads on the system. The above problem contains the vibration of strings and bars as well as of membranes and plates. The aim of this section is to determine the probabilistic distribution of the solution $u(t, x, \omega)$ or the stresses in the considered system, respectively, to give results concerning the expected number of threshold crossings of these quantities. Similar investigations can be found among others in the papers of Bliven, Soong [1]; Bogdanoff, Goldberg [1]; Boyce [1]; Boyce, Goodwin [1]; Caughey [1, 2]; Eringen [1, 2]; Goodwin, Boyce [1]; Haines [1]; Heinrich, Hennig [1]; Herbert [1]; Lyon [1]; Piszczek, Niziol [1]; Samuels, Eringen [1, 2]; Sobczyk [1]; Thomas [1]; van Lear, Uhlenbeck [1]. The results of the first part of this chapter base on the papers Meusel, vom Scheidt [1]; Remke [1, 2]; vom Scheidt [3, 7, 12].

First weakly correlated excitations $p(t, x, \omega)$ are assumed and limit theorems of Chapter 1 are applied to the solution of (0.9). It can be seen that the normalized solution of (0.9) converges to a Gaussian function as $\varepsilon \downarrow 0$; i.e. expansions with respect to ε are only used up to terms of first order. Properties of the random limit function are discussed, in particular, properties of correlation functions and spectral densities. The results are applied to the motion of a string surrounded by a gas (see also Uhlenbeck, Ornstein [1]; van Lear, Uhlenbeck [1]). The results determined by the limit distribution of weakly correlated functions are compared with results obtained from the vibration problem (0.9) as to a realistic random function $p(t, x, \omega)$. Furthermore, partially weakly correlated excitations $p(t, x, \omega)$ as to the spatial variables and as to the time variable are considered. Limit theorems for displacements and stresses can be applied and, for example, the expected number of threshold crossings for these quantities are calculated. As performed for weakly correlated loads concerning the time variable and spatial one some comparing results are given in the case of partially weakly correlated functions as to t.

The second part of Chapter 3 deals with systems of ordinary differential equations having random inhomogeneous terms. Vibrations of multi-mass systems with stochastic excitations are considered. The vibration behaviour of mechanical mass-spring-damper systems with random excitation is investigated in many papers, e.g. Anand, Richard [1]; Bouc, Defilippi [1]; Bunke [1, 2]; Caughey [3, 4, 5, 6]; Crandall [1, 2, 3, 4, 5, 6, 7]; Crandall, Mark [1]; El Madany, Dokainish [1]; Fabian [1]; Friedrich [1]; Friedrich, Heimann, Martins, Renger [1]; Gasparini, Deb Chaudhury [1]; Gossmann [1]; Gossmann, Waller [1]; Heimann [1]; Heinrich, Hennig [1]; Hennig [1]; Helms [1]; Iwan, Lang [1]; Kreuzer, Rill [1]; Lingener [1]; Liu [1]; Macvean [1]; Mitschke [1, 2, 3, 4, 5]; Mühe [1]; Müller [1]; Müller, Popp, Schiehlen [1]; Piszczek, Niziol [1]; Renger [1]; Renger, Gupta [1]; Schiehlen [1]; Schmidt [1]; Sobczyk, Macvean, Robson [1]; Socha [1]; Soong [1]; Szopa [1]; Wallentowitz, Balasubramanian, Biesinger, Meier [1]; Wallrapp, Schwarz [1]; Wedig [1, 2].

Noise processes or processes with given correlation functions or spectral densities are assumed as excitation. The correlation method, spectral method, method of Fokker-Planck equation and Kolmogorov equation, and the Ito theory are mainly applied. From the knowledge of the statistical properties of solutions information is obtained on the suitable dimensioning of the mechanical systems or on the reaction of existing systems excited by stochastic inputs.

The results on vibrations of n-mass systems represented in the second part on vibrations assume weakly correlated input functions. This section is based essentially on the papers DRECHSEL, HASE, NEUMANN, VOM SCHEIDT, WÖHRL [1]; PURKERT, VOM SCHEIDT [4]; VOM SCHEIDT [1]; VOM SCHEIDT, WÖHRL [1, 2, 3, 4, 5, 6]; WÖHRL [2]; WÖHRL, VOM SCHEIDT [1, 2, 3, 4, 5, 6]. If expansions up to the first order as to ε are considered then it follows a normal distribution for the solution. In the case of expansions with higher orders as to ε deviations of the normal distribution can be studied. Such deviations of the normality have been investigated in recent years (see e.g. KOTULSKI, SOBCZYK [1]).

First, some simple results of the correlation method with non-weakly correlated input are represented where the linear case is assumed. In order to use the theory of weakly correlated functions the approximation of random road surfaces and their derivations is investigated. Then expansions of characteristics of solutions as to systems of the form

$$Ax'' + Bx' + Cx = g(t, \omega); \qquad x(t_0) = x_0, \qquad x'(t_0) = x_1, \tag{0.10}$$

are considered. First of all, it is assumed that $g(t, \omega)$ can be written as

$$g(t, \omega) = h(t) + Pf_\varepsilon(t, \omega) \tag{0.11}$$

where $f_\varepsilon(t, \omega)$ denotes a weakly correlated connected vector function. By means of limit theorems (expansions of first order) the stationary solution of (0.10) with (0.11) as inhomogeneous term is studied where, in particular, correlation relations and spectral densities are calculated. With respect to vibrations of cars random excitations (0.11) of the form

$$f_\varepsilon(t, \omega) = \big(f_\varepsilon(t + s_1, \omega), f_\varepsilon(t + s_2, \omega), \ldots, f_\varepsilon(t + s_n, \omega)\big)^\mathsf{T}$$

are considered. Higher approximations of moments or density functions as to ε are given. The examples of a one-mass and a two-mass vibration system are added to these considerations.

Furthermore, an indirect weakly correlated input function of the form

$$g(t, \omega) = h(t) + \int_{-\infty}^{t} \sum_{l=0}^{2} P_l Q^{(l)}(t - s)\, f_\varepsilon(s, \omega)\, \mathrm{d}s \tag{0.12}$$

is assumed. The reason for these investigations consists in the possibility of expanding characteristic quantities of the solution of (0.10) and their derivatives used often in applications. The one-mass vibration system with inhomogeneous term (0.12) is discussed.

Finally, nonlinear systems of the form

$$Ax'' + Bx' + Cx + \eta \sum_{k=2}^{m} B_k(x, x') = Pf_\varepsilon(t, \omega); \qquad x(t_0) = x_0, \qquad x'(t_0) = x_1,$$

are considered where η denotes a small parameter and B_k is assumed to be a homogeneous polynomial of the k-th order as to the components of x and x'. Correlation quantities of x and x' are expanded as to η and ε. These expansions are exact as to η up to terms of the order $O(\varepsilon^3)$. In a similar way the expectations $\langle x(t)\rangle$ and $\langle x'(t)\rangle$ are investigated as expansions with respect to η and ε. This way an approximative solution of the so-called "averaging problem" can be obtained. This averaging problem requires the calculation of the difference between the mean of the random solution and the solution of the averaged equation. As an example the case of one vibration equation with a cubic nonlinearity is presented.

Chapter 4 is concerned with problems of heat conduction influenced by random sources of heat. An adequate random boundary-initial value problem for a parabolic differential equation is considered. Many problems of natural and technological sciences lead to such a problem, e.g. the temperature propagation in brakes of vehicles. For the description of the random influences the weakly correlated functions are used. Advantages in the application of weakly correlated functions are the good interpretation of real physical phenomena and the use of results of limit theorems for the random solutions. Investigations of random temperature propagation in such a way can be found in Fellenberg, vom Scheidt [1, 2, 4, 6, 7]; vom Scheidt [9]; vom Scheidt, Fellenberg, Wöhrl [1]; vom Scheidt, Fellenberg [3]. For another approach applying white noise on the right-hand side of the stochastic parabolic equation to the description of problems mentioned above the papers of Jetschke [1] and Manthey [1, 2, 3, 4] are referred to. Problems in connection with the heat equation with random influences are also contained in Adomian [4]; Bécus [1, 2, 3]; Bécus, Cozzarelli [1, 2, 3]; Parkus [1, 2]; Samuels [1].

First, the simple random boundary-initial value problem

$$\frac{\partial u}{\partial t} = \alpha \Delta u; \qquad u(0, x, y) = u_0(x, y), \qquad \frac{\partial u}{\partial y}(t, x, 0) = P(t, \omega)$$

for a half-plane is considered. The solution can be given by a linear functional as to $P(t, \omega)$ and by this way a limit theorem of Chapter 1 can be applied to this solution. Statistical characteristics as to the solution are determined; e.g. variances and correlation functions, expected rates of crossings, temperature domains with given probabilities.

The following considerations deal with the random boundary-initial value problem

$$\frac{\partial u}{\partial t} = \alpha \Delta u,$$

$$\begin{aligned} &\text{initial condition:} && u(0, x, y) = u_0(x, y), \\ &\text{boundary conditions:} && \frac{\partial u}{\partial y}(t, x, y)\Big|_{(\partial\mathcal{G})_1} = P(t, x, \omega), \\ &&& \left[\lambda \frac{\partial u}{\partial n}(t, x, y) + \alpha_i\{u(t, x, y) - u_R(t, x, y)\}\right]_{(\partial\mathcal{G})_i} = 0 \\ &&& \text{for} \quad i = 2, 3, 4 \end{aligned} \tag{0.13}$$

as to a rectangular domain where $(\partial\mathcal{G})_i$ for $i = 1, 2, 3, 4$ denote the sides of the rectangle. A formula for the "averaged" solution $U(t) = (u - w, \psi)_{\mathcal{G}}$ is derived where w denotes the solution of the averaged boundary-initial value problem. With respect to the probabilistic evaluation of the solution using limit theorems weakly correlated temperature gradients $P(t, x, \omega)$ as to (t, x) are assumed. The cases of partially weakly correlated functions $P(t, x, \omega)$ with respect to (t) or (x) are also treated. In these considerations both short times t and stationary behaviour are investigated. Furthermore, numerical results are also given in the case of small depths y, since this case could be of interest in the considerations of brakes of vehicles. This chapter is concluded by the demonstrating case of a sliding temperature gradient

$$P(t, x, \omega) = P_0(x - vt, \omega)$$

where $P_0(u, \omega)$ is a given weakly correlated process. This sliding temperature gradient is excited again by the procedure in brakes and couplings.

Chapter 5 is concerned with random boundary value problems in which the differential operator can be both ordinary and partial. The case of a Sturm-Liouville operator with a random inhomogeneous term (weakly correlated) which was dealt with by BOYCE [2] is included in the results of this chapter. The papers of MEUSEL, VOM SCHEIDT [3]; PURKERT, VOM SCHEIDT [1, 2, 3, 4]; VOM SCHEIDT [4, 5] have generalized this result to random differential operators in connection with boundary-value problems and eigenvalue problems. These investigations contain expansions up to the first order as to the correlation length ε. The idea of obtaining expansions of higher order as to ε for statistical characteristics of the solutions of boundary value problems and eigenvalue problems is contained in BOYCE, NING-MAO XIA [1]. These considerations are applied in this chapter. Random boundary value problems and problems connected with them are also contained in ADOMIAN [1]; BARRY, BOYCE [1]; BENASSI [1]; BOYCE, DAY [1]; DAY [1].

First, this chapter on random boundary value problems,

$$L(\omega)\, u = g(x, \omega); \qquad U_k[u] = 0, \qquad k = 1, 2, \ldots, 2n, \tag{0.14}$$

deals with ordinary differential operators $L(\omega)$ of the form

$$L(\omega)\, u \doteq (-1)^n \,[f_n(x)\, u^{(n)}]^{(n)} + \sum_{i=0}^{n-1} (-1)^i \,[f_i(x, \omega)\, u^{(i)}]^{(i)}.$$

The solution $u(x, \omega)$ is expanded in a series with terms which are homogeneous as to

$$\bar{g}_\varepsilon(x, \omega) \doteq g(x, \omega) - \langle g(x)\rangle; \quad \bar{f}_{i\varepsilon}(x, \omega) \doteq f_i(x, \omega) - \langle f_i(x)\rangle, \quad i = 0, 1, \ldots, n-1,$$

where explicit formulas for the terms of the series are derived and questions in connection with the convergence of this series are discussed. By application of results of Chapter 1 the convergence in distribution of the solution follows as $\varepsilon \downarrow 0$ to a Gaussian process. The correlation function of this Gaussian limit process can be written by means of the Green's function belonging to the averaged boundary value problem or by means of the eigenvalues and eigenfunctions of this averaged problem.

The n-th Ritz approximation $u_n(x, \omega)$ of $u(x, \omega)$ can be obtained from a system of linear equations. Statistical characteristics of Ritz approximations are investigated where orders as to ε are considered which are of higher order than first one. Examples are contained in the sections.

In the case of partial differential operators $L(\omega)$ in (0.14), first, the special problem

$$L(\omega)\, u = -\Delta u = -\sum_{i=1}^{n} \frac{\partial^2 u}{\partial x_i^2} \tag{0.15}$$

is dealt with. The convergence of $u_\varepsilon(x, \omega)$ as to (0.14) with (0.15) as $\varepsilon \downarrow 0$ depends essentially on the singularity of the Green's function belonging to the averaged boundary value problem. Limit theorems as $\varepsilon \downarrow 0$ can only be obtained under the condition $n \leqq 3$. The reason is that the second moment $\langle u_\varepsilon(x)^2 \rangle$ only converges if the singularity of the Green's function of the averaged boundary value problem is less than the half of the space dimension n.

Furthermore, the problem (0.14) is considered with

$$L(\omega)\, u = Lu + f(x, \omega)\, u$$

and

$$U_k[u] = \left.\frac{\partial^k u}{\partial n^k}\right|_{\partial \mathcal{D}} = 0 \quad \text{for} \quad k = 0, 1, \ldots, m - 1 \quad \text{on} \quad \mathcal{D} \subset \mathbb{R}^n$$

where L is defined by

$$Lu \doteq \sum_{k=1}^{m} \sum_{|\alpha|=|\beta|=k} (-1)^k D^\alpha\bigl(f_{\alpha\beta}(x)\, D^\beta u\bigr).$$

Assuming

$$n \leqq 3 \quad \text{and} \quad |\bar{f}_\varepsilon(x, \omega)| \leqq \eta \quad \text{with} \quad \bar{f}_\varepsilon(x, \omega) \doteq f(x, \omega) - \langle f(x) \rangle$$

where η is a small positive number limit theorems of the usual kind are derived. The example of small transverse vibrations of a rectangular, fixedly, elastically embedded membrane finishes this chapter.

Random eigenvalue problems for differential operators are not considered in this book. Such problems, also in connection with limit theorems contained in Chapter 1, are investigated for example in PURKERT, VOM SCHEIDT [1, 3, 4]; VOM SCHEIDT [4, 5, 6]; VOM SCHEIDT, PURKERT [1, 3]. A list of further publications concerning with random eigenvalue problems is contained in VOM SCHEIDT, PURKERT [3]. Limit theorems in connection with hyperbolic differential equations are investigated in MEUSEL [2, 3]; MEUSEL, VOM SCHEIDT [4]. A good survey on stochastic wave propagation is given by the book of SOBCZYK [2]. The main result of this book on limit theorems was also applied to polynomials with random coefficients in VOM SCHEIDT, BHARUCHA-REID [1] and on functions of random variables as to a concrete case in VOM SCHEIDT, SEIFERT, BAUMGÄRTEL [1]. In the applications of stochastic characteristics of functionals with weakly correlated functions random integral equations were excluded. Results with this subject can be found in FELLENBERG, VOM SCHEIDT [3, 5].

1. Expansions of moments and distributions

1.1. ε-neighbouring point sets

We consider random functions $f(x, \omega)$, $g(x, \omega)$, ..., where $x \in \mathcal{D}$, and $\mathcal{D}$ denotes an arbitrary subset of $\mathbb{R}^m$. $|.|$ is the norm on $\mathbb{R}^m$.

Definition 1.1. Let $\varepsilon > 0$ be an arbitrary real number. A set of points from $\mathbb{R}^m$

$$\{x_i : x_i \in \mathbb{R}^m;\ i \in \mathcal{I}\}, \qquad \mathcal{I} = \{1, 2, \ldots, k\},$$

is said to be *ε-neighbouring* if for every separation $(\mathcal{I}_1, \mathcal{I}_2)$ of $\mathcal{I}$ the relation

$$\mathcal{J}(\mathcal{I}_1) \cap \mathcal{J}(\mathcal{I}_2) \neq \emptyset$$

yields where $\mathcal{J}(\mathcal{I}_p)$ is defined by

$$\mathcal{J}(\mathcal{I}_p) \doteq \bigcup_{i \in \mathcal{I}_p} \{j \in \mathcal{I} : |x_i - x_j| \leqq \varepsilon\}, \qquad p = 1, 2.$$

A single element set is always called ε-neighbouring. Furthermore, the subset $\{x_i : i \in \bar{\mathcal{I}} \subset \mathcal{I}\}$ is said to be maximally ε-neighbouring with respect to $\{x_i : i \in \mathcal{I}\}$ if this subset is ε-neighbouring but the subset $\{x_i : i \in \bar{\mathcal{I}} \subset \mathcal{I}\} \cup \{x_r\}$ is not ε-neighbouring for every point $x_r \in \{x_i, i \in \mathcal{I} \setminus \bar{\mathcal{I}}\}$.

Example. Consider the set of points in Fig. 1.1. Then it follows that the set $\{x_1, x_2, x_3, x_4\}$ is ε-neighbouring and the set $\{x_1, x_2, x_3\}$ is also ε-neighbouring but this set is not maximally ε-neighbouring as to $\{x_1, x_2, x_3, x_4\}$. The set $\{x_1, x_2, x_3, x_4, x_5, x_6\}$ is not ε-neighbouring. The proof follows from $\mathcal{I}_1 = \{1, 2, 3, 4\}$, $\mathcal{I}_2 = \{5, 6\}$ and $\mathcal{J}(\mathcal{I}_1) = \mathcal{I}_1$, $\mathcal{J}(\mathcal{I}_2) = \mathcal{I}_2$.

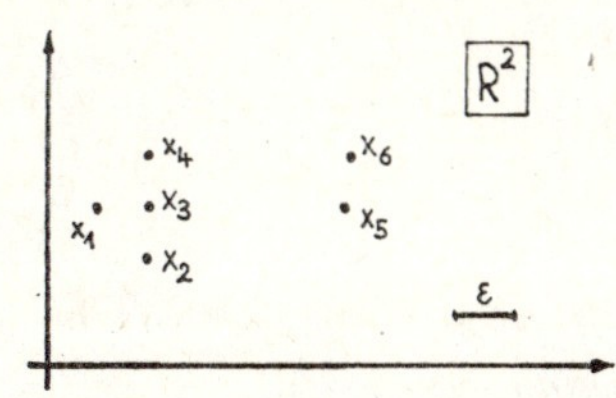

Fig. 1.1. Example of ε-neighbouring sets

Lemma 1.1. *Supposed that $\{x_i : x_i \in \mathbb{R}^m;\ i \in \mathcal{I}\}$, $\mathcal{I} = \{1, 2, \ldots, k\}$ with $k \geqq 2$ is ε-neighbouring then for every $i \in \mathcal{I}$ an element $j \in \mathcal{I}$, $j \neq i$, exists so that $|x_i - x_j| \leqq \varepsilon$.*

Proof. Let $\mathcal{J}_1 = \{i\}$ and $\mathcal{J}_2 = \mathcal{J} \setminus \mathcal{J}_1$. Because of $\mathcal{J}(\mathcal{J}_1) \cap \mathcal{J}(\mathcal{J}_2) \neq \emptyset$ it exists a $p \in \mathcal{J}(\mathcal{J}_1) \cap \mathcal{J}(\mathcal{J}_2)$. Furthermore, we have $|x_i - x_p| \leqq \varepsilon$ and $|x_j - x_p| \leqq \varepsilon$ for an element $j \in \mathcal{J}_2$. In the case of $i = p$ the second inequality leads to the proof and in the case of $i \neq p$ the first inequality. ◀

Lemma 1.2. *For $k \geqq 2$ the set $\{x_i : x_i \in \mathbb{R}^m; i \in \mathcal{J}\}$, $\mathcal{J} = \{1, 2, \ldots, k\}$, is ε-neighbouring if for two arbitrary points x_i, x_j, $i, j \in \mathcal{J}$, indices $i_1, \ldots, i_r \in \mathcal{J}$, $i_1 = i$, $i_r = j$, $r \leqq k$, exist so that*

$$|x_{i_p} - x_{i_{p+1}}| \leqq \varepsilon \quad \textit{for} \quad p = 1, 2, \ldots, r - 1.$$

Proof. Let $\{x_i : x_i \in \mathbb{R}^m; i \in \mathcal{J}\}$ be ε-neighbouring. For an arbitrary $i \in \mathcal{J}$ we define

$$\tilde{\mathcal{J}}_i \doteq \bigcup_{r=2}^{k} \{j \in \mathcal{J} : i_1, \ldots, i_r \in \mathcal{J}, i_1 = i, i_r = j \text{ exist with } |x_{i_p} - x_{i_{p+1}}| \leqq \varepsilon$$

$$\text{for } p = 1, \ldots, r - 1\} \cup \{i\}$$

and show $\tilde{\mathcal{J}}_i = \mathcal{J}$. If $\mathcal{J} \setminus \tilde{\mathcal{J}}_i \neq \emptyset$ then $\mathcal{J}(\tilde{\mathcal{J}}_i) \cap \mathcal{J}(\mathcal{J} \setminus \tilde{\mathcal{J}}_i) \neq \emptyset$ and it exists an element p from the intersection. Hence we have an element $s \in \tilde{\mathcal{J}}_i$ with $|x_p - x_s| \leqq \varepsilon$ and an element $t \in \mathcal{J} \setminus \tilde{\mathcal{J}}_i$ with $|x_p - x_t| \leqq \varepsilon$. Now we can go by these inequation from x_t to x_p then to x_s and to x_i. This is a contradiction to the assumption $\mathcal{J} \setminus \tilde{\mathcal{J}}_i \neq \emptyset$ and $\tilde{\mathcal{J}}_i = \mathcal{J}$ is proved.

We assume the relation $\tilde{\mathcal{J}}_i = \mathcal{J}$. Let $(\mathcal{J}_1, \mathcal{J}_2)$ be a separation of $\mathcal{J}$ and $i \in \mathcal{J}_1$, $j \in \mathcal{J}_2$. Hence indices $i_1, \ldots, i_r \in \mathcal{J}$, $i_1 = i$, $i_r = j$, $r \leqq k$, exist so that $|x_{i_p} - x_{i_{p+1}}| \leqq \varepsilon$ for $p = 1, 2, \ldots, r - 1$. This relation implies $\mathcal{J}(\mathcal{J}_1) \cap \mathcal{J}(\mathcal{J}_2) \neq \emptyset$. The set $\{x_i : x_i \in \mathbb{R}^m; i \in \mathcal{J}\}$ is ε-neighbouring. ◀

Lemma 1.3. *Supposed that $\{x_i : x_i \in \mathbb{R}^m; i \in \mathcal{J}\}$, $\mathcal{J} = \{1, 2, \ldots, k\}$, is ε-neighbouring then the inequation*

$$|x_i - x_j| \leqq (k - 1)\, \varepsilon$$

is fulfilled for all $i, j \in \mathcal{J}$.

Proof. Let $i, j \in \mathcal{J}$ be arbitrary. Using Lemma 1.2 indices $i_1, \ldots, i_r \in \mathcal{J}$, $i_1 = i$, $i_r = j$, $r \leqq k$, exist where $|x_{i_p} - x_{i_{p+1}}| \leqq \varepsilon$ for $p = 1, \ldots, r - 1$. Hence

$$|x_i - x_j| = |x_{i_1} - x_{i_r}| = \left| \sum_{p=1}^{r-1} (x_{i_p} - x_{i_{p+1}}) \right| \leqq \sum_{p=1}^{r-1} |x_{i_p} - x_{i_{p+1}}| \leqq (r - 1)\, \varepsilon \leqq (k - 1)\, \varepsilon. \blacktriangleleft$$

Lemma 1.4. *The set $\{x_i : x_i \in \mathbb{R}^m; i \in \mathcal{J}\}$, $\mathcal{J} = \{1, 2, \ldots, k\}$, is ε-neighbouring if the set $\bigcup_{i \in \mathcal{J}} K_{\varepsilon/2}(x_i)$ is connected where $K_\eta(x)$ is defined by*

$$K_\eta(x) \doteq \{y \in \mathbb{R}^m : |y - x| \leqq \eta\}.$$

Proof. The proof is given immediately by Lemma 1.2. ◀

Lemma 1.5. *Every finite set $\{x_i : x_i \in \mathbb{R}^m; i \in \mathcal{J}\}$ of points from $\mathbb{R}^m$ can be separated into disjoint maximally ε-neighbouring subsets in a unique way.*

Proof. We define the relation $x \sim y$ in the set $\{x_i : x_i \in \mathbb{R}^m; i \in \mathcal{J}\}$ by the fact that $x \sim y$ if a ε-neighbouring subset $\{x_i : x_i \in \mathbb{R}^m; i \in \bar{\mathcal{J}} \subset \mathcal{J}\}$ exists with $x, y \in \{x_i : x_i \in \mathbb{R}^m;$

$i \in \bar{\mathcal{J}}\}$. It easily follows that this relation is an equivalence relation and the equivalence classes are the maximally ε-neighbouring subsets of $\{x_i : x_i \in \mathbb{R}^m; i \in \mathcal{J}\}$. ◀

Let $\mathcal{D}_i$, $i = 1, \ldots, k$, be measurable subsets from $\mathbb{R}^m$. The set $\mathcal{E}$ is defined by

$$\mathcal{E} \doteq \left\{(x_1, \ldots, x_k) \in \mathop{\times}_{p=1}^{k} \mathcal{D}_p : \{x_1, \ldots, x_k\} \text{ } \varepsilon\text{-neighbouring}\right\}.$$

Lemma 1.6. *The measurability of $\mathcal{E}$ follows from the relation*

$$\mathcal{E} = \bigcap_{(\mathcal{J}_1, \mathcal{J}_2)} \bigcup_{i \in \mathcal{J}_1, j \in \mathcal{J}_2} \left\{(x_1, \ldots, x_k) \in \mathop{\times}_{p=1}^{k} \mathcal{D}_p : |x_i - x_j| \leqq \varepsilon\right\}$$

where the intersection is taken over all separations $(\mathcal{J}_1, \mathcal{J}_2)$ of $\mathcal{J} = \{1, \ldots, k\}$.

Proof. We assume $(x_1, \ldots, x_k) \in \mathcal{E}$; i.e. $\{x_1, \ldots, x_k\}$ is ε-neighbouring. Using Definition 1.1 it is $\mathcal{J}(\mathcal{J}_1) \cap \mathcal{J}(\mathcal{J}_2) \neq \emptyset$ for every separation $(\mathcal{J}_1, \mathcal{J}_2)$ of $\mathcal{J}$. Let $p \in \mathcal{J}(\mathcal{J}_1) \cap \mathcal{J}(\mathcal{J}_2)$. Hence we can find an index $s \in \mathcal{J}_1$ with $|x_s - x_p| \leqq \varepsilon$ and an index $t \in \mathcal{J}_2$ with $|x_t - x_p| \leqq \varepsilon$. Also it exists an index $i \in \mathcal{J}_1$ and an index $j \in \mathcal{J}_2$ with $|x_i - x_j| \leqq \varepsilon$. This leads to

$$(x_1, \ldots, x_k) \in \bigcup_{i \in \mathcal{J}_1, j \in \mathcal{J}_2} \left\{(x_1, \ldots, x_k) \in \mathop{\times}_{p=1}^{k} \mathcal{D}_p : |x_i - x_j| \leqq \varepsilon\right\}$$

and since this consideration is correct for all separations $(\mathcal{J}_1, \mathcal{J}_2)$ of $\mathcal{J}$ we have shown

$$\mathcal{E} \subset \bigcap_{(\mathcal{J}_1, \mathcal{J}_2)} \bigcup_{i \in \mathcal{J}_1, j \in \mathcal{J}_2} \left\{(x_1, \ldots, x_k) \in \mathop{\times}_{p=1}^{k} \mathcal{D}_p : |x_i - x_j| \leqq \varepsilon\right\} \doteq \tilde{\mathcal{E}}.$$

Now we assume $(x_1, \ldots, x_k) \in \tilde{\mathcal{E}}$. Let $(\mathcal{J}_1, \mathcal{J}_2)$ be an arbitrary separation of $\mathcal{J}$. Hence we have

$$(x_1, \ldots, x_k) \in \bigcup_{i \in \mathcal{J}_1, j \in \mathcal{J}_2} \left\{(x_1, \ldots, x_k) \in \mathop{\times}_{p=1}^{k} \mathcal{D}_p : |x_i - x_j| \leqq \varepsilon\right\}$$

and it exists an index $i \in \mathcal{J}_1$ and an index $j \in \mathcal{J}_2$ with $|x_i - x_j| \leqq \varepsilon$ so that $i \in \mathcal{J}(\mathcal{J}_1) \cap \mathcal{J}(\mathcal{J}_2)$. We have proved that $\mathcal{J}(\mathcal{J}_1) \cap \mathcal{J}(\mathcal{J}_2) \neq \emptyset$ for every separation $(\mathcal{J}_1, \mathcal{J}_2)$ of $\mathcal{J}$; i.e. $\{x_1, \ldots, x_k\}$ is ε-neighbouring and $(x_1, \ldots, x_k) \in \mathcal{E}$. ◀

1.2. Weakly correlated functions

This section deals with the definition of weakly correlated functions and some properties of such functions. Many facts are contained already in vom Scheidt, Purkert [3].

Definition 1.2. A random function $f_\varepsilon(x, \omega)$, $x \in \mathcal{D} \subset \mathbb{R}^m$, with $\langle f_\varepsilon(x) \rangle = 0$, is called *weakly correlated* with correlation length ε if the relation

$$\left\langle \prod_{i \in I} f_\varepsilon(x_i) \right\rangle = \prod_{j=1}^{p} \left\langle \prod_{i \in I_j} f_\varepsilon(x_i) \right\rangle$$

is satisfied for all k-th moments, $k = 2, 3, \ldots$, where $\mathcal{J} = \{1, \ldots, k\}$ and

$$\{x_i \in \mathcal{D} : i \in \mathcal{J}_1\}, \qquad \{x_i \in \mathcal{D} : i \in \mathcal{J}_2\}, \ldots, \{x_i \in \mathcal{D} : i \in \mathcal{J}_p\}$$

with $\bigcup_{j=1}^{p} \mathcal{J}_j = \mathcal{J}$ denotes the separation of $\{x_i : i \in \mathcal{J}\}$ in the maximally ε-neighbouring subsets.

Let $f_\varepsilon(x, \omega)$ be a weakly correlated random function with correlation length ε. Then, for example, we obtain for the correlation function

$$\langle f_\varepsilon(x_1)\, f_\varepsilon(x_2)\rangle = \begin{cases} R_\varepsilon(x_1, x_2) & \text{for } |x_1 - x_2| \leqq \varepsilon, \\ 0 & \text{for } |x_1 - x_2| > \varepsilon. \end{cases}$$

Examples for correlation functions with this property:

1. $$\langle f_\varepsilon(x_1)\, f_\varepsilon(x_2)\rangle = \begin{cases} \sigma^2(1 - |x_1 - x_2|/\varepsilon) & \text{for } |x_1 - x_2| \leqq \varepsilon, \\ 0 & \text{for } |x_1 - x_2| > \varepsilon. \end{cases}$$

2. We can deduce correlation functions from functions $g_{\varepsilon/2}(z)$ with the property $g_{\varepsilon/2}(z) = 0$ for $|z| > \varepsilon/2$ by the operation of the convolution. Using such a function $g_{\varepsilon/2}(z)$ then we obtain by means of

$$(g_{\varepsilon/2} * g_{\varepsilon/2})\,(z) = \int_{\mathbb{R}^m} g_{\varepsilon/2}(x)\, g_{\varepsilon/2}(z - x)\,\mathrm{d}x = \int_{\mathcal{K}_{\varepsilon/2}(0) \cap \mathcal{K}_{\varepsilon/2}(z)} g_{\varepsilon/2}(x)\, g_{\varepsilon/2}(z - x)\,\mathrm{d}x$$

a correlation function of a weakly correlated function. This statement follows from

$$F[g_{\varepsilon/2} * g_{\varepsilon/2}] = (2\pi)^{m/2}\, \{F[g_{\varepsilon/2}]\}^2 \geqq 0$$

where

$$F[g]\,(x) \doteq (2\pi)^{-m/2} \int_{\mathbb{R}^m} \mathrm{e}^{-\mathrm{i}(x,y)}\, g(y)\,\mathrm{d}y$$

denotes the Fourier transform of g. An example is obtained by

$$g_\eta(x) = g_\eta(|x|) = \begin{cases} \dfrac{d}{\sqrt{\eta}} \exp\left(\dfrac{\eta^2}{|x|^2 - \eta^2}\right) & \text{for } |x| < \eta, \\ 0 & \text{for } |x| \geqq \eta. \end{cases}$$

The function

$$(g_{\varepsilon/2} * g_{\varepsilon/2})\,(z) = \frac{2d^2}{\varepsilon} \int_{\mathbb{R}^m} \exp\left(\frac{(\varepsilon/2)^2}{|x|^2 - (\varepsilon/2)^2} + \frac{(\varepsilon/2)^2}{|z - x|^2 - (\varepsilon/2)^2}\right) \mathrm{d}x$$

possesses the properties of a correlation function of a wide-sense homogeneous random function. This correlation function belongs to $\mathbf{C}_0^\infty(\mathbb{R}^m)$.

3. Using the method written in 2. with the function

$$g_\eta(x) = \begin{cases} \dfrac{d}{\sqrt{\eta^3}} \sqrt{\dfrac{3}{2}}\, (\mathrm{e}^{-|x|} - \mathrm{e}^{-\eta}) & \text{for } |x| < \eta, \\ 0 & \text{for } |x| \geqq \eta \end{cases}$$

for $x \in \mathbb{R}^1$ then we can obtain the correlation function

$$(g_{\varepsilon/2} * g_{\varepsilon/2})\,(z) = \frac{12d^2}{\varepsilon^3} \begin{cases} (1 + z)\,\mathrm{e}^{-z} + \mathrm{e}^{z-\varepsilon} + (2 + \varepsilon - z)\,\mathrm{e}^{-\varepsilon} - 4\mathrm{e}^{-\varepsilon/2} & \text{for } 0 \leqq z \leqq \varepsilon/2, \\ (\varepsilon - z + 2)\,\mathrm{e}^{-z} + (\varepsilon - z + 2)\,\mathrm{e}^{-\varepsilon} & \text{for } \varepsilon/2 \leqq z \leqq \varepsilon, \\ 0 & \text{for } \varepsilon \leqq z \end{cases}$$

of a wide-sense stationary random process.

Theorem 1.1. *Let $f_\varepsilon(x, \omega)$ be a weakly correlated function, and let ε denote the correlation length. A further assumption is*

$$\sum_{k=0}^{\infty} \langle |f_\varepsilon(x)|^k \rangle / k! < \infty$$

for every $x \in \mathcal{D}$. Let the set $\{x_i \in \mathcal{D} : i \in \mathcal{J}\}$ have the separation

$$\{x_i \in \mathcal{D} : i \in \mathcal{J}_1\}, \qquad \{x_i \in \mathcal{D} : i \in \mathcal{J}_2\}, \ldots, \{x_i \in \mathcal{D} : i \in \mathcal{J}_p\}$$

in the maximally ε-neighbouring subsets. Then the random vectors

$$\big(f_\varepsilon(x_i, \omega)\big)_{i \in \mathcal{J}_r} \quad \text{and} \quad \big(f_\varepsilon(x_i, \omega)\big)_{i \in \mathcal{J}_s}$$

are independent for $r \neq s$, $r, s \in \{1, 2, \ldots, p\}$.

In particular, the random variables $f_\varepsilon(x, \omega)$ and $f_\varepsilon(y, \omega)$ are independent if $|x - y| > \varepsilon$.

Theorem 1.2 contains the existence of weakly correlated functions and the proof can be found in VOM SCHEIDT, PURKERT [3].

Theorem 1.2. *For any real number $\varepsilon > 0$ a weakly correlated Gaussian function exists with correlation length ε. A Gaussian function $u(x, \omega)$ satisfying $\langle u(x) \rangle = 0$ and $\langle u(x_1) u(x_2) \rangle = 0$ for $|x_1 - x_2| > \varepsilon$ is a weakly correlated function with correlation length ε', $\varepsilon' \leqq \varepsilon$.*

For applications we need weakly correlated functions where the sample functions are bounded and differentiable.

Theorem 1.3. *Let ε be any real number, $\varepsilon > 0$. Then weakly correlated functions exis possessing sample continuous partial derivatives of any order. Then random function $v(x, \omega) = h\big(u(x, \omega)\big) - \big\langle h\big(u(x, \omega)\big)\big\rangle$ is a weakly correlated function if $\bar{u}(x, \omega) = u(x, \omega) - \langle u(x, \omega) \rangle$ is weakly correlated and*

$$\sum_{k=0}^{\infty} \langle |u(x)|^k \rangle / k!$$

is convergent for every $x \in \mathcal{D} \subset \mathbb{R}^m$ where $h(t)$ is a real function on $\mathbb{R}^1$ and the moments $\langle h\big(u(x)\big)^k \rangle$, $k = 1, 2, \ldots$, are to exist for every $x \in \mathcal{D} \subset \mathbb{R}^m$.

Proof. The correlation function constructed in the above second example belongs to $C_0^\infty(\mathbb{R}^m)$. Thus, the first part of this theorem follows from the connection between the derivatives of a random function and the derivatives of the correlation function. We obtain the second part of this theorem by means of Theorem 1.1. ◀

Now we give the definition of a weakly correlated connected random vector function.

Definition 1.3. A random vector function

$$\big(f_{1\varepsilon}(x, \omega), f_{2\varepsilon}(x, \omega), \ldots, f_{l\varepsilon}(x, \omega)\big), \qquad x \in \mathcal{D} \subset \mathbb{R}^m,$$

with $\langle f_{i\varepsilon}(x, \omega) \rangle = 0$ for $i = 1, 2, \ldots, l$ is said to be *weakly correlated connected* with correlation length ε if

$$\Big\langle \prod_{i \in I} f_{r_{i\varepsilon}}(x_i) \Big\rangle = \prod_{j=1}^{p} \Big\langle \prod_{i \in I_j} f_{r_{i\varepsilon}}(x_i) \Big\rangle$$

is satisfied for k-th moments, $k = 2, 3, \ldots$, where $\mathcal{J} = \{1, 2, \ldots, k\}$, $r_i \in \{1, 2, \ldots, l\}$

and

$$\{x_i \in \mathcal{D}: i \in \mathcal{J}_1\}, \quad \{x_i \in \mathcal{D}: i \in \mathcal{J}_2\}, \ldots, \{x_i \in \mathcal{D}: i \in \mathcal{J}_p\}$$

with $\bigcup_{j=1}^{p} \mathcal{J}_j = \mathcal{J}$ denotes the separation of $\{x_i: i \in \mathcal{J}\}$ in the maximally ε-neighbouring subsets.

The definition of a weakly correlated connected vector function implies the weak correlation of the components of the vector function (e.g. $r_i = 1$ for $i = 1, 2, \ldots, k$). The following theorem is easy to prove (see VOM SCHEIDT, PURKERT [3]).

Theorem 1.4. *A random vector function $\big(f_{1\varepsilon}(x, \omega), f_{2\varepsilon}(x, \omega), \ldots, f_{l\varepsilon}(x, \omega)\big)$ is weakly correlated connected if the random functions $f_{i\varepsilon}(x, \omega)$, $i = 1, 2, \ldots, l$, are independent and weakly correlated.*

Finally, we give the definition of a partially weakly correlated function (see also MEUSEL, VOM SCHEIDT [2]).

Definition 1.4. A random function $f_\varepsilon(x, \omega)$, $x = (y, z) \in \mathcal{D}_1 \times \mathcal{D}_2 \subset \mathbb{R}^r \times \mathbb{R}^s$, with $\langle f_\varepsilon(x) \rangle = 0$, is called *partially weakly correlated* as to y with correlation length ε if the relation

$$\left\langle \prod_{i \in \mathcal{J}} f_\varepsilon(y_i, z_i) \right\rangle = \prod_{j=1}^{p} \left\langle \prod_{i \in \mathcal{J}_j} f_\varepsilon(y_i, z_i) \right\rangle$$

is satisfied for all k-th moments, $k = 2, 3, \ldots$, where $\mathcal{J} = \{1, 2, \ldots, k\}$ and

$$\{y_i \in \mathcal{D}_1: i \in \mathcal{J}_1\}, \ \{y_i \in \mathcal{D}_1: i \in \mathcal{J}_2\}, \ldots, \{y_i \in \mathcal{D}_1: i \in \mathcal{J}_p\}$$

with $\bigcup_{i=1}^{p} \mathcal{J}_j = \mathcal{J}$ denotes the separation of $\{y_i \in \mathcal{D}_1: i \in \mathcal{J}\}$ in the maximally ε-neighbouring subsets of points on $\mathbb{R}^r$.

In particular, the correlation function of such a function can be written as

$$\langle f_\varepsilon(x_1)\, f_\varepsilon(x_2) \rangle = \begin{cases} R_\varepsilon(x_1, x_2) & \text{for} \quad |y_1 - y_2| \leqq \varepsilon, \\ 0 & \text{for} \quad |y_1 - y_2| > \varepsilon \end{cases}$$

where $|y|$ denotes the norm of $\mathbb{R}^r$.

Example. Let $f_\varepsilon(y, z, \omega)$ be a random function on $\mathbb{R}^1 \times \mathbb{R}^1$. In order to explain Definition 1.4 and to make clear the distinction to the (completely) weakly correlated function

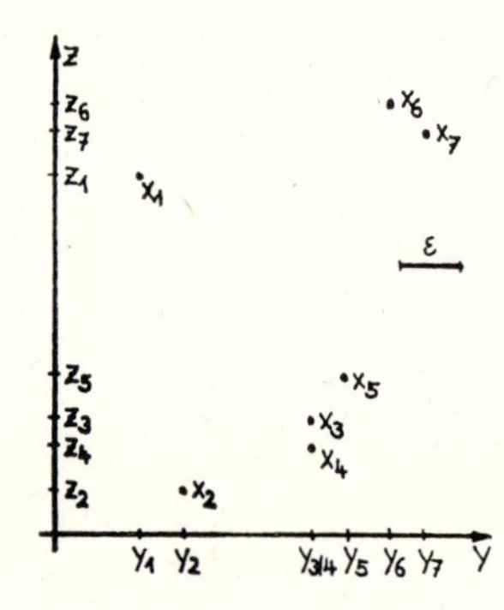

Fig. 1.2. Points in connection with a weakly correlated function $f_\varepsilon(y, z, \omega)$

we consider $f_\varepsilon(y, z, \omega)$ in the seven observation points plotted in Fig. 1.2. First, let $f_\varepsilon(y, z, \omega)$ be weakly correlated; i.e. $f_\varepsilon(y, z, \omega)$ is assumed to be weakly correlated with respect to y and z. Then it follows

$$\left\langle \prod_{i=1}^{7} f_\varepsilon(x_i) \right\rangle = \langle f_\varepsilon(x_1)\rangle \langle f_\varepsilon(x_2)\rangle \langle f_\varepsilon(x_3)\, f_\varepsilon(x_4)\, f_\varepsilon(x_5)\rangle \langle f_\varepsilon(x_6)\, f_\varepsilon(x_7)\rangle .$$

If $f_\varepsilon(y, z, \omega)$ is supposed to be weakly correlated as to y we obtain

$$\left\langle \prod_{i=1}^{7} f_\varepsilon(x_i) \right\rangle = \langle f_\varepsilon(x_1)\, f_\varepsilon(x_2)\rangle \langle f_\varepsilon(x_3)\, f_\varepsilon(x_4)\, f_\varepsilon(x_5)\, f_\varepsilon(x_6)\, f_\varepsilon(x_7)\rangle .$$

Finally, assuming that $f_\varepsilon(y, z, \omega)$ is weakly correlated with respect to z then we have

$$\left\langle \prod_{i=1}^{7} f_\varepsilon(x_i) \right\rangle = \langle f_\varepsilon(x_1)\, f_\varepsilon(x_6)\, f_\varepsilon(x_7)\rangle \langle f_\varepsilon(x_2)\, f_\varepsilon(x_3)\, f_\varepsilon(x_4)\, f_\varepsilon(x_5)\rangle .$$

The random function

$$f_\varepsilon(y, z, \omega) \doteq f_{1\varepsilon}(y, \omega)\, f_2(z)$$

is partially weakly correlated as to y with correlation length ε if $f_{1\varepsilon}(y, \omega)$ is weakly correlated with correlation length ε and $f_2(z)$ is a deterministic function.

1.3. Expansions of moments of random linear functionals

In the following considerations we shall deal with the calculation of moments of random variables

$$r_{ij\varepsilon}(\omega) \doteq \int_{\mathcal{D}_i} F_{ij}(x)\, f_{j\varepsilon}(x, \omega)\, dx, \qquad i = 1, 2, \ldots, n; \quad j = 1, 2, \ldots, l, \tag{1.1}$$

where $\big(f_{1\varepsilon}(x, \omega), f_{2\varepsilon}(x, \omega), \ldots, f_{l\varepsilon}(x, \omega)\big)$ denotes a weakly correlated connected random vector with $x \in \mathcal{D} \subset \mathbb{R}^m$ and correlation length ε. $\mathcal{D}$ is assumed to be a domain with $\mathcal{D} \supset \bigcup_{i=1}^{n} \mathcal{D}_i$ and $\mathcal{D}_i$, $i = 1, 2, \ldots, n$, domains from $\mathbb{R}^m$ with piecewise smooth boundaries. The sample functions of these random functions $f_{j\varepsilon}(x, \omega)$, $j = 1, 2, \ldots, l$, are supposed to be continuous a.s. and

$$\langle |f_{j\varepsilon}(x)|^p \rangle \leqq c_p < \infty \quad \text{for} \quad j = 1, 2, \ldots, l \quad \text{and} \quad p = 1, 2, \ldots$$

Let $F_{ij}(x)$ be integrable functions on $\mathcal{D}_i$ with the properties

$$\sup_{\substack{x \in \mathcal{D}_i \\ i=1,2,\ldots,n \\ j=1,2,\ldots,l}} |F_{ij}(x)| \leqq c < \infty; \qquad \int_{\mathcal{D}_i} |F_{ij}(x)|^p\, dx \leqq d < \infty \quad \text{for} \quad p = 1, 2 .$$

The goal of this section is the expansion of a k-th moment

$$\langle r_{i_1 j_1 \varepsilon} r_{i_2 j_2 \varepsilon} \ldots r_{i_k j_k \varepsilon} \rangle = \left\langle \prod_{p=1}^{k} r_{i_p j_p \varepsilon} \right\rangle; \qquad \begin{array}{l} i_p \in \{1, 2, \ldots, n\}, \\ j_p \in \{1, 2, \ldots, l\}, \end{array} \tag{1.2}$$

with respect to the correlation length ε.

For abbreviation we put

$$\bar{f}_{p\varepsilon}(y_p) \doteq f_{i_p\varepsilon}(y_p), \qquad \bar{F}_p(y_p) \doteq F_{i_pj_p}(y_p), \qquad \bar{r}_{p\varepsilon} \doteq r_{i_pj_p\varepsilon}, \qquad \bar{\mathcal{D}}_p \doteq \mathcal{D}_{i_p}$$

for $p = 1, 2, \ldots, k$. Then (1.2) leads to

$$\left\langle \prod_{q=1}^{k} r_{i_qj_q\varepsilon} \right\rangle = \left\langle \prod_{q=1}^{k} \bar{r}_{q\varepsilon} \right\rangle = \int\limits_{\bar{\mathcal{D}}_1} \cdots \int\limits_{\bar{\mathcal{D}}_k} \prod_{q=1}^{k} \bar{F}_q(y_q) \left\langle \prod_{q=1}^{k} \bar{f}_{q\varepsilon}(y_q) \right\rangle \mathrm{d}y_1 \ldots \mathrm{d}y_k. \tag{1.3}$$

It is useful to introduce the sets $\mathcal{B}(\mathcal{J}_1, \ldots, \mathcal{J}_s)$ by

$$\mathcal{B}(\mathcal{J}_1, \ldots, \mathcal{J}_s) \doteq \left\{(y_1, \ldots, y_k) \in \mathop{\times}_{q=1}^{k} \bar{\mathcal{D}}_q \colon \{y_i \colon i \in \mathcal{J}_q\},\, q = 1, 2, \ldots, s, \right.$$
$$\left. \text{maximally } \varepsilon\text{-neighbouring}\right\}$$

for $s = 1, 2, \ldots, k$ where

$$\mathcal{J}_q = \{i_{q1}, \ldots, i_{qk_q}\} \subset \mathcal{J} = \{1, 2, \ldots, k\}, \; \mathcal{J}_q \cap \mathcal{J}_p = \emptyset \quad \text{for} \quad p \neq q, \; \bigcup_{q=1}^{s} \mathcal{J}_q = \mathcal{J}.$$

The relation

$$\sum_{q=1}^{s} k_q = k$$

is obvious. Using Lemma 1.5 it can be seen that two sets $\mathcal{B}$ are disjoint. Furthermore, let

$$\mathcal{E}(\mathcal{J}_q) \doteq \left\{(y_{i_{q1}}, \ldots, y_{i_{qk_q}}) \in \mathop{\times}_{p=1}^{k_q} \bar{\mathcal{D}}_{i_{qp}} \colon \{y_{i_{q1}}, \ldots, y_{i_{qk_q}}\} \; \varepsilon\text{-neighbouring}\right\}.$$

These sets are measurable (see Lemma 1.6). Now it holds the relation

$$\mathcal{B}(\mathcal{J}_1, \ldots, \mathcal{J}_s) = \mathcal{E}(\mathcal{J}_1) \times \mathcal{E}(\mathcal{J}_2) \times \cdots \times \mathcal{E}(\mathcal{J}_s) \setminus \mathcal{F}(\mathcal{J}_1, \ldots, \mathcal{J}_s) \tag{1.4}$$

where $\mathcal{F}(\mathcal{J}_1, \ldots, \mathcal{J}_s)$ is defined by

$$\mathcal{F}(\mathcal{J}_1, \ldots, \mathcal{J}_s) \doteq \bigcup_{\{q_1,q_2\},\{q_3,\ldots,q_s\}} \mathcal{E}(\mathcal{J}_{q_1} \cup \mathcal{J}_{q_2}) \times \mathcal{E}(\mathcal{J}_{q_3}) \times \mathcal{E}(\mathcal{J}_{q_4}) \times \cdots \times \mathcal{E}(\mathcal{J}_{q_s}) \tag{1.5}$$

for $s \geqq 2$ and $\mathcal{F}(\mathcal{J}_1) \doteq \emptyset$. The above union is taken over all separations $\{q_1, q_2\}, \{q_3, \ldots, q_s\}$ of $\{1, 2, \ldots, s\}$ where two separations are equal if the first and the second set are only distinguished by a permutation. From the relations (1.4) and (1.5) the measurability of $\mathcal{B}(\mathcal{J}_1, \ldots, \mathcal{J}_s)$ follows since $\mathcal{E}(\mathcal{J}_{q_j})$ is measurable.

We obtain

$$\mathop{\times}_{q=1}^{k} \bar{\mathcal{D}}_q = \bigcup_{s=1}^{k} \bigcup_{\{\mathcal{J}_1,\ldots,\mathcal{J}_s\}} \mathcal{B}(\mathcal{J}_1, \ldots, \mathcal{J}_s) \tag{1.6}$$

where the inner union is taken over all nonequivalent separations $\{\mathcal{J}_1, \ldots, \mathcal{J}_s\}$ of $\{1, 2, \ldots, k\}$. Two separations are said to be equivalent if they are only distinguished by a permutation of the sets $\mathcal{J}_q$ and a permutation of the elements in the sets. With the help of (1.6) from (1.3) we obtain

$$\left\langle \prod_{q=1}^{k} \bar{r}_{q\varepsilon} \right\rangle = \sum_{s=1}^{k} \sum_{\{\mathcal{J}_1,\dots,\mathcal{J}_s\}} \int_{\mathcal{B}(\mathcal{J}_1,\dots,\mathcal{J}_s)} \prod_{q=1}^{s} \left\{ \prod_{i\in\mathcal{J}_q} \bar{F}_i(y_i) \left\langle \prod_{i\in\mathcal{J}_q} \bar{f}_{i\varepsilon}(y_i) \right\rangle \right\} dy_1 \dots dy_k$$

$$= \sum_{s=1}^{k} \sum_{\{\mathcal{J}_1,\dots,\mathcal{J}_s\}} \left[\prod_{q=1}^{s} \int_{\mathcal{E}(\mathcal{J}_q)} \prod_{i\in\mathcal{J}_q} \bar{F}_i(y_i) \left\langle \prod_{i\in\mathcal{J}_q} \bar{f}_{i\varepsilon}(y_i) \right\rangle \prod_{i\in\mathcal{J}_q} dy_i \right.$$

$$\left. - \int_{\mathcal{F}(\mathcal{J}_1,\dots,\mathcal{J}_s)} \prod_{q=1}^{s} \left\{ \prod_{i\in\mathcal{J}_q} \bar{F}_i(y_i) \left\langle \prod_{i\in\mathcal{J}_q} \bar{f}_{i\varepsilon}(y_i) \right\rangle \right\} dy_1 \dots dy_k \right]$$

$$= \sum_{s=1}^{k} \sum_{\{\mathcal{J}_1,\dots,\mathcal{J}_s\}} [C_1(\mathcal{J}_1,\dots,\mathcal{J}_s) - C_2(\mathcal{J}_1,\dots,\mathcal{J}_s)] \tag{1.7}$$

where C_1 and C_2 are defined by

$$C_1(\mathcal{J}_1,\dots,\mathcal{J}_s) \doteq \prod_{q=1}^{s} \int_{\mathcal{E}(\mathcal{J}_q)} \prod_{i\in\mathcal{J}_q} \bar{F}_i(y_i) \left\langle \prod_{i\in\mathcal{J}_q} \bar{f}_{i\varepsilon}(y_i) \right\rangle \prod_{i\in\mathcal{J}_q} dy_i$$

$$C_2(\mathcal{J}_1,\dots,\mathcal{J}_s) \doteq \int_{\mathcal{F}(\mathcal{J}_1,\dots,\mathcal{J}_s)} \prod_{q=1}^{s} \left\{ \prod_{i\in\mathcal{J}_q} \bar{F}_i(y_i) \left\langle \prod_{i\in\mathcal{J}_q} \bar{f}_{i\varepsilon}(y_i) \right\rangle \right\} dy_1 \dots dy_k. \tag{1.8}$$

By means of (1.5) the integral C_2 leads to

$$C_2(\mathcal{J}_1,\dots,\mathcal{J}_s) = \sum_{\{q_1,q_2\},\{q_3,\dots,q_s\}} \int_{\mathcal{E}(\mathcal{J}_{q_1}\cup\mathcal{J}_{q_2})} \prod_{q=q_1,q_2}$$

$$\times \left\{ \prod_{i\in\mathcal{J}_q} \bar{F}_i(y_i) \left\langle \prod_{i\in\mathcal{J}_q} \bar{f}_{i\varepsilon}(y_i) \right\rangle \prod_{i\in\mathcal{J}_q} dy_i \right\} \prod_{q=q_3,\dots,q_s}$$

$$\times \int_{\mathcal{E}(\mathcal{J}_q)} \prod_{i\in\mathcal{J}_q} \bar{F}_i(y_i) \left\langle \prod_{i\in\mathcal{J}_q} \bar{f}_{i\varepsilon}(y_i) \right\rangle \prod_{i\in\mathcal{J}_q} dy_i + \hat{C}_2(\mathcal{J}_1,\dots,\mathcal{J}_s). \tag{1.9}$$

The summand $\hat{C}_2$ has to be added since the sets $\mathcal{E}(\mathcal{J}_{q_1}\cup\mathcal{J}_{q_2})\times\mathcal{E}(\mathcal{J}_{q_3})\times\cdots\times\mathcal{E}(\mathcal{J}_{q_s})$ are not disjoint and $\hat{C}_2$ contains integrals over sets of the form $\mathcal{E}(\mathcal{J}_{q_1}\cup\mathcal{J}_{q_2}\cup\mathcal{J}_{q_3})\times\mathcal{E}(\mathcal{J}_{q_4})\times\cdots\times\mathcal{E}(\mathcal{J}_{q_s})$ or $\mathcal{E}(\mathcal{J}_{q_1}\cup\mathcal{J}_{q_2})\times\mathcal{E}(\mathcal{J}_{q_3}\cup\mathcal{J}_{q_4})\times\mathcal{E}(\mathcal{J}_{q_5})\times\cdots\times\mathcal{E}(\mathcal{J}_{q_s})$.

In order to calculate an expansion with respect to the correlation length ε we verify the following estimations for $\mathcal{J} = \{1, 2, \dots, p\}$:

$$\left| \int_{\mathcal{E}(\mathcal{J})} \prod_{i\in\mathcal{J}} \bar{F}_i(y_i) \left\langle \prod_{i\in\mathcal{J}} \bar{f}_{i\varepsilon}(y_i) \right\rangle dy_1 \dots dy_p \right|$$

$$\leq \int_{\mathcal{E}(\mathcal{J})} \prod_{i\in\mathcal{J}} |\bar{F}_i(y_i)| \left| \left\langle \prod_{i\in\mathcal{J}} \bar{f}_{i\varepsilon}(y_i) \right\rangle \right| dy_1 \dots dy_p$$

$$\leq \int_{\mathcal{E}(\mathcal{J})} \prod_{i\in\mathcal{J}} |\bar{F}_i(y_i)| \prod_{i\in\mathcal{J}} \langle |\bar{f}_{i\varepsilon}(y_i)|^p \rangle^{1/p} dy_1 \dots dy_p$$

$$\leq c_p \int_{\bar{\mathcal{D}}_1} |\bar{F}_1(y_1)| \int_{\bar{\mathcal{D}}_2 \cap \mathcal{K}_{(p-1)\varepsilon}(y_1)} |\bar{F}_2(y_2)| dy_2 \dots \int_{\bar{\mathcal{D}}_p \cap \mathcal{K}_{(p-1)\varepsilon}(y_1)} |\bar{F}_p(y_p)| dy_p$$

$$\leq c_p c^{p-1} \int_{\bar{\mathcal{D}}_1} |\bar{F}_1(y_1)| dy_1 \left[v\left(\mathcal{K}_{(p-1)\varepsilon}(0)\right) \right]^{p-1}$$

$$\leq c_p c^{p-1} d (C\varepsilon^m)^{p-1} = O(\varepsilon^{m(p-1)}) \tag{1.10}$$

using Hölder's inequality and Lemma 1.3. This investigation leads to expansions of the form

$$\int_{\mathcal{E}(\mathcal{J})} \prod_{i\in\mathcal{J}} \bar{F}_i(y_i) \left(\prod_{i\in\mathcal{J}} \bar{f}_{i\varepsilon}(y_i) \right) \mathrm{d}y_1 \ldots \mathrm{d}y_p$$

$$= {}^pA_{p-1}(\bar{F}_1, \ldots, \bar{F}_p)\, \varepsilon^{m(p-1)} + {}^pA_p(\bar{F}_1, \ldots, \bar{F}_p)\, \varepsilon^{m(p-1)+1} + o(\varepsilon^{m(p-1)+1}) \tag{1.11}$$

and

$$\int_{\mathcal{E}(\mathcal{J}_1\cup\mathcal{J}_2)} \prod_{i\in\mathcal{J}_1} \bar{F}_i(y_i) \prod_{i\in\mathcal{J}_2} \bar{F}_i(y_i) \left(\prod_{i\in\mathcal{J}_1} \bar{f}_{i\varepsilon}(y_i) \right) \left(\prod_{i\in\mathcal{J}_2} \bar{f}_{i\varepsilon}(y_i) \right) \prod_{i\in\mathcal{J}_1\cup\mathcal{J}_2} \mathrm{d}y_i$$

$$= A_{p_1p_2}(\bar{F}_1, \ldots, \bar{F}_{p_1}; \bar{F}_{p_1+1}, \ldots, \bar{F}_{p_1+p_2})\, \varepsilon^{m(p_1+p_2-1)} + o(\varepsilon^{m(p_1+p_2-1)}) \tag{1.12}$$

where $\mathcal{J}_1 = \{1, 2, \ldots, p_1\}$, $\mathcal{J}_2 = \{p_1 + 1, p_1 + 2, \ldots, p_1 + p_2\}$.

These expansions are introduced in (1.8) and it is

$$C_1(\mathcal{J}_1, \ldots, \mathcal{J}_s)$$

$$= \prod_{q=1}^{s} \left\{ {}^{k_q}A_{k_q-1}\left(\bar{F}_{i_{q1}}, \ldots, \bar{F}_{i_{qk_q}}\right) \varepsilon^{m(k_q-1)} + {}^{k_q}A_{k_q}\left(\bar{F}_{i_{q1}}, \ldots, \bar{F}_{i_{qk_q}}\right) \varepsilon^{m(k_q-1)+1} + \cdots \right\}$$

$$= \left\{ \prod_{q=1}^{s} {}^{k_q}\bar{A}_{k_q-1}(\mathcal{J}_q) \right\} \varepsilon^{m(k-s)} + \sum_{j=1}^{s} \left\{ {}^{k_j}\bar{A}_{k_j}(\mathcal{J}_j) \prod_{\substack{q=1\\ q\neq j}}^{s} {}^{k_q}\bar{A}_{k_q-1}(\mathcal{J}_q) \right\} \varepsilon^{m(k-s)+1} + \cdots \tag{1.13}$$

if we put

$${}^{k_q}\bar{A}_p(\mathcal{J}_q) \doteq {}^{k_q}A_p\left(\bar{F}_{i_{q1}}, \ldots, \bar{F}_{i_{qk_q}}\right) \quad \text{for} \quad p = k_q - 1, k_q.$$

Now we deal with $C_2(\mathcal{J}_1, \ldots, \mathcal{J}_s)$. By means of (1.11) the first summands have the order

$$\varepsilon^{m(k_{q1}+k_{q2}-1)} \prod_{q\in\{q_3,q_4,\ldots,q_s\}} \varepsilon^{m(k_q-1)} = \varepsilon^{m\left(\sum_{j=1}^{s} k_{qj} - (s-1)\right)} = \varepsilon^{m(k-s+1)}$$

and the summands in $\hat{C}_2(\mathcal{J}_1, \ldots, \mathcal{J}_s)$ have an order greater than or equal to

$$\varepsilon^{m(k_{q1}+k_{q2}+k_{q3}-1)} \prod_{q\in\{q_4,q_5,\ldots,q_s\}} \varepsilon^{m(k_q-1)} = \varepsilon^{m(k-s+2)}$$

or

$$\varepsilon^{m(k_{q1}+k_{q2}-1)}\, \varepsilon^{m(k_{q3}+k_{q4}-1)} \prod_{q\in\{q_5,\ldots,q_s\}} \varepsilon^{m(k_q-1)} = \varepsilon^{m(k-s+2)}.$$

The expansion of C_2 can also be written as

$$C_2(\mathcal{J}_1, \ldots, \mathcal{J}_s)$$

$$= \left\{ \sum_{\{q_1,q_2\},\{q_3,\ldots,q_s\}} \bar{A}_{k_{q_1}k_{q_2}}(\mathcal{J}_{q_1}; \mathcal{J}_{q_2}) \prod_{j=3}^{s} {}^{k_{qj}}\bar{A}_{k_{q_j}-1}(\mathcal{J}_{q_j}) \right\} \varepsilon^{m(k-s+1)} + \cdots \tag{1.14}$$

where we have put

$$\bar{A}_{k_qk_p}(\mathcal{J}_q; \mathcal{J}_p) \doteq A_{k_qk_p}(\bar{F}_{q1}, \ldots, \bar{F}_{qk_q}; \bar{F}_{p1}, \ldots, \bar{F}_{pk_p}).$$

First we investigate a moment (1.2) of even order k. It is possible to take $k_q \geqq 2$ for $q = 1, 2, \ldots, s$ since

$$C_1(\mathcal{J}_1, \ldots, \mathcal{J}_s) = C_2(\mathcal{J}_1, \ldots, \mathcal{J}_s) = 0$$

if at least one pertinent k_q is equal 1. The least order of C_1 is obtained for the greatest value of s, also for $s = k/2$. In this case we see that $k_1 = k_2 = \cdots = k_{k/2} = 2$ and (1.13), (1.14) lead to

$$\begin{aligned} &C_1(\mathcal{J}_1, \ldots, \mathcal{J}_{k/2}) - C_2(\mathcal{J}_1, \ldots, \mathcal{J}_{k/2}) \\ &= \left\{\prod_{q=1}^{k/2} {}^2\bar{A}_1(\mathcal{J}_q)\right\} \varepsilon^{mk/2} + \left\{\sum_{j=1}^{k/2} {}^2\bar{A}_2(\mathcal{J}_j) \prod_{\substack{q=1\\q\neq j}}^{k/2} {}^2\bar{A}_1(\mathcal{J}_q)\right\} \varepsilon^{mk/2+1} + \cdots \\ &\quad - \left\{\sum_{\{q_1,q_2\},\{q_3,\ldots,q_{k/2}\}} \bar{A}_{22}(\mathcal{J}_{q_1}; \mathcal{J}_{q_2}) \prod_{j=3}^{k/2} {}^2\bar{A}_1(\mathcal{J}_{q_j})\right\} \varepsilon^{m(k/2+1)} + \cdots \end{aligned}$$

The k-th moment (1.7) contains the terms

$$\begin{aligned} &\sum_{\{s_1,t_1\},\ldots,\{s_{k/2},t_{k/2}\}} \left[\left\{\prod_{q=1}^{k/2} {}^2\bar{A}_1(\{s_q, t_q\})\right\} \varepsilon^{mk/2}\right. \\ &+ \left\{\sum_{j=1}^{k/2} {}^2\bar{A}_2(\{s_j, t_j\}) \prod_{\substack{q=1\\q\neq j}}^{k/2} {}^2\bar{A}_1(\{s_q, t_q\})\right\} \varepsilon^{mk/2+1} + \cdots \\ &\left.- \left\{\sum_{\{q_1,q_2\},\{q_3,\ldots,q_{k/2}\}} \bar{A}_{22}(\{s_{q_1}, t_{q_1}\}; \{s_{q_2}, t_{q_2}\}) \prod_{j=3}^{k/2} {}^2\bar{A}_1(\{s_{q_j}, t_{q_j}\})\right\} \varepsilon^{m(k/2+1)} + \cdots\right]. \end{aligned} \quad (1.15)$$

Now we put $s = k/2 - 1$. Two possibilities follow in this case. First, we have $k_1 = k_2 = 3$, $k_3 = k_4 = \cdots = k_{k/2-1} = 2$ and consequently

$$C_1(\mathcal{J}_1, \ldots, \mathcal{J}_{k/2-1}) - C_2(\mathcal{J}_1, \ldots, \mathcal{J}_{k/2-1}) = {}^3\bar{A}_2(\mathcal{J}_1)\,{}^3\bar{A}_2(\mathcal{J}_2) \prod_{q=3}^{k/2-1} {}^2\bar{A}_1(\mathcal{J}_q)\, \varepsilon^{m(k/2+1)} + \cdots$$

Second, we can take $k_1 = 4$, $k_2 = k_3 = \cdots = k_{k/2-1} = 2$ and obtain

$$C_1(\mathcal{J}_1, \ldots, \mathcal{J}_{k/2-1}) - C_2(\mathcal{J}_1, \ldots, \mathcal{J}_{k/2-1}) = {}^4\bar{A}_3(\mathcal{J}_1) \prod_{q=2}^{k/2-1} {}^2\bar{A}_1(\mathcal{J}_q)\, \varepsilon^{m(k/2+1)} + \cdots$$

Hence, the k-th moment (1.7) has to contain the terms

$$\sum_{\{s_1,t_1,r_1\},\{s_2,t_2,r_2\},\{s_3,t_3\},\ldots,\{s_{k/2-1},t_{k/2-1}\}} \prod_{q=1}^{2} {}^3\bar{A}_2(\{s_q, t_q, r_q\}) \prod_{q=3}^{k/2-1} {}^2\bar{A}_1(\{s_q, t_q\})\, \varepsilon^{m(k/2+1)} + \cdots \quad (1.16)$$

in the first case and

$$\sum_{\{s_1,t_1,r_1,u_1\},\{s_2,t_2\},\ldots,\{s_{k/2-1},t_{k/2-1}\}} {}^4\bar{A}_3(\{s_1, t_1, r_1, u_1\}) \prod_{q=2}^{k/2-1} {}^2\bar{A}_1(\{s_q, t_q\})\, \varepsilon^{m(k/2+1)} + \cdots \quad (1.17)$$

in the second case. The summation containing in (1.15), (1.16), and (1.17) is taken over all adequate nonequivalent separations of $\{1, 2, \ldots, k\}$. For $s < k/2 - 1$ the least order of ε in the term $C_1(\mathcal{J}_1, \ldots, \mathcal{J}_s) - C_2(\mathcal{J}_1, \ldots, \mathcal{J}_s)$ is greater than $m(k/2 + 1)$ and these summands can be neglected.

Now we investigate a moment (1.2) of odd order k. For $s > (k - 1)/2$ the difference $C_1(\mathcal{J}_1, \ldots, \mathcal{J}_s) - C_2(\mathcal{J}_1, \ldots, \mathcal{J}_s)$ is equal to zero since at least one of the numbers k_j is equal to one. We take $s = (k - 1)/2$ and obtain as a necessity $k_1 = 3$, $k_2 = k_3 = \cdots$

$= k_{(k-1)/2} = 2$. The use of (1.13) and (1.14) leads to

$$C_1\left(\mathcal{J}_1, \ldots, \mathcal{J}_{\frac{k-1}{2}}\right) - C_2\left(\mathcal{J}_1, \ldots, \mathcal{J}_{\frac{k-1}{2}}\right) = {}^3\bar{A}_2(\mathcal{J}_1) \prod_{q=2}^{\frac{k-1}{2}} {}^2\bar{A}_1(\mathcal{J}_q)\, \varepsilon^{m\frac{k+1}{2}}$$
$$+ \left\{ {}^3\bar{A}_3(\mathcal{J}_1) \prod_{q=2}^{\frac{k-1}{2}} {}^2\bar{A}_1(\mathcal{J}_q) + {}^3\bar{A}_2(\mathcal{J}_1) \sum_{j=2}^{\frac{k-1}{2}} {}^2\bar{A}_2(\mathcal{J}_j) \prod_{\substack{q=2\\ q\neq j}}^{\frac{k-1}{2}} {}^2\bar{A}_1(\mathcal{J}_q) \right\} \varepsilon^{m\frac{k+1}{2}+1} + \cdots$$
$$- \left\{ \sum_{j=2}^{\frac{k-1}{2}} \bar{A}_{32}(\mathcal{J}_1; \mathcal{J}_j) \prod_{\substack{q=2\\ q\neq j}}^{\frac{k-1}{2}} {}^2\bar{A}_1(\mathcal{J}_q) \right.$$
$$\left. + {}^3\bar{A}_2(\mathcal{J}_1) \sum_{\substack{\{q_1,q_2\} \text{ from}\\ \left\{2,3,\ldots,\frac{k-1}{2}\right\}}} \bar{A}_{22}(\mathcal{J}_{q_1}; \mathcal{J}_{q_2}) \prod_{\substack{q=2\\ q\neq q_1,q_2}}^{\frac{k-1}{2}} {}^2\bar{A}_1(\mathcal{J}_q) \right\} \varepsilon^{m\frac{k+3}{2}} + \cdots \tag{1.18}$$

Without difficulties we can see that the k-th moment contains the terms

$$\sum_{\{s_1,t_1,r_1\},\{s_2,t_2\},\ldots,\left\{s_{\frac{k-1}{2}},t_{\frac{k-1}{2}}\right\}} \left[{}^3\bar{A}_2(\{s_1, t_1, r_1\}) \prod_{q=2}^{\frac{k-1}{2}} {}^2\bar{A}_1(\{s_q, t_q\})\, \varepsilon^{m\frac{k+1}{2}} \right.$$
$$+ \left\{ {}^3\bar{A}_3(\{s_1, t_1, r_1\}) \prod_{q=2}^{\frac{k-1}{2}} {}^2\bar{A}_1(\{s_q, t_q\}) + {}^3\bar{A}_2(\{s_1, t_1, r_1\}) \right.$$
$$\left. \times \sum_{j=2}^{\frac{k-1}{2}} {}^2\bar{A}_2(\{s_j, t_j\}) \prod_{\substack{q=2\\ q\neq j}}^{\frac{k-1}{2}} {}^2\bar{A}_1(\{s_q, t_q\}) \right\} \varepsilon^{m\frac{k+1}{2}+1}$$
$$- \left\{ \sum_{j=2}^{\frac{k-1}{2}} \bar{A}_{32}(\{s_1, t_1, r_1\}; \{s_j, t_j\}) \prod_{\substack{q=2\\ q\neq j}}^{\frac{k-1}{2}} {}^2\bar{A}_1(\{s_q, t_q\}) \right.$$
$$\left.\left. + {}^3\bar{A}_2(\{s_1, t_1, r_1\}) \sum_{\substack{\{q_1,q_2\} \text{ from}\\ \{2,\ldots,(k-1)/2\}}} \bar{A}_{22}(\{s_{q_1}, t_{q_1}\}; \{s_{q_2}, t_{q_2}\}) \prod_{\substack{q=2\\ q\neq q_1,q_2}}^{\frac{k-1}{2}} {}^2\bar{A}_1(\{s_q, t_q\}) \right\} \varepsilon^{m\frac{k+3}{2}} \right]. \tag{1.19}$$

For $s = (k-3)/2$ the following cases have to be considered:

– $k_1 = k_2 = k_3 = 3$, $k_4 = \cdots = k_{(k-3)/2} = 2$ leads to

$$C_1\left(\mathcal{J}_1, \ldots, \mathcal{J}_{\frac{k-3}{2}}\right) - C_2\left(\mathcal{J}_1, \ldots, \mathcal{J}_{\frac{k-3}{2}}\right) = \left\{ \prod_{q=1}^{3} {}^3\bar{A}_2(\mathcal{J}_q) \prod_{q=4}^{\frac{k-3}{2}} {}^2\bar{A}_1(\mathcal{J}_q) \right\} \varepsilon^{m\frac{k+3}{2}} + \cdots; \tag{1.20}$$

– $k_1 = 4$, $k_2 = 3$, $k_3 = \cdots = k_{(k-3)/2} = 2$ leads to

$$C_1\left(\mathcal{J}_1, \ldots, \mathcal{J}_{\frac{k-3}{2}}\right) - C_2\left(\mathcal{J}_1, \ldots, \mathcal{J}_{\frac{k-3}{2}}\right) = \left\{ {}^4\bar{A}_3(\mathcal{J}_1)\, {}^3\bar{A}_2(\mathcal{J}_2) \prod_{q=3}^{\frac{k-3}{2}} {}^2\bar{A}_1(\mathcal{J}_q) \right\} {}^{m\frac{k+3}{2}} + \cdots; \tag{1.21}$$

– $k_1 = 5, k_2 = \cdots = k_{(k-3)/2} = 2$ leads to

$$C_1\left(\mathcal{J}_1, \ldots, \mathcal{J}_{\frac{k-3}{2}}\right) - C_2\left(\mathcal{J}_1, \ldots, \mathcal{J}_{\frac{k-3}{2}}\right) = \left\{{}^5\bar{A}_4(\mathcal{J}_1) \prod_{q=2}^{\frac{k-3}{2}} {}^2\bar{A}_1(\mathcal{J}_q)\right\} \varepsilon^{m\frac{k+3}{2}} + \cdots \tag{1.22}$$

For $s < (k-3)/2$ it follows $k - s > (k+3)/2$ and it is possible to neglett these terms.
Summarizing we obtain the expansion of the k-th moment (1.7) for k even

$$\left\langle \prod_{p=1}^{k} r_{i_p j_p \varepsilon} \right\rangle = H_0(m; i_1, \ldots, i_k; j_1, \ldots, j_k)\, \varepsilon^{m\frac{k}{2}} + H_2(m; i_1, \ldots, i_k; j_1, \ldots, j_k)\, \varepsilon^{m\frac{k}{2}+1} + o\left(\varepsilon^{m\frac{k}{2}+1}\right) \tag{1.23}$$

where

$$H_0(m; i_1, \ldots, i_k; j_1, \ldots, j_k) = \sum_{\{s_1,t_1\},\ldots,\{s_{k/2},t_{k/2}\}} \prod_{q=1}^{k/2} {}^2\bar{A}_1(\{s_q, t_q\}) \tag{1.24}$$

for arbitrary $m \geqq 1$,

$$\begin{aligned} &H_2(m; i_1, \ldots, i_k; j_1, \ldots, j_k) \\ &= \sum_{\{s_1,t_1\},\ldots,\{s_{k/2},t_{k/2}\}} \left\{ \sum_{j=1}^{k/2} {}^2\bar{A}_2(\{s_j, t_j\}) \prod_{\substack{q=1\\q\neq j}}^{k/2} {}^2\bar{A}_1(\{s_q, t_q\}) \right. \\ &\quad - \sum_{\{q_1,q_2\},\{q_3,\ldots,q_{k/2}\}} \bar{A}_{22}(\{s_{q_1}, t_{q_1}\}; \{s_{q_2}, t_{q_2}\}) \prod_{j=3}^{k/2} {}^2\bar{A}_1(\{s_{q_j}, t_{q_j}\}) \\ &\quad + \sum_{\mathcal{J}_1} \prod_{q=1}^{2} {}^3\bar{A}_2(\{s_q, t_q, r_q\}) \prod_{q=3}^{k/2-1} {}^2\bar{A}_1(\{s_q, t_q\}) \\ &\quad + \sum_{\mathcal{J}_2} {}^4\bar{A}_3(\{s_1, t_1, r_1, u_1\}) \prod_{q=2}^{k/2-1} {}^2\bar{A}_1(\{s_q, t_q\}) \end{aligned} \tag{1.25}$$

for $m = 1$, and for $m \geqq 2$

$$H_2(m; i_1, \ldots, i_k; j_1, \ldots, j_k) = \sum_{\{s_1,t_1\},\ldots,\{s_{k/2},t_{k/2}\}} \sum_{j=1}^{k/2} {}^2\bar{A}_2(\{s_j, t_j\}) \prod_{\substack{q=1\\q\neq j}}^{k/2} {}^2\bar{A}_1(\{s_q, t_q\}),$$

where

$$\mathcal{J}_1 = \left\{\{s_1, t_1, r_1\}, \{s_2, t_2, r_2\}, \{s_3, t_3\}, \ldots, \{s_{k/2-1}, t_{k/2-1}\}\right\},$$

$$\mathcal{J}_2 = \left\{\{s_1, t_1, r_1, u_1\}, \{s_2, t_2\}, \ldots, \{s_{k/2-1}, t_{k/2-1}\}\right\}$$

and for k odd

$$\left\langle \prod_{p=1}^{k} r_{i_p j_p \varepsilon} \right\rangle = H_1(m; i_1, \ldots, i_k; j_1, \ldots, j_k)\, \varepsilon^{m\frac{k+1}{2}} + H_3(m; i_1, \ldots, i_k; j_1, \ldots, j_k)\, \varepsilon^{m\frac{k+1}{2}+1} + o\left(\varepsilon^{m\frac{k+1}{2}+1}\right) \tag{1.26}$$

where

$$\begin{aligned} &H_1(m; i_1, \ldots, i_k; j_1, \ldots, j_k) \\ &= \sum_{\{s_1,t_1,r_1\},\{s_2,t_2\},\ldots,\left\{s_{\frac{k-1}{2}},t_{\frac{k-1}{2}}\right\}} {}^3\bar{A}_2(\{s_1, t_1, r_1\}) \prod_{q=2}^{\frac{k-1}{2}} {}^2\bar{A}_1(\{s_q, t_q\}) \end{aligned} \tag{1.27}$$

for arbitrary $m \geqq 1$,

$$
\begin{aligned}
&H_3(m; i_1, \ldots, i_k; j_1, \ldots, j_k) \\
&= \sum_{\mathcal{J}_1} \Bigg\{ {}^3\bar{A}_3(\{s_1, t_1, r_1\}) \prod_{q=2}^{\frac{k-1}{2}} {}^2\bar{A}_1(\{s_q, t_q\}) \\
&\quad + {}^3\bar{A}_2(\{s_1, t_1, r_1\}) \sum_{j=2}^{\frac{k-1}{2}} {}^2\bar{A}_2(\{s_j, t_j\}) \prod_{\substack{q=2\\ q\neq j}}^{\frac{k-1}{2}} (\{s_q, t_q\}) \\
&\quad - \sum_{j=2}^{\frac{k-1}{2}} \bar{A}_{32}(\{s_1, t_1, r_1\}; \{s_j, t_j\}) \prod_{\substack{q=2\\ q\neq j}}^{\frac{k-1}{2}} {}^2\bar{A}_1(\{s_q, t_q\}) \\
&\quad - {}^3\bar{A}_2(\{s_1, t_1, r_1\}) \sum_{\substack{\{q_1,q_2\} \text{ from} \\ \left\{2,3,\ldots,\frac{k-1}{2}\right\}}} \bar{A}_{22}(\{s_{q_1}, t_{q_1}\}; \{s_{q_2}, t_{q_2}\}) \prod_{\substack{q=2\\ q\neq q_1,q_2}}^{\frac{k-1}{2}} {}^2\bar{A}_1(\{s_q, t_q\}) \\
&\quad + \sum_{\mathcal{J}_2} \prod_{q=1}^{3} {}^3\bar{A}_2(\{s_q, t_q, r_q\}) \prod_{q=4}^{\frac{k-3}{2}} {}^2\bar{A}_1(\{s_q, t_q\}) \\
&\quad + \sum_{\mathcal{J}_3} {}^4\bar{A}_3(\{s_1, t_1, r_1, u_1\})\, {}^3\bar{A}_2(\{s_2, t_2, r_2\}) \prod_{q=3}^{\frac{k-3}{2}} {}^2\bar{A}_1(\{s_q, t_q\}) \\
&\quad + \sum_{\mathcal{J}_4} {}^5\bar{A}_4(\{s_1, t_1, r_1, u_1, v_1\}) \prod_{q=2}^{\frac{k-3}{2}} {}^2\bar{A}_1(\{s_q, t_q\})
\end{aligned} \tag{1.28}
$$

for $m = 1$ and for $m \geqq 2$

$$
\begin{aligned}
&H_3(m; i_1, \ldots, i_k; j_1, \ldots, j_k) \\
&= \sum_{\mathcal{J}_1} \Bigg\{ {}^3\bar{A}_3(\{s_1, t_1, r_1\}) \prod_{q=2}^{\frac{k-1}{2}} {}^2\bar{A}_1(\{s_q, t_q\}) \\
&\quad + {}^3\bar{A}_2(\{s_1, t_1, r_1\}) \sum_{j=2}^{\frac{k-1}{2}} {}^2\bar{A}_2(\{s_j, t_j\}) \prod_{\substack{q=2\\ q\neq j}}^{\frac{k-1}{2}} {}^2\bar{A}_1(\{s_q, t_q\}) \Bigg\}
\end{aligned}
$$

where

$$
\begin{aligned}
\mathcal{J}_1 &= \left\{\{s_1, t_1, r_1\}, \{s_2, t_2\}, \ldots, \left\{s_{\frac{k-1}{2}}, t_{\frac{k-1}{2}}\right\}\right\}, \\
\mathcal{J}_2 &= \left\{\{s_1, t_1, r_1\}, \{s_2, t_2, r_2\}, \{s_3, t_3, r_3\}, \{s_4, t_4\}, \ldots, \left\{s_{\frac{k-3}{2}}, t_{\frac{k-3}{2}}\right\}\right\}, \\
\mathcal{J}_3 &= \left\{\{s_1, t_1, r_1, u_1\}, \{s_2, t_2, r_2\}, \{s_3, t_3\}, \ldots, \left\{s_{\frac{k-3}{2}}, t_{\frac{k-3}{2}}\right\}\right\}, \\
\mathcal{J}_4 &= \left\{\{s_1, t_1, r_1, u_1, v_1\}, \{s_2, t_2\}, \ldots, \left\{s_{\frac{k-3}{2}}, t_{\frac{k-3}{2}}\right\}\right\}
\end{aligned}
$$

if we take into consideration (1.15), (1.16), (1.17) for k even and (1.19), (1.20), (1.21), (1.22) for k odd.

For later investigations we define the random variables

$$r_{i\varepsilon}(\omega) \doteq \sum_{j=1}^{l} r_{ij\varepsilon}(\omega) \tag{1.29}$$

and determine the moments of these $r_{i\varepsilon}(\omega)$, $i = 1, 2, \ldots, n$. Thus

$$\left\langle \prod_{q=1}^{k} r_{i_q\varepsilon} \right\rangle = \sum_{j_1,\ldots,j_k=1}^{l} \langle r_{i_1j_1\varepsilon} r_{i_2j_2\varepsilon} \ldots r_{i_kj_k\varepsilon} \rangle .$$

Assuming k even and using (1.23) it follows

$$\left\langle \prod_{q=1}^{k} r_{i_q\varepsilon} \right\rangle = \hat{H}_0(m; i_1, \ldots, i_k)\, \varepsilon^{\frac{mk}{2}} + \hat{H}_2(m; i_1, \ldots, i_k)\, \varepsilon^{\frac{mk}{2}+1} + \cdots \tag{1.30}$$

where

$$\hat{H}_p(m; i_1, \ldots, i_k) = \sum_{j_1,\ldots,j_k=1}^{l} H_p(m; i_1, \ldots, i_k; j_1, \ldots, j_k) \quad \text{for} \quad p = 0, 2 \tag{1.31}$$

and H_0, H_2 are given by (1.24), (1.25). Furthermore, for k odd and with the help of (1.26) we find

$$\left\langle \prod_{q=1}^{k} r_{i_q\varepsilon} \right\rangle = \hat{H}_1(m; i_1, \ldots, i_k)\, \varepsilon^{m\frac{k+1}{2}} + \hat{H}_3(m; i_1, \ldots, i_k)\, \varepsilon^{m\frac{k+1}{2}+1} + \cdots \tag{1.32}$$

where

$$\hat{H}_p(m; i_1, \ldots, i_k) = \sum_{j_1,\ldots,j_k=1}^{l} H_p(m; i_1, \ldots, i_k; j_1, \ldots, j_k) \tag{1.33}$$

and H_1, H_3 are written in (1.27) and (1.28).

Now we want to determine some special cases. Let $l = n = 1$; i.e., we deal with the moments of the random variable

$$r_\varepsilon(\omega) \doteq \int_{\mathcal{D}} F(x)\, f_\varepsilon(x, \omega)\, dx .$$

In preparation of these calculations we remark that e.g.

$$Q(\{s_1, t_1\}, \ldots, \{s_{k/2}, t_{k/2}\})$$

denotes the number of separations

$$\{s_1, t_1\}, \{s_2, t_2\}, \ldots, \{s_{k/2}, t_{k/2}\}$$

of the set $\{1, 2, \ldots, k\}$. Then it is

$$Q(\{s_1, t_1\}, \ldots, \{s_{2/k}, t_{k/2}\}) = \frac{1}{\left(\frac{k}{2}\right)!} \binom{k}{2} \binom{k-2}{2} \cdots \binom{2}{2} = \frac{k!}{2^{k/2} \left(\frac{k}{2}\right)!},$$

$$Q(\{s_1, t_1, r_1\}, \{s_2, t_2, r_2\}, \{s_3, t_3\}, \ldots, \{s_{k/2-1}, t_{k/2-1}\})$$
$$= \frac{1}{2} \binom{k}{3} \binom{k-3}{3} \frac{(k-6)!}{2^{(k-6)/2} \left(\frac{k-6}{2}\right)!} = \frac{k!}{9 \cdot 2^{k/2} \left(\frac{k}{2} - 3\right)!},$$

$$Q(\{s_1, t_1, r_1, u_1\}, \{s_2, t_2\}, \ldots, \{s_{k/2-1}, t_{k/2-1}\})$$

$$= \binom{k}{4} \frac{(k-4)!}{2^{(k-4)/2}\left(\frac{k-4}{2}\right)!} = \frac{k!}{6 \cdot 2^{k/2}\left(\frac{k}{2}-2\right)!},$$

$$Q(\{s_1, t_1, r_1\}, \{s_2, t_2\}, \ldots, \{s_{(k-1)/2}, t_{(k-1)/2}\})$$

$$= \binom{k}{3} \frac{(k-3)!}{2^{(k-3)/2}\left(\frac{k-3}{2}\right)!} = \frac{k!}{3 \cdot 2^{(k-1)/2}\left(\frac{k-3}{2}\right)!},$$

$$Q(\{s_1, t_1, r_1\}, \ldots, \{s_3, t_3, r_3\}, \{s_4, t_4\}, \ldots, \{s_{(k-3)/2}, t_{(k-3)/2}\})$$

$$= \frac{1}{3!}\binom{k}{3}\binom{k-3}{3}\binom{k-6}{3} \frac{(k-9)!}{2^{(k-9)/2}\left(\frac{k-9}{2}\right)!} = \frac{k!}{81 \cdot 2^{(k-1)/2}\left(\frac{k-9}{2}\right)!},$$

$$Q(\{s_1, t_1, r_1, u_1\}, \{s_2, t_2, r_2\}, \{s_3, t_3\}, \ldots, \{s_{(k-3)/2}, t_{(k-3)/2}\})$$

$$= \binom{k}{4}\binom{k}{3} \frac{(k-7)!}{2^{(k-7)/2}\left(\frac{k-7}{2}\right)!} = \frac{k!}{18 \cdot 2^{(k-1)/2}\left(\frac{k-7}{2}\right)!},$$

$$Q(\{s_1, t_1, r_1, u_1, v_1\}, \{s_2, t_2\}, \ldots, \{s_{(k-3)/2}, t_{(k-3)/2}\})$$

$$= \binom{k}{5} \frac{(k-5)!}{2^{(k-5)/2}\left(\frac{k-5}{2}\right)!} = \frac{k!}{30 \cdot 2^{(k-1)/2}\left(\frac{k-5}{2}\right)!}.$$

Assuming k even we obtain

$$\langle r_\varepsilon^k \rangle = \begin{cases} ({}^2A_1\varepsilon)^{k/2} \dfrac{k!}{2^{k/2}\left(\frac{k}{2}\right)!} \left\{1 + \dfrac{k}{2}\left[\dfrac{{}^2A_2}{{}^2A_1} + \dfrac{\left(\frac{k}{2}-1\right)}{6} \dfrac{{}^4A_3 - 3A_{22}}{({}^2A_1)^2} \right.\right. \\ \left.\left. + \dfrac{\left(\frac{k}{2}-1\right)\left(\frac{k}{2}-2\right)}{9} \dfrac{({}^3A_2)^2}{({}^2A_1)^3}\right]\varepsilon + \cdots\right\} \quad \text{for} \quad m = 1, \\ ({}^2A_1\varepsilon^m)^{k/2} \dfrac{k!}{2^{k/2}\left(\frac{k}{2}\right)!} \left\{1 + \dfrac{k}{2}\dfrac{{}^2A_2}{{}^2A_1}\varepsilon + \cdots\right\} \quad \text{for} \quad m \geqq 2 \end{cases} \tag{1.34}$$

and for k odd

$$\langle r_\varepsilon^k \rangle = \begin{cases} ({}^2A_1)^{(k-3)/2}\, \varepsilon^{(k+1)/2} \dfrac{k!}{3 \cdot 2^{(k-1)/2} \left(\dfrac{k-3}{2}\right)!} \\ \times \left\{ {}^3A_2 + \left[{}^3A_3 + \dfrac{k-3}{2} \dfrac{{}^3A_2{}^2A_2 - A_{32}}{{}^2A_1} - \dfrac{(k-3)(k-5)}{8} \dfrac{{}^3A_2A_{22}}{({}^2A_1)^2} \right.\right. \\ + \dfrac{k-3}{20} \dfrac{{}^5A_4}{{}^2A_1} + \dfrac{(k-3)(k-5)(k-7)}{216} \left(\dfrac{{}^3A_2}{{}^2A_1} \right)^3 \\ \left.\left. + \dfrac{(k-3)(k-5)}{24} \dfrac{{}^4A_3{}^3A_2}{({}^2A_1)^2} \right] \varepsilon + \cdots \right\} \quad \text{for } m = 1, \\ ({}^2A_1)^{(k-3)/2}\, \varepsilon^{m(k+1)/2} \dfrac{k!}{3 \cdot 2^{(k-1)/2} \left(\dfrac{k-3}{2}\right)!} \\ \times \left\{ {}^3A_2 + \left[{}^3A_3 + \dfrac{k-3}{2} \dfrac{{}^3A_2{}^2A_2}{{}^2A_1} \right] \varepsilon + \cdots \right\} \quad \text{for } m \geqq 2 \end{cases} \tag{1.35}$$

where ${}^pA_q(F, \ldots, F) \doteq {}^pA_q$ and $A_{pq}(F, \ldots, F; F, \ldots, F) \doteq A_{pq}$. From (1.34) and (1.35) we obtain

$$\langle r_\varepsilon \rangle = 0,$$

$$\langle r_\varepsilon^2 \rangle = {}^2A_1 \varepsilon^m \left\{ 1 + \frac{{}^2A_2}{{}^2A_1} \varepsilon + \cdots \right\} \quad \text{for arbitrary } m,$$

$$\langle r_\varepsilon^3 \rangle = \varepsilon^{2m} \{ {}^3A_2 + {}^3A_3 \varepsilon + \cdots \} \quad \text{for arbitrary } m,$$

$$\langle r_\varepsilon^4 \rangle = \begin{cases} ({}^2A_1\varepsilon)^2 \left\{ 3 + \left[6 \dfrac{{}^2A_2}{{}^2A_1} + \dfrac{{}^4A_3 - 3A_{22}}{({}^2A_1)^2} \right] \varepsilon + \cdots \right\} & \text{for } m = 1, \\ ({}^2A_1\varepsilon^m)^2 \left\{ 3 + 6 \dfrac{{}^2A_2}{{}^2A_1} \varepsilon + \cdots \right\} & \text{for } m \geqq 2, \end{cases}$$

$$\langle r_\varepsilon^5 \rangle = \begin{cases} 10\varepsilon^3 \left\{ {}^3A_2{}^2A_1 + \left[{}^3A_3{}^2A_1 + {}^3A_2{}^2A_2 - A_{32} + \dfrac{1}{10} {}^5A_4 \right] \varepsilon + \cdots \right\} & \text{for } m = 1, \\ 10\varepsilon^{3m} \{ {}^3A_2{}^2A_1 + [{}^3A_3{}^2A_1 + {}^3A_2{}^2A_2] \varepsilon + \cdots \} & \text{for } m \geqq 2. \end{cases}$$

Using (1.23) $k = 2$ leads to

$$\langle r_{i_1j_1\varepsilon} r_{i_2j_2\varepsilon} \rangle = {}^2A_1(F_{i_1j_1}, F_{i_2j_2})\, \varepsilon^m + {}^2A_2(F_{i_1j_1}, F_{i_2j_2})\, \varepsilon^{m+1} + \cdots \tag{1.36}$$

for m arbitrary and with the help of (1.26) it follows

$$\langle r_{i_1j_1\varepsilon} r_{i_2j_2\varepsilon} r_{i_3j_3\varepsilon} \rangle = {}^3A_2(\bar{F}_1, \bar{F}_2, \bar{F}_3)\, \varepsilon^{2m} + {}^3A_3(\bar{F}_1, \bar{F}_2, \bar{F}_3)\, \varepsilon^{2m+1} + \cdots$$

where

$$\bar{F}_q \doteq F_{i_qj_q}.$$

Furthermore, for $k = 4$ we have

$$\begin{aligned}
&\langle r_{i_1j_1\varepsilon} r_{i_2j_2\varepsilon} r_{i_3j_3\varepsilon} r_{i_4j_4\varepsilon}\rangle \\
&= [{}^2A_1(\bar{F}_1, \bar{F}_2)\, {}^2A_1(\bar{F}_3, \bar{F}_4) + {}^2A_1(\bar{F}_1, \bar{F}_3)\, {}^2A_1(\bar{F}_2, \bar{F}_4) \\
&\quad + {}^2A_1(\bar{F}_1, \bar{F}_4)\, {}^2A_1(\bar{F}_2, \bar{F}_3)]\, \varepsilon^{2m} \\
&\quad + [{}^2A_2(\bar{F}_1, \bar{F}_2)\, {}^2A_1(\bar{F}_3, \bar{F}_4) + {}^2A_2(\bar{F}_3, \bar{F}_4)\, {}^2A_1(\bar{F}_1, \bar{F}_2) \\
&\quad + {}^2A_2(\bar{F}_1, \bar{F}_3)\, {}^2A_1(\bar{F}_2, \bar{F}_4) + {}^2A_2(\bar{F}_2, \bar{F}_4)\, {}^2A_1(\bar{F}_1, \bar{F}_3) \\
&\quad + {}^2A_2(\bar{F}_1, \bar{F}_4)\, {}^2A_1(\bar{F}_2, \bar{F}_3) + {}^2A_2(\bar{F}_2, \bar{F}_3)\, {}^2A_1(\bar{F}_1, \bar{F}_4) \\
&\quad + \{{}^4A_3(\bar{F}_1, \bar{F}_2, \bar{F}_3, \bar{F}_4) - A_{22}(\bar{F}_1, \bar{F}_2; \bar{F}_3, \bar{F}_4) - A_{22}(\bar{F}_1, \bar{F}_3; \bar{F}_2, \bar{F}_4) \\
&\quad - A_{22}(\bar{F}_1, \bar{F}_4; \bar{F}_2, \bar{F}_3)\}\, \delta_{1m}]\, \varepsilon^{2m+1} + \cdots
\end{aligned}$$

This section is closed by a summarizing theorem.

Theorem 1.5. *The moments* $\left\langle \prod_{p=1}^{k} r_{i_pj_p\varepsilon} \right\rangle$ *have the expansion*

$$\left\langle \prod_{p=1}^{k} r_{i_pj_p\varepsilon} \right\rangle = \begin{cases} H_0(m; i_1, \ldots, i_k; j_1, \ldots, j_k)\, \varepsilon^{mk/2} \\ \quad + H_2(m; i_1, \ldots, i_k; j_1, \ldots, j_k)\, \varepsilon^{mk/2+1} + \cdots & \textit{for } k \textit{ even}, \\ H_1(m; i_1, \ldots, i_k; j_1, \ldots, j_k)\, \varepsilon^{m(k+1)/2} \\ \quad + H_3(m; i_1, \ldots, i_k; j_1, \ldots, j_k)\, \varepsilon^{m(k+1)/2+1} + \cdots & \textit{for } k \textit{ odd} \end{cases}$$

where H_0 *is given by* (1.24), H_2 *by* (1.25), H_1 *by* (1.27), *and* H_3 *by* (1.28).

1.4. Expansions of distributions of linear functionals

We use the same notations as in the foregoing section. The aim of this section consists in the determination of an approximation of the distribution density of $\big(r_{1\varepsilon}(\omega), r_{2\varepsilon}(\omega), \ldots, r_{n\varepsilon}(\omega)\big)$ where $r_{i\varepsilon}(\omega)$ is defined by (1.29).

In a first step the transformation

$$\tilde{r}_{i\varepsilon}(\omega) = \frac{1}{\sqrt{\varepsilon^m}} \sum_{p=1}^{i} b_{ip} r_{p\varepsilon}, \qquad i = 1, 2, \ldots, n, \tag{1.37}$$

is carried out and the coefficients b_{ip} are determined by the equations

$$\sum_{p=1}^{i} \sum_{q=1}^{j} b_{ip} b_{jq} B_{pq} = \delta_{ij}, \qquad i, j = 1, 2, \ldots, n, \tag{1.38}$$

where

$$B_{pq} \doteq \hat{H}_0(m; p, q) = \sum_{j_1,j_2=1}^{l} H_0(m; p, q; j_1, j_2) = \sum_{j_1,j_2=1}^{l} {}^2A_1(F_{pj_1}, F_{qj_2}). \tag{1.39}$$

Let $r_{i\varepsilon}(\omega)$, $i = 1, 2, \ldots, n$, be linear independent, i.e.

$$\sum_{i=1}^{n} r_{i\varepsilon}(\omega)\, z_i \neq 0 \qquad \text{a.s.}$$

for arbitrary real numbers z_i, $i = 1, 2, \ldots, n$. Then it follows

$$0 < \left\langle \left(\sum_{p=1}^{n} r_{p\varepsilon} z_p \right)^2 \right\rangle = \sum_{p,q=1}^{n} \langle r_{p\varepsilon} r_{q\varepsilon} \rangle z_p z_q$$
$$= \sum_{p,q=1}^{n} \left(\hat{H}_0(m; p, q)\, \varepsilon^m + \hat{H}_2(m; p, q)\, \varepsilon^{m+1} + \cdots \right) z_p z_q$$

and furthermore

$$0 < \sum_{p,q=1}^{n} \hat{H}_0(m; p, q)\, z_p z_q .$$

From this property we have

$$\det \left(\hat{H}_0(m; p, q) \right)_{p,q=1,2,\ldots,s} > 0 \qquad \text{for} \quad s = 1, 2, \ldots, n \tag{1.40}$$

(see e.g. EISENREICH [1]). With the help of the given condition it is possible to calculate the coefficients b_{ip} without difficulties. First, we take (1.37) for $i = j = 1$ and obtain

$$b_{11}^2 B_{11} = 1 \quad \text{or} \quad b_{11} = \sqrt{\frac{1}{B_{11}}}$$

since $B_{11} = \hat{H}_0(m; 1, 1) > 0$ using (1.40). Now we assume that the coefficients b_{ip} for $p \leqq i$, $i < k$, have been calculated and that b_{ii} is greater than zero for $i = 1, 2, \ldots, k - 1$. We will determine the coefficients b_{kp} for $p = k$ and show that $b_{kk} > 0$. Setting $i = k$ and $j = 1, 2, \ldots, k$ in (1.38) it follows

$$\sum_{p=1}^{k} b_{kp}(b_{11} B_{p1}) = 0,$$

$$\sum_{p=1}^{k} b_{kp}(b_{21} B_{p1} + b_{22} B_{p2}) = 0,$$

$$\ldots$$

$$\sum_{p=1}^{k} b_{k-1p}(b_{k-11} B_{p1} + b_{k-12} B_{p2} + \cdots + b_{k-1k-1} B_{pk-1}) = 0,$$

$$\sum_{p=1}^{k} b_{kp}(b_{k1} B_{p1} + b_{k2} B_{p2} + \cdots + b_{kk} B_{pk}) = 1$$

and using $b_{ii} > 0$, $i = 1, 2, \ldots, k - 1$,

$$\sum_{p=1}^{k} b_{kp} B_{pi} = 0, \qquad i = 1, 2, \ldots, k - 1,$$

$$b_{kk} \sum_{p=1}^{k} b_{kp} B_{pk} = 1.$$

Since $b_{kk} \neq 0$ we obtain the system of linear equations

$$\sum_{p=1}^{k} \bar{b}_{kp} B_{pi} = \delta_{ki}, \qquad i = 1, 2, \ldots, k,$$

where $\bar{b}_{kp} \doteq b_{kk} b_{kp}$. The solution has the form

$$\bar{b}_{kp} = b_{kk} b_{kp} = \frac{(-1)^{k+p} \det (B_{ij})_{i=1,\ldots,k-1;\, j=1,\ldots,p-1,p+1,\ldots,k}}{\det (B_{ij})_{i,j=1,2,\ldots,k}} \tag{1.41}$$

for $p = 1, 2, \ldots, k$. In particular, $p = k$ leads to

$$b_{kk} = \sqrt{\frac{\det (B_{ij})_{i,j=1,2,\ldots,k-1}}{\det (B_{ij})_{i,j=1,2,\ldots,k}}} \tag{1.42}$$

if (1.39) is taken into consideration. The inequation $b_{kk} > 0$ is fulfilled and all coefficients b_{kp}, $p = 1, 2, \ldots, k$, can be calculated from (1.41) and (1.42).

For example, taking $k = 2$ it follows

$$b_{21} = -\frac{B_{12}}{B_{11}} \sqrt{\frac{B_{11}}{B_{11}B_{22} - B_{12}^2}}; \qquad b_{22} = \sqrt{\frac{B_{11}}{B_{11}B_{22} - B_{12}^2}}. \tag{1.43}$$

Now the transformed random vector

$$\left(\tilde{r}_{1\varepsilon}(\omega), \tilde{r}_{2\varepsilon}(\omega), \ldots, \tilde{r}_{n\varepsilon}(\omega)\right)$$

is considered. For the second moments we obtain

$$\begin{aligned} \langle \tilde{r}_{i\varepsilon}\tilde{r}_{j\varepsilon} \rangle &= \frac{1}{\varepsilon^m} \sum_{p=1}^{i} \sum_{q=1}^{j} b_{ip}b_{jq} \langle r_{p\varepsilon} r_{q\varepsilon} \rangle \\ &= \frac{1}{\varepsilon^m} \sum_{p=1}^{i} \sum_{q=1}^{j} b_{ip}b_{jq} \{H_1(m; p, q)\, \varepsilon^m + o(\varepsilon^m)\} \\ &= \delta_{ij} + O(\varepsilon) \end{aligned} \tag{1.44}$$

where (1.30) for $k = 2$ and (1.38) have to be taken into consideration; i.e. these second moments are orthonormal in the lowest order. We define

$$\begin{aligned} &\tilde{F}_{ij}(x) = \sum_{p=1}^{i} b_{ip}F_{pj}(x) \quad \text{for} \quad x \in \tilde{\mathcal{D}} \doteq \bigcup_{p=1}^{n} \mathcal{D}_p \\ \text{with} \quad &F_{pj}(x) \doteq 0 \quad \text{for} \quad x \in \tilde{\mathcal{D}} \setminus \mathcal{D}_p \end{aligned} \tag{1.45}$$

since

$$\tilde{r}_{i\varepsilon}(\omega) = \frac{1}{\sqrt{\varepsilon^m}} \sum_{p=1}^{i} b_{ip} \sum_{j=1}^{l} r_{pj\varepsilon}(\omega) = \frac{1}{\sqrt{\varepsilon^m}} \sum_{j=1}^{l} \sum_{p=1}^{i} \int_{\mathcal{D}_p} b_{ip}F_{pj}(x)\, f_{j\varepsilon}(x, \omega)\, \mathrm{d}x$$

and

$$\tilde{r}_{ij\varepsilon}(\omega) \doteq \int_{\tilde{\mathcal{D}}} \tilde{F}_{ij}(x)\, f_{j\varepsilon}(x, \omega)\, \mathrm{d}x \tag{1.46}$$

so that we obtain

$$\tilde{r}_{i\varepsilon}(\omega) = \frac{1}{\sqrt{\varepsilon^m}} \sum_{p=1}^{l} \tilde{r}_{ip\varepsilon}(\omega). \tag{1.47}$$

It is possible to apply the summarized results of Theorem 1.5 and these of (1.30), (1.32) to the random variables $\tilde{r}_{ij\varepsilon}(\omega)$, $\tilde{r}_{i\varepsilon}(\omega)$ substituted F_{ij} by $\tilde{F}_{ij}$. The coefficients relating to $\tilde{F}_{ij}$ are denoted by the symbol "$\sim$". In particular, by means of (1.30) and (1.44) we have

$$\tilde{H}_0(m; i, j) = \sum_{p,q=1}^{l} \tilde{H}_0(m; i, j; p, q) = \sum_{p,q=1}^{l} {}^2A_1(\tilde{F}_{ip}, \tilde{F}_{jq}) = \delta_{ij}. \tag{1.48}$$

Let $\tilde{p}(\tilde{u}_1, \ldots, \tilde{u}_n)$ denote the density of the distribution function of the random vector $(\tilde{r}_{1\varepsilon}(\omega), \ldots, \tilde{r}_{n\varepsilon}(\omega))$. We put

$$\tilde{p}(\tilde{u}_1, \ldots, \tilde{u}_n) = \frac{1}{\sqrt{2\pi}^n} \sum_{k_1, \ldots, k_n = 0}^{\infty} (-1)^k \frac{c_{k_1 \ldots k_n}}{k_1! \ldots k_n!}$$
$$\times \exp\left(-\frac{1}{2}(\tilde{u}_1^2 + \cdots + \tilde{u}_n^2)\right) H_{k_1}(\tilde{u})_1 \ldots H_{k_n}(\tilde{u}_n) \tag{1.49}$$

where $H_p(u)$, $p = 1, 2, \ldots$, denote the Chebyshev-Hermite polynomials and k is defined by $k \doteq k_1 + \cdots + k_n$.

These polynomials are defined by

$$H_p(u) \doteq (-1)^p \exp\left(\frac{1}{2} u^2\right) \frac{d^p}{du^p} \exp\left(-\frac{1}{2} u^2\right).$$

In the following considerations we need these polynomials up to the degree 6:

$$H_0(u) = 1, \quad H_1(u) = u, \quad H_2(u) = u^2 - 1,$$
$$H_3(u) = u^3 - 3u, \quad H_4(u) = u^4 - 6u^2 + 3, \quad H_5(u) = u^5 - 10u^3 + 15u, \tag{1.50}$$
$$H_6(u) = u^6 - 15u^4 + 45u^2 - 15.$$

The relation

$$\int_{-\infty}^{\infty} \exp\left(-\frac{1}{2} u^2\right) H_p(u)\, H_q(u)\, du = \sqrt{2\pi} p!\, \delta_{pq}, \qquad p, q = 1, 2, \ldots, \tag{1.51}$$

is fulfilled. The proof follows for $p < q$ from

$$\int_{-\infty}^{\infty} \exp\left(-\frac{1}{2} u^2\right) H_p(u)\, H_q(u)\, du = (-1)^q \int_{-\infty}^{\infty} H_p(u) \left(\exp\left(-\frac{1}{2} u^2\right)\right)^{(q)} du$$
$$= (-1)^{q+p} \int_{\infty-}^{\infty} H_p^{(p)}(u) \left(\exp\left(-\frac{1}{2} u^2\right)\right)^{(q-p)} du$$
$$= (-1)^{p+q} H_p^{(p)}(u) \left(\exp\left(-\frac{1}{2} u^2\right)\right)^{(q-p-1)} \Bigg|_{-\infty}^{\infty} = 0$$

since $H_p^{(p)} = \text{const}$, $q - p - 1 \geqq 0$ and for $p = q$ from

$$\int_{-\infty}^{\infty} \exp\left(-\frac{1}{2} u^2\right) H_p(u)\, H_p(u)\, du$$
$$= (-1)^{2p} \int_{-\infty}^{\infty} H_p^{(p)}(u) \exp\left(-\frac{1}{2} u^2\right) du = H_p^{(p)} \sqrt{2\pi}$$

if the relation $H_p^{(p)} = p!$ is shown. Using the definition of $H_p(u)$ we have

$$H_p'(u) = (-1)^p \exp\left(\frac{1}{2} u^2\right) \left\{u \frac{d^p}{du^p} + \frac{d^{p+1}}{du^{p+1}}\right\} \exp\left(-\frac{1}{2} u^2\right) = uH_p(u) - H_{p+1}(u).$$

For $p = 1$ it follows $H_1' = 1$ and assuming $H_p^{(p)} = p!$ then

$$H_{p+1}^{(p+1)} = \left(uH_p(u) - H_p'(u)\right)^{(p+1)} = uH_p^{(p+1)} + (p + 1)\, H_p^{(p)} - H_p^{(p+2)} = (p + 1)!.$$

From (1.49) and (1.51) it follows that

$$\int_{-\infty}^{\infty} \cdots \int_{-\infty}^{\infty} \tilde{p}(\tilde{u}_1, \ldots, \tilde{u}_n)\, H_{q_1}(\tilde{u}_1) \ldots H_{q_n}(\tilde{u}_n)\, \mathrm{d}\tilde{u}_1 \ldots \mathrm{d}\tilde{u}_n$$

$$= \frac{1}{\sqrt{2\pi}^n} \sum_{k_1,\ldots,k_n=0}^{\infty} (-1)^k \frac{c_{k_1\ldots k_n}}{k_1! \ldots k_n!} \prod_{s=1}^{n} \int_{-\infty}^{\infty} \exp\left(-\frac{1}{2}\, \tilde{u}_s^2\right) H_{k_s}(\tilde{u}_s)\, H_{q_s}(\tilde{u}_s)\, \mathrm{d}\tilde{u}_s$$

$$= (-1)^{q_1+\cdots+q_n}\, c_{q_1\ldots q_n}$$

and

$$c_{k_1\ldots k_n} = (-1)^k \int_{-\infty}^{\infty} \cdots \int_{-\infty}^{\infty} \tilde{p}(\tilde{u}_1, \ldots, \tilde{u}_n)\, H_{k_1}(\tilde{u}_1) \ldots H_{k_n}(\tilde{u}_n)\, \mathrm{d}\tilde{u}_1 \ldots \mathrm{d}\tilde{u}_n. \tag{1.52}$$

In particular,

$$c_{0\ldots0} = \int_{-\infty}^{\infty} \cdots \int_{-\infty}^{\infty} \tilde{p}(\tilde{u}_1, \ldots, \tilde{u}_n)\, \mathrm{d}\tilde{u}_1 \ldots \mathrm{d}\tilde{u}_n = 1,$$

$$c_{\underset{q}{0\ldots010\ldots0}} = -\int_{-\infty}^{\infty} \cdots \int_{-\infty}^{\infty} \tilde{u}_q \tilde{p}(\tilde{u}_1, \ldots, \tilde{u}_n)\, \mathrm{d}\tilde{u}_1 \ldots \mathrm{d}\tilde{u}_n = -\langle \tilde{r}_{q\varepsilon} \rangle = 0;$$

consequently

$$\tilde{p}(\tilde{u}_1, \ldots, \tilde{u}_n) = \frac{1}{\sqrt{2\pi}^n} \exp\left(-\frac{1}{2}\, (\tilde{u}_1^2 + \cdots + \tilde{u}_n^2)\right)$$

$$\times \left\{1 + \sum_{k_1,\ldots,k_n=0,k\geqq 2}^{\infty} (-1)^k \frac{c_{k_1,\ldots,k_n}}{k_1! \ldots k_n!} H_{k_1}(\tilde{u}_1) \ldots H_{k_n}(\tilde{u}_n)\right\}. \tag{1.53}$$

The characteristic function $\tilde{\psi}(t_1, \ldots, t_n)$ of $\left(\tilde{r}_{1\varepsilon}(\omega), \ldots, \tilde{r}_{n\varepsilon}(\omega)\right)$ is defined by

$$\tilde{\psi}(t_1, \ldots, t_n) = \int_{-\infty}^{\infty} \cdots \int_{-\infty}^{\infty} \exp\left(\mathrm{i} \sum_{s=1}^{n} t_s \tilde{u}_s\right) \tilde{p}(\tilde{u}_1, \ldots, \tilde{u}_n)\, \mathrm{d}\tilde{u}_1 \ldots \mathrm{d}\tilde{u}_n.$$

Hence, from (1.53),

$$\tilde{\psi}(t_1, \ldots, t_n) = \frac{1}{\sqrt{2\pi}^n} \left\{ \int_{-\infty}^{\infty} \cdots \int_{-\infty}^{\infty} \exp\left(\mathrm{i} \sum_{s=1}^{n} t_s \tilde{u}_s - \frac{1}{2} \sum_{s=1}^{n} \tilde{u}_s^2\right) \mathrm{d}\tilde{u}_1 \ldots \mathrm{d}\tilde{u}_n \right.$$

$$+ \sum_{k_1,\ldots,k_n=0,k\geqq 2}^{\infty} (-1)^k \frac{c_{k_1\ldots k_n}}{k_1! \ldots k_n!}$$

$$\left. \times \int_{-\infty}^{\infty} \cdots \int_{-\infty}^{\infty} \exp\left(\mathrm{i} \sum_{s=1}^{n} t_s \tilde{u}_s - \sum_{s=1}^{n} \frac{\tilde{u}_s^2}{2}\right) H_{k_1}(\tilde{u}_1) \ldots H_{k_n}(\tilde{u}_n)\, \mathrm{d}\tilde{u}_1 \ldots \mathrm{d}\tilde{u}_n \right\}.$$

Using

$$\frac{1}{\sqrt{2\pi}} \int_{-\infty}^{\infty} \exp\left(itu - \frac{1}{2}u^2\right) H_q(u)\, du = \frac{(-1)^q}{\sqrt{2\pi}} \int_{-\infty}^{\infty} \exp(itu) \left(\exp\left(-\frac{1}{2}u^2\right)\right)^{(q)} du$$

$$= \frac{1}{\sqrt{2\pi}} (it)^q \int_{-\infty}^{\infty} \exp\left(itu - \frac{1}{2}u^2\right) du = (it)^q \exp\left(-\frac{1}{2}t^2\right)$$

then we find that

$$\tilde{\psi}(t_1, \ldots, t_n) = \exp\left(-\frac{1}{2}\sum_{s=1}^{n} t_s^2\right)$$

$$\times \left\{1 + \sum_{k_1,\ldots,k_n=0,k\geqq 2}^{\infty} (-1)^k \frac{c_{k_1\ldots k_n}}{k_1!\ldots k_n!} (it_1)^{k_1} \ldots (it_n)^{k_n}\right\}. \qquad (1.54)$$

On the other hand, we also have the expression

$$\tilde{\psi}(t_1, \ldots, t_n) = \sum_{k_1,\ldots,k_n=0}^{\infty} \frac{1}{k_1!\ldots k_n!} \tilde{\alpha}_{k_1\ldots k_n} (it_1)^{k_1} \ldots (it_n)^{k_n}$$

where

$$\tilde{\alpha}_{k_1\ldots k_n} = \frac{\partial^{k_1+\cdots+k_n}}{\partial(it_1)^{k_1}\ldots\partial(it_n)^{k_n}} \tilde{\psi}(t_1, \ldots, t_n)\Big|_{t_1=\cdots=t_n=0}$$

$$= \int_{-\infty}^{\infty} \cdots \int_{-\infty}^{\infty} \tilde{u}_1^{k_1} \ldots \tilde{u}_n^{k_n} \tilde{p}(\tilde{u}_1, \ldots, \tilde{u}_n)\, d\tilde{u}_1 \ldots d\tilde{u}_n = \langle \tilde{r}_{1\varepsilon}^{k_1} \ldots \tilde{r}_{n\varepsilon}^{k_n} \rangle$$

and

$$\tilde{\psi}(t_1, \ldots, t_n) = 1 + \sum_{k_1,\ldots,k_n=0,k\geqq 2}^{\infty} \frac{1}{k_1!\ldots k_n!} \tilde{\alpha}_{k_1\ldots k_n} (it_1)^{k_1} \ldots (it_n)^{k_n} \qquad (1.55)$$

because of

$$\tilde{\alpha}_{0\ldots 0} = 1 \quad \text{and} \quad \tilde{\alpha}_{0\ldots 0\underset{q}{1}0\ldots 0} = \langle \tilde{r}_{q\varepsilon} \rangle = 0.$$

Taking into consideration (1.30) and (1.32) we can write

$$\tilde{\alpha}_{k_1\ldots k_n} = \tilde{\alpha}^{(1/2)}_{k_1\ldots k_n} \varepsilon^{m/2} + O(\varepsilon^{m/2+1}) \quad \text{for} \quad k \geqq 3 \text{ odd}$$

and

$$\tilde{\alpha}_{k_1\ldots k_n} = \tilde{\alpha}^{(0)}_{k_1\ldots k_n} + \tilde{\alpha}^{(1)}_{k_1\ldots k_n} \varepsilon + O(\varepsilon^2) \quad \text{for} \quad k \geqq 2 \text{ even}.$$

Substituting these terms into the characteristic function gives the result

$$\tilde{\psi}(t_1, \ldots, t_n) = 1 + \sum_{\substack{k_1,\ldots,k_n=0 \\ k\geqq 2,\ \text{even}}}^{\infty} \frac{1}{k_1!\ldots k_n!} \tilde{\alpha}^{(0)}_{k_1\ldots k_n} (it_1)^{k_1} \ldots (it_n)^{k_n}$$

$$+ \sum_{\substack{k_1,\ldots,k_n=0 \\ k\geqq 2,\ \text{even}}}^{\infty} \frac{1}{k_1!\ldots k_n!} \tilde{\alpha}^{(1)}_{k_1\ldots k_n} (it_1)^{k_1} \ldots (it_n)^{k_n} \varepsilon + O(\varepsilon^2)$$

$$+ \sum_{\substack{k_1,\ldots,k_n=0 \\ k\geqq 3,\ \text{odd}}}^{\infty} \frac{1}{k_1!\ldots k_n!} \tilde{\alpha}^{(1/2)}_{k_1\ldots k_n} (it_1)^{k_1} \ldots (it_n)^{k_n} \varepsilon^{m/2} + O(\varepsilon^{m/2+1}). \qquad (1.56)$$

In order to short the following considerations we introduce the notations

$$ {}^p\hat{\bar{A}}_q(\{1, \ldots, p\}) \doteq \sum_{j_1,\ldots,j_p=1}^{l} {}^p\bar{A}_q(\{1, \ldots, p\}) = \sum_{j_1,\ldots,j_p=1}^{l} {}^pA_q(F_{i_1j_1}, \ldots, F_{i_pj_p}), \tag{1.57} $$

$$ \begin{aligned} &\hat{\bar{A}}_{pq}(\{1, \ldots, p\}, \{p+1, \ldots, p+q\}) \\ &\doteq \sum_{j_1,\ldots,j_p,j_{p+1},\ldots,j_{p+q}=1}^{l} \bar{A}_{pq}(\{1, \ldots, p\}, \{p+1, \ldots, p+q\}) \\ &= \sum_{j_1,\ldots,j_{p+q}=1}^{l} A_{pq}(F_{i_1j_1}, \ldots, F_{i_pj_p}; F_{i_{p+1}j_{p+1}}, \ldots, F_{i_{p+q}j_{p+q}}). \end{aligned} \tag{1.58} $$

It follows

$$ \begin{aligned} \hat{H}_0(m; i_1, \ldots, i_k) &= \sum_{j_1,\ldots,j_k=1}^{l} H_0(m; i_1, \ldots, i_k; j_1, \ldots, j_k) \\ &= \sum_{j_1,\ldots,j_k=1}^{l} \sum_{\{s_1,t_1\},\ldots,\{s_{k/2},t_{k/2}\}} \prod_{q=1}^{k/2} {}^2A_1(\bar{F}_{s_q}, \bar{F}_{t_q}) \\ &= \sum_{\{s_1,t_1\},\ldots,\{s_{k/2},t_{k/2}\}} \prod_{q=1}^{k/2} \sum_{j_{s_q},j_{t_q}=1}^{l} {}^2A_1(F_{i_{s_q}j_{s_q}}, F_{i_{t_q}j_{t_q}}) \\ &= \sum_{\{s_1,t_1\},\ldots,\{s_{k/2},t_{k/2}\}} \prod_{q=1}^{k/2} {}^2\hat{\bar{A}}_1(\{s_q, t_q\}) \end{aligned} \tag{1.59} $$

using (1.24). For $\hat{H}_2(m; i_1, \ldots, i_k)$, $\hat{H}_1(m; i_1, \ldots, i_k)$ we obtain similar formulas as (1.25) and (1.27) where the terms ${}^p\bar{A}_q$, $\bar{A}_{pq}$ only have to be substituted by ${}^p\hat{\bar{A}}_q$, $\hat{\bar{A}}_{pq}$, respectively. By means of these notations (1.48) can be written as

$$ \tilde{\hat{H}}_0(m; i_1, i_2) = {}^2\tilde{\hat{\bar{A}}}_1(\{1, 2\}) = \sum_{j_1,j_2=1}^{l} {}^2A_1(\tilde{\bar{F}}_1, \tilde{\bar{F}}_2) = \delta_{i_1i_2}. \tag{1.60} $$

Now we deal with the summand in (1.55) containing $\alpha^{(0)}_{k_1\ldots k_n}$. Using the above remarks $\tilde{\alpha}^{(0)}_{k_1\ldots k_n}$ has the form

$$ \begin{aligned} \tilde{\alpha}^{(0)}_{k_1\ldots k_n} &= \tilde{\hat{H}}_0(m; \mathcal{V}_{k_1\ldots k_n}) \\ &= \sum_{\{s_1,t_1\},\ldots,\{s_{k/2},t_{k/2}\}} \prod_{q=1}^{k/2} {}^2\tilde{\hat{\bar{A}}}_1(\{s_q, t_q\}) = \sum_{\{s_1,t_1\},\ldots,\{s_{k/2},t_{k/2}\}} \prod_{q=1}^{k/2} \delta_{s_qt_q} \end{aligned} $$

where the sum is taken over all nonequivalent separations

$$ \{s_1, t_1\}, \ldots, \{s_{k/2}, t_{k/2}\} \quad \text{of} \quad \{1, \ldots, k\} \quad \text{as to} \quad (i_1, \ldots, i_k) \doteq \mathcal{V}_{k_1\ldots k_n} $$

with $\quad \mathcal{V}_{k_1\ldots k_n} = \Bigl(1, \ldots, \overset{k_1}{1}, 2, \ldots, \overset{k_2}{2}, \ldots, n, \ldots, \overset{k_n}{n}\Bigr).$

Furthermore, defining

$$ e_p \doteq \begin{cases} 0 & \text{for } p \text{ odd}, \\ \dfrac{p!}{2^{p/2}(p/2)!} & \text{for } p \text{ even} \end{cases} \tag{1.61} $$

we obtain

$$\tilde{\alpha}^{(0)}_{k_1 \ldots k_n} = \begin{cases} 0 & \text{if not all } k_s \text{ even}, \\ \prod_{s=1}^{n} e_{k_s} & \text{if all } k_s,\ s = 1, 2, \ldots, n, \text{ even} \end{cases}$$

and thus

$$1 + \sum_{\substack{k_1, \ldots, k_n = 0 \\ k \geq 2,\ \text{even}}}^{\infty} \frac{1}{k_1! \ldots k_n!} \tilde{\alpha}^{(0)}_{k_1 \ldots k_n} (it_1)^{k_1} \ldots (it_n)^{k_n}$$

$$= \sum_{\substack{p_1, \ldots, p_n = 0 \\ \sum_{s=1}^{n} p_s \geq 0}}^{\infty} \frac{1}{p_1! \ldots p_n!} \left(-\frac{1}{2} t_1^2\right)^{p_1} \ldots \left(-\frac{1}{2} t_n^2\right)^{p_n} = \exp\left(-\frac{1}{2} \sum_{s=1}^{n} t_s^2\right). \tag{1.62}$$

We turn to the coefficient $\tilde{\alpha}^{(1/2)}_{k_1 \ldots k_n}$. (1.32) and (1.27) lead to

$$\tilde{\alpha}^{(1/2)}_{k_1 \ldots k_n} = \tilde{H}_1(m; \mathcal{V}_{k_1 \ldots k_n})$$

$$= \sum_{\{s_1, t_1, r_1\}, \{s_2, t_2\}, \ldots, \left\{s_{\frac{k-1}{2}}, t_{\frac{k-1}{2}}\right\}} {}^3\tilde{\hat{A}}_2(\{s_1, t_1, r_1\}) \prod_{q=2}^{\frac{k-1}{2}} \delta_{i_{s_q} i_{t_q}} \tag{1.63}$$

where we have to summarize over all nonequivalent separations

$$\{s_1, t_1, r_1\}, \{s_2, t_2\}, \ldots, \left\{s_{\frac{k-1}{2}}, t_{\frac{k-1}{2}}\right\} \quad \text{of} \quad \{1, \ldots, k\} \text{ as to } (i_1, \ldots, i_k) = \mathcal{V}_{k_1 \ldots k_n}$$

and k is odd. First, we select $\{s_1, t_1, r_1\}$ from $\{1, \ldots, k_s\}$ as to $(i_1, \ldots, i_{k_s}) \doteq \mathcal{V}_{k_s}$, $\mathcal{V}_{k_s} = (\overset{k_s}{s, \ldots, s})$ and obtain

$$\sum_{s=1}^{n} \binom{k_s}{3} e_{k_s - 3} \prod_{\substack{q=1 \\ q \neq s}}^{n} e_{k_q} \, {}^3\tilde{A}_2' (s, s, s)$$

from (1.63). The terms ${}^p\hat{A}_q'(i_1, \ldots, i_p)$ are defined by ${}^p\hat{A}_q'(i_1, \ldots, i_p) = {}^p\hat{\hat{A}}_q(\{1, \ldots, p\})$. If two elements of $\{s_1, t_1, r_1\}$ are selected from $\{1, \ldots, k_s\}$ as to $(i_1, \ldots, i_{k_s}) = \mathcal{V}_{k_s}$ and one element from $\{1, \ldots, k_t\}$ as to $(i_1, \ldots, i_{k_t}) = \mathcal{V}_{k_t}$ then it follows from (1.63)

$$\sum_{\substack{s,t=1 \\ s \neq t}}^{n} \binom{k_s}{2} k_t e_{k_s - 2} e_{k_t - 1} \prod_{\substack{q=1 \\ q \neq s,t}}^{n} e_{k_q} \, {}^3\tilde{A}_2'(s, s, t)$$

and for s_1 from $\{1, \ldots, k_s\}$ as to $(i_1, \ldots, i_{k_s}) = \mathcal{V}_{k_s}$, t_1 from $\{1, \ldots, k_t\}$ as to $(i_1, \ldots, i_{k_t})$ $= \mathcal{V}_{k_t}$, r_1 from $\{1, \ldots, k_r\}$ as to $(i_1, \ldots, i_{k_r}) = \mathcal{V}_{k_r}$ the term

$$\sum_{\substack{s,t,r=1 \\ s<t<r}}^{n} \prod_{q=s,t,r} k_q e_{k_q - 1} \prod_{\substack{q=1 \\ q \neq s,t,r}}^{n} e_{k_q} \, {}^3\tilde{A}_2'(s, t, r).$$

We find that

$$G \doteq \sum_{\substack{k_1,\dots,k_n=0\\ k\geqq 3,\ \text{odd}}}^{\infty} \frac{1}{k_1!\dots k_n!} \tilde{\alpha}^{(1/2)}_{k_1\dots k_n} (it_1)^{k_1} \dots (it_n)^{k_n}$$

$$= \sum_{s=1}^{n} {}^3\tilde{\bar{A}}_2'(s, s, s) \sum_{\substack{k_1,\dots,k_n=0\\ k\geqq 3,\ \text{odd}}}^{\infty} \binom{k_s}{3} e_{k_s-3} \prod_{\substack{q=1\\ q\neq s}}^{n} e_{k_q} \prod_{q=1}^{n} \frac{1}{k_q!} (it_q)^{k_q}$$

$$+ \sum_{\substack{s,t=1\\ s\neq t}}^{n} {}^3\tilde{\bar{A}}_2'(s, s, t) \sum_{\substack{k_1,\dots,k_n=0\\ k\geqq 3,\ \text{odd}}}^{\infty} \binom{k_s}{2} k_t e_{k_s-2} e_{k_t-1} \prod_{\substack{q=1\\ q\neq s,t}}^{n} e_{k_q} \prod_{q=1}^{n} \frac{(it_q)^{k_q}}{k_q!}$$

$$+ \sum_{\substack{s,t,r=1\\ s<t<r}}^{n} {}^3\tilde{\bar{A}}_2(s, t, r) \sum_{\substack{k_1,\dots,k_n=0\\ k\geqq 3,\ \text{odd}}}^{\infty} \prod_{q=s,t,r} k_q e_{k_q-1} \prod_{\substack{q=1\\ q\neq s,t,r}}^{n} e_{k_q} \prod_{q=1}^{n} \frac{1}{k_q!} (it_q)^{k_q}.$$

With the help of $e_j = 0$ for j odd we can put

$$k_q = 2p_q \quad \text{for} \quad q = 1, 2, \dots, n, \quad q \neq s; \quad k_s - 3 = 2p_s$$

as to the first summand and adequately as to the other summands. Then it follows

$$G = \left[\frac{1}{6} \sum_{s=1}^{n} {}^3\tilde{\bar{A}}_2'(s, s, s)\, (it_s)^3 + \frac{1}{2} \sum_{\substack{s,t=1\\ s\neq t}}^{n} {}^3\tilde{\bar{A}}_2'(s, s, t)\, (it_s)^2\, (it_t)\right.$$

$$\left. + \sum_{\substack{s,t,r=1\\ s<t<r}}^{n} {}^3\tilde{\bar{A}}_2'(s, t, r)\, (it_s)\, (it_t)\, (it_r)\right] \sum_{p_1,\dots,p_n=0}^{\infty} \prod_{q=1}^{n} \frac{1}{p_q!} \left(-\frac{1}{2} t_q^2\right)^{p_q}$$

and

$$\sum_{\substack{k_1,\dots,k_n=0\\ k\geqq 3,\ \text{odd}}}^{\infty} \frac{1}{k_1!\dots k_n!} \tilde{\alpha}^{(1/2)}_{k_1\dots k_n} (it_1)^{k_1} \dots (it_n)^{k_n}$$

$$= \exp\left(-\frac{1}{2} \sum_{q=1}^{n} t_q^2\right) \frac{1}{6} \sum_{s,t,r=1}^{n} {}^3\tilde{\bar{A}}_2'(s, t, r)\, (it_s)\, (it_t)\, (it_r) \tag{1.64}$$

where for symmetric terms a_{str} as to (s, t, r) the relation

$$\sum_{s,t,r=1}^{n} a_{str} = 6 \sum_{\substack{s,t,r=1\\ s<t<r}}^{n} a_{str} + 3 \sum_{\substack{s,t=1\\ s\neq t}}^{n} a_{sst} + \sum_{s=1}^{n} a_{sss}$$

is taken into consideration.

Now we turn to the summand in (1.55) which contains $\tilde{\alpha}^{(1)}_{k_1\dots k_n}$. Using (1.23) and (1.25) we have for $m = 1$

$$\tilde{\alpha}^{(1)}_{k_1\dots k_n} = \tilde{\bar{H}}_2(\mathcal{V}_{k_1\dots k_n})$$

$$= \sum_{\{s_1,t_1\},\dots,\{s_{k/2},t_{k/2}\}} \left\{\sum_{j=1}^{k/2} {}^2\tilde{\hat{\bar{A}}}_2(\{s_j, t_j\}) \prod_{\substack{q=1\\ q\neq j}}^{k/2} \delta_{i_{s_q} i_{t_q}} - \sum_{\{q_1,q_2\},\{q_3,\dots,q_{k/2}\}} \tilde{\hat{\bar{A}}}_{22}(\{s_{q_1}, t_{q_1}\}; \{s_{q_2}, t_{q_2}\}) \prod_{j=3}^{k/2} \delta_{i_{s_q} i_{t_q}}\right\}$$

$$+ \sum_{\mathcal{J}_1} \prod_{q=1}^{2} {}^3\tilde{\hat{\bar{A}}}_2(\{s_q, t_q, r_q\}) \prod_{q=3}^{k/2-1} \delta_{i_{s_q} i_{t_q}} + \sum_{\mathcal{J}_2} {}^4\tilde{\hat{\bar{A}}}_3(\{s_1, t_1, r_1, u_1\}) \prod_{q=2}^{k/2-1} \delta_{i_{s_q} i_{t_q}}; \tag{1.65}$$

$$\mathcal{J}_1 = \{\{s_1, t_1, r_1\}, \{s_2, t_2, r_2\}, \{s_3, t_3\}, \dots, \{s_{k/2-1}, t_{k/2-1}\}\},$$

$$\mathcal{J}_2 = \{\{s_1, t_1, r_1, u_1\}, \{s_2, t_2\}, \dots, \{s_{k/2-1}, t_{k/2-1}\}\}$$

where (1.60) was used and the sums have to be taken over all nonequivalent separations of $\{1, \ldots, k\}$ as to $(i_1, \ldots, i_k) = \mathcal{V}_{k_1\ldots k_n}$. Let us deal with the first summand in (1.65). Thus

$$\tilde{\alpha}^{(1,1)}_{k_1\ldots k_n} \doteq \sum_{\{s_1,t_1\},\ldots,\{s_{k/2},t_{k/2}\}} \sum_{j=1}^{k/2} {}^2\tilde{\hat{A}}_2(\{s_j, t_j\}) \prod_{\substack{q=1\\q\neq j}}^{k/2} \delta_{i_{s_q} i_{t_q}} = \sum_{j=1}^{k/2} \sum_{\{s_1,t_1\},\ldots,\{s_{k/2},t_{k/2}\}} {}^2\tilde{\hat{A}}_2(\{s_j, t_j\}) \prod_{\substack{q=1\\q\neq j}}^{k/2} \delta_{i_{s_q} i_{t_q}}$$

and consequently

$$\tilde{\alpha}^{(1,1)}_{k_1\ldots k_n} = \sum_{s=1}^{n} {}^2\tilde{A}_2'(s, s) \binom{k_s}{2} e_{k_s-2} \prod_{\substack{q=1\\q\neq s}}^{n} e_{k_q} + \sum_{\substack{s,t=1\\s<t}}^{n} {}^2\tilde{A}_2'(s, t)\, k_s k_t e_{k_s-1} e_{k_t-1} \prod_{\substack{q=1\\q\neq s,t}}^{n} e_{k_q}.$$

Furthermore, we obtain

$$\sum_{\substack{k_1,\ldots,k_n=0\\k\geq 2,\text{ even}}} \frac{1}{k_1!\ldots k_n!} \tilde{\alpha}^{(1,1)}_{k_1\ldots k_n} (\mathrm{i}t_1)^{k_1} \ldots (\mathrm{i}t_n)^{k_n}$$

$$= \left[\frac{1}{2} \sum_{s=1}^{n} {}^2\tilde{A}_2'(s, s)\, (\mathrm{i}t_s)^2 + \sum_{\substack{s,t=1\\s<t}}^{n} {}^2\tilde{A}_2'(s, t)\, (\mathrm{i}t_s)\, (\mathrm{i}t_t)\right] \exp\left(-\frac{1}{2} \sum_{q=1}^{n} t_q^2\right)$$

$$= \exp\left(-\frac{1}{2} \sum_{q=1}^{n} t_q^2\right) \frac{1}{2} \sum_{s,t=1}^{n} {}^2\tilde{A}_2'(s, t)\, (\mathrm{i}t_s)\, (\mathrm{i}t_t). \tag{1.66}$$

The second and fourth summand in (1.65) have the form

$$\tilde{\alpha}^{(1,24)}_{k_1\ldots k_n} = \sum_{\mathcal{J}_2} {}^4\tilde{\hat{A}}_3(\{s_1, t_1, r_1, u_1\}) \prod_{q=2}^{k/2-1} \delta_{i_{s_q} i_{t_q}}$$

$$- \sum_{\{s_1,t_1\},\ldots,\{s_{k/2},t_{k/2}\}} \sum_{\{q_1,q_2\},\{q_3,\ldots,q_{k/2}\}} \prod_{j=3}^{k/2} \delta_{i_{s_{q_j}} i_{t_{q_j}}} \tilde{\hat{A}}_{22}(\{s_{q_1}, t_{q_1}\}; \{s_{q_2}, t_{q_2}\}) \tag{1.67}$$

and we define

$${}^4\tilde{\hat{\underline{A}}}_3(\{r, s, t, u\}) \doteq {}^4\tilde{\hat{A}}_3(\{r, s, t, u\}) - \tilde{\hat{A}}_{22}(\{r, s\}, \{t, u\})$$

$$- \tilde{\hat{A}}_{22}(\{r, t\}; \{s, u\}) - \tilde{\hat{A}}_{22}(\{r, u\}; \{s, t\}). \tag{1.68}$$

It is clear that ${}^4\tilde{\hat{\underline{A}}}_3$ is symmetric with respect to r, s, t, u; i.e.

$${}^4\tilde{\hat{\underline{A}}}_3(\{\pi(r), \pi(s), \pi(t), \pi(u)\}) = {}^4\tilde{\hat{\underline{A}}}_3(\{r, s, t, u\})$$

for an arbitrary permutation π of the indices r, s, t, u. Thus

$$\tilde{\alpha}^{(1,23)}_{k_1\ldots k_n} = \sum_{\mathcal{J}} {}^4\tilde{\hat{\underline{A}}}_3(\{s_1, t_1, r_1, u_1\}) \prod_{q=2}^{k/2-1} \delta_{i_{s_q} i_{t_q}} \tag{1.69}$$

with

$$\mathcal{J} = \{\{s_1, t_1, r_1, u_1\}, \{s_2, t_2\}, \ldots, \{s_{k/2-1}, t_{k/2-1}\}\}.$$

The proof of this formula follows from the following considerations:

The separations

$$\{\{s_1, t_1\}, \{s_2, t_2\}, \{s_3, t_3\}, \ldots, \{s_{k/2}, t_{k/2}\}\};$$
$$\{\{s_1, s_2\}, \{t_1, t_2\}, \{s_3, t_3\}, \ldots, \{s_{k/2}, t_{k/2}\}\};$$
$$\{\{s_1, t_2\}, \{t_1, s_2\}, \{s_3, t_3\}, \ldots, \{s_{k/2}, t_{k/2}\}\}$$

are assigned as the separation

$$\{\{s_1, t_1, s_2, t_2\}, \{s_3, t_3\}, \ldots, \{s_{k/2}, t_{k/2}\}\}.$$

For these separations we have the summand ${}^4\hat{\tilde{\underline{A}}}_3(\{s_1, t_1, s_2, t_2\})$. So we have taken into consideration one summand of the first sum in (1.67) and three different summands of the second sum. The first sum in (1.67) has

$$\binom{k}{4} \frac{(k-4)!}{2^{(k-4)/2}\left(\dfrac{k-4}{2}\right)!} = \frac{k!}{3 \cdot 2^{k/2+1}\left(\dfrac{k}{2}-2\right)!}$$

summands and the second sum

$$\frac{k!}{2^{k/2}\left(\dfrac{k}{2}\right)!}\binom{k/2}{2} = 3 \cdot \frac{k!}{3 \cdot 2^{k/2+1}\left(\dfrac{k}{2}-2\right)!}$$

summands so that (1.69) is proved.

For the investigation of the sum in (1.69) we find that

$$\tilde{\alpha}^{(1,24)}_{k_1 \ldots k_n} = \sum_{s=1}^{n} {}^4\tilde{\tilde{\underline{A}}}{}_3'(s, s, s, s) \binom{k_s}{4} e_{k_s-4} \prod_{\substack{q=1\\ q\neq s}}^{n} e_{k_q}$$
$$+ \sum_{\substack{s,t=1\\ s\neq t}}^{n} {}^4\tilde{\tilde{\underline{A}}}{}_3'(s, s, s, t) \binom{k_s}{3} k_t e_{k_s-3} e_{k_t-1} \prod_{\substack{q=1\\ q\neq t,s}}^{n} e_{k_q}$$
$$+ \sum_{\substack{s,t=1\\ s<t}}^{n} {}^4\tilde{\tilde{\underline{A}}}{}_3'(s, s, t, t) \binom{k_s}{2}\binom{k_t}{2} e_{k_s-2} e_{k_t-2} \prod_{\substack{q=1\\ q\neq s,t}}^{n} e_{k_q}$$
$$+ \sum_{\substack{s,t,r=1\\ s\neq t,r\\ t<r}}^{n} {}^4\tilde{\tilde{\underline{A}}}{}_3'(s, s, t, r) \binom{k_s}{2} k_t k_r e_{k_s-1} e_{k_t-1} e_{k_r-1} \prod_{\substack{q=1\\ q\neq s,t,r}}^{n} e_{k_q}$$
$$+ \sum_{\substack{s,t,r,u=1\\ s<t<r<u}}^{n} {}^4\tilde{\tilde{\underline{A}}}{}_3'(s, t, r, u) \prod_{q=s,t,r,u} k_q e_{k_q-1} \prod_{\substack{q=1\\ q\neq s,t,r,u}}^{n} e_{k_q}$$

if in the first summand s_1, t_1, r_1, u_1 is taken from $\{1, \ldots, k_s\}$ as to $(i_1, \ldots, i_{k_s}) = \mathcal{V}_{k_s}$, in the second summand s_1, t_1, r_1 is taken from $\{1, \ldots, k_s\}$ as to $(i_1, \ldots, i_{k_s}) = \mathcal{V}_{k_s}$ and u_1 from $\{1, \ldots, k_t\}$ as to $(i_1, \ldots, i_{k_t}) = \mathcal{V}_{k_t}$ in the third summand s_1, t_1 is taken from $\{1, \ldots, k_s\}$ as to $(i_1, \ldots, i_{k_s}) = \mathcal{V}_{k_s}$ and r_1, u_1 from $\{1, \ldots, k_t\}$ as to $(i_1, \ldots, i_{k_t}) = \mathcal{V}_{k_t}$ and in the other summands we have a similar situation. The calculation of the corresponding term in

(1.55) leads to

$$\sum_{\substack{k_1,\ldots,k_n=1\\ k\geq 2,\ \text{even}}}^{\infty} \frac{1}{k_1!\ldots k_n!}\, \tilde{\alpha}^{(1,24)}_{k_1\ldots k_n}(\mathrm{i}t_1)^{k_1}\ldots(\mathrm{i}t_n)^{k_n}$$

$$= \exp\left(-\frac{1}{2}\sum_{q=1}^{n} t_q^2\right)\left\{\frac{1}{24}\sum_{s=1}^{n} {}^4\underline{\tilde{\tilde{A}}}'_3(s,s,s,s)\,(\mathrm{i}t_s)^4 + \frac{1}{6}\sum_{\substack{s,t=1\\ s\neq t}}^{n} {}^4\underline{\tilde{\tilde{A}}}'_3(s,s,s,t)\,(\mathrm{i}t_s)^3\,(\mathrm{i}t_t)\right.$$

$$+\frac{1}{4}\sum_{\substack{s,t=1\\ s<t}}^{n} {}^4\underline{\tilde{\tilde{A}}}'_3(s,s,t,t)\,(\mathrm{i}t_s)^2\,(\mathrm{i}t_t)^2 + \frac{1}{2}\sum_{\substack{s,t,r=1\\ s\neq t,r;\,t<r}}^{n} {}^4\underline{\tilde{\tilde{A}}}'_3(s,s,t,r)\,(\mathrm{i}t_s)^2\,(\mathrm{i}t_t)\,(\mathrm{i}t_r)$$

$$\left.+\sum_{\substack{s,t,r,u=1\\ s<t<r<u}}^{n} {}^4\underline{\tilde{\tilde{A}}}'_3(s,t,r,u)\,(\mathrm{i}t_s)\,(\mathrm{i}t_t)\,(\mathrm{i}t_r)\,(\mathrm{i}t_u)\right\}$$

$$= \exp\left(-\frac{1}{2}\sum_{q=1}^{n} t_q^2\right)\frac{1}{24}\sum_{s,t,r,u=1}^{n} {}^4\underline{\tilde{\tilde{A}}}'_3(s,t,r,u)\,(\mathrm{i}t_s)\,(\mathrm{i}t_t)\,(\mathrm{i}t_r)\,(\mathrm{i}t_u) \tag{1.70}$$

if the relation

$$\sum_{s,t,r,u=1}^{n} a_{stru} = 24\sum_{\substack{s,t,r,u=1\\ s<t<r<u}}^{n} a_{stru} + 12\sum_{\substack{s,t,r=1\\ s\neq t,r;\,t<r}}^{n} a_{sstr} + 6\sum_{\substack{s,t=1\\ s<t}}^{n} a_{sstt} + 4\sum_{\substack{s,t=1\\ s\neq t}}^{n} a_{ssst} + \sum_{s=1}^{n} a_{ssss}$$

is used for symmetric terms a_{stru} as to (s, t, r, u).

Finally, we turn to the third summand in (1.65)

$$\tilde{\alpha}^{(1,3)}_{k_1\ldots k_n} \doteq \sum_{\mathcal{J}} {}^3\tilde{\tilde{A}}_2(\{s_1,t_1,r_1\})\,{}^3\tilde{\tilde{A}}_2(\{s_2,t_2,r_2\})\prod_{q=3}^{\frac{k}{2}-1}\delta_{i_{s_q}i_{t_q}}$$

where

$$\mathcal{J} \doteq \left\{\{s_1,t_1,r_1\},\{s_2,t_2,r_2\},\{s_3,t_3\},\ldots,\{s_{k/2-1},t_{k/2-1}\}\right\}.$$

$\mathcal{W}_{s_1\ldots s_p}$ is defined by

$$\mathcal{W}_{\underbrace{s\ldots s}_{p}} = \mathcal{V}_{k_1\ldots k_s-p\ldots k_n};\qquad \mathcal{W}_{\underbrace{s\ldots s}_{p}\underbrace{t\ldots t}_{q}} = \mathcal{V}_{k_1\ldots k_s-p\ldots k_t-q\ldots k_n}$$

and analogously in the other cases. k'_q denotes the number of "q" contained in $\mathcal{W}_{s_1\ldots s_p}$. Furthermore, we define

$${}^2P_1(i_1,\ldots,i_k) = \sum_{\{s_1,t_1\},\ldots,\{s_{k/2},t_{k/2}\}}\ \prod_{q=1}^{k/2}\delta_{i_{s_q}i_{t_q}}$$

for k even. Thus

$$\tilde{\alpha}^{(1,3)}_{k_1\ldots k_n} = \frac{1}{2}\left[\sum_{\substack{s,t,r=1\\ s<t<r}}^{n} {}^3\tilde{\tilde{A}}'_2(s,t,r)\,k_sk_tk_r\left(\sum_{\substack{u,v,w=1\\ u<v<w}}^{n} {}^3\tilde{\tilde{A}}'_2(u,v,w)\,k'_uk'_vk'_w\,{}^2P_1(\mathcal{W}_{struvw})\right.\right.$$

$$\left.+\sum_{\substack{u,v=1\\ u\neq v}}^{n} {}^3\tilde{\tilde{A}}'_2(u,u,v)\binom{k'_u}{2}k'_v\,{}^2P_1(\mathcal{W}_{struuv}) + \sum_{u=1}^{n} {}^3\tilde{\tilde{A}}'_2(u,u,u)\binom{k'_u}{3}\,{}^2P_1(\mathcal{W}_{struuu})\right)$$

$$+ \sum_{\substack{s,t=1\\ s\neq t}}^{n} {}^3\tilde{\tilde{A}}_2'(s,s,t) \binom{k_s}{2} k_t \Bigg(\sum_{\substack{u,v,w=1\\ u<v<w}}^{n} {}^3\tilde{\tilde{A}}_2'(u,v,w)\, k_u' k_v' k_w' \, {}^2P_1(\mathscr{W}_{sstuvw})$$

$$+ \sum_{\substack{u,v=1\\ u\neq v}}^{n} {}^3\tilde{\tilde{A}}_2'(u,u,v) \binom{k_u'}{2} k_v'^2 P_1(\mathscr{W}_{sstuuv}) + \sum_{u=1}^{n} {}^3\tilde{\tilde{A}}_2'(u,u,u) \binom{k_u'}{3} {}^2P_1(\mathscr{W}_{sstuuu}) \Bigg)$$

$$+ \sum_{s=1}^{n} {}^3\tilde{\tilde{A}}_2'(s,s,s) \binom{k_s}{3} \Bigg(\sum_{\substack{u,v,w=1\\ u<v<w}}^{n} {}^3\tilde{\tilde{A}}_2'(u,v,w)\, k_u' k_v' k_w' \, {}^2P_1(\mathscr{W}_{sssuvw})$$

$$+ \sum_{\substack{u,v=1\\ u\neq v}}^{n} {}^3\tilde{\tilde{A}}_2'(u,u,v) \binom{k_u'}{2} k_v'^2 P_1(\mathscr{W}_{sssuuv}) + \sum_{u=1}^{n} {}^3\tilde{\tilde{A}}_2'(u,u,u) \binom{k_u'}{3} {}^2P_1(\mathscr{W}_{sssuuu}) \Bigg) \Bigg]. \tag{1.71}$$

It follows

$$\sum_{\substack{k_1,\ldots,k_n=0\\ k\geqq 2;\ \text{even}}}^{\infty} \frac{1}{k_1!\ldots k_n!} (it_1)^{k_1} \ldots (it_n)^{k_n}$$

$$\times \Bigg\{ \sum_{\substack{s,t,r=1\\ s<t<r}}^{n} {}^3\tilde{\tilde{A}}_2'(s,t,r)\, k_s k_t k_r \sum_{\substack{u,v,w=1\\ u<v<w}}^{n} {}^3\tilde{\tilde{A}}_2'(u,v,w)\, k_u' k_v' k_w' \, {}^2P_1(\mathscr{W}_{struvw}) \Bigg\}$$

$$= \sum_{\substack{s,t,r=1\\ s<t<r}}^{n} \sum_{\substack{u,v,w=1\\ u<v<w}}^{n} {}^3\tilde{\tilde{A}}_2'(s,t,r)\, {}^3\tilde{\tilde{A}}_2'(u,v,w)$$

$$\times \sum_{\substack{k_1,\ldots,k_n=0\\ k\geqq 2;\ \text{even}}} k_s k_t k_r k_u' k_v' k_w' \frac{(it_1)^{k_1} \ldots (it_n)^{k_n}}{k_1!\ldots k_n!} \, {}^2P_1(\mathscr{W}_{struvw}).$$

In the case of different numbers s, t, r and u, v, w we have

$$\sum_{\substack{k_1,\ldots,k_n=0\\ k\geqq 2;\ \text{even}}}^{\infty} k_s k_t k_r k_u' k_v' k_w' \frac{1}{k_1!\ldots k_n!} (it_1)^{k_1} \ldots (it_n)^{k_n} \, {}^2P_1(\mathscr{W}_{struvw})$$

$$= \sum_{p_1,\ldots,p_n=0}^{\infty} \prod_{q=1}^{n} \frac{1}{p_q!} \left(-\frac{1}{2} t_q^2\right)^{p_q} (it_s)(it_t)(it_r)(it_u)(it_v)(it_w)$$

$$= \exp\left(-\frac{1}{2} \sum_{q=1}^{n} t_q^2\right) (it_s)(it_t)(it_r)(it_u)(it_v)(it_w)$$

where $k_q - 1 = 2p_q$ for $q = s, t, r, u, v, w$; $k_q = 2p_q$ for $q \neq s, t, r, u, v, w$ are put and e.g. in the case of $s = u$ and different numbers t, r and v, w

$$\sum_{\substack{k_1,\ldots,k_n=0\\ k\geqq 2;\ \text{even}}}^{\infty} k_s k_t k_r k_s' k_v' k_w' \frac{1}{k_1!\ldots k_n!} (it_1)^{k_1} \ldots (it_n)^{k_n} \, {}^2P_1(\mathscr{W}_{strsvw})$$

$$= \exp\left(-\frac{1}{2} \sum_{q=1}^{n} t_q^2\right) (it_s)^2 (it_t)(it_r)(it_v)(it_w)$$

($k_s - 2 = 2p_s$, $k_q - 1 = 2p_q$ for $q = t, r, u, w$; $k_q = 2p_q$ for $q \neq s, t, r, v, w$). By means of these considerations we find from (1.71) that

$$\sum_{\substack{k_1,\ldots,k_n=0\\ k\geqq 2;\ \text{even}}}^{\infty} \frac{1}{k_1!\ldots k_n!}\,\tilde{\alpha}^{(1,3)}_{k_1\ldots k_n}(\mathrm{i}t_1)^{k_1}\ldots(\mathrm{i}t_n)^{k_n}$$

$$= \frac{1}{2}\exp\left(-\frac{1}{2}\sum_{q=1}^{n} t_q^2\right)\left\{\sum_{\substack{s,t,r=1\\ s<t<r}}^{n} {}^3\tilde{\tilde{A}}{}_2'(s,t,r)\,(\mathrm{i}t_s)\,(\mathrm{i}t_t)\,(\mathrm{i}t_r)\right.$$

$$\times\left(\sum_{\substack{u,v,w=1\\ u<v<w}}^{n} {}^3\tilde{\tilde{A}}{}_2'(u,v,w)\,(\mathrm{i}t_u)\,(\mathrm{i}t_v)\,(\mathrm{i}t_w) + \frac{1}{2}\sum_{\substack{u,v=1\\ u\neq v}}^{n} {}^3\tilde{\tilde{A}}{}_2'(u,u,v)\,(\mathrm{i}t_u)^2\,(\mathrm{i}t_v)\right.$$

$$\left.+ \frac{1}{6}\sum_{u=1}^{n} {}^3\tilde{\tilde{A}}{}_2'(u,u,u)\,(\mathrm{i}t_u)^3\right) + \frac{1}{2}\sum_{\substack{s,t=1\\ s\neq t}}^{n} {}^3\tilde{\tilde{A}}{}_2'(s,s,t)\,(\mathrm{i}t_s)^2\,(\mathrm{i}t_t)$$

$$\times\left(\sum_{\substack{u,v,w=1\\ u<v<w}}^{n} {}^3\tilde{\tilde{A}}{}_2'(u,v,w)\,(\mathrm{i}t_u)\,(\mathrm{i}t_v)\,(\mathrm{i}t_w) + \frac{1}{2}\sum_{\substack{u,v=1\\ u\neq v}}^{n} {}^3\tilde{\tilde{A}}{}_2'(u,u,v)\,(\mathrm{i}t_u)^2\,(\mathrm{i}t_v)\right.$$

$$\left.+ \frac{1}{6}\sum_{u=1}^{n} {}^3\tilde{\tilde{A}}{}_2'(u,u,u)\,(\mathrm{i}t_u)^3\right) + \frac{1}{6}\sum_{s=1}^{n} {}^3\tilde{\tilde{A}}{}_2'(s,s,s)\,(\mathrm{i}t_s)^3$$

$$\times\left(\sum_{\substack{u,v,w=1\\ u<v<w}}^{n} {}^3\tilde{\tilde{A}}{}_2'(u,v,w)\,(\mathrm{i}t_u)\,(\mathrm{i}t_v)\,(\mathrm{i}t_w)\right.$$

$$\left.\left.+ \frac{1}{2}\sum_{\substack{u,v=1\\ u\neq v}}^{n} {}^3\tilde{\tilde{A}}{}_2'(u,u,v)\,(\mathrm{i}t_u)^2\,(\mathrm{i}t_v) + \frac{1}{6}\sum_{u=1}^{n} {}^3\tilde{\tilde{A}}{}_2'(u,u,u)\,(\mathrm{i}t_u)^3\right)\right\}$$

$$= \frac{1}{72}\exp\left(-\frac{1}{2}\sum_{q=1}^{n} t_q^2\right)\left[\sum_{s,t,r=1}^{n} {}^3\tilde{\tilde{A}}{}_2'(s,t,r)\,(\mathrm{i}t_s)\,(\mathrm{i}t_t)\,(\mathrm{i}t_r)\right]^2. \tag{1.72}$$

Summarizing we obtain from (1.65)

$$\sum_{\substack{k_1,\ldots,k_n=0\\ k\geqq 2;\ \text{even}}}^{\infty} \frac{1}{k_1!\ldots k_n!}\,\tilde{\alpha}^{(1)}_{k_1\ldots k_n}(\mathrm{i}t_1)^{k_1}\ldots(\mathrm{i}t_n)^{k_n}$$

$$= \exp\left(-\frac{1}{2}\sum_{q=1}^{n} t_q^2\right)\left\{\frac{1}{2}\sum_{u,v=1}^{n} {}^2\tilde{\tilde{A}}{}_2'(u,v)\,(\mathrm{i}t_u)\,(\mathrm{i}t_v)\right.$$

$$+ \frac{1}{24}\sum_{u,v,r,s=1}^{n} {}^4\tilde{\underline{A}}{}_3'(u,v,r,s)\,(\mathrm{i}t_u)\,(\mathrm{i}t_v)\,(\mathrm{i}t_r)\,(\mathrm{i}t_s)$$

$$\left.+ \frac{1}{72}\left[\sum_{u,v,r=1}^{n} {}^3\tilde{\tilde{A}}{}_2'(u,v,r)\,(\mathrm{i}t_u)\,(\mathrm{i}t_v)\,(\mathrm{i}t_r)\right]^2\right\} \tag{1.73}$$

for $m = 1$ using (1.66), (1.70), and (1.72). The equation

$$\tilde{\alpha}^{(1)}_{k_1\ldots k_n} = \hat{\tilde{H}}_2(m;\, v_{k_1\ldots k_n}) = \sum_{\{s_1,t_1\},\ldots,\{s_{k/2},t_{k/2}\}}\ \sum_{j=1}^{k/2} {}^2\tilde{\tilde{A}}{}_2'(s_j,t_j)\prod_{\substack{q=1\\ q\neq j}}^{k/2}\delta_{s_q t_q}$$

for $m \geqq 2$ leads to

$$\sum_{\substack{k_1,\ldots,k_n=0\\ k\geqq 2;\text{even}}}^{\infty} \frac{1}{k_1!\ldots k_n!}\,\tilde{\alpha}^{(1)}_{k_1\ldots k_n}(\mathrm{i}t_1)^{k_1}\ldots(\mathrm{i}t_n)^{k_n}$$

$$= \exp\left(-\frac{1}{2}\sum_{q=1}^{n} t_q^2\right)\frac{1}{2}\sum_{u,v=1}^{n} {}^2\tilde{\tilde{A}}_2'(u,v)\,(\mathrm{i}t_u)\,(\mathrm{i}t_v) \tag{1.74}$$

if (1.66) is applied.

Now it is possible to give the characteristic function $\tilde{\psi}(t_1, \ldots, t_n)$ in the form

$$\tilde{\psi}(t_1,\ldots,t_n)$$

$$= \begin{cases} \exp\left(-\frac{1}{2}\sum\limits_{q=1}^{n} t_q^2\right)\left\{1 + \frac{1}{6}\sum\limits_{u,v,r=1}^{n} {}^3\tilde{\tilde{A}}_2'(u,v,r)\,(\mathrm{i}t_u)\,(\mathrm{i}t_v)\,(\mathrm{i}t_r)\sqrt{\varepsilon}\right. \\ + \left[\frac{1}{2}\sum\limits_{u,v=1}^{n} {}^2\tilde{\tilde{A}}_2'(u,v)\,(\mathrm{i}t_u)\,(\mathrm{i}t_v) + \frac{1}{24}\sum\limits_{u,v,r,s=1}^{n} {}^4\underline{\tilde{\tilde{A}}}_3'(u,v,r,s)\,(\mathrm{i}t_u)\,(\mathrm{i}t_v)\,(\mathrm{i}t_r)\,(\mathrm{i}t_s)\right. \\ \left.\left. + \frac{1}{72}\left[\sum\limits_{u,v,r=1}^{n} {}^3\tilde{\tilde{A}}_2'(u,v,r)\,(\mathrm{i}t_u)\,(\mathrm{i}t_v)\,(\mathrm{i}t_r)\right]^2\right]\varepsilon\right\} + o(\varepsilon) \quad \text{for } m = 1, \\ \exp\left(-\frac{1}{2}\sum\limits_{q=1}^{n} t_q^2\right)\left\{1 + \frac{1}{2}\sum\limits_{u,v=1}^{n} {}^2\tilde{\tilde{A}}_2'(u,v)\,(\mathrm{i}t_u)\,(\mathrm{i}t_v)\,\varepsilon\right. \\ \left. + \frac{1}{6}\sum\limits_{u,v,r=1}^{n} {}^3\tilde{\tilde{A}}_2'(u,v,r)\,(\mathrm{i}t_u)\,\mathrm{i}(t_v)\,(\mathrm{i}t_r)\,\delta_{2m}\varepsilon\right\} + o(\varepsilon) \quad \text{for } m \geqq 2 \end{cases} \tag{1.75}$$

where we start from (1.55) and use (1.62), (1.64), (1.73), and (1.74).

The goal is the determination of the coefficients $c_{k_1\ldots k_n}$ contained in the equation of the distribution density (1.49). In order to obtain these coefficients the expansions (1.54) and (1.75) are compared where the relation

$$\sum_{u_1,\ldots,u_p=1}^{n} \tilde{\tilde{G}}'(u_1,\ldots,u_p)\,(\mathrm{i}t_{u_1})\ldots(\mathrm{i}t_{u_p})$$

$$= \sum_{\substack{k_1,\ldots,k_n=0\\ k=p}}^{\infty} \frac{p!}{k_1!\ldots k_n!}\,\tilde{\tilde{G}}'(\mathcal{V}_{k_1\ldots k_n})\,(\mathrm{i}t_1)^{k_1}\ldots(\mathrm{i}t_n)^{k_n}$$

is used. Thus

for $k = 2$: $c_{k_1\ldots k_n} = {}^2\tilde{\tilde{A}}_2'(\mathcal{V}_{k_1\ldots k_n})\,\varepsilon$ for arbitrary m,

for $k = 3$: $c_{k_1\ldots k_n} = -{}^3\tilde{\tilde{A}}_2'(\mathcal{V}_{k_1\ldots k_n})\,\varepsilon^{m/2}$ for arbitrary m,

for $k = 4$: $c_{k_1\ldots k_n} = \begin{cases} {}^4\underline{\tilde{\tilde{A}}}_3'(\mathcal{V}_{k_1\ldots k_n})\,\varepsilon & \text{for } m = 1, \\ O(\varepsilon)^2 & \text{for } m \geqq 2, \end{cases}$

for $k = 5$: $c_{k_1\ldots k_n} = \begin{cases} O(\varepsilon^{3/2}) & \text{for } m = 1, \\ O(\varepsilon)^2 & \text{for } m \geqq 2, \end{cases}$

$$\text{for } k = 6\colon\ c_{k_1\ldots k_n} = \begin{cases} \dfrac{1}{2} k_1! \ldots k_n! \sum\limits_{\substack{p_1,\ldots,p_n=0 \\ \sum_{s=1}^n p_s = 3;\, 0 \leqq k_s - p_s \leqq 3}}^{\infty} \prod\limits_{q=1}^{n} \dfrac{1}{(k_q - p_q)!\, p_q!} \\ \times\, {}^3\tilde{\tilde{A}}_2'(\mathcal{V}_{k_1-p_1\ldots k_n-p_n}) \\ \times\, {}^3\tilde{\tilde{A}}_2'(\mathcal{V}_{p_1\ldots p_n})\, \varepsilon & \text{for } m = 1, \\ O(\varepsilon^2) & \text{for } m \geqq 2, \end{cases}$$

$$\text{for } k \geqq 7\colon\ c_{k_1\ldots k_n} = \begin{cases} O(\varepsilon^{3/2}) & \text{for } m = 1, \\ O(\varepsilon^2) & \text{for } m \geqq 2. \end{cases}$$

In particular, for $k = 6$ we have

$$\left[\frac{1}{6} \sum_{u,v,r=1}^{n} {}^3\tilde{\tilde{A}}_2'(u, v, r)\, (\mathrm{i}t_u)\, (\mathrm{i}t_v)\, (\mathrm{i}t_r)\right]^2$$

$$= \left[\sum_{\substack{k_1,\ldots,k_n=0, \\ k=3}}^{\infty} \prod_{q=1}^{n} \frac{1}{k_q!} (\mathrm{i}t_q)^{k_q}\, {}^3\tilde{\tilde{A}}_2'(\mathcal{V}_{k_1\ldots k_n})\right]^2$$

$$= \sum_{\substack{k_1,\ldots,k_n=0, \\ k=6}}^{\infty} \sum_{\substack{p_1,\ldots,p_n=0 \\ \sum_{s=1}^n p_s = 3;\, 0 \leqq k_s - p_s \leqq 3}}^{\infty} \prod_{q=1}^{n} \frac{1}{(k_q - p_q)!\, p_q!} (\mathrm{i}t_q)^{k_q}$$

$$\times\, {}^3\tilde{\tilde{A}}_2'(\mathcal{V}_{k_1-p_1\ldots k_n-p_n})\, {}^3\tilde{\tilde{A}}_2'(\mathcal{V}_{p_1\ldots p_n})$$

and from this the above result can be obtained. Finally, from (1.53) it follows the distribution density

$$\tilde{p}(\tilde{u}_1, \ldots, \tilde{u}_n) = \frac{1}{\sqrt{2\pi}^n} \exp\left(-\frac{1}{2} \sum_{q=1}^{n} \tilde{u}_q^2\right)$$

$$\times \left\{1 + \sum_{\substack{k_1,\ldots,k_n=0, \\ k=3}}^{\infty} \prod_{q=1}^{n} \frac{1}{k_q!} H_{k_q}(\tilde{u}_q)\, {}^3\tilde{\tilde{A}}_2'(\mathcal{V}_{k_1\ldots k_n}) \sqrt{\varepsilon}\right.$$

$$+ \left[\sum_{\substack{k_1,\ldots,k_n=0, \\ k=2}}^{\infty} \prod_{q=1}^{n} \frac{1}{k_q!} H_{k_q}(\tilde{u}_q)\, {}^2\tilde{\tilde{A}}_2'(\mathcal{V}_{k_1\ldots k_n})\right.$$

$$+ \sum_{\substack{k_1,\ldots,k_n=0 \\ k=4}}^{\infty} \prod_{q=1}^{n} \frac{1}{k_q!} H_{k_q}(\tilde{u}_q)\, {}^4\tilde{\tilde{A}}_3'(\mathcal{V}_{k_1\ldots k_n})$$

$$+ \frac{1}{2} \sum_{\substack{k_1,\ldots,k_n=0, \\ k=6}}^{\infty} \sum_{\substack{p_1,\ldots,p_n=0 \\ \sum_{s=1}^n p_s = 3;\, 0 \leqq k_s - p_s \leqq 3}}^{\infty} \prod_{q=1}^{n} \frac{1}{(k_q - p_q)!\, p_q!} H_{k_q}(\tilde{u}_q)$$

$$\left.\left.\times\, {}^3\tilde{\tilde{A}}_2'(\mathcal{V}_{k_1-p_1\ldots k_n-p_n})\, {}^3\tilde{\tilde{A}}_2'(\mathcal{V}_{p_1\ldots p_n})\right] \varepsilon\right\} + o(\varepsilon) \tag{1.76}$$

for $m = 1$ and for $m \geqq 2$

$$\tilde{p}(\tilde{u}_1, \ldots, \tilde{u}_n) = \frac{1}{\sqrt{2\pi}^n} \exp\left(-\frac{1}{2}\sum_{q=1}^{n} \tilde{u}_q^2\right) \times \left\{1 + \left[\sum_{\substack{k_1,\ldots,k_n=0,\\ k=2}}^{\infty} \prod_{q=1}^{n} \frac{1}{k_q!} H_{k_q}(\tilde{u}_q)\, {}^2\tilde{\tilde{A}}_2'(\mathcal{V}_{k_1\ldots k_n}) + \sum_{\substack{k_1,\ldots,k_n=0,\\ k=3}}^{\infty} \prod_{q=1}^{n} \frac{1}{k_q!} H_{k_q}(\tilde{u}_q)\, {}^3\tilde{\tilde{A}}_2'(\mathcal{V}_{k_1\ldots k_n})\, \delta_{2m}\right] \varepsilon\right\} + o(\varepsilon).$$

By means of (1.76) the distribution function of the random vector $(\tilde{r}_{1\varepsilon}, \tilde{r}_{2\varepsilon}, \ldots, \tilde{r}_{n\varepsilon})$ is given up to terms of the order ε for arbitrary m.

In particular, in the case of $n = 1$ we obtain from (1.76) the relation

$$\tilde{p}(\tilde{u}_1) = \begin{cases} \dfrac{1}{\sqrt{2\pi}} \exp\left(-\dfrac{1}{2}\tilde{u}_1^2\right)\left\{1 + \dfrac{1}{6}\,{}^3\tilde{\tilde{A}}_2' H_3(\tilde{u}_1)\sqrt{\varepsilon} + \left[\dfrac{1}{2}\,{}^2\tilde{\tilde{A}}_2' H_2(\tilde{u}_1) + \dfrac{1}{24}\,{}^4\underline{\tilde{\tilde{A}}}_3' H_4(\tilde{u}_1) + \dfrac{1}{72}\,({}^3\tilde{\tilde{A}}_2')^2 H_6(\tilde{u}_1)\right]\varepsilon\right\} + o(\varepsilon) & \text{for } m = 1, \\ \dfrac{1}{\sqrt{2\pi}} \exp\left(-\dfrac{1}{2}\tilde{u}_1^2\right)\left\{1 + \dfrac{1}{2}\,{}^2\tilde{\tilde{A}}_2' H_2(\tilde{u}_1)\,\varepsilon + \dfrac{1}{6}\,{}^3\tilde{\tilde{A}}_2' H_3(\tilde{u}_1)\,\varepsilon^{m/2}\right\} + o(\varepsilon) & \text{for } m \geqq 2 \end{cases} \tag{1.77}$$

where

$${}^p\tilde{\tilde{A}}_q'(\mathcal{V}_p) = \sum_{j_1,\ldots,j_p=1}^{l} {}^p\tilde{A}_q'(\mathcal{V}_p) = \sum_{j_1,\ldots,j_p=1}^{l} {}^pA_q(\tilde{F}_{1j_1}, \ldots, \tilde{F}_{1j_p}) \doteq {}^p\tilde{\tilde{A}}_q'$$

and

$$\tilde{F}_{1j}(x) = b_{11}F_{1j}(x) = \sqrt{\frac{1}{B_{11}}}\, F_{1j}(x),$$

$$\tilde{u} = \frac{1}{\sqrt{\varepsilon}^m} b_{11}u = \sqrt{\frac{1}{\varepsilon^m B_{11}}}\, u$$

where

$$B_{11} = \hat{H}_0(m; 1, 1) = \sum_{j_1,j_2=1}^{l} {}^2A_1(F_{1j_1}, F_{1j_2}).$$

Setting $m = 1$, $l = 1$ it follows

$$\tilde{p}(\tilde{u}) = \frac{1}{\sqrt{2\pi}} \exp\left(-\frac{1}{2}\tilde{u}^2\right)\left\{1 + \frac{1}{6}\,{}^3\tilde{A}_2' H_3(\tilde{u})\sqrt{\varepsilon} + \left[\frac{1}{2}\,{}^2\tilde{A}_2' H_2(\tilde{u}) + \frac{1}{24}\,{}^4\underline{\tilde{A}}_3' H_4(\tilde{u}) + \frac{1}{72}\,({}^3\tilde{A}_2')^2 H_6(\tilde{u})\right]\varepsilon\right\} + o(\varepsilon) \tag{1.78}$$

where

$${}^p\tilde{A}_q' \doteq {}^pA_q(\tilde{F}^p); \qquad \tilde{F} = \frac{F}{\sqrt{{}^2A_1(F^2)}}.$$

Neglecting terms of order ε this formula is obtained in BOYCE, XIA [1] and in the case of ${}^3A_2 = 0$ neglecting terms of a higher order than ε the formula resulting from (1.78) is also written in BOYCE, XIA [1].

Now we put $n = 2$. (1.76) leads to

$$\begin{aligned}
\tilde{p}(\tilde{u}_1, \tilde{u}_2) &= \frac{1}{2\pi} \exp\left(-\frac{1}{2} (\tilde{u}_1^2 + \tilde{u}_2^2)\right) \\
&\times \Bigg\{1 + \frac{1}{6} [{}^3\tilde{\bar{A}}_2'(1, 1, 1) H_3(\tilde{u}_1) + 3\, {}^3\tilde{\bar{A}}_2'(1, 1, 2) H_2(\tilde{u}_1) H_1(\tilde{u}_2) \\
&+ 3\, {}^3\tilde{\bar{A}}_2'(1, 2, 2) H_1(\tilde{u}_1) H_2(\tilde{u}_2) + {}^3\tilde{\bar{A}}_2'(2, 2, 2) H_3(\tilde{u}_2)] \sqrt{\varepsilon} \\
&+ \Bigg[\frac{1}{2} \big({}^2\tilde{\bar{A}}_2'(1, 1) H_2(\tilde{u}_1) + 2\, {}^2\tilde{\bar{A}}_2'(1, 2) H_1(\tilde{u}_1) H_1(\tilde{u}_2) \\
&+ {}^2\tilde{\bar{A}}_2'(2, 2) H_2(\tilde{u}_2)\big) + \frac{1}{24} \big({}^4\tilde{\underline{\bar{A}}}_3'(1, 1, 1, 1) H_4(\tilde{u}_1) \\
&+ 4\, {}^4\tilde{\underline{\bar{A}}}_3'(1, 1, 1, 2) H_3(\tilde{u}_1) H_1(\tilde{u}_2) + 6\, {}^4\tilde{\underline{\bar{A}}}_3'(1, 1, 2, 2) H_2(\tilde{u}_1) H_2(\tilde{u}_2) \\
&+ 4\, {}^4\tilde{\underline{\bar{A}}}_3'(1, 2, 2, 2) H_1(\tilde{u}_1) H_3(\tilde{u}_2) + {}^4\tilde{\underline{\bar{A}}}_3'(2, 2, 2, 2) H_4(\tilde{u}_2)\big) \\
&+ \frac{1}{72} \big(({}^3\tilde{\bar{A}}_2'(1, 1, 1))^2 H_6(\tilde{u}_1) + ({}^3\tilde{\bar{A}}_2'(2, 2, 2))^2 H_6(\tilde{u}_2) \\
&+ 6\, {}^3\tilde{\bar{A}}_2'(1, 1, 1)\, {}^3\tilde{\bar{A}}_2'(1, 1, 2) H_5(\tilde{u}_1) H_1(\tilde{u}_2) \\
&+ 6\, {}^3\tilde{\bar{A}}_2'(2, 2, 2)\, {}^3\tilde{\bar{A}}_2'(1, 2, 2) H_1(\tilde{u}_1) H_5(\tilde{u}_2) \\
&+ \big(6\, {}^3\tilde{\bar{A}}_2'(1, 1, 1)\, {}^3\tilde{\bar{A}}_2'(1, 2, 2) + 9({}^3\tilde{\bar{A}}_2'(1, 1, 2))^2\big) H_4(\tilde{u}_1) H_2(\tilde{u}_2) \\
&+ \big(6\, {}^3\tilde{\bar{A}}_2'(2, 2, 2)\, {}^3\tilde{\bar{A}}_2'(1, 1, 2) + 9({}^3\tilde{\bar{A}}_2'(1, 2, 2))^2\big) H_2(\tilde{u}_1) H_4(\tilde{u}_2) \\
&+ \big(2\, {}^3\tilde{\bar{A}}_2'(1, 1, 1)\, {}^3\tilde{\bar{A}}_2'(2, 2, 2) \\
&+ 18\, {}^3\tilde{\bar{A}}_2'(1, 1, 2)\, {}^3\tilde{\bar{A}}_2'(1, 2, 2)\big) H_3(\tilde{u}_1) H_3(\tilde{u}_2)\Bigg] \varepsilon\Bigg\} + o(\varepsilon) \qquad (1.79)
\end{aligned}$$

for $m = 1$ and

$$\begin{aligned}
\tilde{p}(\tilde{u}_1, \tilde{u}_2) &= \frac{1}{2\pi} \exp\left(-\frac{1}{2} (\tilde{u}_1^2 + \tilde{u}_2^2)\right) \\
&\times \Bigg\{1 + \frac{1}{2} [{}^2\tilde{\bar{A}}_2'(1, 1) H_2(\tilde{u}_1) + 2\, {}^2\tilde{\bar{A}}_2'(1, 2) H_1(\tilde{u}_1) H_1(\tilde{u}_2) \\
&+ {}^2\tilde{\bar{A}}_2'(2, 2) H_2(\tilde{u}_2)] \varepsilon + \frac{1}{6} [{}^3\tilde{\bar{A}}_2'(1, 1, 1) H_3(\tilde{u}_1) \\
&+ 3\, {}^3\tilde{\bar{A}}_2'(1, 1, 2) H_2(\tilde{u}_1) H_1(\tilde{u}_2) + 3\, {}^3\tilde{\bar{A}}_2'(1, 2, 2) H_1(\tilde{u}_1) H_2(\tilde{u}_2) \\
&+ {}^3\tilde{\bar{A}}_2'(2, 2, 2) H_3(\tilde{u}_2)] \varepsilon^{m/2}\Bigg\} + o(\varepsilon) \qquad (1.80)
\end{aligned}$$

for $m \geqq 2$.

Now we deal with some special cases. First, let $\big(f_{1\varepsilon}(x), \ldots, f_{l\varepsilon}(x)\big)$ be symmetrically distributed; i.e. for the density functions

$$\varphi_{x_{11}\dots x_{1n_1}\dots x_{l1}\dots x_{ln_l}}(t_{11}, \dots, t_{1n_1}, \dots, t_{l1}, \dots, t_{ln_l})$$

of the vectors

$$\left(f_{1\varepsilon}(x_{11}), \dots, f_{1\varepsilon}(x_{1n_1}), \dots, f_{l\varepsilon}(x_{l1}), \dots, f_{l\varepsilon}(x_{ln_l})\right)^{\mathsf{T}}$$

we have the relation

$$\varphi_{x_{11}\dots x_{1n_1}\dots x_{l1}\dots x_{ln_l}}(-t_{11}, \dots, -t_{1n_1}, \dots, -t_{l1}, \dots, -t_{ln_l})$$
$$= \varphi_{x_{11}\dots x_{1n_1}\dots x_{l1}\dots x_{ln_l}}(t_{11}, \dots, t_{1n_1}, \dots, t_{l1}, \dots, t_{ln_l}).$$

From this assumption it follows that

$$\left\langle \prod_{p=1}^{l} \left(\prod_{q=1}^{n_p} f_{p\varepsilon}(x_{pq}) \right) \right\rangle = 0 \quad \text{for} \quad \sum_{p=1}^{l} n_p \text{ odd}$$

and furthermore

$$ {}^{p}\bar{A}_q(\{1, \dots, p\}) = 0 \quad \text{for} \quad q = p-1, p \text{ and } p \text{ odd using (1.11)},$$
$$ {}^{p}\hat{\bar{A}}_q(\{1, \dots, p\}) = 0 \quad \text{for} \quad q = p-1, p \text{ and } p \text{ odd using (1.57)},$$
$$ {}^{p}\tilde{\bar{A}}_q(\{1, \dots, p\}) = 0 \quad \text{for} \quad q = p-1, p \text{ and } p \text{ odd using (1.46)}.$$

In particular, we obtain

$$ {}^{3}\tilde{\bar{A}}_2(\{1, 2, 3\}) = 0$$

and the terms in (1.76) are zero which are multiplied by $\sqrt{\varepsilon}$. Hence, the first approximation (normal distribution) is for an assumed symmetric distribution of $\left(f_{1\varepsilon}(x), \dots, f_{l\varepsilon}(x)\right)$ a better approximation of the actual distribution than for an assumed non-symmetric distribution of $\left(f_{1\varepsilon}(x), \dots, f_{l\varepsilon}(x)\right)$. In this case for $n = 1$ the relation

$$\tilde{p}(\tilde{u}) = \begin{cases} \dfrac{1}{\sqrt{2\pi}} \exp\left(-\dfrac{1}{2}\tilde{u}^2\right)\left\{1 + \left[\dfrac{1}{2}\,{}^{2}\tilde{\bar{A}}_2(1, 1)\, H_2(\tilde{u}) \right.\right. \\ \left.\left. + \dfrac{1}{24}\,{}^{4}\underline{\tilde{\bar{A}}}_3(1, 1, 1, 1)\, H_4(\tilde{u})\right] \varepsilon\right\} + o(\varepsilon) \quad \text{for } m = 1, \\ \dfrac{1}{\sqrt{2\pi}} \exp\left(-\dfrac{1}{2}\tilde{u}^2\right)\left\{1 + \left[\dfrac{1}{2}\,{}^{2}\tilde{\bar{A}}_2(1, 1)\, H_2(\tilde{u})\right] \varepsilon\right\} + o(\varepsilon) \quad \text{for } m \geqq 2 \end{cases} \tag{1.81}$$

can be found and in a similar way for $n = 2$ from (1.79), (1.80). An example of a symmetrically distributed vector field (weakly correlated) is a Gaussian vector field $\left(f_{1\varepsilon}(x), \dots, f_{l\varepsilon}(x)\right)$. Furthermore, we have for a Gaussian vector field

$$ {}^{4}\underline{\hat{\bar{A}}}_3(\{1, 2, 3, 4\}) = {}^{4}\hat{\bar{A}}_3(\{1, 2, 3, 4\}) - \hat{\bar{A}}_{22}(\{1, 2\}; \{3, 4\})$$
$$ - \hat{\bar{A}}_{22}(\{1, 3\}; \{2, 4\}) - \hat{\bar{A}}_{22}(\{1, 4\}; \{2, 3\})$$
$$ = \sum_{j_1, j_2, j_3, j_4 = 1}^{l} [{}^{4}A_3(F_{i_1 j_1}, F_{i_2 j_2}, F_{i_3 j_3}, F_{i_4 j_4}) - A_{22}(F_{i_1 j_1}, F_{i_2 j_2}; F_{i_3 j_3}, F_{i_4 j_4})$$
$$ - A_{22}(F_{i_1 j_1}, F_{i_3 j_3}; F_{i_2 j_2}, F_{i_4 j_4}) - A_{22}(F_{i_1 j_1}, F_{i_4 j_4}; F_{i_2 j_2}, F_{i_3 j_3})] = 0$$

since

$$
{}^4A_3(\bar{F}_1, \bar{F}_2, \bar{F}_3, \bar{F}_4) = \lim_{\varepsilon \downarrow 0} \frac{1}{\varepsilon^{3m}} \int\limits_{\mathscr{E}(\{1,2,3,4\})} \prod_{p=1}^{4} \bar{F}_p(y_p) \left\langle \prod_{p=1}^{4} \bar{f}_p(y_p) \right\rangle \mathrm{d}y_1 \, \mathrm{d}y_2 \, \mathrm{d}y_3 \, \mathrm{d}y_4
$$

$$
= \lim_{\varepsilon \downarrow 0} \frac{1}{\varepsilon^{3m}} \int\limits_{\mathscr{E}(\{1,2,3,4\})} \prod_{p=1}^{4} \bar{F}_p(y_p) \, [\langle \bar{f}_1(y_1) \, \bar{f}_2(y_2) \rangle \langle \bar{f}_3(y_3) \, \bar{f}_4(y_4) \rangle
$$

$$
+ \langle \bar{f}_1(y_1) \, \bar{f}_3(y_3) \rangle \langle \bar{f}_2(y_2) \, \bar{f}_4(y_4) \rangle + \langle \bar{f}_1(y_1) \, \bar{f}_4(y_4) \rangle \langle \bar{f}_2(y_2) \, \bar{f}_3(y_3) \rangle] \, \mathrm{d}y_1 \, \mathrm{d}y_2 \, \mathrm{d}y_3 \, \mathrm{d}y_4 .
$$

In this special case (1.76) leads to

$$
\tilde{p}(\tilde{u}_1, \ldots, \tilde{u}_n) = \frac{1}{\sqrt{2\pi}^n} \exp\left(-\frac{1}{2} \sum_{q=1}^{n} \tilde{u}_q^2 \right)
$$

$$
\times \left\{ 1 + \sum_{\substack{k_1, \ldots, k_n = 0 \\ k = 2}} \prod_{q=1}^{n} \frac{1}{k_q!} H_{k_q}(\tilde{u}_q) \, {}^2\tilde{\bar{A}}_2'(v_{k_1 \ldots k_n}) \, \varepsilon \right\} + o(\varepsilon)
$$

for arbitrary $m \geqq 1$, in the case of $n = 1$ to

$$
\tilde{p}(\tilde{u}) = \frac{1}{\sqrt{2\pi}} \exp\left(-\frac{1}{2} \tilde{u}^2 \right) \left\{ 1 + \frac{1}{2} \, {}^2\tilde{\bar{A}}_2'(1, 1) \, H_2(\tilde{u}_2) \, \varepsilon \right\} + o(\varepsilon) \tag{1.82}
$$

and of $n = 2$ to

$$
\tilde{p}(\tilde{u}_1, \tilde{u}_2) = \frac{1}{2\pi} \exp\left(-\frac{1}{2} (\tilde{u}_1^2 + \tilde{u}_2^2) \right) \left\{ 1 + \frac{1}{2} \, [{}^2\tilde{\bar{A}}_2'(1, 1) \, H_2(\tilde{u}_1) \right.
$$

$$
\left. + 2 \, {}^2\tilde{\bar{A}}_2'(1, 2) \, H_1(\tilde{u}_1) \, H_1(\tilde{u}_2) + {}^2\tilde{\bar{A}}_2'(2, 2) \, H_2(\tilde{u}_2)] \, \varepsilon \right\} + o(\varepsilon) . \tag{1.83}
$$

Now we determine the density function $p(u_1, \ldots, u_n)$ of $(r_{1\varepsilon}, \ldots, r_{n\varepsilon})$. The transformation relation between $(r_{1\varepsilon}, \ldots, r_{n\varepsilon})$ and $(\tilde{r}_{1\varepsilon}, \ldots, \tilde{r}_{n\varepsilon})$ is given by

$$
\tilde{r}_{i\varepsilon} = \frac{1}{\sqrt{\varepsilon}^m} \sum_{p=1}^{i} b_{ip} r_{p\varepsilon} , \qquad i = 1, 2, \ldots, n ,
$$

(see (1.37)). The density function $p(u_1, \ldots, u_n)$ is obtained by

$$
p(u) = \tilde{p}\left(\frac{1}{\sqrt{\varepsilon}^m} \, Tu \right) \frac{|\det (T)|}{\sqrt{\varepsilon}^{nm}} \tag{1.84}
$$

where $u = (u_1, \ldots, u_n)^\mathsf{T}$, $T = (b_{ip})_{1 \leqq i,p \leqq n}$ and $b_{ip} \doteq 0$ for $i < p$. We consider

$$
\tilde{u}^\mathsf{T} \tilde{u} = \frac{1}{\varepsilon^m} \, u^\mathsf{T} T^\mathsf{T} T u
$$

and by means of $B \doteq (B_{ij})_{1 \leqq i,j \leqq n}$ from (1.38) it follows that

$$
TBT^\mathsf{T} = E \quad \text{or} \quad B^{-1} = T^\mathsf{T} T
$$

and

$$
\tilde{u}^\mathsf{T} \tilde{u} = \frac{1}{\varepsilon^m} \, u^\mathsf{T} B^{-1} u .
$$

Hence, the first approximation $p_0(u)$ of the density function of $(r_{1\varepsilon}, \ldots, r_{n\varepsilon})$ is given by

$$p_0(u) = \frac{1}{\sqrt{2\pi}^n} \frac{1}{\sqrt{\varepsilon^{nm} \det(B)}} \exp\left(-\frac{1}{2} \frac{1}{\varepsilon^m} u^\top B^{-1} u\right) \tag{1.85}$$

(see (1.76)) if $\det(B)$ is calculated from

$$\det(B^{-1}) = \frac{1}{\det(B)} = \left(\det(T)\right)^2.$$

The formula (1.85) describes the density of a Gaussian vector $(\xi_1, \ldots, \xi_n)$ with mean zero where the correlation relations are given by

$$\langle \xi_p \xi_q \rangle = \varepsilon^m B_{pq} = \varepsilon^m \hat{H}_0(m; p, q) = \varepsilon^m \sum_{j_1, j_2 = 1}^{l} {}^2A_1(F_{pj_1}, F_{qj_2}) \tag{1.86}$$

(see (1.39)). Hence the vector $(r_{1\varepsilon}, \ldots, r_{n\varepsilon})$ is a normally distributed vector as to the first approximation with

$$\langle r_{p\varepsilon} r_{q\varepsilon} \rangle = \langle \xi_p \xi_q \rangle .$$

This represents the same result deduced in vom Scheidt, Purkert [3]. The general formula of the transformation is given by (1.84). We want to derive explicit formulas for $n = 1$ and $n = 2$ in the case of $m = 1$. We have

$$b_{11} = \sqrt{\frac{1}{B_{11}}}; \quad \tilde{u}_1 = \frac{1}{\sqrt{\varepsilon^m}} b_{11} u_1 = \frac{u_1}{\sqrt{\varepsilon^m B_{11}}}; \quad \tilde{F}_{1j} = b_{11} F_{1j} = \frac{F_{1j}}{\sqrt{B_{11}}}$$

and B_{11} is given by (1.86). Using these relations (1.77) leads for $m = 1$ to

$$\begin{aligned} p(u_1) = {} & \frac{1}{\sqrt{2\pi\varepsilon B_{11}}} \exp\left(-\frac{1}{2} \frac{u_1^2}{\varepsilon B_{11}}\right) \left\{1 + \frac{1}{6} \frac{{}^3\hat{A}_2}{\sqrt{B_{11}}^3} H_3\left(\frac{u_1}{\sqrt{\varepsilon B_{11}}}\right) \sqrt{\varepsilon} \right. \\ & + \left[\frac{1}{2} \frac{1}{B_{11}} {}^2\hat{A}_2 H_2\left(\frac{u_1}{\sqrt{\varepsilon B_{11}}}\right) + \frac{1}{24} \frac{1}{B_{11}^2} {}^4\underline{\hat{A}}_3 H_4\left(\frac{u_1}{\sqrt{\varepsilon B_{11}}}\right)\right. \\ & \left.\left. + \frac{1}{72} \frac{1}{B_{11}^3} ({}^3\hat{A}_2)^2 H_6\left(\frac{u_1}{\sqrt{\varepsilon B_{11}}}\right)\right] \varepsilon \right\} + o(\varepsilon) \end{aligned} \tag{1.87}$$

where ${}^p\hat{A}_q$ can be calculated by

$${}^p\hat{A}_q = \sum_{j_1, \ldots, j_p = 1}^{l} {}^pA_q(F_{1j_1}, \ldots, F_{1j_p}) \quad ({}^p\underline{\hat{A}}_q \text{ resp.})$$

and for $m \geqq 2$ to

$$\begin{aligned} p(u_1) = {} & \frac{1}{\sqrt{2\pi\varepsilon^m B_{11}}} \exp\left(-\frac{u_1^2}{2\varepsilon^m B_{11}}\right) \left\{1 + \frac{{}^2\hat{A}_2}{2B_{11}} H_2\left(\frac{u_1}{\sqrt{\varepsilon^m B_{11}}}\right) \varepsilon \right. \\ & \left. + \frac{{}^3\hat{A}_2}{6\sqrt{B_{11}}^3} H_3\left(\frac{u_1}{\sqrt{\varepsilon^m B_{11}}}\right) \varepsilon^{m/2} \right\} + o(\varepsilon) . \end{aligned}$$

In the case $n = 2$ it is

$$b_{11} = \sqrt{\frac{1}{B_{11}}}; \quad b_{21} = -\frac{B_{12}}{B_{11}} \sqrt{\frac{B_{11}}{B_{11}B_{22} - B_{12}^2}}; \quad b_{22} = \sqrt{\frac{B_{11}}{B_{11}B_{22} - B_{12}^2}};$$

$$\tilde{u}_1 = \frac{1}{\sqrt{\varepsilon}^m} b_{11}u_1; \quad \tilde{u}_2 = \frac{1}{\sqrt{\varepsilon}^m} (b_{21}u_1 + b_{22}u_2) \tag{1.88}$$

$$\tilde{F}_{1j} = b_{11}F_{1j}; \quad \tilde{F}_{2j} = b_{21}F_{1j} + b_{22}F_{2j}$$

and B^{-1} can be determined by

$$B^{-1} = \frac{1}{B_{11}B_{22} - B_{12}^2} \begin{pmatrix} B_{22} & -B_{12} \\ -B_{12} & B_{11} \end{pmatrix}.$$

Thus

$$p_0(u_1, u_2) = \frac{1}{2\pi\varepsilon^m} \frac{1}{\sqrt{B_{11}B_{22} - B_{12}^2}} \exp\left(-\frac{B_{22}u_1^2 - 2B_{12}u_1u_2 + B_{11}u_2^2}{2\varepsilon^m(B_{11}B_{22} - B_{12}^2)}\right)$$

and with the help of (1.79) the density $p(u_1, u_2)$ can be given. For a Gaussian vector field $(f_{1\varepsilon}(x), \ldots, f_{l\varepsilon}(x))$ using (1.83) we obtain

$$\begin{aligned} p(u_1, u_2) &= \frac{1}{2\pi\varepsilon^m} b_{11}b_{22} \exp\left(-\frac{1}{2\varepsilon^m} \left((b_{11}u_1)^2 + (b_{21}u_1 + b_{22}u_2)^2\right)\right) \\ &\times \left\{1 + \frac{1}{2}\left[b_{11}^2\, {}^2\hat{A}_2'(1,1)\, H_2\left(\frac{1}{\sqrt{\varepsilon}^m} b_{11}u_1\right)\right.\right. \\ &+ 2b_{11}\left({}^2\hat{A}_2'(1,1)\, b_{21} + {}^2\hat{A}_2'(1,2)\, b_{22}\right) H_1\left(\frac{1}{\sqrt{\varepsilon}^m} b_{11}u_1\right) H_1\left(\frac{1}{\sqrt{\varepsilon}^m} (b_{21}u_1 + b_{22}u_2)\right) \\ &+ \left({}^2\hat{A}_2'(1,1)\, b_{21}^2 + 2\, {}^2\hat{A}_2'(1,2)\, b_{21}b_{22}\right. \\ &\left.\left.\left. + {}^2\hat{A}_2'(2,2)\, b_{22}^2\right) H_2\left(\frac{1}{\sqrt{\varepsilon}^m} (b_{21}u_1 + b_{22}u_2)\right)\right] \varepsilon\right\} + o(\varepsilon) \end{aligned} \tag{1.89}$$

where

$${}^2\hat{A}_2'(p, q) = \sum_{j_1, j_2 = 1}^{l} {}^2A_2(F_{pj_1}, F_{qj_2}).$$

We still remark that the density function (1.76) leads naturally to the formulas of the moments containing in Theorem 1.5; i.e. it yields

$$\int_{-\infty}^{\infty} \cdots \int_{-\infty}^{\infty} u_1^{p_1} \ldots u_n^{p_n} p(u_1, \ldots, u_n)\, du_1 \ldots du_n = \langle r_{1\varepsilon}^{p_1} \ldots r_{n\varepsilon}^{p_n} \rangle$$

up to the calculated orders with respect to ε. In order to indicate this statement we restrict to the special case

$$\int_{-\infty}^{\infty} \int_{-\infty}^{\infty} u_1u_2 p(u_1, u_2)\, du_1\, du_2 = \langle r_{1\varepsilon} r_{2\varepsilon} \rangle.$$

Using the relations

$$u_1 = \frac{\sqrt{\varepsilon}^m}{b_{11}} \tilde{u}_1; \qquad u_2 = \frac{\sqrt{\varepsilon}^m}{b_{22}} \left(-\frac{b_{21}}{b_{11}} \tilde{u}_1 + \tilde{u}_2\right)$$

(see (1.88)) we obtain

$$\int_{-\infty}^{\infty}\int_{-\infty}^{\infty} u_1 u_2 p(u_1, u_2)\, \mathrm{d}u_1\, \mathrm{d}u_2 = \frac{\varepsilon^m}{b_{11}b_{22}} \int_{-\infty}^{\infty}\int_{-\infty}^{\infty} \tilde{u}_1 \left(\tilde{u}_2 - \frac{b_{21}}{b_{11}}\,\tilde{u}_1\right) \tilde{p}(\tilde{u}_1, \tilde{u}_2)\, \mathrm{d}\tilde{u}_1\, \mathrm{d}\tilde{u}_2$$

where $\tilde{p}(\tilde{u}_1, \tilde{u}_2)$ is given for $m = 1$ by (1.79) and for $m = 2$ by (1.80). By means of

$$\tilde{p}_0(\tilde{u}_1, \tilde{u}_2) = \frac{1}{2\pi} \exp\left(-\frac{1}{2}\,(\tilde{u}_1^2 + \tilde{u}_2^2)\right)$$

we have

$$\int_{-\infty}^{\infty}\int_{-\infty}^{\infty} \tilde{u}_1^{k_1} \tilde{u}_2^{k_2} \tilde{p}_0(\tilde{u}_1, \tilde{u}_2)\, \mathrm{d}\tilde{u}_1\, \mathrm{d}\tilde{u}_2$$

$$= \frac{1}{\sqrt{2\pi}} \int_{-\infty}^{\infty} \tilde{u}_1^{k_1} \exp\left(-\frac{1}{2}\,\tilde{u}_1^2\right) \mathrm{d}\tilde{u}_1 \frac{1}{\sqrt{2\pi}} \int_{-\infty}^{\infty} \tilde{u}_2^{k_2} \exp\left(-\frac{1}{2}\,\tilde{u}_2^2\right) \mathrm{d}\tilde{u}_2 = a_{k_1} a_{k_2}$$

where a_k denotes the adequate moment of the normal distribution:

$$a_{2k} = \frac{(2k)!}{2^k k!}; \qquad a_{2k+1} = 0 \quad \text{for} \quad k = 0, 1, 2, \ldots$$

Furthermore, we find that

$$\int_{-\infty}^{\infty}\int_{-\infty}^{\infty} u_1^{k_1} u_2^{k_2} \tilde{p}_0(u_1, u_2)\, H_{p_1}(u_1)\, H_{p_2}(u_2)\, \mathrm{d}u_1\, \mathrm{d}u_2 = 0 \qquad \text{for} \quad k_1 + p_1 \quad \text{odd or} \quad k_2 + p_2 \quad \text{odd},$$

$$\int_{-\infty}^{\infty} u^2 \exp\left(-\frac{1}{2}\,u^2\right) H_p(u)\, \mathrm{d}u = 0 \qquad \text{for} \quad k = 4, 6,$$

$$\int_{-\infty}^{\infty} u \exp\left(-\frac{1}{2}\,u^2\right) H_p(u)\, \mathrm{d}u = 0 \qquad \text{for} \quad k = 3, 5,$$

$$\int_{\infty-}^{\infty} \exp\left(-\frac{1}{2}\,u^2\right) H_p(u)\, \mathrm{d}u = 0 \qquad \text{for} \quad k = 2, 4, 6.$$

For arbitrary m we now obtain

$$\frac{\varepsilon^m}{b_{11}b_{22}} \int_{-\infty}^{\infty}\int_{-\infty}^{\infty} \tilde{u}_1 \left(\tilde{u}_2 - \frac{b_{21}}{b_{11}}\,\tilde{u}_1\right) \tilde{p}(\tilde{u}_1, \tilde{u}_2)\, \mathrm{d}\tilde{u}_1\, \mathrm{d}\tilde{u}_2$$

$$= \frac{\varepsilon^m}{b_{11}b_{22}} \left\{-\frac{b_{21}}{b_{11}}\,[1 + {}^2\tilde{\tilde{A}}_2'(1, 1)\,\varepsilon] + {}^2\tilde{\tilde{A}}_2'(1, 2)\,\varepsilon + o(\varepsilon)\right\}$$

$$= B_{12}\varepsilon^m \left[1 + \frac{1}{B_{11}}\,{}^2\hat{A}_2'(1, 1)\,\varepsilon\right] + \varepsilon^{m+1}\left[-\frac{B_{12}}{B_{11}}\,{}^2\hat{A}_2'(1, 1) + {}^2\hat{A}_2'(1, 2)\right] + o(\varepsilon^{m+1})$$

$$= {}^2\hat{A}_1'(1, 2)\,\varepsilon^m + {}^2\hat{A}_2'(1, 2)\,\varepsilon^{m+1} + o(\varepsilon^{m+1}) = \langle r_{1\varepsilon} r_{2\varepsilon}\rangle.$$

1.5. A special case of weakly correlated processes

The case of a normally distributed field $f_\varepsilon(x, \omega)$ is considered. Then the random vector $\big(r_{1\varepsilon}(\omega), r_{2\varepsilon}(\omega)\big)$ is also normally distributed and the density function is given by

$$p(u_1, u_2) = \frac{1}{2\pi\sqrt{\det(m_{ij})}} \exp\left(-\frac{1}{2\det(m_{ij})}\left[m_{22}u_1^2 - 2m_{12}u_1u_2 + m_{11}u_2^2\right]\right)$$

where $m_{ij} = \langle r_{i\varepsilon} r_{j\varepsilon}\rangle$ for $i, j = 1, 2$. The second moments m_{ij} can be calculated from

$$m_{ij} = \int\limits_{\mathcal{D}_i}\int\limits_{\mathcal{D}_j} F_i(x)\, F_j(y)\, \langle f_\varepsilon(x)\, f_\varepsilon(y)\rangle\, \mathrm{d}x\, \mathrm{d}y$$

where the correlation function $\langle f_\varepsilon(x)\, f_\varepsilon(y)\rangle$ has to be known. Using (1.36) we have

$$m_{ij} = {}^2A_1(F_i, F_j)\, \varepsilon^m + {}^2A_2(F_i, F_j)\, \varepsilon^{m+1} + o(\varepsilon^{m+1}) \quad \text{for} \quad i, j = 1, 2.$$

Now we consider the random vector $\big(\tilde r_{1\varepsilon}(\omega), \tilde r_{2\varepsilon}(\omega)\big)$ using (1.37). The quantities refering to the transformed vector are denoted by "$\sim$". (1.48) shows

$${}^2A_1(\tilde F_i, \tilde F_j) = \delta_{ij}$$

and we obtain

$$\tilde p(\tilde u_1, \tilde u_2) = \frac{1}{2\pi\sqrt{\det(\tilde m_{ij})}} \exp\left(-\frac{1}{2\det(\tilde m_{ij})}\left[\tilde m_{22}\tilde u_1^2 - 2\tilde m_{12}\tilde u_1\tilde u_2 + \tilde m_{11}\tilde u_2^2\right]\right)$$

and

$$\tilde m_{ij} = \delta_{ij} + {}^2A_2(\tilde F_i, \tilde F_j)\, \varepsilon + o(\varepsilon) \quad \text{for} \quad i, j = 1, 2.$$

Hence, expansions leads to

$$\det(\tilde m_{ij}) = \tilde m_{11}\tilde m_{22} - \tilde m_{12}^2 = 1 + \big({}^2A_2(\tilde F_1, \tilde F_1) + {}^2A_2(\tilde F_2, \tilde F_2)\big)\, \varepsilon + o(\varepsilon),$$

$$\frac{1}{\sqrt{\det(\tilde m_{ij})}} = 1 - \frac{1}{2}\big({}^2A_2(\tilde F_1, \tilde F_1) + {}^2A_2(\tilde F_2, \tilde F_2)\big)\, \varepsilon + o(\varepsilon),$$

$$\frac{\tilde m_{ij}}{\sqrt{\det(\tilde m_{ij})}} = \delta_{ij} + \left[{}^2A_2(\tilde F_i, \tilde F_j) - \frac{1}{2}\,\delta_{ij}\big({}^2A_2(\tilde F_1, \tilde F_1) + {}^2A_2(\tilde F_2, \tilde F_2)\big)\right] \varepsilon + o(\varepsilon).$$

Now we expand the density function $\tilde p(\tilde u_1, \tilde u_2)$ as to ε:

$$\begin{aligned}
\tilde p(\tilde u_1, \tilde u_2) &= \frac{1}{2\pi\sqrt{\det(\tilde m_{ij})}} \exp\left(-\frac{1}{2}(\tilde u_1^2 + \tilde u_2^2)\right) \\
&\quad \times \exp\left(\frac{\varepsilon}{2}\big({}^2A_2(\tilde F_1, \tilde F_1)\, \tilde u_1^2 + 2\, {}^2A_2(\tilde F_1, \tilde F_2)\, \tilde u_1\tilde u_2 + {}^2A_2(\tilde F_2, \tilde F_2)\, \tilde u_2^2\big) + O(\varepsilon^2)\right) \\
&= \frac{1}{2\pi}\left[1 - \frac{1}{2}\big({}^2A_2(\tilde F_1, \tilde F_1) + {}^2A_2(\tilde F_2, \tilde F_2)\big)\, \varepsilon + O(\varepsilon^2)\right] \exp\left(-\frac{1}{2}(\tilde u_1^2 + \tilde u_2^2)\right) \\
&\quad \times \left[1 + \big({}^2A_2(\tilde F_1, \tilde F_1)\, \tilde u_1^2 + 2\, {}^2A_2(\tilde F_1, \tilde F_2)\, \tilde u_1\tilde u_2 + {}^2A_2(\tilde F_2, \tilde F_2)\, \tilde u_2^2\big)\frac{\varepsilon}{2} + O(\varepsilon)\right] \\
&= \frac{1}{2\pi} \exp\left(-\frac{1}{2}(\tilde u_1^2 + \tilde u_2^2)\right)\left[1 + \left(\frac{1}{2}\, {}^2A_2(\tilde F_1, \tilde F_1)\, (\tilde u_1^2 - 1)\right.\right. \\
&\quad \left.\left. + {}^2A_2(\tilde F_1, \tilde F_2)\, \tilde u_1\tilde u_2 + \frac{1}{2}\, {}^2A_2(\tilde F_2, \tilde F_2)\, (\tilde u_2^2 - 1)\right)\varepsilon + O(\varepsilon)\right].
\end{aligned}$$

Using $H_1(u) = u$, $H_2(u) = u^2 - 1$ we have obtained the same result by expansion of the Gaussian density with respect to the correlation length as (1.83) contains this.

In the case of a normally distributed process $f_\varepsilon(x, \omega)$ the exact density function $p(u)$ of the random variable

$$r_\varepsilon(\omega) = \int_0^1 F(x)\, f_\varepsilon(x, \omega)\, dx$$

is to be compared with the approximative density function following from the above theory. The density $p(u)$ is the density of a normally distributed random variable.

First we calculate $\langle r_\varepsilon^2 \rangle$ exactly. It is

$$\langle r_\varepsilon^2 \rangle = \int_0^1 \int_0^1 F(x_1)\, F(x_2)\, \langle f_\varepsilon(x_1)\, f_\varepsilon(x_2) \rangle\, dx_1\, dx_2$$

and a correlation function of the form

$$R_\varepsilon(|x_2 - x_1|) \doteq \langle f_\varepsilon(x_1)\, f_\varepsilon(x_2) \rangle = \begin{cases} \sigma^2 \left(1 - \dfrac{1}{\varepsilon} |x_2 - x_1|\right) & \text{for } |x_2 - x_1| \leqq \varepsilon, \\ 0 & \text{otherwise} \end{cases}$$

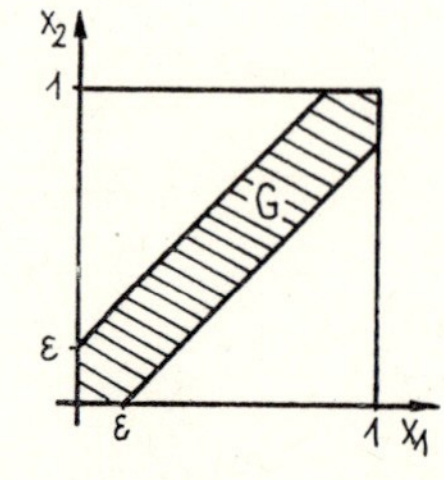

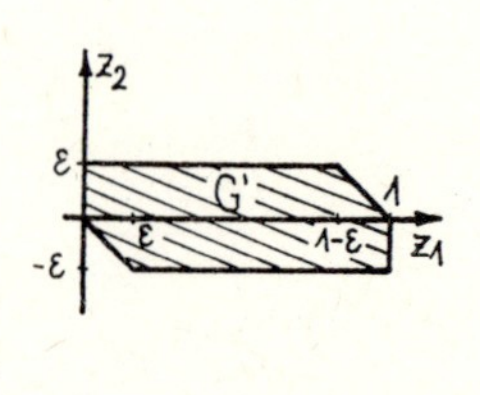

Fig. 1.3. ε-neighbouring points (x_1, x_2) as to $[0, 1] \times [0, 1]$

is assumed. With the help of the transformation $z_1 = x_1$, $z_2 = x_2 - x_1$ it follows

$$\langle r_\varepsilon^2 \rangle = \iint_{\mathscr{G}} F(x_1)\, F(x_2)\, R_\varepsilon(x_2 - x_1)\, dx_1\, dx_2 = \iint_{\mathscr{G}'} F(z_1)\, F(z_1 + z_2)\, R_\varepsilon(z_2)\, dz_1\, dz_2$$

(see Fig. 1.3) and

$$\langle r_\varepsilon^2 \rangle = \int_0^\varepsilon F(z_1) \int_{-z_1}^{\varepsilon} F(z_1 + z_2)\, R_\varepsilon(z_2)\, dz_2\, dz_1 + \int_\varepsilon^{1-\varepsilon} F(z_1) \int_{-\varepsilon}^{\varepsilon} F(z_1 + z_2)\, R_\varepsilon(z_2)\, dz_2\, dz_1$$
$$+ \int_{1-\varepsilon}^{1} F(z_1) \int_{-\varepsilon}^{1-z_1} F(z_1 + z_2)\, R_\varepsilon(z_2)\, dz_2\, dz_1 .$$

Particularly, after some calculations we have

$$\langle r_\varepsilon^2 \rangle = \sigma^2 \left(\varepsilon - \frac{\varepsilon^2}{3}\right) \qquad \text{for } F(x) = 1,$$

$$\langle r_\varepsilon^2 \rangle = \sigma^2 \left(\frac{\varepsilon}{3} - \frac{\varepsilon^2}{6} + \frac{\varepsilon^4}{60}\right) \qquad \text{for } F(x) = x,$$

$$\langle r_\varepsilon^2 \rangle = \sigma^2 \left[\frac{1}{4\pi^2 \varepsilon} \left(1 - \cos(2\pi\varepsilon)\right) - \frac{3}{8\pi^3 \varepsilon} \sin(2\pi\varepsilon) + \frac{1}{4\pi^2} \left(\cos(2\pi\varepsilon) + 2\right)\right] \qquad (1.90)$$
$$\text{for } F(x) = \sin(2\pi x),$$

$$\langle r_\varepsilon^2 \rangle = \sigma^2 \left[\frac{13}{15} \varepsilon - \frac{5}{6} \varepsilon^2 + \frac{1}{18} \varepsilon^3 + \frac{2}{15} \varepsilon^4 - \frac{1}{630} \varepsilon^6\right] \qquad \text{for } F(x) = x^2 + 2x - 1 .$$

By means of this second moment the density function has the form

$$p(u) = \frac{1}{\sqrt{2\pi\langle r_\varepsilon^2\rangle}} \exp\left(-\frac{u^2}{2\langle r_\varepsilon^2\rangle}\right). \tag{1.91}$$

On the other hahd using (1.35) for $k = 2$ the second moment can be written as

$$\langle r_\varepsilon^2\rangle = {}^2A_1\varepsilon + {}^2A_2\varepsilon^2 + O(\varepsilon^3)$$

and the terms ${}^2A_k$, $k = 1, 2$, can be determined from

$${}^2A_1 = \int_0^1 F^2(x)\, a(x)\, \mathrm{d}x,$$

$${}^2A_2 = \int_0^1 F^2(x)\, b(x)\, \mathrm{d}x + F^2(0)\left(\int_0^1 a_0(u)\, \mathrm{d}u - a(0)\right) + F^2(1)\left(\int_0^1 a_1(u)\, \mathrm{d}u - a(1)\right)$$

where $a(x)$ and $a_0(u)$, $a_1(u)$ are defined by

$$a(x) = \lim_{\varepsilon\downarrow 0} \frac{1}{\varepsilon} \int_{-\varepsilon}^{\varepsilon} R_\varepsilon(t)\, \mathrm{d}t,$$

$$a_0(u) = a_1(u) = \lim_{\varepsilon\downarrow 0} \int_{-u}^{1} R_\varepsilon(\varepsilon t)\, \mathrm{d}t$$

(see Section 1.6.2). In the considered case we can calculate easily

$$a(x) = \sigma^2; \qquad b(x) = 0; \qquad \int_0^1 a_0(u)\, \mathrm{d}u = \frac{5}{6}\,\sigma^2$$

and thus

$$\langle r_\varepsilon^2\rangle = \sigma^2 \int_0^1 F^2(x)\, \mathrm{d}x\, \varepsilon - \frac{1}{6}\,\sigma^2\big(F^2(0) + F^2(1)\big)\, \varepsilon^2 + O(\varepsilon^3).$$

Applied to the considered functions we have

$$\langle r_\varepsilon^2\rangle = \sigma^2\left(\varepsilon - \frac{\varepsilon^2}{3}\right) + O(\varepsilon^3) \qquad \text{for} \quad F(x) = 1,$$

$$\langle r_\varepsilon^2\rangle = \sigma^2\left(\frac{\varepsilon}{3} - \frac{1}{6}\,\varepsilon^2\right) + O(\varepsilon^3) \qquad \text{for} \quad F(x) = x,$$

$$\langle r_\varepsilon^2\rangle = \sigma^2\,\frac{\varepsilon}{2} + O(\varepsilon^3) \qquad \text{for} \quad F(x) = \sin(2\pi x),$$

$$\langle r_\varepsilon^2\rangle = \sigma^2\left(\frac{13}{15}\,\varepsilon - \frac{5}{6}\,\varepsilon^2\right) + O(\varepsilon^3) \quad \text{for} \quad F(x) = x^2 + 2x - 1$$

and these expansions coincide with the exact second moments from (1.90) up to terms of second order. Fig. 1.4 shows some examples of this coincidence.

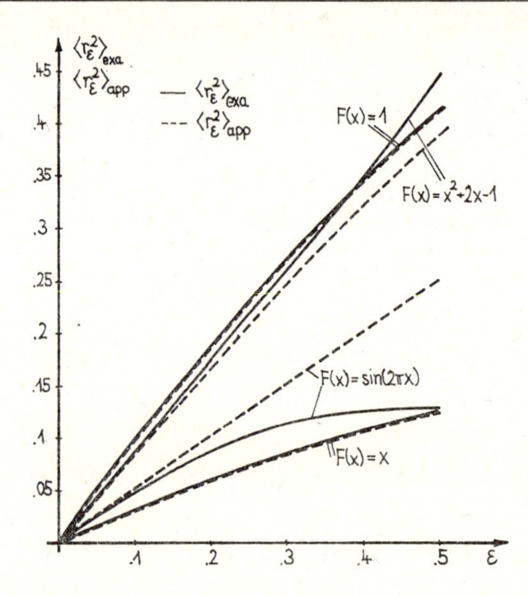

Fig. 1.4. Comparison between the exact second moments $\langle r_\varepsilon^2 \rangle_{\text{exa}}$ and the approximative second moments $\langle r_\varepsilon^2 \rangle_{\text{app}}$ for different functions $F(x)$

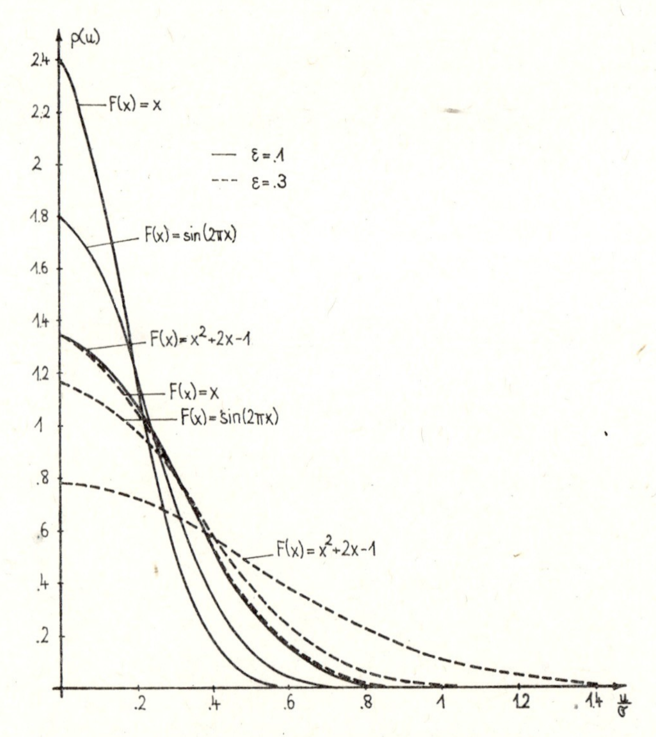

Fig. 1.5. Density functions for normally distributed random variables $r_\varepsilon(\omega)$ for different functions $F(x)$ and correlation lengths ε

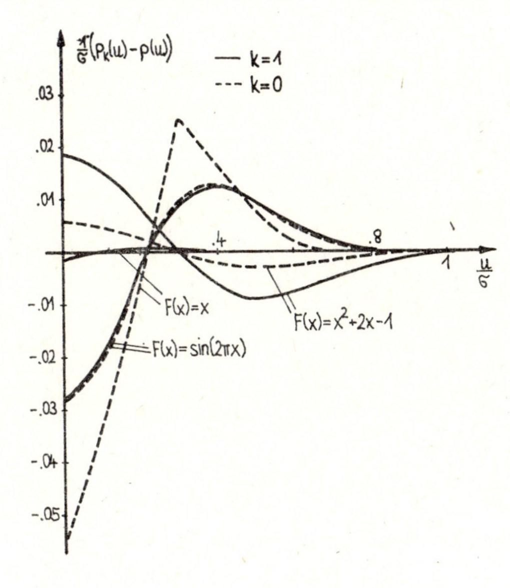

Fig. 1.6a. Deviations between the exact density $p(u)$ and the approximative densities $p_0(u)$, $p_1(u)$ for different functions $F(x)$ and the correlation length $\varepsilon = 0.1$

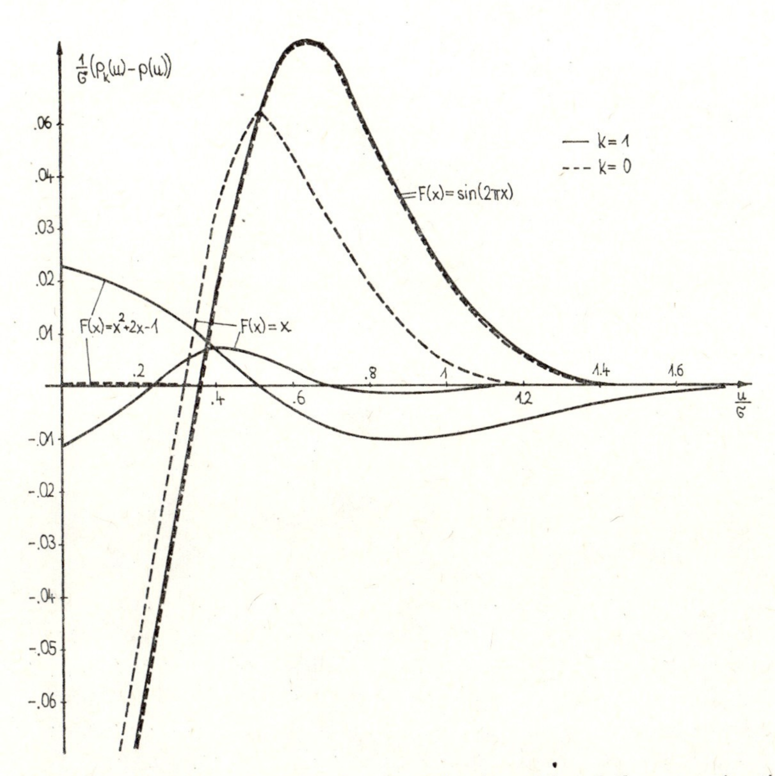

Fig. 1.6b. Deviations between the exact density $p(u)$ and the approximative densities $p_0(u)$, $p_1(u)$ for different functions $F(x)$ and the correlation length $\varepsilon = 0.3$

The density function

$$p(u) = p_1(u) + O(\varepsilon^{3/2})$$

is obtained from (1.87),

$$p_1(u) = \frac{1}{\sqrt{2\pi\varepsilon\,{}^2A_1}} \exp\left(-\frac{u^2}{2\varepsilon\,{}^2A_1}\right) \left\{1 + \frac{{}^2A_2}{2\,{}^2A_1} H_2\left(\frac{u}{\sqrt{\varepsilon\,{}^2A_1}}\right)\varepsilon\right\}.$$

$p_0(u)$ denotes the first approximation of the density $p(u)$:

$$p_0(u) = \frac{1}{\sqrt{2\pi\varepsilon\,{}^2A_1}} \exp\left(-\frac{u^2}{2\varepsilon\,{}^2A_1}\right).$$

The terms ${}^2A_k = {}^2A_k(F, F)$, $k = 1, 2$, are given by

$${}^2A_1(F, F) = \begin{cases} \sigma^2 & \text{for} \quad F(x) = 1, \\ \frac{1}{3}\sigma^2 & \text{for} \quad F(x) = x, \\ \frac{1}{2}\sigma^2 & \text{for} \quad F(x) = \sin(2\pi x), \\ \frac{13}{15}\sigma^2 & \text{for} \quad F(x) = x^2 + 2x - 1, \end{cases}$$

$${}^2A_2(F, F) = \begin{cases} -\frac{1}{3}\sigma^2 & \text{for} \quad F(x) = 1, \\ -\frac{1}{6}\sigma^2 & \text{for} \quad F(x) = x, \\ 0 & \text{for} \quad F(x) = \sin(2\pi x), \\ -\frac{5}{6}\sigma^2 & \text{for} \quad F(x) = x^2 + 2x - 1. \end{cases}$$

The exact density functions $p(u)$ are shown in Fig. 1.5 for different functions $F(x)$ and the correlation lengths $\varepsilon = 0.1$; $\varepsilon = 0.3$. The deviations between the exact density $p(u)$ and the approximative densities $p_0(u)$, $p_1(u)$ are illustrated in Fig. 1.6a and Fig. 1.6b for $\varepsilon = 0.1$ and $\varepsilon = 0.3$, respectively. The larger deviations for $\varepsilon = 0.3$ in comparison with $\varepsilon = 0.1$ can be seen very well. The maximum deviations are observed for $u = 0$. In general, $p_1(u)$ represents a better approximation of $p(u)$ than $p_0(u)$.

The exact calculation of the density function $p(u)$ needs the second moment of $r_\varepsilon(\omega)$. The determination of $\langle r_\varepsilon^2 \rangle$ in an exact way leads to extensive calculations which depend essentially on the function $F(x)$. The above theory allows easily to obtain an approximative second moment and an approximative density function, too.

1.6. Determination of statistical characteristics

1.6.1. Determination of $^2A_1$

First the term $^2A_1(F_1, F_2)$ is to be calculated. VOM SCHEIDT, PURKERT [3] have proved the relation

$$^2A_1(F_1, F_2) = \int_{\mathcal{D}_1 \cap \mathcal{D}_2} F_1(x)\, F_2(x)\, a(x)\, \mathrm{d}x$$

for bounded domains $\mathcal{D}_1$, $\mathcal{D}_2$. We will repeat this proof and we will widen simultaneously this calculation to the case of unbounded domains $\mathcal{D}_i$, $i = 1, 2$.

Using (1.11) the statistical characteristic $^2A_1(F_1, F_2)$ is determined by

$$^2A_1(F_1, F_2) = \lim_{\varepsilon \downarrow 0} \frac{1}{\varepsilon^m} \int_{\mathcal{E}(\{1,2\})} F_1(y_1)\, F_2(y_2)\, \langle f_{1\varepsilon}(y_1)\, f_{2\varepsilon}(y_2)\rangle\, \mathrm{d}y_1\, \mathrm{d}y_2$$

and $\mathcal{E}(\{1, 2\})$ is given by

$$\mathcal{E}(\{1, 2\}) = \big\{(y_1, y_2) \in \mathcal{D}_1 \times \mathcal{D}_2 \colon \{y_1, y_2\}\ \varepsilon\text{-neighbouring}\big\}.$$

In order to calculate $^2A_1$ it is sufficient to prove the following theorem.

Theorem 1.6. *Let $\big(f_{1\varepsilon}(x, \omega), f_{2\varepsilon}(x, \omega)\big)$ be a weakly correlated connected vector function on $\mathcal{D} \subset \mathbb{R}^m$ with correlation length ε and a.s. continuous sample functions and*

$$\langle f_{i\varepsilon}^2(x)\rangle \leqq c_2 < \infty, \qquad x \in \mathcal{D}.$$

Defining

$$r_{i\varepsilon}(\omega) = \frac{1}{\sqrt{\varepsilon^m}} \int_{\mathcal{D}} F_i(x)\, f_{i\varepsilon}(x, \omega)\, \mathrm{d}x, \qquad i = 1, 2,$$

the second moment $\langle r_{1\varepsilon} r_{2\varepsilon}\rangle$ fulfils the relation

$$\lim_{\varepsilon \downarrow 0} \langle r_{1\varepsilon} r_{2\varepsilon}\rangle = \int_{\mathcal{D}} F_1(x)\, F_2(x)\, a_{12}(x)\, \mathrm{d}x \tag{1.92}$$

where

- *$\mathcal{D}$ denotes a domain with piecewise smooth boundary,*
- *$F_i(x)$, $i = 1, 2$, are functions from $\mathbf{L}_2(\mathcal{D})$,*
- *$a_{12}(x)$ is defined by*

$$a_{12}(x) = \lim_{\varepsilon \downarrow 0} \frac{1}{\varepsilon^m} \int_{\mathcal{K}_\varepsilon(0)} \langle f_{1\varepsilon}(x)\, f_{2\varepsilon}(x + y)\rangle\, \mathrm{d}y$$

with $\mathcal{K}_\varepsilon(x) = \{y \in \mathbb{R}^m \colon |x - y| \leqq \varepsilon\}$ and is denoted as the intensity between $f_{1\varepsilon}$ and $f_{2\varepsilon}$.

Remark 1.1. *Let $\big(f_{1\varepsilon}(x, \omega), f_{2\varepsilon}(x, \omega)\big)$ be homogeneously connected, i.e.*

$$R_{12\varepsilon}(x, y) = \tilde{R}_{12\varepsilon}(y - x),$$

then the intensity is constant, $a_{12}(x) = a_{12}$.

Remark 1.2. *The intensity is bounded,*

$$|a_{12}(x)| \leqq V_1 c_2,$$

where V_1 denotes the volume of the sphere with unit radius.

Remark 1.3. *The intensity $a_{12}(x)$ is a continuous function if $R_{12\varepsilon}(x, y)$ satisfies the following condition: For an arbitrary $\eta > 0$ it exists a $\delta > 0$ so that*

$$|R_{12\varepsilon}(x, x+z) - R_{12\varepsilon}(y, y+z)| \leqq \eta$$

for all x, y with $|x - y| < \delta$ and $z \in \mathcal{K}_\varepsilon(0)$. This condition is fulfilled obviously for homogeneously connected fields $f_{1\varepsilon}$ and $f_{2\varepsilon}$.

The constancy of a_{12} can be deduced immediately from the definition.

Proof of Theorem 1.6. This proof is given in four sections. First we give the proof for $F_i(x) = \mathbf{1}_{\mathcal{B}_i}(y)$, $i = 1, 2$, where the Lebesgue measure λ of the boundary $\partial\mathcal{B}_i$ of $\mathcal{B}_i$ is assumed to be zero and $\mathbf{1}_{\mathcal{B}}(y)$ is defined by

$$\mathbf{1}_{\mathcal{B}}(y) \doteq \begin{cases} 1 & \text{for} \quad y \in \mathcal{B}, \\ 0 & \text{for} \quad y \notin \mathcal{B}. \end{cases}$$

Then we establish the assertion of this theorem for continuous functions; and finally for $F_i(x) \in \mathbf{L}_2(\mathcal{D})$. These first three steps assume that $\mathcal{D}$ is bounded. Now, in the fourth step we widen the result of Theorem 1.6 to the case of unbounded domains $\mathcal{D}$.

Step (1). We prove

$$\lim_{\varepsilon \downarrow 0} \frac{1}{\varepsilon^m} \int\limits_{\mathcal{B}_1} \int\limits_{\mathcal{B}_2} \langle f_{1\varepsilon}(y_1) \, f_{2\varepsilon}(y_2) \rangle \, \mathrm{d}y_1 \, \mathrm{d}y_2 = \int\limits_{\mathcal{B}_1 \cap \mathcal{B}_2} a(y) \, \mathrm{d}y \tag{1.93}$$

where $\mathcal{B}_1$ and $\mathcal{B}_2$ are subsets of the bounded set $\mathcal{D}$ with $\lambda(\partial\mathcal{B}_i) = 0$, $i = 1, 2$. It is easy to see that

$$\frac{1}{\varepsilon^m} \int\limits_{\mathcal{B}_1} \int\limits_{\mathcal{B}_2} R_{12\varepsilon}(x, y) \, \mathrm{d}x \, \mathrm{d}y = \frac{1}{\varepsilon^m} \int\limits_{\mathcal{B}_1 \setminus (\partial\mathcal{B}_1 \cup \partial\mathcal{B}_2)} \left(\int\limits_{\mathcal{B}_2 \cap \mathcal{K}_\varepsilon(x)} R_{12\varepsilon}(x, y) \, \mathrm{d}y \right) \mathrm{d}x .$$

With the aid of

$$\lim_{\varepsilon \downarrow 0} \frac{1}{\varepsilon^m} \int\limits_{\mathcal{B}_2 \cap \mathcal{K}_\varepsilon(x)} R_{12\varepsilon}(x, y) \, \mathrm{d}y = \begin{cases} a_{12}(x) & \text{for} \quad x \in \mathcal{B}_1 \cap \mathcal{B}_2 \setminus (\partial\mathcal{B}_1 \cap \partial\mathcal{B}_2), \\ 0 & \text{for} \quad x \in \mathcal{B}_1 \setminus (\mathcal{B}_2 \cap \partial\mathcal{B}_1 \cap \partial\mathcal{B}_2) \end{cases}$$

we obtain

$$\lim_{\varepsilon \downarrow 0} \frac{1}{\varepsilon^m} \int\limits_{\mathcal{B}_2 \cap \mathcal{K}_\varepsilon(x)} R_{12\varepsilon}(x, y) \, \mathrm{d}y = a_{12}(x) \, \mathbf{1}_{\mathcal{B}_1 \cap \mathcal{B}_2}(x) \qquad \text{a.s.}$$

for all $x \in \mathcal{B}_1$. Furthermore, we have

$$\left| \frac{1}{\varepsilon^m} \int\limits_{\mathcal{B}_2 \cap \mathcal{K}_\varepsilon(x)} R_{12\varepsilon}(x, y) \, \mathrm{d}y \right| \leqq V_1 c_2 .$$

Applying Lebesgue's theorem we can show that

$$\lim_{\varepsilon\downarrow 0}\int\limits_{\mathcal{B}_1}\left(\frac{1}{\varepsilon^m}\int\limits_{\mathcal{B}_2\cap\mathcal{K}_\varepsilon(x)} R_{12\varepsilon}(x,y)\,dy\right)dx = \int\limits_{\mathcal{B}_1\cap\mathcal{B}_2} a_{12}(x)\,dx.$$

The first part of this theorem is proved.

Step (2). We now prove (1.92) for continuous functions $F_i(x)$, $i = 1, 2$, on the bounded domain $\mathcal{D}$. Let $t_i(x)$, $i = 1, 2$, be step functions on $\mathcal{D}$, i.e.

$$t_i(y) = \sum_{p=1}^{p_i} c_{ip}\mathbf{1}_{\mathcal{E}_{ip}}(y)$$

where $\mathcal{E}_{ip} \subset \mathcal{D}$, $\lambda(\partial\mathcal{E}_{ip}) = 0$. Using (1.93) we obtain the relation (1.92) for step functions from

$$\lim_{\varepsilon\downarrow 0}\frac{1}{\varepsilon^m}\int\limits_{\mathcal{D}}\int\limits_{\mathcal{D}} t_1(x)\,t_2(y)\,\langle f_{1\varepsilon}(x)\,f_{2\varepsilon}(y)\rangle\,dx\,dy$$

$$= \lim_{\varepsilon\downarrow 0}\sum_{p,q=1}^{p_1,p_2} c_{1p}c_{2q}\frac{1}{\varepsilon^m}\int\limits_{\mathcal{E}_{1p}}\int\limits_{\mathcal{E}_{2q}} R_{12\varepsilon}(x,y)\,dx\,dy$$

$$= \sum_{p,q=1}^{p_1,p_2} c_{1p}c_{2q}\int\limits_{\mathcal{E}_{1p}\cap\mathcal{E}_{2q}} a_{12}(x)\,dx = \int\limits_{\mathcal{D}} t_1(y)\,t_2(y)\,a_{12}(y)\,dy.$$

Now we consider step functions $t_i(x)$ with respect to the continuous functions $F_i(x)$ where

$$|t_i(x) - F_i(x)| < \eta \quad \text{for all} \quad x \in \mathcal{D}, \quad i = 1, 2.$$

Let

$$\bar{r}_{i\varepsilon}(\omega) = \frac{1}{\sqrt{\varepsilon}}\int\limits_{\mathcal{D}} t_i(x)\,f_{i\varepsilon}(x,\omega)\,dx$$

then we find

$$\left|\langle r_{1\varepsilon}r_{2\varepsilon}\rangle - \int\limits_{\mathcal{D}} F_1(x)\,F_2(x)\,a_{12}(x)\,dx\right|$$

$$\leq |\langle r_{1\varepsilon}r_{2\varepsilon}\rangle - \langle\bar{r}_{1\varepsilon}\bar{r}_{2\varepsilon}\rangle| + \left|\langle\bar{r}_{1\varepsilon}\bar{r}_{2\varepsilon}\rangle - \int\limits_{\mathcal{D}} t_1(x)\,t_2(x)\,a_{12}(x)\,dx\right|$$

$$+ \left|\int\limits_{\mathcal{D}} t_1(x)\,t_2(x)\,a_{12}(x)\,dx - \int\limits_{\mathcal{D}} F_1(x)\,F_2(x)\,a_{12}(x)\,dx\right|.$$

The first summand on the right-hand side of the inequality above can be estimated by

$$|\langle r_{1\varepsilon}r_{2\varepsilon}\rangle - \langle\bar{r}_{1\varepsilon}\bar{r}_{2\varepsilon}\rangle| \leq \eta(A_1 + A_2 + \eta)\frac{1}{\varepsilon^m}\int\limits_{\mathcal{D}}\left(\int\limits_{\mathcal{D}\cap\mathcal{K}_\varepsilon(x)} |R_{12\varepsilon}(x,y)|\,dy\right)dx$$

$$\leq V_1 V(\mathcal{D})\,c_2\eta(A_1 + A_2 + \eta)$$

where $|F_i(x)| \leq A_i$ for all $x \in \mathcal{D}$, $i = 1, 2$, and $V(\mathcal{D})$ denotes the volume of $\mathcal{D}$ and the third summand by

$$\left|\int\limits_{\mathcal{D}} \big(t_1(x)\,t_2(x) - F_1(x)\,F_2(x)\big)\,a_{12}(x)\,dx\right| \leq V_1 V(\mathcal{D})\,c_2\eta(A_1 + A_2 + \eta)$$

using the relation $|a_{12}(x)| \leqq V_1 c_2$. The second summand converges to zero as $\varepsilon \downarrow 0$ and finally we obtain

$$\lim_{\varepsilon \downarrow 0} \left| \langle r_{1\varepsilon} r_{2\varepsilon} \rangle - \int_{\mathcal{D}} F_1(x)\, F_2(x)\, a_{12}(x)\, \mathrm{d}x \right| \leqq 2 V_1 V(\mathcal{D})\, c_2 \eta (A_1 + A_2 + \eta).$$

Since η can be an arbitrary positive real number the limit relation (1.92) is proved for continuous functions $F_i(x)$, $i = 1, 2$.

Step (3). In this part we assume that $F_i(x) \in \mathbf{L}_2(\mathcal{D})$ and $\mathcal{D}$ is bounded. Then a sequence $\{F_{ip}(x)\}_{p=1,2,\ldots}$ of continuous functions $F_{ip}(x)$ exists so that

$$\lim_{p \to \infty} \|F_i - F_{ip}\| = 0 \quad \text{for} \quad i = 1, 2$$

where $\|.\|$ denotes the norm in $\mathbf{L}_2(\mathcal{D})$. Defining

$$r_{ip\varepsilon}(\omega) \doteq \frac{1}{\sqrt{\varepsilon^m}} \int_{\mathcal{D}} F_{ip}(x)\, f_{i\varepsilon}(x, \omega)\, \mathrm{d}x$$

it follows that

$$\begin{aligned} &\left| \langle r_{1\varepsilon} r_{2\varepsilon} \rangle - \int_{\mathcal{D}} F_1(x)\, F_2(x)\, a_{12}(x)\, \mathrm{d}x \right| \\ &\leqq |\langle r_{1\varepsilon} r_{2\varepsilon} \rangle - \langle r_{1p\varepsilon} r_{2p\varepsilon} \rangle| + \left| \langle r_{1p\varepsilon} r_{2p\varepsilon} \rangle - \int_{\mathcal{D}} F_{1p}(x)\, F_{2p}(x)\, a_{12}(x)\, \mathrm{d}x \right| \\ &\quad + \left| \int_{\mathcal{D}} \big(F_{1p}(x)\, F_{2p}(x) - F_1(x)\, F_2(x)\big)\, a_{12}(x)\, \mathrm{d}x \right|. \end{aligned} \tag{1.94}$$

The second summand of the right-hand side of (1.94) converges to zero as $\varepsilon \downarrow 0$ if p is a fixed number. For the third summand we can estimate

$$\begin{aligned} &\left| \int_{\mathcal{D}} \big(F_{1p}(x)\, F_{2p}(x) - F_1(x)\, F_2(x)\big)\, a_{12}(x)\, \mathrm{d}x \right| \\ &\leqq V_1 c_2 (\|F_{1p}\|\, \|F_{2p} - F_2\| + \|F_2\|\, \|F_{1p} - F_1\|). \end{aligned}$$

The first summand leads to the following estimation:

$$\begin{aligned} &|\langle r_{1\varepsilon} r_{2\varepsilon} \rangle - \langle r_{1p\varepsilon} r_{2p\varepsilon} \rangle| \\ &\leqq \frac{1}{\varepsilon^m} \left| \int_{\mathcal{D}} \int_{\mathcal{D}} \big(F_1(x)\, F_2(y) - F_{1p}(x)\, F_{2p}(y)\big)\, R_{12\varepsilon}(x, y)\, \mathrm{d}x\, \mathrm{d}y \right| \\ &\leqq \frac{1}{\varepsilon^m} \int_{\mathcal{D}} \int_{\mathcal{D}} |F_1(x)|\, |F_2(y) - F_{2p}(y)|\, |R_{12}(x, y)|\, \mathrm{d}x\, \mathrm{d}y \\ &\quad + \frac{1}{\varepsilon^m} \int_{\mathcal{D}} \int_{\mathcal{D}} |F_{2p}(y)|\, |F_1(x) - F_{1p}(x)|\, |R_{12\varepsilon}(x, y)|\, \mathrm{d}x\, \mathrm{d}y \\ &\leqq \frac{1}{\varepsilon^m} \int_{\mathcal{D}} |F_2(x) - F_{2p}(x)| \int_{\mathcal{D} \cap \mathcal{K}_\varepsilon(x)} |F_1(y)|\, |R_{12\varepsilon}(y, x)|\, \mathrm{d}y\, \mathrm{d}x \\ &\quad + \frac{1}{\varepsilon^m} \int_{\mathcal{D}} |F_1(x) - F_{1p}(x)| \int_{\mathcal{D} \cap \mathcal{K}_\varepsilon(x)} |F_{2p}(y)|\, |R_{12\varepsilon}(x, y)|\, \mathrm{d}y\, \mathrm{d}x \end{aligned}$$

$$\leq \frac{1}{\varepsilon^m} \|F_2 - F_{2p}\| \left\| \int\limits_{\mathcal{D}\cap\mathcal{K}_\varepsilon(.)} |F_1(y)|\, |R_{12\varepsilon}(y, .)|\, \mathrm{d}y \right\|$$

$$+ \frac{1}{\varepsilon^m} \|F_1 - F_{1p}\| \left\| \int\limits_{\mathcal{D}\cap\mathcal{K}_\varepsilon(.)} |F_{2p}(y)|\, |R_{12\varepsilon}(., y)|\, \mathrm{d}y \right\|. \tag{1.95}$$

A further estimation shows that

$$\frac{1}{\varepsilon^m} \left\| \int\limits_{\mathcal{D}\cap\mathcal{K}_\varepsilon(.)} |F_{2p}(y)|\, |R_{12\varepsilon}(., y)|\, \mathrm{d}y \right\| \leq \frac{1}{\varepsilon^m} \left\| \left[\int\limits_{\mathcal{D}\cap\mathcal{K}_\varepsilon(.)} (F_{2p}(y))^2\, \mathrm{d}y \int\limits_{\mathcal{D}\cap\mathcal{K}_\varepsilon(.)} (R_{12\varepsilon}(., y))^2\, \mathrm{d}y \right]^{\frac{1}{2}} \right\|$$

$$\leq \frac{1}{\varepsilon^{m/2}} c_2 \sqrt{V_1} \left\| \left[\int\limits_{\mathcal{D}\cap\mathcal{K}_\varepsilon(.)} (F_{2p}(y))^2\, \mathrm{d}y \right]^{1/2} \right\| \leq \frac{1}{\varepsilon^{m/2}} c_2 \sqrt{V_1} \left[\int\limits_{\mathcal{D}} \int\limits_{\mathcal{D}\cap\mathcal{K}_\varepsilon(x)} (F_{2p}(y))^2\, \mathrm{d}y\, \mathrm{d}x \right]^{1/2}$$

$$\leq \frac{1}{\varepsilon^{m/2}} c_2 \sqrt{V_1} \left[\int\limits_{\mathcal{D}} (F_{2p}(y))^2\, \mathrm{d}y \int\limits_{\mathcal{D}\cap\mathcal{K}_\varepsilon(y)} \mathrm{d}x \right]^{1/2} \leq c_2 V_1 \|F_{2p}\| \tag{1.96}$$

if Fubini's theorem is applied. Thus

$$\lim_{p\to\infty} |\langle r_{1\varepsilon} r_{2\varepsilon}\rangle - \langle r_{1p\varepsilon} r_{2p\varepsilon}\rangle| \leq V_1 c_2 \lim_{p\to\infty} [\|F_2 - F_{2p}\|\, \|F_1\| + \|F_1 - F_{1p}\|\, \|F_{2p}\|] = 0$$

and therefore

$$\lim_{\varepsilon\downarrow 0} \left| \langle r_{1\varepsilon} r_{2\varepsilon}\rangle - \int\limits_{\mathcal{D}} F_1(x)\, F_2(x)\, a_{12}(x)\, \mathrm{d}x \right|$$

$$\leq V_1 c_2 [(\|F_{1p}\| + \|F_1\|)\, \|F_2 - F_{2p}\| + (\|F_{2p}\| + \|F_2\|)\, \|F_1 - F_{1p}\|].$$

By means of this inequality the proof is complete as to step (3).

Step (4). Let $F_i \in \mathbf{L}_2(\mathcal{D})$ and $\mathcal{D}$ unbounded. We define

$$\mathcal{D}_\varrho = \mathcal{K}_\varrho(0) \cap \mathcal{D}$$

and

$$\bar{r}_{i\varepsilon}(\omega) \doteq \frac{1}{\varepsilon^m} \int\limits_{\mathcal{D}_\varrho} F_i(x)\, f_{i\varepsilon}(x)\, \mathrm{d}x; \qquad \bar{\bar{r}}_{i\varepsilon}(\omega) \doteq \frac{1}{\varepsilon^m} \int\limits_{\mathcal{D}\setminus\mathcal{D}_\varrho} F_i(x)\, f_{i\varepsilon}(x)\, \mathrm{d}x$$

and obtain

$$r_{i\varepsilon}(\omega) = \bar{r}_{i\varepsilon}(\omega) + \bar{\bar{r}}_{i\varepsilon}(\omega) \quad \text{for} \quad i = 1, 2.$$

The estimations (1.95) and (1.96) lead to

$$\left| \frac{1}{\varepsilon^m} \int\limits_{\mathcal{G}_1}\int\limits_{\mathcal{G}_2} G_1(x)\, G_2(y)\, R_{12\varepsilon}(x, y)\, \mathrm{d}x\, \mathrm{d}y \right| \leq c_2 V_1 \|G_1\|_{\mathcal{G}_1} \|G_2\|_{\mathcal{G}} \tag{1.97}$$

where $\|G\|_{\mathcal{G}}$ is defined by

$$\|G\|_{\mathcal{G}} \doteq \left[\int\limits_{\mathcal{G}} G^2(x)\, \mathrm{d}x \right]^{1/2}.$$

Now we estimate

$$\left|\langle r_{1\varepsilon} r_{2\varepsilon}\rangle - \int\limits_{\mathscr{D}} F_1(x)\, F_2(x)\, a_{12}(x)\, \mathrm{d}x\right|$$
$$\leqq \left|\langle \bar{r}_{1\varepsilon} \bar{r}_{2\varepsilon}\rangle - \int\limits_{\mathscr{D}_\varrho} F_1(x)\, F_2(x)\, a_{12}(x)\, \mathrm{d}x\right| + |\langle \bar{r}_{1\varepsilon} \bar{\bar{r}}_{2\varepsilon}\rangle| + |\langle \bar{\bar{r}}_{1\varepsilon} \bar{r}_{2\varepsilon}\rangle|$$
$$+ |\langle \bar{\bar{r}}_{1\varepsilon} \bar{\bar{r}}_{2\varepsilon}\rangle| + \left|\int\limits_{\mathscr{D}\setminus\mathscr{D}_\varrho} F_1(x)\, F_2(x)\, a_{12}(x)\, \mathrm{d}x\right|.$$

The first summand of the right-hand side converges to zero as $\varepsilon \downarrow 0$ since $\mathscr{D}_\varrho$ is bounded. Furthermore, by means of (1.97) we have

$$|\langle \bar{r}_{1\varepsilon} \bar{\bar{r}}_{2\varepsilon}\rangle| + |\langle \bar{\bar{r}}_{1\varepsilon} \bar{r}_{2\varepsilon}\rangle| + |\langle \bar{\bar{r}}_{1\varepsilon} \bar{\bar{r}}_{2\varepsilon}\rangle|$$
$$\leqq c_2 V_1 [\|F_1\|_{\mathscr{D}_\varrho} \|F_2\|_{\mathscr{D}\setminus\mathscr{D}_\varrho} + \|F_1\|_{\mathscr{D}\setminus\mathscr{D}_\varrho} \|F_2\|_{\mathscr{D}_\varrho} + \|F_1\|_{\mathscr{D}\setminus\mathscr{D}_\varrho} \|F_2\|_{\mathscr{D}\setminus\mathscr{D}_\varrho}]$$

and the last summand leads to the estimation

$$\left|\int\limits_{\mathscr{D}\setminus\mathscr{D}_\varrho} F_1(x)\, F_2(x)\, a_{12}(x)\, \mathrm{d}x\right| \leqq c_2 V_1 \|F_1\|_{\mathscr{D}\setminus\mathscr{D}_\varrho} \|F_2\|_{\mathscr{D}\setminus\mathscr{D}_\varrho}.$$

Summarizing we obtain

$$\lim_{\varepsilon\downarrow 0} \left|\langle r_{1\varepsilon} r_{2\varepsilon}\rangle - \int\limits_{\mathscr{D}} F_1(x)\, F_2(x)\, a_{12}(x)\, \mathrm{d}x\right|$$
$$\leqq c_2 V_1 [\|F_1\|_{\mathscr{D}} \|F_2\|_{\mathscr{D}\setminus\mathscr{D}_\varrho} + \|F_1\|_{\mathscr{D}\setminus\mathscr{D}_\varrho} \|F_2\|_{\mathscr{D}} + 2\, \|F_1\|_{\mathscr{D}\setminus\mathscr{D}_\varrho} \|F_2\|_{\mathscr{D}\setminus\mathscr{D}_\varrho}]$$

and because of

$$\lim_{\varrho\to 0} \|F_i\|_{\mathscr{D}\setminus\mathscr{D}_\varrho} = 0$$

the statement of this theorem. ◀

Theorem 1.6 contains the determination of ${}^2A_1(F_1, F_2)$ assuming weak conditions to the functions $F_1(x)$, $F_2(x)$. We use this result in order to give the first approximation of the density function of linear functionals with the aid of these weak conditions (see also vom Scheidt, Purkert [3]).

Lemma 1.7. *Let* $(f_{1\varepsilon}(x, \omega), \ldots, f_{l\varepsilon}(x, \omega))$ *be a weakly correlated connected vector function on* $\mathscr{D} \subset \mathbb{R}^m$ *with* $\langle |f_{i\varepsilon}(x)|^k\rangle \leqq c_k < \infty$ *for all* $x \in \mathscr{D}$, $i = 1, 2, \ldots, l$. *Then the convergence relation*

$$\lim_{\varepsilon\downarrow 0} \frac{1}{\sqrt{\varepsilon^{mk}}} \int\limits_{\mathscr{E}(\{1,2,\ldots,k\})} \prod_{p=1}^{k} \bar{F}_p(y_p) \left\langle \prod_{p=1}^{k} \bar{f}_{p\varepsilon}(y_p)\right\rangle \mathrm{d}y_1 \ldots \mathrm{d}y_p = 0$$

is true for $k \geqq 3$ *assuming* $\bar{F}_p = F_{i_p j_p} \in \mathbf{L}_2(\mathscr{D})$ *(see Section 1.3).*

Proof. An easy inequality shows

$$\frac{1}{\varepsilon^m} \iint\limits_{\mathscr{E}(\{1,2\})} |\bar{F}_1(y_1)\, \bar{F}_2(y_2)|\, \mathrm{d}y_1\, \mathrm{d}y_2$$
$$= \frac{1}{\varepsilon^m} \int\limits_{\mathscr{D}} \int\limits_{\mathscr{D}\cap\mathscr{K}_\varepsilon(y_1)} |\bar{F}_1(y_1)\, \bar{F}_2(y_2)|\, \mathrm{d}y_2\, \mathrm{d}y_1 \leqq V_1 \|\bar{F}_1\|\, \|\bar{F}_2\|$$

taking into consideration (1.96). Furthermore, we have

$$\int_{\mathcal{D}\cap\mathcal{K}_\varepsilon(x)} |\bar{F}_i(y)|\,\mathrm{d}y \leqq \sqrt{V_1\varepsilon^m \int_{\mathcal{D}\cap\mathcal{K}_\varepsilon(x)} \bar{F}_i^2(y)\,\mathrm{d}y} \leqq \sqrt{V_1\varepsilon^m A_{i\varepsilon}} \tag{1.98}$$

where

$$A_{i\varepsilon} = \sup_{x\in\mathcal{D}} \int_{\mathcal{D}\cap\mathcal{K}_\varepsilon(x)} \bar{F}_i^2(y)\,\mathrm{d}y\,.$$

The relation $\lim_{\varepsilon\downarrow 0} A_{i\varepsilon} = 0$ follows from the absolute continuity of the Lebesgue integral. Using these estimations we can show that

$$\frac{1}{\sqrt{\varepsilon}^{mk}} \int_{\mathcal{D}} \int_{\mathcal{D}\cap\mathcal{K}_{(k-1)\varepsilon}(y_1)} \cdots \int_{\mathcal{D}\cap\mathcal{K}_{(k-1)\varepsilon}(y_{k-1})} \left| \prod_{p=1}^{k} \bar{F}_p(y_p) \left\langle \prod_{p=1}^{k} \bar{f}_{p\varepsilon}(y_p) \right\rangle \right| \mathrm{d}y_k\,\mathrm{d}y_{k-1}\ldots\mathrm{d}y_1$$

$$\leqq \frac{1}{\sqrt{\varepsilon}^{m(k-2)}} c_k \sqrt{V_1((k-1)\,\varepsilon)^m}^{\,k-2} \sqrt{\prod_{p=3}^{k} A_{p\varepsilon}}\, \frac{1}{\varepsilon^m} \int_{\mathcal{D}} \int_{\mathcal{D}\cap\mathcal{K}_{(k-1)\varepsilon}(y_1)} |\bar{F}_1(y_1)|\,|\bar{F}_2(y_2)|\,\mathrm{d}y_2\,\mathrm{d}y_1$$

$$\leqq c_k V_1^{k/2}(k-1)^{mk/2} \sqrt{\prod_{p=3}^{k} A_{p\varepsilon}}\, \|\bar{F}_1\|\,\|\bar{F}_2\|$$

which completes the proof. ◀

The following theorem includes the convergence of the k-th moments of random variables (1.1) where $k > 2$. The convergence for $k = 2$ was proved in Theorem 1.6.

Theorem 1.7. *Let $(f_{1\varepsilon}(x,\omega), \ldots, f_{l\varepsilon}(x,\omega))$ be a weakly correlated connected vector function on $\mathcal{D} \subset \mathbb{R}^m$ with*

$$\langle |f_{i\varepsilon}(x)|^k \rangle \leqq c_k < \infty \quad \textit{for all} \quad x \in \mathcal{D}, \quad i = 1, 2, \ldots, l.$$

The limit value of a k-th moment

$$\left\langle \prod_{p=1}^{k} r_{i_p j_p} \right\rangle$$

is given by

$$\lim_{\varepsilon\downarrow 0} \left\langle \prod_{p=1}^{k} r_{i_p j_p \varepsilon} \right\rangle = \lim_{\varepsilon\downarrow 0} \left\langle \prod_{p=1}^{k} \bar{r}_{p\varepsilon} \right\rangle = \begin{cases} \sum\limits_{\{p_1,p_2\},\ldots,\{p_{k-1},p_k\}} A_{p_1p_2} \cdots A_{p_{k-1}p_k} & \textit{for } k \textit{ even}, \\ 0 & \textit{for } k \textit{ odd} \end{cases}$$

where

$$A_{pq} \doteq \lim_{\varepsilon\downarrow 0} \langle \bar{r}_{p\varepsilon} \bar{r}_{q\varepsilon} \rangle\,.$$

The sum above is taken over all nonequivalent separations of $\{1, 2, \ldots, k\}$ in pairs. Besides, the same conditions as in Theorem 1.6 are assumed.

Proof. Using (1.6) we have

$$\mathcal{D}^k = \bigcup_{p=1}^{k} \bigcup_{\{\mathcal{J}_1,\ldots,\mathcal{J}_p\}} \mathcal{B}(\mathcal{J}_1, \ldots, \mathcal{J}_p)$$

and obtain

$$\left\langle \prod_{p=1}^{k} \bar{r}_{p\varepsilon} \right\rangle = \sum_{s=1}^{k} \sum_{\mathcal{J}_1,\dots,\mathcal{J}_s} \frac{1}{\sqrt{\varepsilon}^{km}} \int_{\mathcal{B}(\mathcal{J}_1,\dots,\mathcal{J}_s)} \cdots \int \prod_{p=1}^{s} \left\{ \prod_{q\in\mathcal{J}_p} \bar{F}_q(y_q) \left\langle \prod_{q\in\mathcal{J}_q} \bar{f}_{q\varepsilon}(y_q) \right\rangle \right\} \mathrm{d}y_1 \dots \mathrm{d}y_k . \tag{1.99}$$

It is possible to restrict to $k_p \geqq 2$, $p = 1, 2, \dots, s$, (k_p denotes the number of elements of $\mathcal{J}_p$) because all the other summands vanish since $\langle f_{i\varepsilon}(x) \rangle = 0$, $i = 1, 2, \dots, l$. By means of (1.4) we obtain for a summand $A_s = A_s(\mathcal{J}_1, \dots, \mathcal{J}_s)$ of (1.99)

$$A_s = A_{s1} - A_{s2}$$

where

$$A_{s1} = \frac{1}{\sqrt{\varepsilon}^{km}} \int_{\mathcal{E}(\mathcal{J}_1)\times\cdots\times\mathcal{E}(\mathcal{J}_s)} \cdots \int \prod_{p=1}^{s} \left\{ \prod_{q\in\mathcal{J}_q} \bar{F}_q(y_q) \left\langle \prod_{q\in\mathcal{J}_q} \bar{f}_{q\varepsilon}(y_q) \right\rangle \right\} \mathrm{d}y_1 \dots \mathrm{d}y_k ,$$

$$A_{s2} = \frac{1}{\sqrt{\varepsilon}^{km}} \int_{\mathcal{F}(\mathcal{J}_1,\dots,\mathcal{J}_s)} \cdots \int \prod_{p=1}^{s} \left\{ \prod_{q\in\mathcal{J}_q} \bar{F}_q(y_q) \left\langle \prod_{q\in\mathcal{J}_q} \bar{f}_{q\varepsilon}(y_q) \right\rangle \right\} \mathrm{d}y_1 \dots \mathrm{d}y_k .$$

Furthermore, it follows that

$$A_{s1} = \prod_{p=1}^{s} \frac{1}{\sqrt{\varepsilon}^{k_p m}} \int_{\mathcal{E}(\mathcal{J}_p)} \prod_{q\in\mathcal{J}_q} \bar{F}_q(y_q) \left\langle \prod_{q\in\mathcal{J}_q} \bar{f}_{q\varepsilon}(y_q) \right\rangle \prod_{q\in\mathcal{J}_q} \mathrm{d}y_q .$$

The relation

$$\lim_{\varepsilon\downarrow 0} A_{s1} = 0 \text{ for } k \text{ odd or for } k \text{ even and at least one } k_p \geqq 3$$

can be obtained from Lemma 1.7. Now it remains to consider the case k even and $k_p = 2$ for $p = 1, 2, \dots, k/2$. Then, applying Theorem 1.6 it follows

$$\begin{aligned}\lim_{\varepsilon\downarrow 0} A_{k/2,1} &= \prod_{p=1}^{k/2} \lim_{\varepsilon\downarrow 0} \frac{1}{\varepsilon^m} \iint_{\mathcal{E}(\{s_p,t_p\})} \bar{F}_{s_p}(y_{s_p}) \bar{F}_{t_p}(y_{t_p}) \langle \bar{f}_{s_p\varepsilon}(y_{s_p}) \bar{f}_{t_p\varepsilon}(y_{t_p}) \rangle \, \mathrm{d}y_{s_p} \, \mathrm{d}y_{t_p} \\ &= \prod_{p=1}^{k/2} \lim_{\varepsilon\downarrow 0} \langle \bar{r}_{s_p\varepsilon} \bar{r}_{t_p\varepsilon} \rangle .\end{aligned}$$

We can estimate the term A_{s2} by

$$\begin{aligned}|A_{s2}| \leqq \sum_{\{p_1,p_2\},\{p_3,\dots,p_s\}} &\frac{1}{\sqrt{\varepsilon}^{m(k_{p_1}+k_{p_2})}} \int_{\mathcal{E}(\mathcal{J}_{p_1}\cup\mathcal{J}_{p_2})} \cdots \int \left| \prod_{\substack{q\in\mathcal{J}_{p_1} \\ q\in\mathcal{J}_{p_2}}} \bar{F}_q(y_q) \prod_{p=p_1,p_2} \left\langle \prod_{q\in\mathcal{J}_p} \bar{f}_{q\varepsilon}(y_q) \right\rangle \right| \prod_{\substack{q\in\mathcal{J}_{p_1} \\ q\in\mathcal{J}_{p_2}}} \mathrm{d}y_q \\ &\times \prod_{i=3}^{s} \frac{1}{\sqrt{\varepsilon}^{mk_{p_i}}} \int_{\mathcal{E}(\mathcal{J}_{p_i})} \cdots \int \left| \prod_{q\in\mathcal{J}_{p_i}} \bar{F}_q(y_q) \left\langle \prod_{q\in\mathcal{J}_{p_i}} \bar{f}_{q\varepsilon}(y_q) \right\rangle \right| \prod_{q\in\mathcal{J}_{p_i}} \mathrm{d}y_q\end{aligned}$$

using (1.5) if $\mathcal{J}_p = \{s_p, t_p\}$ is put. The first factor of every summand converges to zero as $\varepsilon \downarrow 0$ because $k_{p_1} + k_{p_2} \geqq 4$ and the remaining factors also converge to zero as $\varepsilon \downarrow 0$ for $k_{p_i} \geqq 3$, or they are bounded for $k_{p_i} = 2$. It follows that

$$\lim_{\varepsilon\downarrow 0} A_{s2}(\mathcal{J}_1, \dots, \mathcal{J}_s) = 0 .$$

Hence we obtain

$$\lim_{\varepsilon\downarrow 0} A_s(\mathcal{J}_1, \ldots, \mathcal{J}_s) = \begin{cases} \prod\limits_{p=1}^{k/2} \lim\limits_{\varepsilon\downarrow 0} \langle \bar{r}_{s_p\varepsilon} \bar{r}_{t_p\varepsilon} \rangle & \text{for} \quad k \text{ even and } s = k/2,\ k_p = 2 \text{ for all } p, \\ 0 & \text{otherwise} \end{cases}$$

and finally from (1.99)

$$\lim_{\varepsilon\downarrow 0} \left\langle \prod_{p=1}^{k} \bar{r}_{p\varepsilon} \right\rangle = \begin{cases} \sum\limits_{\{p_1,q_1\},\ldots,\{p_{k/2},q_{k/2}\}} \prod\limits_{i=1}^{k/2} \lim\limits_{\varepsilon\downarrow 0} \langle \bar{r}_{p_i\varepsilon} \bar{r}_{q_i\varepsilon} \rangle & \text{for} \quad k \text{ even}, \\ 0 & \text{for} \quad k \text{ odd}. \end{cases}$$ ◀

The next theorem establishes convergence in distribution of the random vector $(r_{1\varepsilon}, \ldots, r_{n\varepsilon})$ to a Gaussian random vector $(\xi_1, \ldots, \xi_n)$ where

$$r_{i\varepsilon}(\omega) = \frac{1}{\sqrt{\varepsilon^m}} \sum_{p=1}^{l} r_{ip\varepsilon}(\omega); \quad r_{ip\varepsilon}(\omega) = \int_{\mathcal{D}} F_{ip}(x)\, f_{p\varepsilon}(x, \omega)\, \mathrm{d}x.$$

Theorem 1.8. *Let $\big(f_{1\varepsilon}(x, \omega), \ldots, f_{l\varepsilon}(x, \omega)\big)$ be a weakly correlated connected vector function on $\mathcal{D} \subset \mathbb{R}^m$ with*

$$\langle |f_{i\varepsilon}(x)|^k \rangle \leqq c_k < \infty \quad \text{for all } x \in \mathcal{D}, \quad i = 1, 2, \ldots, l, \quad \text{and for all } k \geqq 1,$$

where the sample functions of these random fields are continuous a.s. Furthermore,

- *$\mathcal{D}$ denotes a domain with piecewise smooth boundary,*
- *$F_{ip}(x)$, $i = 1, 2, \ldots, n$; $p = 1, 2, \ldots, l$ are functions from $\mathbf{L}_2(\mathcal{D})$*
- *the intensity $a_{ij}(x)$ between $f_{i\varepsilon}(x, \omega)$ and $f_{j\varepsilon}(x, \omega)$ is defined by*

$$a_{ij}(x) = \lim_{\varepsilon\downarrow 0} \frac{1}{\varepsilon^m} \int_{\mathcal{K}_\varepsilon(0)} \langle f_{i\varepsilon}(x)\, f_{j\varepsilon}(x + y) \rangle\, \mathrm{d}y \quad \text{for} \quad i, j = 1, 2, \ldots, l.$$

Then the convergence in distribution

$$\lim_{\varepsilon\downarrow 0} \big(r_{1\varepsilon}(\omega), \ldots, r_{n\varepsilon}(\omega)\big) = \big(\xi_1(\omega), \ldots, \xi_n(\omega)\big)$$

is obtained. The random vector $\big(\xi_1(\omega), \ldots, \xi_n(\omega)\big)$ is a Gaussian vector with mean

$$\langle \xi_i \rangle = 0 \quad \text{for} \quad i = 1, 2, \ldots, n$$

and correlation relations

$$\langle \xi_i \xi_j \rangle = \sum_{p,q=1}^{l} \int_{\mathcal{D}} F_{ip}(x)\, F_{jq}(x)\, a_{pq}(x)\, \mathrm{d}x \quad \text{for} \quad i, j = 1, 2, \ldots, n.$$

Proof. In order to obtain some facts of this proof we refer to BILLINGSLEY [1]. Using Theorem 1.6 and Theorem 1.7 for $i_q \in \{1, 2, \ldots, n\}$ we have

$$\lim_{\varepsilon\downarrow 0} \left\langle \prod_{q=1}^{k} r_{i_q\varepsilon} \right\rangle = \left\langle \prod_{q=1}^{k} \xi_{i_q} \right\rangle \tag{1.100}$$

if we take into consideration the relations

$$\lim_{\varepsilon\downarrow 0} \langle r_{i_1\varepsilon} r_{i_2\varepsilon}\rangle = \sum_{p,q=1}^{l} \lim_{\varepsilon\downarrow 0} \langle r_{i_1p\varepsilon} r_{i_2q\varepsilon}\rangle$$

$$= \sum_{p,q=1}^{l} \int_{\mathcal{D}} F_{i_1p}(x)\, F_{i_2q}(x)\, a_{pq}(x)\, \mathrm{d}x = \langle \xi_{i_1}\xi_{i_2}\rangle,$$

for k odd

$$\lim_{\varepsilon\downarrow 0} \left\langle \prod_{q=1}^{k} r_{i_q\varepsilon}\right\rangle = \sum_{j_1,\dots,j_k=1}^{l} \lim_{\varepsilon\downarrow 0} \left\langle \prod_{q=1}^{k} r_{i_qj_q\varepsilon}\right\rangle = 0,$$

and for k even

$$\lim_{\varepsilon\downarrow 0} \left\langle \prod_{q=1}^{k} r_{i_q\varepsilon}\right\rangle = \sum_{j_1,\dots,j_k=1}^{l} \lim_{\varepsilon\downarrow 0} \left\langle \prod_{q=1}^{k} r_{i_qj_q\varepsilon}\right\rangle = \sum_{j_1,\dots,j_k=1}^{l} \sum_{\{q_1,q_2\},\dots,\{q_{k-1},q_k\}} A_{q_1q_2} \dots A_{q_{k-1}q_k}$$

$$= \sum_{\{q_1,q_2\},\dots,\{q_{k-1},q_k\}} \langle \xi_{i_{q_1}}\xi_{i_{q_2}}\rangle \dots \langle \xi_{i_{q_{k-1}}}\xi_{i_{q_k}}\rangle = \left\langle \prod_{q=1}^{k} \xi_{i_q}\right\rangle$$

where

$$\sum_{j_p,j_q=1}^{l} A_{pq} = \sum_{j_p,j_q=1}^{l} \int_{\mathcal{D}} F_{i_pj_p}(x)\, F_{i_qj_q}(x)\, a_{j_pj_q}(x)\, \mathrm{d}x = \langle \xi_{i_p}\xi_{i_q}\rangle.$$

The convergence of the distribution function follows from the convergence of the characteristic functions

$$\lim_{\varepsilon\downarrow 0} \left\langle \exp\left(\mathrm{i}\sum_{p=1}^{n} t_p r_{p\varepsilon}\right)\right\rangle = \left\langle \exp\left(\mathrm{i}\sum_{p=1}^{n} t_p\xi_p\right)\right\rangle \quad \text{for every } t_p \in \mathbb{R}^1, \quad p = 1, 2, \dots, n. \tag{1.101}$$

Since $(\xi_1(\omega), \dots, \xi_n(\omega))$ is a Gaussian random vector the right-hand side of (1.101) is a continuous function of $(t_1, \dots, t_n)$. Furthermore, we note that

$$\left| \mathrm{e}^{\mathrm{i}h} - \sum_{q=0}^{2k-1} \frac{1}{q!} (\mathrm{i}h)^q \right| \leqq \frac{h^{2k}}{(2k)!}$$

and obtain the inequality

$$\overline{\lim_{\varepsilon\downarrow 0}} \left| \left\langle \exp\left(\mathrm{i}\sum_{p=1}^{n} t_p r_{p\varepsilon}\right)\right\rangle - \sum_{q=0}^{2k-1} \frac{1}{q!} \left\langle \left(\mathrm{i}\sum_{p=1}^{n} t_p\xi_p\right)^q\right\rangle \right| \leqq \frac{1}{(2k)!} \left\langle \left(\sum_{p=1}^{n} t_p\xi_p\right)^{2n}\right\rangle$$

applying the convergence of the moments of $r_{p\varepsilon}(\omega)$ to the corresponding moments of $\xi_p(\omega)$. It is easy to see that

$$\overline{\lim_{\varepsilon\downarrow 0}} \left| \left\langle \exp\left(\mathrm{i}\sum_{p=1}^{n} t_p r_{p\varepsilon}\right)\right\rangle - \sum_{q=0}^{\infty} \frac{1}{q!} \left\langle \left(\mathrm{i}\sum_{p=1}^{n} t_p\xi_p\right)^q\right\rangle \right|$$

$$\leqq \frac{1}{(2k)!} \left\langle \left(\sum_{p=1}^{n} t_p\xi_p\right)^{2k}\right\rangle + \left| \sum_{q=2k} \frac{1}{q!} \left\langle \left(\mathrm{i}\sum_{p=1}^{n} t_p\xi_p\right)^q\right\rangle \right|.$$

The series $\sum_{q=0}^{\infty} \frac{1}{q!} \left\langle \left(\mathrm{i} \sum_{p=1}^{n} t_p \xi_p \right)^q \right\rangle$ converges since $\sum_{p=1}^{n} t_p \xi_p$ is a Gaussian random variable.
The relation

$$\lim_{k\to\infty} \left[\frac{1}{(2k)!} \left\langle \left(\sum_{p=1}^{n} t_p \xi_p \right)^{2k} \right\rangle + \left| \sum_{q=2k} \frac{1}{q!} \left\langle \left(\mathrm{i} \sum_{p=1}^{n} t_p \xi_p \right)^q \right\rangle \right| \right] = 0$$

completes the proof. ◀

Comparing these results we have shown that a random vector $(r_{1\varepsilon}, \ldots, r_{n\varepsilon})$ with

$$r_{i\varepsilon}(\omega) = \frac{1}{\sqrt{\varepsilon^m}} \sum_{p=1}^{l} r_{ip\varepsilon}(\omega); \qquad r_{ip\varepsilon}(\omega) = \int_{\mathcal{D}} F_{ip}(x)\, f_{p\varepsilon}(x, \omega)\, \mathrm{d}x$$

also for unbounded domains $\mathcal{D}$ and $F_{ip}(x) \in \mathbf{L}_2(\mathcal{D})$ is normally distributed where the first moments are given by

$$\langle r_{i\varepsilon} \rangle \approx 0,$$

$$\langle r_{i\varepsilon} r_{j\varepsilon} \rangle \approx \sum_{p,q=1}^{l} \int_{\mathcal{D}} F_{ip}(x)\, F_{jq}(x)\, a_{pq}(x)\, \mathrm{d}x\,.$$

The convergence of all moments is included in this investigation. In the case of

$$F_1(x) \in \mathbf{C}^0(\mathcal{D}_1); \qquad F_2(x) \in \mathbf{C}^0(\mathcal{D}_2)$$

we apply (1.92) to the functions $F_1(x)\, \mathbf{1}_{\mathcal{D}_1}(x)$ and $F_2(x)\, \mathbf{1}_{\mathcal{D}_2}(x)$ and obtain

$$^2A_1(F_1, F_2) = \int_{\mathcal{D}_1 \cap \mathcal{D}_2} F_1(x)\, F_2(x)\, a_{12}(x)\, \mathrm{d}x\,. \tag{1.102}$$

Hence the statistical characteristic $^2A_1(F_1, F_2)$ is determined completely having the intensity $a_{12}(x)$ between the fields $f_{1\varepsilon}(x, \omega)$ and $f_{2\varepsilon}(x, \omega)$.

1.6.2. Determination of $^2A_2$ in the one-dimensional case

Using the conditions written at the beginning of Section 1.3 (1.11) leads to the equation defining $^2A_2$:

$$J(F_1, F_2) \doteq \iint_{\mathcal{E}(\mathcal{J}_2)} F_1(x)\, F_2(y)\, \langle f_{1\varepsilon}(x)\, f_{2\varepsilon}(y) \rangle\, \mathrm{d}x\, \mathrm{d}y$$

$$= {}^2A_1(F_1, F_2)\, \varepsilon^m + {}^2A_2(F_1, F_2)\, \varepsilon^{m+1} + o(\varepsilon^{m+1})\,. \tag{1.103}$$

In the case of $m = 1$ the goal consists in the calculation of $^2A_2$ for F_1 on $\mathcal{D}_1$ and F_2 on $\mathcal{D}_2$ having the properties:

- $F_i \in \mathbf{C}^1(\mathcal{D}_i)$ for $i = 1, 2$,
- $\sup_{x \in \mathcal{D}_i} |F_i'(x)| < \infty$ for $i = 1, 2$,
- $F_i \in \mathbf{L}_1(\mathcal{D}_i)$; $F_i \in \mathbf{L}_2(\mathcal{D}_i)$ for $i = 1, 2$.

Let the intensity $a_{12}(x)$ be a continuous function.

In a first step we put $\mathcal{D}_1 = \mathcal{D}_2 = [a, b]$. We write

$$R_{12\varepsilon}(x, y) \doteq \langle f_{1\varepsilon}(x) f_{2\varepsilon}(y) \rangle$$

and obtain

$$\begin{aligned} J^{12}(F_1, F_2) &\doteq \int_{\mathcal{D}} \int_{\mathcal{D}} F_1(x) F_2(y) R_{12\varepsilon}(x, y) \, \mathrm{d}x \, \mathrm{d}y \\ &= \int_a^{a+\varepsilon} F_1(x) \left(\int_a^{x+\varepsilon} F_2(y) R_{12\varepsilon}(x, y) \, \mathrm{d}y \right) \mathrm{d}x \\ &\quad + \int_{a+\varepsilon}^{b-\varepsilon} F_1(x) \left(\int_{\mathcal{K}_\varepsilon(x)} F_2(y) R_{12\varepsilon}(x, y) \, \mathrm{d}y \right) \mathrm{d}x \\ &\quad + \int_{b-\varepsilon}^{b} F_1(x) \left(\int_{x-\varepsilon}^{b} F_2(y) R_{12\varepsilon}(x, y) \, \mathrm{d}y \right) \mathrm{d}x. \end{aligned} \tag{1.104}$$

The summands of the right-hand side of (1.104) are denoted by J_1^{12}, J_2^{12}, and J_3^{12}. For J_1^{12} it follows

$$\begin{aligned} J_1^{12} &= \int_a^{a+\varepsilon} \big(F_1(a) + F_1(x) - F_1(a)\big) \\ &\quad \times \left(\int_a^{x+\varepsilon} \big(F_2(a) + F_2(y) - F_2(a)\big) R_{12\varepsilon}(x, y) \, \mathrm{d}y \right) \mathrm{d}x \\ &= F_1F_2(a) \int_a^{a+\varepsilon} \left(\int_a^{x+\varepsilon} R_{12\varepsilon}(x, y) \, \mathrm{d}y \right) \mathrm{d}x \\ &\quad + F_1(a) \int_a^{a+\varepsilon} \left(\int_a^{x+\varepsilon} \big(F_2(y) - F_2(a)\big) R_{12\varepsilon}(x, y) \, \mathrm{d}y \right) \mathrm{d}x \\ &\quad + F_2(a) \int_a^{a+\varepsilon} \big(F_1(x) - F_1(a)\big) \left(\int_a^{x+\varepsilon} R_{12\varepsilon}(x, y) \, \mathrm{d}y \right) \mathrm{d}x \\ &\quad + \int_a^{a+\varepsilon} \big(F_1(x) - F_1(a)\big) \left(\int_a^{x+\varepsilon} \big(F_2(y) - F_2(a)\big) R_{12\varepsilon}(x, y) \, \mathrm{d}y \right) \mathrm{d}x \end{aligned}$$

and furthermore

$$\begin{aligned} J_{11}^{12} &= F_1F_2(a) \int_0^1 \left(\int_0^{u+1} R_{12\varepsilon}(a + \varepsilon u, a + \varepsilon v) \, \mathrm{d}v \right) \mathrm{d}u \cdot \varepsilon^2 \\ &= F_1F_2(a) \int_0^1 \underline{a}_{12}(u; a) \, \mathrm{d}u \cdot \varepsilon^2 + o(\varepsilon^2) \end{aligned} \tag{1.105}$$

where $\underline{a}_{12}$ is defined by

$$\underline{a}_{12}(u; a) \doteq \lim_{\varepsilon \downarrow 0} \left[\frac{1}{\varepsilon} \int_{\mathcal{K}_\varepsilon(x) \cap \mathcal{D}} R_{12\varepsilon}(x, y) \, \mathrm{d}y \right]_{x = a + \varepsilon u} \quad \text{for} \quad 0 \leq u \leq 1. \tag{1.106}$$

If $M_\varepsilon(F; x)$ is defined by

$$M_\varepsilon(F; x) = \sup_{y \in \mathcal{K}_{2\varepsilon}(x) \cap \mathcal{D}} |F(y) - F(x)|$$

then

$$\lim_{\varepsilon \downarrow 0} M_\varepsilon(F; x) = 0 \quad \text{and} \quad |M_\varepsilon(F; x)| \leqq C(F)$$

and we have

$$\begin{aligned} &|J_{12}^{12} + J_{13}^{12} + J_{14}^{12}| \\ &\leqq \frac{3}{2} c_2[|F_1(a)| \, M_\varepsilon(F_2; a) + |F_2(a)| \, M_\varepsilon(F_1; a) + M_\varepsilon(F_1; a) \, M_\varepsilon(F_2; a)] \cdot \varepsilon^2 \end{aligned}$$

or

$$J_{12}^{12} + J_{13}^{12} + J_{14}^{12} = o(\varepsilon^2).$$

In a similar way the term J_3^{12} leads to

$$J_3^{12} = F_1F_2(b) \int_{-1}^{0} \bar{a}_{12}(u; b) \, \mathrm{d}u \cdot \varepsilon^2 + o(\varepsilon^2) \tag{1.107}$$

and $\bar{a}_{12}$ is given by

$$\bar{a}_{12}(u; b) = \lim_{\varepsilon \downarrow 0} \left[\frac{1}{\varepsilon} \int_{\mathcal{K}_\varepsilon(x) \cap \mathcal{D}} R_{12\varepsilon}(x, y) \, \mathrm{d}y \right]_{x = b + \varepsilon u} \quad \text{for} \quad -1 \leqq u \leqq 0. \tag{1.108}$$

Now, J_2^{12} is investigated. It is

$$\begin{aligned} J_2^{12} &= \int_{a+\varepsilon}^{b-\varepsilon} F_1(x) \left(\int_{\mathcal{K}_\varepsilon(x)} \big(F_2(x) + F_2(y) - F_2(x)\big) \, R_{12\varepsilon}(x, y) \, \mathrm{d}y \right) \mathrm{d}x \\ &= \int_{a+\varepsilon}^{b-\varepsilon} F_1F_2(x) \left(\int_{\mathcal{K}_\varepsilon(x)} R_{12\varepsilon}(x, y) \, \mathrm{d}y \right) \mathrm{d}x \\ &\quad + \int_{a+\varepsilon}^{b-\varepsilon} F_1(x) \left(\int_{\mathcal{K}_\varepsilon(x)} \big(F_2(y) - F_2(x)\big) \, R_{12\varepsilon}(x, y) \, \mathrm{d}y \right) \mathrm{d}x. \end{aligned}$$

An expansion of

$$\int_{\mathcal{K}_\varepsilon(x)} R_{12\varepsilon}(x, y) \, \mathrm{d}y$$

as to ε is assumed:

$$\int_{\mathcal{K}_\varepsilon(x)} R_{12\varepsilon}(x, y) \, \mathrm{d}y = a_{12}(x) \, \varepsilon + b_{12}(x) \, \varepsilon^2 + o(\varepsilon^2) \tag{1.109}$$

where $a_{12}(x)$ is continuous, $b_{12}(x)$ bounded, and $o(\varepsilon^2)$ is demanded to be uniform as to x. Then we see that

$$\begin{aligned}
J_{21}^{12} &= \int_{a+\varepsilon}^{b-\varepsilon} F_1F_2(x)\, a_{12}(x)\, \mathrm{d}x \cdot \varepsilon + \int_{a+\varepsilon}^{b-\varepsilon} F_1F_2(x)\, b_{12}(x)\, \mathrm{d}x \cdot \varepsilon^2 + o(\varepsilon^2) \\
&= \int_a^b F_1F_2a_{12}(x)\, \mathrm{d}x\, \varepsilon - [F_1F_2a_{12}(b - \tilde{\varepsilon}) + F_1F_2a_{12}(a + \tilde{\varepsilon})]\, \varepsilon^2 \\
&\quad + \int_a^b F_1F_2b_{12}(x)\, \mathrm{d}x\, \varepsilon^2 + \left[\int_{a+\varepsilon}^{b-\varepsilon} F_1F_2b_{12}(x)\, \mathrm{d}x - \int_a^b F_1F_2b_{12}(x)\, \mathrm{d}x\right] \varepsilon^2 \\
&= \int_a^b F_1F_2a_{12}(x)\, \mathrm{d}x\, \varepsilon + \left[\int_a^b F_1F_2b_{12}(x)\, \mathrm{d}x - F_1F_2a_{12}(a) - F_1F_2a_{12}(b)\right] \varepsilon^2 \\
&\quad + o(\varepsilon^2).
\end{aligned} \tag{1.110}$$

In order to deal with J_{22}^{12} we define

$$\alpha_{12}(x; F) \doteq \lim_{\varepsilon \downarrow 0} \frac{1}{\varepsilon^2} \int_{\mathcal{K}_\varepsilon(x)} \big(F(y) - F(x)\big)\, R_{12\varepsilon}(x, y)\, \mathrm{d}y. \tag{1.111}$$

This limit value exists for $F \in \mathbf{C}^1\big(\mathcal{K}_\eta(x)\big)$ if we have the relation

$$\int_{\mathcal{K}_\varepsilon(x)} (y - x)\, R_{12\varepsilon}(x, y)\, \mathrm{d}y = \alpha_{12}(x)\, \varepsilon^2 + o(\varepsilon^2) \tag{1.112}$$

where $o(\varepsilon^2)$ is demanded to be uniform as to x. This statement follows from

$$\begin{aligned}
&\int_{\mathcal{K}_\varepsilon(x)} \big(F(y) - F(x)\big)\, R_{12\varepsilon}(x, y)\, \mathrm{d}y = \int_{\mathcal{K}_\varepsilon(x)} F'(\tilde{y})\, (y - x)\, R_{12\varepsilon}(x, y)\, \mathrm{d}y \\
&= F'(x) \int_{\mathcal{K}_\varepsilon(x)} (y - x)\, R_{12\varepsilon}(x, y)\, \mathrm{d}y \\
&\quad + \int_{\mathcal{K}_\varepsilon(x)} \big(F'(\tilde{y}) - F'(x)\big)\, (y - x)\, R_{12\varepsilon}(x, y)\, \mathrm{d}y \\
&= F'(x)\, \alpha_{12}(x)\, \varepsilon^2 + o(\varepsilon^2)
\end{aligned}$$

where we take into consideration

$$\left| \int_{\mathcal{K}_\varepsilon(x)} \big(F'(\tilde{y}) - F'(x)\big)\, (y - x)\, R_{12\varepsilon}(x, y)\, \mathrm{d}y \right| = c_2 M_\varepsilon(F'; x)\, \varepsilon^2 = o(\varepsilon^2).$$

Therefore, we have

$$\alpha_{12}(x; F) = F'(x)\, \alpha_{12}(x)\, \varepsilon^2 + o(\varepsilon^2). \tag{1.113}$$

The existence of the limit relation (1.111) for $F \in \mathbf{C}\big(\mathcal{K}_\eta(x)\big)$ cannot be proved. For this a counter-example is given: We take $F(x) = \sqrt{|x|}$ and

$$R_{12\varepsilon}(x, y) = \begin{cases} \sigma^2 \left(1 - \dfrac{1}{\varepsilon}\, |x - y|\right) & \text{for} \quad |x - y| \leqq \varepsilon, \\ 0 & \text{otherwise} \end{cases}$$

and obtain for $x = 0$

$$\int\limits_{\mathcal{K}_\varepsilon(0)} \sqrt{|y|}\, R_{12\varepsilon}(0, y)\, \mathrm{d}y = \sigma^2 \int\limits_{-\varepsilon}^{\varepsilon} \sqrt{|y|} \left(1 - \frac{1}{\varepsilon}|y|\right) \mathrm{d}y = \frac{8}{15}\, \sigma^2 \varepsilon^{3/2}.$$

Now J_{22}^{12} can be expanded as to ε in the form

$$\begin{aligned} J_{22}^{12} &= \int\limits_{a+\varepsilon}^{b-\varepsilon} F_1(x) \left(\int\limits_{\mathcal{K}_\varepsilon(x)} \left(F_2(y) - F_2(x)\right) R_{12\varepsilon}(x, y)\, \mathrm{d}y \right) \mathrm{d}x \\ &= \int\limits_{a+\varepsilon}^{b-\varepsilon} F_1(x)\, F_2'(x)\, \alpha_{12}(x)\, \mathrm{d}x \cdot \varepsilon^2 + o(\varepsilon^2) \\ &= \int\limits_a^b F_1 F_2' \alpha_{12}(x)\, \mathrm{d}x \cdot \varepsilon^2 + o(\varepsilon^2). \end{aligned} \tag{1.114}$$

Using (1.105), (1.107), (1.110), and (1.114) a summary leads to

$$\begin{aligned} J^{12}(F_1, F_2) = \int\limits_a^b F_1 F_2 a_{12}(x)\, \mathrm{d}x \cdot \varepsilon \\ + \Bigg[\int\limits_a^b F_1 F_2 b_{12}(x)\, \mathrm{d}x + \int\limits_a^b F_1 F_2' \alpha_{12}(x)\, \mathrm{d}x \\ + F_1 F_2(a) \left(\int\limits_0^1 \underline{a}_{12}(u; a)\, \mathrm{d}u - a_{12}(a) \right) \\ + F_1 F_2(b) \left(\int\limits_{-1}^0 \bar{a}_{12}(u; b)\, \mathrm{d}u - a_{12}(b) \right) \Bigg] \varepsilon^2 + o(\varepsilon^2). \end{aligned} \tag{1.115}$$

Using (1.103) and (1.115) we have

$$^2A_1(F_1, F_2) = \int\limits_{\mathcal{D} \cap \mathcal{D}} F_1 F_2 a_{12}(x)\, \mathrm{d}x, \tag{1.116}$$

$$\begin{aligned} ^2A_2(F_1, F_2) = \int\limits_{\mathcal{D} \cap \mathcal{D}} F_1 F_2 b_{12}(x)\, \mathrm{d}x + \int\limits_{\mathcal{D} \cap \mathcal{D}} F_1 F_2' \alpha_{12}(x)\, \mathrm{d}x \\ + F_1 F_2(a) \left(\int\limits_0^1 \underline{a}_{12}(u; a)\, \mathrm{d}u - a_{12}(a) \right) \\ + F_1 F_2(b) \left(\int\limits_{-1}^0 \bar{a}_{12}(u; b)\, \mathrm{d}u - a_{12}(b) \right). \end{aligned} \tag{1.117}$$

In order to obtain some properties we see from the relations

$$J^{21}(F_2, F_1) = \int\limits_{\mathcal{D}} \int\limits_{\mathcal{D}} F_2(x)\, F_1(y)\, \langle f_{2\varepsilon}(x)\, f_{1\varepsilon}(y) \rangle\, \mathrm{d}y\, \mathrm{d}x = J^{12}(F_1, F_2)$$

and

$$J^{21}(F_2, F_1) = \int_a^b F_1F_2a_{21}(x)\,dx \cdot \varepsilon$$

$$+ \left[\int_a^b F_1F_2b_{21}(x)\,dx + \int_a^b F_1'F_2\alpha_{21}(x)\,dx\right.$$

$$+ F_1F_2(a)\left(\int_0^1 \underline{a}_{21}(u;a)\,du - a_{21}(a)\right)$$

$$\left.+ F_1F_2(b)\left(\int_{-1}^0 \bar{a}_{21}(u;b)\,du - a_{21}(b)\right)\right]\varepsilon^2 + o(\varepsilon^2)$$

that

$$\int_a^b F_1F_2a_{12}(x)\,dx = \int_a^b F_1F_2a_{21}(x)\,dx\,,$$

$$\int_a^b F_1F_2(b_{12} - b_{21})(x)\,dx + \int_a^b \big(F_1F_2'\alpha_{12}(x) - F_1'F_2\alpha_{21}(x)\big)\,dx$$
$$+ F_1F_2(a)\big(\underline{S}_{12}(a;0,1) - \underline{S}_{21}(a;0,1)\big)$$
$$+ F_1F_2(b)\big(\bar{S}_{12}(b;-1,0) - \bar{S}_{21}(b;-1,0)\big) = 0 \tag{1.118}$$

where we have put

$$\underline{S}_{ij}(a;0,1) \doteq \int_0^1 \underline{a}_{ij}(u;a)\,du - a_{ij}(a)\,,$$

$$\bar{S}_{ij}(b;-1,0) \doteq \int_{-1}^0 \bar{a}_{ij}(u;b)\,du - a_{ij}(b)\,.$$

Since a_{12} is continuous from the above relation it follows

$$a_{12}(x) = a_{21}(x) \quad \text{for all} \quad x \in (a,b)\,. \tag{1.119}$$

Furthermore, we take (1.118) and subtract (1.118) where F_1 and F_2 are changed. Hence

$$\int_a^b (F_1F_2' - F_1'F_2)(\alpha_{12} + \alpha_{21})\,dx = 0$$

and with $F_1 = 1$, $F_2 = \int^x F(t)\,dt$ then

$$\int_a^b F\alpha_{12}(x)\,dx = -\int_a^b F\alpha_{21}(x)\,dx \quad \text{for all} \quad F \in \mathbf{C}\big((a,b)\big)\,. \tag{1.120}$$

Applying this relation and $F \doteq F_1F_2$ we obtain from (1.118)

$$\int_a^b F(b_{12} - b_{21})\,dx + \int_a^b F'\alpha_{12}(x)\,dx$$
$$+ F(a)\big(\underline{S}_{12}(a;0,1) - \underline{S}_{21}(a;0,1)\big)$$
$$+ F(b)\big(\bar{S}_{12}(b;-1,0) - \bar{S}_{21}(b;-1,0)\big) = 0 \tag{1.121}$$

for all $F \in C^1((a, b))$. In the special case $f_{1\varepsilon}(x, \omega) = f_{2\varepsilon}(x, \omega)$ it follows

$$\int_a^b F\alpha_{11}(x)\,dx = 0.$$

Now we assume that $a = -\infty$. Then we can write

$$J^{12}(F_1, F_2) = \int_{-\infty}^{b-\varepsilon} F_1(x) \left(\int_{\mathcal{K}_\varepsilon(x)} F_2(y)\, R_{12\varepsilon}(x, y)\,dy \right) dx + \int_{b-\varepsilon}^{b} F_1(x) \left(\int_{x-\varepsilon}^{b} F_2(y)\, R_{12\varepsilon}(x, y)\,dy \right) dx$$

and hence with the help of (1.107)

$$J_2^{12} = F_1F_2(b) \int_{-1}^{0} \bar{a}_{12}(u; b)\,du \cdot \varepsilon^2 + o(\varepsilon^2).$$

Furthermore,

$$J_1^{12} = \int_{-\infty}^{b-\varepsilon} F_1(x)\, F_2(x) \left(\int_{\mathcal{K}_\varepsilon(x)} R_{12\varepsilon}(x, y)\,dy \right) dx + \int_{-\infty}^{b-\varepsilon} F_1(x) \left(\int_{\mathcal{K}_\varepsilon(x)} (F_2(y) - F_2(x))\, R_{12\varepsilon}(x, y)\,dy \right) dx$$

$$= \int_{-\infty}^{b} F_1F_2a_{12}(x)\,dx \cdot \varepsilon + \left[\int_{-\infty}^{b} F_1F_2b_{12}(x)\,dx + \int_{-\infty}^{b} F_1F_2'\alpha_{12}(x)\,dx - F_1F_2a_{12}(b) \right] \varepsilon^2 + o(\varepsilon^2).$$

Thus,

$$J^{12}(F_1, F_2) = \int_{-\infty}^{b} F_1F_2a_{12}(x)\,dx \cdot \varepsilon + \left[\int_{-\infty}^{b} F_1F_2b_{12}(x)\,dx + \int_{-\infty}^{b} F_1F_2'\alpha_{12}(x)\,dx + F_1F_2(b)\, \bar{S}_{12}(b; -1, 0) \right] \varepsilon^2 + o(\varepsilon^2).$$

Similar considerations as in the case $a > -\infty$ lead to

$$\int_{-\infty}^{b} F(x)\, \alpha_{12}(x)\,dx = - \int_{-\infty}^{b} F(x)\, \alpha_{21}(x)\,dx \quad \text{for all} \quad F \in C((-\infty, b)),$$

$$a_{12}(x) = a_{21}(x) \quad \text{for all} \quad x \in (-\infty, b),$$

$$\int_{-\infty}^{b} F(b_{21} - b_{12})(x)\,dx + \int_{-\infty}^{b} F'\alpha_{12}(x)\,dx + F(b)\left(\bar{S}_{12}(b; -1, 0) - \bar{S}_{21}(b; -1, 0)\right) = 0.$$

Summarizing we have

$$ {}^2A_1(F_1, F_2) = \int_{\mathscr{D}} F_1(x)\, F_2(x)\, a_{12}(x)\, \mathrm{d}x , \tag{1.122} $$

$$ {}^2A_2(F_1, F_2) = \int_{\mathscr{D}} F_1(x)\, F_2(x)\, b_{12}(x)\, \mathrm{d}x + \int_{\mathscr{D}} F_1(x)\, F_2'(x)\, \alpha_{12}(x)\, \mathrm{d}x $$

$$ + \begin{cases} F_1F_2(a)\left(\int\limits_0^1 \underline{a}_{12}(u;a)\,\mathrm{d}u - a_{12}(a)\right) & \\ +F_1F_2(b)\left(\int\limits_{-1}^0 \bar{a}_{12}(u;b)\,\mathrm{d}u - a_{12}(b)\right) & \text{for } \mathscr{D} = [a,b], \\ F_1F_2(b)\left(\int\limits_{-1}^0 \bar{a}_{12}(u;b)\,\mathrm{d}u - a_{12}(b)\right) & \text{for } \mathscr{D} = (-\infty,b], \\ F_1F_2(a)\left(\int\limits_0^1 \underline{a}_{12}(u;a)\,\mathrm{d}u - a_{12}(a)\right) & \text{for } \mathscr{D} = [a,\infty), \\ 0 & \text{for } \mathscr{D} = (-\infty,\infty) \end{cases} \tag{1.123} $$

where $\underline{a}_{12}(u;a)$ is given by (1.106) and $\bar{a}_{12}(u;b)$ by (1.108).

Now we turn to the case $\mathscr{D}_1 \neq \mathscr{D}_2$. First, the case

$$ \mathscr{D}_1 = [a_1, b_1], \qquad \mathscr{D}_2 = [a_2, b_2] \quad \text{with} \quad a_1 < a_2 < b_1 < b_2 $$

is investigated. It follows the separation of $J^{12}(F_1, F_2)$ into J_i^{12} for $i = 1, 2$ with

$$ J_1^{12} = \int\limits_{a_2-\varepsilon}^{a_2+\varepsilon} F_1(x)\left(\int\limits_{a_2}^{x+\varepsilon} F_2(y)\, R_{12\varepsilon}(x,y)\,\mathrm{d}y\right)\mathrm{d}x , $$

$$ J_2^{12} = \int\limits_{a_2+\varepsilon}^{b_1} F_1(x)\left(\int\limits_{\mathscr{K}\varepsilon(x)} F_2(y)\, R_{12\varepsilon}(x,y)\,\mathrm{d}y\right)\mathrm{d}x . $$

J_1^{12} can be determined from

$$ J_1^{12} = F_1F_2(a_2) \int\limits_{a_2-\varepsilon}^{a_2+\varepsilon} \int\limits_{a_2}^{x+\varepsilon} R_{12\varepsilon}(x,y)\,\mathrm{d}y\,\mathrm{d}x $$

$$ + F_1(a_2) \int\limits_{a_2-\varepsilon}^{a_2+\varepsilon} \left(\int\limits_{a_2}^{x+\varepsilon} \big(F_2(y) - F_2(a_2)\big) R_{12\varepsilon}(x,y)\,\mathrm{d}y\right)\mathrm{d}x $$

$$ + F_2(a_2) \int\limits_{a_2-\varepsilon}^{a_2+\varepsilon} \big(F_1(x) - F_1(a_2)\big)\left(\int\limits_{a_2}^{x+\varepsilon} R_{12\varepsilon}(x,y)\,\mathrm{d}y\right)\mathrm{d}x $$

$$ + \int\limits_{a_2-\varepsilon}^{a_2+\varepsilon} \big(F_1(x) - F_1(a_2)\big)\left(\int\limits_{a_2}^{x+\varepsilon} \big(F_2(y) - F_2(a_2)\big) R_{12\varepsilon}(x,y)\,\mathrm{d}y\right)\mathrm{d}x $$

by

$$ J_{11}^{12} = F_1F_2(a_2) \int\limits_{-1}^{1} \underline{a}_{12}(u;a_2)\,\mathrm{d}u\;\varepsilon^2 + o(\varepsilon^2) $$

and

$$|J_{12}^{12} + J_{13}^{12} + J_{14}^{12}| \leqq 2c_2[|F_1(a_2)|\, M_\varepsilon(F_2; a_2) + |F_2(a_2)|\, M_\varepsilon(F_1; a_2) + M_\varepsilon(F_1; a_2)\, M_\varepsilon(F_2; a_2)]\, \varepsilon^2 = o(\varepsilon^2)$$

that is

$$J_1^{12} = F_1F_2(a_2) \int_{-1}^{1} \underline{a}_{12}(u; a_2)\, \mathrm{d}u\, \varepsilon^2 + o(\varepsilon^2).$$

Furthermore, applying (1.109) it is

$$\begin{aligned}
J_{21}^{12} &= \int_{a_2+\varepsilon}^{b_1} F_1F_2(x) \Big(\int_{\mathcal{K}\varepsilon(x)} R_{12\varepsilon}(x, y)\, \mathrm{d}y \Big)\, \mathrm{d}x \\
&= \int_{a_2+\varepsilon}^{b_1} F_1F_2a_{12}(x)\, \mathrm{d}x \cdot \varepsilon + \int_{a_2+\varepsilon}^{b_1} F_1F_2b_{12}(x)\, \mathrm{d}x \cdot \varepsilon^2 + o(\varepsilon^2) \\
&= \int_{a_2}^{b_1} F_1F_2a_{12}(x)\, \mathrm{d}x \cdot \varepsilon - F_1F_2a_{12}(a_2)\, \varepsilon^2 - [F_1F_2a_{12}(a_2 + \tilde{\varepsilon}) - F_1F_2a_{12}(a_2)]\, \varepsilon^2 \\
&\quad + \int_{a_2}^{b_1} F_1F_2b_{12}(x)\, \mathrm{d}x \cdot \varepsilon^2 + \left[\int_{a_2+\varepsilon}^{b_1} F_1F_2b_{12}(x)\, \mathrm{d}x - \int_{a_2}^{b_1} F_1F_2b_{12}(x)\, \mathrm{d}x \right] \varepsilon^2 \\
&= \int_{a_2}^{b_1} F_1F_2a_{12}(x)\, \mathrm{d}x \cdot \varepsilon + \left[\int_{a_2}^{b_1} F_1F_2b_{12}(x)\, \mathrm{d}x - F_1F_2a_{12}(a_2) \right] \varepsilon^2 + o(\varepsilon^2)
\end{aligned}$$

and

$$\begin{aligned}
J_{22}^{12} &= \int_{a_2+\varepsilon}^{b_1} F_1(x) \Big(\int_{\mathcal{K}\varepsilon(x)} \big(F_2(y) - F_2(x)\big)\, R_{12\varepsilon}(x, y)\, \mathrm{d}y \Big)\, \mathrm{d}x \\
&= \int_{a_2+\varepsilon}^{b_1} F_1(x)\, F_2'(x) \Big(\int_{\mathcal{K}\varepsilon(x)} (y - x)\, R_{12\varepsilon}(x, y)\, \mathrm{d}y \Big)\, \mathrm{d}x \\
&\quad + \int_{a_2+\varepsilon}^{b_1} F_1(x) \Big(\int_{\mathcal{K}\varepsilon(x)} \big(F_2'(\tilde{y}) - F_2'(x)\big)\, (y - x)\, R_{12\varepsilon}(x, y)\, \mathrm{d}y \Big)\, \mathrm{d}x .
\end{aligned}$$

Because of (1.112) we have

$$\begin{aligned}
J_{221}^{12} &= \int_{a_2+\varepsilon}^{b_1} F_1F_2'\alpha_{12}(x)\, \mathrm{d}x \cdot \varepsilon^2 + o(\varepsilon^2) \\
&= \int_{a_2}^{b_1} F_1F_2'\alpha_{12}(x)\, \mathrm{d}x \cdot \varepsilon^2 + \left[\int_{a_2+\varepsilon}^{b_1} F_1F_2'\alpha_{12}(x)\, \mathrm{d}x - \int_{a_2}^{b_1} F_1F_2'\alpha_{12}(x)\, \mathrm{d}x \right] \varepsilon^2 + o(\varepsilon^2) \\
&= \int_{a_2}^{b_1} F_1F_2'\alpha_{12}(x)\, \mathrm{d}x \cdot \varepsilon^2 + o(\varepsilon^2)
\end{aligned}$$

and

$$|J_{222}^{12}| \leqq \int_{a_2}^{b_1} |F_1(x)|\, M_\varepsilon(F_2'; x) \left(\int_{\mathcal{K}_\varepsilon(x)} |y - x|\, |R_{12\varepsilon}(x, y)|\, \mathrm{d}y \right) \mathrm{d}x$$

$$\leqq c_2 \int_{a_2}^{b_1} |F_1(x)|\, M_\varepsilon(F_2'; x)\, \mathrm{d}x \cdot \varepsilon^2 = o(\varepsilon^2).$$

With this by means of (1.116) we obtain

$$J^{12}(F_1, F_2) = \int_{\mathcal{D}_1 \cap \mathcal{D}_2} F_1 F_2 a_{12}(x)\, \mathrm{d}x \cdot \varepsilon + \left[\int_{\mathcal{D}_1 \cap \mathcal{D}_2} F_1 F_2 b_{12}(x)\, \mathrm{d}x + \int_{\mathcal{D}_1 \cap \mathcal{D}_2} F_1 F_2' \alpha_{12}(x)\, \mathrm{d}x + F_1 F_2(a_2) \left(\int_{-1}^{1} \underline{a}_{12}(u; a_2)\, \mathrm{d}u - a_{12}(a_2) \right) \right] \varepsilon^2 + o(\varepsilon^2). \tag{1.124}$$

Now we repeat these considerations where we first integrate over $\mathcal{D}_2$:

$$J_1^{12} = \int_{a_2}^{b_1 - \varepsilon} F_2(y) \left(\int_{\mathcal{K}_\varepsilon(y)} F_1(x)\, R_{12\varepsilon}(x, y)\, \mathrm{d}x \right) \mathrm{d}y$$

$$= \int_{a_2}^{b_1 - \varepsilon} F_2(x) \left(\int_{\mathcal{K}_\varepsilon(x)} F_1(y)\, R_{12\varepsilon}(x, y)\, \mathrm{d}y \right) \mathrm{d}x,$$

$$J_2^{12} = \int_{b_1 - \varepsilon}^{b_1 + \varepsilon} F_2(y) \left(\int_{y - \varepsilon}^{b_1} F_1(x)\, R_{12\varepsilon}(x, y)\, \mathrm{d}x \right) \mathrm{d}y$$

$$= \int_{b_1 - \varepsilon}^{b_1 + \varepsilon} F_2(x) \left(\int_{x - \varepsilon}^{b_1} F_1(y)\, R_{21\varepsilon}(x, y)\, \mathrm{d}y \right) \mathrm{d}x.$$

We obtain

$$J_1^{12} = \int_{a_2}^{b_1} F_1 F_2 a_{21}\, \mathrm{d}x \cdot \varepsilon + \left[\int_{a_2}^{b_1} F_1 F_2 b_{21}(x)\, \mathrm{d}x + \int_{a_2}^{b_1} F_1 F_1' \alpha_{21}(x)\, \mathrm{d}x - F_1 F_2 a_{21}(b_1) \right] \varepsilon^2 + o(\varepsilon^2),$$

$$J_2^{12} = F_1 F_2(b_1) \int_{-1}^{1} \bar{a}_{21}(u; b_1)\, \mathrm{d}u \cdot \varepsilon^2 + o(\varepsilon^2)$$

and then

$$J^{12}(F_1, F_2) = \int_{\mathcal{D}_1 \cap \mathcal{D}_2} F_1 F_2 a_{21}(x)\, \mathrm{d}x \cdot \varepsilon + \left[\int_{\mathcal{D}_1 \cap \mathcal{D}_2} F_1 F_2 b_{21}(x)\, \mathrm{d}x + \int_{\mathcal{D}_1 \cap \mathcal{D}_2} F_1' F_2 \alpha_{21}(x)\, \mathrm{d}x + F_1 F_2(b_1) \left(\int_{-1}^{1} \bar{a}_{21}(u; b_1)\, \mathrm{d}u - a_{21}(b_1) \right) \right] \varepsilon^2 + o(\varepsilon^2). \tag{1.125}$$

The difference between (1.124) and (1.125) leads to

$$L^{12}(a_2, b_1) \doteq \int_{\mathscr{D}_1\cap\mathscr{D}_2} F_1F_2(b_{12} - b_{21})(x)\,dx + \int_{\mathscr{D}_1\cap\mathscr{D}_2} \big(F_1F_2'\alpha_{12}(x) - F_1'F_2\alpha_{21}(x)\big)\,dx$$
$$+ F_1F_2(a_2)\,\underline{S}_{12}(a_2; -1, 1) - F_1F_2(b_1)\,\bar{S}_{21}(b_1; -1, 1) = 0 \quad (1.126)$$

where we have put

$$\underline{S}_{ij}(x; -1, 1) \doteq \int_{-1}^{1} \underline{a}_{ij}(u; x)\,du - a_{ij}(x),$$

$$\bar{S}_{ij}(x; -1, 1) = \int_{-1}^{1} \bar{a}_{ij}(u; x)\,du - a_{ij}(x).$$

It is

$$L^{12}(a_2, b_1) + L^{21}(a_2, b_1) = F_1F_2(a_2)\big(\underline{S}_{12}(a_2; -1, 1) + \underline{S}_{21}(a_2; -1, 1)\big)$$
$$- F_1F_2(b_1)\big(\bar{S}_{12}(b_1; -1, 1) + \bar{S}_{21}(b_1; -1, 1)\big) = 0$$

and hence

$$\begin{aligned}&\underline{S}_{12}(a_2; -1, 1) + \underline{S}_{21}(a_2; -1, 1) = 0,\\ &\bar{S}_{12}(b_1; -1, 1) + \bar{S}_{21}(b_1; -1, 1) = 0.\end{aligned} \quad (1.127)$$

In this case ${}^2A_2(F_1, F_2)$ has the form

$${}^2A_2(F_1, F_2) = \int_{\mathscr{D}_1\cap\mathscr{D}_2} F_1F_2b_{12}(x)\,dx + \int_{\mathscr{D}_1\cap\mathscr{D}_2} F_1F_2'\alpha_{12}(x)\,dx + F_1F_2(a_2)\,\underline{S}_{12}(a_2; -1, 1).$$

Using (1.126) we find that

$$L^{12}(a, b) - L^{12}(a, c) - L^{12}(c, b) = F(c)\big(\bar{S}_{21}(c; -1, 1) - \underline{S}_{12}(c; -1, 1)\big) = 0$$

for $a < c < b$ and then

$$\underline{S}_{12}(c; -1, 1) + \bar{S}_{12}(c; -1, 1) = 0. \quad (1.128)$$

Finally, we investigate the case

$$\mathscr{D}_1 = [a_1, b_1]; \qquad \mathscr{D}_2 = [a_2, b_2]; \qquad a_1 < b_1 = a_2 < b_2.$$

We have

$$J^{12}(F_1, F_2) = \int_{b_1-\varepsilon}^{b_1} F_1(x)\left(\int_{b_1}^{x+\varepsilon} F_2(y)\,R_{12\varepsilon}(x, y)\,dy\right)dx$$
$$= F_1F_2(b_1)\int_{b_1-\varepsilon}^{b_1}\left(\int_{b_1}^{x+\varepsilon} R_{12\varepsilon}(x, y)\,dy\right)dx$$
$$+ F_1(b_1)\int_{b_1-\varepsilon}^{b_1}\left(\int_{b_1}^{x+\varepsilon} \big(F_2(y) - F_2(b)\big)\,R_{12\varepsilon}(x, y)\,dy\right)dx$$
$$+ F_2(b_1)\int_{b_1-\varepsilon}^{b_1}\big(F_1(x) - F_1(b_1)\big)\left(\int_{b_1}^{x+\varepsilon} R_{12\varepsilon}(x, y)\,dy\right)dx$$
$$+ \int_{b_1-\varepsilon}^{b_1}\big(F_1(x) - F_1(b_1)\big)\left(\int_{b_1}^{x+\varepsilon}\big(F_2(x) - F_2(b_1)\big)\,R_{12\varepsilon}(x, y)\,dy\right)dx$$

and

$$J_1^{12} = F_1F_2(b_1)\,\varepsilon^2 \int_{-1}^{0} \left(\frac{1}{\varepsilon} \int_{\mathcal{K}_\varepsilon(x)\cap\mathcal{D}_2} R_{12\varepsilon}(x, y)\,\mathrm{d}y\right)_{x=b_1+\varepsilon u} \mathrm{d}u$$

$$= F_1F_2(b_1) \int_{-1}^{0} \underline{a}_{12}(u; a_2)\,\mathrm{d}u\,\varepsilon^2 + o(\varepsilon^2),$$

$$|J_2^{12} + J_3^{12} + J_4^{12}| \leqq \frac{1}{2}\,c_2\varepsilon^2[|F_1(b_1)|\,M_\varepsilon(F_2; b_1) + |F_2(b_1)|\,M_\varepsilon(F_1; b_1)$$
$$+ M_\varepsilon(F_1; b_1)\,M_\varepsilon(F_2; b_1)] = o(\varepsilon^2),$$

so that

$$J^{12}(F_1, F_2) = F_1F_2(b_1) \int_{-1}^{0} \underline{a}_{12}(u; a_2)\,\mathrm{d}u\,\varepsilon^2 + o(\varepsilon^2).$$

On the other hand, it is

$$J^{12}(F_1, F_2) = \int_{a_2}^{a_2+\varepsilon} F_2(y) \left(\int_{y-\varepsilon}^{a_2} F_1(x)\,R_{12\varepsilon}(x, y)\,\mathrm{d}x\right) \mathrm{d}y$$

$$= F_1F_2(b_1) \int_{0}^{1} \bar{a}_{21}(u; b_1)\,\mathrm{d}u\,\varepsilon^2 + o(\varepsilon^2)$$

and thus

$$\int_{-1}^{0} \underline{a}_{12}(u; a_2)\,\mathrm{d}u = \int_{0}^{1} \bar{a}_{21}(u; a_2)\,\mathrm{d}u,$$

i.e.

$$\underline{S}_{12}(a_2; -1, 0) = \bar{S}_{21}(a_2; 0, 1).$$

In the case of $f_{1\varepsilon}(x, \omega) = f_{2\varepsilon}(x, \omega)$ we can write

$$\underline{S}_{11}(c; -1, 1) = \bar{S}_{11}(c; -1, 1) = 0$$

using (1.127) and hence

$$\int_{-1}^{0} \underline{a}_{11}(u; c)\,\mathrm{d}u + \int_{0}^{1} \underline{a}_{11}(u; c)\,\mathrm{d}u = a_{11}(c),$$

$$\int_{-1}^{0} \bar{a}_{11}(u; c)\,\mathrm{d}u + \int_{0}^{1} \bar{a}_{11}(u; c)\,\mathrm{d}u = a_{11}(c)$$

or

$$\begin{aligned} \underline{S}_{11}(c; -1, 0) + \underline{S}_{11}(c; 0, 1) &= -a_{11}(c), \\ \bar{S}_{11}(c; -1, 0) + \bar{S}_{11}(c; 0, 1) &= -a_{11}(c). \end{aligned} \tag{1.128'}$$

The behaviour of ${}^2A_2(F_1, F_2)$ is analogous if infinite intervals $\mathcal{D}_1$ or/and $\mathcal{D}_2$ are considered. We obtain for

$$\mathcal{D}_1 = (-\infty, b_1], \quad \mathcal{D}_2 = (-\infty, b_2] \quad \text{with} \quad b_1 < b_2$$

from

$$J^{12}(F_1, F_2) = \int_{-\infty}^{b_1} F_1(x) \left(\int_{\mathcal{K}_\varepsilon(x)} F_2(y)\, R_{12\varepsilon}(x, y)\, dy \right) dx$$

the relation

$${}^2A_2(F_1, F_2) = \int_{\mathcal{D}_1 \cap \mathcal{D}_2} F_1 F_2 b_{12}(x)\, dx + \int_{\mathcal{D}_1 \cap \mathcal{D}_2} F_1 F_2' \alpha_{12}(x)\, dx$$

and from

$$J^{12}(F_1, F_2) = \int_{-\infty}^{b_1 - \varepsilon} F_2(y) \left(\int_{\mathcal{K}_\varepsilon(y)} F_1(x)\, R_{12\varepsilon}(x, y)\, dx \right) dy + \int_{b_1 - \varepsilon}^{b_1 + \varepsilon} F_2(y) \left(\int_{y - \varepsilon}^{b_1} F_1(x)\, R_{12\varepsilon}(x, y)\, dx \right) dy$$

then

$${}^2A_2(F_1, F_2) = \int_{-\infty}^{b_1} F_1 F_2 b_{21}\, dx + \int_{-\infty}^{b_1} F_1' F_2 \alpha_{21}\, dx + F_1 F_2(b_1)\, \bar{S}_{21}(b_1; -1, 1).$$

We add these both terms for ${}^2A_2(F_1, F_2)$ and can write

$${}^2A_2(F_1, F_2) = \frac{1}{2} \int_{\mathcal{D}_1 \cap \mathcal{D}_2} F_1 F_2 (b_{12} + b_{21})\,(x)\, dx + \frac{1}{2} \int_{\mathcal{D}_1 \cap \mathcal{D}_2} (F_1 F_2' - F_1' F_2)\, \alpha_{12}(x)\, dx + \frac{1}{2} \operatorname{sgn}(b_1 - b_2)\, F_1 F_2(\underline{b})\, \bar{S}_{12}(\underline{b}; -1, 1)$$

where $\underline{b} \doteq \min\{b_1, b_2\}$. For $b_2 < b_1$ we have the same formula as (1.128).

Finally, we want to summarize the results of this section in Theorem 1.9.

Theorem 1.9. *Let F_i, $i = 1, 2$, be functions from $\mathbf{C}^1(\mathcal{D}_i)$ with*

$$\sup_{x \in \mathcal{D}_i} |F_i'(x)| < \infty; \quad F_i \in \mathbf{L}_1(\mathcal{D}_i), \quad F_i \in \mathbf{L}_2(\mathcal{D}_i)$$

and $\mathcal{D}_i$ finite or infinite intervals, $\mathcal{D}_i = (a_i, b_i)$. $\overline{\mathcal{D}} = (\underline{a}, \bar{b})$ denotes the intersection of $\mathcal{D}_1$ and $\mathcal{D}_2$ and $\varkappa(\mathcal{D}_1, \mathcal{D}_2)$ is defined by

$$\varkappa(\mathcal{D}_1, \mathcal{D}_2) = \begin{cases} -1 & \text{for } (\bar{b}, \infty) \cap \mathcal{D}_2 \neq \emptyset \text{ or } (-\infty, \underline{a}) \cap \mathcal{D}_1 \neq \emptyset, \\ 1 & \text{for } (\bar{b}, \infty) \cap \mathcal{D}_1 \neq \emptyset \text{ or } (-\infty, \underline{a}) \cap \mathcal{D}_2 \neq \emptyset. \end{cases}$$

Let the function $R_{12\varepsilon}(x, y) \doteq \langle f_{1\varepsilon}(x)\, f_{2\varepsilon}(y) \rangle$ have the following properties:

– $\displaystyle\int_{\mathcal{K}_\varepsilon(x)} R_{12\varepsilon}(x, y)\, dy = a_{12}(x)\, \varepsilon + b_{12}(x)\, \varepsilon^2 + o(\varepsilon^2)$

with the continuous intensity $a_{12}(x)$,

– $\displaystyle\lim_{\varepsilon \downarrow 0} \frac{1}{\varepsilon^2} \int_{\mathcal{K}_\varepsilon(x)} (y - x)\, R_{12\varepsilon}(x, y)\, dy = \alpha_{12}(x),$

– $\displaystyle\lim_{\varepsilon \downarrow 0} \left[\frac{1}{\varepsilon} \int_{\mathcal{K}_\varepsilon(x) \cap \mathcal{D}} R_{12\varepsilon}(x, y)\, dy \right]_{x = a + \varepsilon u} = \underline{a}_{12}(u; a)$ *for* $u \in [-1, 1]$,

– $\displaystyle\lim_{\varepsilon \downarrow 0} \left[\frac{1}{\varepsilon} \int_{\mathcal{K}_\varepsilon(x) \cap \mathcal{D}} R_{12\varepsilon}(x, y)\, dy \right]_{x = b + \varepsilon u} = \bar{a}_{12}(u; b)$ *for* $u \in [-1, 1]$ *and* $\mathcal{D} = (a, b)$.

Then the statistical characteristic $^2A_2(F_1, F_2)$ *has the form*

$$^2A_2(F_1, F_2) = \frac{1}{2} \int_{\bar{\mathcal{D}}} F_1F_2(b_{12} + b_{21})(x)\,\mathrm{d}x + \frac{1}{2} \int_{\bar{\mathcal{D}}} (F_1F_2' - F_1'F_2)\,\alpha_{12}(x)\,\mathrm{d}x + \frac{1}{2} B(\underline{a}, \bar{b})$$

and the boundary terms $B(\underline{a}, \bar{b})$ *can be written as*

$$B(\underline{a}, \bar{b}) = \begin{cases} 0 & \text{for } \bar{\mathcal{D}} = \emptyset, \\ F_1F_2(\underline{a})\left(\underline{S}_{12}(\underline{a}; -1, 0) + \bar{S}_{21}(\bar{b}; 0, 1) + 2a_{12}(\underline{a})\right) & \text{for } \bar{\mathcal{D}} = \{\underline{a}\}, \\ \varkappa\left(F_1F_2(\bar{b})\,\bar{S}_{12}(\bar{b}; -1, 1) - F_1F_2(\underline{a})\,\underline{S}_{12}(\underline{a}; -1, 1)\right) & \text{for } \bar{\mathcal{D}} \neq \mathcal{D}_1, \mathcal{D}_2, \\ F_1F_2(\bar{b})\left(\bar{S}_{12}(\bar{b}; -1, 0) + \bar{S}_{21}(\bar{b}; -1, 0)\right) - \varkappa F_1F_2(\underline{a})\,\underline{S}_{12}(\underline{a}; -1, 1) & \\ & \text{for } b_1 = b_2 = \bar{b}, \\ \varkappa F_1F_2(\bar{b})\,\bar{S}_{12}(\bar{b}; -1, 1) + F_1F_2(\underline{a})\left(\underline{S}_{12}(\underline{a}; 0, 1) + \underline{S}_{21}(\underline{a}; 0, 1)\right) & \\ & \text{for } a_1 = a_2 = \underline{a}, \\ F_1F_2(\bar{b})\left(\bar{S}_{12}(\bar{b}; -1, 0) + \bar{S}_{21}(\bar{b}; -1, 0)\right) & \\ \quad + F_1F_2(\underline{a})\left(\underline{S}_{12}(\underline{a}; 0, 1) + \underline{S}_{21}(\underline{a}; 0, 1)\right) & \text{for } \mathcal{D}_1 = \mathcal{D}_2 = \bar{\mathcal{D}}, \\ F_1F_2(\bar{b})\,\bar{S}_{12}(\bar{b}; -1, 1) + F_1F_2(\underline{a})\,\underline{S}_{12}(\underline{a}; -1, 1) & \text{for } \mathcal{D}_2 \subset \mathcal{D}_1. \end{cases}$$

The terms $\underline{S}_{ij}$, $\bar{S}_{ij}$ *are defined by*

$$\underline{S}_{ij}(x; y_1, y_2) \doteq \int_{y_1}^{y_2} \underline{a}_{ij}(u; x)\,\mathrm{d}u - a_{ij}(x),$$

$$\bar{S}_{ij}(x; y_1, y_2) \doteq \int_{y_1}^{y_2} \bar{a}_{ij}(u; x)\,\mathrm{d}u - a_{ij}(x).$$

In the case of $f_{1\varepsilon}(x, \omega) = f_{2\varepsilon}(x, \omega)$ *the relations*

$$\int_{\mathcal{D}} F\alpha_{11}(x)\,\mathrm{d}x = 0; \quad \underline{S}_{11}(a; -1, 1) = \bar{S}_{11}(b; -1, 1) = 0$$

are fulfilled.

1.6.3. Examples

Now some correlation functions are considered. First, the correlation function

$$R_{1\varepsilon}(x, y) = \begin{cases} \sigma^2\left(1 - \dfrac{1}{\varepsilon}|x - y|\right) & \text{for } |x - y| < \varepsilon, \\ 0 & \text{otherwise} \end{cases}$$

is investigated. From the relation

$$\int_{\mathcal{K}_\varepsilon(x)} R_{1\varepsilon}(x, y)\,\mathrm{d}y = \sigma^2 \int_{-\varepsilon}^{\varepsilon} \left(1 - \frac{1}{\varepsilon}|x|\right) \mathrm{d}x = \sigma^2 \varepsilon$$

it follows

$$a_{11}(x) = \sigma^2 \quad \text{and} \quad b_{11}(x) \equiv 0.$$

Furthermore, we have for $\mathcal{D}_1 = (a_1, b_1)$

$$\left[\frac{1}{\varepsilon} \int_{\mathcal{K}_\varepsilon(x) \cap \mathcal{D}_1} R_{1\varepsilon}(x, y)\,\mathrm{d}y\right]_{x = a_1 + \varepsilon u} = \begin{cases} \frac{1}{2}\sigma^2(1 + u)^2 & \text{for} \quad -1 \leqq u \leqq 0, \\ \frac{1}{2}\sigma^2(1 + 2u - u^2) & \text{for} \quad 0 \leqq u \leqq 1, \end{cases}$$

$$\left[\frac{1}{\varepsilon} \int_{\mathcal{K}_\varepsilon(x) \cap \mathcal{D}_1} R_{1\varepsilon}(x, y)\,\mathrm{d}y\right]_{x = b_1 + \varepsilon u} = \begin{cases} \frac{1}{2}\sigma^2(1 - 2u - u^2) & \text{for} \quad -1 \leqq u \leqq 0, \\ \frac{1}{2}\sigma^2(1 - u)^2 & \text{for} \quad 0 \leqq u \leqq 1 \end{cases}$$

and hence

$$\underline{a}_{11}(u; a_1) = \begin{cases} \frac{1}{2}\sigma^2(1 + u)^2 & \text{for} \quad -1 \leqq u \leqq 0, \\ \frac{1}{2}\sigma^2(1 + 2u - u^2) & \text{for} \quad 0 \leqq u \leqq 1, \end{cases}$$

$$\bar{a}_{11}(u; b_1) = \begin{cases} \frac{1}{2}\sigma^2(1 - 2u - u^2) & \text{for} \quad -1 \leqq u \leqq 0, \\ \frac{1}{2}\sigma^2(1 - u)^2 & \text{for} \quad 0 \leqq u \leqq 1. \end{cases}$$

Thus,

$$\underline{S}_{11}(a_1; -1, 0) = \int_{-1}^{0} \underline{a}_{11}(u; a_1)\,\mathrm{d}u - a_{11}(a_1) = -\frac{5}{6}\sigma^2,$$

$$\bar{S}_{11}(b_1; -1, 0) = \int_{-1}^{0} \bar{a}_{11}(u; b_1)\,\mathrm{d}u - a_{11}(b_1) = -\frac{1}{6}\sigma^2,$$

$$\underline{S}_{11}(a_1; 0, 1) = \int_{0}^{1} \underline{a}_{11}(u; a_1)\,\mathrm{d}u - a_{11}(a_1) = -\frac{1}{6}\sigma^2,$$

$$\bar{S}_{11}(b_1; 0, 1) = \int_{0}^{1} \bar{a}_{11}(u; b_1)\,\mathrm{d}u - a_{11}(b_1) = -\frac{5}{6}\sigma^2$$

and the relations (1.129) are fulfilled. The function $\underline{a}_{11}(u; a_1)$ is plotted in Fig. 1.7 for this treated correlation function. The above correlation function is not suitable for many cases since the differentiability of the sample functions cannot be shown in general.

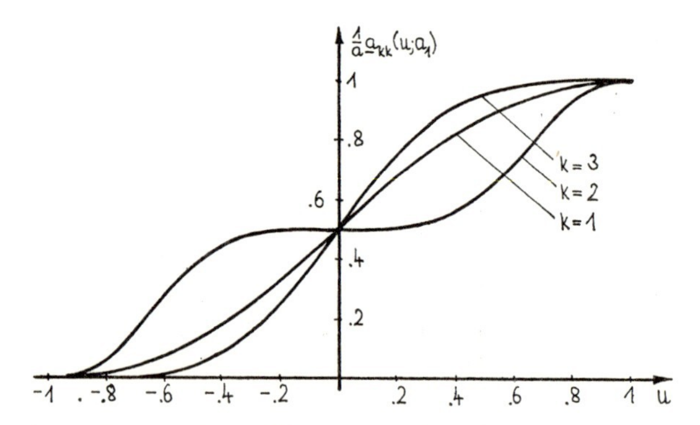

Fig. 1.7. $\underline{a}(u; a_1)$ for different correlation functions

Therefore, we want to consider another correlation function for which the sample functions are arbitrary smooth. Let

$$\overline{R}_{2\varepsilon}(z) \doteq \int_{-\infty}^{\infty} g_{\varepsilon/2}(x)\, g_{\varepsilon/2}(z - x)\, \mathrm{d}x$$

where

$$g_\eta(x) \doteq \begin{cases} \dfrac{d}{\sqrt{\eta}} \exp\left(\dfrac{\eta^2}{x^2 - \eta^2}\right) & \text{for} \quad |x| < \eta\,, \\ 0 & \text{for} \quad |x| \geqq \eta \end{cases}$$

and

$$R_\varepsilon(x, y) = \overline{R}_\varepsilon(|y - x|)\,.$$

Using the definition of $g_\eta(x)$ the determination of the integration domain leads to

$$(-\eta, \eta) \cap (z - \eta, z + \eta) = \begin{cases} (z - \eta, \eta) & \text{for} \quad 0 \leqq z \leqq 2\eta\,, \\ (-\eta, z + \eta) & \text{for} \quad -2\eta \leqq z \leqq 0\,, \\ \emptyset & \text{otherwise} \end{cases}$$

and then for $z \in [0, \varepsilon)$ to

$$\overline{R}_{2\varepsilon}(z) = \frac{2d^2}{\varepsilon} \int_{z-\varepsilon/2}^{\varepsilon/2} \exp\left(\frac{\varepsilon^2}{4x^2 - \varepsilon^2} + \frac{\varepsilon^2}{4(x - z)^2 - \varepsilon^2}\right) \mathrm{d}x$$

$$= \frac{2d^2}{\varepsilon} \int_{-\varepsilon/2}^{\varepsilon/2-z} \exp\left(\frac{\varepsilon^2}{4x^2 - \varepsilon^2} + \frac{\varepsilon^2}{4(x + z)^2 - \varepsilon^2}\right) \mathrm{d}x = \overline{R}_{2\varepsilon}(-z)\,,$$

$$\overline{R}_{2\varepsilon}(z) = d^2 \int_{2z/\varepsilon-1}^{1} \exp\left(\frac{1}{x^2 - 1} + \frac{1}{(2z/\varepsilon - x)^2 - 1}\right) \mathrm{d}x\,.$$

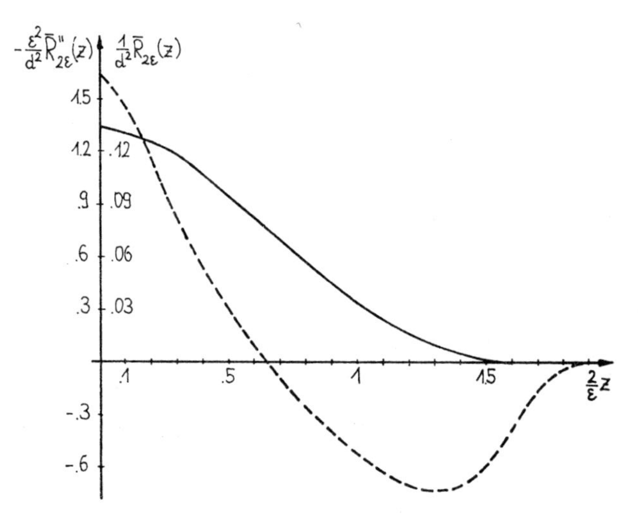

Fig. 1.8. Correlation functions

The function $\frac{1}{d^2}\,\overline{R}_{2\varepsilon}(z)$ is plotted in Fig. 1.8. For the intensity we calculate

$$\frac{1}{\varepsilon}\int\limits_{-\varepsilon}^{\varepsilon}\overline{R}_{2\varepsilon}(z)\,\mathrm{d}z = \frac{2}{\varepsilon}\int\limits_{0}^{\varepsilon}\overline{R}_{2\varepsilon}(z)\,\mathrm{d}z$$

$$= d^2\int\limits_0^2\left(\int\limits_{s-1}^{1}\exp\left(\frac{1}{x^2-1}+\frac{1}{(s-x)^2-1}\right)\mathrm{d}x\right)\mathrm{d}s$$

and obtain

$$a_{22} = d^2\int\limits_{-1}^{1}\exp\left(\frac{1}{x^2-1}\right)\int\limits_{-x}^{1}\exp\left(\frac{1}{s^2-1}\right)\mathrm{d}s\,\mathrm{d}x = 0.09856d^2$$

and

$$b_{22}(x) = 0\,.$$

Setting $\mathcal{D} = (a_1, b_1)$ and $u \in [0, 1]$ it follows

$$\left[\frac{1}{\varepsilon}\int\limits_{\mathcal{K}_\varepsilon(x)\cap\mathcal{D}} R_{2\varepsilon}(x, y)\,\mathrm{d}y\right]_{x=a_1+\varepsilon u} = \int\limits_0^{u+1}\overline{R}_{2\varepsilon}\big(\varepsilon(v-u)\big)\,\mathrm{d}v = \int\limits_0^{u}\overline{R}_{2\varepsilon}(\varepsilon v)\,\mathrm{d}v + \int\limits_0^{1}\overline{R}_{2\varepsilon}(\varepsilon v)\,\mathrm{d}v$$

and then

$$\underline{a}_{22}(u; a_1) = T_2(u) + T_2(1) \quad \text{for} \quad u \in [0, 1]$$

where

$$T_2(u) = \int\limits_0^u \overline{R}_{2\varepsilon}(\varepsilon v)\,\mathrm{d}v = \frac{1}{2}\,d^2\left\{\int\limits_{-1}^{2u-1}\exp\left(\frac{1}{x^2-1}\right)\int\limits_{-x}^{1}\exp\left(\frac{1}{v^2-1}\right)\mathrm{d}v\,\mathrm{d}x\right.$$

$$\left.+\int\limits_{2u-1}^{1}\exp\left(\frac{1}{x^2-1}\right)\int\limits_{-x}^{2u-x}\exp\left(\frac{1}{v^2-1}\right)\mathrm{d}v\,\mathrm{d}x\right\}.$$

This function $\underline{a}_{22}(u; a_1)$ is also illustrated in Fig. 1.7. In the case of $u \in [-1, 0]$ we obtain

$$\left[\frac{1}{\varepsilon} \int\limits_{\mathcal{K}_\varepsilon(x) \cap \mathcal{D}} R_{2\varepsilon}(x, y) \, dy\right]_{x=a_1+\varepsilon u} = \int\limits_{-u}^{1} \overline{R}_{2\varepsilon}(\varepsilon t) \, dt$$

and

$$\underline{a}_{22}(u; a_1) = T_2(1) - T_2(-u) \quad \text{for} \quad u \in [-1, 0].$$

Furthermore, it is

$$\bar{a}_{22}(u; b_1) = \underline{a}_{22}(-u; a_1) \quad \text{for} \quad u \in [-1, 1].$$

A simple consequence shows

$$\int\limits_{-1}^{1} \underline{a}_{22}(u; a_1) \, du = \int\limits_{-1}^{0} \big(T_2(1) - T_2(-u)\big) \, du + \int\limits_{0}^{1} \big(T_2(u) + T_2(1)\big) \, du = 2T_2(1) = a_{22}$$

and

$$\underline{S}_{22}(a_1; -1, 0) = -\int\limits_{0}^{1} T_2(u) \, du - \frac{1}{2} a_{22} = -0.11886 d^2,$$

$$\underline{S}_{22}(a_1; 0, 1) = \int\limits_{0}^{1} T_2(u) \, du - \frac{1}{2} a_{22} = 0.02030 d^2,$$

$$\overline{S}_{22}(b_1; -1, 0) = \underline{S}_{22}(a_1; 0, 1), \quad \overline{S}_{22}(b_1; 0, 1) = \underline{S}_{22}(a_1; -1, 0).$$

The correlation function of the process $f_\varepsilon'(x, \omega)$ can be calculated as

$$-\overline{R}_{2\varepsilon}''(z) = -\frac{16}{\varepsilon^2} d^2 \int\limits_{2z/\varepsilon - 1}^{1} \exp\left(\frac{1}{x^2 - 1} + \frac{1}{\left(\frac{2}{\varepsilon} z - x\right)^2 - 1}\right) \frac{x}{(x^2 - 1)^2}$$

$$\times \frac{\left(\frac{2}{\varepsilon} z - x\right)}{\left(\left(\frac{2}{\varepsilon} z - x\right)^2 - 1\right)^2} \, dx$$

and $-\frac{\varepsilon^2}{d^2} \overline{R}_{2\varepsilon}''(z)$ was plotted in Fig. 1.8 as a hatched line. From the relation

$$\langle f_\varepsilon'^2(x) \rangle = -\overline{R}_{2\varepsilon}''(0) = \frac{16}{\varepsilon^2} d^2 \int\limits_{-1}^{1} \exp\left(\frac{2}{x^2 - 1}\right) \frac{x^2}{(x^2 - 1)^4} \, dx = \frac{d^2}{\varepsilon^2}$$

it follows that the second moments of f_ε' are not uniformly bounded as to ε. Hence, the theory of the weakly correlated process represented can not be applied to $f_\varepsilon'(x, \omega)$. This is also clear from

$$\int\limits_{-\varepsilon}^{\varepsilon} \overline{R}_{2\varepsilon}''(x) \, dx = \overline{R}_{2\varepsilon}'(\varepsilon) - \overline{R}_{2\varepsilon}'(-\varepsilon) = 0.$$

Finally, a third example is to be investigated. We take

$$\overline{R}_{3\varepsilon}(z) = \frac{12d^2}{\varepsilon^3} \begin{cases} (1+z)\,\mathrm{e}^{-z} + \mathrm{e}^{z-\varepsilon} + (2+\varepsilon-z)\,\mathrm{e}^{-\varepsilon} - 4\mathrm{e}^{-\varepsilon/2} & \text{for } 0 \leqq z \leqq \frac{\varepsilon}{2}, \\ (\varepsilon - z - 2)\,\mathrm{e}^{-z} + (\varepsilon - z + 2)\,\mathrm{e}^{-\varepsilon} & \text{for } \frac{\varepsilon}{2} \leqq z \leqq \varepsilon, \\ 0 & \text{otherwise}. \end{cases}$$

It is

$$\overline{R}_{3\varepsilon}(0) = d^2 \left(1 - \frac{5}{8}\,\varepsilon + O(\varepsilon^2)\right)$$

and this correlation function is represented in Fig. 1.9 for different ε. In order to compare the correlation functions we have plotted in Fig. 1.9 also $\overline{R}_{1\varepsilon}$ and $\overline{R}_{2\varepsilon}$ for $\varepsilon = 0.25$. Some numerical calculations lead to

$$\int\limits_{-\varepsilon}^{\varepsilon} \overline{R}_{3\varepsilon}(z)\,\mathrm{d}z = 2\varepsilon - \frac{9}{8}\,\varepsilon^2 + O(\varepsilon^3)$$

so that we obtain

$$a_{33} = \frac{3}{4}\,d^2 \quad \text{and} \quad b_{33} = -\frac{1}{2}\,d^2.$$

For $u \in [0, 1]$ it follows

$$\left[\frac{1}{\varepsilon} \int\limits_{\mathcal{K}_\varepsilon(x)\cap\mathcal{D}} \overline{R}_{3\varepsilon}(x, y)\,\mathrm{d}y\right]_{x=a_1+\varepsilon u} = T_{3\varepsilon}(u) + T_{3\varepsilon}(1)$$

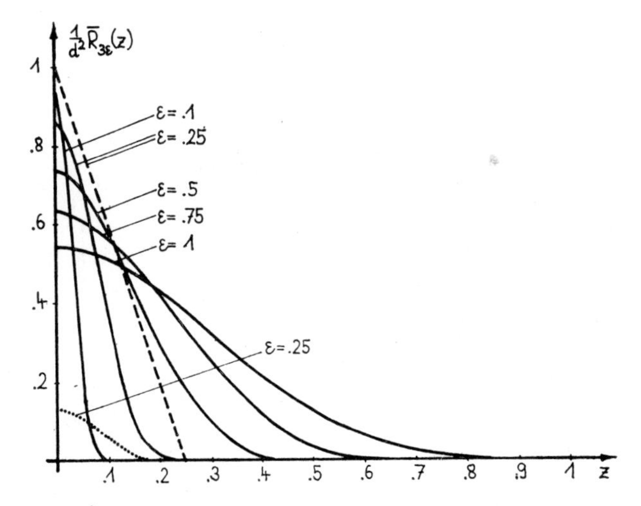

Fig. 1.9. Correlation functions of weakly correlated processes

where

$$T_{3\varepsilon}(u) = \int_0^u \bar{R}_{3\varepsilon}(\varepsilon v)\,dv = \begin{cases} \frac{1}{\varepsilon}\int_0^{\varepsilon u} \bar{R}_{3\varepsilon}(z)\,dz & \text{for } 0 \leqq u \leqq \frac{1}{2}, \\ \frac{1}{\varepsilon}\int_0^{\varepsilon/2} \bar{R}_{3\varepsilon}(z)\,dz + \frac{1}{\varepsilon}\int_{\varepsilon/2}^{\varepsilon u} \bar{R}_{3\varepsilon}(z)\,dz & \text{for } \frac{1}{2} \leqq u \leqq 1. \end{cases}$$

We obtain

$$T_3(u) = \lim_{\varepsilon\downarrow 0} T_{3\varepsilon}(u) = \begin{cases} d^2\left(\frac{3}{2}u^4 - 2u^3 + u\right) & \text{for } 0 \leqq u \leqq \frac{1}{2}, \\ d^2\left(-\frac{1}{2}u^4 + 2u^3 - 3u^2 + 2u - \frac{1}{8}\right) & \text{for } \frac{1}{2} \leqq u \leqq 1 \end{cases}$$

and hence

$$\underline{a}_{33}(u; a_1) = T_3(u) + T_3(1) = T_3(u) + \frac{3}{8}d^2 \qquad \text{for } u \in [0, 1],$$

$$\underline{a}_{33}(u; a_1) = T_3(1) - T_3(-u) = \frac{3}{8}d^2 - T_3(-u) \quad \text{for } u \in [-1, 0],$$

$$\bar{a}_{33}(u; b_1) = \underline{a}_{33}(-u; a_1) \qquad \text{for } u \in [-1, 1].$$

$\underline{a}_{33}(u; a_1)$ is plotted in Fig. 1.7. Furthermore, it is

$$\underline{S}_{33}(a_1; -1, 0) = -\int_0^1 T_3(u)\,du - \frac{1}{2}a_{33} = -\frac{53}{80}d^2,$$

$$\underline{S}_{33}(a_1; 0, 1) = \int_0^1 T_3(u)\,du - \frac{1}{2}a_{33} = -\frac{7}{80}d^2$$

and

$$\bar{S}_{33}(b_1; -1, 0) = -\frac{7}{80}d^2, \quad \bar{S}_{33}(b_1; 0, 1) = -\frac{53}{80}d^2.$$

1.6.4. Determination of ${}^2A_2$ in the multi-dimensional case

In this section the multi-dimensional case $m > 1$ is considered. Let the domains $\mathcal{D}_i$ be bounded with piecewise smooth boundaries. The functions $F_i(x)$ are assumed to be from $\mathbf{C}^1(\mathcal{D}_i)$, $i = 1, 2$.

First, let $\mathcal{D}_1 = \mathcal{D}_2 = \mathcal{D}$. $\mathcal{D}_\varepsilon$ denotes the set of points from $\mathcal{D}$ which have a distance as to $\partial\mathcal{D}$ less than or equal to ε,

$$\mathcal{D}_\varepsilon \doteq \{x \in \mathcal{D}: \text{dist}\,(x, \partial\mathcal{D}) \leqq \varepsilon\}.$$

The smooth pieces of the boundary of $\mathcal{D}$ are denoted by $(\partial\mathcal{D})_1, \ldots, (\partial\mathcal{D})_k$,

$$\bigcup_{p=1}^{k} (\partial\mathcal{D})_p = \partial\mathcal{D}.$$

$(\partial\mathcal{D})_p$ are supposed to have a parametric representation of the form

$$y_i = g_i^p(s_1^p, \ldots, s_{m-1}^p), \quad i = 1, 2, \ldots, m; \quad (s_1^p, \ldots, s_{m-1}^p) \in \mathcal{S}^p.$$

The set

$$\mathcal{D}_{p\varepsilon} = \{y : y_i = g_i^p(s_1^p, \ldots, s_{m-1}^p) + t n_i^p(s_1^p, \ldots, s_{m-1}^p);$$
$$(s_1^p, \ldots, s_{m-1}^p) \in \mathcal{S}^p; 0 \leqq t \leqq \varepsilon\}$$

is considered where $n^p = (n_1^p, \ldots, n_m^p)$ denotes the normal vector on $(\partial\mathcal{D})_p$ which is to show to the interior of the domain $\mathcal{D}$. Then it follows that the Lebesgue measure of

$$\mathcal{D}_\varepsilon \setminus \bigcup_{p=1}^{k} \mathcal{D}_{p\varepsilon}$$

has the order $O(\varepsilon^2)$. The difference of two sets $\mathcal{A}$ and $\mathcal{B}$ is defined by $\mathcal{A} \setminus \mathcal{B} = \mathcal{A} \cup \mathcal{B} - \mathcal{A} \cap \mathcal{B}$. Then, for a continuous function on $\mathcal{D}$ we have

$$\int_{\mathcal{D}_\varepsilon} F(y) \, dy = \sum_{p=1}^{k} \int_{\mathcal{D}_{p\varepsilon}} F(y) \, dy + O(\varepsilon^2)$$

where the function is prolonged to $\bigcup_{p=1}^{k} \mathcal{D}_{p\varepsilon}$ in a continuous way. Introducing the coordinates $(s_1^p, \ldots, s_{m-1}^p, t)$ on $\mathcal{D}_{p\varepsilon}$ we get

$$\int_{\mathcal{D}_{p\varepsilon}} F(y) \, dy = \int_0^\varepsilon \int_{\mathcal{S}^p} F\big(g^p(s_1^p, \ldots, s_{m-1}^p) + t n^p(s_1^p, \ldots, s_{m-1}^p)\big)$$
$$\times \left| \frac{\partial(y_1, \ldots, y_m)}{\partial(s_1^p, \ldots, s_{m-1}^p, t)} \right| ds_1^p \ldots ds_{m-1}^p \, dt$$

and the functional determinant has the form

$$\frac{\partial(y_1, \ldots, y_m)}{\partial(s_1, \ldots, s_{m-1}, t)} = \begin{vmatrix} \frac{\partial g_1}{\partial s_1} + t \frac{\partial n_1}{\partial s_1} & \frac{\partial g_1}{\partial s_2} + t \frac{\partial n_1}{\partial s_2} & \ldots & \frac{\partial g_1}{\partial s_{m-1}} + t \frac{\partial n_1}{\partial s_{m-1}} & n_1 \\ \frac{\partial g_2}{\partial s_1} + t \frac{\partial n_2}{\partial s_1} & \frac{\partial g_2}{\partial s_2} + t \frac{\partial n_2}{\partial s_2} & \ldots & \frac{\partial g_2}{\partial s_{m-1}} + t \frac{\partial n_2}{\partial s_{m-1}} & n_2 \\ \vdots & \vdots & & \vdots & \\ \frac{\partial g_m}{\partial s_1} + t \frac{\partial n_m}{\partial s_1} & \frac{\partial g_m}{\partial s_2} + t \frac{\partial n_m}{\partial s_2} & \ldots & \frac{\partial g_m}{\partial s_{m-1}} + t \frac{\partial n_m}{\partial s_{m-1}} & n_m \end{vmatrix}.$$

It is possible to evaluate the limit value

$$\lim_{t \to 0} \frac{\partial(y_1, \ldots, y_m)}{\partial(s_1, \ldots, s_{m-1}, t)} = \sum_{q=1}^{m} (-1)^{m+q} \frac{\partial(g_1, \ldots, g_{q-1}, g_{q+1}, \ldots, g_m)}{\partial(s_1, \ldots, s_{m-1})} n_q.$$

Because of

$$n_p = \pm \frac{(-1)^p \dfrac{\partial(g_1, \ldots, g_{p-1}, g_{p+1}, \ldots, g_m)}{\partial(s_1, \ldots, s_{m-1})}}{\sqrt{\sum_{q=1}^{m} \left(\dfrac{\partial(g_1, \ldots, g_{q-1}, g_{q+1}, \ldots, g_m)}{\partial(s_1, \ldots, s_{m-1})} \right)^2}}$$

then it follows

$$\lim_{t\to 0} \frac{\partial(y_1, \ldots, y_m)}{\partial(s_1, \ldots, s_{m-1}, t)} = \pm(-1)^m \sqrt{\sum_{q=1}^{m} \left(\frac{\partial(g_1, \ldots, g_{q-1}, g_{q+1}, \ldots, g_m)}{\partial(s_1, \ldots, s_{m-1})}\right)^2}.$$

The vector $(n_1, \ldots, n_m)$ given above is the normed normal vector since, because of

$$\sum_{p=1}^{m} n_p \frac{\partial g_p}{\partial s_i} = \frac{\pm 1}{\sqrt{\sum_{q=1}^{m} \left(\frac{\partial(g_1, \ldots, g_{q-1}, g_{q+1}, \ldots, g_m)}{\partial(s_1, \ldots, s_{m-1})}\right)^2}} \begin{vmatrix} \frac{\partial g_1}{\partial s_i} & \frac{\partial g_1}{\partial s_1} & \cdots & \frac{\partial g_1}{\partial s_{m-1}} \\ \frac{\partial g_2}{\partial s_i} & \frac{\partial g_2}{\partial s_1} & \cdots & \frac{\partial g_2}{\partial s_{m-1}} \\ \vdots & \vdots & & \vdots \\ \frac{\partial g_m}{\partial s_i} & \frac{\partial g_m}{\partial s_1} & \cdots & \frac{\partial g_m}{\partial s_{m-1}} \end{vmatrix} = 0$$

for $i = 1, 2, \ldots, m - 1$, this vector is perpendicular to all tangent vectors

$$\left(\frac{\partial g_1}{\partial s_i}, \frac{\partial g_2}{\partial s_i}, \ldots, \frac{\partial g_m}{\partial s_i}\right), \quad i = 1, 2, \ldots, m - 1.$$

Hence, we obtain

$$\begin{aligned} \int_{\mathcal{D}_{p\varepsilon}} F(y)\,\mathrm{d}y &= \varepsilon \int_0^1 \int_{\mathcal{S}^p} F\big(g^p(s_1^p, \ldots, s_{m-1}^p) + \varepsilon u n(s_1^p, \ldots, s_{m-1}^p)\big) \\ &\quad \times \left|\frac{\partial(g_1^p, \ldots, g_m^p)}{\partial(s_1^p, \ldots, s_{m-1}^p, \varepsilon u)}\right| \mathrm{d}s_1^p \ldots \mathrm{d}s_{m-1}^p\,\mathrm{d}u = \int_{\mathcal{S}^p} F\big(g^p(s_1^p, \ldots, s_{m-1}^p)\big) \\ &\quad \times \sqrt{\sum_{q=1}^{m} \left(\frac{\partial(g_1^p, \ldots, g_{q-1}^p, g_{q+1}^p, \ldots, g_m^p)}{\partial(s_1^p, \ldots, s_{m-1}^p)}\right)^2}\,\mathrm{d}s_1^p \ldots \mathrm{d}s_{m-1}^p \\ &= \int_{(\partial\mathcal{D})_p} F(y)\,\mathrm{d}S\varepsilon + o(\varepsilon). \end{aligned} \tag{1.129}$$

In a similar way it can be easily verified that

$$\begin{aligned} &\int_{\mathcal{D}_{p\varepsilon}} F(y) \left(\int_{\mathcal{D}\cap\mathcal{K}_\varepsilon(y)} R_{12\varepsilon}(y, z)\,\mathrm{d}z\right) \mathrm{d}y \\ &= \varepsilon \int_0^1 \int_{\mathcal{S}^p} F(g^p + \varepsilon u n) \left(\int_{\mathcal{D}\cap\mathcal{K}_\varepsilon(g^p+\varepsilon u n)} R_{12\varepsilon}(g^p + \varepsilon u n, z)\,\mathrm{d}z\right) \\ &\quad \times \left|\frac{\partial(y_1, \ldots, y_m)}{\partial(s_1, \ldots, s_{m-1}, \varepsilon u)}\right| \mathrm{d}s_1 \ldots \mathrm{d}s_{m-1}\,\mathrm{d}u \\ &= \varepsilon^{m+1} \int_{(\partial\mathcal{D})_p} F(y) \int_0^1 a_{12}(u; y)\,\mathrm{d}u\,\mathrm{d}S_y + o(\varepsilon^{m+1}) \end{aligned} \tag{1.130}$$

where $a_{12}(u; g)$ is defined by

$$a_{12}(u; g) \doteq \lim_{\varepsilon\downarrow 0} \left[\frac{1}{\varepsilon^m} \int_{\mathcal{D}\cap\mathcal{K}_\varepsilon(x)} R_{12\varepsilon}(x, z)\,\mathrm{d}z\right]_{x=g+\varepsilon u n} \quad \text{for} \quad g \in \partial\mathcal{D}. \tag{1.131}$$

Now we deal with the calculation of $J^{12}(F_1, F_2)$. The intensity is assumed to be continuous. Thus

$$J^{12}(F_1, F_2) = \int_{\mathcal{D}} \int_{\mathcal{D}} F_1(x)\, F_2(y)\, R_{12\varepsilon}(x, y)\, dx\, dy$$

$$= \int_{\mathcal{D}\setminus\mathcal{D}_\varepsilon} F_1(x) \Big(\int_{\mathcal{K}_\varepsilon(x)} F_2(y)\, R_{12\varepsilon}(x, y)\, dy \Big)\, dx$$

$$+ \int_{\mathcal{D}_\varepsilon} F_1(x) \Big(\int_{\mathcal{D}\cap\mathcal{K}_\varepsilon(x)} F_2(y)\, R_{12\varepsilon}(x, y)\, dy \Big)\, dx .$$

We have

$$J_2^{12}(F_1, F_2) = \int_{\mathcal{D}_\varepsilon} F_1F_2(x) \Big(\int_{\mathcal{D}\cap\mathcal{K}_\varepsilon(x)} R_{12\varepsilon}(x, y)\, dy \Big)\, dx$$

$$+ \int_{\mathcal{D}_\varepsilon} F_1(x) \Big(\int_{\mathcal{D}\cap\mathcal{K}_\varepsilon(x)} \big(F_2(y) - F_2(x)\big)\, R_{12\varepsilon}(x, y)\, dy \Big)\, dx$$

and then

$$|J_{22}^{12}(F_1, F_2)| \leqq \int_{\mathcal{D}_\varepsilon} |F_1(x)|\, M_\varepsilon(F_2; x)\, dx \cdot \varepsilon^m v\big(\mathcal{K}_1(0)\big)\, c_2 = o(\varepsilon^{m+1}),$$

$$J_{21}^{12}(F_1, F_2) = \varepsilon^{m+1} \int_{\partial\mathcal{D}} F_1F_2(x) \int_0^1 a_{12}(u; x)\, du\, dS_x + o(\varepsilon^{m+1})$$

if (1.130) is taken into consideration. As in the case of $m = 1$ an expansion of $\int_{\mathcal{K}_\varepsilon(x)} R_{12\varepsilon}(x,y)\, dy$ is assumed,

$$\int_{\mathcal{K}_\varepsilon(x)} R_{12\varepsilon}(x, y)\, dy = a_{12}(x)\, \varepsilon^m + b_{12}(x)\, \varepsilon^{m+1} + o(\varepsilon^{m+1}). \tag{1.132}$$

With the help of (1.129) and

$$J_1^{12}(F_1, F_2) = \int_{\mathcal{D}\setminus\mathcal{D}_\varepsilon} F_1F_2(x) \Big(\int_{\mathcal{K}_\varepsilon(x)} R_{12\varepsilon}(x, y)\, dy \Big)\, dx$$

$$+ \int_{\mathcal{D}\setminus\mathcal{D}_\varepsilon} F_1(x) \Big(\int_{\mathcal{K}_\varepsilon(x)} \big(F_2(y) - F_2(x)\big)\, R_{12\varepsilon}(x, y)\, dy \Big)\, dx$$

we obtain

$$J_{11}^{12}(F_1, F_2) = \int_{\mathcal{D}\setminus\mathcal{D}_\varepsilon} F_1F_2(x)\, a_{12}(x)\, dx \cdot \varepsilon^m + \int_{\mathcal{D}\setminus\mathcal{D}_\varepsilon} F_1F_2(x)\, b_{12}(x)\, dx \cdot \varepsilon^{m+1} + o(\varepsilon^{m+1})$$

$$= \int_{\mathcal{D}} F_1F_2a_{12}(x)\, dx \cdot \varepsilon^m - \int_{\mathcal{D}_\varepsilon} F_1F_2a_{12}(x)\, dx \cdot \varepsilon^m$$

$$+ \int_{\mathcal{D}} F_1F_2b_{12}(x)\, dx \cdot \varepsilon^{m+1} + o(\varepsilon^{m+1})$$

$$= \int_{\mathcal{D}} F_1F_2a_{12}(x)\, dx \cdot \varepsilon^m + \Big\{ \int_{\mathcal{D}} F_1F_2b_{12}(x)\, dx$$

$$- \int_{\partial\mathcal{D}} F_1F_2a_{12}(x)\, dS_x \Big\}\, \varepsilon^{m+1} + o(\varepsilon^{m+1}).$$

J_{12}^{12} can be calculated by

$$J_{12}^{12}(F_1, F_2) = \sum_{i=1}^{m} \int_{\mathcal{D}\setminus\mathcal{D}_\varepsilon} F_1(x) \left(\int_{\mathcal{K}_\varepsilon(x)} F_{2,y_i}(\tilde{y}) (y_i - x_i) R_{12\varepsilon}(x, y) \, dy \right) dx$$

$$= \sum_{i=1}^{m} \int_{\mathcal{D}\setminus\mathcal{D}_\varepsilon} F_1 F_{2,x_i}(x) \left(\int_{\mathcal{K}_\varepsilon(x)} (y_i - x_i) R_{12\varepsilon}(x, y) \, dy \right) dx$$

$$+ \sum_{i=1}^{m} \int_{\mathcal{D}\setminus\mathcal{D}_\varepsilon} F_1(x) \left(\int_{\mathcal{K}_\varepsilon(x)} \left(F_{2,y_i}(\tilde{y}) - F_{2,y_i}(x)\right) (y_i - x_i) R_{12\varepsilon}(x,y) \, dy \right) dx$$

and

$$J_{121}^{12}(F_1, F_2) = \sum_{i=1}^{m} \int_{\mathcal{D}\setminus\mathcal{D}_\varepsilon} F_1 F_{2,x_i} \alpha_{12,i}(x) \, dx \varepsilon^{m+1} + o(\varepsilon^{m+1})$$

$$= \sum_{i=1}^{m} \int_{\mathcal{D}} F_1 F_{2,x_i} \alpha_{12,i}(x) \, dx \varepsilon^{m+1} + o(\varepsilon^{m+1}),$$

$$|J_{122}^{12}(F_1, F_2)| \leqq c_2 v\left(\mathcal{K}_1(0)\right) \sum_{i=1}^{m} \int_{\mathcal{D}} |F_1| \, M_\varepsilon(F_{2,x_i}; x) \, dx \varepsilon^{m+1} = o(\varepsilon^{m+1})$$

where the limit value $\alpha_{12,i}(x)$ is determined by

$$\alpha_{12,i}(x) \doteq \lim_{\varepsilon \downarrow 0} \frac{1}{\varepsilon^{m+1}} \int_{\mathcal{K}_\varepsilon(x)} (y_i - x_i) R_{12\varepsilon}(x, y) \, dy, \quad i = 1, 2, \ldots, m. \tag{1.133}$$

A summary leads to

$$J^{12}(F_1, F_2) = \int_{\mathcal{D}} F_1 F_2 a_{12}(x) \, dx \varepsilon^m$$

$$+ \left[\int_{\mathcal{D}} F_1 F_2 b_{12}(x) \, dx + \int_{\mathcal{D}} F_1 \operatorname{grad} F_2 \cdot \alpha_{12}(x) \, dx \right.$$

$$\left. + \int_{\partial\mathcal{D}} F_1 F_2(x) \left\{ \int_0^1 a_{12}(u; x) \, du - a_{12}(x) \right\} dS_x \right] \varepsilon^{m+1} + o(\varepsilon^{m+1})$$

if

$$\operatorname{grad} F \cdot \alpha_{12}(x) = \sum_{i=1}^{m} F_{x_i} \alpha_{12,i}(x)$$

is put. On the other hand using $J^{12}(F_1, F_2) = J^{21}(F_2, F_1)$ and

$$J^{21}(F_2, F_1) = \int_{\mathcal{D}} F_1 F_2 a_{21}(x) \, dx \varepsilon^m$$

$$+ \left[\int_{\mathcal{D}} F_1 F_2 b_{21}(x) \, dx + \int_{\mathcal{D}} F_2 \operatorname{grad} F_1 \cdot \alpha_{21}(x) \, dx \right.$$

$$\left. + \int_{\partial\mathcal{D}} F_1 F_2(x) \left\{ \int_0^1 a_{21}(u; x) \, du - a_{21}(x) \right\} dS_x \right] \varepsilon^{m+1} + o(\varepsilon^{m+1})$$

the difference leads to

$$a_{12}(x) = a_{21}(x) \quad \text{for all} \quad x \in \mathcal{D}$$

and

$$\int_{\mathcal{D}} F_1F_2(b_{12} - b_{21})(x)\,dx + \int_{\mathcal{D}} \big(F_1 \operatorname{grad} F_2 \cdot \alpha_{12}(x) - F_2 \operatorname{grad} F_1 \cdot \alpha_{21}(x)\big)\,dx$$
$$+ \int_{\partial\mathcal{D}} F_1F_2(x)\big(S_{12}(x; 0, 1) - S_{21}(x; 0, 1)\big)\,dS_x = 0. \tag{1.134}$$

$S_{ij}(x; 0, 1)$ was introduced by

$$S_{ij}(x; 0, 1) = \int_0^1 a_{ij}(u; x)\,du - a_{ij}(x).$$

Now we apply (1.134) and subtract (1.134) where F_1 and F_2 are changed. The result can be written as

$$\int_{\mathcal{D}} (F_1 \operatorname{grad} F_2 - F_2 \operatorname{grad} F_1)(\alpha_{12} + \alpha_{21})\,dx = 0$$

or

$$\int_{\mathcal{D}} F\alpha_{12,i}(x)\,dx = -\int_{\mathcal{D}} F\alpha_{21,i}(x)\,dx, \quad i = 1, 2, \ldots, m, \quad F \in \mathrm{C}(\mathcal{D}). \tag{1.135}$$

The statistical characteristic ${}^2A_2(F_1, F_2)$ has in this case the form

$${}^2A_2(F_1, F_2) = \frac{1}{2}\int_{\mathcal{D}} F_1F_2(b_{12} + b_{21})(x)\,dx$$
$$+ \frac{1}{2}\int_{\mathcal{D}} (F_1 \operatorname{grad} F_2 - F_2 \operatorname{grad} F_1)\,\alpha_{12}(x)\,dx$$
$$+ \frac{1}{2}\int_{\partial\mathcal{D}} F_1F_2\big(S_{12}(x; 0, 1) + S_{21}(x; 0, 1)\big)\,dS_x.$$

Now, we investigate the general case. For the separation of $\mathcal{D}_1$ we introduce the following sets:

$$\mathcal{D}_{12,i\varepsilon} \doteq \{x \in \mathcal{D}_{12} : \operatorname{dist}(x, \partial\mathcal{D}_i) \leqq \varepsilon\}, \quad i = 1, 2,$$
$$\mathcal{D}_{i,j\varepsilon} \doteq \{x \in \mathcal{D}_i : \operatorname{dist}(x, \partial\mathcal{D}_j) \leqq \varepsilon\}, \quad (i, j) = (1, 2), (2, 1)$$

where $\mathcal{D}_{12} = \mathcal{D}_1 \cap \mathcal{D}_2$. Then we have

$$J^{12}(F_1, F_2) = \int_{\mathcal{D}_1}\int_{\mathcal{D}_2} F_1(x)\,F_2(y)\,R_{12\varepsilon}(x, y)\,dx\,dy$$
$$= \int_{\mathcal{D}_{12}\setminus\mathcal{D}_{12,2\varepsilon}} F_1(x)\left(\int_{\mathcal{K}_\varepsilon(x)} F_2(y)\,R_{12\varepsilon}(x, y)\,dy\right)dx$$
$$+ \int_{\mathcal{D}_{1,2\varepsilon}} F_1(x)\left(\int_{\mathcal{D}_2\cap\mathcal{K}_\varepsilon(x)} F_2(y)\,R_{12\varepsilon}(x, y)\,dy\right)dx.$$

First, we deal with $J_1^{12}(F_1, F_2)$ and obtain

$$J_1^{12}(F_1, F_2) = \int_{\mathcal{D}_{12}\setminus\mathcal{D}_{12,2\varepsilon}} F_1(x)\,F_2(x)\left(\int_{\mathcal{K}_\varepsilon(x)} R_{12\varepsilon}(x, y)\,dy\right)dx$$
$$+ \int_{\mathcal{D}_{12}\setminus\mathcal{D}_{12,2\varepsilon}} F_1(x)\left(\int_{\mathcal{K}_\varepsilon(x)} \big(F_2(y) - F_2(x)\big)\,R_{12\varepsilon}(x, y)\,dy\right)dx$$

and

$$J_{11}^{12}(F_1, F_2) = \int_{\mathcal{D}_{12} \setminus \mathcal{D}_{12,2\varepsilon}} F_1 F_2 a_{12}(x) \, dx \varepsilon^m + \int_{\mathcal{D}_{12} \setminus \mathcal{D}_{12,2\varepsilon}} F_1 F_2 b_{12}(x) \, dx \varepsilon^{m+1} + o(\varepsilon^{m+1})$$

$$= \int_{\mathcal{D}_{12}} F_1 F_2 a_{12}(x) \, dx \varepsilon^m - \int_{\mathcal{D}_{12} \setminus \mathcal{D}_{12,2\varepsilon}} F_1 F_2 a_{12}(x) \, dx \varepsilon^m$$

$$+ \int_{\mathcal{D}_{12}} F_1 F_2 b_{12}(x) \, dx \varepsilon^{m+1} + o(\varepsilon^{m+1})$$

if the expansion

$$\int_{\mathcal{K}_\varepsilon(x)} R_{12\varepsilon}(x, y) \, dy = a_{12}(x) \, \varepsilon^m + b_{12}(x) \, \varepsilon^{m+1} + o(\varepsilon^{m+1})$$

is applied. Using (1.129) we find

$$\int_{\mathcal{D}_{12,2\varepsilon}} F_1 F_2 a_{12}(x) \, dx = \int_{(\partial \mathcal{D}_2)_1^{\mathrm{a}}} F_1 F_2 a_{12}(x) \, dS \cdot \varepsilon + o(\varepsilon)$$

where $(\partial \mathcal{D}_2)_1^{\mathrm{a}}$ denotes the points of $\partial \mathcal{D}_2$ which are accumulation points of $\mathcal{D}_{12} \setminus (\partial \mathcal{D}_1 \cup \partial \mathcal{D}_2)$:

$$(\partial \mathcal{D}_2)_1^{\mathrm{a}} \doteq \{x \in \partial \mathcal{D}_2 \colon x \text{ accumulation point of } \mathcal{D}_{12} \setminus (\partial \mathcal{D}_1 \cup \partial \mathcal{D}_2)\}. \tag{1.136}$$

Thus,

$$J_{11}^{12}(F_1, F_2) = \int_{\mathcal{D}_{12}} F_1 F_2 a_{12}(x) \, dx \varepsilon^m$$

$$+ \left[\int_{\mathcal{D}_{12}} F_1 F_2 b_{12}(x) \, dx - \int_{(\partial \mathcal{D}_2)_1^{\mathrm{a}}} F_1 F_2 a_{12}(x) \, dS_x \right] \varepsilon^{m+1} + o(\varepsilon^{m+1}).$$

Furthermore,

$$J_{12}^{12} = \sum_{i=1}^{m} \int_{\mathcal{D}_{12} \setminus \mathcal{D}_{12,2\varepsilon}} F_1(x) \, F'_{2,x_i}(x) \left(\int_{\mathcal{K}_\varepsilon(x)} (y_i - x_i) \, R_{12\varepsilon}(x, y) \, dy \right) dx$$

$$+ \sum_{i=1}^{m} \int_{\mathcal{D}_{12} \setminus \mathcal{D}_{12,2\varepsilon}} F_1(x) \left(\int_{\mathcal{K}_\varepsilon(x)} \left(F_{2,y_i}(\tilde{y}) - F_{2,y_i}(x) \right) (y_i - x_i) \, R_{12\varepsilon}(x, y) \, dy \right) dx$$

and with the help of

$$\lim_{\varepsilon \downarrow 0} \frac{1}{\varepsilon^{m+1}} \int_{\mathcal{K}_\varepsilon(x)} (y_i - x_i) \, R_{12\varepsilon}(x, y) \, dy = \alpha_{12,i}(x)$$

then

$$J_{121}^{12} = \sum_{i=1}^{m} \int_{\mathcal{D}_{12}} F_1 F_{2,x_i} \alpha_{12,i}(x) \, dx \varepsilon^{m+1} + o(\varepsilon^{m+1})$$

$$= \int_{\mathcal{D}_{12}} F_1 \operatorname{grad} F_2 \cdot \alpha_{12}(x) \, dx \varepsilon^{m+1} + o(\varepsilon^{m+1}),$$

$$|J_{122}^{12}| \leq c_2 v(\mathcal{K}_1(0)) \sum_{i=1}^{m} \int_{\mathcal{D}_{12}} |F_1(x)| \, M_\varepsilon(F_{2,x_i}; x) \, dx \varepsilon^{m+1} = o(\varepsilon^{m+1}).$$

With this, the expansion of $J_1^{12}(F_1, F_2)$ is given by

$$J_1^{12}(F_1, F_2) = \int_{\mathcal{D}_{12}} F_1F_2a_{12}(x)\,\mathrm{d}x\varepsilon^m + \Bigg[\int_{\mathcal{D}_{12}} F_1F_2b_{12}(x)\,\mathrm{d}x + \int_{\mathcal{D}_{12}} F_1 \operatorname{grad} F_2\cdot \alpha_{12}(x)\,\mathrm{d}x - \int_{(\partial\mathcal{D}_2)_1^a} F_1F_2a_{12}(x)\,\mathrm{d}S_x\Bigg]\varepsilon^{m+1} + o(\varepsilon^{m+1}).$$

Now we turn to J_2^{12}. It is

$$\begin{aligned} J_2^{12} &= \int_{\mathcal{D}_{1,2\varepsilon}} F_1(x)\Bigg(\int_{\mathcal{D}_2\cap\mathcal{K}_\varepsilon(x)} F_2(y)\,R_{12\varepsilon}(x, y)\,\mathrm{d}y\Bigg)\,\mathrm{d}x \\ &= \int_{\mathcal{D}_{1,2\varepsilon}} F_1F_2(x)\Bigg(\int_{\mathcal{D}_2\cap\mathcal{K}_\varepsilon(x)} R_{12\varepsilon}(x, y)\,\mathrm{d}y\Bigg)\,\mathrm{d}x \\ &\quad + \int_{\mathcal{D}_{1,2\varepsilon}} F_1(x)\Bigg(\int_{\mathcal{D}_2\cap\mathcal{K}_\varepsilon(x)} \big(F_2(y) - F_2(x)\big)\,R_{12\varepsilon}(x, y)\,\mathrm{d}y\Bigg)\,\mathrm{d}x \end{aligned}$$

and we have the inequality

$$|J_{22}^{12}| \leqq c_2v\big(\mathcal{K}_1(0)\big)\int_{\mathcal{D}_{1,2\varepsilon}} |F_1(x)|\,M_\varepsilon(F_2; x)\,\mathrm{d}x\varepsilon^m = o(\varepsilon^{m+1}).$$

In order to deal with J_{21}^{12} we have to separate $(\partial\mathcal{D}_2)_1$:

$$(\partial\mathcal{D}_2)_1 = (\partial\mathcal{D}_2)_1^{i1} \cup (\partial\mathcal{D}_2)_1^{b1}$$

where

$$\begin{aligned} (\partial\mathcal{D}_2)_1^i &= \{x \in \partial\mathcal{D}_2 : x \in \mathcal{D}_1,\, x \notin \partial\mathcal{D}_1\}, \\ (\partial\mathcal{D}_2)_1^b &= \{x \in \partial\mathcal{D}_2 : x \in \partial\mathcal{D}_1\}. \end{aligned} \tag{1.137}$$

Applying (1.130) and an analogous formula for $(\partial\mathcal{D}_2)_1^{b1}$ we see that

$$J_{21}^{12} = \int_{(\partial\mathcal{D}_2)_1^i} F_1F_2(x)\int_{-1}^{1} a_{12}^2(u; x)\,\mathrm{d}u\,\mathrm{d}S_x + \int_{(\partial\mathcal{D}_2)_1^b} F_1F_2(x)\int_{\mathcal{J}_{12}(x)} a_{12}^2(u; x)\,\mathrm{d}u\,\mathrm{d}S_x$$

where $a_{12}^i(u; x)$ is given by

$$a_{12}^i(u; x) = \lim_{\varepsilon\downarrow 0}\Bigg[\frac{1}{\varepsilon^m}\int_{\mathcal{D}_i\cap\mathcal{K}_\varepsilon(z)} R_{12\varepsilon}(z, y)\,\mathrm{d}y\Bigg]_{z=x+\varepsilon un} \qquad \text{for} \quad x \in \partial\mathcal{D}_i$$

and the normal vector shows to the interior of $\mathcal{D}_i$. The interval $\mathcal{J}_{12}(x)$ is defined by

$$\mathcal{J}_{12}(x) = \begin{cases} [0, 1] & \text{if} \quad n(x, \mathcal{D}_1) = n(x, \mathcal{D}_2), \\ [-1, 0] & \text{if} \quad n(x, \mathcal{D}_2) = -n(x, \mathcal{D}_1), \\ \emptyset & \text{otherwise}. \end{cases}$$

Finally, we have obtained

$$\begin{aligned} {}^2A_2(F_1, F_2) &= \int_{\mathcal{D}_{12}} F_1F_2b_{12}(x)\,\mathrm{d}x + \int_{\mathcal{D}_{12}} F_1 \operatorname{grad} F_2\cdot\alpha_{12}(x)\,\mathrm{d}x \\ &\quad + \int_{(\partial\mathcal{D}_2)_1^i} F_1F_2(x)\int_{-1}^{1} a_{12}^2(u; x)\,\mathrm{d}u\,\mathrm{d}S_x - \int_{(\partial\mathcal{D}_2)_1^a} F_1F_2a_{12}(x)\,\mathrm{d}S_x \\ &\quad + \int_{(\partial\mathcal{D}_2)_1^b} F_1F_2(x)\int_{\mathcal{J}_{12}(x)} a_{12}^2(u; x)\,\mathrm{d}u\,\mathrm{d}S_x. \end{aligned}$$

From $J^{12}(F_1, F_2) = J^{21}(F_2, F_1)$ it follows

$$
\begin{aligned}
{}^2A_2(F_1, F_2) = &\int_{\mathcal{D}_{12}} F_1F_2b_{21}(x)\,\mathrm{d}x + \int_{\mathcal{D}_{12}} F_2 \operatorname{grad} F_1 \cdot \alpha_{21}(x)\,\mathrm{d}x \\
&+ \int_{(\partial\mathcal{D}_1)_2^{\mathrm{i}}} F_1F_2(x) \int_{-1}^{1} a_{21}^1(u; x)\,\mathrm{d}u\,\mathrm{d}S_x \\
&+ \int_{(\partial\mathcal{D}_1)_2^{\mathrm{b}}} F_1F_2(x) \int_{\mathcal{J}_{21}(x)} a_{21}^1(u; x)\,\mathrm{d}u\,\mathrm{d}S_x - \int_{(\partial\mathcal{D}_1)_2^{\mathrm{a}}} F_1F_2a_{21}(x)\,\mathrm{d}S_x
\end{aligned}
$$

and then a symmetric form

$$
\begin{aligned}
2\cdot{}^2A_2(F_1, F_2) = &\int_{\mathcal{D}_{12}} F_1F_2(b_{12} + b_{21})\,(x)\,\mathrm{d}x \\
&+ \int_{\mathcal{D}_{12}} \big(F_1 \operatorname{grad} F_2 \cdot \alpha_{12}(x) + F_2 \operatorname{grad} F_1 \cdot \alpha_{21}(x)\big)\,\mathrm{d}x \\
&- \left\{\int_{(\partial\mathcal{D}_2)_1^{\mathrm{a}}} F_1F_2a_{12}(x)\,\mathrm{d}S_x + \int_{(\partial\mathcal{D}_1)_2^{\mathrm{a}}} F_1F_2a_{21}(x)\,\mathrm{d}S_x\right\} \\
&+ \left\{\int_{(\partial\mathcal{D}_2)_1^{\mathrm{i}}} F_1F_2(x) \int_{-1}^{1} a_{12}^2(u; x)\,\mathrm{d}u\,\mathrm{d}S_x \right. \\
&\left. + \int_{(\partial\mathcal{D}_1)_2^{\mathrm{i}}} F_1F_2(x) \int_{-1}^{1} a_{21}^1(u; x)\,\mathrm{d}u\,\mathrm{d}S_x\right\} \\
&+ \left\{\int_{(\partial\mathcal{D}_2)_1^{\mathrm{b}}} F_1F_2(x) \int_{\mathcal{J}_{12}(x)} a_{12}^2(u; x)\,\mathrm{d}u\,\mathrm{d}S_x \right. \\
&\left. + \int_{(\partial\mathcal{D}_1)_2^{\mathrm{b}}} F_1F_2(x) \int_{\mathcal{J}_{21}(x)} a_{21}^1(u; x)\,\mathrm{d}u\,\mathrm{d}S_x\right\}.
\end{aligned}
$$

The difference between the calculated terms ${}^2A_2(F_1, F_2)$ leads to

$$
\begin{aligned}
&\int_{\mathcal{D}_{12}} F_1F_2(b_{12} - b_{21})\,\mathrm{d}x + \int_{\mathcal{D}_{12}} \operatorname{grad}(F_1F_2) \cdot \alpha_{12}\,\mathrm{d}x \\
&- \left\{\int_{(\partial\mathcal{D}_2)_1^{\mathrm{a}}} F_1F_2a_{12}\,\mathrm{d}S - \int_{(\partial\mathcal{D}_1)_2^{\mathrm{a}}} F_1F_2a_{21}\,\mathrm{d}S\right\} \\
&+ \left\{\int_{(\partial\mathcal{D}_2)_1^{\mathrm{i}}} F_1F_2 \int_{-1}^{1} a_{12}^2\,\mathrm{d}u\,\mathrm{d}S - \int_{(\partial\mathcal{D}_1)_2^{\mathrm{i}}} F_1F_2 \int_{-1}^{1} a_{21}^1\,\mathrm{d}u\,\mathrm{d}S\right\} \\
&+ \left\{\int_{(\partial\mathcal{D}_2)_1^{\mathrm{b}}} F_1F_2 \int_{\mathcal{J}_{12}} a_{12}^2\,\mathrm{d}u\,\mathrm{d}S - \int_{(\partial\mathcal{D}_1)_2^{\mathrm{b}}} F_1F_2 \int_{\mathcal{J}_{21}} a_{21}^1\,\mathrm{d}u\,\mathrm{d}S\right\} = 0.
\end{aligned}
$$

The result of these considerations is summarized by Theorem 1.10.

Theorem 1.10. *Let F_i, $i = 1, 2$, be functions from $\mathbf{C}^1(\mathcal{D}_i)$ and $\mathcal{D}_i$ bounded domains with piecewise smooth boundaries. $\mathcal{D}_{12}$ denotes the intersection of $\mathcal{D}_1$ and $\mathcal{D}_2$ and*

$$(\partial\mathcal{D}_2)_1^{\mathrm{a}} \doteq \{x \in \partial\mathcal{D}_2 : x \text{ accumulation point of } \mathcal{D}_{12} \setminus (\partial\mathcal{D}_1 \cup \partial\mathcal{D}_2)\},$$

$$(\partial\mathcal{D}_2)_1^{\mathrm{i}} \doteq \{x \in \partial\mathcal{D}_2 : x \in \mathcal{D}_1, x \notin \partial\mathcal{D}_1\},$$

$$(\partial\mathcal{D}_2)_1^{\mathrm{b}} \doteq \{x \in \partial\mathcal{D}_2 : x \in \partial\mathcal{D}_1\}.$$

The normal vector n as to $x \in \partial\mathcal{D}_1$ shows in the interior of $\mathcal{D}_1$ and is denoted by $n(x, \mathcal{D}_1)$. Then

$$\mathcal{J}_{12}(x) = \left\{\begin{matrix} [0, 1] & \text{if} & n(x, \mathcal{D}_2) = n(x, \mathcal{D}_1) \\ [-1, 0] & \text{if} & n(x, \mathcal{D}_2) = -n(x, \mathcal{D}_1) \\ \emptyset & & \text{otherwise} \end{matrix}\right\} \quad \text{for} \quad x \in (\partial\mathcal{D}_2)_1^{\mathrm{b}}.$$

Let the function

$$R_{12\varepsilon}(x, y) = \langle f_{1\varepsilon}(x)\, f_{2\varepsilon}(y)\rangle$$

have the following properties:

$$-\int\limits_{\mathcal{K}_\varepsilon(x)} R_{12\varepsilon}(x, y)\, \mathrm{d}y = a_{12}(x)\, \varepsilon^m + b_{12}(x)\, \varepsilon^{m+1} + o(\varepsilon^{m+1})$$

with the continuous intensity $a_{12}(x)$,

$$-\lim_{\varepsilon\downarrow 0} \frac{1}{\varepsilon^{m+1}} \int\limits_{\mathcal{K}_\varepsilon(x)} (y_i - x_i)\, R_{12\varepsilon}(x, y)\, \mathrm{d}y = \alpha_{12,i}(x), \quad i = 1, 2, \ldots, m,$$

$$-\lim_{\varepsilon\downarrow 0} \left[\frac{1}{\varepsilon^m} \int\limits_{\mathcal{D}_i\cap\mathcal{K}_\varepsilon(z)} R_{12\varepsilon}(z, y)\, \mathrm{d}y\right]_{z=x+\varepsilon u n} = a_{12}^{i}(u; x) \quad \text{for} \quad x \in \partial\mathcal{D}_i.$$

Then the statistical characteristic ${}^2A_2(F_1, F_2)$ can be written as

$$\begin{aligned} {}^2A_2(F_1, F_2) = & \int\limits_{\mathcal{D}_{12}} F_1F_2b_{12}(x)\, \mathrm{d}x + \int\limits_{\mathcal{D}_{12}} F_1 \operatorname{grad} F_2 \cdot \alpha_{12}(x)\, \mathrm{d}x \\ & + \int\limits_{(\partial\mathcal{D}_2)_1^{\mathrm{i}}} F_1F_2(x) \int\limits_{-1}^{1} a_{12}^2(u; x)\, \mathrm{d}u\, \mathrm{d}S_x - \int\limits_{(\partial\mathcal{D}_2)_1^{\mathrm{a}}} F_1F_2a_{12}(x)\, \mathrm{d}S_x \\ & + \int\limits_{(\partial\mathcal{D}_2)_1^{\mathrm{b}}} F_1F_2(x) \int\limits_{\mathcal{J}_{12}(x)} a_{12}^2(u; x)\, \mathrm{d}u\, \mathrm{d}S_x \end{aligned}$$

or by

$$\begin{aligned} {}^2A_2(F_1, F_2) = & \int\limits_{\mathcal{D}_{12}} F_1F_2b_{21}(x)\, \mathrm{d}x + \int\limits_{\mathcal{D}_{12}} F_2 \operatorname{grad} F_1 \cdot \alpha_{21}(x)\, \mathrm{d}x \\ & + \int\limits_{(\partial\mathcal{D}_1)_2^{\mathrm{i}}} F_1F_2(x) \int\limits_{-1}^{1} a_{21}^2(u; x)\, \mathrm{d}u\, \mathrm{d}S_x - \int\limits_{(\partial\mathcal{D}_1)_2^{\mathrm{a}}} F_1F_2a_{21}(x)\, \mathrm{d}S_x \\ & + \int\limits_{(\partial\mathcal{D}_1)_2^{\mathrm{b}}} F_1F_2(x) \int\limits_{\mathcal{J}_{21}(x)} a_{21}^1(u; x)\, \mathrm{d}u\, \mathrm{d}S_x. \end{aligned}$$

We give some examples for the calculation of ${}^2A_2$. First we deal with the case $\mathcal{D}_1 = \mathcal{D}_2 = \mathcal{D}$ and it follows $(\partial\mathcal{D}_2)_1^{\mathrm{a}} = \partial\mathcal{D}$, $(\partial\mathcal{D}_2)_1^{\mathrm{i}} = \emptyset$, $(\partial\mathcal{D}_2)_1^{\mathrm{b}} = \partial\mathcal{D}$, $\mathcal{J}_{12}(x) = [0, 1]$ for all $x \in \partial\mathcal{D}$. Then from Theorem 1.10 we obtain the same results as in the first part of this section.

In the case A) of Fig. 1.10 we have

$$(\partial\mathcal{D}_2)_1^{\mathrm{a}} = a \cup b \cup c,$$

$$(\partial\mathcal{D}_2)_1^{\mathrm{i}} = a \cup b, \quad (\partial\mathcal{D}_2)_1^{\mathrm{b}} = c$$

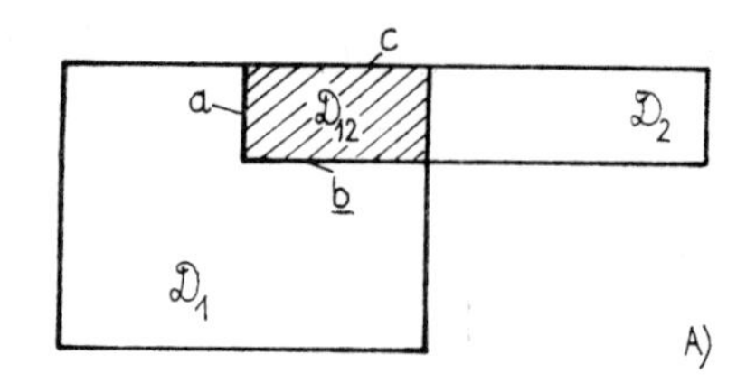

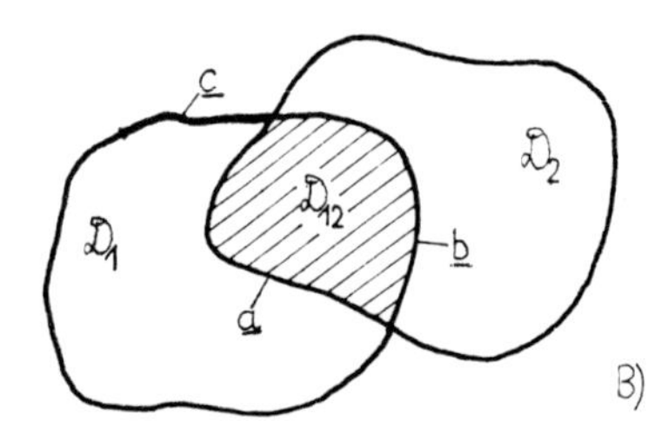

Fig. 1.10. Special domains for the evaluation of ${}^2A_2$

and $\mathcal{J}_{12}(x) = [0, 1]$ for all $x \in c$. From this we can write

$$
\begin{aligned}
{}^2A_2(F_1, F_2) &= \int_{\mathcal{D}_{12}} F_1F_2b_{12}(x)\, dx + \int_{\mathcal{D}_{12}} F_1 \operatorname{grad} F_2 \cdot \alpha_{12}(x)\, dx \\
&+ \int_{a \cup b} F_1F_2(x) \left(\int_{-1}^{1} a_{12}^2(u; x)\, du - a_{12}(x) \right) dS_x \\
&+ \int_{c} F_1F_2(x) \left(\int_{0}^{1} a_{12}^2(u; x)\, du - a_{12}(x) \right) dS_x .
\end{aligned}
$$

In the case B) of Fig. 1.10 the set c is to belong to $\mathcal{D}_1$ and $\mathcal{D}_2$ hence also to $\mathcal{D}_{12}$. Then we can see that

$$(\partial\mathcal{D}_2)_1^{\mathrm{a}} = a; \quad (\partial\mathcal{D}_2)_1^{\mathrm{i}} = a; \quad (\partial\mathcal{D}_2)_1^{\mathrm{b}} = c; \quad \mathcal{J}_{12}(x) = \emptyset \quad \text{for} \quad \text{all } x \in c .$$

Theorem 1.10 leads to

$$
\begin{aligned}
{}^2A_2(F_1, F_2) &= \int_{\mathcal{D}_{12}} F_1F_2b_{12}(x)\, dx + \int_{\mathcal{D}_{12}} F_1 \operatorname{grad} F_2 \cdot \alpha_{12}(x)\, dx \\
&+ \int_{a} F_1F_2(x) \left(\int_{-1}^{1} a_{12}^2(u; x)\, du - a_{12}(x) \right) dS_x .
\end{aligned}
$$

1.6.5. Examples

First, ${}^2A_2$ is calculated for a rectangular domain (see Fig. 1.11a) $\mathcal{D} = \mathcal{D}_1 = \mathcal{D}_2$ and a correlation function of the form

$$
R_{1\varepsilon}(x, y) = \begin{cases} \sigma^2 \left(1 - \dfrac{1}{\tilde{\varepsilon}} |y_1 - x_1|\right) \left(1 - \dfrac{1}{\tilde{\varepsilon}} |y_2 - x_2|\right) \\ \qquad \text{for} \quad \max\{|y_1 - x_1|, |y_2 - x_2|\} \leqq \tilde{\varepsilon}, \\ 0 \quad \text{otherwise}. \end{cases}
$$

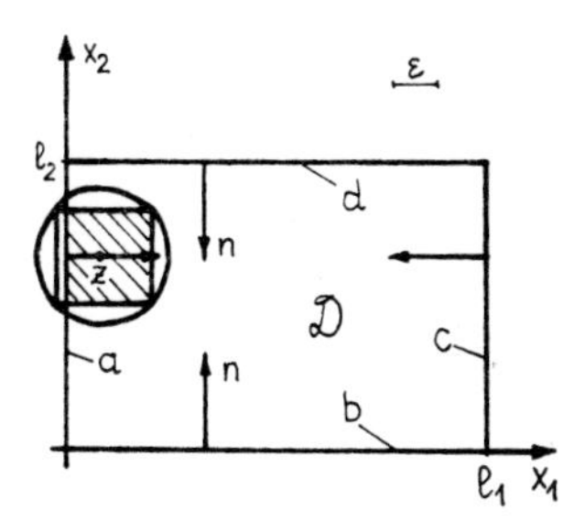

Fig. 1.11a. Domain $\mathscr{D}$ for the calculation of ${}^2A_2$

$R_{1\varepsilon}(x, y)$ is a correlation function using

$$\mathscr{F}(R_{1\varepsilon}) = \sigma^2 \int_{-\tilde{\varepsilon}}^{\tilde{\varepsilon}} \int_{-\tilde{\varepsilon}}^{\tilde{\varepsilon}} \exp\left(\mathrm{i}(t_1 z_1 + t_2 z_2)\right) \left(1 - \frac{1}{\tilde{\varepsilon}} |z_1|\right) \left(1 - \frac{1}{\tilde{\varepsilon}} |z_2|\right) \mathrm{d}z_1 \, \mathrm{d}z_2$$

$$= \frac{4\sigma^2}{\tilde{\varepsilon}^2 t_1^2 t_2^2} \left(1 - \cos(t_1 \tilde{\varepsilon})\right) \left(1 - \cos(t_2 \tilde{\varepsilon})\right) \geqq 0 .$$

We want to be able to compute the necessary values. Thus, using $\varepsilon = \sqrt{2}\tilde{\varepsilon}$

$$\int_{\mathscr{K}_\varepsilon(x)} R_{1\varepsilon}(x, y) \, \mathrm{d}y = \int_{x_1 - \tilde{\varepsilon}}^{x_1 + \tilde{\varepsilon}} \left(\int_{x_2 - \tilde{\varepsilon}}^{x_2 + \tilde{\varepsilon}} R_{1\varepsilon}(x, y) \, \mathrm{d}y_2 \right) \mathrm{d}y_1$$

$$= \sigma^2 \left(\int_{-\tilde{\varepsilon}}^{\tilde{\varepsilon}} \left(1 - \frac{1}{\tilde{\varepsilon}} |z_1|\right) \mathrm{d}z_1 \right)^2 = \sigma^2 \tilde{\varepsilon}^2$$

and we have

$$a_{11}(x) = \frac{1}{2} \sigma^2; \quad b_{11}(x) = 0$$

as to an expansion with respect to ε. Obviously, it is

$$\alpha_{11}(x) = 0 .$$

Now $a_{11}^1(u; x)$ has to be calculated. We take $x \in a$ $(0 < x_2 < l_2)$ and obtain

$$\left[\int_{\mathscr{D} \cap \mathscr{K}_\varepsilon(z)} R_{1\varepsilon}(z, y) \, \mathrm{d}y \right]_{z = x + \varepsilon u n}$$

$$= \sigma^2 \int_{x_2 - \tilde{\varepsilon}}^{x_2 + \tilde{\varepsilon}} \left(1 - \frac{1}{\tilde{\varepsilon}} |y_2 - x_2|\right) \mathrm{d}y_2 \cdot \begin{cases} \displaystyle\int_0^{\varepsilon u + \tilde{\varepsilon}} \left(1 - \frac{1}{\tilde{\varepsilon}} |y_1 - \varepsilon u|\right) \mathrm{d}y_1 & \text{for } 0 \leqq u \leqq \dfrac{1}{\sqrt{2}}, \\ \displaystyle\int_{\varepsilon u - \tilde{\varepsilon}}^{\varepsilon u + \tilde{\varepsilon}} \left(1 - \frac{1}{\tilde{\varepsilon}} |y_1 - \varepsilon u|\right) \mathrm{d}y_1 & \text{for } \dfrac{1}{\sqrt{2}} \leqq u \leqq 1 \end{cases}$$

where $u \in [0, 1]$, $n = (1, 0)$, $x + \varepsilon u n = (\varepsilon u, x_2)$ and

$$\mathscr{D} \cap \mathscr{K}_\varepsilon(y + \varepsilon u n) = \{(y_1, y_2) \colon (y_1 - \varepsilon u)^2 + (y_2 - x_2)^2 \leqq \varepsilon^2\} .$$

Hence

$$a_{11}^1(u;x) = \begin{cases} \frac{1}{4}\,\sigma^2\left(1 + 2\sqrt{2}u - 2u^2\right) & \text{for } 0 \leqq u \leqq \frac{1}{\sqrt{2}}, \\ \frac{1}{2}\,\sigma^2 & \text{for } \frac{1}{\sqrt{2}} \leqq u \leqq 1 \end{cases}$$

for $x = (0, x_2)$, $0 < x_2 < l_2$ and

$$\int_0^1 a_{11}^1(u;x)\,\mathrm{d}u = \frac{1}{2}\,\sigma^2\left(1 - \frac{\sqrt{2}}{12}\right).$$

For $x \in \mathscr{b}, \mathscr{c}, \mathscr{d}$ it follows the same result as in the case $x \in \mathscr{a}$. These considerations show

$$\begin{aligned} {}^2A_2(F_1, F_2) &= \int_{\partial\mathscr{D}} F_1F_2(x)\left(\int_0^1 a_{11}^1(u;x)\,\mathrm{d}u - a_{11}(x)\right)\mathrm{d}S_x \\ &= -\frac{\sqrt{2}}{24}\,\sigma^2 \int_{\mathscr{a}\cup\mathscr{b}\cup\mathscr{c}\cup\mathscr{d}} F_1F_2(x)\,\mathrm{d}S_x. \end{aligned}$$

In a second example we deal with

$$\mathscr{D} = \{(x_1, x_2) : x_1^2 + x_2^2 \leqq l^2\}$$

(see Fig. 1.11b) and use the same correlation function as in the first example.

Let $x = (l\cos\varphi, l\sin\varphi)$, $0 \leqq \varphi < \frac{\pi}{4}$.
Then it follows

$$n = -(\cos\varphi, \sin\varphi),$$

$$\hat{\mathscr{K}}_\varepsilon(x + \varepsilon u n) = \left\{(y_1, y_2) : -1 \leqq \frac{1}{\varepsilon}\,(y_1 - x_1 + \varepsilon u\cos\varphi) \leqq 1,\right.$$
$$\left. -1 \leqq \frac{1}{\varepsilon}\,(y_2 - x_2 + \varepsilon u\sin\varphi) \leqq 1\right\}, \qquad u \in [0, 1]$$

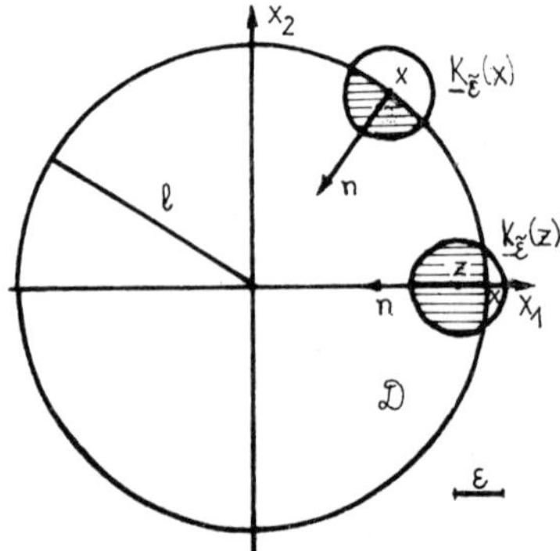

Fig. 1.11b. Domain $\mathscr{D}$ for the calculation of ${}^2A_2$

$R_{1\varepsilon}(x + \varepsilon un, y)$ has to be integrated over $\mathcal{D} \cap \hat{\mathcal{K}}_\varepsilon(x + \varepsilon un)$. Now the new coordinates

$$s_1 = \frac{1}{\varepsilon}\left(y_1 - (l - \varepsilon u)\cos\varphi\right); \quad s_2 = \frac{1}{\varepsilon}\left(y_2 - (l - \varepsilon u)\sin\varphi\right)$$

are introduced where

$$\hat{\mathcal{K}}_\varepsilon(x + \varepsilon un) \quad \text{turns to} \quad \tilde{\mathcal{K}}_0 = \{(s_1, s_2): -1 \leqq s_1 \leqq 1;\ -1 \leqq s_2 \leqq 1\}$$

and $\mathcal{D}$ to

$$\tilde{\mathcal{D}}_\varepsilon = \left\{(s_1, s_2): \left(\tilde{\varepsilon}s_1 + (l - \varepsilon u)\cos\varphi\right)^2 + \left(\tilde{\varepsilon}s_2 + (l - \varepsilon u)\sin\varphi\right)^2 \leqq l^2\right\}.$$

Thus,

$$\frac{1}{\varepsilon^2}\left[\iint\limits_{\mathcal{D}\cap\mathcal{K}_\varepsilon(z)} R_{1\varepsilon}(z, y)\,\mathrm{d}y\right]_{z=x+\varepsilon un} = \sigma^2 \iint\limits_{\tilde{\mathcal{D}}_\varepsilon\cap\tilde{\mathcal{K}}_0} (1 - |s_1|)\,(1 - |s_2|)\,\mathrm{d}s_1\,\mathrm{d}s_2$$

and

$$a_{11}^1(u; x) = \lim_{\varepsilon\downarrow 0} \frac{1}{2}\,\sigma^2 \iint\limits_{\tilde{\mathcal{D}}_{\tilde{\varepsilon}}\cap\tilde{\mathcal{K}}_0} (1 - |s_1|)\,(1 - |s_2|)\,\mathrm{d}s_1\,\mathrm{d}s_2\,.$$

We assume that

$$\mathcal{C}_\varphi(u; \tilde{\varepsilon}) = \left\{(s_1, s_2): \left(s_1 + \left(\frac{l}{\tilde{\varepsilon}} - \sqrt{2}u\right)\cos\varphi\right)^2 + \left(s_2 + \left(\frac{l}{\tilde{\varepsilon}} - \sqrt{2}u\right)\sin\varphi\right)^2 = \left(\frac{l}{\tilde{\varepsilon}}\right)^2\right\}$$

and compute

$$\mathcal{C}_\varphi(u; \tilde{\varepsilon}) \cap \{(s_1, s_2): s_2 = 1\} = \{s_{11\varphi}(u; \tilde{\varepsilon})\},$$

$$s_{11\varphi}(u; \tilde{\varepsilon}) = \frac{1}{\tilde{\varepsilon}}\left(\sqrt{l^2 - \left(\tilde{\varepsilon} + \left(l - \sqrt{2}\tilde{\varepsilon}u\right)\sin\varphi\right)^2} - \left(l - \sqrt{2}\tilde{\varepsilon}u\right)\cos\varphi\right)$$

$$= \frac{\sqrt{2}u - \sin\varphi}{\cos\varphi} + o(\tilde{\varepsilon})$$

where

$$s_{11\varphi}(0; 0) = -\tan\varphi \geqq -1;$$

$$s_{11\varphi}(u_2; 0) = 1 \quad \text{for} \quad u_2 = \frac{1}{\sqrt{2}}(\sin\varphi + \cos\varphi),$$

$$\mathcal{C}_\varphi(u; \tilde{\varepsilon}) \cap \{(s_1, s_2): s_2 = -1\} = \{s_{12\varphi}(u; \tilde{\varepsilon})\},$$

$$s_{12\varphi}(u; \tilde{\varepsilon}) = \frac{1}{\tilde{\varepsilon}}\left(\sqrt{l^2 - \left(-\tilde{\varepsilon} + \left(l - \sqrt{2}\tilde{\varepsilon}u\right)\sin\varphi\right)^2} - \left(l - \sqrt{2}\tilde{\varepsilon}u\right)\cos\varphi\right)$$

$$= \frac{\sqrt{2}u + \sin\varphi}{\cos\varphi} + o(\tilde{\varepsilon})$$

where

$$s_{12\varphi}(0;0) = \tan\varphi \geqq 0,$$

$$s_{12\varphi}(u_1;0) = 1 \quad \text{for} \quad u_1 = \frac{1}{\sqrt{2}}(\cos\varphi - \sin\varphi),$$

$$s_{12\varphi}(u;0) \geqq 1 \quad \text{for} \quad u \geqq u_1,$$

and

$$\mathscr{C}_\varphi(u;\tilde\varepsilon) \cap \{(s_1, s_2): s_1 = 1\} = \{s_{21\varphi}(u;\tilde\varepsilon)\},$$

$$s_{21\varphi}(u;\tilde\varepsilon) = \frac{1}{\tilde\varepsilon}\left(\sqrt{l^2 - \left(\tilde\varepsilon + \left(l - \sqrt{2}\tilde\varepsilon u\right)\cos\varphi\right)^2} - \left(l - \sqrt{2}\tilde\varepsilon u\right)\sin\varphi\right)$$

$$= \frac{\sqrt{2}u - \cos\varphi}{\sin\varphi} + o(\tilde\varepsilon)$$

(see Fig. 1.12) where

$$s_{21}(u;0) \leqq -1 \quad \text{for} \quad u \leqq u_1,$$

$$s_{21}(u_1;0) = -1, \qquad s_{21}(u_2;0) = 1,$$

$$s_{21}(u;0) \geqq 1 \quad \text{for} \quad u \geqq u_2.$$

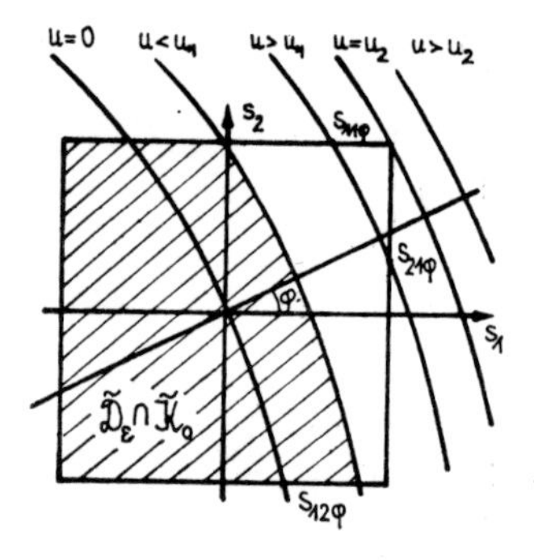

Fig. 1.12. Domain for the integration as to the computation of $a_{12}^2(u;x)$

Now it is

$$a_{11}^1(u;x) = \frac{1}{2}\sigma^2 \iint_{\mathscr{S}} (1 - |s_1|)(1 - |s_2|)\,\mathrm{d}s_1\,\mathrm{d}s_2$$

where

$$\mathscr{S} \doteq \left\{(s_1, s_2) \in \tilde{\mathscr{K}}_0 : s_1 \leqq s_1(s_2);\ s_1(s_2) = -s_2\tan\varphi + \frac{\sqrt{2}u}{\cos\varphi}\right\}.$$

This statement follows from

$$0 = \left[\left(\tilde\varepsilon s_1 + \left(l - \sqrt{2}\tilde\varepsilon u\right)\cos\varphi\right)^2 + \left(\tilde\varepsilon s_2 + \left(l - \sqrt{2}\tilde\varepsilon u\right)\sin\varphi\right)^2 - l^2\right]$$

$$= 2l\tilde\varepsilon\left[s_1\cos\varphi + s_2\sin\varphi - \sqrt{2}u\right] + 2\tilde\varepsilon^2\left[(s_1 - \sqrt{2}u\cos\varphi)^2 + \left(s_2 - \sqrt{2}u\sin\varphi\right)^2\right].$$

Thus,

$$a_{11}^1(u;x) = \frac{1}{2}\sigma^2 \int\limits_{-1}^{1} ds_2 \int\limits_{-1}^{s_1(s_2)} ds_1 (1 - |s_1|)(1 - |s_2|) \quad \text{for} \quad u \leqq u_1,$$

$$a_{11}^1(u;x) = \frac{1}{2}\sigma^2 \int\limits_{-1}^{s_{21\varphi}} ds_2 (1 - |s_2|) + \int\limits_{s_{21\varphi}}^{1} ds_2 \int\limits_{-1}^{s_1(s_2)} ds_1 (1 - |s_1|)(1 - |s_2|)$$

for $u_1 \leqq u \leqq u_2$,

$$a_{11}^1(u;x) = \frac{1}{2}\sigma^2 \quad \text{for} \quad u \geqq u_2.$$

In particular, for $u \leqq u_1$ we obtain

$$a_{11}^1(u;x) = \frac{1}{2}\sigma^2 \left[\int\limits_{-1}^{0} (1 + s_2) \left(\frac{1}{2} + \int\limits_{0}^{s_1(s_2)} (1 - s_1)\, ds_1 \right) ds_2 \right.$$

$$+ \int\limits_{0}^{s_{20}} (1 - s_2) \left(\frac{1}{2} + \int\limits_{0}^{s_1(s_2)} (1 - s_1)\, ds_1 \right) ds_2$$

$$\left. + \int\limits_{s_{20}}^{1} (1 - s_2) \left(\int\limits_{-1}^{s_1(s_2)} (1 + s_1)\, ds_1 \right) ds_2 \right] \quad \text{for} \quad s_{11\varphi} \leqq 0,$$

$$a_{11}^1(u;x) = \frac{1}{2}\sigma^2 \left[\int\limits_{-1}^{0} (1 + s_2) \left(\frac{1}{2} + \int\limits_{0}^{s_1(s_2)} (1 - s_1)\, ds_1 \right) ds_2 \right.$$

$$\left. + \int\limits_{0}^{1} (1 - s_2) \left(\frac{1}{2} + \int\limits_{0}^{s_1(s_2)} (1 - s_1)\, ds_1 \right) ds_2 \right] \quad \text{for} \quad s_{11\varphi} \geqq 0$$

where $s_{20} = \dfrac{\sqrt{2}u}{\sin\varphi}$ and for $u_1 \leqq u \leqq u_2$

$$a_{12}^2(u;x) = \int\limits_{-1}^{s_{21\varphi}} (1 + s_2)\, ds_2 + \int\limits_{s_{21\varphi}}^{0} (1 + s_2) \left(\frac{1}{2} + \int\limits_{0}^{s_1(s_2)} (1 - s_1)\, ds_1 \right) ds_2$$

$$+ \int\limits_{0}^{s_{20}} (1 - s_2) \left(\frac{1}{2} + \int\limits_{0}^{s_1(s_2)} (1 - s_1)\, ds_1 \right) ds_2$$

$$+ \int\limits_{s_{20}}^{1} (1 - s_2) \int\limits_{-1}^{s_1(s_2)} (1 + s_1)\, ds_1\, ds_2 \quad \text{for} \quad s_{11\varphi} \leqq 0,$$

$$a_{12}^2(u;x) = \int_{-1}^{s_{21}} (1+s_2)\,\mathrm{d}s_2 + \int_{s_{21\varphi}}^{0} (1+s_2)\left(\frac{1}{2} + \int_0^{s_1(s_2)} (1-s_1)\,\mathrm{d}s_1\right)\mathrm{d}s_2$$

$$+ \int_0^1 (1-s_2)\left(\frac{1}{2} + \int_0^{s_1(s_2)} (1-s_1)\,\mathrm{d}s_1\right)\mathrm{d}s_2 \quad \text{for} \quad s_{21\varphi} \leqq 0;\ s_{11\varphi} \geqq 0,$$

$$a_{12}^2(u;x) = \frac{1}{2} + \int_0^{s_{21}} (1-s_2)\,\mathrm{d}s_2 + \int_{s_{21\varphi}}^{1} (1-s_2)\left(\frac{1}{2} + \int_0^{s_1(s_2)} (1-s_1)\,\mathrm{d}s_1\right)\mathrm{d}s_2$$

for $s_{21\varphi} \geqq 0;\ s_{11\varphi} \geqq 0$.

Assuming $\varphi = 0$ it follows

$$u_1 = u_2 = \frac{1}{\sqrt{2}}, \quad s_1(s_2) = \sqrt{2}u$$

and

$$a_{12}^2(u;\ (1,0)) = \begin{cases} \dfrac{1}{2}\,\sigma^2\left(\dfrac{1}{2} + \sqrt{2}u - u^2\right) & \text{for} \quad u \leqq \dfrac{1}{2}\sqrt{2}, \\[2ex] \dfrac{1}{2}\,\sigma^2 & \text{for} \quad u \geqq \dfrac{1}{2}\sqrt{2}, \end{cases}$$

$$\int^1 a_{12}^2(u;\ (1,0))\,\mathrm{d}u = \frac{1}{2}\,\sigma^2\left(1 - \frac{\sqrt{2}}{12}\right) = 0.441074\sigma^2.$$

The results from the above formulas are plotted in Fig. 1.13 with respect to $a_{11}^1(u;x)$, $x = (l\cos\varphi, l\sin\varphi)$ and in Fig. 1.14 as to the function

$$\int_0^1 a_{11}^1(u;x)\,\mathrm{d}u - a_{11}$$

which is contained in the statistical characteristic ${}^2A_2$.

For the considered case the considerations above lead to the expansion

$$\int_{\mathcal{D}}\int_{\mathcal{D}} F_1(x)\,F_2(y)\,R_{1\varepsilon}(x,y)\,\mathrm{d}x\,\mathrm{d}y$$

$$= \frac{1}{2}\,\sigma^2 \int_{\mathcal{D}} F_1F_2(x)\,\mathrm{d}x\varepsilon^2 + \int_{\partial\mathcal{D}} F_1F_2(x)\left(\int_0^1 a_{11}^1(u;x)\,\mathrm{d}u - a_{12}(x)\right)\mathrm{d}S_x\varepsilon^3 + o(\varepsilon^3)$$

$$\approx \sigma^2\left[\frac{1}{2}\int_{\mathcal{D}} F_1F_2(x)\,\mathrm{d}x\varepsilon^2 - 0.058926\int_{\partial\mathcal{D}} F_1F_2(x)\,\mathrm{d}S_x\varepsilon^3\right] + o(\varepsilon^3).$$

It is also possible to deal with ${}^2A_2(F_1, F_2)$ as to other domains and other correlation functions.

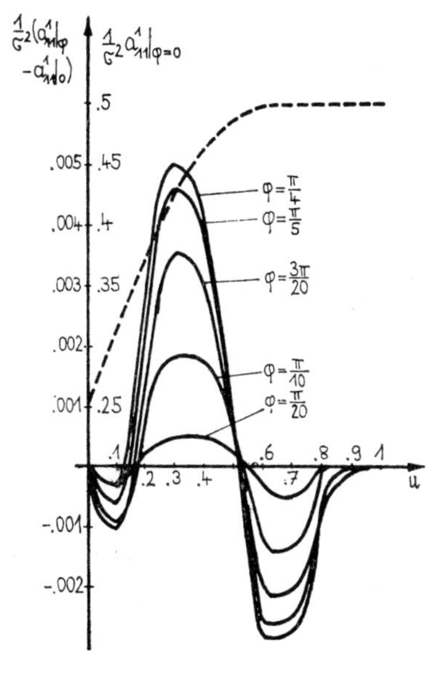

Fig. 1.13. a_{11}^1 as a function of u for different values of φ

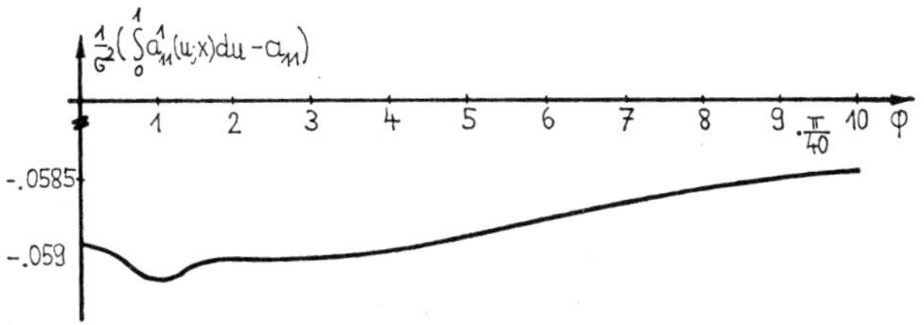

Fig. 1.14. Integral over $a_{11}^1(u; x)$ as to u as a function of φ

1.6.6. Determination of ${}^pA_{p-1}$ for $p = 3, 4$

The coefficients ${}^pA_{p-1}$ of the expansion of the p-th moments as to the correlation length follow from

$$ {}^pA_{p-1}(F_1, \ldots, F_p) = \lim_{\varepsilon \downarrow 0} \frac{1}{\varepsilon^{m(p-1)}} \int\limits_{\mathscr{E}(\mathscr{J})} \prod_{q=1}^{p} F_q(y_q) \left\langle \prod_{q=1}^{p} f_{q\varepsilon}(y_q) \right\rangle \mathrm{d}y_1 \ldots \mathrm{d}y_p $$

where $\mathscr{E}(\mathscr{J})$ is given by

$$ \mathscr{E}(\mathscr{J}) = \{(y_1, \ldots, y_p) \in \mathop{\times}_{q=1}^{p} \mathscr{D}_q \colon \{y_1, \ldots, y_p\} \ \varepsilon\text{-neighbouring}\} . $$

The functions $F_q(x)$, $q = 1, 2, \ldots, p$, are assumed to be continuous and the domains $\mathscr{D}_q$ bounded with piecewise smooth boundaries.

Now we deal with the case $p = 3$. The sets $\mathscr{M}$ and $\mathscr{M}_{ij}$ are defined by

$$ \mathscr{M} \doteq \left\{(y_1, y_2, y_3) \in \mathop{\times}_{q=1}^{3} \mathscr{D}_q \colon |y_1 - y_2| \leqq \varepsilon, |y_1 - y_3| \leqq \varepsilon, |y_2 - y_3| \leqq \varepsilon\right\}, $$

$$ \mathscr{M}_{ij} \doteq \left\{(y_1, y_2, y_3) \in \mathop{\times}_{q=1}^{3} \mathscr{D}_q \colon |y_i - y_j| > \varepsilon; \text{ else } |y_k - y_l| \leqq \varepsilon\right\} \quad \text{for} \quad i, j = 1, 2, 3; i \neq j $$

and we can write

$$\mathcal{E}(\mathcal{J}_3) = \mathcal{M} \cup \mathcal{M}_{12} \cup \mathcal{M}_{13} \cup \mathcal{M}_{23}$$

and $\mathcal{M}$, $\mathcal{M}_{12}$, $\mathcal{M}_{13}$, $\mathcal{M}_{23}$ are pairwise disjoint sets. The integral over $\mathcal{M}$ can be written as

$$J(\mathcal{M}) = \int_{\mathcal{M}} \prod_{q=1}^{3} F_q(y_q) \left\langle \prod_{q=1}^{3} f_{q\varepsilon}(y_q) \right\rangle \mathrm{d}y_1 \, \mathrm{d}y_2 \, \mathrm{d}y_3$$
$$= \int_{\mathcal{D}_1} \mathrm{d}y_1 \int_{\mathcal{D}_2 \cap \mathcal{K}_\varepsilon(y_1)} \mathrm{d}y_2 \int_{\mathcal{D}_3 \cap \mathcal{K}_\varepsilon(y_2) \cap \mathcal{K}_\varepsilon(y_1)} \mathrm{d}y_3 \prod_{q=1}^{3} F_q(y_q) \left\langle \prod_{q=1}^{3} f_{q\varepsilon}(y_q) \right\rangle .$$

The integrand is zero for all points y_1 which fulfil the relation

$$\operatorname{dist} \left(y_1, \bigcap_{q=1}^{3} \mathcal{D}_q \right) > \varepsilon .$$

The volume of the integration domain possesses the order $O(\varepsilon^{2m})$ and hence the consideration of ε-boundaries as in the case ${}^2A_2$ can be omitted. Put

$$\mathcal{D}_{123} \doteq \mathcal{D}_1 \cap \mathcal{D}_2 \cap \mathcal{D}_3; \quad \left\langle \prod_{q=1}^{3} f_{q\varepsilon}(y_q) \right\rangle = R_{123,\varepsilon}(y_1, y_2, y_3);$$

and

$$\mathcal{M}(y_1) \doteq \left\{ (y_2, y_3) \in \bigtimes_{q=2}^{3} \mathcal{D}_q : y_2 \in \mathcal{K}_\varepsilon(y_1), \, y_3 \in \mathcal{K}_\varepsilon(y_1) \cap \mathcal{K}_\varepsilon(y_2) \right\}$$

then

$$J(\mathcal{M}) = \int_{\mathcal{D}_{123}} \iint_{\mathcal{M}(y_1)} \prod_{q=1}^{3} F_q(y_q) \, R_{123,\varepsilon}(y_1, y_2, y_3) \, \mathrm{d}y_1 \, \mathrm{d}y_2 \, \mathrm{d}y_3 + o(\varepsilon^{2m})$$
$$= \int_{\mathcal{D}_{123}} \prod_{q=1}^{3} F_q(y_1) \left(\iint_{\mathcal{M}(y_1)} R_{123,\varepsilon}(y_1, y_2, y_3) \, \mathrm{d}y_2 \, \mathrm{d}y_3 \right) \mathrm{d}y_1$$
$$+ \int_{\mathcal{D}_{123}} \left(\iint_{\mathcal{M}(y_1)} \left(\prod_{q=1}^{3} F_q(y_q) - \prod_{q=1}^{3} F_q(y_1) \right) R_{123,\varepsilon}(y_1, y_2, y_3) \, \mathrm{d}y_2 \, \mathrm{d}y_3 \right) \mathrm{d}y_1 + o(\varepsilon^{2m})$$
$$= \int_{\mathcal{D}_{123}} \prod_{q=1}^{3} F_q(y_1) \left(\iint_{\mathcal{M}(y_1)} R_{123,\varepsilon}(y_1, y_2, y_3) \, \mathrm{d}y_2 \, \mathrm{d}y_3 \right) \mathrm{d}y_1 + o(\varepsilon^{2m})$$

because of the continuity of F_q. In a similar way we obtain

$$J(\mathcal{M}_{12}) = \int_{\mathcal{M}_{ij}} \prod_{q=1}^{3} F_q(y_q) \left\langle \prod_{q=1}^{3} f_{q\varepsilon}(y_q) \right\rangle \mathrm{d}y_1 \, \mathrm{d}y_2 \, \mathrm{d}y_3$$
$$= \int_{\mathcal{D}_{123}} \prod_{q=1}^{3} F_q(y_1) \left(\iint_{\mathcal{M}_{ij}(y_1)} R_{123,\varepsilon}(y_1, y_2, y_3) \, \mathrm{d}y_2 \, \mathrm{d}y_3 \right) \mathrm{d}y_1 + o(\varepsilon^{2m})$$

for $(i, j) = (1, 2), (1, 3), (2, 3)$ where $\mathcal{M}_{ij}(y_1)$ is given by

$$\mathcal{M}_{ij}(y_1) = \left\{ (y_2, y_3) \in \bigtimes_{q=2}^{3} \mathcal{D}_q : |y_i - y_j| > \varepsilon; \text{ else } |y_k - y_l| \leq \varepsilon \right\}.$$

Furthermore, we have

$$J\big(\mathscr{E}(\mathscr{I}_3)\big) = \int\limits_{\mathscr{D}_{123}} \prod_{q=1}^{3} F_q(y_1) \Big(\iint\limits_{\tilde{\mathscr{M}}(y_1)} R_{123,\varepsilon}(y_1, y_2, y_3)\, \mathrm{d}y_2\, \mathrm{d}y_3 \Big)\, \mathrm{d}y_1 + o(\varepsilon^{2m})$$

and defining

$$a_{123}(y) = \lim_{\varepsilon \downarrow 0} \frac{1}{\varepsilon^{2m}} \int\limits_{\tilde{\mathscr{M}}(y)} R_{123,\varepsilon}(y, y_2, y_3)\, \mathrm{d}y_2\, \mathrm{d}y_3 \tag{1.138}$$

then

$${}^3A_2(F_1, F_2, F_3) = {}^3A_2 \left(\prod_{q=1}^{3} F_q \right) = \int\limits_{\mathscr{D}_{123}} \prod_{q=1}^{3} F_q(y)\, a_{123}(y)\, \mathrm{d}y \tag{1.139}$$

where we have put

$$\tilde{\mathscr{M}}(y) \doteq \mathscr{M}(y) \cup \mathscr{M}_{12}(y) \cup \mathscr{M}_{13}(y) \cup \mathscr{M}_{23}(y).$$

Introducing the coordinates

$$s_1 \doteq y - y_2; \quad s_2 \doteq y_2 - y_3 \quad \text{and hence} \quad y_2 = y - s_1; \quad y_3 = y - s_1 - s_2$$

it follows

$$\left| \frac{\partial(y_2, y_3)}{\partial(s_1, s_2)} \right| = 1$$

and

$$\iint\limits_{\mathscr{M}(y)} R_{123,\varepsilon}(y, y_2, y_3)\, \mathrm{d}y_2\, \mathrm{d}y_3 = \iint\limits_{\mathscr{S}} R_{123,\varepsilon}(y, y - s_1, y - s_1 - s_2)\, \mathrm{d}s_1\, \mathrm{d}s_2$$

with

$$\mathscr{S} = \{(s_1, s_2) \colon |s_1| \leqq \varepsilon,\ |s_2| \leqq \varepsilon,\ |s_1 + s_2| \leqq \varepsilon\},$$

$$\iint\limits_{\mathscr{M}_{ij}(y)} R_{123,\varepsilon}(y, y_2, y_3)\, \mathrm{d}y_2\, \mathrm{d}y_3 = \iint\limits_{\mathscr{S}_{ij}} R_{123,\varepsilon}(y, y - s_1, y - s_1 - s_2)\, \mathrm{d}s_1\, \mathrm{d}s_2$$

with

$$\mathscr{S}_{12} = \{(s_1, s_2) \colon |s_1| > \varepsilon,\ |s_2| \leqq \varepsilon,\ |s_1 + s_2| \leqq \varepsilon\},$$

$$\mathscr{S}_{13} = \{(s_1, s_2) \colon |s_1| \leqq \varepsilon,\ |s_2| \leqq \varepsilon,\ |s_1 + s_2| > \varepsilon\},$$

$$\mathscr{S}_{23} = \{(s_1, s_2) \colon |s_1| \leqq \varepsilon,\ |s_2| > \varepsilon,\ |s_1 + s_2| \leqq \varepsilon\}.$$

Hence, $a_{123}(y)$ can be calculated by

$$a_{123}(y) = \lim_{\varepsilon \downarrow 0} \frac{1}{\varepsilon^{m}} \int\limits_{\mathscr{S}_0} R_{123,\varepsilon}(y, y - s_1, y - s_1 - s_2)\, \mathrm{d}s_1\, \mathrm{d}s_2 \tag{1.138'}$$

with

$$\mathscr{S}_0 = \mathscr{S} \cup \mathscr{S}_{12} \cup \mathscr{S}_{13} \cup \mathscr{S}_{23}.$$

In the case of $m = 1$ the integration domain is drawn in Fig. 1.15. Assuming $m = 1$ we obtain

$$v(\mathscr{S}) = 3\varepsilon^2,$$

$$v(\mathscr{S}_{12}) = v(\mathscr{S}_{13}) = v(\mathscr{S}_{23}) = \varepsilon^2$$

and

$$v(\mathcal{G}_0) = 6\varepsilon^2.$$

For $m = 1$ supposing additionally

$$F_q \in L_k(\mathcal{D}_q), \qquad k = 1, 2, 3,$$

we can also deduce the same result (1.139) if the domains are unbounded.

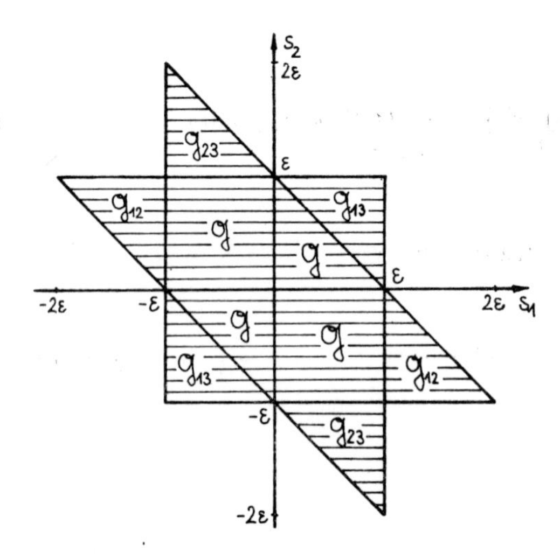

Fig. 1.15. Integration domain for the evaluation of a_{123} in the case of $m = 1$

We turn to the case ${}^4\underline{A}_3$ defined by

$$\begin{aligned}{}^4\underline{A}_3(F_1, F_2, F_3, F_4) &= {}^4A_3(F_1, F_2, F_3, F_4) - A_{22}(F_1, F_2; F_3, F_4)\\ &\quad - A_{22}(F_1, F_3; F_2, F_4) - A_{22}(F_1, F_4; F_2, F_3)\end{aligned}$$

if we take into consideration (1.11), (1.12), and (1.68) where

$${}^4A_3(F_1, F_2, F_3, F_4) = \lim_{\varepsilon\downarrow 0} \frac{1}{\varepsilon^{3m}} \int\limits_{\mathcal{E}(\mathcal{J}_4)} \prod_{q=1}^{4} F_q(y_q) \left\langle \prod_{q=1}^{4} f_{q\varepsilon}(y_q) \right\rangle \mathrm{d}y_1\, \mathrm{d}y_2\, \mathrm{d}y_3\, \mathrm{d}y_4,$$

$$\begin{aligned}&A_{22}(F_1, F_2; F_3, F_4)\\ &= \lim_{\varepsilon\downarrow 0} \frac{1}{\varepsilon^{3m}} \int\limits_{\mathcal{E}(\mathcal{J}_4)} \prod_{q=1}^{4} F_q(y_q) \left\langle f_{1\varepsilon}(y_1) f_{2\varepsilon}(y_2)\right\rangle \left\langle f_{3\varepsilon}(y_3) f_{4\varepsilon}(y_4)\right\rangle \mathrm{d}y_1\, \mathrm{d}y_2\, \mathrm{d}y_3\, \mathrm{d}y_4.\end{aligned}$$

For abbreviation we put

$$\begin{aligned}\underline{R}_{1234,\varepsilon}(y_1, y_2, y_3, y_4) &\doteq \left\langle \prod_{q=1}^{4} f_{q\varepsilon}(y_q) \right\rangle - \left\langle \prod_{q=1,2} f_{q\varepsilon}(y_q)\right\rangle \left\langle \prod_{q=3,4} f_{q\varepsilon}(y_q)\right\rangle\\ &\quad - \left\langle \prod_{q=1,3} f_{q\varepsilon}(y_q)\right\rangle \left\langle \prod_{q=2,4} f_{q\varepsilon}(y_q)\right\rangle - \left\langle \prod_{q=1,4} f_{q\varepsilon}(y_q)\right\rangle \left\langle \prod_{q=2,3} f_{q\varepsilon}(y_q)\right\rangle\end{aligned}$$

and obtain

$${}^4\underline{A}_3(F_1, F_2, F_3, F_4) = \lim_{\varepsilon\downarrow 0} \frac{1}{\varepsilon^{3m}} \int\limits_{\mathcal{E}(\mathcal{J}_4)} \prod_{q=1}^{4} F_q(y_q)\, \underline{R}_{1234,\varepsilon}(y_1, y_2, y_3, y_4)\, \mathrm{d}y_1\, \mathrm{d}y_2\, \mathrm{d}y_3\, \mathrm{d}y_4.$$

Analogously to the case ${}^3A_2$ the sets $\mathcal{M}$, $\mathcal{M}_{i_1j_1,\dots,i_sj_s}$ are defined by

$$\mathcal{M} \doteq \left\{(y_1, y_2, y_3, y_4) \in \mathop{\times}_{q=1}^{4} \mathcal{D}_q : |y_i - y_j| \leqq \varepsilon \quad \text{for} \quad i, j = 1, 2, 3, 4; i < j\right\}.$$

$$\mathcal{M}_{i_1j_1,\dots,i_sj_s} \doteq \left\{(y_1, y_2, y_3, y_4) \in \mathop{\times}_{q=1}^{4} \mathcal{D}_q : |y_{i_q} - y_{j_q}| > \varepsilon \quad \text{for} \quad q = 1, 2, \dots, s; \right.$$
$$\left. \text{else} \quad |y_i - y_j| \leqq \varepsilon\right\}.$$

It is easy to see that

$$\mathcal{M} \subset \mathcal{E}(\mathcal{J}_4); \qquad \mathcal{M}_{i_1j_1} \subset \mathcal{E}(\mathcal{J}_4); \qquad \mathcal{M}_{i_1j_1, i_2j_2} \subset \mathcal{E}(\mathcal{J}_4);$$

and

$$\mathcal{M}_{i_1j_1, i_2j_2, i_3j_3} \subset \mathcal{E}(\mathcal{J}_4)$$

if $\mathcal{M}_{i_1j_1, i_2j_2, i_3j_3}$ is not equal to a set of the form $\mathcal{M}_{ik_1, ik_2, ik_3}$. Furthermore, we have

$$\mathcal{M}_{i_1j_1, i_2j_2,\dots,i_sj_s} \not\subset \mathcal{E}(\mathcal{J}_4) \quad \text{for} \quad s \geqq 4.$$

From this it follows

$$\mathcal{E}(\mathcal{J}_4) = \mathcal{M} \cup \bigcup_{\substack{i,j=1\\ i<j}}^{4} \mathcal{M}_{ij} \cup \bigcup_{\substack{i,j,k,l=1\\ i<j,k<l\\ (i,j)\neq(k,l)}}^{4} \mathcal{M}_{ij,kl} \cup \bigcup_{\substack{i,j,k,l,p,q=1\\ i<j,k<l,p<q\\ (i,j),(k,l),(p,q)\text{different}\\ \neq(s,k_1),(s,k_2),(s,k_3)}}^{4} \mathcal{M}_{ij,kl,pq}.$$

Using similar considerations as in the case ${}^3A_2$ we obtain

$$J(\mathcal{M}) = \int_{\mathcal{M}} \prod_{q=1}^{4} F_q(y_q)\, \underline{R}_{1234,\varepsilon}(y_1, y_2, y_3, y_4)\, dy_1\, dy_2\, dy_3\, dy_4$$
$$= \int_{\mathcal{D}_1} dy_1 \int_{\mathcal{D}_2 \cap \mathcal{K}_\varepsilon(y_1)} dy_2 \int_{\mathcal{D}_3 \cap \mathcal{K}_\varepsilon(y_1) \cap \mathcal{K}_\varepsilon(y_2)} dy_3$$
$$\times \int_{\mathcal{D}_4 \cap \mathcal{K}_\varepsilon(y_1) \cap \mathcal{K}_\varepsilon(y_2) \cap \mathcal{K}_\varepsilon(y_3)} dy_4 \prod_{q=1}^{4} F_q(y_q)\, \underline{R}_{1234,\varepsilon}(y_1, y_2, y_3, y_4)$$
$$= \int_{\mathcal{D}_{1234}} \prod_{q=1}^{4} F_q(y_1) \left(\iiint_{\mathcal{M}(y_1)} \underline{R}_{1234,\varepsilon}(y_1, y_2, y_3, y_4)\, dy_2\, dy_3\, dy_4\right) dy_1$$
$$+ \int_{\mathcal{D}_{1234}} \left(\iiint_{\mathcal{M}(y_1)} \left(\prod_{q=1}^{4} F_q(y_q) - \prod_{q=1}^{4} F_q(y_1)\right) \underline{R}_{1234,\varepsilon}(y_1, y_2, y_3, y_4)\right.$$
$$\left. \times\, dy_2\, dy_3\, dy_4\right) dy_1 + o(\varepsilon^{3m})$$
$$= \int_{\mathcal{D}_{1234}} \prod_{q=1}^{4} F_q(y_1) \left(\iiint_{\mathcal{M}(y_1)} \underline{R}_{1234,\varepsilon}(y_1, y_2, y_3, y_4)\, dy_2\, dy_3\, dy_4\right) dy_1 + o(\varepsilon^{3m})$$

as to $\mathcal{M}$ where we have put

$$\mathcal{M}(y_1) = \left\{(y_2, y_3, y_4) : y_2 \in \mathcal{K}_\varepsilon(y_1),\ y_3 \in \bigcap_{q=1}^{2} \mathcal{K}_\varepsilon(y_q),\ y_4 \in \bigcap_{q=1}^{3} \mathcal{K}_\varepsilon(y_q)\right\}.$$

Analogous considerations lead to

$$J(\mathscr{M}_{ij}) = \int\limits_{\mathscr{D}_{1234}} \prod_{q=1}^{4} F_q(y_1) \left(\iiint\limits_{\mathscr{M}_{ij}(y_1)} \underline{R}_{1234,\varepsilon}(y_1, y_2, y_3, y_4)\, \mathrm{d}y_2\, \mathrm{d}y_3\, \mathrm{d}y_4 \right) \mathrm{d}y_1 + o(\varepsilon^{3m})$$

with

$$\mathscr{M}_{ij}(y) = \{(y_2, y_3, y_4) \colon |y_i - y_j| > \varepsilon;\ \text{else } |y_k - y_l| \leqq \varepsilon$$
$$\text{for}\quad k, l = 1, 2, 3, 4,\ k < l\}$$

and $J(\mathscr{M}_{i_1j_1, i_2j_2})$, $J(\mathscr{M}_{i_1j_1, i_2j_2, i_3j_3})$ with the adequate sets $\mathscr{M}_{i_1j_1, i_2j_2}(y_1)$, $\mathscr{M}_{i_1j_1, i_2j_2, i_3j_3}(y_1)$. Summarizing we have

$$J\big(\mathscr{E}(\mathscr{I}_4)\big) = \int\limits_{\mathscr{D}_{1234}} \prod_{q=1}^{4} F_q(y_1) \left(\int\limits_{\tilde{\mathscr{M}}(y_1)} \underline{R}_{1234,\varepsilon}(y_1, y_2, y_3, y_4)\, \mathrm{d}y_2\, \mathrm{d}y_3\, \mathrm{d}y_4 \right) \mathrm{d}y_1 + o(\varepsilon^{3m})$$

where $\tilde{\mathscr{M}}(y_1)$ is defined by the union of the sets $\mathscr{M}(y_1)$, $\mathscr{M}_{ij}(y_1)$, $\mathscr{M}_{i_1j_1, i_2j_2}(y_1)$, and $\mathscr{M}_{i_1j_1, i_2j_2, i_3j_3}(y_1)$. Then it results

$$^4\underline{A}_3(F_1, F_2, F_3, F_4) = \int\limits_{\mathscr{D}_{1234}} \prod_{q=1}^{4} F_q(y)\, \underline{a}_{1234}(y)\, \mathrm{d}y \tag{1.140}$$

where

$$\underline{a}_{1234}(y) \doteq \lim_{\varepsilon \downarrow 0} \frac{1}{\varepsilon^{3m}} \int\limits_{\tilde{\mathscr{M}}(y)} \underline{R}_{1234,\varepsilon}(y, y_2, y_3, y_4)\, \mathrm{d}y_2\, \mathrm{d}y_3\, \mathrm{d}y_4 . \tag{1.141}$$

We investigate the integration domain $\tilde{\mathscr{M}}(y)$ for $m = 1$. Introducing the new coordinates

$$s_1 \doteq y - y_2; \qquad s_2 \doteq y_2 - y_3; \qquad s_3 \doteq y_3 - y_4$$

or

$$y_2 = y - s_1; \qquad y_3 = y - s_1 - s_2; \qquad y_4 = y - s_1 - s_2 - s_3$$

it follows

$$\left| \frac{\partial(y_2, y_3, y_4)}{\partial(s_1, s_2, s_3)} \right| = 1 .$$

The domain $\mathscr{M}(y)$ turns to

$$\mathscr{S} = \{(s_1, s_2, s_3) \colon |s_1| \leqq \varepsilon,\ |s_2| \leqq \varepsilon,\ |s_3| \leqq \varepsilon,\ |s_1 + s_2| \leqq \varepsilon,\ |s_2 + s_3| \leqq \varepsilon,$$
$$|s_1 + s_2 + s_3| \leqq \varepsilon\}$$

and by simple calculation for the volume of $\mathscr{S}$ we obtain

$$\begin{aligned} v(\mathscr{S}) = {} & \iint\limits_{\mathscr{S}_1} \left(\int\limits_{-\varepsilon}^{\varepsilon - s_1 - s_2} \mathrm{d}s_3 \right) \mathrm{d}s_1\, \mathrm{d}s_2 + \iint\limits_{\mathscr{S}_2} \left(\int\limits_{-\varepsilon}^{\varepsilon - s_2} \mathrm{d}s_3 \right) \mathrm{d}s_1\, \mathrm{d}s_2 \\ & + \iint\limits_{\mathscr{S}_3} \left(\int\limits_{-s_1 - s_2 - \varepsilon}^{\varepsilon - s_2} \mathrm{d}s_3 \right) \mathrm{d}s_1\, \mathrm{d}s_2 + \iint\limits_{\mathscr{S}_4} \left(\int\limits_{-s_1 - s_2 - \varepsilon}^{\varepsilon} \mathrm{d}s_3 \right) \mathrm{d}s_1\, \mathrm{d}s_2 \\ & + \iint\limits_{\mathscr{S}_5} \left(\int\limits_{-s_2 - \varepsilon}^{\varepsilon} \mathrm{d}s_3 \right) \mathrm{d}s_1\, \mathrm{d}s_2 + \iint\limits_{\mathscr{S}_6} \left(\int\limits_{-s_2 - \varepsilon}^{\varepsilon - s_1 - s_2} \mathrm{d}s_3 \right) \mathrm{d}s_1\, \mathrm{d}s_2 \end{aligned}$$

where the domains $\bar{\mathscr{G}}_i$ are plotted in Fig. 1.16a and

$$\bar{\mathscr{G}} = \bigcup_{i=1}^{6} \bar{\mathscr{G}}_i = \{(s_1, s_2) : |s_1| \leqq \varepsilon, |s_2| \leqq \varepsilon, |s_1 + s_2| \leqq \varepsilon\}.$$

Thus

$$v(\mathscr{M}(y)) = v(\mathscr{G}) = 4\varepsilon^3.$$

The transformation of the domains

$$\mathscr{M}_{14}(y); \quad \mathscr{M}_{24}(y); \quad \mathscr{M}_{34}(y); \quad \mathscr{M}_{14,24}; \quad \mathscr{M}_{14,34}; \quad \mathscr{M}_{24,34}$$

leads also to integrals over $\bar{\mathscr{G}}$ as to (s_1, s_2):

$$\begin{aligned}
&v(\mathscr{M}) + v(\mathscr{M}_{14}) + v(\mathscr{M}_{24}) + v(\mathscr{M}_{34}) + v(\mathscr{M}_{14,24}) + v(\mathscr{M}_{14,34}) + v(\mathscr{M}_{24,34}) \\
&= \iint\limits_{\bar{\mathscr{G}}_1} \left(\int\limits_{-\varepsilon - s_1 - s_2}^{\varepsilon} ds_3 \right) ds_1\, ds_2 + \iint\limits_{\bar{\mathscr{G}}_2} \left(\int\limits_{-\varepsilon - s_2}^{\varepsilon} ds_3 \right) ds_1\, ds_2 \\
&\quad + \iint\limits_{\bar{\mathscr{G}}_3} \left(\int\limits_{-\varepsilon - s_2}^{\varepsilon - s_1 - s_2} ds_3 \right) ds_1\, ds_2 + \iint\limits_{\bar{\mathscr{G}}_4} \left(\int\limits_{-\varepsilon}^{\varepsilon - s_1 - s_2} ds_3 \right) ds_1\, ds_2 \\
&\quad + \iint\limits_{\bar{\mathscr{G}}_5} \left(\int\limits_{-\varepsilon}^{\varepsilon - s_2} ds_3 \right) ds_1\, ds_2 + \iint\limits_{\bar{\mathscr{G}}_6} \left(\int\limits_{-\varepsilon - s_1 - s_2}^{\varepsilon - s_2} ds_3 \right) ds_1\, ds_2 \\
&= 6 \left(\frac{4}{3} \varepsilon^3 \right) = 8\varepsilon^3.
\end{aligned}$$

Furthermore, we obtain

$$\bar{\mathscr{G}}_{12} = \bigcup_{i=1}^{2} \bar{\mathscr{G}}_{12;i} = \{(s_1, s_2) : |s_1| > \varepsilon, |s_2| \leqq \varepsilon, |s_1 + s_2| \leqq \varepsilon\},$$

$$\bar{\mathscr{G}}_{13} = \bigcup_{i=1}^{2} \bar{\mathscr{G}}_{13;i} = \{(s_1, s_2) : |s_1| \leqq \varepsilon, |s_2| \leqq \varepsilon, |s_1 + s_2| > \varepsilon\},$$

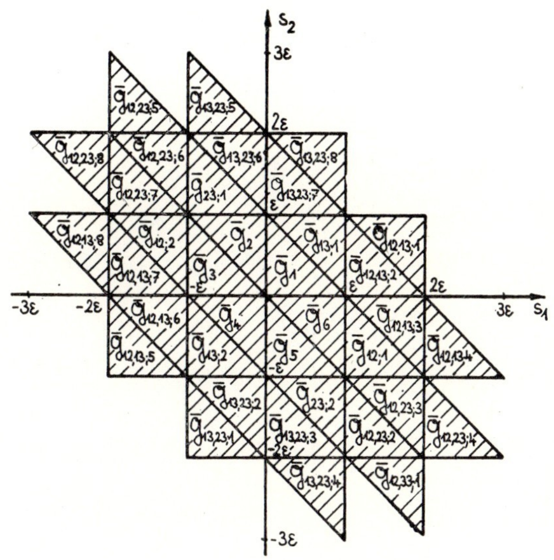

Fig. 1.16a. Integration domain as to the calculation of $\underline{a}_{1234}(y)$ in the case of $m = 1$

$$\overline{\mathscr{G}}_{23} = \bigcup_{i=1}^{2} \overline{\mathscr{G}}_{23;i} = \{(s_1, s_2) \colon |s_1| \leqq \varepsilon, |s_2| > \varepsilon, |s_1 + s_2| \leqq \varepsilon\}$$

and

$$v(\mathscr{M}_{ij}) + v(\mathscr{M}_{ij,34}) + v(\mathscr{M}_{ij,24}) + v(\mathscr{M}_{ij,14}) + v(\mathscr{M}_{ij,34,24}) + v(\mathscr{M}_{ij,34,14}) + v(\mathscr{M}_{ij,24,14})$$

$$= \begin{cases} \iint\limits_{\overline{\mathscr{G}}_{12,4}} \left(\int\limits_{-\varepsilon-s_1-s_2}^{\varepsilon-s_2} ds_3 \right) ds_1\, ds_2 + \iint\limits_{\overline{\mathscr{G}}_{12,2}} \left(\int\limits_{-\varepsilon-s_2}^{\varepsilon-s_1-s_2} ds_3 \right) ds_1\, ds_2 & \text{for} \quad i = 1, j = 2 \\ \iint\limits_{\overline{\mathscr{G}}_{13,1}} \left(\int\limits_{-\varepsilon-s_1-s_2}^{\varepsilon} ds_3 \right) ds_1\, ds_2 + \iint\limits_{\overline{\mathscr{G}}_{13,2}} \left(\int\limits_{-\varepsilon}^{\varepsilon-s_1-s_2} ds_3 \right) ds_1\, ds_2 & \text{for} \quad i = 1, j = 3 \\ \iint\limits_{\overline{\mathscr{G}}_{23,1}} \left(\int\limits_{-\varepsilon-s_2}^{\varepsilon} ds_3 \right) ds_1\, ds_2 + \iint\limits_{\overline{\mathscr{G}}_{23,2}} \left(\int\limits_{-\varepsilon}^{\varepsilon-s_2} ds_3 \right) ds_1\, ds_2 & \text{for} \quad i = 2, j = 3 \end{cases}$$

$$= 2\left(\frac{5}{3}\,\varepsilon^3\right) = \frac{10}{3}\,\varepsilon^3 .$$

Now we consider the sets

$$\overline{\mathscr{G}}_{12,13} = \bigcup_{i=1}^{8} \overline{\mathscr{G}}_{12,13;i} = \{(s_1, s_2) \colon |s_1| > \varepsilon, |s_2| \leqq \varepsilon, |s_1 + s_2| > \varepsilon\},$$

$$\overline{\mathscr{G}}_{12,23} = \bigcup_{i=1}^{8} \overline{\mathscr{G}}_{12,23;i} = \{(s_1, s_2) \colon |s_1| > \varepsilon, |s_2| > \varepsilon, |s_1 + s_2| \leqq \varepsilon\},$$

$$\overline{\mathscr{G}}_{13,23} = \bigcup_{i=1}^{8} \overline{\mathscr{G}}_{13,23;i} = \{(s_1, s_2) \colon |s_1| \leqq \varepsilon, |s_2| > \varepsilon, |s_1 + s_2| > \varepsilon\}$$

and obtain

$$v(\mathscr{M}_{1i,j3}) + v(\mathscr{M}_{1i,j3,34}) + v(\mathscr{M}_{1i,j3,24}) + v(\mathscr{M}_{1i,j3,14})$$

$$= \iint\limits_{\overline{\mathscr{G}}_{1i,j3,1} \cup \overline{\mathscr{G}}_{1i,j3,2}} \begin{Bmatrix} \int\limits_{-\varepsilon-s_2}^{\varepsilon-s_1-s_2} ds_3 \\ \int\limits_{-\varepsilon-s_2}^{\varepsilon-s_1-s_2} ds_3 \\ \int\limits_{-\varepsilon-s_2}^{\varepsilon} ds_3 \end{Bmatrix} ds_1\, ds_2 + \iint\limits_{\overline{\mathscr{G}}_{1i,j3,3} \cup \overline{\mathscr{G}}_{1i,j3,4}} \begin{Bmatrix} \int\limits_{-\varepsilon}^{\varepsilon-s_1-s_2} ds_3 \\ \int\limits_{-\varepsilon-s_2}^{\varepsilon} ds_3 \\ \int\limits_{-\varepsilon-s_1-s_2}^{\varepsilon} ds_3 \end{Bmatrix} ds_1\, ds_2$$

$$+ \iint\limits_{\overline{\mathscr{G}}_{1i,j3,5} \cup \overline{\mathscr{G}}_{1i,j3,6}} \begin{Bmatrix} \int\limits_{-\varepsilon-s_1-s_2}^{\varepsilon-s_2} ds_3 \\ \int\limits_{-\varepsilon-s_1-s_2}^{\varepsilon-s_2} ds_3 \\ \int\limits_{-\varepsilon}^{\varepsilon-s_1-s_2} ds_3 \end{Bmatrix} ds_1\, ds_2 + \iint\limits_{\overline{\mathscr{G}}_{1i,j3,7} \cup \overline{\mathscr{G}}_{1i,j3,8}} \begin{Bmatrix} \int\limits_{-\varepsilon-s_1-s_2}^{\varepsilon} ds_3 \\ \int\limits_{-\varepsilon}^{\varepsilon-s_2} ds_3 \\ \int\limits_{-\varepsilon}^{\varepsilon-s_2} ds_3 \end{Bmatrix} ds_1\, ds_2 \quad \begin{matrix} \text{for } i = 2, j = 1 \\ \text{for } i = 2, j = 2 \\ \text{for } i = 3, j = 2 \end{matrix}$$

$$= 4\left(\frac{1}{2}\,\varepsilon^3\right) = 2\varepsilon^3 .$$

Define $\overline{\mathcal{H}}_k$, $k = 1, 2, \ldots, 6$, by

$$\overline{\mathcal{H}}_1 \doteq \overline{\mathcal{G}}_1 \cup \overline{\mathcal{G}}_{13;1} \cup \bigcup_{i=7}^{8} \overline{\mathcal{G}}_{12,13;i} \cup \bigcup_{i=3}^{4} \overline{\mathcal{G}}_{13,23;i},$$

$$\overline{\mathcal{H}}_2 \doteq \overline{\mathcal{G}}_2 \cup \overline{\mathcal{G}}_{23;1} \cup \bigcup_{i=1}^{2} \overline{\mathcal{G}}_{13,23;i} \cup \bigcup_{i=3}^{4} \overline{\mathcal{G}}_{12,23;i},$$

$$\overline{\mathcal{H}}_3 \doteq \overline{\mathcal{G}}_3 \cup \overline{\mathcal{G}}_{12;2} \cup \bigcup_{i=1}^{2} \overline{\mathcal{G}}_{12,13;i} \cup \bigcup_{i=1}^{2} \overline{\mathcal{G}}_{12,23;i},$$

$$\overline{\mathcal{H}}_4 \doteq \overline{\mathcal{G}}_4 \cup \overline{\mathcal{G}}_{13;2} \cup \bigcup_{i=3}^{4} \overline{\mathcal{G}}_{12,13;i} \cup \bigcup_{i=5}^{6} \overline{\mathcal{G}}_{13,23;i},$$

$$\overline{\mathcal{H}}_5 \doteq \overline{\mathcal{G}}_5 \cup \overline{\mathcal{G}}_{23;2} \cup \bigcup_{i=7}^{8} \overline{\mathcal{G}}_{12,23;i} \cup \bigcup_{i=7}^{8} \overline{\mathcal{G}}_{13,23;i},$$

$$\overline{\mathcal{H}}_6 \doteq \overline{\mathcal{G}}_6 \cup \overline{\mathcal{G}}_{12;1} \cup \bigcup_{i=5}^{6} \overline{\mathcal{G}}_{12,23;i} \cup \bigcup_{i=5}^{6} \overline{\mathcal{G}}_{12,13;i}$$

then

$$\begin{aligned} v(\tilde{\mathcal{M}}(y)) &= \iint_{\overline{\mathcal{H}}_1} \left(\int_{-\varepsilon-s_1-s_2}^{\varepsilon} \mathrm{d}s_3 \right) \mathrm{d}s_1 \, \mathrm{d}s_2 + \iint_{\overline{\mathcal{H}}_2} \left(\int_{-\varepsilon-s_2}^{\varepsilon} \mathrm{d}s_3 \right) \mathrm{d}s_1 \, \mathrm{d}s_2 \\ &+ \iint_{\overline{\mathcal{H}}_3} \left(\int_{-\varepsilon-s_2}^{\varepsilon-s_1-s_2} \mathrm{d}s_3 \right) \mathrm{d}s_1 \, \mathrm{d}s_2 + \iint_{\overline{\mathcal{H}}_4} \left(\int_{-\varepsilon}^{\varepsilon-s_1-s_2} \mathrm{d}s_3 \right) \mathrm{d}s_1 \, \mathrm{d}s_2 \\ &+ \iint_{\overline{\mathcal{H}}_5} \left(\int_{-\varepsilon}^{\varepsilon-s_2} \mathrm{d}s_3 \right) \mathrm{d}s_1 \, \mathrm{d}s_2 + \iint_{\overline{\mathcal{H}}_6} \left(\int_{-\varepsilon-s_1-s_2}^{\varepsilon-s_2} \mathrm{d}s_3 \right) \mathrm{d}s_1 \, \mathrm{d}s_2 \\ &= 6\left(\frac{4}{3}\,\varepsilon^3 + \frac{5}{3}\,\varepsilon^3 + \varepsilon^3\right) = 24\varepsilon^3. \end{aligned}$$

The domains $\overline{\mathcal{H}}_k$ are illustrated in Fig. 1.16b. An example for the computation of ${}^3A_2$ and ${}^4\underline{A}_3$ can be found in Section 2.1.

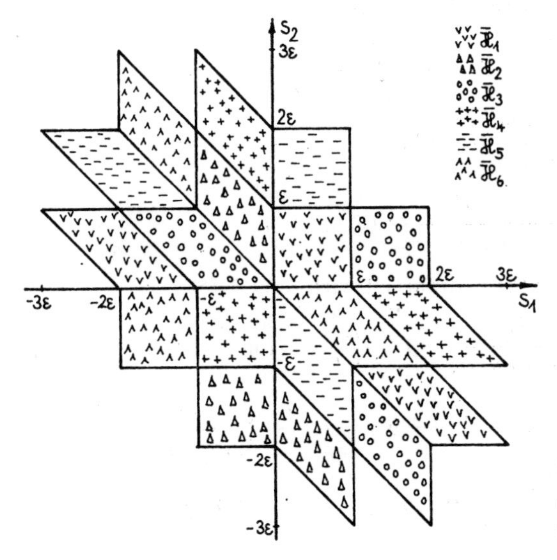

Fig. 1.16b. Integration domain as to the evaluation of $\underline{a}_{1234}(y)$ in the case of $m = 1$

1.7. Expansions of moments of random functional-functions

In this section we will expanse moments of random functions of linear functionals with respect to the correlation length ε. These results will be used for expansions of distribution functions which will be investigated in a following section. We will refer to the notations of Section 1.3.

Let d_k, $k = 1, 2, \ldots, s$, be functions of $y \doteq (y_1, \ldots, y_n)$. These functions are to have the following properties:

(1) d_k possesses the representation

$$d_k(y_1, \ldots, y_n) = d_{k,0} + \sum_{a=1}^{n} d_{k,a} y_a + \sum_{a,b=1}^{n} d_{k,ab} y_a y_b + \sum_{a,b,c=1}^{n} d_{k,abc} y_a y_b y_c + \left(\sum_{a=1}^{n} y_a^2\right)^{\alpha} g_k(y_1, \ldots, y_n) \tag{1.142}$$

where $3/2 < \alpha$ and the real function g_k is bounded on $\mathcal{K}_\delta(0)$ for a $\delta > 0$, $|g_k(y_1, \ldots, y_n)| \leqq c_0$.

(2) All moments of

$$d_{k\varepsilon}(\omega) \doteq d_k\big(r_{1\varepsilon}(\omega), \ldots, r_{n\varepsilon}(\omega)\big)$$

exist and

$$\langle |d_{k\varepsilon}|^p \rangle \leqq c_p < \infty \quad \text{for} \quad k = 1, 2, \ldots, n;\ \varepsilon > 0$$

where $r_{i\varepsilon}(\omega)$, $i = 1, 2, \ldots, n$, denote linear functionals of the form

$$r_{i\varepsilon}(\omega) = \sum_{j=1}^{l} r_{ij\varepsilon}(\omega); \qquad r_{ij\varepsilon}(\omega) = \int_{\mathcal{D}_t} F_{ij}(x)\, f_{j\varepsilon}(x, \omega)\, dx .$$

Especially, condition (1) is satisfied if the function

$$d_k \in \mathbf{C}^4\big(\mathcal{K}_\delta(0)\big) \quad \text{for} \quad k = 1, 2, \ldots, s .$$

In this case we have by means of the Taylor expansion

$$\begin{aligned} d_k(y) = d_k(0) &+ \sum_{a=1}^{n} \frac{\partial d_k}{\partial y_a}(0)\, y_a + \frac{1}{2} \sum_{a,b=1}^{n} \frac{\partial^2 d_k}{\partial y_a\, \partial y_b}(0)\, y_a y_b \\ &+ \frac{1}{6} \sum_{a,b,c=1}^{n} \frac{\partial^3 d_k}{\partial y_a\, \partial y_b\, \partial y_c}(0)\, y_a y_b y_c \\ &+ \frac{1}{24} \sum_{a,b,c,d=1}^{n} \frac{\partial^4 d_k}{\partial y_a\, \partial y_b\, \partial y_c\, \partial y_d}(\bar{y})\, y_a y_b y_c y_d \end{aligned}$$

where $y \in \mathcal{K}_\delta(0)$. From this, (1.142) is obtained with $\alpha = 2$ if

$$g_k(y) \doteq \frac{1}{24} \sum_{a,b,c,d=1}^{n} \frac{\partial^4 d_k}{\partial y_a\, \partial y_b\, \partial y_c\, \partial y_d}(\bar{y})\, \frac{y_a y_b y_c y_d}{|y|^4} \quad \text{for} \quad y \in \mathcal{K}_\delta(0) \setminus \{0\},$$

$$\begin{aligned} g_k(y) \doteq \frac{1}{|y|^4} \Bigg[d_k(y) - d_k(0) &- \sum_{a=1}^{n} \frac{\partial d_k}{\partial y_a}(0)\, y_a - \frac{1}{2} \sum_{a,b=1}^{n} \frac{\partial^2 d_k}{\partial y_a\, \partial y_b}(0)\, y_a y_b \\ &- \frac{1}{6} \sum_{a,b,c=1}^{n} \frac{\partial^3 d_k}{\partial y_a\, \partial y_b\, \partial y_c}(0)\, y_a y_b y_c \Bigg] \quad \text{for} \quad y \notin \mathcal{K}_\delta(0) \end{aligned}$$

and $g_k(0) \doteq 0$. The condition

$$|g_k(y)| \leqq c_0 \quad \text{for} \quad y \in \mathcal{K}_\delta(0)$$

follows from the boundedness of the fourth derivatives of d_k on $\mathcal{K}_\delta(0)$ and

$$\left|\frac{y_a y_b y_c y_d}{|y|^4}\right| \leqq 1 .$$

First, the transformation

$$\tilde{d}_k = \frac{1}{\sqrt{\varepsilon}^m} \sum_{p=1}^{n} b_{kp}(d_p - d_{p,0}), \qquad k = 1, 2, \ldots, s,$$

is carried out and the coefficients b_{kp} are determined by the relation

$$\langle \tilde{d}_k \tilde{d}_h \rangle = \delta_{kh} + O(\varepsilon) .$$

Thus,

$$\begin{aligned}
\langle \tilde{d}_k \tilde{d}_h \rangle &= \frac{1}{\varepsilon^m} \sum_{p=1}^{k} \sum_{q=1}^{h} b_{kp} b_{hq} \langle (d_p - d_{p,0})\,(d_q - d_{q,0}) \rangle = \frac{1}{\varepsilon^m} \sum_{p=1}^{k} \sum_{q=1}^{h} b_{kp} b_{hq} \\
&\quad \times \Bigg\langle \Bigg(\sum_{a=1}^{n} d_{p,a} r_{a\varepsilon} + \sum_{a,b=1}^{n} d_{p,ab} r_{a\varepsilon} r_{b\varepsilon} + \sum_{a,b,c=1}^{n} d_{p,abc} r_{a\varepsilon} r_{b\varepsilon} r_{c\varepsilon} \\
&\quad + \left(\sum_{a=1}^{n} r_{a\varepsilon}^2 \right)^{\alpha} g_p(r_{1\varepsilon}, \ldots, r_{n\varepsilon}) \Bigg) \\
&\quad \times \Bigg(\sum_{a=1}^{n} d_{q,a} r_{a\varepsilon} + \sum_{a,b=1}^{n} d_{q,ab} r_{a\varepsilon} r_{b\varepsilon} + \sum_{a,b,c=1}^{n} d_{q,abc} r_{a\varepsilon} r_{b\varepsilon} r_{c\varepsilon} \\
&\quad + \left(\sum_{a=1}^{n} r_{a\varepsilon}^2 \right)^{\alpha} g_q(r_{1\varepsilon}, \ldots, r_{n\varepsilon}) \Bigg) \Bigg\rangle \\
&= \frac{1}{\varepsilon^m} \left(\sum_{p=1}^{k} \sum_{q=1}^{h} b_{kp} b_{hq} \langle \tilde{r}_{p\varepsilon} \tilde{r}_{q\varepsilon} \rangle + O(\varepsilon^{m+1}) \right)
\end{aligned}$$

where $\tilde{r}_{k\varepsilon}(\omega)$ are defined by

$$\tilde{r}_{k\varepsilon} = \sum_{a=1}^{n} \tilde{d}_{k,a} r_{a\varepsilon} \quad \text{for} \quad k = 1, 2, \ldots, s \tag{1.143}$$

and the following relations are taken into consideration:

$$\begin{aligned}
&\left\langle \prod_{i=1}^{p} r_{a_i\varepsilon} \right\rangle = O(\varepsilon^{(p-1)m}), \\
&\left\langle \prod_{i=1}^{p} r_{a_i\varepsilon} \left(\sum_{a=1}^{n} r_{a\varepsilon}^2 \right)^{\alpha} g_k(r_{1\varepsilon}, \ldots, r_{n\varepsilon}) \right\rangle = o\left(\varepsilon^{\left(\frac{p}{2}+\frac{3}{2}\right)m} \right).
\end{aligned} \tag{1.144}$$

The first relation follows from (1.11). In order to prove the second relation we estimate

$$\begin{aligned}
&\left\langle \prod_{i=1}^{p} r_{a_i\varepsilon} \left(\sum_{a=1}^{n} r_{a\varepsilon}^2 \right)^{\alpha} g_k(r_{1\varepsilon}, \ldots, r_{n\varepsilon}) \right\rangle \\
&\leqq \left[\left\langle \left(\prod_{i=1}^{p} r_{a_i\varepsilon} \right)^4 \right\rangle \left\langle \left(\sum_{a=1}^{n} r_{a\varepsilon}^2 \right)^{4\alpha} \right\rangle \left\langle g_k^4(r_{1\varepsilon}, \ldots, r_{n\varepsilon}) \right\rangle \right]^{1/4}
\end{aligned}$$

and furthermore with the help of (1.23)

$$\left\langle\left(\prod_{i=1}^{p} r_{a_i\varepsilon}\right)^4\right\rangle = O(\varepsilon^{2pm}),$$

$$\left\langle\left(\sum_{a=1}^{n} r_{a\varepsilon}^2\right)^{4\alpha}\right\rangle \leq \left\langle\left(\sum_{a=1}^{n} r_{a\varepsilon}^2\right)^{[4\alpha]} + \left(\sum_{a=1}^{n} r_{a\varepsilon}^2\right)^{[4\alpha]+1}\right\rangle = O(\varepsilon^{[4\alpha]m}),$$

$$\langle g_k^{2q}(r_{1\varepsilon}, \ldots, r_{n\varepsilon})\rangle = \int\limits_{\{y:\,|y|<\delta\}} g_k^{2q}(y)\,\mathrm{d}P_\varepsilon(y) + \int\limits_{\{y:\,|y|>\delta\}} g_k^{2q}(y)\,\mathrm{d}P_\varepsilon(y)$$

$$\leq c_0^{2q} + \frac{1}{\delta^{4\alpha q}}\left\langle\left(d_k(r_{1\varepsilon}, \ldots, r_{n\varepsilon}) - d_{k,0} - \sum_{a=1}^{n} d_{k,a} r_{a\varepsilon}\right.\right.$$

$$\left.\left. - \sum_{a,b=1}^{n} d_{k,ab} r_{a\varepsilon} r_{b\varepsilon} - \sum_{a,b,c=1}^{n} d_{k,abc} r_{a\varepsilon} r_{b\varepsilon} r_{c\varepsilon}\right)^{2q}\right\rangle.$$

We return to $\langle \tilde{d}_k\, \tilde{d}_h\rangle$ and with

$$\tilde{B}_{pq} = \lim_{\varepsilon\downarrow 0} \frac{1}{\varepsilon^m} \langle \tilde{r}_{p\varepsilon} \tilde{r}_{q\varepsilon}\rangle = \sum_{a,b=1}^{n} d_{p,a} d_{q,b} B_{ab},$$

$$B_{ab} = \lim_{\varepsilon\downarrow 0} \frac{1}{\varepsilon^m} \langle r_{a\varepsilon} r_{b\varepsilon}\rangle = \hat{H}_1(m; a, b) = \sum_{j_1,j_2=1}^{l} {}^2A_1(F_{aj_1}, F_{bj_2})$$

we obtain the equations for the determination of b_{kp}:

$$\sum_{p=1}^{k} \sum_{q=1}^{h} b_{kp} b_{hq} \tilde{B}_{pq} = \delta_{kh}, \qquad k, h = 1, 2, \ldots, s. \tag{1.145}$$

The linear independence of

$$\tilde{r}_{k\varepsilon} = \sum_{a=1}^{n} \tilde{d}_{k,a} r_{a\varepsilon}, \qquad k = 1, 2, \ldots, s,$$

leads to the determination of the coefficients b_{kp} as in (1.41), (1.42)

$$b_{kp} b_{kk} = (-1)^{k+p} \frac{\det (\tilde{B}_{ij})_{i=1,\ldots,k-1;\,j=1,\ldots,p-1,p+1,\ldots,k}}{\det (\tilde{B}_{ij})_{i,j=1,2,\ldots,k}}$$

for $p = 1, 2, \ldots, k$,

$$b_{kk} = \sqrt{\frac{\det (\tilde{B}_{ij})_{i,j=1,2,\ldots,k-1}}{\det (\tilde{B}_{ij})_{i,j=1,2,\ldots,k}}}.$$

Put

$$\tilde{d}_k = \frac{1}{\sqrt{\varepsilon^m}} \sum_{p=1}^{k} b_{kp}(d_p - d_{p,0})$$

$$= \frac{1}{\sqrt{\varepsilon^m}} \sum_{p=1}^{k} b_{kp} \left(\sum_{a=1}^{n} d_{p,a} y_a + \sum_{a,b=1}^{n} d_{p,ab} y_a y_b + \sum_{a,b,c=1}^{n} d_{p,abc} y_a y_b y_c + |y|^{2\alpha} g_p\right)$$

$$= \frac{1}{\sqrt{\varepsilon^m}} \left\{\sum_{a=1}^{n} \tilde{d}_{k,a} y_a + \sum_{a,b=1}^{n} \tilde{d}_{k,ab} y_a y_b + \sum_{a,b,c=1}^{n} \tilde{d}_{k,abc} y_a y_b y_c + |y|^{2\alpha} \tilde{g}_k\right\} \tag{1.146}$$

where

$$\tilde{d}_{k,a_1\ldots a_t} \doteq \sum_{p=1}^{k} b_{kp} d_{p,a_1\ldots a_t} \quad \text{for} \quad t = 1, 2, 3,$$
$$\tilde{g}_k \doteq \sum_{p=1}^{k} b_{kp} g_p . \tag{1.147}$$

The computation of a k-th moment leads to

$$\left\langle \prod_{i=1}^{s} \tilde{d}_i^{k_i} \right\rangle = \frac{1}{\sqrt{\varepsilon}^{mk}} \left\langle \prod_{i=1}^{s} \left\{ \sum_{a=1}^{n} \tilde{d}_{i,a} r_{a\varepsilon} + \sum_{a,b=1}^{n} \tilde{d}_{i,ab} r_{a\varepsilon} r_{b\varepsilon} + \sum_{a,b,c=1}^{n} \tilde{d}_{i,abc} r_{a\varepsilon} r_{b\varepsilon} r_{c\varepsilon} + |r_\varepsilon|^{2\alpha} \tilde{g}_i \right\}^{k_i} \right\rangle$$
$$= \frac{1}{\sqrt{\varepsilon}^{mk}} \left\langle \prod_{i=1}^{s} \left\{ \left[\sum_{a=1}^{n} \tilde{d}_{i,a} r_{a\varepsilon} \right]^{k_i} + k_i \left[\sum_{a=1}^{n} \tilde{d}_{i,a} r_{a\varepsilon} \right]^{k_i - 1} \left[\sum_{a,b=1}^{n} \tilde{d}_{i,ab} r_{a\varepsilon} r_{b\varepsilon} \right] \right.\right.$$
$$+ k_i \left[\sum_{a=1}^{n} \tilde{d}_{i,a} r_{a\varepsilon} \right]^{k_i - 1} \left[\sum_{a,b,c=1}^{n} \tilde{d}_{i,abc} r_{a\varepsilon} r_{b\varepsilon} r_{c\varepsilon} \right]$$
$$\left.\left. + \binom{k_i}{2} \left[\sum_{a=1}^{n} \tilde{d}_{i,a} r_{a\varepsilon} \right]^{k_i - 2} \left[\sum_{a,b=1}^{n} \tilde{d}_{i,ab} r_{a\varepsilon} r_{b\varepsilon} \right]^2 + \cdots \right\} \right\rangle$$
$$= \frac{1}{\sqrt{\varepsilon}^{mk}} \left\{ \left\langle \prod_{i=1}^{s} \left[\sum_{a=1}^{n} \tilde{d}_{i,a} r_{a\varepsilon} \right]^{k_i} \right\rangle \right.$$
$$+ \left\langle \sum_{p=1}^{s} k_p \left[\sum_{a=1}^{n} \tilde{d}_{p,a} r_{a\varepsilon} \right]^{k_p - 1} \left[\sum_{a,b=1}^{n} \tilde{d}_{p,ab} r_{a\varepsilon} r_{b\varepsilon} \right] \prod_{\substack{i=1 \\ i \neq p}}^{s} \left[\sum_{a=1}^{n} \tilde{d}_{i,a} r_{a\varepsilon} \right]^{k_i} \right\rangle$$
$$+ \left\langle \sum_{p=1}^{s} k_p \left[\sum_{a=1}^{n} \tilde{d}_{p,a} r_{a\varepsilon} \right]^{k_p - 1} \left[\sum_{a,b,c=1}^{n} \tilde{d}_{p,abc} r_{a\varepsilon} r_{b\varepsilon} r_{c\varepsilon} \right] \prod_{\substack{i=1 \\ i \neq p}}^{s} \left[\sum_{a=1}^{n} \tilde{d}_{i,a} r_{a\varepsilon} \right]^{k_i} \right\rangle$$
$$+ \left\langle \sum_{p=1}^{s} \binom{k_p}{2} \left[\sum_{a=1}^{n} \tilde{d}_{p,a} r_{a\varepsilon} \right]^{k_p - 2} \left[\sum_{a,b=1}^{n} \tilde{d}_{p,ab} r_{a\varepsilon} r_{b\varepsilon} \right]^2 \prod_{\substack{i=1 \\ i \neq p}}^{s} \left[\sum_{a=1}^{n} \tilde{d}_{i,a} r_{a\varepsilon} \right]^{k_i} \right\rangle$$
$$+ \left\langle \sum_{\substack{p,q=1 \\ p<q}}^{s} k_p k_q \left[\sum_{a=1}^{n} \tilde{d}_{p,a} r_{a\varepsilon} \right]^{k_p - 1} \left[\sum_{q=1}^{n} \tilde{d}_{q,a} r_{a\varepsilon} \right]^{k_q - 1} \right.$$
$$\left.\left. \times \left[\sum_{a,b=1}^{n} \tilde{d}_{p,ab} r_{a\varepsilon} r_{b\varepsilon} \right] \left[\sum_{a,b=1}^{n} \tilde{d}_{q,ab} r_{a\varepsilon} r_{b\varepsilon} \right] \prod_{\substack{i=1 \\ i \neq p,q}}^{s} \left[\sum_{a=1}^{n} \tilde{d}_{i,a} r_{a\varepsilon} \right]^{k_i} \right\rangle + \cdots \right\}$$

where $k \doteq \sum_{i=1}^{s} k_i$ and the points denote terms having an order greater than $(k + 2)\, m/2$ as to ε. Introduce the terms

$$R_{i,t}(\omega) \doteq \sum_{a_1, a_2, \ldots, a_t = 1}^{n} \tilde{d}_{i,a_1\ldots a_t} r_{a_1\varepsilon} r_{a_2\varepsilon} \cdots r_{a_t\varepsilon} . \tag{1.148}$$

Thus,

$$\left\langle \prod_{i=1}^{s} \tilde{d}_i^{k_i} \right\rangle = \frac{1}{\sqrt{\varepsilon}^{mk}} \left\{ \left[\left\langle \prod_{i=1}^{s} R_{i,1}^{k_i} \right\rangle \right] + \left[\left\langle \sum_{j=1}^{s} k_j R_{j,1}^{k_j - 1} R_{j,2} \prod_{\substack{p=1 \\ p \neq j}}^{s} R_{p,1}^{k_p} \right\rangle \right] \right.$$
$$+ \left[\left\langle \sum_{j=1}^{s} k_j R_{j,1}^{k_j - 1} R_{j,3} \prod_{\substack{p=1 \\ p \neq j}}^{s} R_{p,1}^{k_p} \right\rangle + \left\langle \sum_{j=1}^{s} \binom{k_j}{2} R_{j,1}^{k_j - 2} R_{j,2}^2 \prod_{\substack{p=1 \\ p \neq j}}^{s} R_{p,1}^{k_p} \right\rangle \right.$$
$$\left.\left. + \left\langle \sum_{\substack{i,j=1 \\ j<i}}^{s} k_j k_i R_{j,1}^{k_j - 1} R_{i,1}^{k_i - 1} R_{j,2} R_{i,2} \prod_{\substack{p=1 \\ p \neq j,i}}^{s} R_{p,1}^{k_p} \right\rangle \right] + \cdots \right\} . \tag{1.149}$$

Now, the adequate moments are determined. As in Section 1.4 we introduce

$$\tilde{F}_{i,j}(x) \doteq \sum_{a=1}^{n} \tilde{d}_{i,a} F_{aj}(x) \quad \text{for} \quad x \in \tilde{\mathcal{D}} \doteq \bigcup_{i=1}^{n} \mathcal{D}_i \tag{1.150}$$

with

$$F_{aj}(x) \doteq 0 \quad \text{for} \quad x \in \tilde{\mathcal{D}} \setminus \mathcal{D}_a$$

and

$$\tilde{r}_i(\omega) \doteq \sum_{j=1}^{l} \tilde{r}_{ij}(\omega), \qquad \tilde{r}_{ij}(\omega) \doteq \int_{\tilde{\mathcal{D}}} \tilde{F}_{i,j}(x)\, f_{j\varepsilon}(x, \omega)\, \mathrm{d}x . \tag{1.151}$$

Then it follows

$$R_{i,1}(\omega) = \sum_{a=1}^{n} \tilde{d}_{i,a} r_{a\varepsilon}(\omega) = \tilde{r}_{i\varepsilon}(\omega)$$

and the moments

$$\left\langle \prod_{i=1}^{s} R_{i,1}^{k_i} \right\rangle = \left\langle \prod_{i=1}^{s} \tilde{r}_{i\varepsilon}^{k_i} \right\rangle = \left\langle \prod_{p=1}^{k} \tilde{r}_{i_p\varepsilon} \right\rangle; \qquad i_p \in \{1, 2, \ldots, s\},$$

have been calculated in (1.30) for k even and in (1.32) for k odd. We can write

$$\left\langle \prod_{i=1}^{s} R_{i,1}^{k_i} \right\rangle = \begin{cases} \tilde{C}_0^0(k_1, \ldots, k_s)\, \varepsilon^{km/2} + \tilde{C}_0^1(k_1, \ldots, k_s)\, \varepsilon^{km/2+1} \\ + o(\varepsilon^{km/2+1}) \quad \text{for } k \text{ even} \\ \tilde{C}_0^0(k_1, \ldots, k_s)\, \varepsilon^{m(k+1)/2} + O(\varepsilon^{m(k+1)/2+1}) \quad \text{for } k \text{ odd.} \end{cases} \tag{1.152}$$

Moments of the form

$$\left\langle R_{j,2} \prod_{i=1}^{s} R_{i,1}^{k_i} \right\rangle$$

are now expanded. We consider

$$R_{i,2} = \sum_{a,b=1}^{n} \tilde{d}_{i,ab} r_{a\varepsilon} r_{b\varepsilon} = \sum_{j_1,j_2=1}^{l} \int_{\tilde{\mathcal{D}}} \int_{\tilde{\mathcal{D}}} \tilde{F}_{i,j_1j_2}(x, y)\, f_{j_1\varepsilon}(x)\, f_{j_2\varepsilon}(y)\, \mathrm{d}x\, \mathrm{d}y$$

where

$$\tilde{F}_{i,j_1j_2}(x, y) \doteq \sum_{a,b=1}^{n} \tilde{d}_{i,ab} F_{aj_1}(x)\, F_{bj_2}(y) .$$

For the investigation of

$$\left\langle R_{p,2} \prod_{i=1}^{s} R_{i,1}^{k_i} \right\rangle$$

we have to expand moments of the form

$$J_1 = \int_{\tilde{\mathcal{D}}} \ldots \int_{\tilde{\mathcal{D}}} G_{12}(y_1, y_2)\, G_3(y_3) \ldots G_k(y_k) \left\langle \prod_{q=1}^{k} f_{i_q\varepsilon}(y_q) \right\rangle \mathrm{d}y_1 \ldots \mathrm{d}y_k .$$

Using the results of Section 1.3 it is

$$J_1 = \sum_{p=1}^{k} \sum_{\{\mathcal{J}_1, \ldots, \mathcal{J}_p\}} \int_{\mathcal{B}(\mathcal{J}_1, \ldots, \mathcal{J}_p)} G_{12}(y_1, y_2)\, G_3(y_3) \ldots G_k(y_k) \prod_{q=1}^{p} \left\langle \prod_{j \in \mathcal{J}_q} f_{i_j\varepsilon}(y_j) \right\rangle \mathrm{d}y_1 \ldots \mathrm{d}y_k \tag{1.153}$$

and for an even k we obtain the lowest order as to ε for $p = k/2$,

$$J_1 = \sum_{\{\mathcal{J}_1,\dots,\mathcal{J}_{k/2}\}} \int_{\mathscr{E}(\mathcal{J}_1)} \dots \int_{\mathscr{E}(\mathcal{J}_{k/2})} G_{12}(y_1, y_2)\, G_3(y_3) \dots G_k(y_k)$$

$$\times \prod_{q=1}^{k/2} \Big\langle \prod_{j\in\mathcal{J}_q} f_{i_j\varepsilon}(y_j)\Big\rangle\, \mathrm{d}y_1 \dots \mathrm{d}y_k + o(\varepsilon^{km/2})$$

$$= \int_{\mathscr{E}(\{1,2\})} G_{12}(y_1, y_2)\, \langle f_{i_1\varepsilon}(y_1)\, f_{i_2\varepsilon}(y_2)\rangle\, \mathrm{d}y_1\, \mathrm{d}y_2$$

$$\times \left[\sum_{\substack{\{s_1,t_1\},\dots,\{s_{(k/2)-1},t_{(k/2)-1}\}\\ \text{separations of } \{3,4,\dots,k\}}} \prod_{q=1}^{(k/2)-1} {}^2A_1(G_{s_q}, G_{t_q})\, \varepsilon^{m\left(\frac{k}{2}-1\right)} + o\left(\varepsilon^{m\left(\frac{k}{2}-1\right)}\right)\right]$$

$$+ \sum_{\substack{p_1,p_2=3\\ p_1\neq p_2}}^{k} \int_{\mathscr{E}(\{1,p_1\})} \int_{\mathscr{E}(\{2,p_2\})} G_{12}(y_1, y_2)\, G_{p_1}(y_{p_1})\, G_{p_2}(y_{p_2})$$

$$\times \langle f_{i_1\varepsilon}(y_1)\, f_{i_{p_1}\varepsilon}(y_{p_1})\rangle \langle f_{i_2\varepsilon}(y_2)\, f_{i_{p_2}\varepsilon}(y_{p_2})\rangle\, \mathrm{d}y_1\, \mathrm{d}y_2\, \mathrm{d}y_{p_1}\, \mathrm{d}y_{p_2}$$

$$\times \left[\sum_{\substack{\{s_1,t_1\},\dots,\{s_{(k/2)-2},t_{(k/2)-2}\}\\ \text{separations of}\\ \{1,\dots,k\}\setminus\{1,2,p_1,p_2\}}} \prod_{q=1}^{(k/2)-2} {}^2A_1(G_{s_q}, G_{t_q})\, \varepsilon^{m\left(\frac{k}{2}-2\right)} + o\left(\varepsilon^{m\left(\frac{k}{2}-2\right)}\right)\right].$$

The terms ${}^2A_1(G_s, G_t)$ can be written as

$${}^2A_1(G_s, G_t) = \int_{\tilde{D}} G_s G_t(x)\, a_{i_s i_t}(x)\, \mathrm{d}x$$

where $a_{i_s i_t}(x)$ denotes the intensity between $f_{i_s\varepsilon}(x, \omega)$ and $f_{i_t\varepsilon}(x, \omega)$. For abbreviation it is defined

$${}^2P_1(\{p_1, \dots, p_k\}) \doteq \sum_{\substack{\{s_1,t_1\},\dots,\{s_{k/2},t_{k/2}\}\\ \text{separations of } \{1,\dots,k\}}} \prod_{q=1}^{k/2} {}^2A_1\big(G_{p_{s_q}}, G_{p_{t_q}}\big). \tag{1.154}$$

In the above formula $G_{12}(y_1, y_2)$ has the special form

$$G_{12}(y_1, y_2) = \sum_{a,b=1}^{n} d_{ab} H_a(y_1)\, K_b(y_2)$$

so that

$$\lim_{\varepsilon\downarrow 0} \frac{1}{\varepsilon^m} \int_{\mathscr{E}(\{1,2\})} G_{12}(y_1, y_2)\, \langle f_{i_1\varepsilon}(y_1)\, f_{i_2\varepsilon}(y_2)\rangle\, \mathrm{d}y_1\, \mathrm{d}y_2$$

$$= \int_{\tilde{D}} G_{12}(y, y)\, a_{i_1 i_2}(y)\, \mathrm{d}y = {}^2A_1(G_{12}),$$

$$\lim_{\varepsilon\downarrow 0} \frac{1}{\varepsilon^{2m}} \int_{\mathscr{E}(\{1,3\})} \int_{\mathscr{E}(\{2,4\})} G_{12}(y_1, y_2)\, G_3(y_3)\, G_4(y_4)$$

$$\times \langle f_{i_1\varepsilon}(y_1)\, f_{i_3\varepsilon}(y_3)\rangle \langle f_{i_2\varepsilon}(y_2)\, f_{i_4\varepsilon}(y_4)\rangle\, \mathrm{d}y_1\, \mathrm{d}y_2\, \mathrm{d}y_3\, \mathrm{d}y_4$$

$$= \lim_{\varepsilon\downarrow 0} \sum_{a,b=1}^{n} d_{ab} \left[\frac{1}{\varepsilon^m} \int_{\mathscr{E}(\{1,3\})} H_a(y_1)\, G_3(y_3)\, \langle f_{i_1\varepsilon}(y_1)\, f_{i_3\varepsilon}(y_3)\rangle\, \mathrm{d}y_1\, \mathrm{d}y_3\right]$$

$$\times \left[\frac{1}{\varepsilon^m} \int\limits_{\mathscr{E}(\{2,4\})} K_b(y_2)\, G_4(y_4)\, \langle f_{i_2\varepsilon}(y_2)\, f_{i_4\varepsilon}(y_4) \rangle \,\mathrm{d}y_2\, \mathrm{d}y_4 \right]$$

$$= \sum_{a,b=1}^{n} d_{ab} \int\limits_{\tilde{D}} H_a(y)\, G_3(y)\, a_{i_1 i_3}(y)\, \mathrm{d}y \int\limits_{\tilde{D}} K_b(z)\, G_4(z)\, a_{i_2 i_4}(z)\, \mathrm{d}z$$

$$= \int\limits_{\tilde{D}} \int\limits_{\tilde{D}} G_{12}(y, z)\, G_3(y)\, G_4(z)\, a_{i_1 i_3}(y)\, a_{i_2 i_4}(z)\, \mathrm{d}y\, \mathrm{d}z$$

$$= {}^{22}A_{11}\big(G_{12}(y, z)\, G_3(y)\, G_4(z)\big).$$

For the new term we can write the relation

$${}^{22}A_{11}\big(G_{12}(y, z)\, G_3(y)\, G_4(z)\big) = \sum_{a,b=1}^{n} d_{ab}\, {}^{2}A_1(H_a, G_3)\, {}^{2}A_1(K_b, G_4)$$

where the adequate intensities are used. In the case of k even we can find

$$\begin{aligned} J_1 = \Big\{ & {}^{2}A_1(G_{12})\, {}^{2}P_1(\{3, 4, \ldots, k\}) \\ & + \sum_{\substack{p_1, p_2 = 3 \\ p_1 \neq p_2}}^{k} {}^{22}A_{11}\big(G_{12}(y, z)\, G_{p_1}(y)\, G_{p_2}(z)\big)\, {}^{2}P_1(\{3, \ldots, k\} \setminus \{p_1, p_2\}) \Big\} \\ & \times \varepsilon^{mk/2} + o(\varepsilon^{mk/2}). \end{aligned} \tag{1.155}$$

Deal with J_1 in the case of k odd. The lowest order as to ε results from $s = (k-1)/2$,

$$\begin{aligned} J_1 = & \sum_{\substack{\mathscr{I}_1 = \{s_1, t_1, r_1\}, \mathscr{I}_q = \{s_q, t_q\}, \\ q = 2, 3, \ldots, (k-1)/2}} \int\limits_{\mathscr{E}(\mathscr{I}_1)} \int\limits_{\mathscr{E}(\mathscr{I}_2)} \cdots \int\limits_{\mathscr{E}(\mathscr{I}_{(k-1)/2})} G_{12}(y_1, y_2) \\ & \times G_3(y_3) \ldots G_k(y_k) \prod_{q=2}^{(k-1)/2} \Big\langle \prod_{j \in \mathscr{I}_q} f_{i_j\varepsilon}(y_j) \Big\rangle\, \mathrm{d}y_1 \ldots \mathrm{d}y_k + o\Big(\varepsilon^{\frac{m}{2}(k+1)}\Big) \\ = & \Bigg[\sum_{p=3}^{k} {}^{3}A_2(G_{12} G_p)\, {}^{2}P_1(\{3, \ldots, k\} \setminus \{p\}) \\ & + \sum_{\substack{\{1, p_1, p_2\}, \{2, p_3\} \\ \text{from } \{3, \ldots, k\}}} {}^{32}A_{21}\big(G_{12}(y, z)\, G_{p_1}(y)\, G_{p_2}(y)\, G_{p_3}(z)\big)\, {}^{2}P_1(\{3, \ldots, k\} \setminus \{p_1, p_2, p_3\}) \\ & + \sum_{\substack{\{2, p_1, p_2\}, \{1, p_3\} \\ \text{from } \{3, \ldots, k\}}} {}^{32}A_{21}\big(G_{12}(y, z)\, G_{p_1}(z)\, G_{p_2}(z)\, G_{p_3}(y)\big)\, {}^{2}P_1(\{3, \ldots, k\} \setminus \{p_1, p_2, p_3\}) \\ & + \sum_{\substack{\{p_1, p_2, p_3\} \\ \text{from } \{3, \ldots, k\}}} {}^{3}A_2(G_{p_1} G_{p_2} G_{p_3}) \Big\{ {}^{2}A_1(G_{12})\, {}^{2}P_1(\{3, \ldots, k\} \setminus \{p_1, p_2, p_3\}) \\ & + \sum_{\substack{\{1, p_4\}, \{2, p_5\} \\ \text{from } \{3, \ldots, k\}}} {}^{22}A_{11}\big(G_{12}(z, y)\, G_{p_4}(z)\, G_{p_5}(y)\big)\, {}^{2}P_1(\{3, \ldots, k\} \setminus \{p_1, p_2, p_3, p_4, p_5\}) \Big\} \Bigg] \\ & \times \varepsilon^{m(k+1)/2} + o(\varepsilon^{m(k+1)/2}) \end{aligned} \tag{1.156}$$

where ${}^{32}A_{21}$ is defined by

$${}^{32}A_{21}\big(G_{12}(y, z)\, G_3(y)\, G_4(y)\, G_5(z)\big) \doteq \sum_{a,b=1}^{n} d_{ab}\, {}^{3}A_2(H_a, G_3, G_4)\, {}^{2}A_1(K_b, G_5).$$

Compute the term

$$\left\langle R_{j,2} R_{j,1}^{k_j-1} \prod_{\substack{p=1\\p\neq j}}^{s} R_{p,1}^{k_p} \right\rangle.$$

First for k odd using (1.154) we obtain

$$\left\langle R_{j,2} R_{j,1}^{k_j-1} \prod_{\substack{p=1\\p\neq j}}^{s} R_{p,1}^{k_p} \right\rangle$$
$$= \sum_{j_1,\ldots,j_{k+1}=1}^{l} \int_{\tilde{D}} \cdots \int_{\tilde{D}} \tilde{F}_{j,j_1j_2}(y_1, y_2) \prod_{\substack{q=3\\(i_3,\ldots,i_{k+1})=\mathscr{W}_j}}^{k+1} \tilde{F}_{i_q,j_q}(y_q) \left\langle \prod_{q=1}^{k+1} f_{j_q\varepsilon}(y_q) \right\rangle \mathrm{d}y_1 \ldots \mathrm{d}y_{k+1},$$

where $\mathscr{W}$ ... is defined in Section 1.4.

It is necessary to introduce further notations,

$${}^p\tilde{\tilde{A}}_q'(i_1, \ldots, i_p) \doteq \sum_{j_1,\ldots,j_p=1}^{l} {}^pA_q(\tilde{F}_{i_1j_1}, \ldots, \tilde{F}_{i_pj_p}),$$

$$\tilde{\tilde{A}}_{pq}'(i_1, \ldots, i_p; i_{p+1}, \ldots, i_{p+q}) = \sum_{j_1,\ldots,j_{p+q}=1}^{l} A_{pq}(\tilde{F}_{i_1j_1}, \ldots, \tilde{F}_{i_pj_p}; \tilde{F}_{i_{p+1}j_{p+1}}, \ldots, \tilde{F}_{i_{p+q}j_{p+q}}).$$

In particular, we write

$${}^2\tilde{\tilde{A}}_1'(i) \doteq \sum_{j_1,j_2=1}^{l} {}^2A_1(\tilde{F}_{i,j_1j_2}),$$

$${}^3\tilde{\tilde{A}}_2'(i, j) \doteq \sum_{j_1,j_2,j_3=1}^{l} {}^3A_2(\tilde{F}_{i,j_1j_2}, \tilde{F}_{j,j_3})$$

and in the case of ${}^{pq}A_{p-1,q-1}$ we have to denote these abbreviations some more accurately; e.g.

$${}^{22}\tilde{\tilde{A}}_{11}'\big(i_1(x, y), i_2(x), i_3(y)\big) = \sum_{j_1,j_2,j_3,j_4=1}^{l} {}^{22}A_{11}\big(\tilde{F}_{i_1,j_1j_2}(x, y)\, \tilde{F}_{i_2,j_3}(x)\, \tilde{F}_{i_3,j_4}(y)\big),$$

$${}^{32}\tilde{\tilde{A}}_{21}'\big(i_1(x, y), i_2(x), i_3(x), i_4(y)\big)$$
$$= \sum_{j_1,j_2,j_3,j_4,j_5=1}^{l} {}^{32}A_{21}\big(\tilde{F}_{i_1,j_1j_2}(x, y)\, \tilde{F}_{i_2,j_3}(x)\, \tilde{F}_{i_3,j_4}(x)\, \tilde{F}_{i_4,j_5}(y)\big).$$

For k odd using (1.154) it follows

$$\left\langle R_{j,2} R_{j,1}^{k_j-1} \prod_{\substack{p=1\\p\neq j}}^{s} R_{p,1}^{k_p} \right\rangle$$
$$= \left[{}^2\tilde{\tilde{A}}_1'(j)\, {}^2\tilde{\tilde{P}}_1(\mathscr{W}_j) + \sum_{\substack{(p,q) \text{ from } \mathscr{W}_j\\ \text{w.a.}}} {}^{22}\tilde{\tilde{A}}_{11}'\big(j(y, z), p(y), q(z)\big)\, {}^2\tilde{\tilde{P}}_1(\mathscr{W}_{jpq}) \right] \varepsilon^{\frac{m}{2}(k+1)} + o\left(\varepsilon^{\frac{m}{2}(k+1)}\right) \tag{1.157}$$

if we take into consideration

$$\sum_{j_1,\ldots,j_k=1}^{l} \sum_{\substack{\{s_1,t_1\},\ldots,\{s_{k/2},t_{k/2}\}\\ \text{separations of } \{1,\ldots,k\}}} \prod_{q=1}^{k/2} {}^2A_1(\tilde{F}_{p_{s_q}j_{s_q}}, \tilde{F}_{p_{t_q}j_{t_q}})$$
$$= \sum_{\substack{\{s_1,t_1\},\ldots,\{s_{k/2},t_{k/2}\}\\ \text{separations of } \{1,\ldots,k\}}} \prod_{q=1}^{k/2} {}^2\tilde{\tilde{A}}_1'(p_{s_q}, p_{t_q}) = {}^2\tilde{\tilde{P}}_1(p_1, \ldots, p_k)$$

and the abbreviation "w.a." means "with arrangement", i.e. the sum is to be taken over all (p, q) from $\mathcal{W}_j$ where the arrangement is important, $(p, q) \neq (q, p)$. Later we use the abbreviation "w.o.a." which says that the choise (p, q) from $\mathcal{W}_j$ equals to (q, p) from $\mathcal{W}_j$. In the case of k even we need (1.155) and obtain

$$\left\langle R_{j,2} R_{j,1}^{k_j-1} \prod_{\substack{p=1 \\ p \neq j}}^{s} R_{p,1}^{k_p} \right\rangle$$
$$= \Bigg[\sum_{p \text{ from } \mathcal{W}_j} {}^3\tilde{A}_2'(j, p)\, {}^2\tilde{\tilde{P}}_1(\mathcal{W}_{jp}) + 2 \sum_{\substack{(p,q) \text{ from } \mathcal{W}_j \\ \text{w.o.a.}}} \sum_{u \text{ from } \mathcal{W}_{jpq}} {}^{32}\tilde{\tilde{A}}_{21}'\big(j(y, z), p(y), q(y), u(z)\big)$$
$$\times\, {}^2\tilde{\tilde{P}}_1(\mathcal{W}_{jpqu}) + \sum_{\substack{(p,q,t) \text{ from } \mathcal{W}_j \\ \text{w.o.a.}}} {}^3\tilde{\tilde{A}}_2'(p, q, t)$$
$$\times \Bigg\{ {}^2\tilde{\tilde{A}}_1'(j)\, {}^2\tilde{\tilde{P}}_1(\mathcal{W}_{jpqt}) + \sum_{\substack{(u,v) \text{ from } \mathcal{W}_{jpqt} \\ \text{w.a.}}} {}^{22}\tilde{\tilde{A}}_{11}'\big(j(y, z), u(y), v(z)\big)$$
$$\times\, {}^2\tilde{\tilde{P}}_1(\mathcal{W}_{jpqtuv}) \Bigg\} \Bigg]\, \varepsilon^{\frac{m}{2}(k+2)} + O\left(\varepsilon^{\frac{m}{2}(k+2)+1}\right) \tag{1.158}$$

where the relation

$${}^{32}\tilde{\tilde{A}}_{21}'\big(j(y, z), p(y), q(y), u(z)\big) = {}^{32}\tilde{\tilde{A}}_{21}'\big(j(y, z), p(z), q(z), u(y)\big)$$

is correct since

$${}^{32}\tilde{\tilde{A}}_{21}'\big(j(y, z), p(y), q(y), u(z)\big) = \sum_{a,b,c,f,g=1}^{s} \tilde{d}_{j,ab} \tilde{d}_{p,c} \tilde{d}_{q,f} \tilde{d}_{u,g}\, {}^3\hat{A}_2'(a, c, f)\, {}^2\hat{A}_1'(b, g).$$

Finally, we summarize the results as to the considered moment

$$\left\langle R_{j,2} R_{j,1}^{k_j-1} \prod_{\substack{p=1 \\ p \neq j}}^{s} R_{p,1}^{k_p} \right\rangle$$
$$= \begin{cases} \tilde{C}_{1j}(k_1, \ldots, k_s)\, \varepsilon^{\frac{m}{2}(k+1)} + O\left(\varepsilon^{\frac{m}{2}(k+1)+1}\right) & \text{for } k \text{ odd,} \\ \tilde{\tilde{C}}_{1j}(k_1, \ldots, k_s)\, \varepsilon^{\frac{m}{2}(k+2)} + o\left(\varepsilon^{\frac{m}{2}(k+2)}\right) & \text{for } k \text{ even} \end{cases} \tag{1.159}$$

where $\tilde{C}_{1j}(k_1, \ldots, k_s)$ is given by (1.157) and $\tilde{\tilde{C}}_{1j}(k_1, \ldots, k_s)$ by (1.158).

Now we deal with

$$\left\langle R_{j,3} R_{j,1}^{k_j-1} \prod_{\substack{p=1 \\ p \neq j}}^{s} R_{p,1}^{k_p} \right\rangle \quad \text{for} \quad \sum_{i=1}^{s} k_i = k \quad \text{even}.$$

For k odd this term has the lowest order $\varepsilon^{\frac{m}{2}(k+3)}$ and terms of this order will later not be used. Hence, terms of the form

$$J_2 = \int_{\tilde{\mathcal{D}}} \cdots \int_{\tilde{\mathcal{D}}} G_{123}(y_1, y_2, y_3)\, G_4(y_4) \ldots G_k(y_k) \left\langle \prod_{q=1}^{k} f_{i_q{}^\varepsilon}(y_q) \right\rangle \mathrm{d}y_1 \ldots \mathrm{d}y_k$$

have to be investigated for k even. The lowest order is obtained for $s = k/2$ considering a representation of J_2 which is analogous to (1.153). In this way J_2 can be written as

$$J_2 = \Bigg[\sum_{p=4}^{k} \{{}^{22}A_{11}(G_{123}(y, y, z)\, G_p(z)) + {}^{22}A_{11}(G_{123}(y, z, y)\, G_p(z))$$

$$+ {}^{22}A_{11}(G_{123}(z, y, y)\, G_p(z))\}\, {}^2P_1(\{4, \ldots, k\} \setminus \{p\})$$

$$+ \sum_{\substack{\{p_1,p_2,p_3\} \subset \{4,\ldots,k\} \\ \text{w.a.}}} {}^{222}A_{111}(G_{123}(x, y, z)\, G_{p_1}(x)\, G_{p_2}(y)\, G_{p_3}(z))$$

$$\times {}^2P_1(\{4, \ldots, k\} \setminus \{p_1, p_2, p_3\})\Bigg]\, \varepsilon^{mk/2} + o(\varepsilon^{mk/2})$$

since we have to deal with J_2 only in the case of

$$G_{123}(y_1, y_2, y_3) = \sum_{a,b,c=1}^{n} d_{abc} H_a(y_1)\, K_b(y_2)\, L_c(y_3).$$

Hence, we notice

$${}^{22}A_{11}(G_{123}(y, y, z)\, G_p(z)) = \sum_{a,b,c=1}^{n} d_{abc}\, {}^2A_1(H_a, K_b)\, {}^2A_1(L_c, G_p),$$

$${}^{222}A_{111}(G_{123}(x, y, z)\, G_p(x)\, G_q(y)\, G_t(z))$$

$$= \sum_{a,b,c=1}^{n} d_{abc}\, {}^2A_1(H_a, G_p)\, {}^2A_1(K_b, G_q)\, {}^2A_1(L_c, G_t).$$

Defining the function

$$\tilde{F}_{i,j_1j_2j_3}(x, y, z) = \sum_{a,b,c=1}^{n} \tilde{d}_{i,abc} F_{aj_1}(x)\, F_{bj_2}(y)\, F_{cj_3}(z),$$

so that

$$R_{i,3} = \sum_{a,b,c=1}^{n} \tilde{d}_{i,abc} r_{a\varepsilon} r_{b\varepsilon} r_{c\varepsilon}$$

$$= \sum_{j_1,j_2,j_3=1}^{l} \int_{\tilde{\mathcal{D}}} \int_{\tilde{\mathcal{D}}} \int_{\tilde{\mathcal{D}}} \tilde{F}_{i,j_1j_2j_3}(x, y, z)\, f_{j_1\varepsilon}(x)\, f_{j_2\varepsilon}(y)\, f_{j_3\varepsilon}(z)\, \mathrm{d}x\, \mathrm{d}y\, \mathrm{d}z,$$

we see that

$$\left\langle R_{j,3} R_{j,1}^{k_j-1} \prod_{\substack{p=1 \\ p\neq j}}^{s} R_{p,1}^{k_p} \right\rangle$$

$$= \sum_{j_1,\ldots,j_{k+2}=1}^{l} \int_{\tilde{\mathcal{D}}} \cdots \int_{\tilde{\mathcal{D}}} \tilde{F}_{j,j_1j_2j_3} \prod_{\substack{q=4 \\ (i_4,\ldots,i_{k+2}) = \mathcal{W}_j}}^{k+2} \tilde{F}_{i_q,j_q}(y_q) \left\langle \prod_{q=1}^{k+2} f_{j_q\varepsilon}(y_q) \right\rangle \mathrm{d}y_1 \ldots \mathrm{d}y_{k+2}$$

$$= \Bigg[3 \sum_{p \text{ from } \mathcal{W}_j} {}^{22}\tilde{\tilde{A}}{}'_{11}(j(y, y, z), p(z))\, {}^2\tilde{\tilde{P}}_1(\mathcal{W}_{jp})$$

$$+ \sum_{\substack{(p,q,t) \text{ from } \mathcal{W}_j \\ \text{w.a.}}} {}^{222}\tilde{\tilde{A}}{}'_{111}(j(x, y, z), p(x), q(y), t(z))\, {}^2\tilde{\tilde{P}}_1(\mathcal{W}_{jpqt})\Bigg]\, \varepsilon^{\frac{m}{2}(k+2)} + o\left(\varepsilon^{\frac{m}{2}(k+2)}\right). \tag{1.160}$$

For the later consideration this moment is written in the form

$$\left\langle R_{j,3} R_{j,1}^{k_j-1} \prod_{\substack{p=1\\p\neq j}}^{s} R_{p,1}^{k_p} \right\rangle = \tilde{C}_{2j}(k_1, \ldots, k_s)\, \varepsilon^{\frac{m}{2}(k+2)} + o\left(\varepsilon^{\frac{m}{2}(k+2)}\right) \tag{1.161}$$

for k even and $\tilde{C}_{2j}(k_1, \ldots, k_s)$ is given by the right hand side of (1.160).

The expansion of

$$\left\langle R_{j,2}^2 R_{j,1}^{k_j-2} \prod_{\substack{p=1\\p\neq j}}^{s} R_{p,1}^{k_p} \right\rangle \quad \text{and} \quad \left\langle R_{j,2} R_{i,2} R_{j,1}^{k_j-1} R_{i,1}^{k_i-1} \prod_{\substack{p=1\\p\neq j,i}}^{s} R_{p,1}^{k_p} \right\rangle$$

for k even demands the consideration of

$$J_3 = \int_{\tilde{D}} \cdots \int_{\tilde{D}} G_{12}(y_1, y_2)\, G_{34}(y_3, y_4)\, G_5(y_5) \ldots G_k(y_k) \left\langle \prod_{q=1}^{k} f_{i_q\varepsilon}(y_q) \right\rangle \mathrm{d}y_1 \ldots \mathrm{d}y_k$$

where a separation in pairs leads to the lowest order as to ε. For this, the following separations with respect to $\{1, 2, 3, 4\}$ have to be taken into consideration:

$$\begin{array}{ll}
\{\{1,2\},\{3,4\}\}, & \{\{1,4\},\{2,3\}\},\\
\{\{1,2\},\{3,p_1\},\{4,p_2\}\}, & \{\{1,4\},\{2,p_1\},\{3,p_2\}\},\\
\{\{1,3\},\{2,4\}\}, & \{\{1,p_1\},\{2,3\},\{4,p_2\}\},\\
\{\{1,3\},\{2,p_1\},\{4,p_2\}\}, & \{\{1,p_1\},\{2,4\},\{3,p_2\}\},\\
\{\{1,p_1\},\{2,p_2\},\{3,4\}\}, & \{\{1,p_1\},\{2,p_2\},\{3,p_3\},\{4,p_4\}\}.
\end{array}$$

Then, similarly to the case of J_1 and J_2 we obtain

$$\begin{aligned}
J_3 = \Big[& \big\{{}^2A_1(G_{12})\, {}^2A_1(G_{34}) + {}^{22}A_{11}\big(G_{12}(y, z)\, G_{34}(y, z)\big)\\
& + {}^{22}A_{11}\big(G_{12}(y, z)\, G_{34}(z, y)\big)\big\}\, {}^2P_1(\{5, \ldots, k\})\\
& + \sum_{\substack{\{p_1, p_2\} \in \{5, \ldots, k\}\\ \text{w.a.}}} \big\{{}^2A_1(G_{12})\, {}^{22}A_{11}\big(G_{34}(y, z)\, G_{p_1}(y)\, G_{p_2}(z)\big)\\
& + {}^2A_1(G_{34})\, {}^{22}A_{11}\big(G_{12}(y, z)\, G_{p_1}(y)\, G_{p_2}(z)\big)\\
& + {}^{222}A_{111}\big(G_{12}(x, y)\, G_{34}(x, z)\, G_{p_1}(y)\, G_{p_2}(z)\big)\\
& + {}^{222}A_{111}\big(G_{12}(x, y)\, G_{34}(z, x)\, G_{p_1}(y)\, G_{p_2}(z)\big)\\
& + {}^{222}A_{111}\big(G_{12}(x, y)\, G_{34}(y, z)\, G_{p_1}(x)\, G_{p_2}(z)\big)\\
& + {}^{222}A_{111}\big(G_{12}(x, y)\, G_{34}(z, y)\, G_{p_1}(x)\, G_{p_2}(z)\big)\big\}\, {}^2P_1(\{5, \ldots, k\} \setminus \{p_1, p_2\})\\
& + \sum_{\{p_1, p_2, p_3, p_4\} \in \{5, \ldots, k\}} {}^{2222}A_{1111}\big(G_{12}(x, y)\, G_{34}(z, v)\, G_{p_1}(x)\, G_{p_2}(y)\, G_{p_3}(z)\, G_{p_4}(v)\big)\\
& \times {}^2P_1(\{5, \ldots, k\} \setminus \{p_1, p_2, p_3, p_4\})\Big]\, \varepsilon^{\frac{m}{2}k} + o\left(\varepsilon^{\frac{m}{2}k}\right).
\end{aligned}$$

The definition of ${}^{2222}A_{1111}$ is clear taking functions G_{12}, G_{34} as given above.

Then, as in the foregoing cases it follows

$$\left\langle R_{j,2}^2 R_{j,1}^{k_j-2} \prod_{\substack{p=1\\p\neq j}}^{s} R_{p,1}^{k_p} \right\rangle$$

$$= \sum_{j_1,\ldots,j_{k+2}=1}^{l} \int_{\tilde{\mathcal{D}}} \cdots \int_{\tilde{\mathcal{D}}} \tilde{F}_{j,j_1j_2}(y_1, y_2)\, \tilde{F}_{j,j_3j_4}(y_3, y_4)$$

$$\times \prod_{\substack{q=5\\(i_5,\ldots,i_{k+2})=\mathscr{W}_{jj}}}^{k+2} \tilde{F}_{i_q,j_q}(y_q) \left\langle \prod_{q=1}^{k+2} f_{j_q\varepsilon}(y_q) \right\rangle \mathrm{d}y_1 \ldots \mathrm{d}y_{k+2}$$

$$= \Bigg[\left\{{}^2\tilde{\tilde{A}}_1'(j)^2 + 2\,{}^{22}\tilde{\tilde{A}}_{11}'\big(j(y,z), j(y,z)\big)\right\} {}^2\tilde{\tilde{P}}_1(\mathscr{W}_{jj})$$

$$+ \sum_{\substack{(p,q)\text{ from }\mathscr{W}_{jj}\\\text{w.a.}}} \left\{2\,{}^2\tilde{\tilde{A}}_1'(j)\,{}^{22}\tilde{\tilde{A}}_{11}'\big(j(y,z), p(y), q(z)\big)\right.$$

$$\left. + 4\,{}^{222}\tilde{\tilde{A}}_{111}'\big(j(x,y), j(z,y), p(x), q(z)\big)\right\} {}^2\tilde{\tilde{P}}_1(\mathscr{W}_{jjpq})$$

$$+ \sum_{\substack{(p,q,t,r)\text{ from }\mathscr{W}_{jj}\\\text{w.a.}}} {}^{2222}\tilde{\tilde{A}}_{1111}'\big(j(x,y), j(z,v), p(x), q(y), t(z), r(v)\big)$$

$$\times {}^2\tilde{\tilde{P}}_1(\mathscr{W}_{jjpqtr})\Bigg]\, \varepsilon^{\frac{m}{2}(k+2)} + o\left(\varepsilon^{\frac{m}{2}(k+2)}\right)$$

and

$$\left\langle R_{j,2} R_{i,2} R_{j,1}^{k_j-1} R_{i,1}^{k_i-1} \prod_{\substack{p=1\\p\neq i,j}}^{s} R_{p,1}^{k_p} \right\rangle$$

$$= \sum_{j_1,\ldots,j_{k+2}=1}^{l} \int_{\tilde{\mathcal{D}}} \cdots \int_{\tilde{\mathcal{D}}} \tilde{F}_{j,j_1j_2}(y_1, y_2)\, \tilde{F}_{i,j_3j_4}(y_3, y_4)$$

$$\times \prod_{\substack{q=5\\(i_5,\ldots,i_{k+2})=\mathscr{W}_{ji}}}^{k+2} \tilde{F}_{i_q,j_q}(y_q) \left\langle \prod_{q=1}^{k+2} f_{j_q\varepsilon}(y_q) \right\rangle \mathrm{d}y_1 \ldots \mathrm{d}y_{k+2}$$

$$= \Bigg[\left\{{}^2\tilde{\tilde{A}}_1'(j)\,{}^2\tilde{\tilde{A}}_1'(i) + 2\,{}^{22}\tilde{\tilde{A}}_{11}'\big(j(y,z), i(y,z)\big)\right\} {}^2\tilde{\tilde{P}}_1(\mathscr{W}_{ji})$$

$$+ \sum_{\substack{(p,q)\text{ from }\mathscr{W}_{ji}\\\text{w.a.}}} \left\{{}^2\tilde{\tilde{A}}_1'(j)\,{}^{22}\tilde{\tilde{A}}_{11}'\big(i(y,z), p(y), q(z)\big) + {}^2\tilde{\tilde{A}}_1'(i)\,{}^{22}\tilde{\tilde{A}}_{11}'\big(j(y,z), p(y), q(z)\big)\right]$$

$$\left. + 4\,{}^{222}\tilde{\tilde{A}}_{111}'\big(j(x,y), i(x,z), p(y), q(z)\big)\right\} {}^2\tilde{\tilde{P}}_1(\mathscr{W}_{jipq})$$

$$+ \sum_{\substack{(p,q,t,r)\text{ from }\mathscr{W}_{ji}\\\text{w.a.}}} {}^{2222}\tilde{\tilde{A}}_{1111}'\big(j(x,y), i(z,v), p(x), q(y), t(z), r(v)\big)$$

$$\times {}^2\tilde{\tilde{P}}_1(\mathscr{W}_{jipqtr})\Bigg]\, \varepsilon^{\frac{m}{2}(k+2)} + o\left(\varepsilon^{\frac{m}{2}(k+2)}\right). \tag{1.162}$$

Finally we put

$$\left\langle R_{j,2} R_{i,2} R_{j,1}^{k_j-1} R_{i,1}^{k_i-1} \prod_{\substack{p=1\\p\neq i,j}}^{s} R_{p,1}^{k_p} \right\rangle$$

$$= \tilde{C}_{3ji}(k_1, \ldots, k_s)\, \varepsilon^{m(k+2)/2} + o(\varepsilon^{m(k+2)/2}) \quad \text{for} \quad k \text{ even} \tag{1.163}$$

where $\tilde{C}_{3ji}$ is defined by the right hand side of (1.162).

Starting from (1.149) a summary shows

$$\left\langle \prod_{i=1}^{s} \tilde{d}_i^{k_i} \right\rangle = \begin{cases} \tilde{C}_0^0(k_1, \ldots, k_s) + \tilde{C}_0^1(k_1, \ldots, k_s)\,\varepsilon + o(\varepsilon) \\ + \left[\sum_{j=1}^{s} k_j \left(\bar{\bar{C}}_{1j}(k_1, \ldots, k_s) + \tilde{C}_{2j}(k_1, \ldots, k_s)\right) \right. \\ + \sum_{j=1}^{s} \binom{k_j}{2} \tilde{C}_{3jj}(k_1, \ldots, k_s) \\ \left. + \sum_{\substack{j,i=1 \\ j<i}}^{s} k_j k_i \tilde{C}_{3ji}(k_1, \ldots, k_s) \right] \varepsilon^m + o(\varepsilon^m) \quad \text{for} \quad k \text{ even}, \\ \left[\bar{\bar{C}}_0^0(k_1, \ldots, k_s) + \sum_{j=1}^{s} k_j \tilde{C}_{1j}(k_1, \ldots, k_s) \right] \varepsilon^{\frac{m}{2}} + O\left(\varepsilon^{\frac{m}{2}+1}\right) \\ \text{for} \quad k \text{ odd}. \end{cases} \tag{1.164}$$

We see that

$$\tilde{B}_{pq} = \sum_{a,b=1}^{n} d_{p,a} d_{q,b} B_{ab} = \sum_{a,b=1}^{n} d_{p,a} d_{q,b} \sum_{j_1,j_2=1}^{l} {}^2A_1(F_{aj_1}, F_{bj_2}) = \sum_{a,b=1}^{n} d_{p,a} d_{q,b}\, {}^2\hat{A}_1'(a, b)$$

and then with the aid of (1.145)

$$\delta_{kh} = \sum_{p=1}^{k} \sum_{q=1}^{h} b_{kp} b_{hq} \tilde{B}_{pq} = \sum_{p=1}^{k} \sum_{q=1}^{h} b_{kp} b_{hq} \sum_{a,b=1}^{n} d_{p,a} d_{q,b}\, {}^2\hat{A}_1'(a, b) = {}^2\tilde{\bar{A}}_1'(k, h). \tag{1.165}$$

We want to consider some special cases. Before, the terms $\tilde{C}_0^0$, $\tilde{C}_0^1$, and $\bar{\bar{C}}_0^0$ are to be written. Using (1.30) and (1.24) it follows

$$\begin{aligned} \tilde{C}_0^0(k_1, \ldots, k_s) &= \tilde{\bar{H}}_0(m; i_1, \ldots, i_k) = \sum_{j_1,\ldots,j_k=1}^{l} \tilde{H}_0(m; i_1, \ldots, i_k; j_1, \ldots, j_k) \\ &= \sum_{j_1,\ldots,j_k=1}^{l} \sum_{\{s_1,t_1\},\ldots,\{s_{k/2},t_{k/2}\}} \prod_{q=1}^{k/2} {}^2A_1\left(\tilde{F}_{i_{s_q} j_{s_q}}, \tilde{F}_{i_{t_q} j_{t_q}}\right) \\ &= \sum_{\{s_1,t_1\},\ldots,\{s_{k/2},t_{k/2}\}} \prod_{q=1}^{k/2} {}^2\tilde{\bar{A}}_1'(i_{s_q}, i_{t_q}) \\ &= {}^2\tilde{\bar{P}}_1(\mathcal{V}) = \prod_{p=1}^{s} e_{k_p} \quad \text{for} \quad k \geqq 2 \quad \text{and} \quad k \text{ even} \end{aligned} \tag{1.166}$$

where $\mathcal{V}$ is defined by $\mathcal{V} \doteq \mathcal{V}_{k_1 \ldots k_s}$ and e_q by

$$e_q = \begin{cases} \dfrac{q!}{2^{q/2}(q/2)!} & \text{for } q \text{ even}, \\ 0 & \text{for } q \text{ odd}. \end{cases}$$

With the aid of (1.30) and (1.25) we have

$$\tilde{C}_0^1(k_1, \ldots, k_s) = \tilde{\bar{H}}_2(m; i_1, \ldots, i_k) = \sum_{j_1,\ldots,j_k=1}^{l} \tilde{H}_2(m; i_1, \ldots, i_k; j_1, \ldots, j_k)$$

and then applying (1.65) and (1.69) for k even

$$\tilde{C}_0^1(k_1, \ldots, k_s) = \sum_{\substack{(p,q)\text{ from } \mathcal{V} \\ \text{w.o.a.}}} {}^2\tilde{\tilde{A}}_2'(p, q)\, {}^2\tilde{\tilde{P}}_1(\mathcal{W}_{pq}) + \sum_{\substack{(p,q,t,s)\text{ from } \mathcal{V} \\ \text{w.o.a.}}} {}^4\underline{\tilde{\tilde{A}}}_3'(p, q, t, s)\, {}^2\tilde{\tilde{P}}_1(\mathcal{W}_{pqts})$$
$$+ \frac{1}{2} \sum_{\substack{(p,q,t)\text{ from } \mathcal{V} \\ \text{w.o.a.}}} \sum_{\substack{(u,v,r)\text{ from } \mathcal{W}_{pqt} \\ \text{w.o.a.}}} {}^3\tilde{\tilde{A}}_2'(p, q, t)\, {}^3\tilde{\tilde{A}}_2'(u, v, r)$$
$$\times\, {}^2\tilde{\tilde{P}}_1(\mathcal{W}_{pqtuvr}) \quad \text{for} \quad m = 1,$$
$$\tilde{C}_0^1(k_1, \ldots, k_s) = \sum_{\substack{(p,q)\text{ from } \mathcal{V} \\ \text{w.o.a.}}} {}^2\tilde{\tilde{A}}_2'(p, q)\, {}^2\tilde{\tilde{P}}_1(\mathcal{W}_{pq}) \quad \text{for} \quad m \geqq 2. \tag{1.167}$$

By means of (1.27) we have

$$\bar{\bar{C}}_0^0(k_1, \ldots, k_s) = \hat{\hat{H}}_1(m; i_1, \ldots, i_k) = \sum_{j_1,\ldots,j_k=1}^{l} H_1(m; i_1, \ldots, i_k; j_1, \ldots, j_k)$$
$$= \sum_{\substack{(p,q,t)\text{ from } \mathcal{V} \\ \text{w.o.a.}}} {}^3\tilde{\tilde{A}}_2'(p, q, t)\, {}^2\tilde{\tilde{P}}_1(\mathcal{W}_{pqt}) \quad \text{for} \quad k \text{ odd}. \tag{1.168}$$

We deal with the case of one function, $s = 1$. Then, from the above considerations we can obtain

$$\tilde{C}_0^0(k) = e_k \quad \text{for} \quad k \text{ even}, \qquad \bar{\bar{C}}_0^0(k) = e_{k-3}\, {}^3\tilde{\tilde{A}}_2'(1, 1, 1) \quad \text{for} \quad k \text{ odd},$$

$$\tilde{C}_0^1(k) = \begin{cases} e_k \left[\frac{k}{2}\, {}^2\tilde{\tilde{A}}_2'(1, 1) + \frac{1}{9}\frac{k}{2}\left(\frac{k}{2} - 1\right)\left(\frac{k}{2} - 2\right)\left({}^3\tilde{\tilde{A}}_2'(1, 1, 1)\right)^2 \right. \\ \left. + \frac{1}{6}\frac{k}{2}\left(\frac{k}{2} - 1\right) {}^4\underline{\tilde{\tilde{A}}}_3'(1, 1, 1, 1)\right] & \text{for} \quad m = 1, k \text{ even}, \\ e_k \frac{k}{2}\, {}^2\tilde{\tilde{A}}_2'(1, 1) & \text{for} \quad m \geqq 2, k \text{ even}, \end{cases}$$

$$\tilde{C}_{11}(k) = e_{k-1}\left[{}^2\tilde{\tilde{A}}_1'(1) + (k-1)\, {}^{22}\tilde{\tilde{A}}_{11}'\big(1(y, z), 1(y), 1(z)\big)\right] \quad \text{for} \quad k \text{ odd},$$

$$\bar{\bar{C}}_{11}(k) = e_{k-2}(k-1)\left[{}^3\tilde{\tilde{A}}_2'(1, 1) + (k-2)\, {}^{32}\tilde{\tilde{A}}_{21}'\big(1(y, z), 1(y), 1(y), 1(z)\big)\right.$$
$$+ \frac{1}{6}(k-2)\, {}^3\tilde{\tilde{A}}_2'(1, 1, 1)\, {}^2\tilde{\tilde{A}}_1'(1) + \frac{2}{3}\left(\frac{k}{2} - 1\right)\left(\frac{k}{2} - 2\right)$$
$$\left.\times\, {}^3\tilde{\tilde{A}}_2'(1, 1, 1)\, {}^{22}\tilde{\tilde{A}}_{11}'\big(1(y, z), 1(y), 1(z)\big)\right] \quad \text{for} \quad k \text{ even},$$

$$\tilde{C}_{21}(k) = e_{k-2}(k-1)\left[3\, {}^{22}\tilde{\tilde{A}}_{11}'\big(1(y, y, z), 1(z)\big)\right.$$
$$\left.+ (k-2)\, {}^{222}\tilde{\tilde{A}}_{111}'\big(1(x, y, z). 1(x), 1(y), 1(z)\big)\right] \quad \text{for} \quad k \text{ even},$$

$$\tilde{C}_{311}(k) = e_{k-2}\left[\left({}^2\tilde{\tilde{A}}_1'(1)\right)^2 + 2\, {}^{22}\tilde{\tilde{A}}_{11}'\big(1(y, z), 1(y, z)\big)\right.$$
$$+ (k-2)\left(2\, {}^2\tilde{\tilde{A}}_1'(1)\, {}^{22}\tilde{\tilde{A}}_{11}'\big(1(y, z), 1(y), 1(z)\big)\right.$$
$$\left.+ 4\, {}^{222}\tilde{\tilde{A}}_{111}'\big(1(x, y), 1(z, y), 1(x), 1(z)\big)\right)$$
$$\left.+ (k-2)(k-4)\, {}^{2222}\tilde{\tilde{A}}_{1111}'\big(1(x, y), 1(z, v), 1(x), 1(y), 1(z), 1(v)\big)\right]$$

for k even.

With these preparations it follows

$$\langle \tilde{d}^k \rangle = \begin{cases} \tilde{C}_0^0(k) + \Big[\tilde{C}_0^1(k) + k\{\bar{\tilde{C}}_{11}(k) + \tilde{C}_{21}(k)\} \\ \quad + \dfrac{1}{2}\, k(k-1)\, \tilde{C}_{311}(k)\Big]\, \varepsilon + o(\varepsilon) & \text{for } k \text{ even}, \\ & \qquad m = 1, \\ [\bar{\tilde{C}}_0^0(k) + k\tilde{C}_{11}(k)]\, \sqrt{\varepsilon} + O(\varepsilon^{3/2}) & \text{for } k \text{ odd}, \\ \tilde{C}_0^0(k) + \tilde{C}_0^1(k)\, \varepsilon + o(\varepsilon) & \text{for } k \text{ even}, \\ & \qquad m \geqq 2, \\ [\bar{\tilde{C}}_0^0(k) + k\tilde{C}_{11}(k)]\, \sqrt{\varepsilon}^m + O(\varepsilon^{m/2+1}) & \text{for } k \text{ odd}, \end{cases}$$

and then for $k = 1$

$$\langle \tilde{d} \rangle = \tilde{C}_{11}(1)\, \sqrt{\varepsilon}^m + o\big(\sqrt{\varepsilon}^m\big) = {}^2\tilde{A}_1'(1)\, \sqrt{\varepsilon}^m + O\left(\varepsilon^{\frac{m}{2}+1}\right)$$

and for $k = 2$

$$\begin{aligned} \langle \tilde{d}^2 \rangle &= \tilde{C}_0^0(2) + [\tilde{C}_0^1(2) + \{2\bar{\tilde{C}}_{11}(2) + 2\tilde{C}_{21}(2) + \tilde{C}_{311}(2)\}\, \delta_{1m}]\, \varepsilon + o(\varepsilon) \\ &= 1 + \big[{}^2\tilde{A}_2'(1, 1) + \big\{2\,\big({}^3\tilde{A}_2'(1, 1) + 3\, {}^{22}\tilde{A}_{11}'(1(y, y, z), 1(z))\big) \\ &\quad + \big({}^2\tilde{A}_1'(1)\big)^2 + 2\, {}^{22}\tilde{A}_{11}'\big(1(y, z), 1(y, z)\big)\big\}\, \delta_{1m}\big]\, \varepsilon + o(\varepsilon). \end{aligned}$$

Use the transformation formulas

$$\tilde{d} = \frac{1}{\sqrt{\varepsilon}^m}\, b_{11}(d - d_{,0}), \qquad b_{11} = \frac{1}{\sqrt{\tilde{B}_{11}}}, \qquad \tilde{B}_{11} = \sum_{a,b=1}^{n} d_{,a} d_{,b} B_{ab},$$

$$B_{ab} = \sum_{j_1, j_2 = 1}^{l} {}^2A_1(F_{aj_1}, F_{bj_2}),$$

$$\tilde{F}_{,j}(x) = \sum_{a=1}^{n} \tilde{d}_{,a} F_{aj}(x),$$

$$\tilde{F}_{,j_1 j_2}(x, y) = \sum_{a,b=1}^{n} \tilde{d}_{,ab} F_{aj_1}(x)\, F_{bj_2}(y),$$

$$\tilde{F}_{,j_1 j_2 j_3}(x, y, z) = \sum_{a,b,c=1}^{n} \tilde{d}_{,abc} F_{aj_1}(x)\, F_{bj_2}(y)\, F_{cj_3}(z),$$

$$\tilde{d}_{,a} = b_{11} d_{,a}, \qquad \tilde{d}_{,ab} = b_{11} d_{,ab}, \qquad \tilde{d}_{,abc} = b_{11} d_{,abc}.$$

Hence, we can write

$$\langle d - d_{,0} \rangle = \sum_{a,b=1}^{n} d_{,ab}\, {}^2\hat{A}_1'(a, b)\, \varepsilon^m + o(\varepsilon^m) \tag{1.169}$$

where

$${}^2\hat{A}_1'(a, b) = \sum_{j_1, j_2 = 1}^{l} {}^2A_1(F_{aj_1}, F_{bj_2}),$$

and

$$\langle (d - d_{,0})^2 \rangle = \sum_{a,b=1}^{n} d_{,a} d_{,b}\, {}^2\hat{A}_1'(a, b)\, \varepsilon^m + \Big[\sum_{a,b=1}^{n} d_{,a} d_{,b}\, {}^2\hat{A}_2'(a, b)$$

$$+ \left\{2 \sum_{a,b,c=1}^{n} d_{,ab} d_{,c}\, {}^3\hat{A}_2'(a, b, c) + \left(\sum_{a,b=1}^{n} d_{,ab}\, {}^2\hat{A}_1'(a, b)\right)^2 \right.$$

$$+ 6 \sum_{a,b,c,f=1}^{n} d_{,abc} d_{,f}\, {}^2\hat{A}_1'(a, b)\, {}^2\hat{A}_1'(c, f)$$

$$\left. \left. + 2 \sum_{a,b,c,f=1}^{n} d_{,ab} d_{,cf}\, {}^2\hat{A}_1'(a, c)\, {}^2\hat{A}_1'(b, f)\right\} \delta_{1m} \right] \varepsilon^{m+1} + o(\varepsilon^{m+1}) \qquad (1.170)$$

where we have applied

$${}^2\tilde{A}_2'(1, 1) = b_{11}^2 \sum_{a,b=1}^{n} d_{,a} d_{,b}\, {}^2\hat{A}_2'(a, b),$$

$${}^3\tilde{A}_2'(1, 1) = b_{11}^2 \sum_{a,b,c=1}^{n} d_{,ab} d_{,c}\, {}^3\hat{A}_2'(a, b, c),$$

$${}^{22}\tilde{A}_{11}'\big(1(y, y, z), 1(z)\big) = b_{11}^2 \sum_{a,b,c,f=1}^{n} d_{,abc} d_{,f}\, {}^2\hat{A}_1'(a, b)\, {}^2\hat{A}_1'(c, f),$$

$${}^2\tilde{A}_1'(1) = b_{11} \sum_{a,b=1}^{n} d_{,ab}\, {}^2\hat{A}_1'(a, b),$$

$${}^{22}\tilde{A}_{11}'\big(1(y, z), 1(y, z)\big) = b_{11}^2 \sum_{a,b,c,f=1}^{n} d_{,ab} d_{,cf}\, {}^2\hat{A}_1'(a, c)\, {}^2\hat{A}_1'(b, f).$$

We still want to investigate the case of two functions, $s = 2$. Using the formulas deduced we have

$$\tilde{C}_0^0(k_1, k_2) = e_{k_1} e_{k_2},$$

$$\bar{\tilde{C}}_0^0(k_1, k_2) = \binom{k_1}{3} e_{k_1-3} e_{k_2}\, {}^3\tilde{A}_2'(1, 1, 1) + \binom{k_2}{3} e_{k_1} e_{k_2-3}\, {}^3\tilde{A}_2'(2, 2, 2)$$

$$+ \binom{k_1}{2} k_2 e_{k_1-2} e_{k_2-1}\, {}^3\tilde{A}_2'(1, 1, 2) + k_1 \binom{k_2}{2} e_{k_1-1} e_{k_2-2}\, {}^3\tilde{A}_2'(1, 2, 2),$$

$$\tilde{C}_0^1(k_1, k_2) = \binom{k_1}{2} e_{k_1-2} e_{k_2}\, {}^2\tilde{A}_2'(1, 1) + k_1 k_2 e_{k_1-1} e_{k_2-1}\, {}^2\tilde{A}_2'(1, 2)$$

$$+ \binom{k_2}{2} e_{k_1} e_{k_2-2}\, {}^2\tilde{A}_2'(2, 2) + \left\{\binom{k_1}{4} e_{k_1-4} e_{k_2}\, {}^4\underline{\tilde{A}}_3'(1, 1, 1, 1)\right.$$

$$+ \binom{k_1}{3} k_2 e_{k_1-3} e_{k_2-1}\, {}^4\underline{\tilde{A}}_3'(1, 1, 1, 2) + \binom{k_1}{2}\binom{k_2}{2} e_{k_1-2} e_{k_2-2}\, {}^4\underline{\tilde{A}}_3'(1, 1, 2, 2)$$

$$+ k_1 \binom{k_2}{3} e_{k_1-1} e_{k_2-3}\, {}^4\underline{\tilde{A}}_3'(1, 2, 2, 2) + \binom{k_2}{4} e_{k_1} e_{k_2-4}\, {}^4\underline{\tilde{A}}_3'(2, 2, 2, 2)$$

$$+ \binom{k_1}{3} {}^3\tilde{A}_2'(1, 1, 1) \left[\binom{k_1 - 3}{3} e_{k_1-6} e_{k_2}\, {}^3\tilde{A}_2'(1, 1, 1)\right.$$

$$+ 2 \binom{k_1 - 3}{2} k_2 e_{k_1-5} e_{k_2-1}\, {}^3\tilde{A}_2'(1, 1, 2)$$

$$+ 2\,(k_1 - 3) \binom{k_2}{2} e_{k_1-4} e_{k_2-2}\, {}^3\tilde{A}_2'(1, 2, 2)$$

$$+ 2\binom{k_2}{3} e_{k_1-3}e_{k_2-3}\, {}^3\tilde{\bar{A}}_2'(2, 2, 2)\Big]$$

$$+ \binom{k_1}{2} k_2\, {}^3\tilde{\bar{A}}_2'(1, 1, 2) \Big[\binom{k_1-2}{2}(k_2-1)\, e_{k_1-4}e_{k_2-2}\, {}^3\tilde{A}_2'(1, 1, 2)$$

$$+ 2(k_1-2)\binom{k_2-1}{2} e_{k_1-3}e_{k_2-3}\, {}^3\tilde{\bar{A}}_2'(1, 2, 2)$$

$$+ 2\binom{k_2-1}{3} e_{k_1-2}e_{k_2-4}\, {}^3\tilde{\bar{A}}_2'(2, 2, 2)\Big]$$

$$+ k_1\binom{k_2}{2}\, {}^3\tilde{\bar{A}}_2'(1, 2, 2)\Big[(k_1-1)\binom{k_2-2}{2} e_{k_1-2}e_{k_2-4}\, {}^3\tilde{\bar{A}}_2(1, 2, 2)$$

$$+ 2\binom{k_2-2}{3} e_{k_1-1}e_{k_2-5}\, {}^3\tilde{\bar{A}}_2'(2, 2, 2)\Big]$$

$$+ \binom{k_2}{3}\binom{k_2-3}{3} e_{k_1}e_{k_2-6}\left({}^3\tilde{\bar{A}}_2'(2, 2, 2)\right)^2\Big\}\, \delta_{1m},$$

$$\tilde{C}_{1j}(k_1, k_2) = e_{k_j-1}e_{k_p}\, {}^2\tilde{\bar{A}}_1'(j)$$
$$+ (k_j-1)(k_j-2)\, e_{k_j-3}e_{k_p}\, {}^{22}\tilde{\bar{A}}_{11}'\big(j(y, z), j(y), j(z)\big)$$
$$+ k_p(k_p-1)\, e_{k_j-1}e_{k_p-2}\, {}^{22}\tilde{\bar{A}}_{11}'\big(j(y, z), p(y), p(z)\big)$$
$$+ 2(k_j-1)\, k_p e_{k_j-2}e_{k_p-1}\, {}^{22}\tilde{\bar{A}}_{11}'\big(j(y, z), j(y), p(z)\big)$$

with $p = 2$ for $j = 1$, $p = 1$ for $j = 2$,

$$\bar{\bar{C}}_{1j}(k_1, k_2) = (k_j-1)\, e_{k_j-2}e_{k_p}\, {}^3\tilde{\bar{A}}_2'(j, j) + k_p e_{k_j-1}e_{k_p-1}\, {}^3\tilde{\bar{A}}_2'(j, p)$$
$$+ 2(k_j-1)(k_j-2)\Big[e_{k_j-4}e_{k_p}\, {}^{32}\tilde{\bar{A}}_{21}'\big(j(y, z), j(y), j(y), j(z)\big)$$
$$+ e_{k_j-3}e_{k_p-1}\, {}^{32}\tilde{\bar{A}}_{21}'\big(j(y, z), j(y), j(y), p(z)\big)\Big]$$
$$+ 2(k_j-1)\, k_p\Big[e_{k_j-3}e_{k_p-1}\, {}^{32}\tilde{\bar{A}}_{21}'\big(j(y, z), j(y), p(y), j(z)\big)$$
$$+ e_{k_j-2}e_{k_p-2}\, {}^{32}\tilde{\bar{A}}_{21}'\big(j(y, z), j(y), p(y), p(z)\big)\Big]$$
$$+ 2k_p(k_p-1)\Big[e_{k_j-2}e_{k_p-2}\, {}^{32}\tilde{\bar{A}}_{21}'\big(j(y, z), p(y), p(y), j(z)\big)$$
$$+ e_{k_j-1}e_{k_p-3}\, {}^{32}\tilde{\bar{A}}_{21}'\big(j(y, z), p(y), p(y), p(z)\big)\Big]$$
$$+ \sum_{q=0}^{3}\binom{k_j-1}{3-q}\binom{k_p}{q}\, {}^3\tilde{\bar{A}}_2(\mathcal{N}_q)\Big[e_{k_j-4+q}e_{k_p-q}\, {}^2\tilde{\bar{A}}_1'(j)$$
$$+ (k_j-4+q)(k_j-5+q)\, e_{k_j-6+q}e_{k_p-q}\, {}^{22}\tilde{\bar{A}}_{11}'\big(j(y, z), j(y), j(z)\big)$$
$$+ (k_j-4+q)(k_p-q)\, e_{k_j-5+q}e_{k_p-1-q}$$
$$\times \left\{{}^{22}\tilde{\bar{A}}_{11}'\big(j(y, z), j(y), p(z)\big) + {}^{22}\tilde{\bar{A}}_{11}'\big(j(y, z), p(y), j(z)\big)\right\}$$
$$+ (k_j-q)(k_j-1-q)\, e_{k_j-4+q}e_{k_p-2-q}\, {}^{22}\tilde{\bar{A}}_{11}'\big(j(y, z), p(y), p(z)\big)\Big]$$

with $p = 2$ for $j = 1$, $p = 1$ for $j = 2$

where

$$\mathcal{N}_q \doteq \big(\overbrace{j, \ldots, j}^{3-q}, \overbrace{p, \ldots, p}^{q}\big),$$

$$\begin{aligned}
\tilde{C}_{2j}(k_1, k_2) = {} & 3(k_j - 1)\, e_{k_j-2} e_{k_p}\, {}^{22}\tilde{\bar{A}}'_{11}\big(j(y, y, z), j(z)\big) \\
& + 3k_p e_{k_j-1} e_{k_p-1}\, {}^{22}\tilde{\bar{A}}'_{11}\big(j(y, y, z), p(z)\big) \\
& + (k_j - 1)(k_j - 2)(k_j - 3)\, e_{k_j-4} e_{k_p} \\
& \times {}^{222}\tilde{\bar{A}}'_{111}\big(j(x, y, z), j(x), j(y), j(z)\big) \\
& + k_p(k_p - 1)(k_p - 2)\, e_{k_j-1} e_{k_p-3} \\
& \times {}^{222}\tilde{\bar{A}}'_{111}\big(j(x, y, z), p(x), p(y), p(z)\big) \\
& + 3(k_j - 1)(k_j - 2)\, k_p e_{k_j-3} e_{k_p} \\
& \times {}^{222}\tilde{\bar{A}}'_{111}\big(j(x, y, z), j(x), j(y), p(z)\big) \\
& + 3(k_j - 1)\, k_p(k_p - 1)\, e_{k_j-2} e_{k_p-2} \\
& \times {}^{222}\tilde{\bar{A}}'_{111}\big(j(x, y, z), j(x), p(y), p(z)\big)
\end{aligned}$$

with $p = 2$ for $j = 1$, $p = 1$ for $j = 2$,

$$\begin{aligned}
\tilde{C}_{3jj}(k_1, k_2) = {} & e_{k_j-2} e_{k_p}\big[\big({}^2\tilde{\bar{A}}'_1(j)\big)^2 + 2\, {}^{22}\tilde{\bar{A}}'_{11}\big(j(y, z), j(y, z)\big)\big] \\
& + 2(k_j - 2)(k_j - 3)\, e_{k_j-4} e_{k_p} \\
& \times \big[2\tilde{S}_{jjjj} + {}^2\tilde{\bar{A}}'_1(j)\, {}^{22}\tilde{\bar{A}}'_{11}\big(j(y, z), j(y), j(z)\big)\big] \\
& + 4(k_j - 2)\, k_p e_{k_j-3} e_{k_p-1} \\
& \times \big[2\tilde{S}_{jjjp} + {}^2\tilde{\bar{A}}'_1(j)\, {}^{22}\tilde{\bar{A}}'_{11}\big(j(y, z), j(y), p(z)\big)\big] \\
& + 2k_p(k_p - 1)\, e_{k_j-2} e_{k_p-2} \\
& \times \big[2\tilde{S}_{jjpp} + {}^2\tilde{\bar{A}}'_1(j)\, {}^{22}\tilde{\bar{A}}'_{11}\big(j(y, z), p(y), p(z)\big)\big] \\
& + (k_j - 2)(k_j - 3)(k_j - 4)(k_j - 5)\, e_{k_j-6} e_{k_p} \tilde{S}_{jjjjjj} \\
& + 4(k_j - 2)(k_j - 3)(k_j - 4)\, k_p e_{k_j-5} e_{k_p-1} \tilde{S}_{jjjjjp} \\
& + 6(k_j - 2)(k_j - 3)\, k_p(k_p - 1)\, e_{k_j-4} e_{k_p-2} \tilde{S}_{jjjjpp} \\
& + 4(k_j - 2)\, k_p(k_p - 1)(k_p - 2)\, e_{k_j-3} e_{k_p-3} \tilde{S}_{jjjppp} \\
& + k_p(k_p - 1)(k_p - 2)(k_p - 3)\, e_{k_j-2} e_{k_p-4} \tilde{S}_{jjpppp}
\end{aligned}$$

with $p = 2$ for $j = 1$, $p = 1$ for $j = 2$,

$$\begin{aligned}
\tilde{C}_{312}(k_1, k_2) = {} & e_{k_1-1} e_{k_2-1}\big[{}^2\tilde{\bar{A}}'_1(1)\, {}^2\tilde{\bar{A}}'_1(2) + 2\, {}^{22}\tilde{\bar{A}}'_{11}\big(1(y, z), 2(y, z)\big)\big] \\
& + (k_1 - 1)(k_1 - 2)\, e_{k_1-3} e_{k_2-1} \\
& \times \big[4\tilde{S}_{1211} + {}^2\tilde{\bar{A}}'_1(1)\, {}^{22}\tilde{\bar{A}}'_{11}\big(2(y, z), 1(y), 1(z)\big) \\
& + {}^2\tilde{\bar{A}}'_1(2)\, {}^{22}\tilde{\bar{A}}'_{11}\big(1(y, z), 1(y), 1(z)\big)\big] \\
& + 2(k_1 - 1)(k_2 - 1)\, e_{k_1-2} e_{k_2-2} \\
& \times \big[4\tilde{S}_{1212} + {}^2\tilde{\bar{A}}'_1(1)\, {}^{22}\tilde{\bar{A}}'_{11}\big(2(y, z), 1(y), 1(z)\big) \\
& + {}^2\tilde{\bar{A}}'_1(2)\, {}^{22}\tilde{\bar{A}}'_{11}\big(1(y, z), 1(y), 2(z)\big)\big] \\
& + (k_2 - 1)(k_2 - 2)\, e_{k_1-1} e_{k_2-3} \\
& \times \big[4\tilde{S}_{1222} + {}^2\tilde{\bar{A}}'_1(1)\, {}^{22}\tilde{\bar{A}}'_{11}\big(2(y, z), 2(y), 2(z)\big) \\
& + {}^2\tilde{\bar{A}}'_1(2)\, {}^{22}\tilde{\bar{A}}'_{11}\big(1(y, z), 2(y), 2(z)\big)\big]
\end{aligned}$$

$$+ (k_1 - 1)(k_1 - 2)(k_1 - 3)(k_1 - 4)\, e_{k_1-5}e_{k_2-1}\tilde{S}_{121111}$$
$$+ 4(k_1 - 1)(k_1 - 2)(k_1 - 3)(k_2 - 1)\, e_{k_1-4}e_{k_2-2}\tilde{S}_{121112}$$
$$+ 6(k_1 - 1)(k_1 - 2)(k_2 - 1)(k_2 - 2)\, e_{k_1-3}e_{k_2-3}\tilde{S}_{121122}$$
$$+ 4(k_1 - 1)(k_2 - 1)(k_2 - 2)(k_2 - 3)\, e_{k_1-2}e_{k_2-4}\tilde{S}_{121222}$$
$$+ (k_2 - 1)(k_2 - 2)(k_2 - 3)(k_2 - 4)\, e_{k_1-1}e_{k_2-5}\tilde{S}_{122222}$$

where

$$\tilde{S}_{jipq} \doteq {}^{222}\tilde{\tilde{A}}'_{111}\big(j(x, y), i(x, z), p(y), q(z)\big),$$
$$\tilde{S}_{jipqtr} \doteq {}^{2222}\tilde{\tilde{A}}'_{1111}\big(j(x, y), i(z, v), p(x), q(y), t(z), r(v)\big).$$

From these formulas and (1.164) we obtain

$$\langle\tilde{d}_i\rangle = {}^2\tilde{\tilde{A}}'_1(i)\sqrt{\varepsilon}^m + O\left(\varepsilon^{\frac{m}{2}+1}\right) \quad \text{for} \quad i = 1, 2,$$
$$\langle\tilde{d}_i^2\rangle = 1 + \big[{}^2\tilde{\tilde{A}}'_2(i, i) + \big\{2\,{}^3\tilde{\tilde{A}}'_2(i, i) + 3\,{}^{22}\tilde{\tilde{A}}'_{11}\big(i(y, y, z), i(z)\big) + \big({}^2\tilde{\tilde{A}}'_1(i)\big)^2$$
$$+ 2\,{}^{22}\tilde{\tilde{A}}'_{11}\big(i(y, z), i(y, z)\big)\big\}\,\delta_{1m}\big]\,\varepsilon + o(\varepsilon) \quad \text{for} \quad i = 1, 2,$$

and

$$\langle\tilde{d}_1\tilde{d}_2\rangle = \big[{}^2\tilde{\tilde{A}}'_2(1, 2) + \big\{{}^3\tilde{\tilde{A}}'_2(1, 2) + {}^3\tilde{\tilde{A}}'_2(2, 1) + 3\,{}^{22}\tilde{\tilde{A}}'_{11}\big(1(y, y, z), 2(z)\big)$$
$$+ 3\,{}^{22}\tilde{\tilde{A}}'_{11}\big(2(y, y, z), 1(z)\big) + {}^2\tilde{\tilde{A}}'_1(1)\,{}^2\tilde{\tilde{A}}'_1(2)$$
$$+ 2\,{}^{22}\tilde{\tilde{A}}'_{11}\big(1(y, z), 2(y, z)\big)\big\}\,\delta_{1m}\big]\,\varepsilon + o(\varepsilon).$$

Summarize the transformation formulas for $s = 2$:

$$\tilde{d}_1 = \frac{1}{\sqrt{\varepsilon}^m}\, b_{11}(d_1 - d_{1,0}), \qquad \tilde{d}_2 = \frac{1}{\sqrt{\varepsilon}^m}\,[b_{21}(d_1 - d_{1,0}) + b_{22}(d_2 - d_{2,0})],$$

$$b_{11} = \sqrt{\frac{1}{\tilde{B}_{11}}}, \quad b_{21} = -\frac{\tilde{B}_{12}}{\tilde{B}_{11}}\sqrt{\frac{\tilde{B}_{11}}{\tilde{B}_{11}\tilde{B}_{22} - \tilde{B}_{12}^2}}, \quad b_{22} = \sqrt{\frac{\tilde{B}_{11}}{\tilde{B}_{11}\tilde{B}_{22} - \tilde{B}_{12}^2}}$$

(see (1.43)),

$$\tilde{B}_{pq} = \sum_{a,b=1}^{n} d_{p,a}d_{q,b}B_{ab} \quad \text{for} \quad p, q = 1, 2,$$

$$B_{ab} = {}^2\hat{A}'_1(a, b) \quad \text{for} \quad a, b = 1, 2, \ldots, n,$$

$$\tilde{F}_{i,j_1}(x) = \sum_{a=1}^{n} \tilde{d}_{i,a}F_{aj_1}(x),$$

$$\tilde{F}_{i,j_1j_2}(x, y) = \sum_{a,b=1}^{n} \tilde{d}_{i,ab}F_{aj_1}(x)\,F_{aj_2}(y),$$

$$\tilde{F}_{i,j_1j_2j_3}(x, y, z) = \sum_{a,b,c=1}^{n} \tilde{d}_{i,abc}F_{aj_1}(x)\,F_{bj_2}(y)\,F_{cj_3}(z) \quad \text{for} \quad i = 1, 2,$$

$$\tilde{d}_{1,a_1\ldots a_t} = b_{11}d_{1,a_1\ldots a_t},$$

$$\tilde{d}_{2,a_1\ldots a_t} = b_{21}d_{1,a_1\ldots a_t} + b_{22}d_{2,a_1\ldots a_t}.$$

Thus

$$
\begin{aligned}
{}^2\tilde{\tilde{A}}_1'(i) &= \sum_{a,b=1}^{n} \tilde{d}_{i,ab}\, {}^2\hat{A}_1'(a,b) \\
&= \begin{cases} b_{11} \sum\limits_{a,b=1}^{n} d_{1,ab}\, {}^2\hat{A}_1'(a,b) & \text{for} \quad i = 1, \\ b_{21} \sum\limits_{a,b=1}^{n} d_{1,ab}\, {}^2\hat{A}_1'(a,b) + b_{22} \sum\limits_{a,b=1}^{n} d_{2,ab}\, {}^2\hat{A}_1'(a,b) & \text{for} \quad i = 2 \end{cases}
\end{aligned}
$$

and

$$
\begin{aligned}
\langle d_1 - d_{1,0} \rangle &= \frac{1}{b_{11}}\, {}^2\tilde{\tilde{A}}_1'(1)\, \varepsilon^m + o(\varepsilon^m) \\
&= \sum_{a,b=1}^{n} d_{1,ab}\, {}^2\hat{A}_1'(a,b)\, \varepsilon^m + o(\varepsilon^m), \\
\langle d_2 - d_{2,0} \rangle &= \frac{1}{b_{22}} \left[{}^2\tilde{\tilde{A}}_1'(2) - \frac{b_{21}}{b_{11}}\, {}^2\tilde{\tilde{A}}_1'(1) \right] \varepsilon^m + o(\varepsilon^m) \\
&= \sum_{a,b=1}^{n} d_{2,ab}\, {}^2\hat{A}_1'(a,b)\, \varepsilon^m + o(\varepsilon^m).
\end{aligned}
\tag{1.171}
$$

For the second moments it follows

$$
\langle \tilde{d}_i \tilde{d}_j \rangle = \frac{1}{\varepsilon^m} \sum_{p=1}^{i} \sum_{q=1}^{j} b_{ip} b_{jq} \langle (d_p - d_{p,0})\, (d_q - d_{q,0}) \rangle
$$

and on the other hand

$$
\begin{aligned}
\langle \tilde{d}_i \tilde{d}_j \rangle = \sum_{p=1}^{i} \sum_{q=1}^{j} b_{ip} b_{jq} \Bigg\{ &\sum_{a,b=1}^{n} d_{p,a} d_{q,b} B_{ab} + \Bigg[\sum_{a,b=1}^{n} d_{p,a} d_{q,b}\, {}^2\hat{A}_2'(a,b) \\
&+ \Bigg\{ \sum_{a,b,c=1}^{n} (d_{p,ab} d_{q,c} + d_{q,ab} d_{p,c})\, {}^3\hat{A}_2'(a,b,c) + \sum_{a,b,c,f=1}^{n} d_{p,ab} d_{q,cf} B_{ab} B_{cf} \\
&+ \sum_{a,b,c,f=1}^{n} (d_{p,abc} d_{q,f} + d_{q,abc} d_{p,f})\, (B_{ab} B_{cf} + B_{ac} B_{bf} + B_{af} B_{bc}) \\
&+ \sum_{a,b,c,f=1}^{n} d_{p,ab} d_{q,cf} (B_{ac} B_{bf} + B_{af} B_{bc}) \Bigg\}\, \delta_{1m} \Bigg]\, \varepsilon + o(\varepsilon) \Bigg\} \doteq \sum_{p=1}^{i} \sum_{q=1}^{j} b_{ip} b_{jq} X_{pq}.
\end{aligned}
\tag{1.172}
$$

Particularly, we have

$$
\langle (d_i - d_{i,0})\, (d_j - d_{j,0}) \rangle = X_{ij} \varepsilon^m \quad \text{for} \quad i, j = 1, 2
$$

and then

$$
\begin{aligned}
&\langle (d_i - d_{i,0})\, (d_j - d_{j,0}) \rangle \\
&= \sum_{a,b=1}^{n} d_{i,a} d_{j,b} B_{ab} \varepsilon^m + \Bigg[\sum_{a,b=1}^{n} d_{i,a} d_{j,b}\, {}^2\hat{A}_2'(a,b) \\
&\quad + \Bigg\{ \sum_{a,b,c=1}^{n} (d_{i,ab} d_{j,c} + d_{j,ab} d_{i,c})\, {}^3\hat{A}_2'(a,b,c)
\end{aligned}
$$

$$+ \sum_{a,b,c,f=1}^{n} (d_{i,ab}d_{j,cf} + d_{i,abc}d_{j,f} + d_{j,abc}d_{i,f})$$

$$\left. \left. \times (B_{ab}B_{cf} + B_{ac}B_{bf} + B_{af}B_{bc}) \right\} \delta_{1m} \right] \varepsilon^{m+1} + o(\varepsilon^{m+1}). \qquad (1.173)$$

The central second moments can be written as

$$\langle (d_i - \langle d_i \rangle)(d_j - \langle d_j \rangle) \rangle = \langle (d_i - d_{i,0})(d_j - d_{j,0}) \rangle - (\langle d_i \rangle - d_{i,0})(\langle d_j \rangle - d_{j,0})$$

$$= \sum_{a,b=1}^{n} d_{i,a}d_{j,b}B_{ab}\varepsilon^m + \left[\sum_{a,b=1}^{n} d_{i,a}d_{j,b}\, {}^2\hat{A}_2'(a,b) \right.$$

$$+ \left\{ \sum_{a,b,c=1}^{n} (d_{i,ab}d_{j,c} + d_{j,ab}d_{i,c})\, {}^3\hat{A}_2'(a,b,c) \right.$$

$$+ \sum_{a,b,c,f=1}^{n} (d_{i,abc}d_{j,f} + d_{j,abc}d_{i,f})(B_{ab}B_{cf} + B_{ac}B_{bf} + B_{af}B_{bc})$$

$$\left. \left. + \sum_{a,b,c,f=1}^{n} d_{i,ab}d_{j,cf}(B_{ac}B_{bf} + B_{af}B_{bc}) \right\} \delta_{1m} \right] \varepsilon^{m+1} + o(\varepsilon^{m+1}). \qquad (1.174)$$

Investigate the correlation coefficient ϱ of d_1 and d_2. We define

$$\langle (d_i - \langle d_i \rangle)(d_j - \langle d_j \rangle) \rangle = Y_{ij}^1 \varepsilon^m + Y_{ij}^2 \varepsilon^{m+1} + o(\varepsilon^{m+1}) \quad \text{for} \quad i, j = 1, 2.$$

Then, ϱ can be written as

$$\varrho = \frac{\langle (d_1 - \langle d_1 \rangle)(d_2 - \langle d_2 \rangle) \rangle}{\sqrt{\langle (d_1 - \langle d_1 \rangle)^2 \rangle \langle (d_2 - \langle d_2 \rangle)^2 \rangle}} = \frac{Y_{12}^1 + Y_{12}^2 \varepsilon + o(\varepsilon)}{\sqrt{\left(Y_{11}^1 + Y_{11}^2 \varepsilon + o(\varepsilon)\right)\left(Y_{22}^1 + Y_{22}^2 \varepsilon + o(\varepsilon)\right)}}$$

$$= \frac{Y_{12}^1}{\sqrt{Y_{11}^1 Y_{22}^1}} + \left(\frac{Y_{12}^2}{\sqrt{Y_{11}^1 Y_{22}^1}} - \frac{Y_{12}^1}{\sqrt{Y_{11}^1 Y_{22}^1}} \frac{Y_{11}^1 Y_{22}^2 + Y_{11}^2 Y_{22}^1}{2 Y_{11}^1 Y_{22}^1} \right) \varepsilon + o(\varepsilon).$$

1.8. Expansions of distributions of random functional-functions

Using the expansion of the moments the expansion of the density function will be deduced. We start with the same notations and assumptions as in Section 1.7.

Let $\tilde{p}(\tilde{u}_1, \ldots, \tilde{u}_s)$ denote the density of the distribution function of the random vector $(\tilde{d}_1, \ldots, \tilde{d}_s)$ and $\tilde{\psi}(t_1, \ldots, t_s)$ the corresponding characteristic function. This characteristic function can be written in the form

$$\tilde{\psi}(t_1, \ldots, t_s) = 1 + \sum_{\substack{k_1, \ldots, k_s = 0 \\ k \geq 1}}^{\infty} \frac{1}{k_1! \ldots k_s!} \tilde{\alpha}_{k_1 \ldots k_s} (\mathrm{i}t_1)^{k_1} \ldots (\mathrm{i}t_s)^{k_s}$$

where

$$\tilde{\alpha}_{k_1 \ldots k_s} = \left\langle \prod_{p=1}^{s} \tilde{d}_p^{k_p} \right\rangle.$$

Using (1.164) the characteristic function $\tilde{\psi}$ is as follows:

$$\tilde{\psi}(t_1, \ldots, t_s) = 1 + \sum_{\substack{k_1, \ldots, k_s = 0 \\ k \geq 2, \text{even}}}^{\infty} \frac{\tilde{C}_0^0(k_1, \ldots, k_s)}{k_1! \ldots k_s!} (\mathrm{i}t_1)^{k_1} \ldots (\mathrm{i}t_s)^{k_s}$$

$$+ \sum_{\substack{k_1,\ldots,k_s=0\\ k\geq 2,\text{even}}}^{\infty} \frac{\tilde{C}_0^1(k_1,\ldots,k_s)}{k_1!\ldots k_s!} (\mathrm{i}t_1)^{k_1}\ldots(\mathrm{i}t_s)^{k_s}\,\varepsilon + o(\varepsilon)$$

$$+ \sum_{\substack{k_1,\ldots,k_s=0\\ k\geq 2,\text{even}}}^{\infty} \frac{(\mathrm{i}t_1)^{k_1}\ldots(\mathrm{i}t_s)^{k_s}}{k_1!\ldots k_s!} \left\{\sum_{j=1}^{s}\left[k_j\bar{\tilde{C}}_{1j}(k_1,\ldots,k_s)\right.\right.$$

$$\left. + k_j\tilde{C}_{2j}(k_1,\ldots,k_s) + \binom{k_j}{2}\tilde{C}_{3jj}(k_1,\ldots,k_s)\right]$$

$$\left. + \sum_{\substack{j,i=1\\ j<i}}^{s} k_jk_i\tilde{C}_{3ji}(k_1,\ldots,k_s)\right\}\varepsilon^m + o(\varepsilon^m) + \sum_{\substack{k_1,\ldots,k_s=0\\ k\geq 1,\text{odd}}}^{s} \frac{(\mathrm{i}t_1)^{k_1}\ldots(\mathrm{i}t_s)^{k_s}}{k_1!\ldots k_s!}$$

$$\times\left\{\bar{\tilde{C}}_0^0(k_1,\ldots,k_s) + \sum_{j=1}^{s} k_j\tilde{C}_{1j}(k_1,\ldots,k_s)\right\}\sqrt{\varepsilon}^m + O\left(\varepsilon^{\frac{m}{2}+1}\right). \tag{1.175}$$

Now, we have to compute the single sums in (1.175). Expression (1.62) leads to

$$1 + \sum_{\substack{k_1,\ldots,k_s=0\\ k\geq 2,\text{even}}}^{\infty} \frac{\tilde{C}_0^0(k_1,\ldots,k_s)}{k_1!\ldots k_s!} (\mathrm{i}t_1)^{k_1}\ldots(\mathrm{i}t_s)^{k_s} = \exp\left(-\frac{1}{2}\sum_{p=1}^{s} t_p^2\right), \tag{1.176}$$

(1.73) to

$$\sum_{\substack{k_1,\ldots,k_s=0\\ k\geq 2,\text{even}}}^{\infty} \frac{\tilde{C}_0^0(k_1,\ldots,k_s)}{k_1!\ldots k_s!} (\mathrm{i}t_1)^{k_1}\ldots(\mathrm{i}t_s)^{k_s}$$

$$= \begin{cases} \exp\left(-\frac{1}{2}\sum_{p=1}^{s} t_p^2\right)\left\{\frac{1}{2}\sum_{a,b=1}^{s} {}^2\tilde{\bar{A}}_2'(a,b)\,(\mathrm{i}t_a)\,(\mathrm{i}t_b)\right. \\ \quad + \frac{1}{24}\sum_{a,b,c,f=1}^{s} {}^4\tilde{\bar{A}}_3'(a,b,c,f)\,(\mathrm{i}t_a)\,(\mathrm{i}t_b)\,(\mathrm{i}t_c)\,(\mathrm{i}t_f) \\ \quad \left. + \frac{1}{72}\left[\sum_{a,b,c=1}^{s} {}^3\tilde{\bar{A}}_2'(a,b,c)\,(\mathrm{i}t_a)\,(\mathrm{i}t_b)\,(\mathrm{i}t_c)\right]^2\right\} & \text{for } m=1, \\ \exp\left(-\frac{1}{2}\sum_{p=1}^{s} t_p^2\right)\frac{1}{2}\sum_{a,b=1}^{s} {}^2\tilde{\bar{A}}_2'(a,b)\,(\mathrm{i}t_a)\,(\mathrm{i}t_b) & \text{for } m=2, \end{cases} \tag{1.177}$$

and (1.64) to

$$\sum_{\substack{k_1,\ldots,k_s=0\\ k\geq 1,\text{odd}}}^{\infty} \frac{\bar{\tilde{C}}_0^0(k_1,\ldots,k_s)}{k_1!\ldots k_s!} (\mathrm{i}t_1)^{k_1}\ldots(\mathrm{i}t_s)^{k_s}$$

$$= \exp\left(-\frac{1}{2}\sum_{p=1}^{s} t_p^2\right)\frac{1}{6}\sum_{a,b,c=1}^{s} {}^3\tilde{\bar{A}}_2'(a,b,c)\,(\mathrm{i}t_a)\,(\mathrm{i}t_b)\,(\mathrm{i}t_c). \tag{1.178}$$

We turn to the term of (1.175) which contains $\tilde{C}_{1j}$. With the aid of (1.157) it follows

$$G_1 \doteq \sum_{\substack{k_1,\ldots,k_s=0\\ k\geq 1,\text{odd}}}^{\infty} \frac{1}{k_1!\ldots k_s!} (\mathrm{i}t_1)^{k_1}\ldots(\mathrm{i}t_s)^{k_s} \sum_{j=1}^{s} k_j\tilde{C}_{1j}(k_1,\ldots,k_s)$$

$$= \sum_{j=1}^{s} \sum_{\substack{k_1,\ldots,k_s=0 \\ k\geq 1,\text{odd}}}^{\infty} k_j \frac{1}{k_1!\ldots k_s!} (it_1)^{k_1} \ldots (it_s)^{k_s}$$

$$\times \left\{ {}^2\tilde{\tilde{A}}_1'(j)\, {}^2\tilde{\tilde{P}}_1(\mathcal{W}_j) + \sum_{\substack{(p,q)\text{ from }\mathcal{W}_j \\ \text{w.a.}}} {}^{22}\tilde{\tilde{A}}_{11}'\big(j(y,z), p(y), q(z)\big)\, {}^2\tilde{\tilde{P}}_1(\mathcal{W}_{jpq}) \right\}$$

$$\doteq G_{11} + G_{12}.$$

First we deal with G_{11}. Applying (1.166) we obtain

$$G_{11} = \sum_{j=1}^{s} {}^2\tilde{\tilde{A}}_1'(j) \sum_{\substack{k_1,\ldots,k_s=0 \\ k\geq 1,\text{odd}}}^{\infty} k_j \frac{1}{k_1!\ldots k_s!} (it_1)^{k_1} \ldots (it_s)^{k_s} e_{k_j-1} \prod_{\substack{q=1 \\ q\neq j}}^{s} e_{k_q}$$

and by means of

$$k_a = 2p_a \quad \text{for} \quad a = 1, 2, \ldots, j-1, j+1, \ldots, s; \quad k_j - 1 = 2p_j,$$

then

$$G_{11} = \sum_{j=1}^{s} {}^2\tilde{\tilde{A}}_1'(j)\, (it_j) \sum_{p_1,\ldots,p_s=0}^{\infty} \prod_{a=1}^{s} e_{2p_a} \frac{1}{(2p_a)!} (it_a)^{2p_a}$$

$$= \exp\left(-\frac{1}{2} \sum_{p=1}^{s} t_p^2\right) \sum_{j=1}^{s} {}^2\tilde{\tilde{A}}_1'(j)\, (it_j).$$

In order to compute G_{12} the term $\tilde{S}_{jpq}$ is defined by

$$\tilde{S}_{jpq} \doteq {}^{22}\tilde{\tilde{A}}_{11}'\big(j(y,z), p(y), q(z)\big). \tag{1.179}$$

Thus

$$G_{12} = \sum_{j=1}^{s} \sum_{\substack{k_1,\ldots,k_s=0 \\ k\geq 1,\text{odd}}}^{\infty} k_j \frac{1}{k_1!\ldots k_s!} (it_1)^{k_1} \ldots (it_s)^{k_s} \sum_{\substack{(p,q)\text{ from }\mathcal{W}_j \\ \text{w.a.}}} \tilde{S}_{jpq}\, {}^2\tilde{\tilde{P}}_1(\mathcal{W}_{jpq});$$

$$\sum_{j=1}^{s} \sum_{\substack{(p,q)\text{ from }\mathcal{W}_j \\ \text{w.a.}}} k_j \tilde{S}_{jpq}\, {}^2\tilde{\tilde{P}}_1(\mathcal{W}_{jpq}) = \sum_{a,b,c=1}^{s} \tilde{S}_{abc} m_{abc}\, {}^2\tilde{\tilde{P}}_1(\mathcal{W}_{abc})$$

where m_{abc} denotes the number of possibilities to choose a, b, c from $\mathcal{V}$. It is easy to see that

$$\frac{1}{k_1!\ldots k_s!} (it_1)^{k_1} \ldots (it_s)^{k_s}\, m_{abc}\, {}^2\tilde{\tilde{P}}_1(\mathcal{W}_{abc})$$

$$= \frac{1}{p_1!\ldots p_s!} (it_1)^{2p_1} \ldots (it_s)^{2p_s} \left(\frac{1}{2}\right)^{p_1+\cdots+p_s} (it_a)\, (it_b)\, (it_c)$$

with

$$k_a - 1 = 2p_a, \quad k_b - 1 = 2p_b, \quad k_c - 1 = 2p_c, \text{ otherwise } k_q = 2p_q$$
$$\text{for} \quad a \neq b,\ a \neq c,\ b \neq c,$$

$$k_a - 2 = 2p_a, \quad k_c - 1 = 2p_c, \quad \text{otherwise } k_q = 2p_q \quad \text{for} \quad a = b,\ a \neq c,$$

$$k_a - 3 = 2p_a, \text{ otherwise } k_q = 2p_q \quad \text{for} \quad a = b = c.$$

Hence, we have

$$G_{12} = \sum_{a,b,c=1}^{s} \tilde{S}_{abc} \sum_{\substack{k_1,\ldots,k_s=0\\ k\geq 1,\text{odd}}}^{\infty} \frac{1}{k_1!\ldots k_s!} (it_1)^{k_1} \ldots (it_s)^{k_s} m_{abc}\, {}^2\tilde{\tilde{P}}_1(\mathcal{W}_{abc})$$

$$= \sum_{a,b,c=1}^{s} \tilde{S}_{abc}(it_a)(it_b)(it_c) \sum_{p_1,\ldots,p_s=0}^{\infty} \frac{1}{p_1!\ldots p_s!} \left(-\frac{1}{2}t_1^2\right)^{p_1} \ldots \left(-\frac{1}{2}t_s^2\right)^{p_s}$$

$$= \exp\left(-\frac{1}{2}\sum_{p=1}^{s} t_p^2\right) \sum_{a,b,c=1}^{s} \tilde{S}_{abc}(it_a)(it_b)(it_c)$$

and G_1 can be written as

$$G_1 = \exp\left(-\frac{1}{2}\sum_{p=1}^{s} t_p^2\right)$$
$$\times \left\{\sum_{a=1}^{s} {}^2\tilde{\tilde{A}}_1'(a)(it_a) + \sum_{a,b,c=1}^{s} {}^{22}\tilde{\tilde{A}}_{11}'\big(a(y,z), b(y), c(z)\big)(it_a)(it_b)(it_c)\right\}. \tag{1.180}$$

In the following we compute the term in (1.175) containing $\bar{\bar{C}}_{1j}$; i.e.

$$G_2 \doteq \sum_{\substack{k_1,\ldots,k_s=0\\ k\geq 2,\text{even}}}^{\infty} \frac{1}{k_1!\ldots k_s!} (it_1)^{k_1} \ldots (it_s)^{k_s} \sum_{j=1}^{s} k_j \bar{\bar{C}}_{1j}(k_1,\ldots,k_s)$$

$$= \sum_{j=1}^{s} \sum_{\substack{k_1,\ldots,k_s=0\\ k\geq 2,\text{even}}}^{\infty} k_j \frac{1}{k_1!\ldots k_s!} (it_1)^{k_1} \ldots (it_s)^{k_s}$$

$$\times \Bigg\{ \sum_{p \text{ from } \mathcal{W}_j} {}^3\tilde{\tilde{A}}_2'(j,p)\, {}^2\tilde{\tilde{P}}_1(\mathcal{W}_{jp}) + 2 \sum_{\substack{(p,q) \text{ from } \mathcal{W}_j\\ \text{w.o.a.}}} \sum_{u \text{ from } \mathcal{W}_{jpq}} {}^2\tilde{\tilde{P}}_1(\mathcal{W}_{jpqu})$$

$$\times {}^{32}\tilde{\tilde{A}}_{21}'\big(j(y,z), p(y), q(y), u(z)\big) + \sum_{\substack{(p,q,t) \text{ from } \mathcal{W}_j\\ \text{w.o.a.}}} {}^3\tilde{\tilde{A}}_2'(p,q,t)$$

$$\times \Bigg[{}^2\tilde{\tilde{A}}_1'(j)\, {}^2\tilde{\tilde{P}}_1(\mathcal{W}_{jpqt}) + \sum_{\substack{(u,v) \text{ from } \mathcal{W}_{jpqt}\\ \text{w.a.}}} {}^{22}\tilde{\tilde{A}}_{11}'\big(j(y,z), u(y), v(z)\big)\, {}^2\tilde{\tilde{P}}_1(\mathcal{W}_{jpqtuv})\Bigg]\Bigg\}$$

$$\doteq G_{21} + G_{22} + G_{23} + G_{24}.$$

First, it is

$$G_{21} = \sum_{j=1}^{s} \sum_{\substack{k_1,\ldots,k_s=0\\ k\geq 2,\text{even}}}^{\infty} k_j \frac{1}{k_1!\ldots k_s!} (it_1)^{k_1} \ldots (it_s)^{k_s} \sum_{p \text{ from } \mathcal{W}_j} {}^3\tilde{\tilde{A}}_2'(j,p)\, {}^2\tilde{\tilde{P}}_1(\mathcal{W}_{jp})$$

and

$$\sum_{j=1}^{s} \sum_{p \text{ from } \mathcal{W}_j} k_j\, {}^3\tilde{\tilde{A}}_2'(j,p)\, {}^2\tilde{\tilde{P}}_1(\mathcal{W}_{jp}) = \sum_{a,b=1}^{s} m_{ab}\, {}^3\tilde{\tilde{A}}_2'(a,b)\, {}^2\tilde{\tilde{P}}_1(\mathcal{W}_{ab})$$

where m_{ab} denotes the number of possibilities to choose a, b from $\mathcal{V}$. Using

$$\frac{1}{k_1!\ldots k_s!} (it_1)^{k_1} \ldots (it_s)^{k_s} m_{ab}\, {}^2\tilde{\tilde{P}}_1(\mathcal{W}_{ab})$$

$$= \frac{1}{p_1!\ldots p_s!} (it_1)^{2p_1} \ldots (it_s)^{2p_s} \left(\frac{1}{2}\right)^{p_1+\cdots+p_s} (it_a)(it_b)$$

with

$$k_a - 1 = 2p_a, \quad k_b - 1 = 2p_b, \quad \text{otherwise} \quad k_q = 2p_q \quad \text{for} \quad a \neq b,$$

$$k_a - 2 = 2p_a, \quad \text{otherwise} \quad k_q = 2p_q \quad \text{for} \quad a = b$$

then we obtain

$$G_{21} = \sum_{a,b=1}^{s} {}^3\tilde{\tilde{A}}_2'(a, b)\,(\mathrm{i}t_a)\,(\mathrm{i}t_b) \sum_{p_1,\ldots,p_s=0}^{\infty} \frac{1}{p_1!\ldots p_s!} \left(-\frac{1}{2}\,t_1^2\right)^{p_1} \ldots \left(-\frac{1}{2}\,t_s^2\right)^{p_s}$$

$$= \exp\left(-\frac{1}{2}\sum_{p=1}^{s} t_p^2\right) \sum_{a,b=1}^{s} {}^3\tilde{\tilde{A}}_2'(a, b)\,(\mathrm{i}t_a)\,(\mathrm{i}t_b).$$

Now, we deal with

$$G_{22} = 2 \sum_{j=1}^{s} \sum_{\substack{k_1,\ldots,k_s=0\\ k\geqq 2,\text{even}}}^{\infty} k_j \frac{1}{k_1!\ldots k_s!} (\mathrm{i}t_1)^{k_1} \ldots (\mathrm{i}t_s)^{k_s} \sum_{\substack{(p,q)\text{ from } \mathscr{W}_j\\ \text{w.o.a.}}} \sum_{u \text{ from } \mathscr{W}_{jpq}} \tilde{S}^{32}_{jpqu}\, {}^2\tilde{\tilde{P}}_1(\mathscr{W}_{jpqu})$$

where

$$\tilde{S}^{32}_{jpqu} \doteq {}^{32}\tilde{\tilde{A}}_{21}'\big(j(y, z), p(y), q(y), u(z)\big) \tag{1.181}$$

is put. We obtain

$$\sum_{j=1}^{s} \sum_{\substack{(p,q)\text{from}\mathscr{W}_j\\ \text{w.o.a.}}} \sum_{u \text{ from } \mathscr{W}_{jpq}} k_j \tilde{S}^{32}_{jpqu}\, {}^2\tilde{\tilde{P}}_1(\mathscr{W}_{jpqu}) = \frac{1}{2} \sum_{a,b,c,f=1}^{s} m_{abcf} \tilde{S}^{32}_{abcf}\, {}^2\tilde{\tilde{P}}_1(\mathscr{W}_{abcf})$$

where m_{abcf} denotes again the number of possibilities to choose a, b, c, f from $\mathscr{V}$. Similar considerations carried out above lead to

$$G_{22} = \exp\left(-\frac{1}{2}\sum_{p=1}^{s} t_p^2\right) \sum_{a,b,c,f=1}^{s} {}^{32}\tilde{\tilde{A}}_{21}'\big(a(y, z), b(y), c(y), f(z)\big)\,(\mathrm{i}t_a)\,(\mathrm{i}t_b)\,(\mathrm{i}t_c)\,(\mathrm{i}t_f).$$

The third summand in G_2 has the form

$$G_{23} \doteq \sum_{j=1}^{s} {}^2\tilde{\tilde{A}}_1'(j) \sum_{\substack{k_1,\ldots,k_s=0\\ k\geqq 2,\text{even}}}^{\infty} k_j \frac{1}{k_1!\ldots k_s!} (\mathrm{i}t_1)^{k_1} \ldots (\mathrm{i}t_s)^{k_s} \sum_{\substack{(p,q,t)\text{ from } \mathscr{W}_j\\ \text{w.o.a.}}} {}^3\tilde{\tilde{A}}_2'(p, q, t)\, {}^2\tilde{\tilde{P}}_1(\mathscr{W}_{jpqt})$$

and we use

$$\sum_{j=1}^{s} \sum_{\substack{(p,q,t)\text{ from } \mathscr{W}_j\\ \text{w.o.a.}}} k_j\, {}^2\tilde{\tilde{A}}_1'(j)\, {}^3\tilde{\tilde{A}}_2'(p, q, t)\, {}^2\tilde{\tilde{P}}_1(\mathscr{W}_{jpqt})$$

$$= \frac{1}{6} \sum_{a,b,c,f=1}^{s} m_{abcf}\, {}^2\tilde{\tilde{A}}_1'(a)\, {}^3\tilde{\tilde{A}}_2'(b, c, f)\, {}^2\tilde{\tilde{P}}_1(\mathscr{W}_{abcf}).$$

Hence, we can easily notice that

$$G_{23} = \exp\left(-\frac{1}{2}\sum_{p=1}^{s} t_p^2\right) \frac{1}{6} \sum_{a,b,c,f=1}^{s} {}^2\tilde{\tilde{A}}_1'(a)\, {}^3\tilde{\tilde{A}}_2'(b, c, f)\,(\mathrm{i}t_a)\,(\mathrm{i}t_b)\,(\mathrm{i}t_c)\,(\mathrm{i}t_f).$$

Finally, the term G_{24},

$$G_{24} \doteq \sum_{j=1}^{s} \sum_{\substack{k_1,\ldots,k_s=0 \\ k\geq 2,\text{even}}}^{\infty} k_j \frac{1}{k_1!\ldots k_s!} \sum_{\substack{(p,q,t)\text{ from }\mathcal{W}_j \\ \text{w.o.a.}}} {}^3\tilde{\tilde{A}}'_2(p,q,t)$$
$$\times \sum_{\substack{(u,v)\text{ from }\mathcal{W}_{jpqt} \\ \text{w.a.}}} {}^{22}\tilde{\tilde{A}}'_{11}\big(j(y,z),u(y),v(z)\big)\,{}^2\tilde{\tilde{P}}_1(\mathcal{W}_{jpqtuv})$$

has to be analysed. In this case we see that

$$\sum_{j=1}^{s} \sum_{\substack{(p,q,t)\text{ from }\mathcal{W}_j \\ \text{w.o.a.}}} \sum_{\substack{(u,v)\text{ from }\mathcal{W}_{jpqt} \\ \text{w.a.}}} k_j\,{}^3\tilde{\tilde{A}}'_2(p,q,t)\,{}^{22}\tilde{\tilde{A}}'_{11}\big(j(y,z),u(y),v(z)\big)\,{}^2\tilde{\tilde{P}}_1(\mathcal{W}_{jpqtuv})$$
$$= \frac{1}{6} \sum_{a,b,c,f,g,h=1}^{s} m_{abcfgh}\,{}^{22}\tilde{\tilde{A}}'_{11}\big(a(y,z),b(y),c(z)\big)\,{}^3\tilde{\tilde{A}}'_2(f,g,h)\,{}^2\tilde{\tilde{P}}_1(\mathcal{W}_{abcfgh})$$

where m_{abcfgh} denotes the number of possibilities to choose a, b, c, f, g, h from the set $\mathcal{V}$. As in the other cases we take into consideration that

$$\frac{1}{k_1!\ldots k_s!}(\mathrm{i}t_1)^{k_1}\ldots(\mathrm{i}t_s)^{k_s}\,m_{abcfgh}\,{}^2\tilde{\tilde{P}}_1(\mathcal{W}_{abcfgh})$$
$$= \frac{1}{p_1!\ldots p_s!}(\mathrm{i}t_1)^{2p_1}\ldots(\mathrm{i}t_s)^{2p_s}\left(\frac{1}{2}\right)^{p_1+\cdots+p_s}(\mathrm{i}t_a)(\mathrm{i}t_b)(\mathrm{i}t_c)(\mathrm{i}t_f)(\mathrm{i}t_g)(\mathrm{i}t_h)$$

where

$k_q - 1 = 2p_q$ for $q = a, b, c, f, g, h$, otherwise $k_q = 2p_q$ if all indices a, b, c, f, g, h are different;

$k_a - 2 = 2p_a$, $k_q - 1 = 2p_q$ for $q = c, f, g, h$, otherwise $k_q = 2p_q$ if $a = b$ and a, c, f, g, h are different;

$k_a - 3 = 2p_a$, $k_q - 1 = 2p_q$ for $q = f, g, h$, otherwise $k_q = 2p_q$ if $a = b = c$ and a, f, g, h are different;

$k_a - 4 = 2p_a$, $k_q - 1 = 2p_q$ for $q = g, h$, otherwise $k_q = 2p_q$ if $a = b = c = f$ and a, g, h are different;

$k_a - 5 = 2p_a$, $k_h - 1 = 2p_h$, otherwise $k_q = 2p_q$ if $a = b = c = f = g$ and a, h are different;

$k_a - 6 = 2p_a$, otherwise $k_q = 2p_q$ if $a = b = c = f = g = h$

and analogous relations for the other cases. With this, it follows

$$G_{24} = \exp\left(-\frac{1}{2}\sum_{q=1}^{s} t_q^2\right)\frac{1}{6}\sum_{a,b,c,f,g,h=1}^{s} {}^{22}\tilde{\tilde{A}}'_{11}\big(a(y,z),b(y),c(z)\big)$$
$$\times\,{}^3\tilde{\tilde{A}}'_2(f,g,h)\,(\mathrm{i}t_a)(\mathrm{i}t_b)(\mathrm{i}t_c)(\mathrm{i}t_f)(\mathrm{i}t_g)(\mathrm{i}t_h)$$
$$= \exp\left(-\frac{1}{2}\sum_{q=1}^{s} t_q^2\right)\frac{1}{6}\sum_{a,b,c=1}^{s} {}^{22}\tilde{\tilde{A}}'_{11}\big(a(y,z),b(y),c(z)\big)$$
$$\times\,(\mathrm{i}t_a)(\mathrm{i}t_b)(\mathrm{i}t_c)\sum_{a,b,c=1}^{s} {}^3\tilde{\tilde{A}}'_2(a,b,c)\,(\mathrm{i}t_a)(\mathrm{i}t_b)(\mathrm{i}t_c).$$

Summarizing G_2 can be written in the form

$$G_2 = \exp\left(-\frac{1}{2}\sum_{q=1}^{s} t_q^2\right)\left\{\sum_{a,b=1}^{s} {}^3\tilde{\bar{A}}_2'(a, b)\,(it_a)\,(it_b)\right.$$

$$+\sum_{a,b,c,f=1}^{s} {}^{32}\tilde{\bar{A}}_{21}'\big(a(y, z), b(y), c(y), f(z)\big)\,(it_a)\,(it_b)\,(it_c)\,(it_f)$$

$$+\frac{1}{6}\sum_{a,b,c=1}^{s} {}^3\tilde{\bar{A}}_2'(a, b, c)\,(it_a)\,(it_b)\,(it_c)\left[\sum_{a=1}^{s} {}^2\tilde{\bar{A}}_1'(a)\,(it_a)\right.$$

$$\left.\left.+\sum_{a,b,c=1}^{s} {}^{22}\tilde{\bar{A}}_{11}'\big(a(y, z), b(y), c(z)\big)\,(it_a)\,(it_b)\,(it_c)\right]\right\}. \tag{1.182}$$

Investigate

$$G_3 \doteq \sum_{\substack{k_1,\dots,k_s=0\\ k\geqq 2,\text{even}}}^{\infty} \frac{1}{k_1!\dots k_s!}\,(it_1)^{k_1}\dots(it_s)^{k_s}\sum_{j=1}^{s} k_j\tilde{C}_{2j}(k_1, \dots, k_s).$$

Applying (1.160) we find

$$G_3 = \sum_{j=1}^{s}\sum_{\substack{k_1,\dots,k_s=0\\ k\geqq 2,\text{even}}}^{\infty} k_j\,\frac{1}{k_1!\dots k_s!}\,(it_1)^{k_1}\dots(it_s)^{k_s}\left[3\sum_{p\,\text{from}\,\mathscr{W}_j}\tilde{S}_{jp}\,{}^2\tilde{P}_1(\mathscr{W}_{jp})\right.$$

$$\left.+\sum_{\substack{(p,q,t)\,\text{from}\,\mathscr{W}_j\\ \text{w.a.}}} {}^{222}\tilde{\bar{A}}_{111}'\big(j(x, y, z), p(x), q(y), t(z)\big)\,{}^2\tilde{P}_1(\mathscr{W}_{jpqt})\right]$$

where $\tilde{S}_{jp}$ is defined by

$$\tilde{S}_{jp} \doteq {}^{22}\tilde{\bar{A}}_{11}'\big(j(y, y, z), p(z)\big). \tag{1.183}$$

Now, by means of the above considerations we find without difficulties

$$G_3 = \exp\left(-\frac{1}{2}\sum_{p=1}^{s} t_p^2\right)\left\{3\sum_{a,b=1}^{s} {}^{22}\tilde{\bar{A}}_{11}'\big(a(y, y, z), b(z)\big)\,(it_a)\,(it_b)\right.$$

$$\left.+\sum_{a,b,c,f=1}^{s} {}^{222}\tilde{\bar{A}}_{111}'\big(a(x, y, z), b(x), c(y), f(z)\big)\,(it_a)\,(it_b)\,(it_c)\,(it_f)\right\}. \tag{1.184}$$

G_4 is given by

$$G_4 = \sum_{\substack{j,h=1\\ j<h}}^{s}\sum_{\substack{k_1,\dots,k_s=1\\ k\geqq 2,\text{even}}}^{\infty} k_jk_h\,\frac{1}{k_1!\dots k_s!}\,(it_1)^{k_1}\dots(it_s)^{k_s}\,\tilde{C}_{3jh}(k_1, \dots, k_s)$$

$$+\sum_{j=1}^{s}\sum_{\substack{k_1,\dots,k_s=1\\ k\geqq 2,\text{even}}}^{\infty}\binom{k_j}{2}\frac{1}{k_1!\dots k_s!}\,(it_1)^{k_1}\dots(it_s)^{k_s}\,\tilde{C}_{3jj}(k_1, \dots, k_s).$$

With the notations

$$\tilde{S}_{jhpq}^{222} \doteq {}^{222}\tilde{\bar{A}}_{111}'\big(j(x, y), h(x, z), p(y), q(z)\big), \tag{1.185}$$

$$\tilde{S}_{jhpqtr} \doteq {}^{2222}\tilde{\bar{A}}_{1111}'\big(j(x, y), h(z, v), p(x), q(y), t(z), r(v)\big) \tag{1.186}$$

and (1.162) we can write

$$G_4 = \sum_{\substack{j,h=1 \\ j<h}}^{s} \sum_{\substack{k_1,\ldots,k_s=1 \\ k\geqq 2,\text{even}}}^{\infty} k_j k_h \frac{1}{k_1!\ldots k_s!} (it_1)^{k_1} \ldots (it_s)^{k_s}$$

$$\times \Big\{\big[{}^2\tilde{\tilde{A}}_1'(j)\, {}^2\tilde{\tilde{A}}_1'(h) + 2\, {}^{22}\tilde{\tilde{A}}_{11}'\big(j(y,z), h(y,z)\big)\big]\, {}^2\tilde{\tilde{P}}_1(\mathscr{W}_{jh})$$

$$+ \sum_{\substack{(p,q)\text{ from }\mathscr{W}_{jh} \\ \text{w.a.}}} [{}^2\tilde{\tilde{A}}_1'(j)\, \tilde{S}_{hpq} + {}^2\tilde{\tilde{A}}_1'(h)\, S_{jpq} + 4\tilde{S}^{222}_{jhpq}]\, {}^2\tilde{\tilde{P}}_1(\mathscr{W}_{jhpq})$$

$$+ \sum_{\substack{(p,q,t,r)\text{ from }\mathscr{W}_{jh} \\ \text{w.a.}}} \tilde{S}_{jhpqtr}\, {}^2\tilde{\tilde{P}}_1(\mathscr{W}_{jhpqtr})\Big\}$$

$$+ \sum_{j=1}^{s} \sum_{\substack{k_1,\ldots,k_s=1 \\ k\geqq 2,\text{even}}}^{\infty} \binom{k_j}{2} \frac{1}{k_1!\ldots k_s!} (it_1)^{k_1} \ldots (it_s)^{k_s}$$

$$\times \Big\{\big[\big({}^2\tilde{\tilde{A}}_1'(j)\big)^2 + {}^{22}\tilde{\tilde{A}}_{11}'\big(j(y,z), j(y,z)\big) + {}^{22}\tilde{\tilde{A}}_{11}'\big(j(y,z)\, j(z,y)\big)\big]\, {}^2\tilde{\tilde{P}}_1(\mathscr{W}_{jj})$$

$$+ 2 \sum_{\substack{(p,q)\text{ from }\mathscr{W}_{jj} \\ \text{w.a.}}} [{}^2\tilde{\tilde{A}}_1'(j)\, \tilde{S}_{jpq} + 2\tilde{S}^{222}_{jjpq}]\, {}^2\tilde{\tilde{P}}_1(\mathscr{W}_{jjpq})$$

$$+ \sum_{\substack{(p,q,t,r)\text{ from }\mathscr{W}_{jj} \\ \text{w.a.}}} \tilde{S}_{jjpqtr}\, {}^2\tilde{\tilde{P}}_1(\mathscr{W}_{jjpqtr})\Big\}.$$

Take the first summands in the terms of G_4 which are characterized by $\tilde{C}_{3jh}$ and $\tilde{C}_{3jj}$, respectively,

$$G_{41} \doteq \frac{1}{2} \sum_{j,h=1}^{s} \sum_{\substack{k_1,\ldots,k_s=1 \\ k\geqq 2,\text{even}}}^{\infty} \big(k_j k_h (1-\delta_{jh}) + k_j(k_j-1)\,\delta_{jh}\big)$$

$$\times \frac{1}{k_1!\ldots k_s!} (it_1)^{k_1} \ldots (it_s)^{k_s} \big[{}^2\tilde{\tilde{A}}_1'(j)\, {}^2\tilde{\tilde{A}}_1'(h)$$

$$+ 2\, {}^{22}\tilde{\tilde{A}}_{11}'\big(j(y,z), h(y,z)\big)\big]\, {}^2\tilde{\tilde{P}}_1(\mathscr{W}_{jh}).$$

Furthermore, we obtain

$$\sum_{j,h=1}^{s} \big(k_j k_h (1-\delta_{jh}) + k_j(k_j-1)\,\delta_{jh}\big)\, \tilde{T}_{jh}\, {}^2\tilde{\tilde{P}}_1(\mathscr{W}_{jh}) = \sum_{a,b=1}^{s} m_{ab} \tilde{T}_{ab}\, {}^2\tilde{\tilde{P}}_1(\mathscr{W}_{ab})$$

and then

$$G_{41} = \exp\left(-\frac{1}{2} \sum_{p=1}^{s} t_p^2\right) \frac{1}{2} \sum_{a,b=1}^{s} \big[{}^2\tilde{\tilde{A}}_1'(a)\, {}^2\tilde{\tilde{A}}_1'(b) + 2\, {}^{22}\tilde{\tilde{A}}_{11}'\big(a(y,z), b(y,z)\big)\big]\, (it_a)\, (it_b).$$

Take the second summands in the term of G_4

$$G_{42} = \frac{1}{2} \sum_{j,h=1}^{s} \sum_{\substack{k_1,\ldots,k_s=1 \\ k\geqq 2,\text{even}}}^{\infty} \big(k_j k_h (1-\delta_{jh}) + k_j(k_j-1)\,\delta_{jh}\big) \frac{1}{k_1!\ldots k_s!}$$

$$\times (it_1)^{k_1} \ldots (it_s)^{k_s} \sum_{\substack{(p,q)\text{ from }\mathscr{W}_{jh} \\ \text{w.a.}}} [{}^2\tilde{\tilde{A}}_1'(j)\, \tilde{S}_{hpq} + {}^2\tilde{\tilde{A}}_1'(h)\, \tilde{S}_{jpq} + 4\tilde{S}^{222}_{jhpq}]\, {}^2\tilde{\tilde{P}}_1(\mathscr{W}_{jhpq}).$$

By the aid of

$$\sum_{j,h=1}^{s} \sum_{\substack{(p,q)\text{ from }\mathscr{W}_{jh}\\ \text{w.a.}}} \big(k_jk_h(1-\delta_{jh}) + k_j(k_j-1)\,\delta_{jh}\big)\,\tilde{T}_{jhpq}\,{}^2\tilde{\tilde{P}}_1(\mathscr{W}_{jhpq})$$

$$= \sum_{a,b,c,f=1}^{s} m_{abcf}\tilde{T}_{abcf}\,{}^2\tilde{\tilde{P}}_1(\mathscr{W}_{abcf})$$

it follows

$$G_{42} = \exp\left(-\frac{1}{2}\sum_{p=1}^{s} t_p^2\right)\frac{1}{2}\sum_{a,b,c,f=1}^{s}\left[{}^2\tilde{\tilde{A}}_1'(a)\,\tilde{S}_{bcf} + {}^2\tilde{\tilde{A}}_1'(b)\,S_{acf} + 4\tilde{S}_{abcf}^{222}\right]$$
$$\times\,(\mathrm{i}t_a)\,(\mathrm{i}t_b)\,(\mathrm{i}t_c)\,(\mathrm{i}t_f)\,.$$

The third summands in the term of G_4 lead to

$$G_{43} = \frac{1}{2}\sum_{j,h=1}^{s}\sum_{\substack{k_1,\ldots,k_s=1\\ k\geqq 2,\text{even}}}^{\infty}\big(k_jk_h(1-\delta_{jh}) + k_j(k_j-1)\,\delta_{jh}\big)\sum_{\substack{(p,q,t,r)\text{ from }\mathscr{W}_{jh}\\ \text{w.a.}}}\tilde{S}_{jhpqtr}\,{}^2\tilde{\tilde{P}}_1(\mathscr{W}_{jhpqtr})\,.$$

Taking into consideration

$$\sum_{j,h=1}^{s}\sum_{\substack{(p,q,t,r)\text{ from }\mathscr{W}_{jh}\\ \text{w.a.}}}\big(k_jk_h(1-\delta_{jh}) + k_j(k_j-1)\,\delta_{jh}\big)\,\tilde{S}_{jhpqtr}\,{}^2\tilde{\tilde{P}}_1(\mathscr{W}_{jhpqtr})$$

$$= \sum_{a,b,c,f,g,h=1}^{s} m_{abcfgh}\tilde{S}_{abcfgh}\,{}^2\tilde{\tilde{P}}_1(\mathscr{W}_{abcfgh})$$

we find

$$G_{43} = \exp\left(-\frac{1}{2}\sum_{p=1}^{s}t_p^2\right)\frac{1}{2}\sum_{a,b,c,f,g,h=1}^{s}\tilde{S}_{abcfgh}(\mathrm{i}t_a)\,(\mathrm{i}t_b)\,(\mathrm{i}t_c)\,(\mathrm{i}t_f)\,(\mathrm{i}t_g)\,(\mathrm{i}t_h)$$

and furthermore

$$G_4 = \exp\left(-\frac{1}{2}\sum_{p=1}^{s}t_p^2\right)\frac{1}{2}\sum_{a,b=1}^{s}\Bigg\{{}^2\tilde{\tilde{A}}_1'(a)\,{}^2\tilde{\tilde{A}}_1'(b) + 2\,{}^{22}\tilde{\tilde{A}}_{11}'\big(a(y,z),b(y,z)\big)$$
$$+\,2\sum_{c,f=1}^{s}\Bigg[{}^2\tilde{\tilde{A}}_1'(a)\,{}^{22}\tilde{\tilde{A}}_{11}'\big(b(y,z),c(y),f(z)\big)$$
$$+\,2\,{}^{222}\tilde{\tilde{A}}_{111}'\big(a(x,y),b(z,x),c(y),f(z)\big)$$
$$+\sum_{g,h=1}^{s}{}^{2222}\tilde{\tilde{A}}_{1111}'\big(a(x,y),b(z,v),c(x),f(y),g(z),h(v)\big)$$
$$\times\,(\mathrm{i}t_g)\,(\mathrm{i}t_h)\Bigg]\,(\mathrm{i}t_c)\,(\mathrm{i}t_f)\Bigg\}\,(\mathrm{i}t_a)\,(\mathrm{i}t_b)\,. \tag{1.187}$$

If we start from (1.175) and use the relations (1.176), (1.177), (1.182), (1.184), (1.187), (1.178), and (1.180) then we obtain the characteristic function

$$\tilde{\psi}(t_1,\ldots,t_s)$$
$$= \exp\left(-\frac{1}{2}\sum_{p=1}^{s}t_p^2\right)\Bigg\{1 + \Bigg[\frac{1}{2}\sum_{a,b=1}^{s}{}^2\tilde{\tilde{A}}_2'(a,b)\,(\mathrm{i}t_a)\,(\mathrm{i}t_b)$$

$$+\frac{1}{24}\sum_{a,b,c,f=1}^{s}{}^{4}\tilde{\bar{A}}'_{3}(a,b,c,f)\,(it_a)\,(it_b)\,(it_c)\,(it_f)\,\delta_{1m}$$

$$+\frac{1}{2}\left[\left[\frac{1}{6}\sum_{a,b,c=1}^{s}{}^{3}\tilde{\bar{A}}'_{2}(a,b,c)\,(it_a)\,(it_b)\,(it_c)\right]^{2}\delta_{1m}\right]\varepsilon+o(\varepsilon)$$

$$+\left[\sum_{a=1}^{s}{}^{2}\tilde{\bar{A}}'_{1}(a)\,(it_a)+\sum_{a,b,c=1}^{s}\left[{}^{22}\tilde{\bar{A}}'_{11}\big(a(y,z),b(y),c(z)\big)\right.\right.$$

$$\left.\left.+\frac{1}{6}\,{}^{3}\tilde{\bar{A}}'_{2}(a,b,c)\right]\,(it_a)\,(it_b)\,(it_c)\right]\sqrt{\varepsilon}^{\,m}+O\left(\varepsilon^{\frac{m}{2}+1}\right)+\left[\sum_{a,b=1}^{s}{}^{3}\tilde{\bar{A}}'_{2}(a,b)\,(it_a)\,(it_b)\right.$$

$$+\sum_{a,b,c,f=1}^{s}{}^{32}\tilde{\bar{A}}'_{21}\big(a(y,z),b(y),c(y),f(z)\big)\,(it_a)\,(it_b)\,(it_c)\,(it_f)$$

$$+\frac{1}{6}\sum_{a,b,c=1}^{s}{}^{3}\tilde{\bar{A}}'_{2}(a,b,c)\,(it_a)\,(it_b)\,(it_c)\left\{\sum_{a=1}^{s}{}^{2}\tilde{\bar{A}}'_{1}(a)\,(it_a)\right.$$

$$\left.+\sum_{a,b,c=1}^{s}{}^{22}\tilde{\bar{A}}'_{11}\big(a(y,z),b(y),c(z)\big)\,(it_a)\,(it_b)\,(it_c)\right\}$$

$$+3\sum_{a,b=1}^{s}{}^{22}\tilde{\bar{A}}'_{11}\big(a(y,y,z),b(z)\big)\,(it_a)\,(it_b)$$

$$+\sum_{a,b,c,f=1}^{s}{}^{222}\tilde{\bar{A}}'_{111}\big(a(x,y,z),b(x),c(y),f(z)\big)\,(it_a)\,(it_b)\,(it_c)\,(it_f)$$

$$+\frac{1}{2}\sum_{a,b=1}^{s}\left[{}^{2}\tilde{\bar{A}}'_{1}(a)\,{}^{2}\tilde{\bar{A}}'_{1}(b)+2\,{}^{22}\tilde{\bar{A}}'_{11}\big(a(y,z),b(y,z)\big)\right](it_a)\,(it_b)$$

$$+\sum_{a,b,c,f=1}^{s}\left[{}^{2}\tilde{\bar{A}}'_{1}(a)\,{}^{22}\tilde{\bar{A}}_{11}\big(b(y,z),c(y),f(z)\big)\right.$$

$$\left.+2\,{}^{222}\tilde{\bar{A}}'_{111}\big(a(x,y),b(z,x),c(y),f(z)\big)\right](it_a)\,(it_b)\,(it_c)\,(it_f)$$

$$+\frac{1}{2}\sum_{a,b,c,f,g,h=1}^{s}{}^{2222}\tilde{\bar{A}}'_{1111}\big(a(x,y),b(z,v),c(x),f(y),g(z),h(v)\big)$$

$$\left.\left.\times\,(it_a)\,(it_b)\,(it_c)\,(it_f)\,(it_g)\,(it_h)\right]\varepsilon^{m}+o(\varepsilon^{m})\right\}. \tag{1.188}$$

On the other hand, comparing with (1.54) the characteristic function can be written as

$$\tilde{\psi}(t_1,\ldots,t_s)$$

$$=\exp\left(-\frac{1}{2}\sum_{p=1}^{s}t_p^2\right)\left\{1+\sum_{\substack{k_1,\ldots,k_s=0\\k\geq1}}^{\infty}(-1)^k\frac{1}{k_1!\ldots k_s!}c_{k_1\ldots k_s}(it_1)^{k_1}\ldots(it_s)^{k_s}\right\}$$

$$=\exp\left(-\frac{1}{2}\sum_{p=1}^{s}t_p^2\right)\left\{1+\sum_{k=1}^{\infty}(-1)^k\sum_{k_1,\ldots,k_s=0}^{\infty}\frac{1}{k_1!\ldots k_s!}c_{k_1\ldots k_s}\,(it_1)^{k_1}\ldots(it_s)^{k_s}\right\}. \tag{1.189}$$

Our aim is the computation of the density function possessing the form

$$\tilde{p}(\tilde{u}_1, \ldots, \tilde{u}_s) = \frac{1}{\sqrt{2\pi}^s} \exp\left(-\frac{1}{2} \sum_{p=1}^{s} \tilde{u}_p^2\right) \times \left\{1 + \sum_{k=1}^{\infty} (-1)^k \sum_{k_1,\ldots,k_s=0}^{\infty} \frac{1}{k_1! \ldots k_s!} c_{k_1,\ldots,k_s} H_{k_1}(\tilde{u}_1) \ldots H_{k_s}(\tilde{u}_s)\right\} \tag{1.190}$$

(see also (1.53)). Now, the characteristic function can be written as

$$\begin{aligned}
&\tilde{\psi}(t_1, \ldots, t_s) \\
&= \exp\left(-\frac{1}{2} \sum_{p=1}^{s} t_p^2\right) \left\{1 + \sum_{a=1}^{s} \tilde{T}_a(it_a) + \sum_{a,b=1}^{s} \tilde{T}_{ab}(it_a)(it_b)\right. \\
&\quad + \sum_{a,b,c=1}^{s} \tilde{T}_{abc}(it_a)(it_b)(it_c) + \sum_{a,b,c,f=1}^{s} \tilde{T}_{abcf}(it_a)(it_b)(it_c)(it_f) \\
&\quad \left. + \sum_{a,b,c,f,g,h=1}^{s} \tilde{T}_{abcfgh}(it_a)(it_b)(it_c)(it_f)(it_g)(it_h)\right\} + o(\varepsilon)
\end{aligned} \tag{1.191}$$

and we obtain from (1.188)

$$\tilde{T}_a = {}^{2}\tilde{\bar{A}}_1'(a) \left[\sqrt{\varepsilon}\delta_{1m} + \varepsilon\delta_{2m}\right], \tag{1.192}$$

$$\begin{aligned}
\tilde{T}_{ab} = \Bigg[\frac{1}{2}\, {}^{2}\tilde{\bar{A}}_2'(a, b) + \Big\{{}^{3}\tilde{\bar{A}}_2'(a, b) + 3\, {}^{22}\tilde{\bar{A}}_{11}'\big(a(y, y, z), b(z)\big) \\
+ \frac{1}{2}\, {}^{2}\tilde{\bar{A}}_1'(a)\, {}^{2}\tilde{\bar{A}}_1'(b) + {}^{22}\tilde{\bar{A}}_{11}'\big(a(y, z), b(y, z)\big)\Big\}\, \delta_{1m}\Bigg]\, \varepsilon,
\end{aligned} \tag{1.193}$$

$$\tilde{T}_{abc} = \left[{}^{22}\tilde{\bar{A}}_{11}'\big(a(y, z), b(y), c(z)\big) + \frac{1}{6}\, {}^{3}\tilde{\bar{A}}_2'(a, b, c)\right] \left[\sqrt{\varepsilon}\delta_{1m} + \varepsilon\delta_{2m}\right], \tag{1.194}$$

$$\begin{aligned}
\tilde{T}_{abcf} = \Bigg[&\frac{1}{24}\, {}^{4}\underline{\tilde{\bar{A}}}_3'(a, b, c, f) + {}^{32}\tilde{\bar{A}}_{21}'\big(a(y, z), b(y), c(y), f(z)\big) \\
&+ \frac{1}{6}\, {}^{2}\tilde{\bar{A}}_1'(a)\, {}^{3}\tilde{\bar{A}}_2'(b, c, f) + {}^{222}\tilde{\bar{A}}_{111}'\big(a(x, y, z), b(x), c(y), f(z)\big) \\
&+ {}^{2}\tilde{\bar{A}}_1'(a)\, {}^{22}\tilde{\bar{A}}_{11}'\big(b(y, z), c(y), f(z)\big) \\
&+ 2\, {}^{222}\tilde{\bar{A}}_{111}'\big(a(x, y), b(z, x), c(y), f(z)\big)\Bigg]\, \delta_{1m}\varepsilon,
\end{aligned} \tag{1.195}$$

$$\begin{aligned}
\tilde{T}_{abcfgh} = \Bigg[&\frac{1}{72}\, {}^{3}\tilde{\bar{A}}_2'(a, b, c)\, {}^{3}\tilde{\bar{A}}_2'(f, g, h) \\
&+ \frac{1}{6}\, {}^{3}\tilde{\bar{A}}_2'(a, b, c)\, {}^{22}\tilde{\bar{A}}_{11}'\big(f(y, z), g(y), h(z)\big) \\
&+ {}^{2222}\tilde{\bar{A}}_{1111}'\big(a(x, y), b(z, v), c(x), f(y), g(z), h(v)\big)\Bigg]\, \delta_{1m}\varepsilon.
\end{aligned} \tag{1.196}$$

We define

$$G_k(\tilde{u}_1, \ldots, \tilde{u}_s) \doteq (-1)^k \sum_{k_1,\ldots,k_s=1}^{\infty} \frac{1}{k_1! \ldots k_s!} c_{k_1\ldots k_s} H_{k_1}(\tilde{u}_1) \ldots H_{k_s}(\tilde{u}_s) \tag{1.197}$$

and this leads to

$$\overset{\bullet}{\tilde{p}}(\tilde{u}_1, \ldots, \tilde{u}_s) = \frac{1}{\sqrt{2\pi}^s} \exp\left(-\frac{1}{2}\sum_{p=1}^{s} \tilde{u}_p^2\right)\left\{1 + \sum_{k=1}^{\infty} G_k(\tilde{u}_1, \ldots, \tilde{u}_s)\right\}. \tag{1.198}$$

A comparison between (1.189) and (1.191) shows

$$(-1)^k \sum_{k_1,\ldots,k_s=1}^{\infty} \frac{1}{k_1! \ldots k_s!} c_{k_1\ldots k_s} (\mathrm{i}t_1)^{k_1} \ldots (\mathrm{i}t_s)^{k_s}$$

$$= \sum_{a_1,a_2\ldots,a_k=1}^{s} \tilde{T}_{a_1a_2\ldots a_k} (\mathrm{i}t_{a_1}) \ldots (\mathrm{i}t_{a_k}) \quad \text{for} \quad k = 1, 2, 3, 4, 5, 6$$

where

$$\tilde{T}_{a_1\ldots a_s} = o(\varepsilon)$$

and then

$$G_1(\tilde{u}_1, \ldots, \tilde{u}_s) = \sum_{a=1}^{s} \tilde{T}_a H_1(\tilde{u}_a), \tag{1.199}$$

$$G_2(\tilde{u}_1, \ldots, \tilde{u}_s) = \sum_{\substack{a,b=1 \\ a,b\,\text{different}}}^{s} \tilde{T}_{ab} H_1(\tilde{u}_a)\, H_1(\tilde{u}_b) + \sum_{a=1}^{s} \tilde{T}_{aa} H_2(\tilde{u}_a), \tag{1.200}$$

$$G_3(\tilde{u}_1, \ldots, \tilde{u}_s) = \sum_{\substack{a,b,c=1 \\ a,b,c\,\text{different}}}^{s} \tilde{T}_{abc} H_1(\tilde{u}_a)\, H_1(\tilde{u}_b)\, H_1(\tilde{u}_c)$$
$$+ \sum_{\substack{a,b=1 \\ a,b\,\text{different}}}^{s} [\tilde{T}_{aab} + \tilde{T}_{aba} + \tilde{T}_{baa}]\, H_2(\tilde{u}_a)\, H_1(\tilde{u}_b) + \sum_{a=1}^{s} \tilde{T}_{aaa} H_3(\tilde{u}_a), \tag{1.201}$$

$$G_4(\tilde{u}_1, \ldots, \tilde{u}_s)$$
$$= \sum_{\substack{a,b,c,f=1 \\ a,b,c,f\,\text{different}}}^{s} \tilde{T}_{abcf} H_1(\tilde{u}_a)\, H_1(\tilde{u}_b)\, H_1(\tilde{u}_c)\, H_1(\tilde{u}_f)$$
$$+ \sum_{\substack{a,b,c=1 \\ a,b,c\,\text{different}}}^{s} [\tilde{T}_{aabc} + \tilde{T}_{abac} + \tilde{T}_{abca} + \tilde{T}_{baac} + \tilde{T}_{baca} + \tilde{T}_{bcaa}]\, H_2(\tilde{u}_a)\, H_1(\tilde{u}_b)\, H_1(\tilde{u}_c)$$
$$+ \sum_{\substack{a,b=1 \\ a,b\,\text{different}}}^{s} [\tilde{T}_{aabb} + \tilde{T}_{abab} + \tilde{T}_{abba}]\, H_2(\tilde{u}_a)\, H_2(\tilde{u}_b)$$
$$+ \sum_{\substack{a,b=1 \\ a,b\,\text{different}}}^{s} [\tilde{T}_{aaab} + \tilde{T}_{aaba} + \tilde{T}_{abaa} + \tilde{T}_{baaa}]\, H_3(\tilde{u}_a)\, H_1(\tilde{u}_b) + \sum_{a=1}^{s} \tilde{T}_{aaaa} H_4(\tilde{u}_a), \tag{1.202}$$

$$G_6(\tilde{u}_1, \ldots, \tilde{u}_s)$$

$$= \sum_{\substack{a,b,c,f,g,h=1 \\ a,b,c,f,g,h \text{ different}}}^{s} \tilde{T}_{abcfgh} H_1(\tilde{u}_a) H_1(\tilde{u}_b) H_1(\tilde{u}_c) H_1(\tilde{u}_f) H_1(\tilde{u}_g) H_1(\tilde{u}_h)$$

$$+ \sum_{\substack{a,b,c,f,g=1 \\ a,b,c,f,g \text{ different}}}^{s} \Bigg[\sum_{(a_1,\ldots,a_6) \text{from} (a,a,b,c,f,g)}' \tilde{T}_{a_1a_2a_3a_4a_5a_6} \Bigg] H_2(\tilde{u}_a) H_1(\tilde{u}_b) H_1(\tilde{u}_c) H_1(\tilde{u}_f) H_1(\tilde{u}_g)$$

$$+ \sum_{\substack{a,b,c,f=1 \\ a,b,c,f \text{ different}}}^{s} \Bigg[\sum_{(a_1,\ldots,a_6) \text{from} (a,a,a,b,c,f)}' \tilde{T}_{a_1a_2a_3a_4a_5a_6} \Bigg] H_3(\tilde{u}_a) H_1(\tilde{u}_b) H_1(\tilde{u}_c) H_1(\tilde{u}_f)$$

$$+ \sum_{\substack{a,b,c,f=1 \\ a,b,c,f \text{ different}}}^{s} \Bigg[\sum_{(a_1,\ldots,a_6) \text{from} (a,a,b,b,c,f)}' \tilde{T}_{a_1a_2a_3a_4a_5a_6} \Bigg] H_2(\tilde{u}_a) H_2(\tilde{u}_b) H_1(\tilde{u}_c) H_1(\tilde{u}_f)$$

$$+ \sum_{\substack{a,b,c=1 \\ a,b,c \text{ different}}}^{s} \Bigg[\sum_{(a_1,\ldots,a_6) \text{from} (a,a,a,a,b,c)}' \tilde{T}_{a_1a_2a_3a_4a_5a_6} \Bigg] H_4(\tilde{u}_a) H_1(\tilde{u}_b) H_1(\tilde{u}_c)$$

$$+ \sum_{\substack{a,b,c=1 \\ a,b,c \text{ different}}}^{s} \Bigg[\sum_{(a_1,\ldots,a_6) \text{from} (a,a,a,b,b,c)}' \tilde{T}_{a_1a_2a_3a_4a_5a_6} \Bigg] H_3(\tilde{u}_a) H_2(\tilde{u}_b) H_1(\tilde{u}_c)$$

$$+ \sum_{\substack{a,b,c=1 \\ a,b,c \text{ different}}}^{s} \Bigg[\sum_{(a_1,\ldots,a_6) \text{from} (a,a,b,b,c,c)}' \tilde{T}_{a_1a_2a_3a_4a_5a_6} \Bigg] H_2(\tilde{u}_a) H_2(\tilde{u}_b) H_2(\tilde{u}_c)$$

$$+ \sum_{\substack{a,b=1 \\ a,b \text{ different}}}^{s} \Bigg[\sum_{(a_1,\ldots,a_6) \text{from} (a,a,a,a,a,b)}' \tilde{T}_{a_1a_2a_3a_4a_5a_6} \Bigg] H_5(\tilde{u}_a) H_1(\tilde{u}_b)$$

$$+ \sum_{\substack{a,b=1 \\ a,b \text{ different}}}^{s} \Bigg[\sum_{(a_1,\ldots,a_6) \text{from} (a,a,a,a,b,b)}' \tilde{T}_{a_1a_2a_3a_4a_5a_6} \Bigg] H_4(\tilde{u}_a) H_2(\tilde{u}_b)$$

$$+ \sum_{\substack{a,b=1 \\ a,b \text{ different}}}^{s} \Bigg[\sum_{(a_1,\ldots,a_6) \text{from} (a,a,a,b,b,b)}' \tilde{T}_{a_1a_2a_3a_4a_5a_6} \Bigg] H_3(\tilde{u}_a) H_3(\tilde{u}_b)$$

$$+ \sum_{a=1}^{s} \tilde{T}_{aaaaaa} H_6(\tilde{u}_a) \tag{1.203}$$

and "$\sum'$" means that the sum is taken over all different selections $(a_1, \ldots, a_6)$ from the given indices. Two selections are different if they are not equal after a redenomination of the indices $a, b, \ldots$; e.g. the selections (a, a, b, c, f, g) and (a, a, c, g, f, b) are equal. In order to obtain an expansion of the density function we introduce $\overline{G}_k$ by the equations

$$G_2(\tilde{u}_1, \ldots, \tilde{u}_s) = \left(\overline{G}_{2,1}(\tilde{u}_1, \ldots, \tilde{u}_s) + \overline{G}_{2,2}(\tilde{u}_1, \ldots, \tilde{u}_s)\, \delta_{1m}\right) \varepsilon,$$

$$G_k(\tilde{u}_1, \ldots, \tilde{u}_s) = \overline{G}_k(\tilde{u}_1, \ldots, \tilde{u}_s)\, \delta_{1m}\varepsilon \quad \text{for} \quad k = 4, 6, \tag{1.204}$$

$$G_k(\tilde{u}_1, \ldots, \tilde{u}_s) = \overline{G}_k(\tilde{u}_1, \ldots, \tilde{u}_s) \left(\sqrt{\varepsilon}\delta_{1m} + \varepsilon\delta_{2m}\right) \quad \text{for} \quad k = 1, 3$$

and denote the term of $\tilde{T}_{a_1\ldots a_k}$ which belongs to $\overline{G}_{k,p}$ by $\overline{\tilde{T}}{}^{p}_{a_1\ldots a_k}$.

Finally, using (1.198) $\tilde{p}(\tilde{u}_1, \ldots, \tilde{u}_s)$ can be expressed by

$$\begin{aligned}\tilde{p}(\tilde{u}_1, \ldots, \tilde{u}_s) &= \frac{1}{\sqrt{2\pi}^s} \exp\left(-\frac{1}{2}\sum_{p=1}^{s} \tilde{u}_p^2\right)\\ &\times \Big\{1 + [\overline{G}_1(\tilde{u}_1, \ldots, \tilde{u}_s) + \overline{G}_3(\tilde{u}_1, \ldots, \tilde{u}_s)]\, \delta_{1m}\sqrt{\varepsilon}\\ &+ \Big[\big(\overline{G}_1(\tilde{u}_1, \ldots, \tilde{u}_s) + \overline{G}_3(\tilde{u}_1, \ldots, \tilde{u}_s)\big)\, \delta_{2m} + \overline{G}_{2,1}(\tilde{u}_1, \ldots, \tilde{u}_s)\\ &+ \{\overline{G}_{2,2}(\tilde{u}_1, \ldots, \tilde{u}_s) + \overline{G}_4(\tilde{u}_1, \ldots, \tilde{u}_s)\\ &+ \overline{G}_6(\tilde{u}_1, \ldots, \tilde{u}_s)\}\, \delta_{1m}\Big]\, \varepsilon\Big\} + o(\varepsilon)\,. \qquad (1.205)\end{aligned}$$

Some special cases are to be considered. First, we put $s = 1$. Then it follows

$$\overline{G}_1(\tilde{u}) = {}^2\tilde{\bar{A}}_1'(1)\, H_1(\tilde{u})\,,$$

$$\overline{G}_{2,1}(\tilde{u}) = \frac{1}{2}\, {}^2\tilde{\bar{A}}_2'(1, 1)\, H_2(\tilde{u})\,,$$

$$\begin{aligned}\overline{G}_{2,2}(\tilde{u}) &= \Big[\frac{1}{2}\left({}^2\tilde{\bar{A}}_1'(1)\right)^2 + {}^3\tilde{\bar{A}}_2'(1, 1) + {}^{22}\tilde{\bar{A}}_{11}'\big(1(y, z), 1(y, z)\big)\\ &+ 3\; {}^{22}\tilde{\bar{A}}_{11}'\big(1(y, y, z), 1(z)\big)\Big]\, H_2(\tilde{u})\,,\end{aligned}$$

$$\overline{G}_3(\tilde{u}) = \Big[\frac{1}{6}\, {}^3\tilde{\bar{A}}_2'(1, 1, 1) + {}^{22}\tilde{\bar{A}}_{11}'\big(1(y, z), 1(y), 1(z)\big)\Big]\, H_3(\tilde{u})\,,$$

$$\begin{aligned}\overline{G}_4(\tilde{u}) &= \frac{1}{24}\, {}^4\tilde{\bar{A}}_3'(1, 1, 1, 1)\, H_4(\tilde{u})\\ &+ \Big[{}^2\tilde{\bar{A}}_1'(1)\left\{\frac{1}{6}\, {}^3\tilde{\bar{A}}_2'(1, 1) + {}^{22}\tilde{\bar{A}}_{11}'\big(1(y, z), 1(y), 1(z)\big)\right\}\\ &+ {}^{32}\tilde{\bar{A}}_{21}'\big(1(y, z), 1(y), 1(y), 1(z)\big) + {}^{222}\tilde{\bar{A}}_{111}'\big(1(x, y, z), 1(x), 1(y), 1(z)\big)\\ &+ 2\; {}^{222}\tilde{\bar{A}}_{111}'\big(1(x, y), 1(z, x), 1(y), 1(z)\big)\Big]\, H_4(\tilde{u})\,,\end{aligned}$$

$$\begin{aligned}\overline{G}_6(\tilde{u}) &= \frac{1}{2}\left(\frac{1}{6}\, {}^3\tilde{\bar{A}}_2'(1, 1, 1)\right)^2 H_6(\tilde{u}) + \Big[\frac{1}{6}\, {}^3\tilde{\bar{A}}_2'(1, 1, 1)\; {}^{22}\tilde{\bar{A}}_{11}'\big(1(y, z), 1(y), 1(z)\big)\\ &+ {}^{2222}\tilde{\bar{A}}_{1111}'\big(1(x, y), 1(z, v), 1(x), 1(y), 1(z), 1(v)\big)\Big]\, H_6(\tilde{u})\end{aligned}$$

and the density function $\tilde{p}(\tilde{u})$ has the form

$$\begin{aligned}\tilde{p}(\tilde{u}) &= \frac{1}{\sqrt{2\pi}} \exp\left(-\frac{1}{2}\tilde{u}^2\right)\Big\{1 + \Big[{}^2\tilde{\bar{A}}_1'(1)\, H_1(\tilde{u}) + \Big(\frac{1}{6}\, {}^3\tilde{\bar{A}}_2'(1, 1, 1)\\ &+ {}^{22}\tilde{\bar{A}}_{11}'\big(1(y, z), 1(y), 1(z)\big)\Big)\, H_3(\tilde{u})\Big]\sqrt{\varepsilon}\\ &+ \Big[\Big\{\frac{1}{2}\, {}^2\tilde{\bar{A}}_2'(1, 1) + \frac{1}{2}\left({}^2\tilde{\bar{A}}_1'(1)\right)^2 + {}^3\tilde{\bar{A}}_2'(1, 1)\end{aligned}$$

$$+ {}^{22}\tilde{A}'_{11}\big(1(y,z), 1(y,z)\big) + 3\, {}^{22}\tilde{A}'_{11}\big(1(y,y,z), 1(z)\big)\Big\} H_2(\tilde{u})$$

$$+ \left\{\frac{1}{24}\, {}^4\underline{\tilde{A}}'_3(1,1,1,1) + {}^2\tilde{A}'_1(1)\left(\frac{1}{6}\, {}^3\tilde{A}'_2(1,1) + {}^{22}\tilde{A}'_{11}\big(1(y,z), 1(y), 1(z)\big)\right)\right.$$

$$+ {}^{32}\tilde{A}'_{21}\big(1(y,z), 1(y), 1(y), 1(z)\big) + {}^{222}\tilde{A}'_{111}\big(1(x,y,z), 1(x), 1(y), 1(z)\big)$$

$$\left. + 2\, {}^{222}\tilde{A}'_{111}\big(1(x,y), 1(z,x), 1(y), 1(z)\big)\right\} H_4(\tilde{u})$$

$$+ \left\{\frac{1}{2}\left(\frac{1}{6}\, {}^3\tilde{A}'_2(1,1,1)\right)^2 + \frac{1}{6}\, {}^3\tilde{A}'_2(1,1,1)\, {}^{22}\tilde{A}'_{11}\big(1(y,z), 1(y), 1(z)\big)\right.$$

$$\left.\left. \left. + {}^{2222}\tilde{A}'_{1111}\big(1(x,y), 1(z,v), 1(x), 1(y), 1(z), 1(v)\big)\right\} H_6(\tilde{u})\right] \varepsilon + o(\varepsilon)\right\}$$

$$\text{for} \quad m = 1, \tag{1.206.1}$$

$$\tilde{p}(\tilde{u}) = \frac{1}{\sqrt{2\pi}} \exp\left(-\frac{1}{2}\,\tilde{u}^2\right)\left\{1 + \left[\left\{{}^2\tilde{A}'_1(1)\, H_1(\tilde{u})\right.\right.\right.$$

$$\left. + \left\{\frac{1}{6}\, {}^3\tilde{A}'_2(1,1,1) + {}^{22}\tilde{A}'_{11}\big(1(y,z), 1(y), 1(z)\big)\right\} H_3(\tilde{u})\right\} \delta_{2m}$$

$$\left.\left. + \frac{1}{2}\, {}^2\tilde{A}'_2(1,1)\, H_2(\tilde{u})\right] \varepsilon + o(\varepsilon)\right\} \quad \text{for} \quad m \geqq 2 \tag{1.206.2}$$

where the statistical characteristics containing in $\tilde{p}(\tilde{u})$ are given by the relations after (1.170) and

$${}^{22}\tilde{A}'_{11}\big(1(y,z), 1(y), 1(z)\big) = b_{11}^3 \sum_{a,b,c,f=1}^{n} d_{,ab} d_{,c} d_{,f}\, {}^2\hat{A}'_1(a,c)\, {}^2\hat{A}'_1(b,f),$$

$${}^3\tilde{A}'_2(1,1,1) = b_{11}^3 \sum_{a,b,c=1}^{n} d_{,a} d_{,b} d_{,c}\, {}^3\hat{A}'_2(a,b,c),$$

$${}^4\underline{\tilde{A}}'_3(1,1,1,1) = b_{11}^4 \sum_{a,b,c,f=1}^{n} d_{,a} d_{,b} d_{,c} d_{,f}\, {}^4\underline{\hat{A}}'_3(a,b,c,f),$$

$${}^{32}\tilde{A}'_{21}\big(1(y,z), 1(y), 1(y), 1(z)\big) = b_{11}^4 \sum_{a,b,c,f,g=1}^{n} d_{,ab} d_{,c} d_{,f} d_{,g}\, {}^3\hat{A}'_2(a,c,f)\, {}^2\hat{A}'_1(b,g),$$

$${}^{222}\tilde{A}'_{111}\big(1(x,y,z), 1(x), 1(y), 1(z)\big)$$

$$= b_{11}^4 \sum_{a,b,c,f,g,h=1}^{n} d_{,abc} d_{,f} d_{,g} d_{,h}\, {}^2\hat{A}'_1(a,f)\, {}^2\hat{A}'_1(b,g)\, {}^2\hat{A}'_1(c,h),$$

$${}^{222}\tilde{A}'_{111}\big(1(x,y), 1(z,x), 1(y), 1(z)\big)$$

$$= b_{11}^4 \sum_{a,b,c,f,g,h=1}^{n} d_{,ab} d_{,cf} d_{,g} d_{,h}\, {}^2\hat{A}'_1(a,f)\, {}^2\hat{A}'_1(b,g)\, {}^2\hat{A}'_1(c,h),$$

$${}^{2222}\tilde{A}'_{1111}\big(1(x,y), 1(z,v), 1(x), 1(y), 1(z), 1(v)\big)$$

$$= b_{11}^6 \sum_{a,b,c,f,g,h,p,q=1}^{n} d_{,ab} d_{,cf} d_{,g} d_{,h} d_{,p} d_{,q}\, {}^2\hat{A}'_1(a,g)\, {}^2\hat{A}'_1(b,h)\, {}^2\hat{A}'_1(c,p)\, {}^2\hat{A}'_1(f,q).$$

Now we want to deal with the case $s = 2$. As functions of $\tilde{T}_{a_1 \ldots a_p}$ the G-values can be written in the following form:

$$\begin{aligned}
\bar{G}_1(\tilde{u}_1, \tilde{u}_2) &= \tilde{\bar{T}}_1 H_1(\tilde{u}_1) + \tilde{\bar{T}}_2 H_1(\tilde{u}_2), \\
\bar{G}_{2,p}(\tilde{u}_1, \tilde{u}_2) &= \tilde{\bar{T}}^p_{(12)} H_1(\tilde{u}_1) H_1(\tilde{u}_2) + \tilde{\bar{T}}^p_{11} H_2(\tilde{u}_1) + \tilde{\bar{T}}^p_{22} H_2(\tilde{u}_2), \\
\bar{G}_3(\tilde{u}_1, \tilde{u}_2) &= \tilde{\bar{T}}_{(112)} H_2(\tilde{u}_1) H_1(\tilde{u}_2) + \tilde{\bar{T}}_{(221)} H_2(\tilde{u}_2) H_1(\tilde{u}_1) \\
&\quad + \tilde{\bar{T}}_{111} H_3(\tilde{u}_1) + \tilde{\bar{T}}_{222} H_3(\tilde{u}_2), \\
\bar{G}_4(\tilde{u}_1, \tilde{u}_2) &= \tilde{\bar{T}}_{(1112)} H_3(\tilde{u}_1) H_1(\tilde{u}_2) + \tilde{\bar{T}}_{(2221)} H_3(\tilde{u}_2) H_1(\tilde{u}_1) \\
&\quad + \tilde{\bar{T}}_{(1122)} H_2(\tilde{u}_1) H_2(\tilde{u}_2) + \tilde{\bar{T}}_{1111} H_4(\tilde{u}_1) + \tilde{\bar{T}}_{2222} H_4(\tilde{u}_2), \\
\bar{G}_6(\tilde{u}_1, \tilde{u}_2) &= \tilde{\bar{T}}_{(111112)} H_5(\tilde{u}_1) H_1(\tilde{u}_2) + \tilde{\bar{T}}_{(222221)} H_5(\tilde{u}_2) H_1(\tilde{u}_1) \\
&\quad + \tilde{\bar{T}}_{(111122)} H_4(\tilde{u}_1) H_2(\tilde{u}_2) + \tilde{\bar{T}}_{(222211)} H_4(\tilde{u}_2) H_2(\tilde{u}_1) \\
&\quad + \tilde{\bar{T}}_{(111222)} H_3(\tilde{u}_1) H_3(\tilde{u}_2) + \tilde{\bar{T}}_{111111} H_6(\tilde{u}_1) + \tilde{\bar{T}}_{222222} H_6(\tilde{u}_2)
\end{aligned}$$

where $\tilde{\bar{T}}_{(a_1 \ldots a_q)}$ denotes the sum over all different permutations of the indices $a_1, \ldots, a_q$; e.g.

$$\tilde{\bar{T}}_{(111112)} = \tilde{\bar{T}}_{111112} + \tilde{\bar{T}}_{111121} + \tilde{\bar{T}}_{111211} + \tilde{\bar{T}}_{112111} + \tilde{\bar{T}}_{121111} + \tilde{\bar{T}}_{211111}.$$

$\tilde{\bar{T}}^{(p)}_{a_1 \ldots a_q}$ are given for $q = 1$ by (1.192), for $q = 2$ by (1.193), for $q = 3$ by (1.194), for $q = 4$ (1.195), and for $q = 6$ by (1.196). Using (1.205) we obtain in our case of $s = 2$

$$\begin{aligned}
\tilde{p}(\tilde{u}_1, \tilde{u}_2) &= \frac{1}{2\pi} \exp\left(-\frac{1}{2}\left(\tilde{u}_1^2 + \tilde{u}_2^2\right)\right) \\
&\quad \times \Big\{1 + [\tilde{\bar{T}}_1 H_1(\tilde{u}_1) + \tilde{\bar{T}}_2 H_1(\tilde{u}_2) + \tilde{\bar{T}}_{(112)} H_2(\tilde{u}_1) H_1(\tilde{u}_2) \\
&\quad + \tilde{\bar{T}}_{(221)} H_1(\tilde{u}_1) H_2(\tilde{u}_2) + \tilde{\bar{T}}_{111} H_3(\tilde{u}_1) + \tilde{\bar{T}}_{222} H_3(\tilde{u}_2)] \left[\delta_{1m} \sqrt{\varepsilon} + \delta_{2m} \varepsilon\right] \\
&\quad + [\tilde{\bar{T}}^1_{(12)} H_1(\tilde{u}_1) H_1(\tilde{u}_2) + \tilde{\bar{T}}^1_{11} H_2(\tilde{u}_1) + \tilde{\bar{T}}^1_{22} H_2(\tilde{u}_2) \\
&\quad + \{\tilde{\bar{T}}^2_{(12)} H_1(\tilde{u}_1) H_1(\tilde{u}_2) + \tilde{\bar{T}}^2_{11} H_2(\tilde{u}_1) + \tilde{\bar{T}}^2_{22} H_2(\tilde{u}_2) \\
&\quad + \tilde{\bar{T}}_{(1112)} H_3(\tilde{u}_1) H_1(\tilde{u}_2) + \tilde{\bar{T}}_{(2221)} H_1(\tilde{u}_1) H_3(\tilde{u}_2) \\
&\quad + \tilde{\bar{T}}_{(1122)} H_2(\tilde{u}_1) H_2(\tilde{u}_2) + \tilde{\bar{T}}_{1111} H_4(\tilde{u}_1) + \tilde{\bar{T}}_{2222} H_4(\tilde{u}_2) \\
&\quad + \tilde{\bar{T}}_{(111112)} H_5(\tilde{u}_1) H_1(\tilde{u}_2) + \tilde{\bar{T}}_{(222221)} H_1(\tilde{u}_1) H_5(\tilde{u}_2) \\
&\quad + \tilde{\bar{T}}_{(111122)} H_4(\tilde{u}_1) H_2(\tilde{u}_2) + \tilde{\bar{T}}_{(222211)} H_2(\tilde{u}_1) H_4(\tilde{u}_2) \\
&\quad + \tilde{\bar{T}}_{(111222)} H_3(\tilde{u}_1) H_3(\tilde{u}_2) \\
&\quad + \tilde{\bar{T}}_{111111} H_6(\tilde{u}_1) + \tilde{\bar{T}}_{222222} H_6(\tilde{u}_2)\} \, \delta_{1m}] \, \varepsilon\Big\} + o(\varepsilon). \qquad (1.207)
\end{aligned}$$

Consider the case of special linear functions d_i, $i = 1, 2, \ldots, n$;

$$d_i = d_{i,0} + y_i; \qquad d_{i,a} = \delta_{ia}, \qquad d_{i,ab} = 0, \qquad d_{i,abc} = 0,$$

in order to compare these results with the results of Section 1.4. Assuming these conditions from (1.192) up to (1.196) it follows

$$\tilde{T}_a = 0,$$

$$\tilde{T}_{ab} = \frac{1}{2} \, {}^2\tilde{A}_2'(a, b) \, \varepsilon,$$

$$\tilde{T}_{abc} = \frac{1}{6}\,{}^3\tilde{\bar{A}}_2'(a, b, c)\left(\sqrt{\varepsilon}\delta_{1m} + \varepsilon\delta_{2m}\right),$$

$$\tilde{T}_{abcf} = \frac{1}{24}\,{}^4\underline{\tilde{\bar{A}}}_3'(a, b, c, f)\,\delta_{1m}\varepsilon,$$

$$\tilde{T}_{abcfgh} = \frac{1}{72}\,{}^3\tilde{\bar{A}}_2'(a, b, c)\,{}^3\tilde{\bar{A}}_2'(f, g, h)\,\delta_{1m}\varepsilon$$

and with this

$$\bar{G}_1(\tilde{u}_1, \ldots, \tilde{u}_n) = 0; \qquad \bar{G}_{2,2}(\tilde{u}_1, \ldots, \tilde{u}_n) = 0,$$

$$\bar{G}_{2,1}(\tilde{u}_1, \ldots, \tilde{u}_n)$$

$$= \sum_{\substack{a,b=1\\a<b}}^{n} {}^2\tilde{\bar{A}}_2'(a, b)\,H_1(\tilde{u}_a)\,H_1(\tilde{u}_b) + \frac{1}{2}\sum_{a=1}^{n} {}^2\tilde{\bar{A}}_2'(a, a)\,H_2(\tilde{u}_a)$$

$$= \sum_{\substack{k_1,\ldots,k_n=0\\k=2}}^{\infty} {}^2\tilde{\bar{A}}_2'(\mathcal{V}_{k_1\ldots k_n})\prod_{q=1}^{n}\frac{1}{k_q!}H_{k_q}(\tilde{u}_q),$$

$$\bar{G}_3(\tilde{u}_1, \ldots, \tilde{u}_n) = \sum_{\substack{a,b,c=1\\a<b<c}}^{n} {}^3\tilde{\bar{A}}_2'(a, b, c)\,H_1(\tilde{u}_a)\,H_1(\tilde{u}_b)\,H_1(\tilde{u}_c)$$

$$+ \frac{1}{2}\sum_{\substack{a,b=1\\a\neq b}}^{n} {}^3\tilde{\bar{A}}_2'(a, a, b)\,H_2(\tilde{u}_a)\,H_1(\tilde{u}_b) + \frac{1}{6}\sum_{a=1}^{n} {}^3\tilde{\bar{A}}_2'(a, a, a)\,H_3(\tilde{u}_a)$$

$$= \sum_{\substack{k_1,\ldots,k_n=0\\k=3}}^{\infty} {}^3\tilde{\bar{A}}_2'(\mathcal{V}_{k_1\ldots k_n})\prod_{q=1}^{n}\frac{1}{k_q!}H_{k_q}(\tilde{u}_q),$$

$$\bar{G}_4(\tilde{u}_1, \ldots, \tilde{u}_n) = \sum_{\substack{a,b,c,f=1\\a<b<c<f}}^{n} {}^4\underline{\tilde{\bar{A}}}_3'(a, b, c, f)\,H_1(\tilde{u}_a)\,H_1(\tilde{u}_b)\,H_1(\tilde{u}_c)\,H_1(\tilde{u}_f)$$

$$+ \frac{1}{2}\sum_{\substack{a,b,c=1\\a\neq b,c;\,b<c}}^{n} {}^4\underline{\tilde{\bar{A}}}_3'(a, a, b, c)\,H_2(\tilde{u}_a)\,H_1(\tilde{u}_b)\,H_1(\tilde{u}_c)$$

$$+ \frac{1}{4}\sum_{\substack{a,b=1\\a\neq b}}^{n} {}^4\underline{\tilde{\bar{A}}}_3'(a, a, b, b)\,H_2(\tilde{u}_a)\,H_2(\tilde{u}_b)$$

$$+ \frac{1}{6}\sum_{\substack{a,b=1\\a\neq b}}^{n} {}^4\underline{\tilde{\bar{A}}}_3'(a, a, a, b)\,H_3(\tilde{u}_a)\,H_1(\tilde{u}_b)$$

$$+ \frac{1}{24}\sum_{a=1}^{n} {}^4\underline{\tilde{\bar{A}}}_3'(a, a, a, a)\,H_4(\tilde{u}_a)$$

$$= \sum_{\substack{k_1,\ldots,k_n=0\\k=4}}^{\infty} {}^4\underline{\tilde{\bar{A}}}_3'(\mathcal{V}_{k_1\ldots k_n})\prod_{q=1}^{n}\frac{1}{k_q!}H_{k_q}(\tilde{u}_q),$$

$$\bar{G}_6(\tilde{u}_1, \ldots, \tilde{u}_n)$$

$$= \frac{1}{72} \Bigg[\sum_{\substack{a,b,c,f,g,h=1 \\ a,b,c,f,g,h \,\text{different}}}^{n} {}^3\bar{\tilde{A}}'_2(a, b, c)\, {}^3\bar{\tilde{A}}'_2(f, g, h)\, H_1(\tilde{u}_a)\, H_1(\tilde{u}_b)\, H_1(\tilde{u}_c)$$

$$\times H_1(\tilde{u}_f)\, H_1(\tilde{u}_g)\, H_1(\tilde{u}_h)$$

$$+ \sum_{\substack{a,b,c,f,g=1 \\ a,b,c,f,g \,\text{different}}}^{n} [6\, {}^3\bar{\tilde{A}}'_2(a, a, b)\, {}^3\bar{\tilde{A}}'_2(f, g, h) + 9\, {}^3\bar{\tilde{A}}'_2(a, b, c)\, {}^3\bar{\tilde{A}}'_2(a, f, g)]$$

$$\times H_2(\tilde{u}_a)\, H_1(\tilde{u}_b)\, H_1(\tilde{u}_c)\, H_1(\tilde{u}_f)\, H_1(\tilde{u}_g)$$

$$+ \sum_{\substack{a,b,c,f=1 \\ a,b,c,f \,\text{different}}}^{n} [2\, {}^3\bar{\tilde{A}}'_2(a, a, a)\, {}^3\bar{\tilde{A}}'_2(b, c, f) + 18\, {}^3\bar{\tilde{A}}'_2(a, a, b)\, {}^3\bar{\tilde{A}}'_2(a, c, f)]$$

$$\times H_3(\tilde{u}_a)\, H_1(\tilde{u}_b)\, H_1(\tilde{u}_c)\, H_1(\tilde{u}_f)$$

$$+ \sum_{\substack{a,b,c,f=1 \\ a,b,c,f \,\text{different}}}^{n} [36\, {}^3\bar{\tilde{A}}'_2(a, a, b)\, {}^3\bar{\tilde{A}}'_2(b, c, f) + 36\, {}^3\bar{\tilde{A}}'_2(a, b, c)\, {}^3\bar{\tilde{A}}'_2(a, b, f)$$

$$+ 18\, {}^3\bar{\tilde{A}}'_2(a, a, c)\, {}^3\bar{\tilde{A}}'_2(b, b, f)]\, H_2(\tilde{u}_a)\, H_2(\tilde{u}_b)\, H_1(\tilde{u}_c)\, H_1(\tilde{u}_f)$$

$$+ \sum_{\substack{a,b,c=1 \\ a,b,c \,\text{different}}}^{n} [6\, {}^3\bar{\tilde{A}}'_2(a, a, a)\, {}^3\bar{\tilde{A}}'_2(a, b, c) + 9\, {}^3\bar{\tilde{A}}'_2(a, a, b)\, {}^3\bar{\tilde{A}}'_2(a, a, c)]$$

$$\times H_4(\tilde{u}_a)\, H_1(\tilde{u}_b)\, H_1(\tilde{u}_c)$$

$$+ \sum_{\substack{a,b,c=1 \\ a,b,c \,\text{different}}}^{n} [6\, {}^3\bar{\tilde{A}}'_2(a, a, a)\, {}^3\bar{\tilde{A}}'_2(b, b, c) + 36\, {}^3\bar{\tilde{A}}'_2(a, a, b)\, {}^3\bar{\tilde{A}}'_2(a, b, c)$$

$$+ 18\, {}^3\bar{\tilde{A}}'_2(a, a, c)\, {}^3\bar{\tilde{A}}'_2(b, b, a)]\, H_3(\tilde{u}_a)\, H_2(\tilde{u}_b)\, H_1(\tilde{u}_c)$$

$$+ \sum_{\substack{a,b,c=1 \\ a,b,c \,\text{different}}}^{n} [9\, {}^3\bar{\tilde{A}}'_2(a, a, b)\, {}^3\bar{\tilde{A}}'_2(b, c, c) + 6\, {}^3\bar{\tilde{A}}'_2(a, b, c)\, {}^3\bar{\tilde{A}}'_2(a, b, c)]$$

$$\times H_2(\tilde{u}_a)\, H_2(\tilde{u}_b)\, H_2(\tilde{u}_c)$$

$$+ \sum_{\substack{a,b=1 \\ a,b \,\text{different}}}^{n} 6\, {}^3\bar{\tilde{A}}'_2(a, a, a)\, {}^3\bar{\tilde{A}}'_2(a, a, b)\, H_5(\tilde{u}_a)\, H_1(\tilde{u}_b)$$

$$+ \sum_{\substack{a,b=1 \\ a,b \,\text{different}}}^{n} [6\, {}^3\bar{\tilde{A}}'_2(a, a, a)\, {}^3\bar{\tilde{A}}'_2(a, b, b) + 9\, {}^3\bar{\tilde{A}}'_2(a, a, b)\, {}^3\bar{\tilde{A}}'_2(a, a, b)]\, H_4(\tilde{u}_a)\, H_2(\tilde{u}_b)$$

$$+ \sum_{\substack{a,b=1 \\ a,b \,\text{different}}}^{n} [{}^3\bar{\tilde{A}}'_2(a, a, a)\, {}^3\bar{\tilde{A}}'_2(b, b, b) + 9\, {}^3\bar{\tilde{A}}'_2(a, a, b)\, {}^3\bar{\tilde{A}}'_2(a, b, b)]\, H_3(\tilde{u}_a)\, H_3(\tilde{u}_b)$$

$$+ \sum_{a=1}^{n} {}^3\bar{\tilde{A}}'_2(a, a, a)\, {}^3\bar{\tilde{A}}'_2(a, a, a)\, H_6(\tilde{u}_a) \Bigg]$$

$$= \frac{1}{2} \sum_{\substack{k_1,\dots,k_n=0 \\ k=6}}^{\infty} \sum_{\substack{p_1,\dots,p_n=0 \\ \sum_{q=1}^{n} p_q=3;\, 0\leqq k_q-p_q\leqq 3}}^{\infty} {}^3\tilde{\tilde{A}}_2'(V_{k_1-p_1\dots k_n-p_n})$$

$$\times\, {}^3\tilde{\tilde{A}}_2(V_{p_1,\dots,p_n}) \prod_{q=1}^{n} \frac{1}{(k_q-p_q)!\,p_q!} H_{k_q}(\tilde{u}_q).$$

Using (1.205) these considerations lead to (1.76).

Compute some moments from the density function $\tilde{p}(\tilde{u}_1, \dots, \tilde{u}_s)$. First, setting $s = 1$ from

$$\frac{1}{\sqrt{2\pi}} \int_{-\infty}^{\infty} \tilde{u} \exp\left(-\frac{1}{2}\tilde{u}^2\right) H_p(\tilde{u})\, d\tilde{u} = \begin{cases} 1 & \text{for} \quad p = 1, \\ 0 & \text{for} \quad p = 0, 2, 3, 4, 6 \end{cases} \tag{1.208}$$

it follows

$$\langle \tilde{d} \rangle = \int_{-\infty}^{\infty} \tilde{u}\tilde{p}(\tilde{u})\, d\tilde{u} = \begin{cases} {}^2\tilde{\tilde{A}}_1'(1) \sqrt{\varepsilon}^m + o(\varepsilon) & \text{for} \quad m = 1, 2, \\ o(\varepsilon) & \text{for} \quad m \geqq 3 \end{cases}$$

or

$$\langle d - d_0 \rangle = \begin{cases} \sum_{a,b=1}^{n} d_{,ab}\, {}^2\hat{A}_1'(a, b)\, \varepsilon^m + o(\varepsilon^{m/2+1}) & \text{for} \quad m = 1, 2, \\ o(\varepsilon^{m/2+1}) & \text{for} \quad m \geqq 3 \end{cases}$$

where the relations (1.206.1), (1.206.2) were needed. For $m = 1, 2$ we have the same results as written in (1.169). Naturally, these results for $m \geqq 3$ are weaker. From

$$\frac{1}{\sqrt{2\pi}} \int_{-\infty}^{\infty} \tilde{u}^2 \exp\left(-\frac{1}{2}\tilde{u}^2\right) H_p(\tilde{u})\, d\tilde{u} = \begin{cases} 1 & \text{for} \quad p = 0, \\ 2 & \text{for} \quad p = 2, \\ 0 & \text{for} \quad p = 1, 3, 4, 6 \end{cases}$$

with the aid of (1.206.1), (1.206.2) it follows

$$\langle \tilde{d}^2 \rangle = \int_{-\infty}^{\infty} \tilde{u}^2\tilde{p}(\tilde{u})\, d\tilde{u} = \begin{cases} 1 + \big[{}^2\tilde{\tilde{A}}_2'(1, 1) + \big({}^2\tilde{\tilde{A}}_1'(1)\big)^2 + 2\, {}^3\tilde{\tilde{A}}_2'(1, 1) \\ \quad + 2\, {}^{22}\tilde{\tilde{A}}_{11}'\big(1(y, z), 1(y, z)\big) \\ \quad + 6\, {}^{22}\tilde{\tilde{A}}_{11}'\big(1(y, y, z), 1(z)\big)\big]\, \varepsilon + o(\varepsilon) & \text{for} \quad m = 1, \\ 1 + {}^2\tilde{\tilde{A}}_2'(1, 1)\, \varepsilon + o(\varepsilon) & \text{for} \quad m \geqq 2 \end{cases}$$

and then the relation (1.170). We take into consideration

$$\frac{1}{\sqrt{2\pi}} \int_{-\infty}^{\infty} \tilde{u}^3 \exp\left(-\frac{1}{2}\tilde{u}^2\right) H_p(\tilde{u})\, d\tilde{u} = \begin{cases} 3 & \text{for} \quad p = 1, \\ 6 & \text{for} \quad p = 3, \\ 0 & \text{for} \quad p = 0, 2, 4, 6 \end{cases}$$

and obtain

$$\langle \tilde{d}^3 \rangle = \int_{-\infty}^{\infty} \tilde{u}^3\tilde{p}(\tilde{u})\, d\tilde{u} = \begin{cases} \big[3\, {}^2\tilde{\tilde{A}}_1'(1) + {}^3\tilde{\tilde{A}}_2'(1, 1, 1) \\ \quad + 6\, {}^{22}\tilde{\tilde{A}}_{11}'\big(1(y, z), 1(y), 1(z)\big)\big] \sqrt{\varepsilon}^m + o(\varepsilon) & \text{for} \quad m = 1, 2, \\ o(\varepsilon) & \text{for} \quad m \geqq 3. \end{cases}$$

In this special case the relation (1.164) contains more information as to the third moment for $m \geqq 3$ than it follows from the density function. Finally, we want to compute $\langle \tilde{d}_1 \tilde{d}_2 \rangle$ from the density. By means of (1.208) it is

$$\frac{1}{2\pi} \int_{-\infty}^{\infty} \int_{-\infty}^{\infty} \tilde{u}_1 \tilde{u}_2 \exp\left(-\frac{1}{2} (\tilde{u}_1^2 + \tilde{u}_2^2)\right) H_p(\tilde{u}_1) H_q(\tilde{u}_2) \, d\tilde{u}_1 \, d\tilde{u}_2 = \delta_{1p}\delta_{1q}$$

and with (1.207) it follows

$$\begin{aligned} \langle \tilde{d}_1 \tilde{d}_2 \rangle &= [\tilde{\bar{T}}^1_{(12)} + \tilde{\bar{T}}^2_{(12)}\delta_{1m}] \, \varepsilon + o(\varepsilon) \\ &= \left[{}^2\tilde{\bar{A}}_2'(1, 2) + \left\{{}^3\tilde{\bar{A}}_2'(1, 2) + {}^3\tilde{\bar{A}}_2'(2, 1) + 3 \, {}^{22}\tilde{\bar{A}}_{11}'\big(1(y, y, z), 2(z)\big) \right.\right. \\ &\quad + 3 \, {}^{22}\tilde{\bar{A}}_{11}'\big(2(y, y, z), 1(z)\big) + {}^2\tilde{\bar{A}}_1'(1) \, {}^2\tilde{\bar{A}}_1'(2) \\ &\quad \left.\left. + 2 \, {}^{22}\tilde{\bar{A}}_{11}'\big(1(y, z), 2(y, z)\big)\right\} \delta_{1m}\right] \varepsilon + o(\varepsilon) . \end{aligned}$$

This relation also leads to (1.174).
The distribution function of $(\tilde{d}_1, \ldots, \tilde{d}_s)$ can be written by

$$\mathsf{P}(\tilde{d}_1 < \tilde{u}_1, \ldots, \tilde{d}_s < \tilde{u}_s) = \int_{-\infty}^{\tilde{u}_1} \cdots \int_{-\infty}^{\tilde{u}_s} \tilde{p}(\tilde{v}_1, \ldots, \tilde{v}_s) \, d\tilde{v}_1 \ldots d\tilde{v}_s .$$

Then the density function of $(d_1 - d_{1,0}, \ldots, d_s - d_{s,0})$ is given by

$$p(u) = \tilde{p}\left(\frac{Tu}{\sqrt{\varepsilon}^m}\right) \frac{|\det(T)|}{\sqrt{\varepsilon}^{sm}}$$

where

$$u = (u_1, \ldots, u_s)^\mathsf{T}, \qquad \tilde{u} = (\tilde{u}_1, \ldots, \tilde{u}_s)^\mathsf{T}$$

and

$$\tilde{u} = \frac{1}{\sqrt{\varepsilon}^m} \, Tu, \quad T = (b_{ik})_{1 \leq i,k \leq s} .$$

The coefficients b_{ik} are determined by (1.145). For further considerations we can use the investigations after (1.84) in Section 1.4.

The lowest order of the density is given by (1.205) as

$$\tilde{p}_0(\tilde{u}) = \frac{1}{\sqrt{2\pi}^s} \exp\left(-\frac{1}{2} \tilde{u}^\mathsf{T}\tilde{u}\right).$$

Hence, $p_0(u)$ has the form

$$p_0(u) = \frac{1}{\sqrt{2\pi}^s} \frac{1}{\sqrt{\varepsilon^{sm} \det(\tilde{B})}} \exp\left(-\frac{1}{2} \frac{1}{\varepsilon^m} u^\mathsf{T} \tilde{B}^{-1} u\right)$$

(see (1.85)) where

$$\tilde{B} = (\tilde{B}_{pq})_{1 \leq p,q \leq s}, \quad \tilde{B}_{pq} = \sum_{a,b=1}^{n} d_{p,a} d_{q,b} B_{ab}, \quad B_{ab} = {}^2\hat{A}_1'(a, b) .$$

This result is contained in vom Scheidt, Purkert [3] (see Theorem 2.9). For the calculation of the lowest order of the density we need only the characteristic ${}^2\hat{A}_1'(a, b)$. This characteristic exists already if the functions $F_{ij}(x)$ are contained in L_2. This special case is summarized by Theorem 1.11.

Theorem 1.11. *Let d_k, $k = 1, 2, \ldots, s$, be real functions satisfying conditions* (1) *and* (2) *written in Section* 1.7. *Assume that*

(i) *the deterministic functions $F_{ij}(x) \in \mathbf{L}_2(\mathcal{D})$ where $\mathcal{D} \subset \mathbb{R}^m$ denotes a domain with piecewise smooth boundary,*

(ii) *the weakly correlated connected vector function $\big(f_{1\varepsilon}(x, \omega), \ldots, f_{l\varepsilon}(x, \omega)\big)$ on $\mathcal{D}$ possesses continuous sample functions and*

$$\langle |f_{i\varepsilon}(x)|^k \rangle \leqq c_k < \infty \quad \text{for all} \quad x \in \mathcal{D},\ i = 1, 2, \ldots, l, \quad \text{and for all} \quad k \geqq 1.$$

The limit theorem refers to

$$d_k(\omega) = d_k\big(r_{1\varepsilon}(\omega), \ldots, r_{n\varepsilon}(\omega)\big), \quad k = 1, 2, \ldots, s,$$

where

$$r_{i\varepsilon}(\omega) = \sum_{j=1}^{l} r_{ij\varepsilon}(\omega), \qquad r_{ij\varepsilon}(\omega) = \int_{\mathcal{D}} F_{ij}(x)\, f_{j\varepsilon}(x, \omega)\, \mathrm{d}x.$$

Then the random vector

$$V_\varepsilon(\omega) = \frac{1}{\sqrt{\varepsilon^m}} \big(d_1(\omega) - d_{1,0}, \ldots, d_s(\omega) - d_{s,0}\big)$$

converges in distribution to a Gaussian vector $\xi(\omega)$ as $\varepsilon \downarrow 0$,

$$\lim_{\varepsilon \downarrow 0} V_\varepsilon(\omega) = \xi(\omega).$$

The mean vector of ξ is zero and the elements of the correlation matrix are given by

$$\langle \xi_p \xi_q \rangle = \sum_{a,b=1}^{n} d_{p,a} d_{q,b}\, {}^2\hat{A}_1'(a, b) \quad \text{for} \quad p, q = 1, \ldots, s$$

where

$${}^2\hat{A}_1'(a, b) = \sum_{i,j=1}^{l} {}^2A_1(F_{ai}, F_{bj}) = \sum_{i,j=1}^{l} \int_{\mathcal{D}} F_{ai}(x)\, F_{bj}(x)\, a_{ij}(x)\, \mathrm{d}x$$

and $a_{ij}(x)$ denotes the intensity between the random fields $f_{i\varepsilon}(x, \omega)$ and $f_{j\varepsilon}(x, \omega)$.

Higher approximations for the density function of

$$\big(d_1(\omega), \ldots, d_s(\omega)\big)$$

have been given in this section where some additional conditions have to be satisfied. The main result is contained in (1.205).

2. Simulation of weakly correlated processes

2.1. Linear functionals

For simulation of weakly correlated processes we use processes of the form

$$f(x, \omega) \doteq \bar{b}_i(x)\, \zeta_i(\omega) + b_i(x)\, \zeta_{i+1}(\omega) \quad \text{for} \quad x \in [a_i, a_{i+1}], \quad i = 0, 1, \ldots, k-1, \tag{2.1}$$

where

$$a_i = \alpha + \frac{i}{k}(\beta - \alpha), \qquad i = 0, 1, \ldots, k \quad (k \text{ even}),$$

$$b_i(x) \doteq \frac{x - a_i}{a_{i+1} - a_i}, \qquad \bar{b}_i(x) \doteq 1 - b_i(x), \qquad i = 0, 1, \ldots, k-1,$$

and ζ_i, $i = 0, 1, \ldots, k$, denote independent random variables for which the moments $\langle \zeta_i^p \rangle$, $p = 1, 2, \ldots$, exist. In particular, using given smooth functions $\sigma_p(x)$ for $x \in [\alpha, \beta]$ we put

$$\langle \zeta_i \rangle = 0, \qquad \langle \zeta_i^p \rangle \doteq \sigma_p(a_i) = \sigma_{pi} \quad \text{for} \quad p = 2, 3, 4, \ldots$$

The definition of the process $f(x, \omega)$ implies that the random variables $f(x, \omega)$ and $f(y, \omega)$ are independent if x and y satisfy the inequality

$$|x - y| \geqq \frac{2}{k}(\beta - \alpha).$$

Hence, the process $f(x, \omega)$ defined by (2.1) is a weakly correlated process having the correlation length $\varepsilon = \frac{2}{k}(\beta - \alpha)$. We write $f_\varepsilon(x, \omega)$.

Assuming $x \in [a_i, a_{i+1}]$, the correlation function is given by

$$R_\varepsilon(x, y) = \langle f_\varepsilon(x)\, f_\varepsilon(y) \rangle$$

$$= \begin{cases} \bar{b}_i(x)\, b_{i-1}(y)\, \sigma_{2i} & \text{for} \quad y \in [a_{i-1}, a_i], \\ \bar{b}_i(x)\, \bar{b}_i(y)\, \sigma_{2i} + b_i(x)\, b_i(y)\, \sigma_{2i+1} & \text{for} \quad y \in [a_i, a_{i+1}], \\ b_i(x)\, \bar{b}_{i+1}(y)\, \sigma_{2i+1} & \text{for} \quad y \in [a_{i+1}, a_{i+2}], \\ 0 & \text{otherwise}. \end{cases} \tag{2.2}$$

The variance of $f_\varepsilon(x, \omega)$ is obtained by

$$\tilde{\sigma}_2(x) \doteq \langle f_\varepsilon^2(x) \rangle = \overline{b}_i^2(x)\, \sigma_{2i} + b_i^2(x)\, \sigma_{2i+1} \quad \text{for} \quad x \in [a_i, a_{i+1}]$$

and we have plotted $\tilde{\sigma}_2(x)$ for

$$(1) \quad \sigma_2(x) = r^2$$

and

$$(2) \quad \sigma_2(x) = r^2 \left[1 - \frac{4}{(\beta - \alpha)^2} \left(x - \frac{\alpha + \beta}{2} \right)^2 \right]$$

in Fig. 2.1. Furthermore, we have

$$\int_\alpha^\beta f_\varepsilon(y)\, \mathrm{d}y = \frac{h}{2} \left(\zeta_0 + \zeta_k + 2 \sum_{p=1}^{k-1} \zeta_p \right)$$

with $h \doteq (\beta - \alpha)/k$ and then

$$\int_\alpha^\beta R_\varepsilon(x, y)\, \mathrm{d}y = \left\langle f_\varepsilon(x) \int_\alpha^\beta f_\varepsilon(y)\, \mathrm{d}y \right\rangle$$

$$= \begin{cases} \dfrac{h}{2} \left(\overline{b}_0(x)\, \sigma_{20} + 2b_0(x)\, \sigma_{21} \right) & \text{for} \quad x \in [a_0, a_1], \\ h\left(\overline{b}_i(x)\, \sigma_{2i} + b_i(x)\, \sigma_{2i+1} \right) & \text{for} \quad x \in [a_i, a_{i+1}], \quad i = 1, \ldots, k-2, \\ \dfrac{h}{2} \left(2\overline{b}_{k-1}(x)\, \sigma_{2k-1} + b_{k-1}(x)\, \sigma_{2k} \right) & \text{for} \quad x \in [a_{k-1}, a_k]. \end{cases} \tag{2.3}$$

Now, we calculate the characteristics ${}^2A_1$, ${}^2A_2$, ${}^3A_2$, and ${}^4\underline{A}_3$ for this process $f_\varepsilon(x, \omega)$.

First, we consider ${}^2A_1$ and ${}^2A_2$ using the results of Section 1.6 and, in particular, formula (1.109). Assuming $(x - \varepsilon, x + \varepsilon) \subset [\alpha, \beta]$ from (2.3) it follows

$$\int_{\mathcal{K}_\varepsilon(x)} R_\varepsilon(x, y)\, \mathrm{d}y = \int_\alpha^\beta R_\varepsilon(x, y)\, \mathrm{d}y = h\left(\overline{b}_i(x)\, \sigma_{2i} + b_i(x)\, \sigma_{2i+1} \right)$$

for $\quad x \in [a_i, a_{i+1}], \quad i = 1, 2, \ldots, k - 2.$

We expanse the function σ_2 as to x and obtain

$$\sigma_{2i} = \sigma_2(a_i) = \sigma_2(x) + \sigma_2'(x)\, (a_i - x) + o(h),$$

$$\sigma_{2i+1} = \sigma_2(a_{i+1}) = \sigma_2(x) + \sigma_2'(x)\, (a_{i+1} - x) + o(h).$$

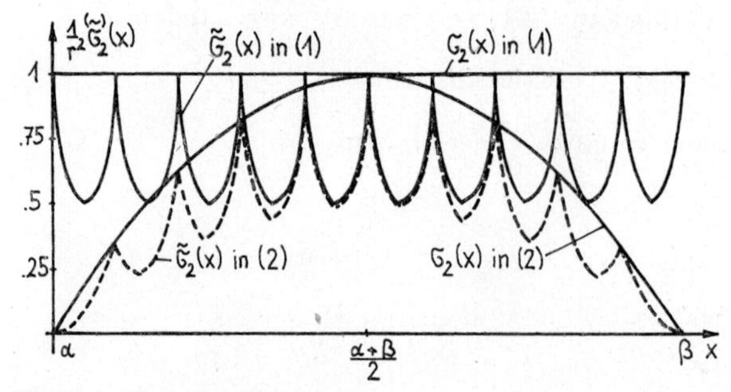

Fig. 2.1. Variance of $f_\varepsilon(x, \omega)$

Then it is

$$\int\limits_{\mathcal{K}_\varepsilon(x)} R_\varepsilon(x, y)\,\mathrm{d}y = (a_{i+1} - x)\,\sigma_{2i} + (x - a_i)\,\sigma_{2i+1} = \sigma_2(x)\,h + o(h^2)$$

and

$$a(x) = \frac{1}{2}\,\sigma_2(x), \qquad b(x) = 0 \quad \text{for} \quad x \in (\alpha, \beta). \tag{2.4}$$

Now we turn to the values

$$\int\limits_0^1 \underline{a}(u; \alpha)\,\mathrm{d}u \quad \text{and} \quad \int\limits_{-1}^0 \bar{a}(u; \beta)\,\mathrm{d}u$$

which have to be computed in our case. Applying the definition which is contained in Theorem 1.9 we obtain

$$\frac{1}{\varepsilon}\left[\int\limits_{\mathcal{K}_\varepsilon(x)\cap[\alpha,\beta]} R_\varepsilon(x, y)\,\mathrm{d}y\right]_{x=\alpha+\varepsilon u} = \frac{1}{2h}\int\limits_\alpha^{\alpha+2h(u+1)} R_\varepsilon(\alpha + \varepsilon u, y)\,\mathrm{d}y$$

$$= \begin{cases} \dfrac{1}{2h}\displaystyle\int\limits_\alpha^{a_2} R_\varepsilon(\alpha + \varepsilon u, y)\,\mathrm{d}y & \text{for} \quad 0 \leqq u \leqq \dfrac{1}{2}, \\ \dfrac{1}{2h}\displaystyle\int\limits_\alpha^{a_3} R_\varepsilon(\alpha + \varepsilon u, y)\,\mathrm{d}y & \text{for} \quad \dfrac{1}{2} \leqq u \leqq 1 \end{cases}$$

and furthermore by means of (2.3)

$$\frac{1}{2h}\int\limits_\alpha^{a_2} R_\varepsilon(\alpha + \varepsilon u, y)\,\mathrm{d}y = \frac{1}{4}\big((1 - 2u)\,\sigma_{20} + 4u\sigma_{21}\big)$$

$$= \frac{1}{4}\big((1 - 2u)\,\sigma_{20} + 4u\big(\sigma_{20} + o(h)\big)\big)$$

$$= \frac{1}{4}\,\sigma_{20}(1 + 2u) + o(h) \quad \text{for} \quad 0 \leqq u \leqq \frac{1}{2},$$

$$\frac{1}{2h}\int\limits_\alpha^{a_3} R_\varepsilon(\alpha + \varepsilon u, y)\,\mathrm{d}y = \frac{1}{2}\big((1 - 2u)\,\sigma_{21} + 2u\sigma_{22}\big) = \frac{1}{2}\,\sigma_{20} + o(h)$$

$$\text{for} \quad \frac{1}{2} \leqq u \leqq 1.$$

The function $\underline{a}(u; \alpha)$ has the form

$$\underline{a}(u; \alpha) = \begin{cases} \dfrac{1}{4}\,(1 + 2u)\,\sigma_{20} & \text{for} \quad 0 \leqq u \leqq \dfrac{1}{2}, \\ \dfrac{1}{2}\,\sigma_{20} & \text{for} \quad \dfrac{1}{2} \leqq u \leqq 1 \end{cases}$$

and it holds

$$\int_0^1 \underline{a}(u\,;\alpha)\,\mathrm{d}u = \frac{7}{16}\,\sigma_{20}$$

and analogously

$$\bar{a}(u\,;\beta) = \begin{cases} \dfrac{1}{4}\,(1 - 2u)\,\sigma_{2k} & \text{for} \quad -\dfrac{1}{2} \leqq u \leqq 0\,, \\ \dfrac{1}{2}\,\sigma_{2k} & \text{for} \quad -1 \leqq u \leqq -\dfrac{1}{2}\,. \end{cases}$$

We have

$$\begin{aligned} \underline{S}(\alpha\,;0,1) &= \int_0^1 \underline{a}(u\,;\alpha)\,\mathrm{d}u - a(\alpha) = -\frac{1}{16}\,\sigma_{20}\,, \\ \bar{S}(\beta\,;-1,0) &= \int_{-1}^0 \bar{a}(u\,;\beta)\,\mathrm{d}u - a(\beta) = -\frac{1}{16}\,\sigma_{2k}\,. \end{aligned} \tag{2.5}$$

Then the following lemma expresses the characteristics ${}^2A_1$ and ${}^2A_2$ for the linear functionals

$$r_{i\varepsilon}(\omega) = \int_\alpha^\beta F_i(x)\,f_\varepsilon(x,\omega)\,\mathrm{d}x \quad \text{for} \quad i = 1, 2$$

using Theorem 1.9 of Section 1.6.2.

Lemma 2.1. *The characteristics*

$${}^2A_1(F_1, F_2) = \frac{1}{2}\int_\alpha^\beta F_1(y)\,F_2(y)\,\sigma_2(y)\,\mathrm{d}y$$

and

$${}^2A_2(F_1, F_2) = -\frac{1}{16}\,[F_1(\alpha)\,F_2(\alpha)\,\sigma_2(\alpha) + F_1(\beta)\,F_2(\beta)\,\sigma_2(\beta)]$$

are connected with the weakly correlated process (2.1).

Secondly, we consider the characteristics ${}^3A_2$ and ${}^4\underline{A}_3$ using the expressions of Section 1.6.6. ${}^3A_2$ can be calculated from (1.139) using (1.138):

$$\begin{aligned} {}^3A_2(F_1, F_2, F_3) &= \int_\alpha^\beta \prod_{p=1}^3 F_p(y)\,a_{111}(y)\,\mathrm{d}y\,, \\ a_{111}(y) &= \lim_{\varepsilon\downarrow 0}\frac{1}{\varepsilon^2}\iint_{\tilde{\mathcal{M}}(y)} \langle f_\varepsilon(y)\,f_\varepsilon(y_2)\,f_\varepsilon(y_3)\rangle\,\mathrm{d}y_2\,\mathrm{d}y_3\,. \end{aligned}$$

Define the sets

$$\mathcal{S}_p \doteq (a_{i+p-3}, a_{i+p-2}) \quad \text{for} \quad p = 1, 2, \ldots, 5$$

and

$$\mathscr{S}_{pq} \doteq \mathscr{S}_p \times \mathscr{S}_q$$

and take $y \in [a_i, a_{i+1}]$ with $i = 3, 4, \ldots, k - 3$ then we obtain

$$\mathscr{M}(y) = [\mathscr{S}_{22} \cup \mathscr{S}_{23}] \cup [\mathscr{S}_{32} \cup \mathscr{S}_{33} \cup \mathscr{S}_{34}] \cup [\mathscr{S}_{43} \cup \mathscr{S}_{44}],$$

$$\mathscr{M}_{12}(y) = \mathscr{S}_{12} \cup \mathscr{S}_{54}, \quad \mathscr{M}_{13}(y) = \mathscr{S}_{21} \cup \mathscr{S}_{45}, \quad \mathscr{M}_{23}(y) = \mathscr{S}_{24} \cup \mathscr{S}_{42}.$$

Taking into consideration the independence of ζ_i, $i = 0, 1, \ldots, k$, we have

$$\iint\limits_{\tilde{\mathscr{M}}(y)} \langle f_\varepsilon(y) f_\varepsilon(y_2) f_\varepsilon(y_3)\rangle \, dy_2 \, dy_3$$
$$= \Bigg[\int\limits_{\mathscr{S}_2} dy_2 \Big(\int\limits_{\mathscr{S}_2} dy_3 + \int\limits_{\mathscr{S}_3} dy_3\Big) + \int\limits_{\mathscr{S}_3} dy_2 \Big(\int\limits_{\mathscr{S}_2} dy_3 + \int\limits_{\mathscr{S}_3} dy_3 + \int\limits_{\mathscr{S}_4} dy_3\Big)$$
$$+ \int\limits_{\mathscr{S}_4} dy_2 \Big(\int\limits_{\mathscr{S}_3} dy_3 + \int\limits_{\mathscr{S}_4} dy_3\Big)\Bigg] \langle f_\varepsilon(y) f_\varepsilon(y_2) f_\varepsilon(y_3)\rangle. \tag{2.6}$$

Easy calculations lead to

$$\int\limits_{a_p}^{a_{p+1}} f_\varepsilon(y, \omega) \, dy = \frac{h}{2} (\zeta_p + \zeta_{p+1})$$

and

$$\langle \zeta_a \zeta_b \zeta_c \rangle = \sigma_{3a} \delta_{abc}$$

where

$$\delta_{abc} \doteq \begin{cases} 1 & \text{for} \quad a = b = c, \\ 0 & \text{otherwise.} \end{cases}$$

Then, the determination of the integrals results in

$$\iint\limits_{\tilde{\mathscr{M}}(y)} \langle f_\varepsilon(y) f_\varepsilon(y_2) f_\varepsilon(y_3)\rangle \, dy_2 \, dy_3$$
$$= \frac{1}{4} h^2 \big\langle \big(\bar{b}_i(y) \zeta_i + b_i(y) \zeta_{i+1}\big) [(\zeta_{i-1} + \zeta_i)(\zeta_{i-1} + 2\zeta_i + \zeta_{i+1})$$
$$+ (\zeta_i + \zeta_{i+1})(\zeta_{i-1} + 2\zeta_i + 2\zeta_{i+1} + \zeta_{i+2}) + (\zeta_{i+1} + \zeta_{i+2})(\zeta_i + 2\zeta_{i+1} + \zeta_{i+2})] \big\rangle$$
$$= h^2 \big(\bar{b}_i(y) \sigma_{3i} + b_i(y) \sigma_{3i+1}\big).$$

We expanse

$$\sigma_{3i} = \sigma_3(y) + O(h), \qquad \sigma_{3i+1} = \sigma_3(y) + O(h).$$

Thus,

$$\frac{1}{\varepsilon^2} \iint\limits_{\tilde{\mathscr{M}}(y)} \langle f_\varepsilon(y) f_\varepsilon(y_2) f_\varepsilon(y_3)\rangle \, dy_2 \, dy_3$$
$$= \frac{1}{4h} \big[(a_{i+1} - y)\big(\sigma_3(y) + O(h)\big) + (y - a_i)\big(\sigma_3(y) + O(h)\big)\big] = \frac{1}{4} \sigma_3(y) + o(1).$$

By this we have obtained

$$a_{111}(y) = \frac{1}{4}\,\sigma_3(y)$$

and

$${}^3A_2(F_1, F_2, F_3) = \frac{1}{4}\int\limits_\alpha^\beta \prod_{p=1}^{3} F_p(y)\,\sigma_3(y)\,\mathrm{d}y.$$

The calculation of the function $\underline{a}_{1111}(y)$ can be carried out in the same manner, of course with more extensive considerations concerning the set $\tilde{\mathcal{M}}(y)$. Let us assume that $y \in [a_i, a_{i+1}]$ for $i = 3, 4, \ldots, k - 4$ and

$$\mathcal{S}_p = (a_{i+p-4}, a_{i+p-3}) \quad \text{for} \quad p = 1, 2, \ldots, 7,$$
$$\mathcal{S}_{pqr} = \mathcal{S}_p \times \mathcal{S}_q \times \mathcal{S}_r.$$

We obtain

$$\mathcal{M}(y) = \mathcal{S}_{333} \cup \mathcal{S}_{334} \cup \mathcal{S}_{343} \cup \mathcal{S}_{344} \cup \mathcal{S}_{433} \cup \mathcal{S}_{434} \cup \mathcal{S}_{443} \cup \mathcal{S}_{444} \cup \mathcal{S}_{445} \cup \mathcal{S}_{454} \cup \mathcal{S}_{455} \cup \mathcal{S}_{544} \cup \mathcal{S}_{545} \cup \mathcal{S}_{554} \cup \mathcal{S}_{555},$$

$$\mathcal{M}_{12}(y) = \mathcal{S}_{233} \cup \mathcal{S}_{655}, \quad \mathcal{M}_{14}(y) = \mathcal{S}_{332} \cup \mathcal{S}_{556}, \quad \mathcal{M}_{24}(y) = \mathcal{S}_{345} \cup \mathcal{S}_{543},$$

$$\mathcal{M}_{13}(y) = \mathcal{S}_{323} \cup \mathcal{S}_{565}, \quad \mathcal{M}_{23}(y) = \mathcal{S}_{354} \cup \mathcal{S}_{534}, \quad \mathcal{M}_{34}(y) = \mathcal{S}_{435} \cup \mathcal{S}_{453},$$

$$\mathcal{M}_{12,13}(y) = \mathcal{S}_{223} \cup \mathcal{S}_{665}, \qquad \mathcal{M}_{13,14}(y) = \mathcal{S}_{322} \cup \mathcal{S}_{566},$$

$$\mathcal{M}_{14,34}(y) = \mathcal{S}_{342} \cup \mathcal{S}_{546}, \qquad \mathcal{M}_{12,14}(y) = \mathcal{S}_{232} \cup \mathcal{S}_{656},$$

$$\mathcal{M}_{13,23}(y) = \mathcal{S}_{423} \cup \mathcal{S}_{465}, \qquad \mathcal{M}_{23,24}(y) = \mathcal{S}_{355} \cup \mathcal{S}_{533},$$

$$\mathcal{M}_{12,23}(y) = \mathcal{S}_{243} \cup \mathcal{S}_{645}, \qquad \mathcal{M}_{13,34}(y) = \mathcal{S}_{324} \cup \mathcal{S}_{564},$$

$$\mathcal{M}_{23,34}(y) = \mathcal{S}_{535} \cup \mathcal{S}_{353}, \qquad \mathcal{M}_{12,24}(y) = \mathcal{S}_{234} \cup \mathcal{S}_{654},$$

$$\mathcal{M}_{14,24}(y) = \mathcal{S}_{432} \cup \mathcal{S}_{456}, \qquad \mathcal{M}_{24,34}(y) = \mathcal{S}_{553} \cup \mathcal{S}_{335},$$

$$\mathcal{M}_{12,13,24}(y) = \mathcal{S}_{123} \cup \mathcal{S}_{765}, \qquad \mathcal{M}_{12,23,34}(y) = \mathcal{S}_{253} \cup \mathcal{S}_{635},$$

$$\mathcal{M}_{13,23,24}(y) = \mathcal{S}_{523} \cup \mathcal{S}_{365}, \qquad \mathcal{M}_{12,13,34}(y) = \mathcal{S}_{213} \cup \mathcal{S}_{675},$$

$$\mathcal{M}_{12,24,34}(y) = \mathcal{S}_{235} \cup \mathcal{S}_{653}, \qquad \mathcal{M}_{13,24,34}(y) = \mathcal{S}_{325} \cup \mathcal{S}_{563},$$

$$\mathcal{M}_{12,14,23}(y) = \mathcal{S}_{132} \cup \mathcal{S}_{756}, \qquad \mathcal{M}_{13,14,23}(y) = \mathcal{S}_{312} \cup \mathcal{S}_{576},$$

$$\mathcal{M}_{14,23,24}(y) = \mathcal{S}_{532} \cup \mathcal{S}_{356}, \qquad \mathcal{M}_{12,14,34}(y) = \mathcal{S}_{231} \cup \mathcal{S}_{657},$$

$$\mathcal{M}_{13,14,24}(y) = \mathcal{S}_{321} \cup \mathcal{S}_{567}, \qquad \mathcal{M}_{14,23,34}(y) = \mathcal{S}_{352} \cup \mathcal{S}_{536}$$

We have to compute the integral over $\mathcal{S}_{abc}$ with the integrand

$$\begin{aligned}
&\underline{R}_{1111,\varepsilon}(y_1, y_2, y_3, y_4) \\
&= \left\langle \prod_{q=1}^{4} f_\varepsilon(y_q) \right\rangle - \left\langle \prod_{q=1,2} f_\varepsilon(y_q) \right\rangle \left\langle \prod_{q=3,4} f_\varepsilon(y_q) \right\rangle \\
&\quad - \left\langle \prod_{q=1,3} f_\varepsilon(y_q) \right\rangle \left\langle \prod_{q=2,4} f_\varepsilon(y_q) \right\rangle - \left\langle \prod_{q=1,4} f_\varepsilon(y_q) \right\rangle \left\langle \prod_{q=2,3} f_\varepsilon(y_q) \right\rangle.
\end{aligned}$$

Using

$$\int_{\mathscr{S}_p} f_\varepsilon(y, \omega)\, \mathrm{d}y = \frac{1}{2}\, h(\bar{\zeta}_p + \bar{\zeta}_{p+1}) \quad \text{with} \quad \bar{\zeta}_q \doteq \zeta_{i+q-4}$$

for $a, b, c \in \{1, 2, \ldots, 7\}$ it follows

$$\iiint_{\mathscr{S}_{abc}} \underline{R}_{1111,\varepsilon}(y, y_2, y_3, y_4)\, \mathrm{d}y_2\, \mathrm{d}y_3\, \mathrm{d}y_4$$
$$= \left(\frac{h}{2}\right)^3 \Big\{\Big\langle\big(\bar{b}_i(y)\, \bar{\zeta}_4 + b_i(y)\, \bar{\zeta}_5\big)\, (\bar{\zeta}_a + \bar{\zeta}_{a+1})\, (\bar{\zeta}_b + \bar{\zeta}_{b+1})\, (\bar{\zeta}_c + \bar{\zeta}_{c+1})\Big\rangle$$
$$- \Big\langle\big(\bar{b}_i(y)\, \bar{\zeta}_4 + b_i(y)\, \bar{\zeta}_5\big)\, (\bar{\zeta}_a + \bar{\zeta}_{a+1})\Big\rangle \langle(\bar{\zeta}_b + \bar{\zeta}_{b+1})\, (\bar{\zeta}_c + \bar{\zeta}_{c+1})\rangle$$
$$- \Big\langle\big(\bar{b}_i(y)\, \bar{\zeta}_4 + b_i(y)\, \bar{\zeta}_5\big)\, (\bar{\zeta}_b + \bar{\zeta}_{b+1})\Big\rangle \langle(\bar{\zeta}_a + \bar{\zeta}_{a+1})\, (\bar{\zeta}_c + \bar{\zeta}_{c+1})\rangle$$
$$- \Big\langle\big(\bar{b}_i(y)\, \bar{\zeta}_4 + b_i(y)\, \bar{\zeta}_5\big)\, (\bar{\zeta}_c + \bar{\zeta}_{c+1})\Big\rangle \langle(\bar{\zeta}_a + \bar{\zeta}_{a+1})\, (\bar{\zeta}_b + \bar{\zeta}_{b+1})\rangle\Big\}.$$

Hence

$$\iiint_{\mathscr{S}_{abc}} \underline{R}_{1111,\varepsilon}(y, y_2, y_3, y_4)\, \mathrm{d}y_2\, \mathrm{d}y_3\, \mathrm{d}y_4$$
$$= \left(\frac{h}{2}\right)^3 \{\bar{b}_i(y)\, \overline{\boldsymbol{M}}(4, a, b, c) + b_i(y)\, \overline{\boldsymbol{M}}(5, a, b, c)\}$$

where

$$\overline{\boldsymbol{M}}(p, a, b, c)$$
$$= \overline{m}\begin{pmatrix} p \\ a \\ b \\ c \end{pmatrix} + \overline{m}\begin{pmatrix} p \\ a \\ b \\ c+1 \end{pmatrix} + \overline{m}\begin{pmatrix} p \\ a \\ b+1 \\ c \end{pmatrix} + \overline{m}\begin{pmatrix} p \\ a \\ b+1 \\ c+1 \end{pmatrix} + \overline{m}\begin{pmatrix} p \\ a+1 \\ b \\ a \end{pmatrix}$$
$$+ \overline{m}\begin{pmatrix} p \\ a+1 \\ b \\ c+1 \end{pmatrix} + \overline{m}\begin{pmatrix} p \\ a+1 \\ b+1 \\ c \end{pmatrix} + \overline{m}\begin{pmatrix} p \\ a+1 \\ b+1 \\ c+1 \end{pmatrix}$$
$$- \left\{\left[\overline{m}\begin{pmatrix} p \\ a \end{pmatrix} + \overline{m}\begin{pmatrix} p \\ a+1 \end{pmatrix}\right]\left[\overline{m}\begin{pmatrix} b \\ c \end{pmatrix} + \overline{m}\begin{pmatrix} b \\ c+1 \end{pmatrix} + \overline{m}\begin{pmatrix} b+1 \\ c \end{pmatrix} + \overline{m}\begin{pmatrix} b+1 \\ c+1 \end{pmatrix}\right]\right.$$
$$+ \left[\overline{m}\begin{pmatrix} p \\ b \end{pmatrix} + \overline{m}\begin{pmatrix} p \\ b+1 \end{pmatrix}\right]\left[\overline{m}\begin{pmatrix} a \\ c \end{pmatrix} + \overline{m}\begin{pmatrix} a \\ c+1 \end{pmatrix} + \overline{m}\begin{pmatrix} a+1 \\ c \end{pmatrix} + \overline{m}\begin{pmatrix} a+1 \\ c+1 \end{pmatrix}\right]$$
$$\left. + \left[\overline{m}\begin{pmatrix} p \\ c \end{pmatrix} + \overline{m}\begin{pmatrix} p \\ c+1 \end{pmatrix}\right]\left[\overline{m}\begin{pmatrix} a \\ b \end{pmatrix} + \overline{m}\begin{pmatrix} a \\ b+1 \end{pmatrix} + \overline{m}\begin{pmatrix} a+1 \\ b \end{pmatrix} + \overline{m}\begin{pmatrix} a+1 \\ b+1 \end{pmatrix}\right]\right\}$$

and we have put

$$\overline{m}\begin{pmatrix} a \\ b \end{pmatrix} \doteq \langle\bar{\zeta}_a \bar{\zeta}_b\rangle, \qquad \overline{m}\begin{pmatrix} p \\ a \\ b \\ c \end{pmatrix} \doteq \langle\bar{\zeta}_p \bar{\zeta}_a \bar{\zeta}_b \bar{\zeta}_c\rangle.$$

The independence of $\bar{\zeta}_a$, $a = 1, 2, \ldots, 7$, implies that

$$\overline{m}\begin{pmatrix} a \\ b \end{pmatrix} = \bar{\sigma}_{2a}\delta_{ab},$$

$$\overline{m}\begin{pmatrix} p \\ a \\ b \\ c \end{pmatrix} = \bar{\sigma}_{4p}\delta_{pabc} + \big(\delta_{pa}\delta_{bc}(1 - \delta_{pb})\,\bar{\sigma}_{2b} + \delta_{pb}\delta_{ac}(1 - \delta_{pa})\,\bar{\sigma}_{2a}$$

$$+ \delta_{pc}\delta_{ab}(1 - \delta_{pa})\,\bar{\sigma}_{2a}\big)\,\bar{\sigma}_{2p}$$

$$= (\bar{\sigma}_{4p} - 3\bar{\sigma}_{2p}^2)\,\delta_{pabc} + \overline{m}\begin{pmatrix} p \\ a \end{pmatrix}\overline{m}\begin{pmatrix} b \\ c \end{pmatrix} + \overline{m}\begin{pmatrix} p \\ b \end{pmatrix}\overline{m}\begin{pmatrix} a \\ c \end{pmatrix} + \overline{m}\begin{pmatrix} p \\ c \end{pmatrix}\overline{m}\begin{pmatrix} a \\ b \end{pmatrix}$$

with

$$\bar{\sigma}_{pq} \doteq \sigma_{pi+q-4} \quad \text{for} \quad p = 2, 4 \quad \text{and} \quad \delta_{pabc} = \begin{cases} 1 & \text{for} \quad p = a = b = c, \\ 0 & \text{otherwise.} \end{cases}$$

We now find that

$$\overline{M}(p, a, b, c) = (\bar{\sigma}_{4p} - 3\bar{\sigma}_{2p}^2)\,(\delta_{pabc} + \delta_{pabc+1} + \delta_{pab+1c} + \delta_{pab+1c+1}$$
$$+ \delta_{pa+1bc} + \delta_{pa+1bc+1} + \delta_{pa+1b+1c} + \delta_{pa+1b+1c+1})$$

and $\overline{M}(p, a, b, c) = 0$ if one index from $\{a, b, c\}$ is greater than p or

$$\max\{p - a, p - b, p - c\} \geqq 2.$$

These facts lead to

$$\iiint\limits_{\tilde{\mathcal{M}}(y)} \underline{R}_{1111,\varepsilon}(y, y_2, y_3, y_4)\,\mathrm{d}y_2\,\mathrm{d}y_3\,\mathrm{d}y_4$$
$$= h^3\{(\bar{\sigma}_{44} - 3\bar{\sigma}_{24}^2)\,\bar{b}_i(y) + (\bar{\sigma}_{45} - 3\bar{\sigma}_{25}^2)\,b_i(y)\}$$
$$= h^3\{(\sigma_{4i} - 3\sigma_{2i}^2)\,\bar{b}_i(y) + (\sigma_{4i+1} - 3\sigma_{2i+1}^2)\,b_i(y)\}\,.$$

Again we expanse σ_{4i}, σ_{4i+1} and σ_{2i}, σ_{2i+1} as to y,

$$\sigma_p(z) = \sigma_p(y) + O(h) \quad \text{for} \quad z = a_i, \quad a_{i+1}; \quad p = 2, 4,$$

and substitute these expressions in the above formula. This results in

$$\frac{1}{\varepsilon^3}\iiint\limits_{\tilde{\mathcal{M}}(y)} \underline{R}_{1111,\varepsilon}(y, y_2, y_3, y_4)\,\mathrm{d}y_2\,\mathrm{d}y_3\,\mathrm{d}y_4$$
$$= \frac{1}{8h}\{(a_{i+1} - y)\,[\sigma_4(y) - 3\sigma_2(y)^2 + O(h)] + (y - a_i)\,[\sigma_4(y) - 3\sigma_2(y)^2 + O(h)]\}$$
$$= \frac{1}{8}\,\big(\sigma_4(y) - 3\sigma_2(y)^2\big) + o(1)\,.$$

The expressions of the characteristics ${}^3A_2$ and ${}^4\underline{A}_3$ are summarized by the following lemma.

Lemma 2.2. *As to the treated case the characteristics* ${}^3A_2$ *and* ${}^4\underline{A}_3$ *are*

$$ {}^3A_2(F_1, F_2, F_3) = \frac{1}{4} \int_\alpha^\beta F_1(y)\, F_2(y)\, F_3(y)\, \sigma_3(y)\, \mathrm{d}y , $$

$$ {}^4\underline{A}_3(F_1, F_2, F_3, F_4) = \frac{1}{8} \int_\alpha^\beta F_1(y)\, F_2(y)\, F_3(y)\, F_4(y) \left(\sigma_4(y) - 3\sigma_2(y)^2\right) \mathrm{d}y . $$

Consider linear functionals of our weakly correlated process $f_\varepsilon(x, \omega)$

$$ r_{i\varepsilon}(\omega) = \int_\alpha^\beta F_i(x)\, f_\varepsilon(x, \omega)\, \mathrm{d}x , \qquad i = 1, 2, \ldots, $$

assuming that $F_i(x) \in \mathbf{C}^1\big((\alpha, \beta)\big)$, $i = 1, 2, \ldots$ Because of the special form of the given process we obtain

$$ \begin{aligned} r_{i\varepsilon}(\omega) &= \sum_{p=0}^{k-1} \int_{a_p}^{a_{p+1}} F_i(x) \left(\bar{b}_p(x)\, \zeta_p + b_p(x)\, \zeta_{p+1}\right) \mathrm{d}x \\ &= \sum_{p=0}^{k-1} [d_{i,p}\zeta_p + c_{i,p}(\zeta_{p+1} - \zeta_p)] \\ &= (f_{i,0} - c_{i,0})\, \zeta_0 + \sum_{p=1}^{k-1} (f_{i,p} + c_{i,p-1} - c_{i,p})\, \zeta_p + c_{i,k-1}\zeta_k \doteq \sum_{p=0}^{k} g_{i,p}\zeta_p \end{aligned} \tag{2.7} $$

where $c_{i,p}$, $f_{i,p}$, and $g_{i,p}$ are defined by

$$ f_{i,p} = \int_{a_p}^{a_{p+1}} F_i(x)\, \mathrm{d}x , $$

$$ c_{i,p} = \int_{a_p}^{a_{p+1}} b_p(x)\, F_i(x)\, \mathrm{d}x = \frac{1}{h} \int_0^h x F_i(x + a_p)\, \mathrm{d}x , $$

$$ g_{i,p} = \begin{cases} f_{i,0} - c_{i,0} & \text{for } p = 0, \\ f_{i,p} + c_{i,p-1} - c_{i,p} & \text{for } p = 1, 2, \ldots, k-1, \\ c_{i,k-1} & \text{for } p = k. \end{cases} $$

Using the independence of ζ_p, $p = 0, 1, \ldots, k$, the moments of $r_{i\varepsilon}$ have the form

$$ \langle r_{i_1\varepsilon} r_{i_2\varepsilon} \ldots r_{i_t\varepsilon} \rangle = \sum_{p=0}^{k} g_{i_1,p} g_{i_2,p} \cdots g_{i_t,p} \sigma_{tp} \quad \text{for} \quad t = 2, 3, $$

$$ \begin{aligned} &\langle r_{i_1\varepsilon} r_{i_2\varepsilon} r_{i_3\varepsilon} r_{i_4\varepsilon} \rangle \\ &= \sum_{p=0}^{k} g_{i_1,p} g_{i_2,p} g_{i_3,p} g_{i_4,p} \sigma_{4p} \\ &\quad + \sum_{\substack{p,q=0 \\ p \neq q}}^{k} (g_{i_1,p} g_{i_2,p} g_{i_3,q} g_{i_4,q} + g_{i_1,p} g_{i_2,q} g_{i_3,p} g_{i_4,q} + g_{i_1,p} g_{i_2,q} g_{i_3,q} g_{i_4,p})\, \sigma_{2p} \sigma_{2q} \end{aligned} \tag{2.8} $$

for $i_1, \ldots, i_t \in \{1, 2, \ldots\}$. On the other hand the above moments can be calculated from the theory of weakly correlated processes using the formulas written at the end of Section 1.3:

$$\langle r_{i_1\varepsilon} r_{i_2\varepsilon} \rangle = {}^2A_1(F_{i_1}, F_{i_2})\, \varepsilon + {}^2A_2(F_{i_1}, F_{i_2})\, \varepsilon^2 + o(\varepsilon^2), \tag{2.9}$$

$$\langle r_{i_1\varepsilon} r_{i_2\varepsilon} r_{i_3\varepsilon} \rangle = {}^3A_2(F_{i_1}, F_{i_2}, F_{i_3})\, \varepsilon^2 + o(\varepsilon^2), \tag{2.10}$$

$$\begin{aligned}
&\langle r_{i_1\varepsilon} r_{i_2\varepsilon} r_{i_3\varepsilon} r_{i_4\varepsilon} \rangle \\
&= [{}^2A_1(F_{i_1}, F_{i_2})\, {}^2A_1(F_{i_3}, F_{i_4}) + {}^2A_1(F_{i_1}, F_{i_3})\, {}^2A_1(F_{i_2}, F_{i_4}) \\
&\quad + {}^2A_1(F_{i_1}, F_{i_4})\, {}^2A_1(F_{i_2}, F_{i_3})]\, \varepsilon^2 \\
&\quad + [{}^2A_2(F_{i_1}, F_{i_2})\, {}^2A_1(F_{i_3}, F_{i_4}) + {}^2A_1(F_{i_1}, F_{i_2})\, {}^2A_2(F_{i_3}, F_{i_4}) \\
&\quad + {}^2A_2(F_{i_1}, F_{i_3})\, {}^2A_1(F_{i_2}, F_{i_4}) + {}^2A_1(F_{i_1}, F_{i_3})\, {}^2A_2(F_{i_2}, F_{i_4}) \\
&\quad + {}^2A_2(F_{i_1}, F_{i_4})\, {}^2A_1(F_{i_2}, F_{i_3}) + {}^2A_1(F_{i_1}, F_{i_4})\, {}^2A_2(F_{i_2}, F_{i_3}) \\
&\quad + {}^4\underline{A}_3(F_{i_1}, F_{i_2}, F_{i_3}, F_{i_4})]\, \varepsilon^3 + o(\varepsilon^3).
\end{aligned} \tag{2.11}$$

Three cases are to be investigated with respect to the distribution of ζ_p:

(i) normal distribution,

(ii) uniform distribution,

(iii) logarithmic normal distribution.

Considering the case of normally distributed random variables ζ_p we deal with two different variances:

(i1) $\sigma_2(x) = r^2$ for $x \in [\alpha, \beta]$,

(i2) $\sigma_2(x) = r^2 \left(1 - \dfrac{4}{(\beta - \alpha)^2} \left(x - \dfrac{\alpha + \beta}{2}\right)^2\right)$ for $x \in [\alpha, \beta]$.

Then we have

$$\sigma_3(x) \equiv 0 \quad \text{and} \quad \sigma_4(x) = 3\sigma_2^2(x).$$

In the case (ii) we assume that ζ_p is uniformly distributed on $[-d(a_p), d(a_p)]$ and $d(x)$ is given by

$$d(x) = d_m - \frac{4d_0}{(\beta - \alpha)^2} \left(x - \frac{\alpha + \beta}{2}\right)^2. \tag{2.12}$$

Then, it follows

$$\begin{aligned}
\langle \zeta_p^2 \rangle &= \frac{1}{2d(a_p)} \int_{-d(a_p)}^{d(a_p)} x^2 \,\mathrm{d}x = \frac{1}{3}\, d^2(a_p), \\
\langle \zeta_p^3 \rangle &= 0, \\
\langle \zeta_p^4 \rangle &= \frac{1}{2d(a_p)} \int_{-d(a_p)}^{d(a_p)} x^4 \,\mathrm{d}x = \frac{1}{5}\, d^4(a_p)
\end{aligned} \tag{2.13}$$

for $p = 0, 1, \ldots, k$; hence

$$\sigma_2(x) = \frac{1}{3} d^2(x), \qquad \sigma_3(x) = 0, \qquad \sigma_4(x) = \frac{1}{5} d^4(x)$$

and the two cases

$$\text{(ii 1)} \quad d_m = \sqrt{3} r, \qquad d_0 = 0,$$

$$\text{(ii 2)} \quad d_m = \sqrt{3} r, \qquad d_0 = \frac{3}{4} \sqrt{3} r$$

are considered in detail.

In the case (iii) we suppose that ξ_p has a logarithmic normal distribution of the form

$$\varphi_p(t) = \frac{1}{\sqrt{2\pi} s_p t} \exp\left(-\frac{(\ln(t))^2}{2s_p^2}\right) \quad \text{for} \quad t > 0$$

and $s^2(x)$ is given by

$$s^2(x) = d_m - \frac{4}{(\beta - \alpha)^2} d_0 \left(x - \frac{\alpha + \beta}{2}\right)^2$$

and $s_p = s(a_p)$. We put

$$\zeta_p = \xi_p - \langle \xi_p \rangle \quad \text{with} \quad \langle \xi_p \rangle = \exp\left(\frac{1}{2} s_p^2\right).$$

Thus,

$$\begin{aligned}
\langle \zeta_p^2 \rangle &= \langle \xi_p^2 \rangle - \langle \xi_p \rangle^2 = \exp(2s_p^2) - \exp(s_p^2) = S_p(S_p - 1), \\
\langle \zeta_p^3 \rangle &= \langle \xi_p^3 \rangle - 3\langle \xi_p^2 \rangle \langle \xi_p \rangle + 2\langle \xi_p \rangle^3 \\
&= \exp\left(\frac{9}{2} s_p^2\right) - 3 \exp\left(\frac{5}{2} s_p^2\right) + 2 \exp\left(\frac{3}{2} s_p^2\right) = \sqrt{S_p}^3 (S_p^3 - 3S_p + 2), \\
\langle \zeta_p^4 \rangle &= \langle \xi_p^4 \rangle - 4\langle \xi_p^3 \rangle \langle \xi_p \rangle + 6\langle \xi_p^2 \rangle \langle \xi_p \rangle^2 - 3\langle \xi_p \rangle^4 \\
&= \exp(8s_p^2) - 4 \exp(5s_p^2) + 6 \exp(3s_p^2) - 3 \exp(2s_p^2) \\
&= S_p^2(S_p^6 - 4S_p^3 + 6S_p - 3)
\end{aligned}$$

where we have used

$$\langle \xi_p^t \rangle = \exp\left(\frac{1}{2} t^2 s_p^2\right) \quad \text{and} \quad S(x) = \exp\left(s^2(x)\right), \qquad S_p = S(a_p)$$

for $p = 0, 1, \ldots, k$. The $\sigma_q(x)$ can be written as

$$\begin{aligned}
\sigma_2(x) &= S(x)\left(S(x) - 1\right), \\
\sigma_3(x) &= \sqrt{S(x)}^3 \left(S(x)^3 - 3S(x) + 2\right), \\
\sigma_4(x) &= S(x)^2 \left(S(x)^6 - 4S(x)^3 + 6S(x) - 3\right).
\end{aligned} \tag{2.14}$$

In particular, we investigate the two special cases

$$\text{(iii 1)} \quad d_m = 0.5r, \qquad d_0 = 0.4r,$$

$$\text{(iii 2)} \quad d_m = 0.1r, \qquad d_0 = -0.4r.$$

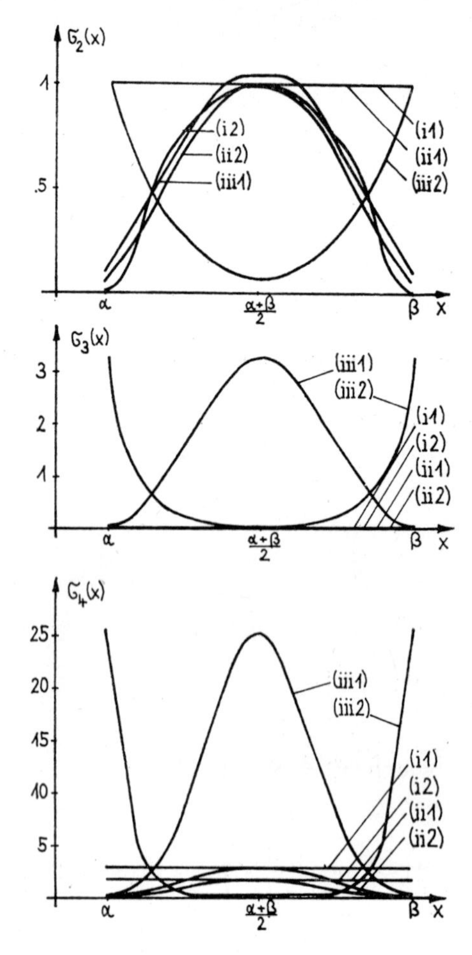

Fig. 2.2. Moments of the order 2, 3, 4 of ζ_p, $p = 0, 1, \ldots, k$

In order to compare the moments of the ζ_p in the cases (i), (ii), and (iii) we have plotted Fig. 2.2 with $r = 1$.

By means of these results we can compute the characteristics ${}^2A_1(F_{i_1}, F_{i_2})$, ${}^2A_2(F_{i_1}, F_{i_2})$ with the aid of Lemma 2.1 and ${}^3A_2(F_{i_1}, F_{i_2}, F_{i_3})$, ${}^4\underline{A}_3(F_{i_1}, F_{i_2}, F_{i_3}, F_{i_4})$ with Lemma 2.2 for the given three cases (i), (ii), (iii). For the further investigations we put

$$F_1(x) = 1,$$

$$F_2(x) = x,$$

$$F_3(x) = \sin(2\pi x),$$

$$F_4(x) = x^2 + 2x - 1$$

using $(\alpha, \beta) = (0, 1)$. These simple functions were chosen in order to demonstrate different behaviour at the boundary of the interval. Table 2.1 contains the values of ${}^2A_1(F_{i_1}, F_{i_2})$ in the various cases for $r = 1$ and Table 2.2 the adequate values of ${}^2A_2(F_{i_1}, F_{i_2})$. The notation ${}^2A_1(F_s^pF_{s+1}^q)$, ${}^2A_2(F_s^pF_{s+1}^q)$ and later ${}^3A_2(F_s^pF_{s+1}^q)$, ${}^4\underline{A}_3(F_s^pF_{s+1}^q)$ is at once clear

Table 2.1

$^2A_1(F_s^pF_{s+1}^q)$			(i1)	(i2)	(ii1)	(ii2)	(iii1)	(iii2)
s	p	q						
1	2	0	0.5000	0.3333	0.5000	0.3063	0.3429	0.1854
1	1	1	0.2500	0.1667	0.2500	0.1530	0.1715	0.0927
1	0	2	0.1667	0.1000	0.1667	0.0909	0.1031	0.0715
3	2	0	0.2500	0.1793	0.2500	0.1599	0.1787	0.0770
3	1	1	−0.2387	−0.1451	−0.2387	−0.1260	−0.1441	−0.0959
3	0	2	0.4333	0.1810	0.4333	0.1557	0.1881	0.2551

Table 2.2

$^2A_2(F_s^pF_{s+1}^q)$			(i1)	(i2)	(ii1)	(ii2)	(iii1)	(iii2)
s	p	q						
1	2	0	−0.1250	0	−0.1250	−0.0078	−0.0146	−0.1337
1	1	1	−0.0625	0	−0.0625	−0.0039	−0.0073	−0.0668
1	0	2	−0.0625	0	−0.0625	−0.0039	−0.0073	−0.0668
3	2	0	0	0	0	0	0	0
3	1	1	0	0	0	0	0	0
3	0	2	−0.3125	0	−0.3125	−0.0195	−0.0363	−0.3342

Table 2.3

$^3A_2(F_s^pF_{s+1}^q)$			(iii1)	(iii2)
s	p	q		
1	3	0	0.4013	0.1481
1	2	1	0.2006	0.0741
1	1	2	0.1139	0.0628
1	0	3	0.0706	0.0572
3	3	0	0	0
3	2	1	0.0587	0.0154
3	1	2	−0.0887	−0.0553
3	0	3	0.1332	0.3227

from the formulas after which these values are calculated. As to $^3A_2(F_{i_1}, F_{i_2}, F_{i_3})$ we obtain immediately

$$^3A_2(F_s^pF_{s+1}^q) = 0$$

in the case (i) and (ii) because of $\sigma_3(y) = 0$ (the distribution assumed in the case (i) and (ii) is symmetric). For logarithmic normally distributed ζ_p the values $^3A_2$ are summarized in Table 2.3 for $r = 1$. Finally, the characteristics $^4\underline{A}_3$ are zero in the case (i) and the other values are contained in Table 2.4.

Table 2.4

${}^4\underline{A}_3(F_s^p F_{s+1}^q)$			(ii1)	(ii2)	(iii1)	(iii2)
s	p	q				
1	4	0	−0.1500	−0.0704	1.0753	0.3077
1	3	1	−0.0750	−0.0352	0.5376	0.1538
1	2	2	−0.0500	−0.0197	0.2949	0.1377
1	1	3	−0.0375	−0.0120	0.1735	0.1296
1	0	4	−0.0300	−0.0078	0.1080	0.1231
3	4	0	−0.0562	−0.0251	0.3407	0.0285
3	3	1	0.0477	0.0180	−0.2358	−0.0381
3	2	2	−0.0556	−0.0170	0.2152	0.0755
3	1	3	0.0899	0.0193	−0.2294	−0.2443
3	0	4	−0.2652	−0.0262	0.2875	1.7354

Table 2.5a

$m_{s,pq}^t$			(i1)		(i2)	(ii1)		(ii2)	
s	p	q	$t = 1$	$t = 2$	$t = 1, 2$	$t = 1$	$t = 2$	$t = 1$	$t = 2$
1	2	0	0.1000	0.0950	0.0667	0.0999	0.0951	0.0612	0.0609
1	1	1	0.0500	0.0475	0.0333	0.0501	0.0474	0.0312	0.0312
1	0	2	0.0333	0.0308	0.0200	0.0333	0.0309	0.0183	0.0180
3	2	0	0.0500	0.0500	0.0359	0.0501	0.0501	0.0321	0.0321
3	1	1	−0.0477	−0.0477	−0.0290	−0.0477	−0.0477	−0.0252	−0.0252
3	0	2	0.0867	0.0742	0.0362	0.0867	0.0741	0.0312	0.0303

			(iii1)		(iii2)	
s	p	q	$t = 1$	$t = 2$	$t = 1$	$t = 2$
1	2	0	0.0686	0.0680	0.0371	0.0317
1	1	1	0.0343	0.0340	0.0185	0.0159
1	0	2	0.0206	0.0203	0.0143	0.0116
3	2	0	0.0357	0.0357	0.0154	0.0154
3	1	1	−0.0288	−0.0288	−0.0192	−0.0192
3	0	2	0.0376	0.0362	0.0510	0.0377

Now, we use $k = 10$ and $k = 20$ and then it follows $\varepsilon = 0.2$ and $\varepsilon = 0.1$, respectively. The moments $\langle r_{s\varepsilon}^p r_{s+1\varepsilon}^q \rangle$ which are computed by the theory of weakly correlated processes are denoted by $m_{s,pq}$ and these moments exactly by use of (2.8) are denoted by $m_{s,pq}^{\text{ex}}$. Later, the estimates of $\langle r_{s\varepsilon}^p r_{s+1\varepsilon}^q \rangle$ are denoted by $\overline{m}_{s,pq}$. By means of these characteristics and (2.9), (2.10), (2.11) it is easy to determine the approximative moments $m_{s,pq}$. Table 2.5a contains these second moments for $k = 10$ in the first and second order, $m_{s,pq}^1$,

Table 2.5b

$m^{ex}_{s,pq}$			(i1)	(i2)	(ii1)	(ii2)	(iii1)	(iii2)
s	p	q						
1	2	0	0.0950	0.0660	0.0950	0.0607	0.0676	0.0327
1	1	1	0.0475	0.0330	0.0475	0.0303	0.0338	0.0164
1	0	2	0.0308	0.0197	0.0308	0.0179	0.0201	0.0121
3	2	0	0.0470	0.0336	0.0470	0.0300	0.0335	0.0146
3	1	1	−0.0461	−0.0281	−0.0461	−0.0244	−0.0279	−0.0185
3	0	2	0.0743	0.0346	0.0743	0.0298	0.0353	0.0402

Table 2.6

		(iii1)			(iii2)		
p	q	$m^1_{3,pq}$	$m^2_{3,pq}$	$m^{ex}_{3,pq}$	$m^1_{3,pq}$	$m^2_{3,pq}$	$m^{ex}_{3,pq}$
2	0	0.0179	0.0179	0.0176	0.0077	0.0077	0.0076
1	1	−0.0144	−0.0144	−0.0143	−0.0096	−0.0096	−0.0095
0	2	0.0188	0.0184	0.0183	0.0255	0.0222	0.0225

Table 2.7

			(iii1)		(iii2)	
s	p	q	$m^2_{s,pq}$	$m^{ex}_{s,pq}$	$m^2_{s,pq}$	$m^{ex}_{s,pq}$
1	3	0	0.0161	0.0160	0.0059	0.0041
1	2	1	0.0080	0.0080	0.0030	0.0020
1	1	2	0.0046	0.0045	0.0025	0.0016
1	0	3	0.0028	0.0028	0.0023	0.0014
3	3	0	0	0	0	0
3	2	1	0.0023	0.0022	0.0006	0.0006
3	1	2	−0.0035	−0.0035	−0.0022	−0.0019
3	0	3	0.0053	0.0052	0.0129	0.0070

$m^2_{s,pq}$. The exact moments $m^{ex}_{s,pq}$ can be calculated from

$$\langle r_{i\varepsilon} r_{j\varepsilon} \rangle = \sum_{p=0}^{k} g_{i,p} g_{j,p} \sigma_{2p} \quad \text{for} \quad i, j = 1, 2$$

(see (2.8)) and are written in Table 2.5b for $k = 10$. We see that $m^2_{s,pq}$ are better than $m^1_{s,pq}$ with respect to $m^{ex}_{s,pq}$ in all cases. This agreement is better for increasing k. An example for this statement is contained in Table 2.6 written for $k = 20$. Now we consider the third moments. The cases (i) and (ii) are not interesting for the third moments

Table 2.8

$m^t_{s,pq}$			(i1)		(i2)	(ii1)		(ii2)	
s	p	q	$t = 1$	$t = 2$	$t = 1, 2$	$t = 1$	$t = 2$	$t = 1$	$t = 2$
1	4	0	0.0300	0.0270	0.0133	0.0300	0.0264	0.0113	0.0105
1	3	1	0.0150	0.0135	0.0067	0.0150	0.0129	0.0056	0.0054
1	2	2	0.0083	0.0074	0.0036	0.0083	0.0070	0.0030	0.0028
1	1	3	0.0050	0.0044	0.0020	0.0050	0.0044	0.0017	0.0016
1	0	4	0.0033	0.0028	0.0012	0.0033	0.0026	0.0010	0.0009
3	4	0	0.0075	0.0075	0.0039	0.0075	0.0071	0.0031	0.0029
3	3	1	−0.0072	−0.0072	−0.0031	−0.0072	−0.0068	−0.0024	−0.0023
3	2	2	0.0089	0.0083	0.0030	0.0089	0.0078	0.0023	0.0021
3	1	3	−0.0124	−0.0106	−0.0032	−0.0124	−0.0099	−0.0024	−0.0021
3	0	4	0.0225	0.0160	0.0039	0.0225	0.0139	0.0029	0.0025

			(iii1)		(iii2)	
s	p	q	$t = 1$	$t = 2$	$t = 1$	$t = 2$
1	4	0	0.0141	0.0225	0.0041	0.0054
1	3	1	0.0071	0.0112	0.0021	0.0027
1	2	2	0.0038	0.0061	0.0012	0.0019
1	1	3	0.0021	0.0035	0.0008	0.0016
1	0	4	0.0013	0.0021	0.0006	0.0014
3	4	0	0.0038	0.0066	0.0007	0.0009
3	3	1	−0.0031	−0.0050	−0.0009	−0.0012
3	2	2	0.0030	0.0047	0.0015	0.0019
3	1	3	−0.0033	−0.0050	−0.0029	−0.0041
3	0	4	0.0042	0.0062	0.0078	0.0176

Table 2.9

$m^{ex}_{s,pq}$			(i1)	(i2)	(ii1)	(ii2)	(iii1)	(iii2)
s	p	q						
1	4	0	0.0271	0.0131	0.0260	0.0105	0.0223	0.0044
1	3	1	0.0135	0.0065	0.0130	0.0052	0.0112	0.0022
1	2	2	0.0074	0.0035	0.0071	0.0028	0.0060	0.0014
1	1	3	0.0044	0.0020	0.0041	0.0015	0.0034	0.0010
1	0	4	0.0028	0.0012	0.0027	0.0009	0.0021	0.0008
3	4	0	0.0066	0.0034	0.0062	0.0025	0.0058	0.0008
3	3	1	−0.0065	−0.0028	−0.0062	−0.0021	−0.0045	−0.0011
3	2	2	0.0078	0.0027	0.0073	0.0020	0.0044	0.0018
3	1	3	−0.0103	−0.0029	−0.0096	−0.0020	−0.0047	−0.0036
3	0	4	0.0166	0.0036	0.0152	0.0024	0.0060	0.0094

since

$$m^{\mathrm{ex}}_{s,pq} = m^1_{s,pq} = m^2_{s,pq} = 0 \quad \text{for} \quad p + q = 3.$$

Table 2.7 contains the results for the case (iii) for $k = 10$. We remember that the third moments are zero with respect to the first order. Now we consider some results as to the fourth moments again for $k = 10$. The approximative moments $m^t_{s,pq}$ for $p + q = 4$ are written in Table 2.8 and the exact moments $m^{\mathrm{ex}}_{s,pq}$ for $p + q = 4$ in Table 2.9. The improvement of the second approximations $m^2_{s,pq}$ upon $m^1_{s,pq}$ as to $m^{\mathrm{ex}}_{s,pq}$ is obvious.

We deal with the density functions of $r_{j\varepsilon}$. Using (1.87) the approximations $p_{j,t}(u)$ of the density $p_j(u)$ of $r_{j\varepsilon}$ can be written as

$$p_{j,1}(u) = \frac{1}{\sqrt{2\pi}} \frac{1}{\sqrt{\varepsilon^2 A_1(F_j^2)}} \exp\left(-\frac{1}{2} \frac{u^2}{\varepsilon^2 A_1(F_j^2)}\right),$$

$$p_{j,2}(u) = p_{j,1}(u) \left\{1 + \frac{{}^3A_2(F_j^3)}{6({}^2A_1(F_j^2))^{3/2}} H_3\left(\frac{u}{\sqrt{\varepsilon^2 A_1(F_j^2)}}\right) \sqrt{\varepsilon}\right\},$$

$$p_{j,3}(u) = p_{j,2}(u) + p_{j,1}(u) \left\{\frac{1}{2} \frac{{}^2A_2(F_j^2)}{{}^2A_1(F_j^2)} H_2\left(\frac{u}{\sqrt{\varepsilon^2 A_1(F_j^2)}}\right)\right.$$

$$\left. + \frac{1}{24} \frac{{}^4A_3(F_j^4)}{({}^2A_1(F^2))^2} H_4\left(\frac{u}{\sqrt{\varepsilon^2 A_1(F_j^2)}}\right) + \frac{1}{72} \frac{({}^3A_2(F_j^3))^2}{({}^2A_1(F_j^2))^3} H_6\left(\frac{u}{\sqrt{\varepsilon^2 A_1(F_j^2)}}\right)\right\} \varepsilon \tag{2.15}$$

and it is

$$p_j(u) = p_{j,3}(u) + o(\varepsilon).$$

In the case (i) we know the exact density function of $r_{i\varepsilon}(\omega)$. Fig. 2.3a shows the density of the normal distribution $p_j(u)$ and the approximations $p_{j,1}(u)$ and $p_{j,3}(u)$ for $j = 1, 4$. Because of ${}^3A_2(F_j^3) = 0$ we have $p_{j,1} = p_{j,2}$. The functions p_j, $p_{j,1}$, and $p_{j,3}$ are symmetric. We have greater deviations for small values of $|u|$. On the other hand the

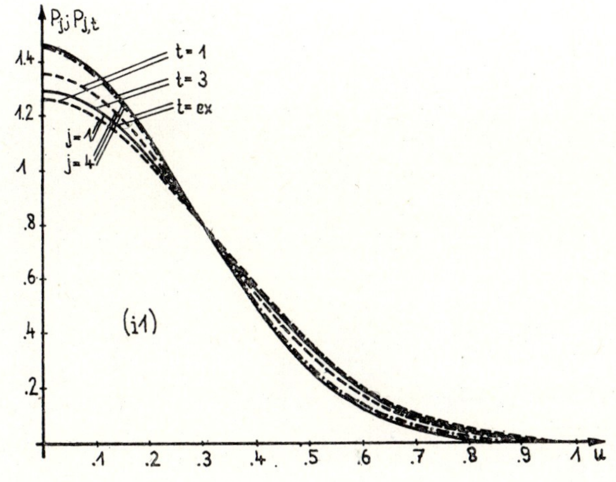

Fig. 2.3a. Density functions p_j and their approximations $p_{j,1}$ and $p_{j,3}$ for $j = 1, 4$

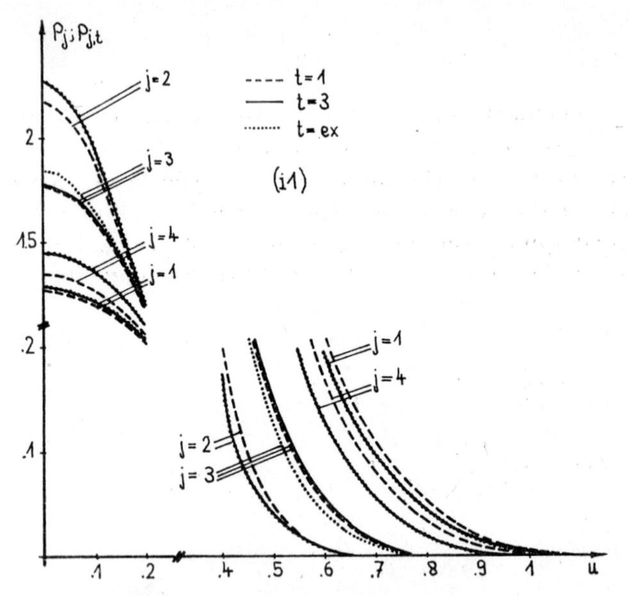

Fig. 2.3b. Density functions p_j and their approximations $p_{j,1}$ and $p_{j,3}$ for $j = 1, 2, 3, 4$ for small values $|u|$ and great values u

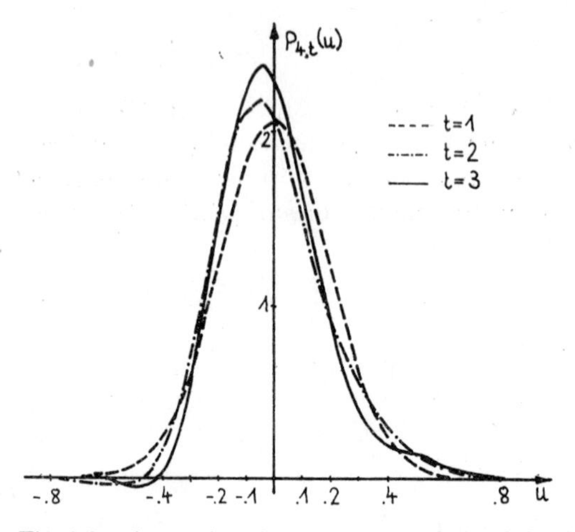

Fig. 2.3c. Approximations $p_{4,t}$, $t = 1, 2, 3$, of the density function p_4 in the case of (iii 1)

deviations between p_j and $p_{j,t}$ for great values of u are interesting as to applications. Fig. 2.3b contains p_j and $p_{j,1}$, $p_{j,3}$ for $j = 1, 2, 3, 4$ where we have only plotted these values for small values of $|u|$ and greater values of u. In all cases, $p_{j,3}$ agrees very well with p_j. The case (i2) shows a similar behaviour.

Now, we turn to the case (ii). Because of ${}^3A_2(F_j^3) = 0$ we obtain $p_{j,1} = p_{j,2}$ and $p_{j,t}$ are symmetric functions. A similar behaviour between $p_{j,1}$ and $p_{j,3}$ can be found.

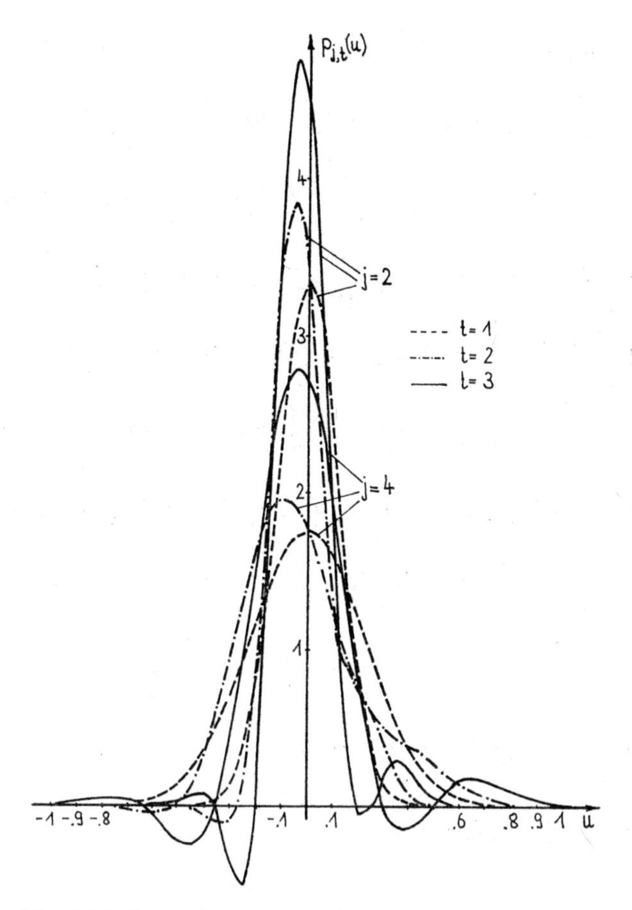

Fig. 2.3d. Approximations $p_{j,t}$, $j = 2, 4$; $t = 1, 2, 3$, of the density functions p_j, $j = 2, 4$, in the case of (iii2)

In the case (iii) $p_{j,1}$ only is a symmetric function and remarkable deviations between the approximations $p_{j,t}$ for $t = 1, 2, 3$ can be established. Fig. 2.3c shows $p_{4,t}$ for $t = 1, 2, 3$ in the case of (iii1) and Fig. 2.3d $p_{j,t}$ for $j = 2, 4$; $t = 1, 2, 3$ in the case of (iii2). In general, $p_{j,t}$ are not density functions but approximations of density functions. This is the cause that $p_{j,t}(u)$ can also have negative values. It is necessary to compute higher approximations in order to have only nonnegative values of $p_{j,t}(u)$.

Now, we discuss the case of $n = 2$. The density function of

$$(r_{j\varepsilon}, r_{j+1\varepsilon}), \qquad j = 1, 3,$$

is given by (1.79) using the relations of (1.88). We obtain

$$p_{j,1}(u_1, u_2) = \frac{b_{11}b_{22}}{2\pi\varepsilon} \exp\left(-\frac{1}{2\varepsilon}\left[(b_{11}u_1)^2 + (b_{21}u_1 + b_{22}u_2)^2\right]\right),$$

$$p_{j,2}(u_1, u_2) = p_{j,1}(u_1, u_2) \left\{1 + \frac{1}{2}\left[\frac{1}{3}\left({}^3\tilde{A}_{2,111}H_3(\tilde{u}_1) + {}^3\tilde{A}_{2,222}H_3(\tilde{u}_2)\right)\right.\right.$$

$$\left.\left. + {}^3\tilde{A}_{2,112}H_2(\tilde{u}_1)\,H_1(\tilde{u}_2) + {}^3\tilde{A}_{2,122}H_1(\tilde{u}_1)\,H_2(\tilde{u}_2)\right]\sqrt{\varepsilon}\right\},$$

$$p_{j,3}(u_1, u_2) = p_{j,2}(u_1, u_2) + p_{j,1}(u_1, u_2)\left\{\frac{1}{2}\left({}^2\tilde{A}_{2,11}H_2(\tilde{u}_1) + {}^2\tilde{A}_{2,22}H_2(\tilde{u}_2)\right)\right.$$

$$+ {}^2\tilde{A}_{2,12}H_1(\tilde{u}_1)\,H_1(\tilde{u}_2) + \frac{1}{24}\left({}^4\underline{\tilde{A}}_{3,1111}H_4(\tilde{u}_1) + {}^4\underline{\tilde{A}}_{3,2222}H_4(\tilde{u}_2)\right)$$

$$+ \frac{1}{6}\left({}^4\underline{\tilde{A}}_{3,1112}H_3(\tilde{u}_1)\,H_1(\tilde{u}_2) + {}^4\underline{\tilde{A}}_{3,1222}H_1(\tilde{u}_1)\,H_3(\tilde{u}_2)\right)$$

$$+ \frac{1}{4}\,{}^4\underline{\tilde{A}}_{3,1122}H_2(\tilde{u}_1)\,H_2(\tilde{u}_2) + \frac{1}{72}\left({}^3\tilde{A}^2_{2,111}H_6(\tilde{u}_1) + {}^3\tilde{A}^2_{2,222}H_6(\tilde{u}_2)\right)$$

$$+ \frac{1}{12}\left({}^3\tilde{A}_{2,111}{}^3\tilde{A}_{2,112}H_5(\tilde{u}_1)H_1(\tilde{u}_2) + {}^3\tilde{A}_{2,222}{}^3\tilde{A}_{2,122}H_1(\tilde{u}_1)\,H_5(\tilde{u}_2)\right)$$

$$+ \frac{1}{12}\left({}^3\tilde{A}_{2,111}{}^3\tilde{A}_{2,122}H_4(\tilde{u}_1)\,H_2(\tilde{u}_2) + {}^3\tilde{A}_{2,222}{}^3\tilde{A}_{2,112}H_2(\tilde{u}_1)\,H_4(\tilde{u}_2)\right)$$

$$+ \frac{1}{8}\left({}^3\tilde{A}^2_{2,112}H_4(\tilde{u}_1)\,H_2(\tilde{u}_2) + {}^3\tilde{A}^2_{2,122}H_2(\tilde{u}_1)\,H_4(\tilde{u}_2)\right)$$

$$+ \frac{1}{36}\,{}^3\tilde{A}_{2,111}{}^3\tilde{A}_{2,222}H_3(\tilde{u}_1)\,H_3(\tilde{u}_2)$$

$$\left. + \frac{1}{4}\,{}^3\tilde{A}_{2,112}{}^3\tilde{A}_{2,122}H_3(\tilde{u}_1)\,H_3(\tilde{u}_2)\right\}\varepsilon \tag{2.16}$$

and

$$p_j(u_1, u_2) = p_{j,3}(u_1, u_2) + o(\varepsilon).$$

We have to take into consideration

$$b_{11} = \sqrt{\frac{1}{B_{jj}}}; \qquad b_{21} = -\frac{B_{jj+1}}{B_{jj}}\sqrt{\frac{B_{jj}}{B_{jj}B_{j+1j+1} - B^2_{jj+1}}};$$

$$b_{22} = \sqrt{\frac{B_{jj}}{B_{jj}B_{j+1j+1} - B^2_{jj+1}}} \quad \text{with} \quad B_{pq} = {}^2A_1(F_p, F_q) \quad \text{for} \quad p, q = j, j+1,$$

$$\tilde{u}_1 = \frac{1}{\sqrt{\varepsilon}}\,b_{11}u_1; \qquad \tilde{u}_2 = \frac{1}{\sqrt{\varepsilon}}\,(b_{21}u_1 + b_{22}u_2),$$

$$\tilde{F}_1 = b_{11}F_j; \qquad \tilde{F}_2 = b_{21}F_j + b_{22}F_{j+1},$$

and

$$
{}^2\tilde{A}_{2,pq} \doteq {}^2A_2(\tilde{F}_p, \tilde{F}_q)
= \begin{cases}
b^2_{11}{}^2A_{2,jj} & \text{for } (p,q) \text{ from } (1,1),\\
b_{11}(b_{21}{}^2A_{2,jj} + b_{22}{}^2A_{2,jj+1}) & \text{for } (p,q) \text{ from } (1,2),\\
b^2_{21}{}^2A_{2,jj} + 2b_{21}b_{22}{}^2A_{2,jj+1} + b^2_{22}{}^2A_{2,j+1j+1} & \text{for } (p,q) \text{ from } (2,2),
\end{cases}
$$

$$
{}^3\tilde{A}_{2,pqs} \doteq {}^3A_2(\tilde{F}_p, \tilde{F}_q, \tilde{F}_s)
$$

$$
= \begin{cases}
b_{11}^3\,{}^3A_{2,jjj} & \text{for } (p,q,s) \text{ from } (1,1,1),\\
b_{11}^2(b_{21}\,{}^3A_{2,jjj} + b_{22}\,{}^3A_{2,jjj+1}) & \text{for } (p,q,s) \text{ from } (1,1,2),\\
b_{11}(b_{21}^2\,{}^3A_{2,jjj} + 2b_{21}b_{22}\,{}^3A_{2,jjj+1} + b_{22}^2\,{}^3A_{2,jj+1j+1}) & \\
 & \text{for } (p,q,s) \text{ from } (1,2,2),\\
b_{21}^3\,{}^3A_{2,jjj} + 3b_{21}^2b_{22}\,{}^3A_{2,jjj+1} + 3b_{21}b_{22}^2\,{}^3A_{2,jj+1j+1} & \\
+ b_{22}^3\,{}^3A_{2,j+1j+1j+1} & \text{for } (p,q,s) \text{ from } (2,2,2),
\end{cases}
$$

$$
{}^4\underline{\tilde{A}}_{3,pqst} \doteq {}^4\underline{A}_3(\tilde{F}_p, \tilde{F}_q, \tilde{F}_s, \tilde{F}_t)
$$

$$
= \begin{cases}
b_{11}^4\,{}^4\underline{A}_{3,jjjj} & \text{for } (p,q,s,t) \text{ from } (1,1,1,1),\\
b_{11}^3(b_{21}\,{}^4\underline{A}_{3,jjjj} + b_{22}\,{}^4\underline{A}_{3,jjjj+1}) & \text{for } (p,q,s,t) \text{ from } (1,1,1,2),\\
b_{11}^2(b_{21}^2\,{}^4\underline{A}_{3,jjjj} + 2b_{21}b_{22}\,{}^4\underline{A}_{3,jjjj+1} + b_{22}^2\,{}^4\underline{A}_{3,jjj+1j+1}) & \\
 & \text{for } (p,q,s,t) \text{ from } (1,1,2,2),\\
b_{11}(b_{21}^3\,{}^4\underline{A}_{3,jjjj} + 3b_{21}^2b_{22}\,{}^4\underline{A}_{3,jjjj+1} + 3b_{21}b_{22}^2\,{}^4\underline{A}_{3,jjj+1j+1} & \\
+ b_{22}^3\,{}^4\underline{A}_{3,jj+1j+1j+1}) & \text{for } (p,q,s,t) \text{ from } (1,2,2,2),\\
b_{21}^4\,{}^4\underline{A}_{3,jjjj} + 4b_{21}^3b_{22}\,{}^4\underline{A}_{3,jjjj+1} + 6b_{21}^2b_{22}^2\,{}^4\underline{A}_{3,jjj+1j+1} & \\
+ 4b_{21}b_{22}^3\,{}^4\underline{A}_{3,jj+1j+1j+1} + b_{22}^4\,{}^4\underline{A}_{3,j+1j+1j+1j+1} & \\
 & \text{for } (p,q,s,t) \text{ from } (2,2,2,2).
\end{cases}
$$

In the case of (iii 1) the approximations $p_{3,t}(u_1, u_2)$ for $t = 1, 2, 3$ are plotted in Fig. 2.4. Relatively great differences between $p_{3,1}(u_1, u_2)$ and $p_{3,3}(u_1, u_2)$ can be observed.

Now we introduce the intervals $\mathcal{J}_p$, $p = -k_0, -k_0 + 1, \ldots, -1, 1, 2, \ldots, k_0 - 1, k_0$ defined by

$$
\begin{aligned}
\mathcal{J}_{-k_0} &\doteq \{x\colon -\infty < x \leqq (-k_0 + 1)\,c\},\\
\mathcal{J}_{k_0} &\doteq \{x\colon (k_0 - 1)\,c < x < \infty\},\\
\mathcal{J}_k &\doteq \{x\colon kc < x \leqq (k+1)\,c\} \quad \text{for} \quad k = -k_0 + 1, -k_0 + 2, \ldots, -1,\\
\mathcal{J}_k &\doteq \{x\colon (k - 1)\,c < x \leqq kc\} \quad \text{for} \quad k = 1, 2, \ldots, k_0 - 1.
\end{aligned}
\tag{2.17}
$$

In general, with respect to the random variable $r(\omega)$ the constant c is chosen that

$$
(k_0 - 1)\,c \approx 2\sqrt{\langle (r - \langle r\rangle)^2\rangle}.
$$

In particular, we take

$$
(k_0 - 1)\,c_j \approx 2\sqrt{{}^2A_1(F_j, F_j)\,\varepsilon}
$$

for the random variable $r_{j,\varepsilon}(\omega)$. Then, the probabilities $P_{jk,t}$ are defined by

$$
P_{jk,t} = \int\limits_{\mathcal{J}_{j,k}} p_{j,t}(u)\,du \quad \text{for} \quad j = 1, 2, 3, 4; \quad t = 1, 2, 3; \quad k = -k_0, \ldots, -1, 1, \ldots, k_0
$$

and these values $P_{jk,t}$ are approximations of the propabilities

$$
\mathsf{P}\big(r_{j,\varepsilon}(\omega) \in \mathcal{J}_{j,k}\big).
$$

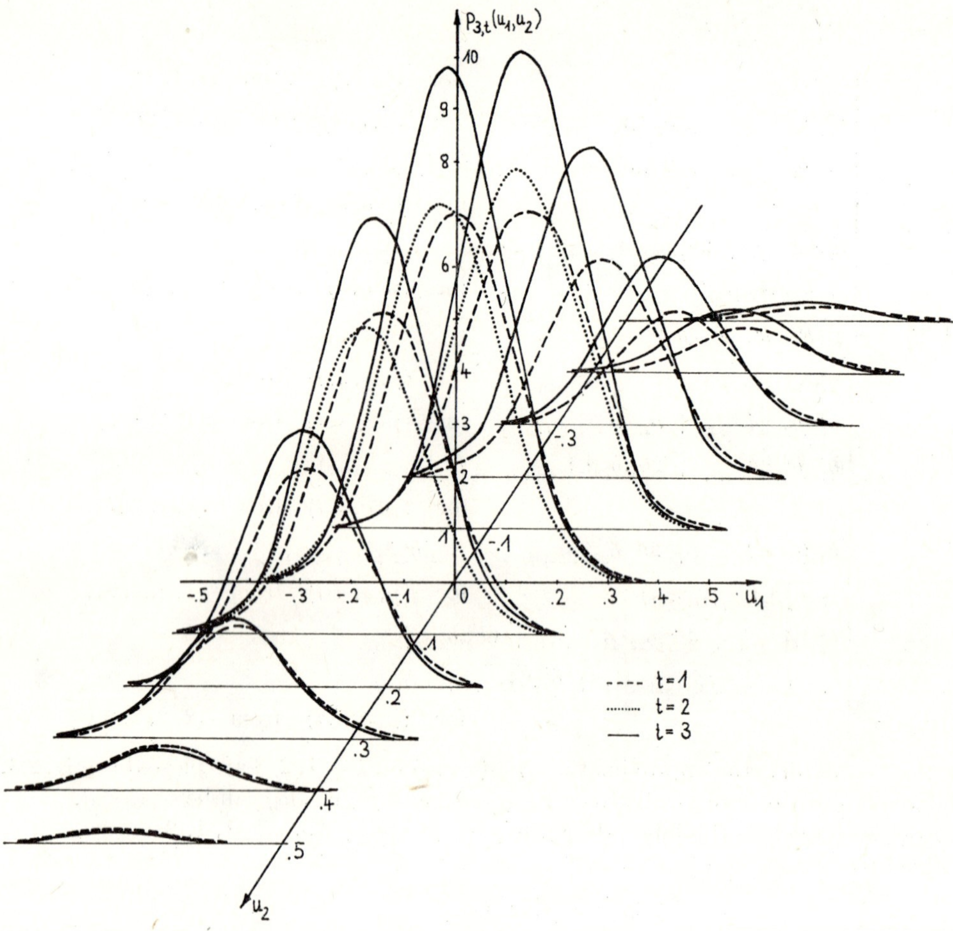

Fig. 2.4. Approximations $p_{3,t}(u_1, u_2)$ of the density function $p_3(u_1, u_2)$ in the case of (iii 1) for $\varepsilon = 0.2$

$\mathcal{J}_{j,k}$ denotes the interval $\mathcal{J}_k$ with the adequate value c_j;

$$\mathcal{J}_{j,k} = (a_{jk}, a_{jk+1}] .$$

Hence, for $t = 1$ it follows

$$P_{jk,1} = \int_{a_{jk}}^{a_{jk+1}} p_{j,1}(u) \,\mathrm{d}u = \frac{1}{\sqrt{2\pi}} \int_{\tilde{a}_{jk}}^{\tilde{a}_{jk+1}} \exp\left(-\frac{1}{2}\, u^2\right) \mathrm{d}u = \Phi(u)|_{\tilde{a}_{jk}}^{\tilde{a}_{jk+1}} \tag{2.18}$$

where

$$\tilde{a}_{jk} \doteq \frac{a_{jk}}{\sqrt{\varepsilon^2 A_{1,j}}}; \qquad {}^2A_{1,j} \doteq {}^2A_1(F_j, F_j);$$

$$\Phi(u) \doteq \frac{1}{\sqrt{2\pi}} \int_{-\infty}^{u} \exp\left(-\frac{1}{2}\, t^2\right) \mathrm{d}t .$$

Thus,

$$\tilde{a}_{j,-k_0} = -\infty; \qquad \tilde{a}_{jk} = \frac{kc_j}{\sqrt{{}^2A_{1,j}\varepsilon}} = \frac{2k}{k_0 - 1} \quad \text{for} \quad k = -k_0 + 1, \ldots, -1,$$

$$\tilde{a}_{j,k-1} = \frac{(k-1)\,c_j}{\sqrt{{}^2A_{1,j}\varepsilon}} = \frac{2(k-1)}{k_0 - 1} \quad \text{for} \quad k = 1, 2, \ldots, k_0; \qquad \tilde{a}_{jk_0} = \infty.$$

Furthermore, setting $t = 2$ we obtain

$$\begin{aligned} P_{jk,2} &= \int_{a_{jk}}^{a_{jk+1}} p_{j,2}(u)\,du \\ &= P_{jk,1} + \frac{{}^3A_{2,j}}{6({}^2A_{1,j})^{3/2}} \frac{\sqrt{\varepsilon}}{\sqrt{2\pi}} \int_{\tilde{a}_{jk}}^{\tilde{a}_{jk+1}} H_3(u) \exp\left(-\frac{1}{2}\,u^2\right) du \\ &= P_{jk,1} + c_{21,j} \sqrt{\frac{\varepsilon}{2\pi}}\,(1 - u^2) \exp\left(-\frac{1}{2}\,u^2\right)\Bigg|_{\tilde{a}_{jk}}^{\tilde{a}_{jk+1}} \end{aligned} \tag{2.19}$$

where $c_{21,j}$ is defined by

$$c_{21,j} \doteq \frac{{}^3A_{2,j}}{6\sqrt{{}^2A_{1,j}}^{\,3}}; \qquad {}^pA_{q,j} \doteq {}^pA_q\Big(F_j, \ldots, \overset{k}{\check{F}_j}\Big).$$

For $t = 3$ it is

$$\begin{aligned} P_{jk,3} = P_{jk,2} &+ \frac{\varepsilon}{\sqrt{2\pi}} \left[\frac{{}^2A_{2,j}}{2^2A_{1,j}} \int_{\tilde{a}_{jk}}^{\tilde{a}_{jk+1}} H_2(u) \exp\left(-\frac{1}{2}\,u^2\right) du \right. \\ &+ \frac{{}^4A_{3,j}}{24({}^2A_{1,j})^2} \int_{\tilde{a}_{jk}}^{\tilde{a}_{jk+1}} H_4(u) \exp\left(-\frac{1}{2}\,u^2\right) du \\ &\left. + \frac{({}^3A_{2,j})^2}{72({}^2A_{1,j})^3} \int_{\tilde{a}_{jk}}^{\tilde{a}_{jk+1}} H_6(u) \exp\left(-\frac{1}{2}\,u^2\right) du \right] \\ = P_{jk,2} &+ \frac{\varepsilon}{\sqrt{2\pi}} \left[-c_{31,j}u + c_{32,j}(3u - u^3)\right. \\ &\left. + c_{33,j}(-15u + 10u^3 - u^5)\right] \exp\left(-\frac{1}{2}\,u^2\right)\Bigg|_{\tilde{a}_{jk}}^{\tilde{a}_{jk+1}} \end{aligned} \tag{2.20}$$

where we have put

$$c_{31,j} \doteq \frac{{}^2A_{2,j}}{2^2A_{1,j}}; \qquad c_{32,j} \doteq \frac{{}^4A_{3,j}}{24({}^2A_{1,j})^2}; \qquad c_{33,j} \doteq \frac{({}^3A_{2,j})^2}{72({}^2A_{1,j})^3}.$$

We still note that from

$$\int_{-\infty}^{\infty} H_k(u) \exp\left(-\frac{1}{2}\,u^2\right) du = (-1)^k \int_{-\infty}^{\infty} \frac{d^k}{du^k} \exp\left(-\frac{1}{2}\,u^2\right) du = 0 \quad \text{for} \quad k \geqq 1$$

it follows

$$\int_{-\infty}^{\infty} p_{j,t}(u)\,du = \int_{-\infty}^{\infty} p_j(u)\,du = 1 \quad \text{for} \quad t = 1, 2, 3.$$

In the cases of (i1) and (i2) we know the exact probabilities $P_{jk,\mathrm{ex}}$. Fig. 2.5 shows the averaged absolute deviation between $P_{jk,t}$ and $P_{jk,\mathrm{ex}}$ for $k_0 = 7$:

$$AD_{j,t} \doteq \frac{1}{2k_0} \sum_{|k|=1}^{k_0} |P_{jk,t} - P_{jk,\mathrm{ex}}|.$$

The values $AD_{j,t}$ for $t = 3$ are less than or equal to the values for $t = 1, 2$ in all considered cases. In general, we have a good correspondence between $P_{jk,t}$ and $P_{jk,\mathrm{ex}}$. Fig. 2.6 contains some probabilities $P_{jk,t}$ as to the investigated cases. For (i1), (ii2) we do not find great differences between $P_{jk,1}$ and $P_{jk,3}$. We note that for (i1) the values $P_{jk,3}$ correspond very well with $P_{jk,\mathrm{ex}}$. The calculations for $t = 3$ show an improvement with respect to the results for $t = 1$. Great differences between $P_{jk,t}$ for $t = 1, 2, 3$ can be seen in the case (iii1). These differences decrease naturally if ε goes to less values. In the cases (i) and (ii) the probabilities $P_{jk,t}$ are symmetric as to k since for these cases ${}^3A_{2,j} = 0$. Considering the case (iii) the probabilities are strongly antisymmetric as to k for $t = 2, 3$ since the coefficient ${}^3A_2$ is not equal to zero. Now we will compute some probabilities as to the two-dimensional distribution. For this we introduce the intervals

$$\mathcal{I}_{kh} \doteq \{(x, y): (2k-1)\,2c_1 < x \leqq (2k+1)\,2c_1,\ (2h-1)\,2c_2 < y \leqq (2h+1)\,2c_2\}$$

for $k, h = 0, \pm 1, \pm 2, \ldots, \pm k_0$ where

$$(2k-1)\,2c_i|_{k=-k_0} \triangleq -\infty; \qquad (2k+1)\,2c_i|_{k=k_0} \triangleq \infty$$

are put. We want to calculate probabilities with respect to $(r_{j\varepsilon}(\omega), r_{j+1\varepsilon}(\omega))$. Now, as in the one-dimensional case, we put

$$2(2(k_0 - 1) + 1)\,c_{1j} \approx 2\sqrt{\langle r_j^2 \rangle}; \qquad 2(2(k_0 - 1) + 1)\,c_{2j} \approx 2\sqrt{\langle r_{j+1}^2 \rangle}$$

and hence

$$c_{1j} = \frac{1}{2k_0 - 1}\sqrt{\varepsilon^2 A_{1,j}}; \qquad c_{2j} = \frac{1}{2k_0 - 1}\sqrt{\varepsilon^2 A_{1,j+1}}.$$

The values $P_{jkh,t}$ are defined by

$$P_{jkh,t} = \int_{\mathcal{I}_{j,kh}} p_{j,t}(u_1, u_2)\,du_1\,du_2 \quad \text{for} \quad j = 1, 3; \qquad t = 1, 2, 3; \quad k, h = 0, \pm 1, \pm 2, \ldots, \pm k_0$$

where these values $P_{jkh,t}$ are approximations of the probabilities

$$\mathsf{P}((r_{j\varepsilon}, r_{j+1\varepsilon}) \in \mathcal{I}_{j,kh}).$$

Using

$$\mathcal{I}_{j,kh} = (a_{jk}^1, a_{jk+1}^1] \times (a_{jh}^2, a_{jh+1}^2]$$

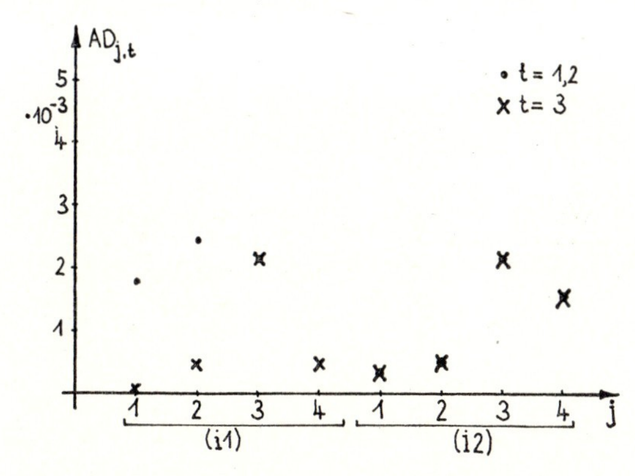

Fig. 2.5. Averaged absolute deviation between $P_{jk,t}$ and $P_{jk,\mathrm{ex}}$

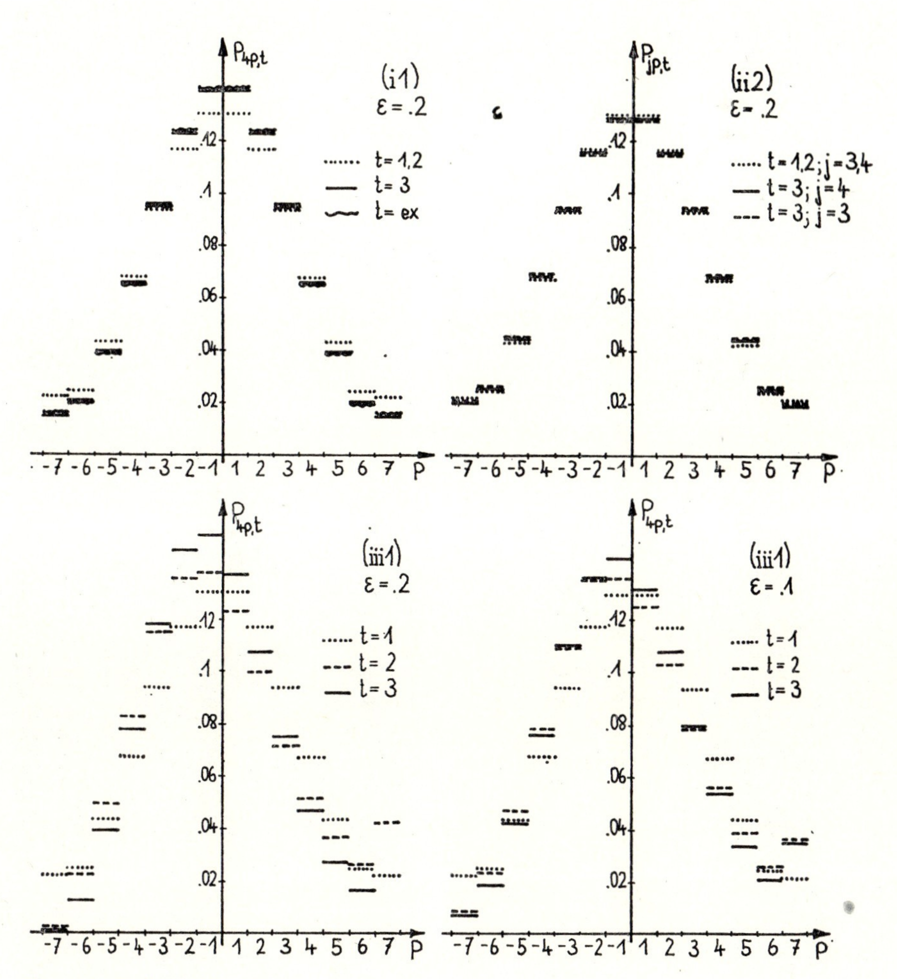

Fig. 2.6. Probabilities $P_{jk,t}$

and (2.16) we obtain

$$K_{j,kh}^{pq} \doteq \iint_{\mathcal{J}_{j,kh}} p_{j,1}(u_1, u_2)\, H_p(b_{11}u_1)\, H_q(b_{21}u_1 + b_{22}u_2)\, du_1\, du_2$$

$$= \frac{b_{11}b_{22}}{2\pi\varepsilon} \int_{a_{jk}^1}^{a_{jk+1}^1} \left(\int_{a_{jh}^2}^{a_{jh+1}^2} \exp\left(-\frac{1}{2\varepsilon}\left[(b_{11}u_1)^2 + (b_{21}u_1 + b_{22}u_2)^2\right]\right) \right.$$

$$\left. \times H_p\left(\frac{1}{\sqrt{\varepsilon}}\, b_{11}u_1\right) H_q\left(\frac{1}{\sqrt{\varepsilon}}\,(b_{21}u_1 + b_{22}u_2)\right) du_2 \right) du_1$$

$$= \frac{b_{22}}{2\pi\sqrt{\varepsilon}} \int_{\tilde{a}_{jk}^1}^{\tilde{a}_{jk+1}^1} \left(\int_{a_{jh}^2}^{a_{jh+1}^2} \exp\left(-\frac{1}{2}\, v_1^2 - \frac{1}{2\varepsilon}\left(\sqrt{\varepsilon}\,\frac{b_{21}}{b_{11}}\, v_1 + b_{22}u_2\right)^2\right) \right.$$

$$\left. \times H_p(v_1)\, H_q\left(\frac{b_{21}}{b_{11}}\, v_1 + \frac{1}{\sqrt{\varepsilon}}\, b_{22}u_2\right) du_2 \right) dv_1$$

$$= \frac{1}{2\pi} \int_{\tilde{a}_{jk}^1}^{\tilde{a}_{jk+1}^1} \left(\exp\left(-\frac{1}{2}\, v_1^2\right) H_p(v_1) \int_{\tilde{a}_{jh}^2(v_1)}^{\tilde{a}_{jh+1}^2(v_1)} \exp\left(-\frac{1}{2}\, v_2^2\right) H_q(v_2)\, dv_2 \right) dv_1$$

where

$$\tilde{a}_{jk}^1 = \frac{1}{\sqrt{\varepsilon}}\, b_{11}a_{jk}^1 = \frac{2(2k-1)}{2k_0 - 1}$$

for $k = -k_0 + 1, -k_0 + 2, \ldots, -1, 0, 1, \ldots, k_0$;

$$\tilde{a}_{j,-k_0}^1 = -\infty; \qquad \tilde{a}_{j,k_0+1}^1 = \infty,$$

$$\tilde{a}_{jh}^2(v_1) = \frac{b_{21}}{b_{11}}\, v_1 + \frac{1}{\sqrt{\varepsilon}}\, b_{22}a_{jh}^2 = b_{22}\left(\frac{b_{21}}{b_{11}b_{22}}\, v_1 + \frac{2(2h-1)}{2k_0 - 1}\sqrt{B_{j+1j+1}}\right)$$

$$= \sqrt{\frac{B_{jj}B_{j+1j+1}}{B_{jj}B_{j+1j+1} - B_{jj+1}^2}}\left(-\frac{B_{jj+1}}{\sqrt{B_{jj}B_{j+1j+1}}}\, v_1 + \frac{2(2h-1)}{2k_0 - 1}\right)$$

for $h = -k_0 + 1, -k_0 + 2, \ldots, -1, 0, 1, \ldots, k_0$;

$$\tilde{a}_{j,-k_0}^2(v_1) = -\infty; \qquad \tilde{a}_{j,k_0+1}^2(v_1) = \infty.$$

Putting

$$T_h(a, b) \doteq \int_a^b \exp\left(-\frac{1}{2}\, v^2\right) H_h(v)\, dv$$

we obtain

$$T_h(a, b) = \begin{cases} \Phi(u)\big|_a^b & \text{for } h = 0, \\ -\exp\left(-\frac{1}{2}v^2\right)\Big|_a^b & \text{for } h = 1, \\ -v\exp\left(-\frac{1}{2}v^2\right)\Big|_a^b & \text{for } h = 2, \\ (1 - v^2)\exp\left(-\frac{1}{2}v^2\right)\Big|_a^b & \text{for } h = 3, \\ v(3 - v^2)\exp\left(-\frac{1}{2}v^2\right)\Big|_a^b & \text{for } h = 4, \\ (-v^4 + 6v^2 - 3)\exp\left(-\frac{1}{2}v^2\right)\Big|_a^b & \text{for } h = 5, \\ v(-v^4 + 10v^2 - 15)\exp\left(-\frac{1}{2}v^2\right)\Big|_a^b & \text{for } h = 6. \end{cases}$$

Now, we find

$$P_{jkh,1} = K^{00}_{j,kh},$$

$$P_{jkh,2} = K^{00}_{j,kh} + \frac{1}{2}\left[\frac{1}{3}\left({}^3\tilde{A}_{2,111}K^{30}_{j,kh} + {}^3\tilde{A}_{2,222}K^{03}_{j,kh}\right) + {}^3\tilde{A}_{2,112}K^{21}_{j,kh} + {}^3\tilde{A}_{2,122}K^{12}_{j,kh}\right]\sqrt{\varepsilon},$$

$$\begin{aligned} P_{jkh,3} = P_{jkh,2} + \Bigg[&\frac{1}{2}\left({}^2\tilde{A}_{2,11}K^{20}_{j,kh} + {}^2\tilde{A}_{2,22}K^{02}_{j,kh}\right) + {}^2\tilde{A}_{2,12}K^{11}_{j,kh} \\ &+ \frac{1}{24}\left({}^4\underline{\tilde{A}}_{3,1111}K^{40}_{j,kh} + {}^4\underline{\tilde{A}}_{3,2222}K^{04}_{j,kh}\right) \\ &+ \frac{1}{6}\left({}^4\underline{\tilde{A}}_{3,1112}K^{31}_{j,kh} + {}^4\underline{\tilde{A}}_{3,1222}K^{13}_{j,kh}\right) + \frac{1}{4}\,{}^4\underline{\tilde{A}}_{3,1122}K^{22}_{j,kh} \\ &+ \frac{1}{72}\left({}^3\tilde{A}^2_{2,111}K^{60}_{j,kh} + {}^3\tilde{A}^2_{2,222}K^{06}_{j,kh}\right) \\ &+ \frac{1}{12}\left({}^3\tilde{A}_{2,111}{}^3\tilde{A}_{2,112}K^{51}_{j,kh} + {}^3\tilde{A}_{2,222}{}^3\tilde{A}_{2,122}K^{15}_{j,kh}\right) \\ &+ \frac{1}{4}\left(\frac{1}{3}\,{}^3\tilde{A}_{2,111}{}^3\tilde{A}_{2,122} + \frac{1}{2}\,{}^3\tilde{A}^2_{2,112}\right)K^{42}_{j,kh} \\ &+ \frac{1}{4}\left(\frac{1}{3}\,{}^3\tilde{A}_{2,222}{}^3\tilde{A}_{2,112} + \frac{1}{2}\,{}^3\tilde{A}^2_{2,122}\right)K^{24}_{j,kh} \\ &+ \frac{1}{4}\left({}^3\tilde{A}_{2,112}{}^3\tilde{A}_{2,122} + \frac{1}{9}\,{}^3\tilde{A}_{2,111}{}^3\tilde{A}_{2,222}\right)K^{33}_{j,kh}\Bigg]\varepsilon \end{aligned}$$

Table 2.10

	$k=-3$	$k=-2$	$k=-1$	$k=0$	$k=1$	$k=2$	$k=3$	
$h=-3$	0	0	0	0.0007	0.0046	0.0092	0.0069	0.0214
	0	0	0	0.0002	0.0030	0.0078	0.0068	0.0178
	0	0	0	0.0003	0.0031	0.0078	0.0067	0.0179
$h=-2$	0	0	0.0014	0.0132	0.0364	0.0315	0.0092	0.0917
	0	0	0.0007	0.0111	0.0341	0.0327	0.0098	0.0884
	0	0	0.0008	0.0107	0.0341	0.0327	0.0095	0.0878
$h=-1$	0	0.0014	0.0190	0.0749	0.0943	0.0364	0.0046	0.2306
	0	0.0009	0.0172	0.0742	0.0981	0.0384	0.0045	0.2333
	0	0.0009	0.0166	0.0743	0.0992	0.0380	0.0042	0.2332
$h=0$	0.0007	0.0132	0.0749	0.1350	0.0749	0.0132	0.0007	0.3126
	0.0005	0.0122	0.0771	0.1414	0.0771	0.0122	0.0005	0.3210
	0.0004	0.0117	0.0770	0.1440	0.0770	0.0117	0.0004	0.3222
$h=1$	0.0046	0.0364	0.0943	0.0749	0.0190	0.0014	0	0.2306
	0.0045	0.0384	0.0981	0.0742	0.0172	0.0009	0	0.2333
	0.0042	0.0380	0.0992	0.0743	0.0166	0.0009	0	0.2332
$h=2$	0.0092	0.0315	0.0364	0.0132	0.0014	0	0	0.0917
	0.0098	0.0327	0.0341	0.0111	0.0007	0	0	0.0884
	0.0095	0.0327	0.0341	0.0107	0.0008	0	0	0.0878
$h=3$	0.0069	0.0092	0.0046	0.0007	0	0	0	0.0214
	0.0068	0.0078	0.0030	0.0002	0	0	0	0.0178
	0.0067	0.0078	0.0031	0.0003	0	0	0	0.0179
	0.0214	0.0917	0.2306	0.3126	0.2306	0.0917	0.0214	1
	0.0216	0.0920	0.2302	0.3124	0.2302	0.0920	0.0216	1
	0.0208	0.0911	0.2308	0.3146	0.2308	0.0911	0.0208	1

where $K^{pq}_{j,kh}$ can be calculated by

$$K^{pq}_{j,kh} = \frac{1}{2\pi} \int\limits_{\tilde{a}^1_{jk}}^{\tilde{a}^1_{jk+1}} \exp\left(-\frac{1}{2}\,v^2\right) H_p(v)\, T_q\big(\tilde{a}^2_{jh}(v), \tilde{a}^2_{jh+1}(v)\big)\, dv\,.$$

In the case of (i) we know the exact distribution function

$$p_{j,\mathrm{ex}}(u_1, u_2) = \frac{1}{2\pi\sqrt{m^{\mathrm{ex}}_{j,20} m^{\mathrm{ex}}_{j,02} - (m^{\mathrm{ex}}_{j,11})^2}}$$

$$\times \exp\left(-\frac{1}{2(1-\varrho_j^2)}\left(\frac{u_1^2}{m^{\mathrm{ex}}_{j,20}} - 2\varrho_j\,\frac{u_1 u_2}{\sqrt{m^{\mathrm{ex}}_{j,20} m^{\mathrm{ex}}_{j,02}}} + \frac{u_2^2}{m^{\mathrm{ex}}_{j,02}}\right)\right)$$

where

$$\varrho_j \doteq \frac{m_{j,11}^{\text{ex}}}{\sqrt{m_{j,20}^{\text{ex}} m_{j,02}^{\text{ex}}}}.$$

It follows

$$P_{jkh,\text{ex}} = \int_{\mathcal{J}_{j,kh}} p_{j,\text{ex}}(u_1, u_2)\,\mathrm{d}u_1\,\mathrm{d}u_2$$

$$= \frac{1}{\sqrt{2\pi}} \int_{\hat{a}_{jk}^1}^{\hat{a}_{jk+1}^1} \exp\left(-\frac{1}{2}v^2\right)\left[\Phi\left(\hat{a}_{jh+1}^2(v)\right) - \Phi\left(\hat{a}_{jh}^2(v)\right)\right]\mathrm{d}v$$

with

$$\hat{a}_{jk}^1 = \frac{1}{\sqrt{m_{j,20}^{\text{ex}}}}\,a_{jk}^1 \quad \text{for} \quad k = -k_0+1, -k_0+2, \ldots, -1, 0, 1, \ldots, k_0;$$

$$\hat{a}_{j,-k_0}^1 = -\infty, \quad \hat{a}_{j,k_0+1}^1 = \infty,$$

$$\hat{a}_{jh}^2(v) = \sqrt{\frac{m_{j,20}^{\text{ex}}}{m_{j,20}^{\text{ex}} m_{j,02}^{\text{ex}} - (m_{j,11}^{\text{ex}})^2}}\left(a_{jh}^2 - \frac{m_{j,12}^{\text{ex}}}{\sqrt{m_{j,20}^{\text{ex}}}}\,v\right)$$

for $h = -k_0+1, -k_0+2, \ldots, -1, 0, 1, \ldots, k_0$;

$$\hat{a}_{j,-k_0}^2(v) = -\infty, \quad \hat{a}_{j,k_0+1}^2(v) = \infty.$$

Table 2.10 shows some values in the case of (i1) for $j = 3$ and $\varepsilon = 0.1$. We have taken this case since this case allows to calculate the exact probabilities $P_{jkh,\text{ex}}$ with the help of the normal distribution. The first number is the value for $P_{3kh,1}$ and the second number for $P_{3kh,3}$. Furthermore, the third number gives the exact probability $P_{3kh,\text{ex}}$. In the case of (i1), $j = 3$ and $\varepsilon = 0.2; 0.1$ Fig. 2.7 shows average deviations

$$\overline{P}_{j..,t} \doteq \frac{1}{N}\sum_{k,h=-3}^{3} |P_{jkh,t} - P_{jkh,\text{ex}}|;$$

$$\overline{P}_{jk.,t} \doteq \frac{1}{N_k}\sum_{h=-3}^{3} |P_{jkh,t} - P_{jkh,\text{ex}}|;$$

$$\overline{P}_{j.h,t} \doteq \frac{1}{N_h}\sum_{k=-3}^{3} |P_{jkh,t} - P_{jkh,\text{ex}}|$$

where N, N_k, N_h denote the number of summands for which $P_{jkh,\text{ex}}$ is not equal to zero. It is easy to see that the results for $t = 3$ are essentially better with respect to the values of $P_{jkh,\text{ex}}$ than the results for $t = 1$. The values for $t = 3$ are smaller than 0.002 in all cases for $\varepsilon = 0.2$ and smaller than 0.0006 for $\varepsilon = 0.1$. For smaller ε we can observe a better coincidence between $P_{jkh,t}$ and $P_{jkh,\text{ex}}$. By the approximations of the density function it is possible that $P_{jkh,t} < 0$.

Table 2.11 contains some results for the case (ii1), $j = 3$, and $\varepsilon = 0.04$. This value of ε corresponds to $k = 50$. The first value gives $P_{3kh,1} = P_{3kh,2}$ and the second one $P_{3kh,3}$. As in Table 2.10 at the margin of the tables the boundary probabilities

$$P_{j.h,t} \doteq \sum_{k=-3}^{3} P_{jkh,t}, \qquad P_{jk.,t} \doteq \sum_{h=-3}^{3} P_{jkh,t}$$

are given.

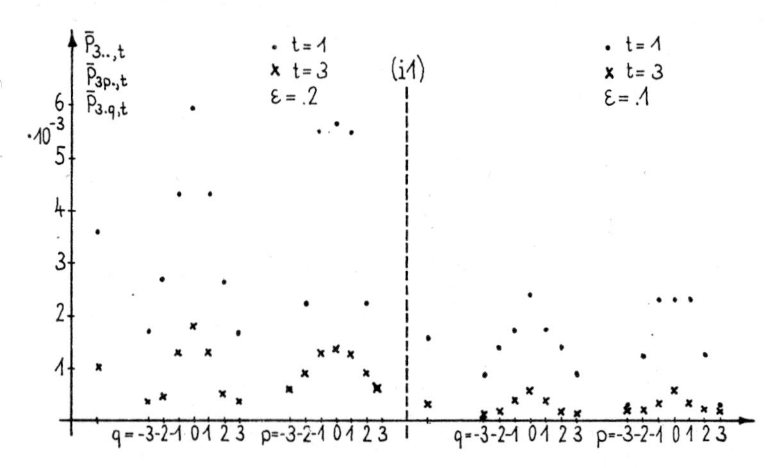

Fig. 2.7. Averaged deviations between approximative probabilities and exact probabilities for two-dimensional distributions

Table 2.11

	$k=-3$	$k=-2$	$k=-1$	$k=0$	$k=1$	$k=2$	$k=3$	
$h=-3$	0 0	0 0	0 0	0.0011 0.0004	0.0049 0.0038	0.0094 0.0088	0.0071 0.0070	0.0225 0.0200
$h=-2$	0 0	0 0	0.0014 0.0011	0.0132 0.0122	0.0364 0.0362	0.0315 0.0329	0.0092 0.0095	0.0917 0.0919
$h=-1$	0 0	0.0014 0.0012	0.0190 0.0183	0.0749 0.0749	0.0931 0.0942	0.0364 0.0373	0.0050 0.0046	0.2298 0.2305
$h=0$	0.0007 0.0011	0.0132 0.0130	0.0749 0.0756	0.1344 0.1358	0.0749 0.0756	0.0132 0.0130	0.0007 0.0011	0.3120 0.3152
$h=1$	0.0050 0.0046	0.0364 0.0373	0.0931 0.0942	0.0749 0.0749	0.0190 0.0183	0.0014 0.0012	0 0	0.2298 0.2305
$h=2$	0.0092 0.0095	0.0315 0.0329	0.0364 0.0362	0.0132 0.0122	0.0014 0.0011	0 0	0 0	0.0917 0.0919
$h=3$	0.0071 0.0070	0.0094 0.0088	0.0049 0.0038	0.0011 0.0004	0 0	0 0	0 0	0.0225 0.0200
	0.0220 0.0222	0.0919 0.0932	0.2297 0.2292	0.3128 0.3108	0.2297 0.2292	0.0919 0.0932	0.0220 0.0222	1 1

2.2. Simulation results of linear functionals

The generation of independent uniformly distributed random variables based on the algorithm

$$\bar{\eta}_{k+1} = (57\bar{\eta} + 645497) \bmod 999997; \qquad \eta_{k+1} = \frac{\bar{\eta}_{k+1}}{999997},$$

where η_{k+1} is the $(k+1)$-th realization. We can put $\bar{\eta}_0$ arbitrary but smaller than 999997. The number

$$\xi = 2\eta s - s \tag{2.21}$$

gives a realization of a uniformly distributed random variable on $[-s, s]$. Furthermore, the number

$$\xi = \sigma_0 \sqrt{-2 \ln (\eta_1)} \cos (2\pi\eta_2) \tag{2.22}$$

is a realization of a mean zero Gaussian random variable with variance σ_0^2. In this formula η_1 and η_2 denote independent realizations of a uniformly distributed random variable on $[0, 1]$. Finally, the formula

$$\xi = \exp\left(\sigma_0 \sqrt{-2 \ln (\eta_1)} \cos (2\pi\eta_2)\right) - \exp\left(\frac{1}{2}\sigma_0^2\right) \tag{2.23}$$

leads to realizations ξ of a mean zero logarithmic normal distribution where η_1 and η_2 are again independent realizations of a uniformly distributed random variable on $[0, 1]$.

Using these possibilities for the generation of realizations of the random variables ζ_p (2.7) leads to realizations of the linear functionals $r_{i\varepsilon}(\omega)$.

Estimates of moments up to the fourth order are obtained from the realizations $(x_i, y_i)_{i=1,2,\ldots,m}$ of the random vector $\left(x(\omega), y(\omega)\right)$ by the following formulas:

$$s_{x_1x_2\ldots x_k} \doteq \frac{1}{m} \sum_{i=1}^{m} x_{1i}x_{2i}\ldots x_{ki};$$

$$\overline{m}_x = s_x,$$

$$\overline{m}_{xx} = \frac{1}{m} \sum_{i=1}^{m} (x_i - \overline{m}_x)^2 = s_{xx} - s_x^2,$$

$$\overline{m}_{xy} = \frac{1}{m} \sum_{i=1}^{m} (x_i - \overline{m}_x)(y_i - \overline{m}_y) = s_{xy} - s_x s_y,$$

$$\overline{m}_{xxx} = \frac{1}{m} \sum_{i=1}^{m} (x_i - \overline{m}_x)^3 = s_{xxx} - 3s_{xx}s_x + 2s_x^3,$$

$$\overline{m}_{xxy} = \frac{1}{m} \sum_{i=1}^{m} (x_i - \overline{m}_x)^2 (y_i - \overline{m}_y) = s_{xxy} - 2s_{xy}s_x - s_{xx}s_y + 2s_x^2 s_y,$$

$$\overline{m}_{xxxx} = \frac{1}{m} \sum_{i=1}^{m} (x_i - \overline{m}_x)^4 = s_{xxxx} - 4s_{xxx}s_x + 6s_{xx}s_x^2 - 3s_x^4,$$

$$\overline{m}_{xxxy} = \frac{1}{m} \sum_{i=1}^{m} (x_i - \overline{m}_x)^3 (y_i - \overline{m}_y) = s_{xxxy} - 3s_{xxy}s_x + 3s_{xy}s_x^2 - s_{xxx}s_y$$
$$+ 3s_{xx}s_xs_y - 3s_x^3s_y,$$
$$\overline{m}_{xxyy} = \frac{1}{m} \sum_{i=1}^{m} (x_i - \overline{m}_x)^2 (y_i - \overline{m}_y)^2 = s_{xxyy} - 2s_{xyy}s_x - 2s_{xxy}s_y + s_{yy}s_x^2$$
$$+ s_{xx}s_y^2 + 4s_{xy}s_xs_y - 3s_x^2s_y^2$$

where the estimates of the moments are denoted by a "–". The estimates of $m_{s,pq} = \langle r_{s\varepsilon}^p r_{s+1\varepsilon}^q \rangle$ are denoted by $\overline{m}_{s,pq}$.

First, estimates of the first moments are contained in Table 2.12 where we note that $m_{s,pq} = m_{s,pq}^t = 0$ for $p + q = 1$, $t = 1, 2, 3$. The values of Table 2.12 are written for $\varepsilon = 0.2$ and for $m = 1000$ realizations. We have put $r = 1$. The estimates for the second moments are contained in Table 2.13 with $\varepsilon = 0.2$, $r = 1$, $m = 1000$. A comparison with Table 2.5a/2.5b shows the good correspondence between calculated results and simulation results. For the considered case ($\varepsilon = 0.2$, $r = 1$, $m = 1000$) Table 2.14 contains estimates for the third moments and Table 2.15 for the fourth moments. Setting $p + q = 3$ then we have

$$m_{s,pq}^{\text{ex}} = m_{s,pq}^1 = m_{s,pq}^2 = 0$$

for the cases (i) and (ii). The values of $m_{s,pq}^2$ and $m_{s,pq}^{\text{ex}}$ in the case (iii) are contained in Table 2.7 and we note that $m_{s,pq}^1 = 0$. The comparison values for $p + q = 4$ can be found in Table 2.8 and Table 2.9.

Table 2.12

$\overline{m}_{s,pq}$

s	p	q	(i1)	(i2)	(ii1)	(ii2)	(iii1)	(iii2)
1	1	0	−0.0088	−0.0053	−0.0085	−0.0064	0.0052	−0.0005
1	0	1	−0.0038	−0.0031	−0.0010	−0.0041	0.0016	0.0002
3	1	0	−0.0017	−0.0035	−0.0095	0.0034	0.0003	−0.0010
3	0	1	0.0069	0.0006	0.0151	−0.0055	0.0013	0.0028

Table 2.13

$\overline{m}_{s,pq}$

s	p	q	(i1)	(i2)	(ii1)	(ii2)	(iii1)	(iii2)
1	2	0	0.0915	0.0661	0.0930	0.0615	0.0658	0.0335
1	1	1	0.0457	0.0341	0.0471	0.0309	0.0328	0.0168
1	0	2	0.0301	0.0207	0.0318	0.0180	0.0193	0.0121
3	2	0	0.0450	0.0324	0.0447	0.0288	0.0317	0.0130
3	1	1	−0.0423	−0.0268	−0.0459	−0.0234	−0.0252	−0.0160
3	0	2	0.0685	0.0339	0.0750	0.0288	0.0321	0.0364

Fig. 2.8 shows a comparison between the calculated moments $m^t_{s,pq}$ for $t = 1, 2$ and simulation results $\overline{m}_{s,pq}$ in dependence of the number of realizations m. The case (ii1), $s = 3$, $\varepsilon = 0.1$ was chosen for this figure. Putting $p + q = 3$ we note $m^1_{s,pq} = m^2_{s,pq} = 0$ and the estimates of the third moments approach very well to zero in dependence of increasing m. For $p + q = 2$ and $p + q = 4$ the estimates coincide better with the second approximations $m^2_{s,pq}$ than with the first approximation $m^1_{s,pq}$ considering greater m. Similar results can be obtained in the other cases (i) and (iii).

In order to value these results as to $r_{i\varepsilon}(\omega)$ we give some simulation results for the $\zeta_i(\omega)$. Independent realizations of the ζ_i were obtained from the given methods. Fig. 2.9 shows estimates of moments of $\zeta_2(\omega)$ in the cases of (i1) and (iii1). Setting $c_i = \frac{1}{3}\sqrt{\langle\zeta_i^2\rangle}$ we obtain the intervals $\mathcal{J}^{\zeta}_{i,k}$ as in the case of the random variable $r_{j\varepsilon}(\omega)$. The corresponding

Table 2.14

$\overline{m}_{s,pq}$

s	p	q	(i1)	(i2)	(ii1)	(ii2)	(iii1)	(iii2)
1	3	0	−0.0005	0.0005	−0.0004	−0.0020	0.0109	0.0038
1	2	1	−0.0002	0.0004	0.0002	−0.0010	0.0055	0.0019
1	1	2	−0.0001	0.0003	0.0004	−0.0005	0.0032	0.0014
1	0	3	−0.0001	0.0002	0.0003	−0.0003	0.0021	0.0011
3	3	0	0	0.0002	0.0001	0.0002	0.0008	0.0001
3	2	1	0.0008	0	−0.0002	−0.0002	0.0015	0.0004
3	1	2	−0.0015	0	0.0005	0.0002	−0.0027	−0.0012
3	0	3	0.0019	0	−0.0009	−0.0003	0.0041	0.0048

Table 2.15

$\overline{m}_{s,pq}$

s	p	q	(i1)	(i2)	(ii1)	(ii2)	(iii1)	(iii2)
1	4	0	0.0237	0.0116	0.0263	0.0113	0.0146	0.0041
1	3	1	0.0119	0.0060	0.0133	0.0057	0.0074	0.0020
1	2	2	0.0066	0.0034	0.0073	0.0030	0.0041	0.0012
1	1	3	0.0039	0.0020	0.0043	0.0017	0.0024	0.0008
1	0	4	0.0025	0.0012	0.0028	0.0010	0.0015	0.0005
3	4	0	0.0064	0.0029	0.0054	0.0022	0.0042	0.0006
3	3	1	−0.0059	−0.0024	−0.0056	−0.0018	−0.0031	−0.0007
3	2	2	0.0068	0.0023	0.0070	0.0017	0.0030	0.0012
3	1	3	−0.0088	−0.0025	−0.0095	−0.0018	−0.0034	−0.0022
3	0	4	0.0144	0.0032	0.0154	0.0023	0.0043	0.0057

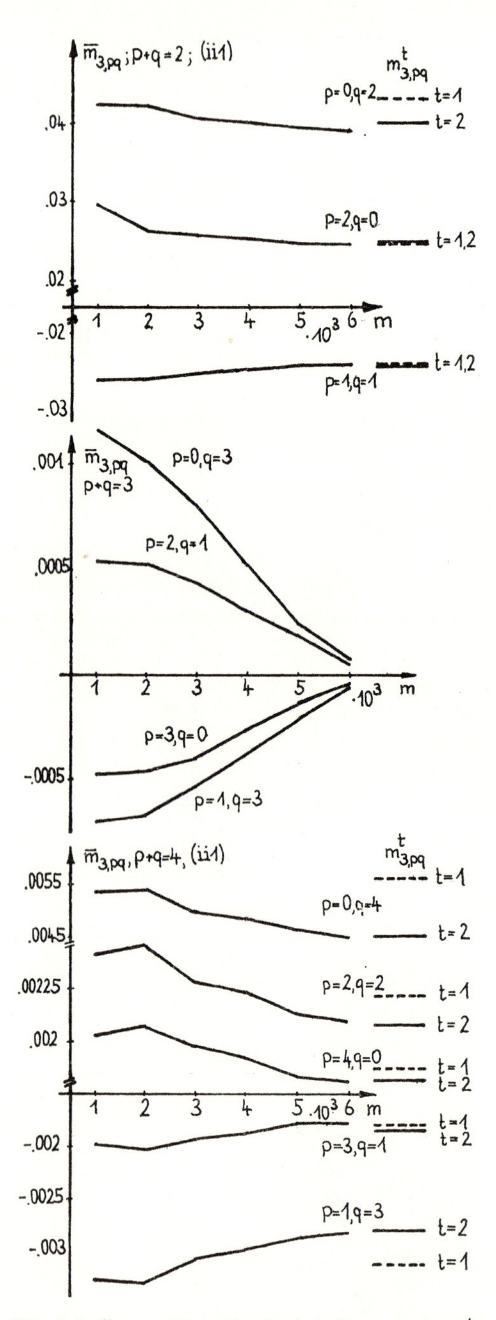

Fig. 2.8. Comparison of calculated moments $m^t_{s,pq}$ and simulation results $\overline{m}_{s,pq}$ in dependence on m in the case of (ii 1), $s = 3$

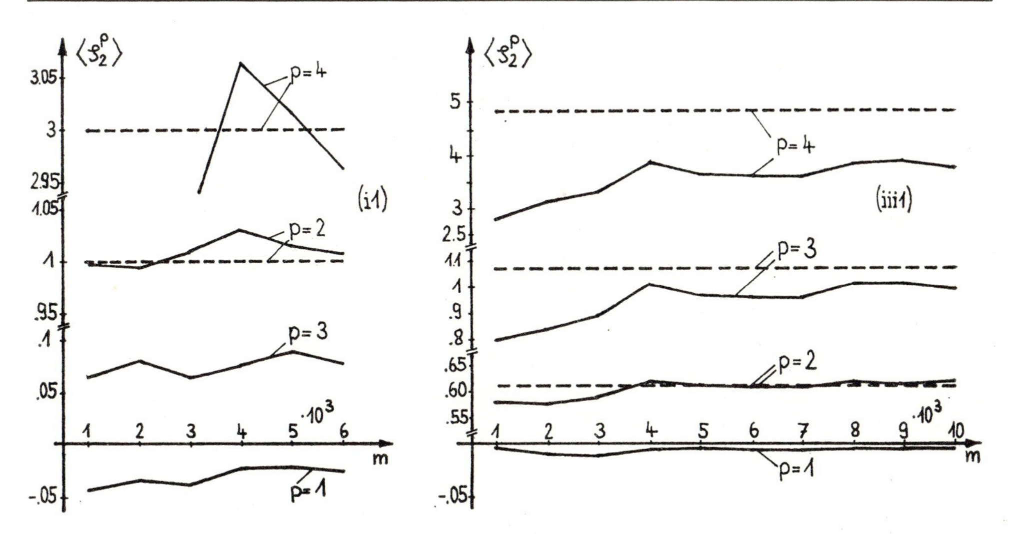

Fig. 2.9. Estimates of moments of ζ_2 in dependence on m

probabilities are denoted by P^{ζ}_{ik}. Fig. 2.10 shows some simulation results of ζ_2 in the cases (i1) and (iii1) in dependence on m. The relative frequencies are denoted by H^{ζ}_{ik}, $H^{\zeta}_{ik} \doteq \frac{1}{m} \cdot$ (number of realizations of $\zeta_i \in \mathcal{J}^{\zeta}_{ik}$). These values H^{ζ}_{ik} correspond to the bold lines in Fig. 2.10 and the probabilities P^{ζ}_{ik} to the hatched lines. This figure also contains the χ^2-values which are defined by

$$\chi_i^{2,\zeta} \doteq m \sum_k \frac{(P^{\zeta}_{ik} - H^{\zeta}_{ik})^2}{P^{\zeta}_{ik}}.$$

We have 14 classes in the case of (i1) and 12 classes for (iii1). These investigations have been given in order to compare with the results for the $r_{j\varepsilon}(\omega)$. Considering the case of (iii1), $\varepsilon = 0.2$, $j = 4$ Fig. 2.11 shows a comparison between the theoretical probabilities $P_{jk,t}$ for $t = 1, 2, 3$ and the relative frequencies H_{jk} for $m = 1000$ and $m = 8000$. Some further results as to $P_{4k,t}$ and H_{4k} are contained in Fig. 2.12. In this figure we have plotted $P_{4k,t}$ for $t = 1, 2, 3$ and $P_{4k,\mathrm{ex}}$ in the case of (i1) as horizontal line and the values H_{4k} were connected in form of a polygon. It can be seen that in general the results for $m = 1000$ differ strongly from the probabilities $P_{4k,3}$. In general, we can note that for great numbers m the probabilities $P_{4k,3}$ agree better with H_{4k} than $P_{4k,t}$ for $t = 1, 2$. The following χ^2-values also show this better agreement.

Fig. 2.13 shows some $\chi^2_{j,t}$-values for $m = 1000$ in all considered cases. $\chi^2_{j,t}$ is defined by

$$\chi^2_{j,t} \doteq m \sum_k \frac{1}{P_{jk,t}} (P_{jk,t} - H_{jk})^2$$

where it is summarized over all possible classes. In the cases (iii1) and (iii2) some classes were joined since the values $P_{jk,t}$ also can be negative. For these cases the value $\varepsilon = 0.2$ seems to be to great. Some $\chi^2_{j,t}$-values in dependence on m are contained in Fig. 2.14. We note that in these cases the best results are obtained for $t = 3$. Since ${}^3A_2 = 0$ for

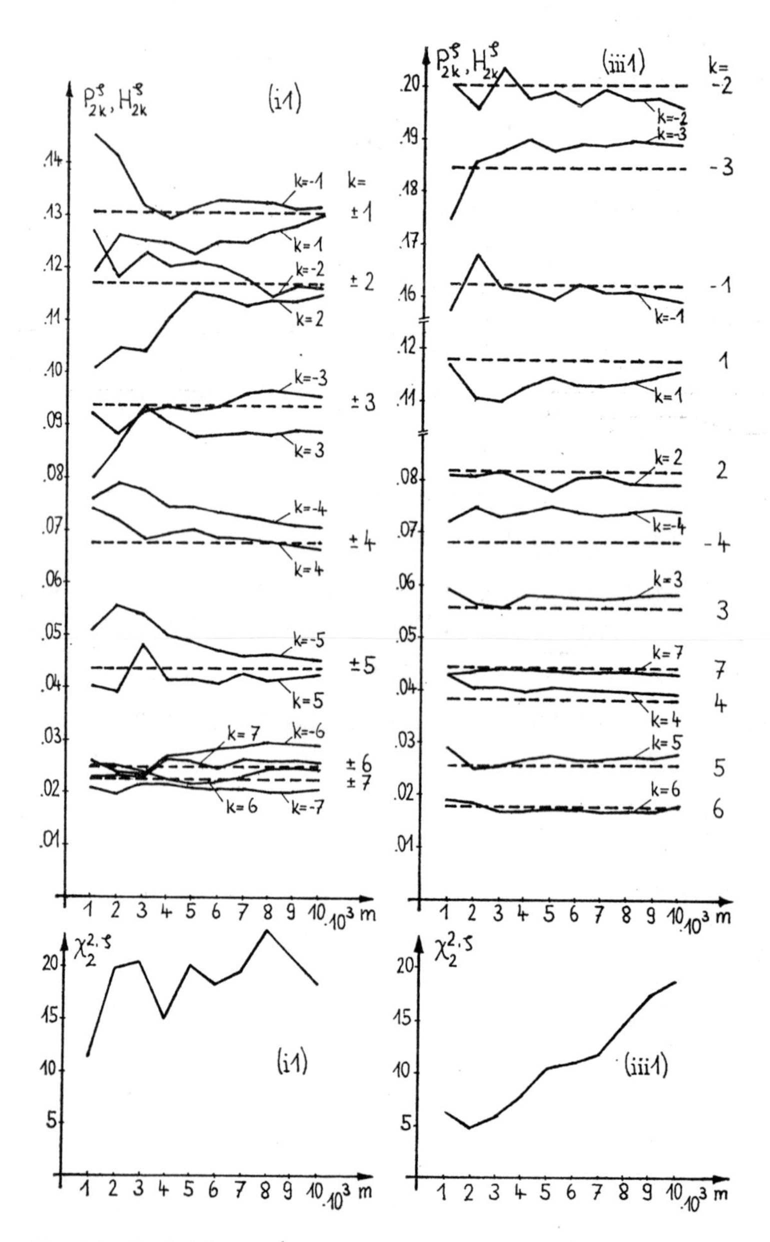

Fig. 2.10. Probabilities P_{2k}^{ζ} and relative frequencies H_{2k}^{ζ} for $\zeta_2(\omega)$ and corresponding χ^2-values

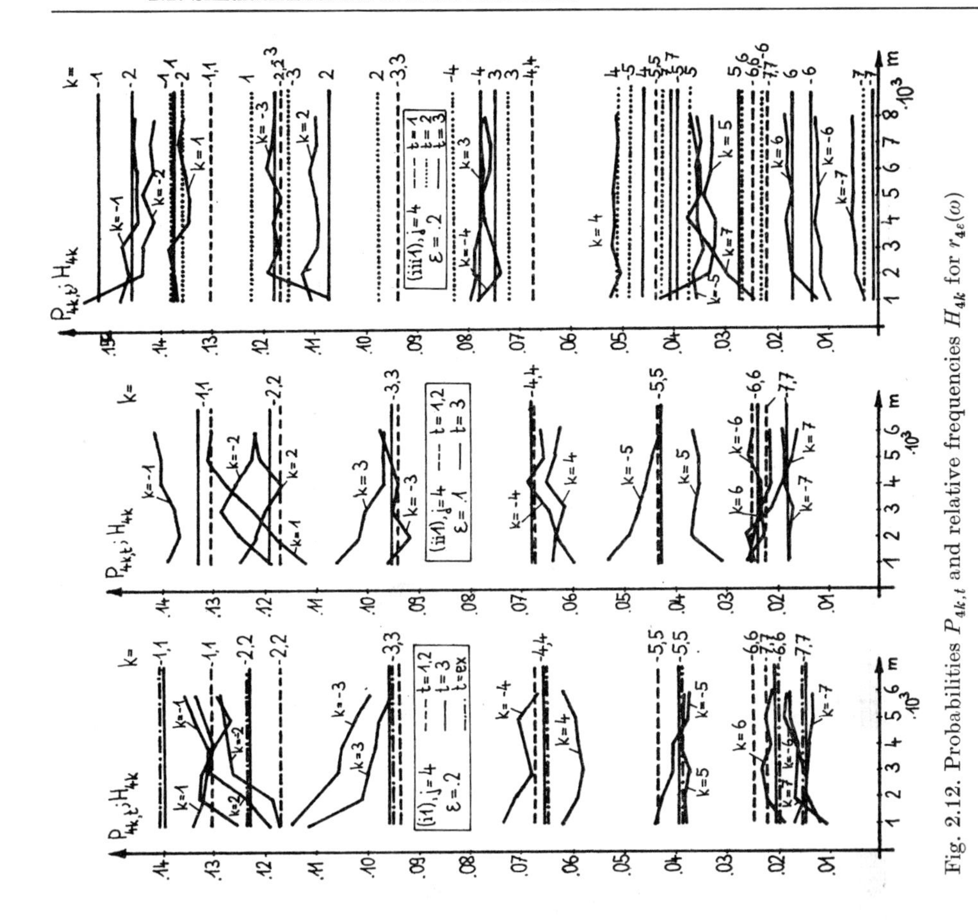

Fig. 2.11. Comparison between the probabilities $P_{jk,t}$ and the relative frequencies H_{jk} for $j = 4$ in the case of (iii1), $\varepsilon = 0.2$

Fig. 2.12. Probabilities $P_{4k,t}$ and relative frequencies H_{4k} for $r_{4\varepsilon}(\omega)$

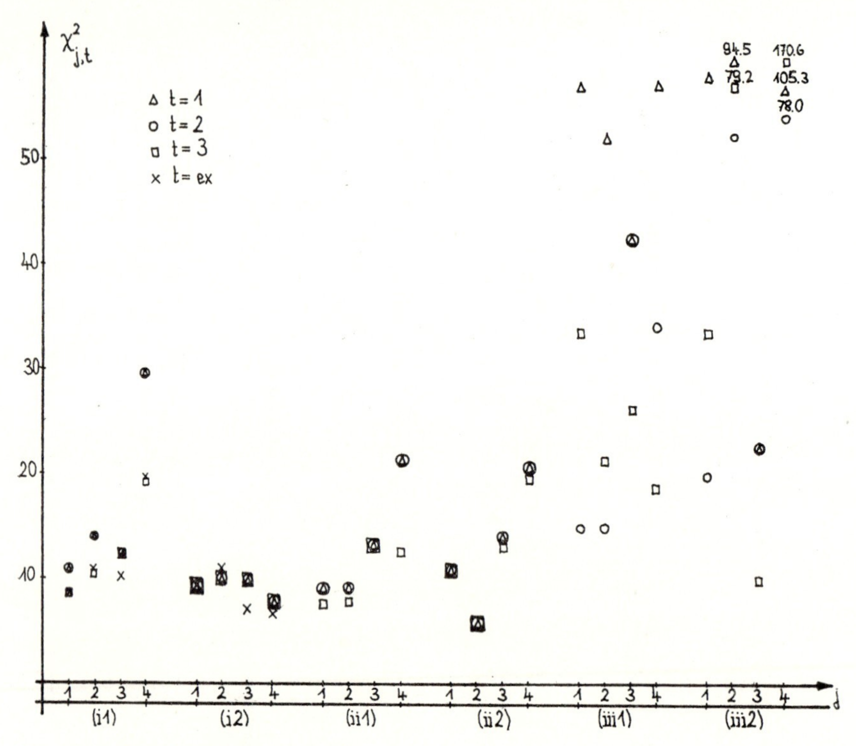

Fig. 2.13. $\chi^2_{j,t}$-values for $m = 1000$

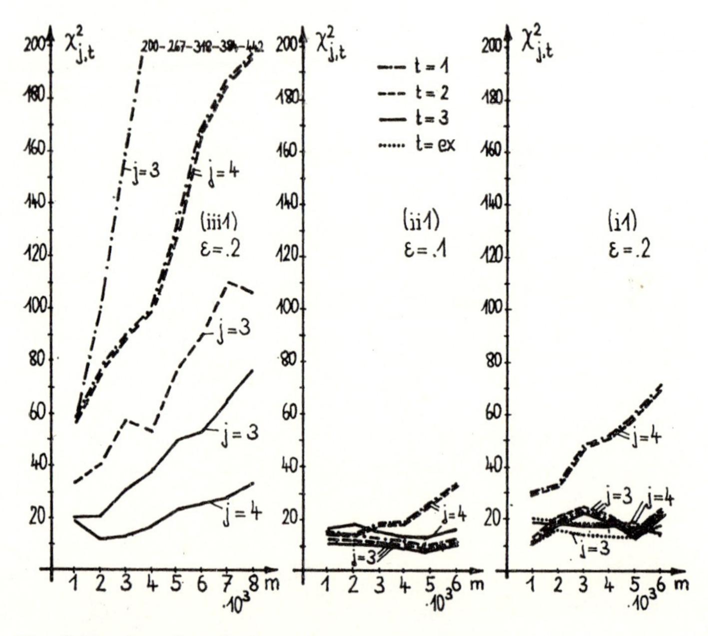

Fig. 2.14. $\chi^2_{j,t}$-values in dependence on m

(i1) and (ii1) it is $\chi^2_{j,1} = \chi^2_{j,2}$ in these cases. But for (iii1) we have the relation

$$\chi^2_{3,3} < \chi^2_{3,2} < \chi^2_{3,1}.$$

Now we give some results as to the two-dimensional probabilities. Table 2.16 contains some simulation results in the case (i1), $j = 3$, $\varepsilon = 0.1$. The first number gives the value for H_{jkh} for $m = 1000$, the second one for $m = 4000$ and the third one for $m = 5000$. These relative frequencies can be compared with the approximative probabilities $P_{jkh,t}$ for $t = 1, 3$ and the exact probabilities $P_{jkh,\text{ex}}$ of Table 2.10. Table 2.17 shows some relative frequencies in the case of (ii1), $s = 3$, and $\varepsilon = 0.04$. The approximative probabilities belonging to Table 2.17 can be found in Table 2.11.

For comparison we have calculated the χ^2-values as to the classes for which $P_{jkh,\text{ex}} \neq 0$

Table 2.16

	$k = -3$	$k = -2$	$k = -1$	$k = 0$	$k = 1$	$k = 2$	$k = 3$	
$h = -3$	0	0	0	0	0.0049	0.0070	0.0060	0.0179
	0	0	0	0	0.0032	0.0062	0.0082	0.0176
	0	0	0	0.0003	0.0027	0.0075	0.0061	0.0166
$h = -2$	0	0	0.0010	0.0160	0.0300	0.0320	0.0099	0.0889
	0	0	0.0015	0.0117	0.0347	0.0347	0.0095	0.0921
	0	0	0.0008	0.0112	0.0322	0.0325	0.0089	0.0856
$h = -1$	0	0.0010	0.0140	0.0719	0.1080	0.0390	0.0020	0.2359
	0	0.0007	0.0147	0.0775	0.0980	0.0395	0.0040	0.2344
	0	0.0006	0.0166	0.0748	0.0986	0.0390	0.0038	0.2334
$h = 0$	0.0020	0.0099	0.0760	0.1466	0.0740	0.0120	0	0.3205
	0.0007	0.0117	0.0752	0.1482	0.0735	0.0110	0.0005	0.3208
	0.0006	0.0108	0.0766	0.1443	0.0771	0.0113	0.0005	0.3212
$h = 1$	0.0040	0.0350	0.0960	0.0710	0.0160	0.0020	0	0.2240
	0.0030	0.0377	0.0935	0.0712	0.0205	0.0022	0	0.2281
	0.0031	0.0382	0.0952	0.0723	0.0202	0.0013	0	0.2303
$h = 2$	0.0060	0.0310	0.0370	0.0160	0	0	0	0.0900
	0.0055	0.0337	0.0340	0.0125	0.0002	0	0	0.0879
	0.0068	0.0350	0.0383	0.0116	0.0010	0	0	0.0927
$h = 3$	0.0049	0.0070	0.0089	0.0020	0	0	0	0.0228
	0.0060	0.0089	0.0035	0.0007	0	0	0	0.0191
	0.0062	0.0097	0.0033	0.0010	0	0	0	0.0202
	0.0169	0.0839	0.2329	0.3235	0.2329	0.0920	0.0179	1
	0.0152	0.0927	0.2244	0.3218	0.2301	0.0936	0.0222	1
	0.0167	0.0943	0.2308	0.3155	0.2318	0.0916	0.0193	1

and the values

$$F_{(j),t} \doteq \frac{1}{M} \sum_{k,h=-3}^{3} |P_{jkh,t} - H_{jkh}|.$$

In this case we obtain $M = 37$. Fig. 2.15 shows $\chi^2_{(j),t}$ and $F_{(j),t}$ for (i1), $j = 3$, $\varepsilon = 0.1$ and for (ii1), $j = 3$, $\varepsilon = 0.04$.

Fig. 2.16 shows the χ^2-values $\chi^2_{j,t}$ and the F-values $F_{j,t}$ for $j = 3, 4$ which follow from Table 2.16 and 2.17 in comparison with Table 2.10 and 2.11 as to the marginal distribution.

Table 2.17

	$k=-3$	$k=-2$	$k=-1$	$k=0$	$k=1$	$k=2$	$k=3$	
$h=-3$	0	0	0	0	0.0040	0.0109	0.0099	0.0248
	0	0	0	0.0002	0.0057	0.0102	0.0067	0.0228
	0	0	0	0.0004	0.0046	0.0087	0.0069	0.0206
$h=-2$	0	0	0.0020	0.0090	0.0369	0.0369	0.0099	0.0947
	0	0	0.0017	0.0107	0.0357	0.0382	0.0092	0.0955
	0	0	0.0014	0.0123	0.0366	0.0371	0.0091	0.0965
$h=-1$	0	0.0010	0.0199	0.0849	0.0940	0.0290	0.0059	0.2347
	0	0.0010	0.0179	0.0817	0.0979	0.0377	0.0049	0.2411
	0	0.0010	0.0170	0.0755	0.0942	0.0391	0.0051	0.2319
$h=0$	0.0010	0.0109	0.0679	0.1309	0.0819	0.0119	0	0.3015
	0.0010	0.0119	0.0732	0.1312	0.0697	0.0135	0.0010	0.3015
	0.0010	0.0126	0.0751	0.1368	0.0726	0.0133	0.0010	0.3124
$h=1$	0.0020	0.0330	0.0890	0.0760	0.0199	0.0029	0	0.2228
	0.0044	0.0339	0.0952	0.0737	0.0202	0.0025	0	0.2299
	0.0044	0.0359	0.0924	0.0719	0.0197	0.0018	0	0.2261
$h=2$	0.0059	0.0279	0.0380	0.0160	0.0010	0	0	0.0888
	0.0105	0.0310	0.0342	0.0117	0.0007	0	0	0.0881
	0.0101	0.0331	0.0350	0.0125	0.0007	0	0	0.0914
$h=3$	0.0059	0.0149	0.0089	0	0	0	0	0.0297
	0.0074	0.0080	0.0047	0.0010	0	0	0	0.0211
	0.0076	0.0089	0.0040	0.0006	0	0	0	0.0211
	0.0148	0.0877	0.2257	0.3168	0.2377	0.0916	0.0257	1
	0.0233	0.0858	0.2269	0.3102	0.2299	0.1021	0.0218	1
	0.0231	0.0915	0.2249	0.3100	0.2284	0.1000	0.0221	1

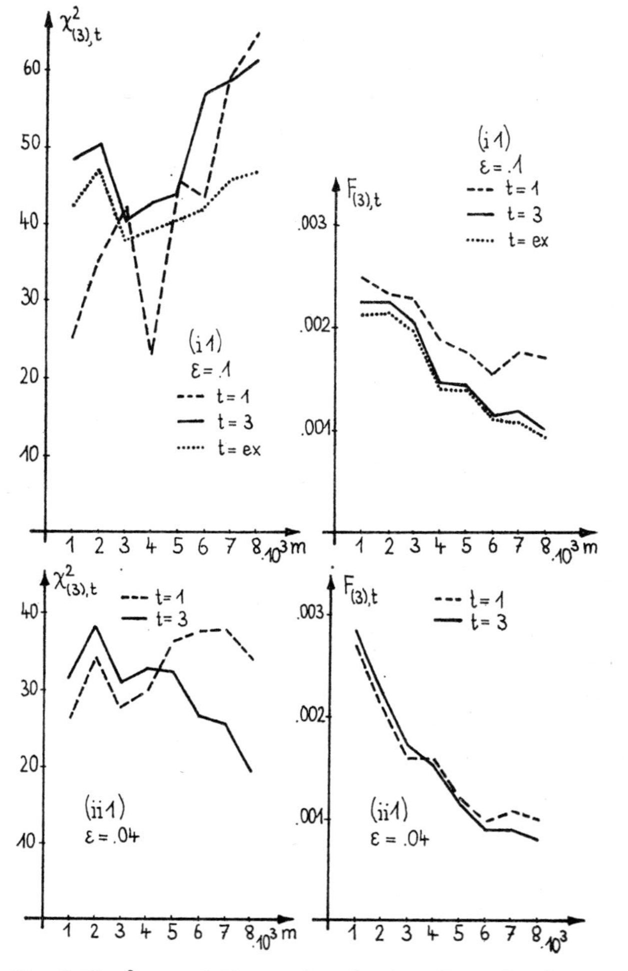

Fig. 2.15. $\chi^2_{(3),t}$- and $F_{(3),t}$-values for two-dimensional distributions

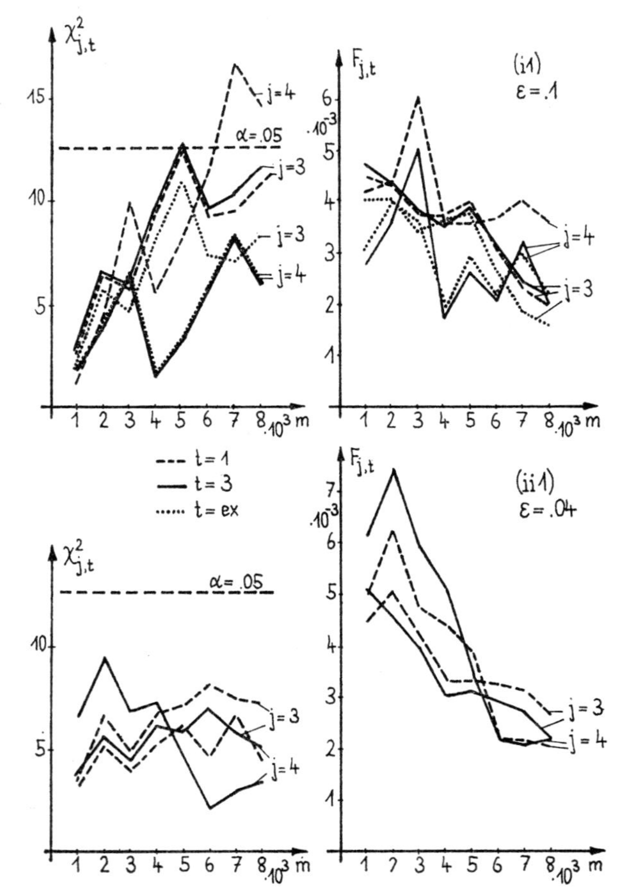

Fig. 2.16. $\chi^2_{j,t}$- and $F_{j,t}$-values for the marginal distribution of two-dimensional distribution

2.3. Functional-functions

In this section we wish to describe an application of the results as to functions of linear functionals of weakly correlated processes to a simple eigenvalue problem.

We consider the eigenvalue problem

$$\big(A_0 + B(\omega)\big)\, U = \Lambda U \tag{2.24}$$

where A_0, $B(\omega)$ have the special form

$$A_0 = \begin{pmatrix} 2 & -2 \\ -2 & -1 \end{pmatrix}, \qquad B(\omega) = \begin{pmatrix} r_s(\omega) & 0 \\ 0 & r_{s+1}(\omega) \end{pmatrix} \quad \text{for} \quad s = 1, 3$$

and $r_{j\varepsilon}(\omega)$, $j = 1, 2, 3, 4$, denote linear functionals

$$r_{j\varepsilon}(\omega) = \int_\alpha^\beta F_j(x)\, f_\varepsilon(x, \omega)\, \mathrm{d}x$$

of a weakly correlated process $f_\varepsilon(x, \omega)$. These linear functionals were investigated. The averaged problem

$$A_0 U_0 = \Lambda_0 U_0$$

has the simple eigenvalues $\Lambda_{10} = -2$, $\Lambda_{20} = 3$ and the pertinent eigenvectors

$$U_{10} = \frac{1}{\sqrt{5}}\,(1, 2)^{\mathsf{T}}, \qquad U_{20} = \frac{1}{\sqrt{5}}\,(2, -1)^{\mathsf{T}}.$$

The eigenvalues of the eigenvalue problem can be calculated as

$$\begin{aligned} \Lambda_{1/2}(\omega) &= d_{1/2}\big(r_{s\varepsilon}(\omega), r_{s+1\varepsilon}(\omega)\big) \\ &= \frac{1}{2}\left(r_{s\varepsilon} + r_{s+1\varepsilon} + 1 \mp \sqrt{(3 + r_{s\varepsilon} - r_{s+1\varepsilon})^2 + 16}\right). \end{aligned} \tag{2.24.1}$$

These functions fulfil the conditions which are written at the beginning of Section 1.7. The expansions of $d_1(y_1, y_2)$ and $d_2(y_1, y_2)$ can be written as

$$\begin{aligned} d_1(y_1, y_2) &= -2 + \frac{1}{5}\, y_1 + \frac{4}{5}\, y_2 - \frac{4}{5^3}\,(y_1 - y_2)^2 + \frac{12}{5^5}\,(y_1 - y_2)^3 + \dots, \\ d_2(y_1, y_2) &= 3 + \frac{4}{5}\, y_1 + \frac{1}{5}\, y_2 + \frac{4}{5^3}\,(y_1 - y_2)^2 - \frac{12}{5^5}\,(y_1 - y_2)^3 + \dots \end{aligned} \tag{2.25}$$

and it follows

$$\begin{pmatrix} d_{1,1} & d_{1,2} \\ d_{2,1} & d_{2,2} \end{pmatrix} = \frac{1}{5} \begin{pmatrix} 1 & 4 \\ 4 & 1 \end{pmatrix},$$

$$\begin{pmatrix} d_{1,11} & d_{1,12} & d_{1,22} \\ d_{2,11} & d_{2,12} & d_{2,22} \end{pmatrix} = \frac{4}{5^3} \begin{pmatrix} -1 & 1 & -1 \\ 1 & -1 & 1 \end{pmatrix},$$

$$\begin{pmatrix} d_{1,111} & d_{1,112} & d_{1,122} & d_{1,222} \\ d_{2,111} & d_{2,112} & d_{2,122} & d_{2,222} \end{pmatrix} = \frac{12}{5^5} \begin{pmatrix} 1 & -1 & 1 & -1 \\ -1 & 1 & -1 & 1 \end{pmatrix}.$$

Now we deal with the first and second moments of $(\Lambda_1(\omega), \Lambda_2(\omega))$. Using (1.169) or (1.171) we obtain

$$\langle \Lambda_i - \Lambda_{i0} \rangle = \sum_{a,b=1}^{2} d_{i,ab}{}^2 A_1(F_a, F_b)\, \varepsilon + o(\varepsilon)$$

where

$${}^2A_1(F_a, F_b) = \frac{1}{2} \int_{\alpha}^{\beta} F_a(x)\, F_b(x)\, \sigma_2(x)\, \mathrm{d}x$$

is calculated as in Section 2.1. Hence, the mean value $\langle \Lambda_i \rangle$ is given approximatively by

$$\langle \Lambda_i \rangle = \Lambda_{i0} + \big(d_{i,11}{}^2A_1(F_1, F_1) + 2d_{i,12}{}^2A_1(F_1, F_2) + d_{i,22}{}^2A_1(F_2, F_2)\big)\, \varepsilon + o(\varepsilon)\,. \tag{2.26}$$

This formula leads to an approximative solution of the averaging problem since it allows to determine the difference between the mean value of the random solution $\langle \Lambda_i \rangle$ and the solution of the averaged problem Λ_{i0}.

The numerical values are given for the cases (i1), (i2), (ii1), (ii2), (iii1), (iii2) as to the random process $f_\varepsilon(x, \omega)$ (see Section 2.1). We introduce the notation $\Lambda_{,is}$ for $s = 1, 3$ and $\Lambda_{i,1}$ denote the eigenvalues of the eigenvalue problem (2.24) with

$$r_{1\varepsilon}(\omega),\ r_{2\varepsilon}(\omega) \quad \text{and} \quad F_1(x) = 1, \qquad F_2(x) = x$$

and $\Lambda_{i,3}$ the eigenvalues with

$$r_{3\varepsilon}(\omega), \qquad r_{4\varepsilon}(\omega) \quad \text{and} \quad F_3(x) = \sin(2\pi x), \qquad F_4(x) = x^2 + 2x - 1$$

on the interval $(\alpha, \beta) = (0, 1)$. The moments

$$\big\langle (\Lambda_{1,s} - \langle \Lambda_{1,s} \rangle)^p\, (\Lambda_{2,s} - \langle \Lambda_{2,s} \rangle)^q \big\rangle$$

which are computed by the theory of weakly correlated processes are denoted by $M^t_{s,pq}$. In particular, we denote the moments for $p + q = 1$, $\langle \Lambda_{i,s} \rangle$, by $M^t_{s,i}$ and the estimates by $\overline{M}_{s,i}$. The index t corresponds to the order of the approximation of the weakly correlated theory. Later, the estimates of the moments $M^t_{s,pq}$ are denoted by $\overline{M}_{s,pq}$. We have

$$M^2_{s,1} = \Lambda_{10} + \frac{4}{125} \big(-{}^2A_1(F_s, F_s) + 2\,{}^2A_1(F_s, F_{s+1}) - {}^2A_1(F_{s+1}, F_{s+1})\big)\, \varepsilon\,,$$

$$M^2_{s,2} = \Lambda_{20} - \frac{4}{125} \big(-{}^2A_1(F_s, F_s) + 2\,{}^2A_1(F_s, F_{s+1}) - {}^2A_1(F_{s+1}, F_{s+1})\big)\, \varepsilon$$

and it is

$$M^2_{s,1} + M^2_{s,2} = \Lambda_{10} + \Lambda_{20} = 1\,.$$

This relation is also fulfilled in a more general way,

$$\langle \Lambda_{1,s} \rangle + \langle \Lambda_{2,s} \rangle = \Lambda_{10} + \Lambda_{20} = 1\,.$$

Table 2.18 contains some numerical values for $M^2_{s,pq}$, $p + q = 1$, in the case of $\varepsilon = 0.2$, $\varepsilon = 0.1$ and $\varepsilon = 0.04$ for which we will add some simulation results.

Table 2.18

$M^2_{s,i}$		$\varepsilon = 0.2$						$\varepsilon = 0.1$	$\varepsilon = 0.04$
s	i	(i1)	(i2)	(ii1)	(ii2)	(iii1)	(iii2)	(i1)	(ii1)
1	1	−2.0011	−2.0006	−2.0011	−2.0006	−2.0007	−2.0005	−2.0005	−2.0002
1	2	3.0011	3.0006	3.0011	3.0006	3.0007	3.0005	3.0005	3.0002
3	1	−2.0074	−2.0042	−2.0074	−2.0036	−2.0042	−2.0034	−2.0037	−2.0015
3	2	3.0074	3.0042	3.0074	3.0036	3.0042	3.0034	3.0037	3.0015

Now we go over to the second moments. With the help of (1.174) we obtain

$$\begin{aligned}
&\langle (d_i - \langle d_i \rangle)(d_j - \langle d_j \rangle) \rangle \\
&= \sum_{a,b=1}^{2} d_{i,a} d_{j,b} {}^2A_1(F_a, F_b)\, \varepsilon + \Bigg[\sum_{a,b=1}^{2} d_{i,a} d_{j,b} {}^2A_2(F_a, F_b) \\
&\quad + \sum_{a,b,c=1}^{2} (d_{i,ab} d_{j,c} + d_{j,ab} d_{i,c})\, {}^3A_2(F_a, F_b, F_c) + \sum_{a,b,c,f=1}^{2} (d_{i,abc} d_{j,f} + d_{j,abc} d_{i,f}) \\
&\quad \times \big({}^2A_1(F_a, F_b)\, {}^2A_1(F_c, F_f) + {}^2A_1(F_a, F_c)\, {}^2A_1(F_b, F_f) + {}^2A_1(F_a, F_f)\, {}^2A_1(F_b, F_c) \big) \\
&\quad + \sum_{a,b,c,f=1}^{2} d_{i,ab} d_{j,cf} \big({}^2A_1(F_a, F_c)\, {}^2A_1(F_b, F_f) \\
&\quad + {}^2A_1(F_a, F_f)\, {}^2A_1(F_b, F_c) \big) \Bigg] \varepsilon^2 + o(\varepsilon^2).
\end{aligned}$$

The application to our case leads to

$$\langle (A_{1,s} - \langle A_{1,s} \rangle)^p (A_{2,s} - \langle A_{2,s} \rangle)^q \rangle = \begin{cases} M^1_{s,pq} + o(\varepsilon), \\ M^2_{s,pq} + o(\varepsilon^2) \end{cases}$$

where

$$\begin{aligned}
&M^1_{s,20} = {}^1N_{s,11}; \qquad M^1_{s,11} = {}^1N_{s,12}; \qquad M^1_{s,02} = {}^1N_{s,22}; \\
&M^2_{s,20} = {}^2N_{s,11} + {}^1N_{s,11}; \qquad M^2_{s,11} = {}^2N_{s,12} + {}^1N_{s,12}; \\
&M^2_{s,02} = {}^2N_{s,22} + {}^1N_{s,22}
\end{aligned}$$

with

$${}^1N_{s,ij} = \left(\sum_{a,b=1}^{2} d_{i,a} d_{j,b} {}^2A_1(F_{s+a-1}, F_{s+b-1}) \right) \varepsilon,$$

$${}^2N_{s,ij} = ({}^1H_{s,ij} + {}^2H_{s,ij} + {}^2H_{s,ji} + {}^3H_{s,ij} + {}^3H_{s,ji} + {}^4H_{s,ij})\, \varepsilon^2$$

and

$${}^1H_{s,ij} = \sum_{a,b=1}^{2} d_{i,a} d_{j,b} {}^2A_2(F_{s+a-1}, F_{s+b-1}),$$

$${}^2H_{s,ij} = \sum_{a,b,c=1}^{2} d_{i,ab} d_{j,c} {}^3A_2(F_{s+a-1}, F_{s+b-1}, F_{s+c-1}),$$

Table 2.19a

$M^t_{s,pq}$			(i1)		(i2)		(ii1)		(ii2)	
s	p	q	$t = 1$	$t = 2$	$t = 1$	$t = 2$	$t = 1$	$t = 2$	$t = 1$	$t = 2$
1	2	0	0.0413	0.0388	0.0261	0.0261	0.0413	0.0388	0.0239	0.0238
1	1	1	0.0553	0.0524	0.0365	0.0366	0.0553	0.0524	0.0335	0.0333
1	0	2	0.0813	0.0770	0.0541	0.0540	0.0813	0.0770	0.0497	0.0494
3	2	0	0.0422	0.0320	0.0153	0.0148	0.0422	0.0320	0.0131	0.0122
3	1	1	−0.0106	−0.0108	−0.0082	−0.0077	−0.0106	−0.0108	−0.0070	−0.0067
3	0	2	0.0202	0.0184	0.0151	0.0146	0.0202	0.0184	0.0136	0.0131

$M^t_{s,pq}$			(iii1)		(iii2)	
s	p	q	$t = 1$	$t = 2$	$t = 1$	$t = 2$
1	2	0	0.0269	0.0265	0.0166	0.0138
1	1	1	0.0376	0.0372	0.0208	0.0177
1	0	2	0.0557	0.0554	0.0302	0.0259
3	2	0	0.0163	0.0142	0.0271	0.0171
3	1	1	−0.0079	−0.0071	−0.0024	−0.0038
3	0	2	0.0152	0.0144	0.0058	0.0052

Table 2.19b

$M^t_{s,pq}$			(i1); $\varepsilon = 0.1$		(ii1); $\varepsilon = 0.04$	
s	p	q	$t = 1$	$t = 2$	$t = 1$	$t = 2$
3	2	0	0.0211	0.0185	0.0084	0.0080
3	1	1	−0.0053	−0.0054	−0.0021	−0.0021
3	0	2	0.0101	0.0097	0.0040	0.0039

$$\begin{aligned}{}^3H_{s,ij} = \sum_{a,b,c,f=1}^{2} d_{i,abc}d_{j,f}[&{}^2A_1(F_{s+a-1}, F_{s+b-1})\,{}^2A_1(F_{s+c-1}, F_{s+f-1}) \\ &+ {}^2A_1(F_{s+a-1}, F_{s+c-1})\,{}^2A_1(F_{s+b-1}, F_{s+f-1}) \\ &+ {}^2A_1(F_{s+a-1}, F_{s+f-1})\,{}^2A_1(F_{s+b-1}, F_{s+c-1})],\end{aligned}$$

$$\begin{aligned}{}^4H_{s,ij} = \sum_{a,b,c,f=1}^{2} d_{i,ab}d_{j,cf}[&{}^2A_1(F_{s+a-1}, F_{s+c-1})\,{}^2A_1(F_{s+b-1}, F_{s+f-1}) \\ &+ {}^2A_1(F_{s+a-1}, F_{s+f-1})\,{}^2A_1(F_{s+b-1}, F_{s+c-1})].\end{aligned}$$

Now, we can obtain by some numerical calculations for example in the case (iii1) and $s = 1$ the values

$$({}^1N_{1,ij}) = \begin{pmatrix} 0.1346 & 0.1880 \\ 0.1880 & 0.2785 \end{pmatrix} \varepsilon,$$

$$({}^1H_{1,ij}) = \begin{pmatrix} -0.0076 & -0.0085 \\ -0.0085 & -0.0120 \end{pmatrix},$$

$$({}^2H_{1,ij}) = \begin{pmatrix} -0.0018 & -0.0032 \\ 0.0018 & 0.0032 \end{pmatrix},$$

$$({}^3H_{1,ij}) = \begin{pmatrix} 0.0005 & 0.0009 \\ -0.0005 & -0.0009 \end{pmatrix},$$

$$({}^4H_{1,ij}) = \begin{pmatrix} 0.00002 & -0.00002 \\ -0.00002 & 0.00002 \end{pmatrix}$$

and furthermore we obtain Table 2.19a for $\varepsilon = 0.2$ and Table 2.19b for other ε-values. Now we deal with the density functions $p_{i,s}(u)$ of $\Lambda_{i,s}(\omega)$. Using (1.206.1) we can write

$$\tilde{p}_{i,s}(\tilde{u}) = \begin{cases} \tilde{p}_{i,s1}(\tilde{u}) + O(\sqrt{\varepsilon}), \\ \tilde{p}_{i,s2}(\tilde{u}) + O(\varepsilon), \\ \tilde{p}_{i,s3}(\tilde{u}) + o(\varepsilon) \end{cases}$$

where

$$\tilde{p}_{i,s1}(\tilde{u}) = \frac{1}{\sqrt{2\pi}} \exp\left(-\frac{1}{2}\,\tilde{u}^2\right),$$

$$\tilde{p}_{i,s2}(\tilde{u}) = \tilde{p}_{i,s1}(\tilde{u}) + \tilde{p}_{i,s1}(\tilde{u}) \Bigg[{}^2\tilde{\tilde{A}}_1'(i)\,H_1(\tilde{u}) + \left\{\frac{1}{6}\,{}^3\tilde{\tilde{A}}_2'(i,i,i) + {}^{22}\tilde{\tilde{A}}_{11}'\big(i(y,z),i(y),i(z)\big)\right\} H_3(\tilde{u})\Bigg]\sqrt{\varepsilon},$$

$$\begin{aligned}
\tilde{p}_{i,s3}(\tilde{u}) = {} & \tilde{p}_{i,s2}(\tilde{u}) + \tilde{p}_{i,s1}(\tilde{u}) \\
& \times \Bigg[\bigg\{\frac{1}{2}\,{}^2\tilde{\tilde{A}}_2'(i,i) + \frac{1}{2}\,({}^2\tilde{\tilde{A}}_1'(i))^2 + {}^3\tilde{\tilde{A}}_2'(i,i) \\
& + {}^{22}\tilde{\tilde{A}}_{11}'\big(i(y,z),i(y,z)\big) + 3\,{}^{22}\tilde{\tilde{A}}_{11}'\big(i(y,y,z),i(z)\big)\bigg\} H_2(\tilde{u}) \\
& + \bigg\{\frac{1}{24}\,{}^4\tilde{\tilde{A}}_3'(i,i,i,i) + {}^2\tilde{\tilde{A}}_1'(i)\left(\frac{1}{6}\,{}^3\tilde{\tilde{A}}_2'(i,i) + {}^{22}\tilde{\tilde{A}}_{11}'\big(i(y,z),i(y),i(z)\big)\right) \\
& + {}^{32}\tilde{\tilde{A}}_{21}'\big(i(y,z),i(y),i(y),i(z)\big) + {}^{222}\tilde{\tilde{A}}_{111}'\big(i(x,y,z),i(x),i(y),i(z)\big) \\
& + 2\,{}^{222}\tilde{\tilde{A}}_{111}'\big(i(x,y),i(z,x),i(y),i(z)\big)\bigg\} H_4(\tilde{u}) \\
& + \bigg\{\frac{1}{72}\,({}^3\tilde{\tilde{A}}_2'(i,i,i))^2 + \frac{1}{6}\,{}^3\tilde{\tilde{A}}_2'(i,i,i)\,{}^{22}\tilde{\tilde{A}}_{11}'\big(i(y,z),i(y),i(z)\big) \\
& + {}^{2222}\tilde{\tilde{A}}_{1111}'\big(i(x,y),i(z,v),i(x),i(y),i(z),i(v)\big)\bigg\} H_6(u)\Bigg]\,\varepsilon.
\end{aligned}$$

In order to compute the statistical characteristics we summarize these ones:

$${}^2\tilde{\tilde{A}}_1'(i) = {}^sb_{ii} \sum_{a,b=1}^{2} d_{i,ab}\,{}^2A_1(F_{a+s-1}, F_{b+s-1}),$$

$$^3\tilde{A}_2'(i, i, i) = {}^sb_{ii}^3 \sum_{a,b,c=1}^{2} d_{i,a}d_{i,b}d_{i,c}{}^3A_2(F_{a+s-1}, F_{b+s-1}, F_{c+s-1}),$$

$$^{22}\tilde{A}_{11}'\big(i(y, z), i(y), i(z)\big)$$

$$= {}^sb_{ii}^3 \sum_{a,b,c,f=1}^{2} d_{i,ab}d_{i,c}d_{i,f}{}^2A_1(F_{a+s-1}, F_{c+s-1})\ {}^2A_1(F_{b+s-1}, F_{f+s-1}),$$

$$^2\tilde{A}_2'(i, i) = {}^sb_{ii}^2 \sum_{a,b=1}^{2} d_{i,a}d_{i,b}{}^2A_2(F_{a+s-1}, F_{b+s-1}),$$

$$^3\tilde{A}_2'(i, i) = {}^sb_{ii}^2 \sum_{a,b,c=1}^{2} d_{i,ab}d_{i,c}{}^3A_2(F_{a+s-1}, F_{b+s-1}, F_{c+s-1}),$$

$$^{22}\tilde{A}_{11}'\big(i(y, z), i(y, z)\big) = {}^sb_{ii}^2 \sum_{a,b,c,f=1}^{2} d_{i,ab}d_{i,cf}{}^2A_1(F_{a+s-1}, F_{c+s-1})\ {}^2A_1(F_{b+s-1}, F_{f+s-1}),$$

$$^{22}\tilde{A}_{11}'\big(i(y, y, z), i(z)\big) = {}^sb_{ii}^2 \sum_{a,b,c,f=1}^{2} d_{i,abc}d_{i,f}{}^2A_1(F_{a+s-1}, F_{b+s-1})\ {}^2A_1(F_{c+s-1}, F_{f+s-1}),$$

$$^4\tilde{\underline{A}}_3'(i, i, i, i) = {}^sb_{ii}^4 \sum_{a,b,c,f=1}^{2} d_{i,a}d_{i,b}d_{i,c}d_{i,f}{}^4\underline{A}_3(F_{a+s-1}, F_{b+s-1}, F_{c+s-1}, F_{f+s-1}),$$

$$^{32}\tilde{A}_{21}'\big(i(y, z), i(y), i(y), i(z)\big)$$

$$= {}^sb_{11}^4 \sum_{a,b,c,f,g=1}^{2} d_{i,ab}d_{i,c}d_{i,f}d_{i,g}{}^3A_2(F_{a+s-1}, F_{c+s-1}, F_{f+s-1})\ {}^2A_1(F_{b+s-1}, F_{g+s-1}),$$

$$^{222}\tilde{A}_{111}'\big(i(x, y, z), i(x), i(y), i(z)\big) = {}^sb_{ii}^4 \sum_{a,b,c,f,g,h=1}^{2} d_{i,abc}d_{i,f}d_{i,g}d_{i,h}$$

$$\times\, {}^2A_1(F_{a+s-1}, F_{f+s-1})\ {}^2A_1(F_{b+s-1}, F_{g+s-1})\ {}^2A_1(F_{c+s-1}, F_{h+s-1}),$$

$$^{222}\tilde{A}_{111}'\big(i(x, y), i(z, x), i(y), i(z)\big) = {}^sb_{ii}^4 \sum_{a,b,c,f,g,h=1}^{2} d_{i,ab}d_{i,cf}d_{i,g}d_{i,h}$$

$$\times\, {}^2A_1(F_{a+s-1}, F_{f+s-1})\ {}^2A_1(F_{b+s-1}, F_{g+s-1})\ {}^2A_1(F_{c+s-1}, F_{h+s-1}),$$

$$^{2222}\tilde{A}_{1111}'\big(i(x, y), i(z, v), i(x), i(y), i(z), i(v)\big)$$

$$= {}^sb_{ii}^6 \sum_{a,b,c,f,g,h,p,q=1}^{2} d_{i,ab}d_{i,cf}d_{i,g}d_{i,h}d_{i,p}d_{i,q}$$

$$\times\, {}^2A_1(F_{a+s-1}, F_{g+s-1})\ {}^2A_1(F_{b+s-1}, F_{h+s-1})$$

$$\times\, {}^2A_1(F_{c+s-1}, F_{p+s-1})\ {}^2A_1(F_{f+s-1}, F_{q+s-1}).$$

The coefficients $d_{i,a_1\ldots a_p}$ are given for $p = 1, 2, 3$ and we have

$$^sb_{ii} = \sqrt{\frac{1}{\tilde{B}_{ii}^s}}; \qquad \tilde{B}_{ii}^s = \sum_{a,b=1}^{2} d_{i,a}d_{i,b}{}^2A_1(F_{a+s-1}, F_{b+s-1})$$

Furthermore, we use

$a_3(y) = 0$ for the cases (i1), (i2), (ii1), (ii2)

and consequently

$$^{3}\tilde{A}_2'(i, i, i) = 0, \qquad ^{3}\tilde{A}_2'(i, i) = 0,$$
$$^{32}\tilde{A}_{21}'\big(i(y, z), i(y), i(y), i(z)\big) = 0.$$

From

$$\underline{a}_4(y) = 0 \quad \text{for the cases (i1), (i2)}$$

it follows

$$^{4}\underline{\tilde{A}}_3'(i, i, i, i) = 0.$$

For (iii1) we get the calculated values of Table 2.20. For abbreviation we have put

$$i_x \doteq i(x); \qquad i_{xy} \doteq i(x, y); \qquad i_{xyz} \doteq i(x, y, z).$$

Now we are able to compute the approximations $p_{i,st}(u)$ of the density functions $p_{i,s}(u)$. We obtain

$$p_{i,s1}(u) = \frac{1}{\sqrt{2\pi\varepsilon\tilde{B}_{ii}^s}} \exp\left(-\frac{u^2}{2\varepsilon\tilde{B}_{ii}^s}\right),$$

$$p_{i,s2}(u) = p_{i,s1}(u) + p_{i,s1}(u)\left[C_{i,s}^1 H_1\left(\frac{u}{\sqrt{\varepsilon\tilde{B}_{ii}^s}}\right) + C_{i,s}^3 H_3\left(\frac{u}{\sqrt{\varepsilon\tilde{B}_{ii}^s}}\right)\right]\sqrt{\varepsilon},$$

$$p_{i,s3}(u) = p_{i,s2}(u) + p_{i,s1}(u)$$
$$\times\left[C_{i,s}^2 H_2\left(\frac{u}{\sqrt{\varepsilon\tilde{B}_{ii}^s}}\right) + C_{i,s}^4 H_4\left(\frac{u}{\sqrt{\varepsilon\tilde{B}_{ii}^s}}\right) + C_{i,s}^6 H_6\left(\frac{u}{\sqrt{\varepsilon\tilde{B}_{ii}^s}}\right)\right]\varepsilon$$

Table 2.20

s	1	1	3	3
i	1	2	1	2
$^{2}\tilde{A}_1'(i)$	−0.008984	0.006246	−0.073465	0.076131
$^{3}\tilde{A}_2'(i, i, i)$	2.072688	2.000318	1.712577	0.723158
$^{22}\tilde{A}_{11}'(i_{yz}, i_y, i_z)$	−0.005133	0.004951	−0.055779	0.056410
$^{2}\tilde{A}_2'(i, i)$	−0.056404	−0.042987	−0.285406	−0.019156
$^{3}\tilde{A}_2'(i, i)$	−0.013675	0.011476	−0.099939	−0.038426
$^{22}\tilde{A}_{11}'(i_{yz}, i_{yz})$	0.000081	0.000039	0.005397	0.005796
$^{22}\tilde{A}_{11}'(i_{yyz}, i_z)$	0.000262	−0.000214	−0.006217	−0.006364
$^{4}\underline{\tilde{A}}_3'(i, i, i, i)$	9.719130	9.160998	7.751515	12.289659
$^{32}\tilde{A}_{21}'(i_{yz}, i_y, i_y, i_z)$	−0.009980	0.009825	−0.094445	0.000773
$^{222}\tilde{A}_{111}'(i_{xyz}, i_x, i_y, i_z)$	0.000149	−0.000170	−0.004720	−0.004716
$^{222}\tilde{A}_{111}'(i_{xy}, i_{zx}, i_y, i_z)$	0.000046	0.000031	0.004098	0.004295
$^{2222}\tilde{A}_{1111}'(i_{xy}, i_{zv}, i_x, i_y, i_z, i_v)$	0.000026	0.000025	0.003111	0.003182
$\tilde{B}_{ii}^s$	0.1346	0.2785	0.0814	0.0758

where the coefficients $C^p_{i,s}$ are given by Table 2.21. For example, we can write for (i1), $s = 1$, $i = 1$ the approximations

$$p_{1,11}(u) = \frac{1}{\sqrt{2\pi \cdot 0.2067\varepsilon}} \exp\left(-\frac{u^2}{0.4134\varepsilon}\right),$$

$$p_{1,12}(u) = p_{1,11}(u) + p_{1,11}(u)\left[-0.011733H_1\left(\frac{u}{\sqrt{0.2067\varepsilon}}\right) - 0.004635H_3\left(\frac{u}{\sqrt{0.2067\varepsilon}}\right)\right]\sqrt{\varepsilon},$$

$$p_{1,13}(u) = p_{1,12}(u) + p_{1,11}(u)\left[-0.155939H_2\left(\frac{u}{\sqrt{0.2067\varepsilon}}\right) + 0.000305H_4\left(\frac{u}{\sqrt{0.2067\varepsilon}}\right) + 0.000021H_6\left(\frac{u}{\sqrt{0.2067\varepsilon}}\right)\right]\varepsilon$$

of the density function $p_{1,1}(u)$. Fig. 2.17 shows approximations of density functions of $\Lambda_{1,3} - \Lambda_{10}$ in the cases of (i1) and (iii1) with $\varepsilon = 0.1$. For (iii1) we note a stronger deviation of the symmetry of the first approximation.

With the help of

$$\frac{1}{\sqrt{2\pi\varepsilon\tilde{B}^s_{ii}}} \int_{-\infty}^{\infty} u \exp\left(-\frac{u^2}{2\varepsilon\tilde{B}^s_{ii}}\right) H_k\left(\frac{u}{\sqrt{\varepsilon\tilde{B}^s_{ii}}}\right) du = \begin{cases} \sqrt{\varepsilon\tilde{B}^s_{ii}} & \text{for } k = 1, \\ 0 & \text{for } k = 2, 3, 4, 5, 6 \end{cases}$$

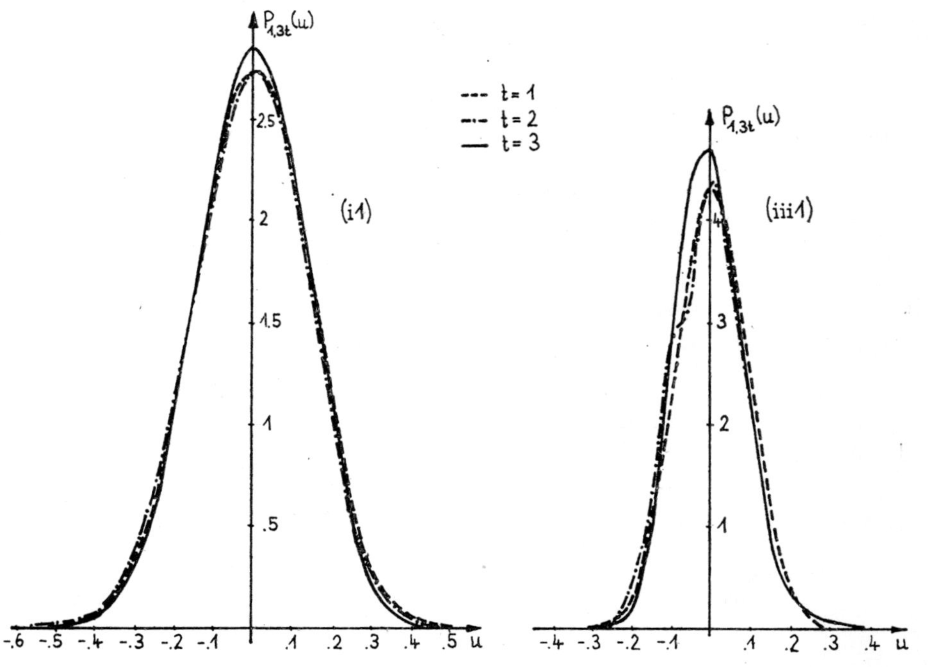

Fig. 2.17. Approximations $p_{i,st}(u)$ of density functions of $\Lambda_{i,s} - \Lambda_{i0}$ for $i = 1$, $s = 3$, $\varepsilon = 0.1$

Table 2.21

$C^p_{i,s}$		(i1)				
s	i	$p = 1$	$p = 2$	$p = 3$	$p = 4$	$p = 6$
1	1	−0.011733	−0.155939	−0.004635	0.000305	0.000021
1	2	0.008365	−0.126939	0.005792	−0.000092	0.000034
3	1	−0.080880	−0.492227	−0.063922	0.008163	0.004086
3	2	0.116931	−0.075431	0.065716	0.016683	0.004319
		(i2)				
1	1	−0.008852	0.000882	−0.005086	0.000281	0.000026
1	2	0.006150	−0.000567	0.004886	−0.000075	0.000024
3	1	−0.075210	−0.010688	−0.057983	0.008156	0.003362
3	2	0.075708	−0.010662	0.058120	0.008271	0.003378
		(ii1)				
1	1	−0.011731	−0.155939	−0.004635	−0.036252	0.000021
1	2	0.008365	−0.126939	0.005792	−0.025498	0.000034
3	1	−0.080880	−0.492227	−0.063922	−0.066037	0.004086
3	2	0.116931	−0.075431	0.065716	−0.024766	0.004319
		(ii2)				
1	1	−0.008446	−0.016170	−0.005006	−0.035244	0.000025
1	2	0.005853	−0.013382	0.004709	−0.031485	0.000022
3	1	−0.070737	−0.104193	−0.053876	−0.041231	0.002889
3	2	0.069428	−0.014998	0.052894	−0.038524	0.002850
		(iii1)				
1	1	−0.008984	−0.040970	0.340315	0.395291	0.057920
1	2	0.006246	−0.010601	0.338337	0.391468	0.057249
3	1	−0.073465	−0.253197	0.229651	0.237332	0.027925
3	2	0.076131	−0.058402	0.176935	0.545812	0.017244
		(iii2)				
1	1	−0.007951	−0.425692	0.429899	0.782881	0.092408
1	2	0.005884	−0.351997	0.317341	0.395070	0.050358
3	1	−0.045510	−0.865339	0.446176	1.314749	0.100290
3	2	0.098788	−0.196511	0.139042	0.212090	0.010128

we see that

$$\langle \Lambda_{i,s} - \Lambda_{i0} \rangle = \int_{-\infty}^{\infty} u\big(p_{i,s3}(u) + o(\varepsilon)\big)\, du = \sqrt{\tilde{B}^s_{ii}}\, C^1_{i,s}\varepsilon + o(\varepsilon)$$

and this formula also leads to the values of Table 2.18. Furthermore, using

$$\frac{1}{\sqrt{2\pi\varepsilon\tilde{B}^s_{ii}}} \int_{-\infty}^{\infty} u^2 \exp\left(-\frac{u^2}{2\varepsilon\tilde{B}^s_{ii}}\right) H_k\left(\frac{u}{\sqrt{\varepsilon\tilde{B}^s_{ii}}}\right) du = \begin{cases} \varepsilon\tilde{B}^s_{ii} & \text{for } k = 0, \\ 2\varepsilon\tilde{B}^s_{ii} & \text{for } k = 2, \\ 0 & \text{for } k = 1, 3, 4, 6 \end{cases}$$

we have

$$\langle (\Lambda_{i,s} - \Lambda_{i0})^2 \rangle = \int_{-\infty}^{\infty} u^2\big(p_{i,s3}(u) + o(\varepsilon)\big)\, du = \tilde{B}^s_{ii}\varepsilon + 2C^2_{i,s}\tilde{B}^s_{ii}\varepsilon^2 + o(\varepsilon^2)$$

and then Table 2.19a and Table 2.19b.

Now we go over to the calculation of

$$\int_{\mathcal{J}_p} p_{i,st}(u)\, du \quad \text{where} \quad \mathcal{J}_p = (a_p, b_p).$$

Introducing

$$\tilde{\mathcal{J}}^s_{i,p} = (\tilde{a}^s_{ip}, \tilde{b}^s_{ip}); \qquad \tilde{a}^s_{ip} = \frac{a_p}{\sqrt{\varepsilon\tilde{B}^s_{ii}}}, \qquad \tilde{b}^s_{ip} = \frac{b_p}{\sqrt{\varepsilon\tilde{B}^s_{ii}}}$$

we obtain the following relations

$$\frac{1}{\sqrt{2\pi\varepsilon\tilde{B}^s_{ii}}} \int_{\mathcal{J}_p} \exp\left(-\frac{u^2}{2\varepsilon\tilde{B}^s_{ii}}\right) H_k\left(\frac{u}{\sqrt{\varepsilon\tilde{B}^s_{ii}}}\right) du$$

$$= \begin{cases} \Phi(u)\Big|_{\tilde{a}^s_{ip}}^{\tilde{b}^s_{ip}} & \text{for } k = 0, \\ -\dfrac{1}{\sqrt{2\pi}} \exp\left(-\dfrac{1}{2}u^2\right)\Big|_{\tilde{a}^s_{ip}}^{\tilde{b}^s_{ip}} & \text{for } k = 1, \\ -\dfrac{u}{\sqrt{2\pi}} \exp\left(-\dfrac{1}{2}u^2\right)\Big|_{\tilde{a}^s_{ip}}^{\tilde{b}^s_{ip}} & \text{for } k = 2, \\ \dfrac{1-u^2}{\sqrt{2\pi}} \exp\left(-\dfrac{1}{2}u^2\right)\Big|_{\tilde{a}^s_{ip}}^{\tilde{b}^s_{ip}} & \text{for } k = 3, \\ \dfrac{3u-u^3}{\sqrt{2\pi}} \exp\left(-\dfrac{1}{2}u^2\right)\Big|_{\tilde{a}^s_{ip}}^{\tilde{b}^s_{ip}} & \text{for } k = 4, \\ \dfrac{-u^5+10u^3-15u}{\sqrt{2\pi}} \exp\left(-\dfrac{1}{2}u^2\right)\Big|_{\tilde{a}^s_{ip}}^{\tilde{b}^s_{ip}} & \text{for } k = 6 \end{cases}$$

where $\Phi(u)$ is defined by

$$\Phi(u) = \frac{1}{\sqrt{2\pi}} \int_{-\infty}^{u} \exp\left(-\frac{1}{2}t^2\right) dt.$$

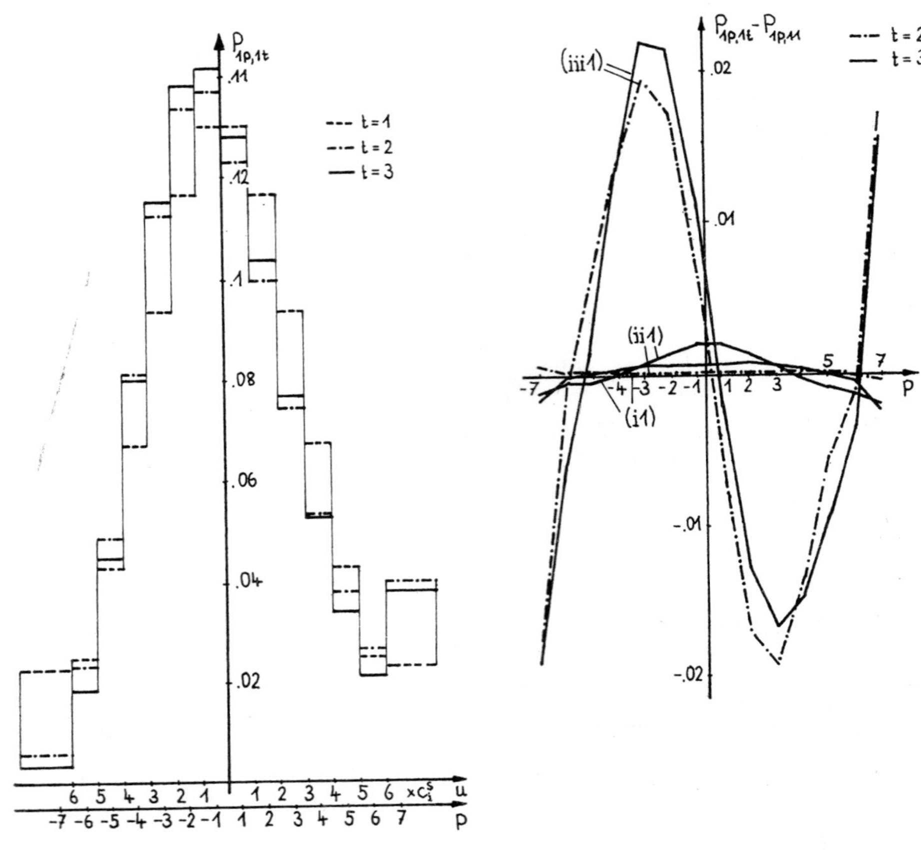

Fig. 2.18a. Approximative probabilities $P_{ip,st}$ for $i = 1$, $s = 1$ in the case of (iii 1), $\varepsilon = 0.1$

Fig. 2.18b. Differences $P_{ip,st} - P_{ip,s1}$ for $i = 1$, $s = 1$, $\varepsilon = 0.1$

Hence, it is

$$P_{ip,s1} \doteq \int_{\mathfrak{I}_p} p_{i,s1}(u)\,\mathrm{d}u = \Phi(u)\Big|_{\tilde{a}_{ip}^s}^{\tilde{b}_{ip}^s},$$

$$P_{ip,s2} \doteq \int_{\mathfrak{I}_p} p_{i,s2}(u)\,\mathrm{d}u$$

$$= P_{ip,s1} + \frac{1}{\sqrt{2\pi}}\left[-C_{i,s}^1 + C_{i,s}^3(1-u^2)\right]\exp\left(-\frac{1}{2}u^2\right)\Big|_{\tilde{a}_{ip}^s}^{\tilde{b}_{ip}^s}\sqrt{\varepsilon},$$

$$P_{ip,s3} \doteq \int_{\mathfrak{I}_p} p_{i,s3}(u)\,\mathrm{d}u$$

$$= P_{ip,s2} + \frac{1}{\sqrt{2\pi}}\left[-C_{i,s}^2 u + C_{i,s}^4(3u-u^3) + C_{i,s}^6(-u^5+10u^3-15u)\right]$$

$$\times \exp\left(-\frac{1}{2}u^2\right)\Big|_{\tilde{a}_{ip}^s}^{\tilde{b}_{ip}^s}\varepsilon.$$

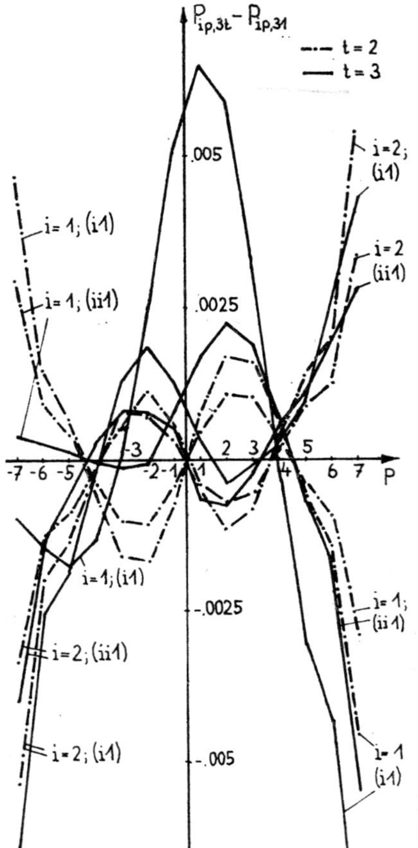

Fig. 2.18c. Differences $P_{ip,st} - P_{ip,s1}$ for $i = 1, 2$, $s = 3$ in the cases of (i1), $\varepsilon = 0.1$ and (ii1), $\varepsilon = 0.04$

In particular, we choose $\mathcal{J}_p$, $p = -k_0, -k_0 + 1, \ldots, -1, 1, 2, \ldots, k_0 - 1, k_0$, in the form

$$\begin{aligned} \mathcal{J}_{-k_0} &\doteq \{x\colon -\infty < x \leqq (-k_0 + 1)\, c\}, \\ \mathcal{J}_p &\doteq \{x\colon pc < x \leqq (p + 1)\, c\} \quad \text{for} \quad p = -k_0 + 1, \ldots, -1, \\ \mathcal{J}_p &\doteq \{x\colon (p - 1)\, c < x \leqq pc\} \quad \text{for} \quad p = 1, 2, \ldots, k_0 - 1, \\ \mathcal{J}_{k_0} &\doteq \{x\colon (k_0 - 1)\, c < x < \infty\} \end{aligned}$$

and c is determined by the equation

$$(k_0 - 1)\, c = 2 \sqrt{\varepsilon \tilde{B}^s_{ii}}.$$

In general, k_0 is taken to be 7. Fig. 2.18a shows some approximative probabilities $P_{jp,st}$ for $i = 1$, $s = 1$ in the case of (iii1), $\varepsilon = 0.1$. The differences $P_{ip,st} - P_{ip,s1}$ for $t = 2, 3$ and $i = 1$, $s = 1$ are plotted in Fig. 2.18b where we have chosen the cases (i1), (ii1), and (iii1) with $\varepsilon = 0.1$. Since we will give some simulation results for the cases (i1), $\varepsilon = 0.1$ and (ii1), $\varepsilon = 0.04$ we have plotted the differences $P_{ip,st} - P_{ip,s1}$ for $i = 1, 2$, $s = 3$ in Fig. 2.18c.

2.4. Simulation results of functional-functions

Simulation results of

$$\left(\Lambda_{1,s}(\omega), \Lambda_{2,s}(\omega)\right)$$

are obtained from simulation results of $\left(r_{s\varepsilon}(\omega), r_{s+1\varepsilon}(\omega)\right)$ by replacing these values in (2.24.1). Then it is easy possible to get statistical characteristics of $\left(\Lambda_{1,s}(\omega), \Lambda_{2,s}(\omega)\right)$ as to the different cases.

First, estimates of the first moments $\overline{M}_{s,i}$ are contained in Table 2.22a and Table 2.22b. The relations

$$\langle \Lambda_{1,s} \rangle \leqq \Lambda_{10} \quad \text{and} \quad \langle \Lambda_{2,s} \rangle \geqq \Lambda_{20}$$

are fulfilled by examples with small ε. Furthermore, the equation

$$\langle \Lambda_{1,s} \rangle + \langle \Lambda_{2,s} \rangle = \Lambda_{10} + \Lambda_{20}$$

with respect to the estimates shows some deviations.

Some simulation results as to the second moments are summarized in Table 2.23. If we compare these values with the values of Table 2.19a we find a better coincidence between the simulation results $\overline{M}_{s,pq}$ and the second approximations $M^2_{s,pq}$ than in

Table 2.22a

$\overline{M}_{s,i}$		$\varepsilon = 0.2$; $m = 1000$					
s	i	(i1)	(i2)	(ii1)	(ii2)	(iii1)	(iii2)
1	1	−2.0058	−2.0042	−2.0053	−2.0082	−1.9982	−2.0003
1	2	2.9932	2.9957	2.9888	2.9903	3.0052	3.0000
3	1	−2.0103	−2.0040	−1.9892	−2.0097	−2.0026	−2.0006
3	2	3.0051	3.0012	2.9989	3.0062	3.0041	3.0024

Table 2.22b

$\overline{M}_{3,i}$	(i1); $\varepsilon = 0.1$		(ii1); $\varepsilon = 0.04$		(iii1); $\varepsilon = 0.2$	
m	$i = 1$	$i = 2$	$i = 1$	$i = 2$	$i = 1$	$i = 2$
1000	−2.0044	3.0048	−2.0029	3.0047	−2.0026	3.0041
2000	−2.0047	3.0074	−2.0032	3.0037	−2.0032	3.0039
3000	−2.0031	3.0063	−2.0036	3.0028	−2.0017	3.0047
4000	−2.0045	3.0060	−2.0038	3.0028	−2.0020	3.0035
5000	−2.0042	3.0057	−2.0030	3.0025	−2.0021	3.0036
6000	−2.0020	3.0046	−2.0023	3.0025	−2.0023	3.0035
7000	−2.0006	3.0045	−2.0024	3.0023	−2.0018	3.0037
8000	−2.0003	3.0043	−2.0025	3.0024	−2.0018	3.0032

comparison with the first approximation $M^1_{s,pq}$. The coincidence between $\overline{M}_{s,pq}$ and $M^t_{s,pq}$ in dependence of m shows Fig. 2.19. For great ε ($\varepsilon = 0.2$) we can see, in general, a great difference between $M^1_{s,pq}$ and $M^2_{s,pq}$ and a very good coincidence between $\overline{M}_{s,pq}$ and $M^2_{s,pq}$ for increasing values of m. Fig. 2.19 shows in all cases a very good approximation of $\overline{M}_{s,pq}$ by $M^2_{s,pq}$. The calculation of the second approximation $M^2_{s,pq}$ leads to

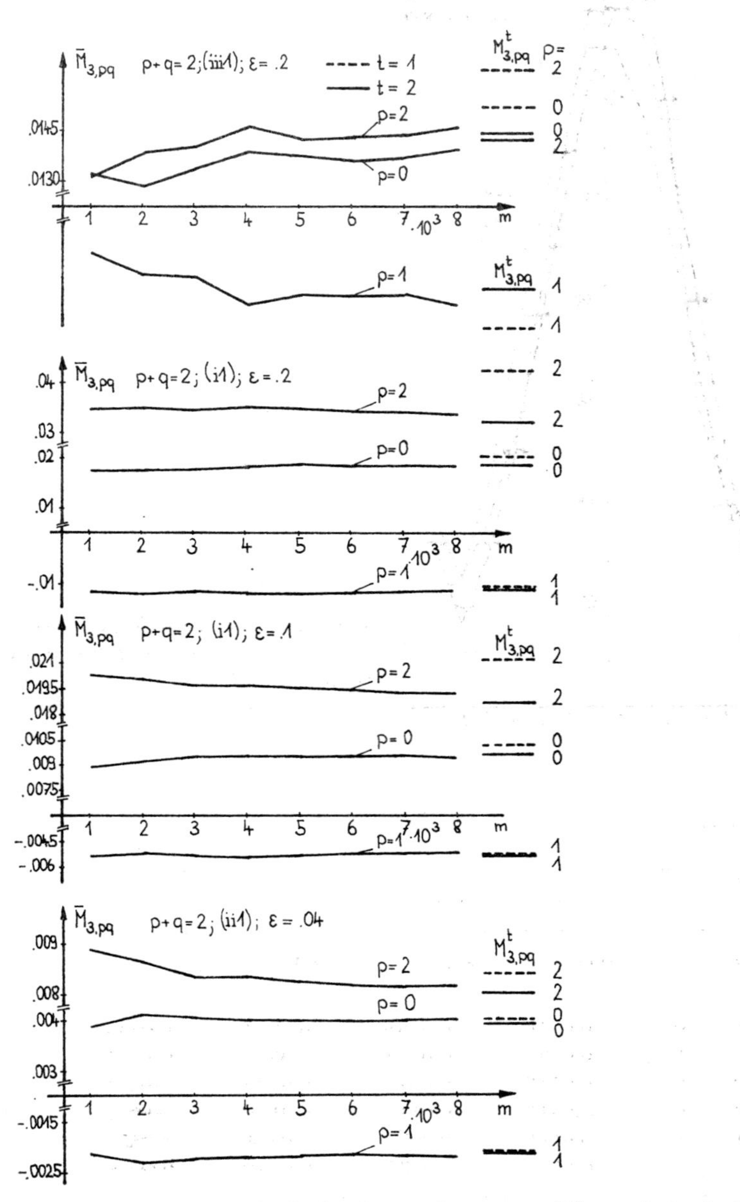

Fig. 2.19. Comparison of calculated second moments $M^t_{s,pq}$ and simulation results $\overline{M}_{s,pq}$ in dependence on m

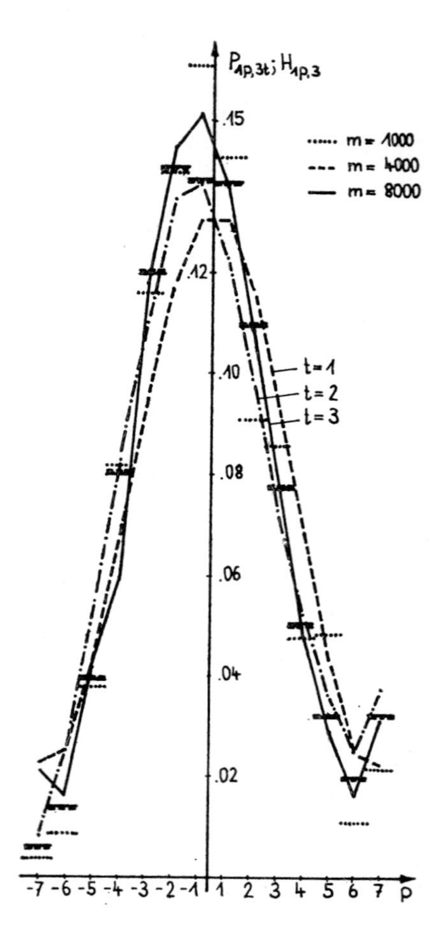

Fig. 2.20. Comparison between the calculated values $P_{ip,st}$ for $t = 1, 2, 3$ and the relative frequencies $H_{ip,s}$ in the case of (iii 1), $i = 1$, $s = 3$, $\varepsilon = 0.2$

Table 2.23

$\overline{M}_{s,pq}$			$\varepsilon = 0.2$; $m = 1000$			
s	p	q	(i1)	(i2)	(iii1)	(iii2)
1	2	0	0.0376	0.0268	0.0254	0.0144
1	1	1	0.0505	0.0371	0.0359	0.0187
1	0	2	0.0743	0.0540	0.0536	0.0273
3	2	0	0.0343	0.0143	0.0130	0.0182
3	1	1	−0.0116	−0.0075	−0.0064	−0.0027
3	0	2	0.0176	0.0135	0.0132	0.0046

Table 2.24

$\chi^2_{1,3t}$	$t = 1$	$t = 2$	$t = 3$
$m = 1000$	68.5	44.9	31.3
$m = 4000$	196.6	58.0	39.6
$m = 8000$	391.0	114.9	80.2

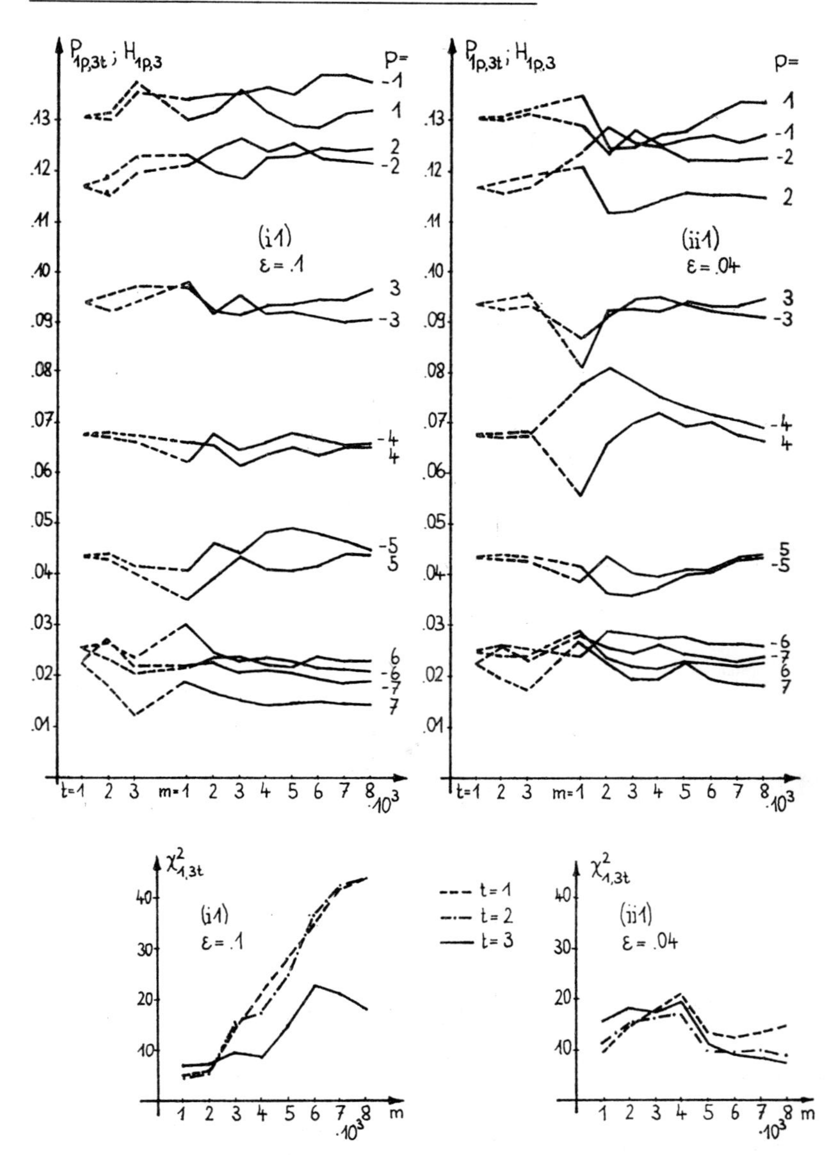

Fig. 2.21. Relative frequencies $H_{ip,s}$ in dependence on m in comparison with $P_{ip,st}$ and the χ^2-values

an essential correction of the investigated values for greater ε. Although we deal with functions of linear functionals there is no difference between the results of Fig. 2.8 for $p + q = 2$ and the results of Fig. 2.19.

Now we investigate the approximative density functions and the relative frequencies. In the case of (iii 1) with the values $i = 1$, $s = 3$, and $\varepsilon = 0.2$ we have plotted the approximative probabilities $P_{ip,st}$ for $t = 1, 2, 3$ and the relative frequencies $H_{ip,s}$ in Fig. 2.20. The best coincidence between $P_{ip,st}$ and $H_{ip,s}$ can be observed for $t = 3$. This also follows from the $\chi^2_{i,st}$-values. The adequate χ^2-values which belong to this figure are included in Table 2.24.

Fig. 2.21 shows some relative frequencies $H_{ip,s}$ in dependence on m in comparison with the approximative probabilities $P_{ip,st}$ where we have chosen the cases (i1), $\varepsilon = 0.1$, $i = 1$, $s = 3$ and (ii1), $\varepsilon = 0.04$, $i = 1$, $s = 3$. We have added to this figure the

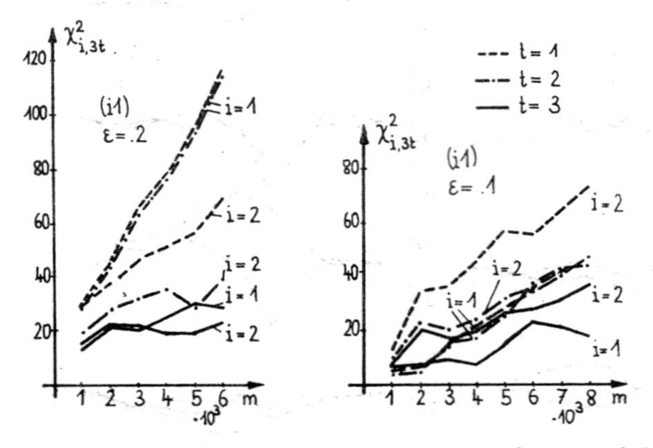

Fig. 2.22a. χ^2-values in dependence on m in the case of (i1)

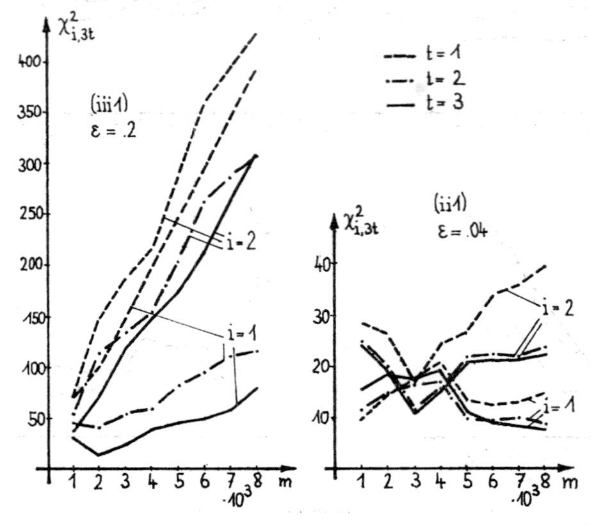

Fig. 2.22b. χ^2-values in dependence on m in the cases of (ii1) and (iii1)

χ^2-values $\chi^2_{i,st}$ belonging to the relative frequencies $H_{ip,s}$ and the approximative probabilities $P_{ip,st}$,

$$\chi^2_{i,st} = m \sum_{p=-k_0}^{k_0} \frac{(H_{ip,s} - P_{ip,st})^2}{P_{ip,st}}.$$

With regard to these χ^2-values we obtain the best results for great numbers m in the case of $t = 3$. The influence of the correlation length ε is obvious.

In Fig. 2.22a and Fig. 2.22b we have plotted some more χ^2-values in the case (i1) for $\varepsilon = 0.2, 0.1$; (ii1) for $\varepsilon = 0.04$, and (iii1) for $\varepsilon = 0.2$. In all cases the smallest values of $\chi^2_{i,st}$ are obtained for $t = 3$. Furthermore, the value $\varepsilon = 0.2$ is to great as to a χ^2-test. Naturally, the χ^2-values are strongly dependent on the correlation length ε. In this case we have dealt with functions of linear functionals nevertheless the comparison between the theoretical results and the simulation results of this chapter are similar to the comparison in the linear case (see Section 2.2).

3. Random vibrations

3.1. Random vibrations of continua

3.1.1. Deterministic solutions

We consider the boundary-initial-value problem on $[0, T] \times \mathscr{D}$

$$\frac{\partial^2 u}{\partial t^2} + 2\beta \frac{\partial u}{\partial t} + Lu = p(t, x, \omega) \tag{3.1}$$

with the initial conditions

$$u(0, x) = u_0(x), \qquad \frac{\partial u}{\partial t}(0, x) = u_1(x)$$

and the boundary conditions on $\partial\mathscr{D}$ according to the order of the elliptic differential operator L. $\mathscr{D}$ denotes a bounded domain of $\mathbb{R}^n$ with piecewise smooth boundaries.

With the help of the Fourier method a formal solution of (3.1) can be found. The eigenvalues and eigenfunctions of the problem

$Lf = \lambda f$
with the boundary conditions of the problem (3.1)

are denoted by λ_k and $f_k(x)$, respectively. Then, the formal solution can be determined in the form of a series

$$u(t, x) = \sum_{k=1}^{\infty} c_k(t)\, f_k(x). \tag{3.2}$$

Using (3.1) the function $c_k(t)$ fulfils the differential equation

$$c_k''(t) + 2\beta c_k'(t) + \lambda_k c_k(t) = P_k(t)$$

where P_k is defined by

$$P_k(t) \doteq \big(p(t, .), f_k\big)$$

and $(.,.)$ denotes the scalar product on $\mathbf{L}_2(\mathscr{D})$. The general solution of this differential equation is obtained by

$$c_k(t) = \mathrm{e}^{-\beta t}\big(a_k \cos(\beta_k t) + b_k \sin(\beta_k t)\big) + \int_0^t h_k(t - s)\, P_k(s)\, \mathrm{d}s$$

where

$$\beta_k \doteq \sqrt{\lambda_k - \beta^2}$$

and

$$h_k(t) \doteq \frac{1}{\beta_k} e^{-\beta t} \sin(\beta_k t).$$

The constants a_k, b_k can be calculated by the initial conditions

$$u_0(x) = u(0, x) = \sum_{k=1}^{\infty} a_k f_k(x),$$

$$u_1(x) = \frac{\partial u}{\partial t}(0, x) = \sum_{k=1}^{\infty} [-\beta a_k + b_k \beta_k] f_k(x)$$

and we have

$$a_k = (u_0, f_k), \qquad b_k = \frac{1}{\beta_k} \big(\beta a_k + (u_1, f_k)\big).$$

Using these constants a_k, b_k the formal solution of the boundary-initial-value problem (3.1) can be written as

$$u(t, x) = \sum_{k=1}^{\infty} \left[e^{-\beta t}\big(a_k \cos(\beta_k t) + b_k \sin(\beta_k t)\big) + \int_0^t h_k(t - s) P_k(s) \, ds \right] f_k(x). \tag{3.3}$$

The solution of the averaged problem associated with the above problem (3.1) is denoted by $w(t, x)$ and we define

$$\tilde{u}(t, x, \omega) \doteq u(t, x, \omega) - w(t, x).$$

The averaged problem is obtained from (3.1) if the random function $p(t, x, \omega)$ is substituted by $\langle p(t, x) \rangle$. From (3.3) it follows

$$\tilde{u}(t, x, \omega) = \sum_{k=1}^{\infty} \int_0^t h_k(t - s) \overline{P}_k(s) \, ds f_k(x) \tag{3.4}$$

where

$$\overline{P}_k(t) \doteq \big(\overline{p}(t, .), f_k\big),$$

$$\overline{p}(t, x, \omega) \doteq p(t, x, \omega) - \langle p(t, x) \rangle,$$

if the boundary conditions and initial conditions are assumed to be non-random.

Now, we deal with the "averaged" solution

$$U(t, \omega) \doteq \big(\tilde{u}(t, ., \omega), g\big) \tag{3.5}$$

where (f, g) denotes the scalar product on $\mathbf{L}_2(\mathcal{D})$ and $g \in \mathbf{C}_0^{\infty}(\mathcal{D})$. It follows

$$U(t, \omega) = \sum_{k=1}^{\infty} \int_0^t h_k(t - s) \overline{P}_k(s) \, ds (f_k, g)$$

assuming the sum in (3.4) is convergent in $\mathbf{L}_2(\mathcal{D})$. Furthermore, we have

$$\frac{\partial}{\partial t} U(t, \omega) = \sum_{k=1}^{\infty} \int_0^t h_k'(t - s) \, \overline{P}_k(s) \, ds (f_k, g)$$

and

$$\frac{\partial^2}{\partial t^2} U(t, \omega) = \sum_{k=1}^{\infty} \left[\overline{P}_k(t) + \int_0^t h_k''(t - s) \, \overline{P}_k(s) \, ds \right] (f_k, g)$$

using the uniform convergence of the series which is based on

$$(f_k, g) = \frac{1}{\lambda_k} (Lf_k, g) = \frac{1}{\lambda_k} (f_k, Lg) .$$

Hence, we obtain

$$\frac{\partial^2}{\partial t^2} (\bar{u}, g) + 2\beta \frac{\partial}{\partial t} (\bar{u}, g) + (\bar{u}, Lg)$$

$$= \sum_{k=1}^{\infty} \int_0^t \{h_k''(t - s) + 2\beta h_k'(t - s) + \lambda_k h_k(t - s)\} \, \overline{P}_k(s) \, ds (f_k, g) + \big(p(t, .), g\big)$$

$$= \big(p(t, .), g\big)$$

because of

$$h_k''(s) + 2\beta h_k'(s) + \lambda_k h_k(s) = 0 \quad \text{for} \quad 0 \leqq s \leqq t .$$

Defining

$$F(t, y) \doteq \sum_{k=1}^{\infty} h_k(t) \, f_k(y) \, (f_k, g)$$

it is

$$U(t, \omega) = \int_0^t \int_{\mathcal{D}} F(t - s, y) \, \overline{p}(s, y, \omega) \, ds \, dy \tag{3.6}$$

since the series determining $F(t, y)$ possesses strong convergence properties. From

$$F(t, y, x) \doteq \sum_{k=1}^{\infty} h_k(t) \, f_k(x) \, f_k(y) \tag{3.7}$$

the relation

$$U(t, \omega) = \left(\int_0^t \int_{\mathcal{D}} F(t - s, y, .) \, \overline{p}(s, y) \, ds \, dy, g(.) \right)$$

is obtained assuming

$$F(t, y, x) \in \mathbf{L}_2([0, T] \times \mathcal{D} \times \mathcal{D}) .$$

Assuming the uniform convergence of

$$\sum_{k=1}^{\infty} h_k^2(t) \, f_k^2(x)$$

as to x the function

$$\int_0^t \int_{\mathcal{D}} F(t-s, y, x)\, \overline{p}(s, y)\, \mathrm{d}s\, \mathrm{d}y$$

is continuous with respect to x. This property follows from the inequalities

$$\left| \int_0^t \int_{\mathcal{D}} \big(F(t-s, y, x) - F(t-s, y, \overline{x})\big)\, \overline{p}(s, y)\, \mathrm{d}s\, \mathrm{d}y \right|$$

$$\leqq \left[\int_0^t \int_{\mathcal{D}} \big(F(t-s, y, x) - F(t-s, y, \overline{x})\big)^2\, \mathrm{d}s\, \mathrm{d}y \int_0^t \int_{\mathcal{D}} \overline{p}^2(s, y)\, \mathrm{d}s\, \mathrm{d}y \right]^{1/2}$$

$$\leqq c(\omega) \left[\int_0^t \sum_{k=1}^{\infty} h_k^2(t-s)\, \big(f_k(x) - f_k(\overline{x})\big)^2\, \mathrm{d}s \right]^{1/2}.$$

Under these conditions the solution of

$$\frac{\partial^2 \overline{u}}{\partial t^2} + 2\beta \frac{\partial \overline{u}}{\partial t} + L\overline{u} = \overline{p}(t, x, \omega) \tag{3.8}$$

with homogeneous boundary and initial conditions can be written as

$$\overline{u}(t, x, \omega) = \int_0^t \int_{\mathcal{D}} F(t-s, y, x)\, \overline{p}(s, y, \omega)\, \mathrm{d}s\, \mathrm{d}y. \tag{3.9}$$

Defining

$$\mathrm{D}_t^p \mathrm{D}_x^q v \doteq \frac{\partial^{p+q} v}{\partial t^p \partial x_1^{q_1} \ldots \partial x_n^{q_n}} \qquad \left(\sum_{i=1}^{n} q_i = q \right)$$

for the derivations we obtain the relation

$$\begin{aligned} \big(\mathrm{D}_t^p \mathrm{D}_x^q \overline{u}(t, x, \omega), g(x)\big) &= (-1)^q \big(\mathrm{D}_t^p \overline{u}(t, x, \omega), \mathrm{D}_x^q g(x)\big) \\ &= (-1)^q \sum_{k=1}^{\infty} \mathrm{D}_t^p \int_0^t h_k(t-s)\, \overline{P}_k(s)\, \mathrm{d}s (f_k, \mathrm{D}^q g) \\ &= \int_0^t \int_{\mathcal{D}} \mathrm{D}_t^p \mathrm{D}_x^q F(t-s, y)\, \overline{p}(s, y)\, \mathrm{d}s\, \mathrm{d}y \end{aligned}$$

where

$$\mathrm{D}_t^p \mathrm{D}_x^q F(t, y) \doteq \sum_{k=1}^{\infty} \frac{\partial^p}{\partial t^p} h_k(t)\, f_k(y)\, (\mathrm{D}^q f_k, g) \quad \text{for} \quad p = 0, 1.$$

By means of the condition

$$\mathrm{D}_t^p \mathrm{D}_x^q F(t, y, x) \doteq \sum_{k=1}^{\infty} \frac{\partial^p}{\partial t^p} h_k(t)\, f_k(y)\, \mathrm{D}_x^q f_k(x) \in \mathbf{L}_2([0, T] \times \mathcal{D} \times \mathcal{D})$$

we can deduce

$$\big(\mathrm{D}_t^p \mathrm{D}_x^q \overline{u}(t, x, \omega), g(x)\big) = \left(\int_0^t \int_{\mathcal{D}} \mathrm{D}_t^p \mathrm{D}_x^q F(t-s, y, .)\, \overline{p}(s, y)\, \mathrm{d}s\, \mathrm{d}y, g \right).$$

From similar considerations as for $\bar{u}(t, x, \omega)$ the continuity of

$$\int_0^t \int_{\mathcal{D}} \mathrm{D}_t^p \mathrm{D}_x^q F(t - s, y, x)\, \bar{p}(s, y)\, \mathrm{d}s\, \mathrm{d}y$$

as to x follows from the uniform convergence of

$$\sum_{k=1}^{\infty} \left[\frac{\partial^p}{\partial t^p} h_k(t)\, \mathrm{D}_x^q f_k(x)\right]^2 \tag{3.10}$$

and we have

$$\mathrm{D}_t^p \mathrm{D}_x^q \bar{u}(t, x, \omega) = \int_0^t \int_{\mathcal{D}} \mathrm{D}_t^p \mathrm{D}_x^q F(t - s, y, x)\, \bar{p}(s, y)\, \mathrm{d}s\, \mathrm{d}y\,. \tag{3.11}$$

A condition of the form (3.10) with $p = 1$, $q = 0$ would demand the convergence of the series

$$\sum_{k=1}^{\infty} \left(h_k'(t)\right)^2$$

which is not fulfilled.

3.1.2. Weakly correlated excitation

The inhomogeneous term $\bar{p}(t, x, \omega)$ of (3.8) is assumed to be weakly correlated. The limit theorem (see Section 1.4) can be applied to terms of the form (3.9) or (3.11). We obtain

$$\lim_{\varepsilon \downarrow 0} \frac{1}{\sqrt{\varepsilon}^{n+1}} \mathrm{D}_x^q \bar{u}(t, x, \omega) = \eta_q(t, x, \omega) \qquad \text{in distribution} \tag{3.12}$$

if the condition $\|\mathrm{D}_x^q F(., ., x)\| < \infty$ for $x \in \mathcal{D}$ is satisfied. The random function $\eta_q(t, x, \omega)$ is Gaussian having the moments $\langle \eta_q(t, x) \rangle = 0$ and

$$\langle \eta_q(t, x)\, \eta_q(r, y) \rangle = \int_0^{\bar{t}} \int_{\mathcal{D}} \mathrm{D}_x^q F(t - s, z, x)\, \mathrm{D}_y^q F(r - s, z, y)\, a(s, z)\, \mathrm{d}s\, \mathrm{d}z \tag{3.13}$$

with $\bar{t} \doteq \min\{t, r\}$. It follows that $\mathrm{D}_x^q u(t, x, \omega)$ is approximately Gaussian distributed for small ε with the first moments

$$\langle \mathrm{D}_x^q u(t, x) \rangle \approx \mathrm{D}_x^q w(t, x)\,,$$

$$\langle \mathrm{D}_x^q \bar{u}(t, x)\, \mathrm{D}_y^q \bar{u}(r, y) \rangle \approx \varepsilon^{n+1} \langle \eta_q(t, x)\, \eta_q(r, y) \rangle\,.$$

A similar limit theorem can be obtained for random vector functions

$$\frac{1}{\sqrt{\varepsilon}^{n+1}} \left(\mathrm{D}_x^{q_1} \bar{u}(t, x, \omega), \ldots, \mathrm{D}_x^{q_k} \bar{u}(t, x, \omega)\right).$$

The given limit theorem can also be applied to the random process $U(t, \omega)$ or to

$$\mathrm{D}_t^p \mathrm{D}_x^q U(t, \omega) \doteq \left(\mathrm{D}_t^p \mathrm{D}_\cdot^q \bar{u}(t, \cdot, \omega), g\right) = \int_0^t \int_{\mathscr{D}} F^{pq}(t-s, y)\, \overline{p}(s, y, \omega) \,\mathrm{d}s \,\mathrm{d}y$$

for $p = 0, 1$; $q = 0, 1, 2, \ldots$, respectively, where

$$F^{pq}(t, x) \doteq \sum_{k=1}^{\infty} \mathrm{D}_t^p h_k(t-s)\, f_k(y)\, (\mathrm{D}_x^q f_k, g).$$

Then we have

$$\lim_{\varepsilon \downarrow 0} \frac{1}{\sqrt{\varepsilon}^{n+1}} \mathrm{D}_t^p \mathrm{D}_x^q U(t, \omega) = \zeta_{pq}(t, \omega) \qquad \text{in distribution} \tag{3.14}$$

and the Gaussian process $\zeta_{pq}(t, \omega)$ possesses the moments $\langle \zeta_{pq}(t) \rangle = 0$ and

$$\langle \zeta_{pq}(t)\, \zeta_{pq}(r) \rangle = \int_0^{\bar{t}} \int_{\mathscr{D}} F^{pq}(t-s, y)\, F^{pq}(r-s, y)\, a(s, y) \,\mathrm{d}s \,\mathrm{d}y. \tag{3.15}$$

Without essential difficulties the limit theorem can be applied to the random vector

$$\frac{1}{\sqrt{\varepsilon}^{n+1}} \left(\mathrm{D}_t^{p_1} \mathrm{D}_x^{q_1} U(t, \omega), \ldots, \mathrm{D}_t^{p_k} \mathrm{D}_x^{q_k} U(t, \omega)\right).$$

Investigate the limit function $\eta(t, x, \omega)$ of $\bar{u}(t, x, \omega)$. Let the field $\overline{p}(t, x, \omega)$ be in the wide-sense homogeneous so that the intensity is constant, $a(t, x) = a = \text{const}$. Hence, from (3.13) we obtain

$$\begin{aligned} \langle \eta(t, x)\, \eta(r, y) \rangle &= a \int_0^{\bar{t}} \sum_{k=1}^{\infty} h_k(t-s)\, h_k(r-s)\, f_k(x)\, f_k(y) \,\mathrm{d}s \\ &= a \sum_{k=1}^{\infty} T_k(t, r)\, f_k(x)\, f_k(y) \end{aligned} \tag{3.16}$$

where

$$T_k(t, r) = \int_0^{\bar{t}} h_k(t-s)\, h_k(r-s) \,\mathrm{d}s$$

if

$$F(t, x, y) \in \mathbf{L}_2([0, T] \times \mathscr{D} \times \mathscr{D}) \quad \text{and} \quad \sum_{k=1}^{\infty} h_k^2(t)\, f_k^2(x)$$

is uniformly convergent as to x. The functions $T_k(t, r)$ can be calculated as

$$\begin{aligned} T_k(t, r) = {} & \frac{1}{4\lambda_k} \mathrm{e}^{-\beta|t-r|} \left\{\frac{1}{\beta} \cos\left(\beta_k |t-r|\right) + \frac{1}{\beta_k} \sin\left(\beta_k |t-r|\right)\right\} \\ & + \frac{1}{4\beta_k^2} \mathrm{e}^{-\beta(t+r)} \left\{-\frac{1}{\beta} \cos\left(\beta_k |t-r|\right) + \frac{\beta}{\beta_k} \cos\left(\beta_k (t+r)\right) \right. \\ & \left. - \frac{\beta_k}{\lambda_k} \sin\left(\beta_k (t+r)\right)\right\} \end{aligned} \tag{3.17}$$

and it is

$$\lim_{\substack{t,r\to\infty \\ r-t=s=\text{const}}} T_k(t, r) = \frac{1}{4\lambda_k} e^{-\beta|s|} \left\{\frac{1}{\beta} \cos(\beta_k |s|) + \frac{1}{\beta_k} \sin(\beta_k |s|)\right\} \doteq \tilde{T}_k(s).$$

Fig. 3.1 illustrates the convergence of

$$R(t, s, x; \beta) \doteq \langle \eta(t, x)\, \eta(t + s, x)\rangle$$

to

$$\tilde{R}(s, x; \beta) \doteq \lim_{t\to\infty} \langle \eta(t, x)\, \eta(t + s, x)\rangle$$

for $t \to \infty$. We can see that the value $\tilde{R}(s, x; \omega)$ already approximates very well the value $R(t, s, x; \beta)$ for $t \geqq 1$ assuming $\beta \geqq 1$. For $\beta = 0.1$ this good approximation property can only be observed for larger values of t. In Fig. 3.1 it was put

$$f_k(x) = \sqrt{2} \sin(\pi k x); \qquad \lambda_k = (k\pi)^2.$$

For example, these eigenvalues and eigenfunctions correspond to an operator $Lu = -u_{xx}$ on $\mathscr{D} = [0, 1]$ with the boundary condition $u(t, 0) = u(t, 1) = 0$.

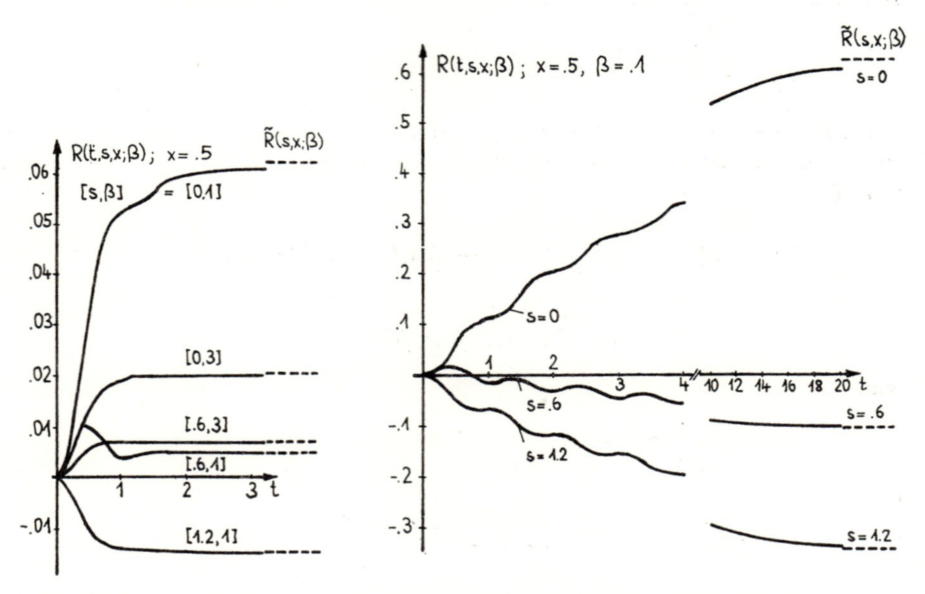

Fig. 3.1. Convergence of the correlation function to the limit correlation function as $t \to \infty$

The random function $\eta(t, x, \omega)$ can be taken for large times t as a stationary Gaussian process $\tilde{\eta}(t, x, \omega)$ as to t having the correlation function

$$\tilde{R}(s, x, y) \doteq \langle \tilde{\eta}(t, x)\, \tilde{\eta}(t + s, y)\rangle$$
$$= \frac{a}{4} e^{-\beta|s|} \sum_{k=1}^{\infty} \frac{1}{\lambda_k} \left\{\frac{1}{\beta} \cos(\beta_k |s|) + \frac{1}{\beta_k} \sin(\beta_k |s|)\right\} f_k(x)\, f_k(y).$$

The spectral density

$$S(\alpha; x, y) \doteq \frac{1}{2\pi} \int_{-\infty}^{\infty} e^{-is\alpha} \tilde{R}(s, x, y) \, ds$$

of the wide-sense stationary process $\tilde{\eta}(t, x, \omega)$ as to t can be calculated. Supposing $\sum_{k=1}^{\infty} 1/\lambda_k < \infty$ we have

$$S(\alpha; x, y) = \frac{a}{4\pi} \sum_{k=1}^{\infty} \frac{1}{\lambda_k} \int_0^{\infty} e^{-\beta s} \cos(s\alpha) \left\{ \frac{1}{\beta} \cos(\beta_k s) + \frac{1}{\beta_k} \sin(\beta_k s) \right\} f_k(x) f_k(y) \, ds$$

$$= \frac{a}{2\pi} \sum_{k=1}^{\infty} \frac{1}{(\lambda_k + \alpha^2)^2 - 4\beta_k^2 \alpha^2} f_k(x) f_k(y).$$

Similar results follow for the limit process $\zeta_{pq}(t, \omega)$ of

$$\frac{1}{\sqrt{\varepsilon^{n+1}}} D_t^p D_x^q U(t, \omega).$$

We can write for the correlation function

$$\langle \zeta_{p_1q_1}(t) \, \zeta_{p_2q_2}(r) \rangle = a \int_0^{\bar{t}} \sum_{k=1}^{\infty} D_t^{p_1} h_k(t - s) \, D_r^{p_2} h_k(r - s) \, (D_x^{q_1} f_k, g) \, (D_x^{q_2} f_k, g) \, ds$$

$$= a \sum_{k=1}^{\infty} T_k^{p_1p_2}(t, r) \, (D_x^{q_1} f_k, g) \, (D_x^{q_2} f_k, g)$$

where

$$T_k^{p_1p_2}(t, r) = \int_0^{\bar{t}} D_t^{p_1} h_k(t - s) \, D_r^{p_2} h_k(r - s) \, ds.$$

There

$$T_k^{00}(t, r) = T_k(t, r)$$

given by (3.16) and furthermore

$$\lim_{\substack{t,r \to \infty \\ r - t = s = \text{const}}} T_k^{10}(t, r) = \frac{1}{4\beta\beta_k} e^{-\beta|s|} \sin(\beta_k |s|),$$

$$\lim_{\substack{t,r \to \infty \\ r - t = s = \text{const}}} T_k^{11}(t, r) = \frac{1}{4} e^{-\beta|s|} \left\{ \frac{1}{\beta} \cos(\beta_k |s|) - \frac{1}{\beta_k} \sin(\beta_k |s|) \right\}.$$

The stationary process corresponding to $\zeta_{pq}(t, \omega)$ is denoted by $\tilde{\zeta}_{pq}(t, \omega)$. Then the vector process

$$(\tilde{\zeta}_{00}(t, \omega), \tilde{\zeta}_{10}(t, \omega))$$

is a stationary Gaussian vector process. It follows for large times

$$\begin{aligned}
\langle U^2(t)\rangle &\approx \varepsilon^{n+1}\langle \xi_{00}^2(t)\rangle = \frac{a\varepsilon^{n+1}}{4\beta}\sum_{k=1}^{\infty}\frac{1}{\lambda_k}(f_k, g)^2 \doteq \frac{a\varepsilon^{n+1}}{4\beta}R(g),\\
\langle U(t)\,U'(t)\rangle &\approx \varepsilon^{n+1}\langle \xi_{00}(t)\,\xi_{10}(t)\rangle = 0,\\
\langle U'(t)^2\rangle &\approx \varepsilon^{n+1}\langle \xi_{10}^2(t)\rangle = \frac{a\varepsilon^{n+1}}{4\beta}\sum_{k=1}^{\infty}(f_k, g)^2 = \frac{a\varepsilon^{n+1}}{4\beta}(g, g).
\end{aligned} \tag{3.18}$$

The spectral density with respect to $\xi_{pq}(t, \omega)$ is denoted by

$$S_{p_1q_1p_2q_2}(\alpha) \doteq \frac{1}{2\pi}\int_{-\infty}^{\infty} e^{-i\alpha s}\langle \xi_{p_1q_1}(t)\,\xi_{p_2q_2}(t+s)\rangle\, ds.$$

Then we have

$$\begin{aligned}
&S_{0q_10q_2}(\alpha)\\
&= \frac{a}{8\pi}\sum_{k=1}^{\infty}\frac{1}{\lambda_k}\int_{-\infty}^{\infty} e^{-is\alpha}\, e^{-\beta|s|}\left\{\frac{1}{\beta}\cos(\beta_k|s|) + \frac{1}{\beta_k}\sin(\beta_k|s|)\right\} ds(D_x^{q_1}f_k, g)\,(D_x^{q_2}f_k, g)\\
&= \frac{a}{2\pi}\sum_{k=1}^{\infty}\frac{1}{(\lambda_k+\alpha^2)^2 - 4\beta_k^2\alpha^2}(D_x^{q_1}f_k, g)\,(D_x^{q_2}f_k, g),\\
&S_{1q_10q_2}(\alpha) = -i\alpha S_{0q_10q_2}(\alpha),\\
&S_{1q_11q_2}(\alpha) = \alpha^2 S_{0q_10q_2}(\alpha).
\end{aligned}$$

As an application the expected rate of crossings $r(l, g)$ of the process $U(t, \omega)$ at the level l can be calculated by

$$r(l, g) = \frac{1}{\pi}\sqrt{\frac{\langle U'(t)^2\rangle}{\langle U(t)^2\rangle}}\exp\left(-\frac{l^2}{2\langle U(t)^2\rangle}\right) = \frac{1}{\pi}\sqrt{\frac{(g, g)}{R(g)}}\exp\left(-\frac{2\beta l^2}{\varepsilon^{n+1}aR(g)}\right) \tag{3.19}$$

(see e.g. SOONG [1]). In order to obtain results as to $\bar{u}(t, x, \omega)$ we use

$$\begin{aligned}
U(t, \omega) &\approx \bar{u}(t, x, \omega)\int_{\mathcal{D}_\delta(x)} g(y)\, dy,\\
U'(t, \omega) &\approx \bar{u}_t(t, x, \omega)\int_{\mathcal{D}_\delta(x)} g(y)\, dy
\end{aligned}$$

for functions $g(y)$ which are zero outside of a small neighbourhood $\mathcal{D}_\delta(x)$ of x. Thus,

$$\langle \bar{u}^2(t, x)\rangle \approx \frac{a\varepsilon^{n+1}}{4\beta}\frac{R(g)}{(g, 1)^2}; \qquad \langle \bar{u}_t^2(t, x)\rangle \approx \frac{a\varepsilon^{n+1}}{4\beta}\frac{(g, g)}{(g, 1)^2}.$$

Selecting a bounded function $g_\delta(y)$ then it follows

$$\langle \bar{u}_t^2(t, x)\rangle \xrightarrow[\delta\to 0]{} \infty.$$

Now, we use

$$g(s) \doteq \begin{cases} \frac{1}{\delta}(\delta - x + s) & \text{for} \quad x - \delta \leqq s \leqq x, \\ \frac{1}{\delta}(\delta + x - s) & \text{for} \quad x \leqq s \leqq x + \delta, \\ 0 & \text{otherwise} \end{cases}$$

and then we have

$$\langle U^2(t) \rangle = \frac{a\varepsilon^{n+1}}{4\beta} \frac{\sqrt{2}}{120} \left(15x\delta^3(1 - x) - \frac{7}{2}\delta^4 \right),$$

$$\langle U'(t)^2 \rangle = \frac{a\varepsilon^{n+1}}{4\beta} \frac{2}{3}\delta \quad \text{for} \quad 0 \leqq x - \delta < x < x + \delta < 1.$$

Now, we will investigate the motion of a string surrounded by a gas. Let the string have a constant density ϱ and a length l (see UHLENBECK, ORNSTEIN [1]; VAN LEAR, UHLENBECK [1]). Furthermore, let the string be elastically bounded at its ends and we assume that the string is under the tension $\tau(x)$. The equation of motion of the string can be written in the form

$$\varrho \frac{\partial^2 u}{\partial t^2} + f \frac{\partial u}{\partial t} - \frac{\partial}{\partial x}\left(\tau \frac{\partial u}{\partial x}\right) = F(t, x, \omega), \qquad 0 \leqq x \leqq l,$$

where f is the friction coefficient and $F(t, x, \omega)$ denotes the fluctuating force which acts on the string by means of the gas. The conditions to be satisfied at the ends are:

$$h_0 u(t, 0) - \frac{\partial u}{\partial x}(t, 0) = 0; \qquad -h_l u(t, l) - \frac{\partial u}{\partial x}(t, l) = 0$$

and the initial conditions are

$$u(0, x) = \frac{\partial u}{\partial t}(0, x) = 0.$$

If $F(t, x, \omega)$ is assumed to be weakly correlated in (t, x) then the given method can be applied to the calculation of the limit distribution. It is possible to prove that $F(t, x, \omega)$ is a homogeneous weakly correlated field with very small $\varepsilon > 0$ (see EINSTEIN [1]). The intensity corresponding to $F(t, x, \omega)$ can be determined with the help of the equipartition of energy (see UHLENBECK, ORNSTEIN [1]):

$$a = \frac{2fkT}{\varrho^2}$$

where T denotes the absolute temperature and $k = R/N$ (R – gas constant; N – Loschmidt number).

For a string with a constant tension τ we have

$$u_{tt} + 2\beta u_t - A u_{xx} = p_\varepsilon(t, x, \omega),$$

with the boundary conditions

$$(h_0 u - u_x)|_{x=0} = 0; \qquad (-h_l u - u_x)|_{x=l} = 0$$

and the initial conditions

$$u(0, x) = u_t(0, x) = 0$$

where

$$2\beta \doteq \frac{f}{\varrho}; \qquad A \doteq \frac{\tau}{\varrho}; \qquad p \doteq \frac{F}{\varrho}.$$

Now, $p_\varepsilon(t, x, \omega)$ is assumed to be a homogeneous weakly correlated field. The eigenvalues λ_k and the eigenfunctions f_k are calculated from

$$-Af_{xx} = \lambda f; \qquad (h_0 f - f_x)|_{x=0} = 0, \qquad (-h_l f - f_x)|_{x=l} = 0.$$

The equation for the eigenvalues λ_k is

$$\left(\sqrt{\frac{\lambda}{A}} - \frac{h_0 h_l}{\sqrt{\frac{\lambda}{A}}}\right) \tan\left(l\sqrt{\frac{\lambda}{A}}\right) = h_0 + h_l$$

and the eigenfunctions $f_k(x)$ are obtained from

$$f(x) = c_1 \cos\left(\sqrt{\frac{\lambda}{A}}\, x\right) + c_2 \sin\left(\sqrt{\frac{\lambda}{A}}\, x\right), \qquad c_2 = h_0 \sqrt{\frac{A}{\lambda}}\, c_1.$$

Now, the normalized solution $\frac{1}{\varepsilon} u_\varepsilon(t, x, \omega)$ converges in distribution as $\varepsilon \downarrow 0$ to the Gaussian function $\eta(t, x, \omega)$ having the correlation function given by (3.16) since in this case the function $F(t, x, y)$ is contained in $\mathbf{L}_2([0, T] \times \mathcal{D} \times \mathcal{D})$. For large times t, the stationary limit function $\tilde{\eta}(t, x, \omega)$ is obtained having the correlation function

$$\tilde{R}(s, x, y) \doteq \langle \tilde{\eta}(t, x)\, \tilde{\eta}(t + s, y) \rangle$$
$$= \frac{a}{4} e^{-\beta|s|} \sum_{k=1}^{\infty} \frac{1}{\lambda_k} \left\{ \frac{1}{\beta} \cos(\beta_k |s|) + \frac{1}{\beta_k} \sin(\beta_k |s|) \right\} f_k(x)\, f_k(y). \tag{3.20}$$

The spectral density can be calculated as

$$S(\alpha; x, y) = \frac{a}{2\pi} \sum_{k=1}^{\infty} \frac{1}{(\lambda_k + \alpha^2)^2 - 4\beta_k^2 \alpha^2} f_k(x)\, f_k(y). \tag{3.21}$$

In the case of $h_0, h_l \to \infty$ we deduce

$$\lambda_k = A \left(\frac{k\pi}{l}\right)^2 \quad \text{and} \quad f_k(x) = \sqrt{\frac{2}{l}} \sin\left(\frac{k\pi x}{l}\right)$$

from the above considerations. From this it follows

$$\tilde{R}(0, x, y) = \frac{a}{2\beta l} \sum_{k=1}^{\infty} \frac{1}{\lambda_k} \sin\left(\frac{k\pi x}{l}\right) \sin\left(\frac{k\pi y}{l}\right) = \frac{al}{4\beta A} \left[\min\left\{\frac{x}{l}, \frac{y}{l}\right\} - \frac{x}{l} \frac{y}{l}\right]$$

and for $x = y$

$$\langle \tilde{\eta}^2(t, x) \rangle = \frac{a}{4\beta A} x \left(1 - \frac{x}{l}\right).$$

This is the same result also determined by UHLENBECK, ORNSTEIN [1] if

$$2\beta = \frac{f}{\varrho}, \quad A = \frac{\tau}{\varrho}, \quad a = \frac{2fkT}{\varrho^2}$$

are put.

Similar results are obtained by ERINGEN [1, 2] relative to $\tilde{\eta}(t, x, \omega)$ by means of a generalized harmonic analysis. All examples given in these two papers can also be treated by the above method of the weakly correlated functions.

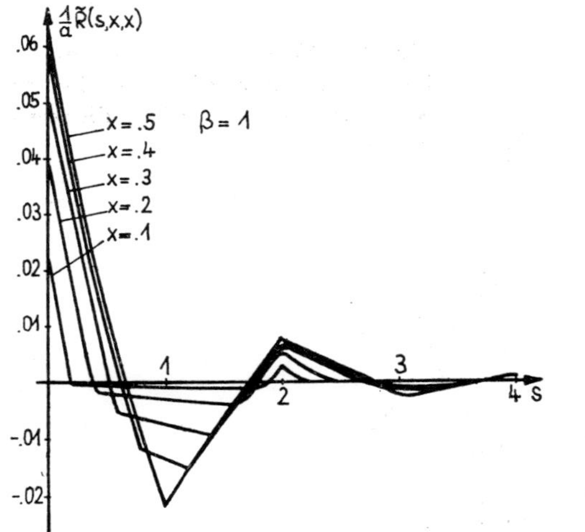

Fig. 3.2. Correlation function $\tilde{R}(s, x, x)$ of the stationary limit function $\tilde{\eta}(t, x, \omega)$

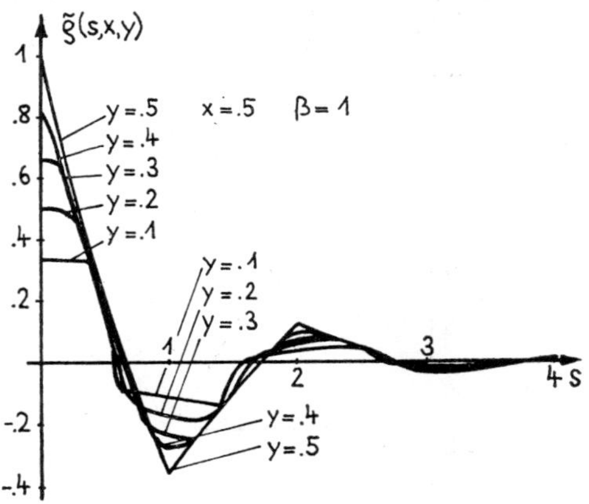

Fig. 3.3. Normalized correlation function $\tilde{\varrho}(s, x, y)$ of the stationary limit function $\tilde{\eta}(t, x, \omega)$

The correlation function in dependence on s and for $x = y$ is plotted in Fig. 3.2. In Fig. 3.3 the normalized correlation function $\tilde{\varrho}(s, x, y)$ of $\tilde{\eta}(t, x, \omega)$

$$\tilde{\varrho}(s, x, y) = \frac{\langle \tilde{\eta}(t, x)\, \tilde{\eta}(t + s, y) \rangle}{\sqrt{\langle \tilde{\eta}^2(t, x) \rangle \langle \tilde{\eta}^2(t + s, y) \rangle}}$$

is illustrated as a function of s for fixed x, y, and $\beta = 1$. For these numerical calculations we have put $A = 1$, $l = 1$. Fig. 3.4 shows the normalized correlation function for diverse values of β. In this figure the function $\tilde{\varrho}(s, x, y)$ is plotted for $x = y = 0.5$ and for various values of β as a function of s. This normalized correlation function shows a strong dependence on β. In Fig. 3.5 the normalized correlation function $\tilde{\varrho}$ is plotted for $\beta = 1$, $x = 0.5$ in dependence on y and for different times s. The correlation function $\tilde{\varrho}(s, x, y)$ is plotted in Fig. 3.6 as a function of y for $s = 0$, $\beta = 1$ and in Fig. 3.7 as a function of y for different values of (s, β) and $x = 0.5$. These figures give a very good

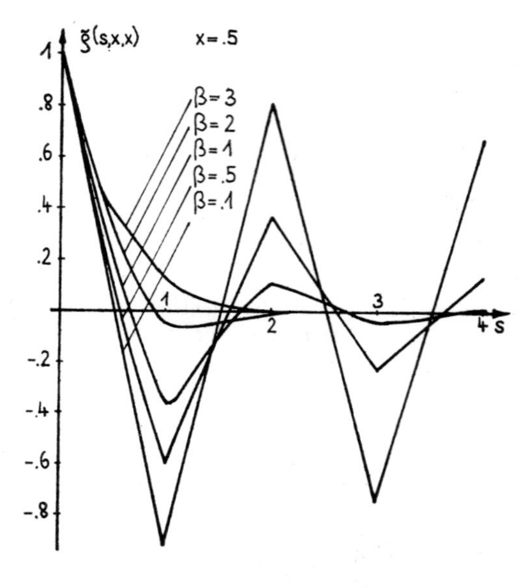

Fig. 3.4. Normalized correlation function $\tilde{\varrho}(s, x, x)$ of the stationary limit function $\tilde{\eta}(t, x, \omega)$ for different values β

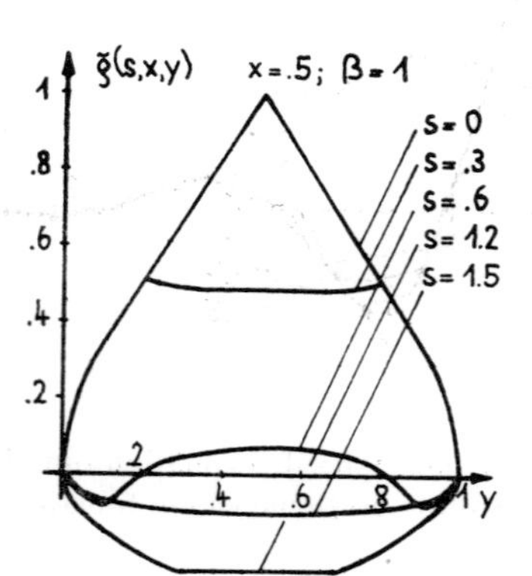

Fig. 3.5. Normalized correlation function for different values of s in dependence on y

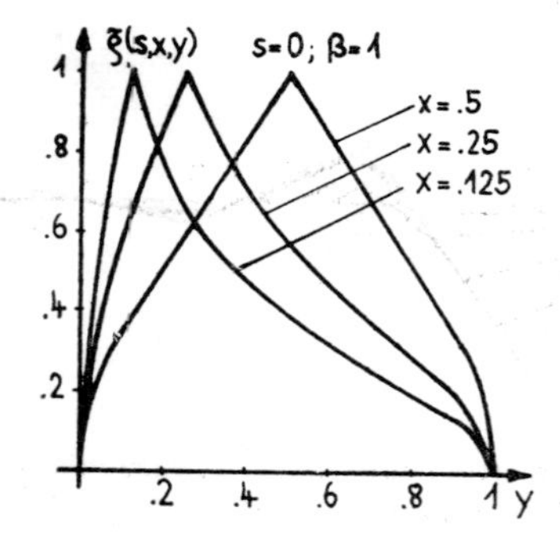

Fig. 3.6. Normalized correlation function for different values of x in dependence on y

survey on the dependence of the normalized correlation function $\tilde{\varrho}(s, x, y)$ on the parameters s, x, y, β.

Fig. 3.8 shows the spectral density of the stationary limit function $\tilde{\eta}(t, x, y)$ as to t in dependence on α for different values of $x = y$ and for $\beta = 1$. Furthermore, Fig. 3.9 illustrates $S(\alpha; x, y)$ for different values of (x, y) and for $\beta = 1$.

Finally, the dependence of the spectral density on β is plotted in Fig. 3.10 where we can establish a strong dependence on β. As is seen the frequencies around $\alpha \approx 2.8$ are very important since these frequencies give a relatively large amount to the variance of the function $\tilde{\eta}$.

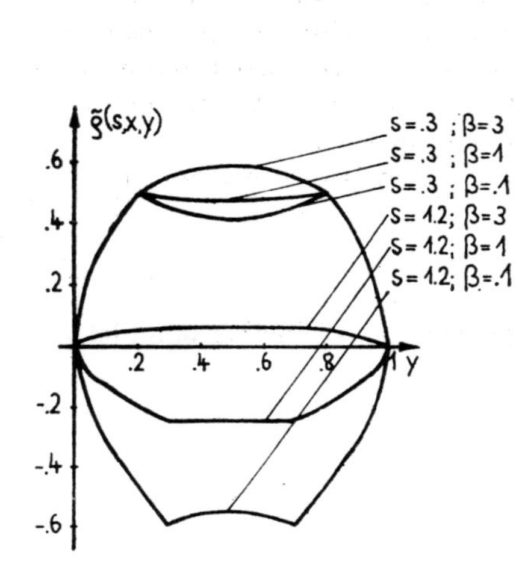

Fig. 3.7. Normalized correlation function for different values of β in dependence on y

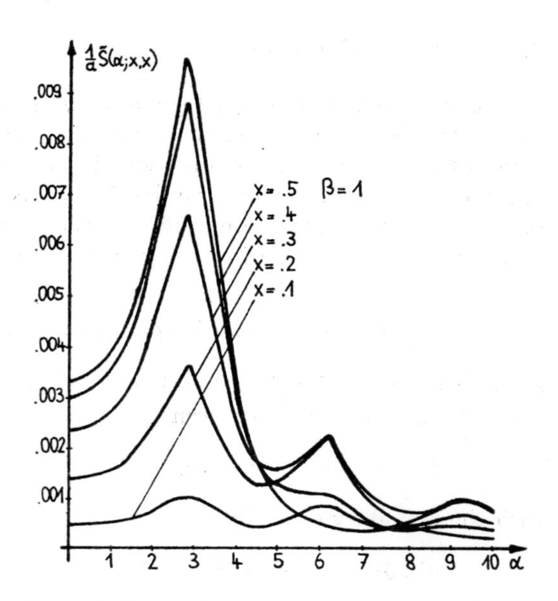

Fig. 3.8. Spectral density of the stationary limit function for different values of $x = y$

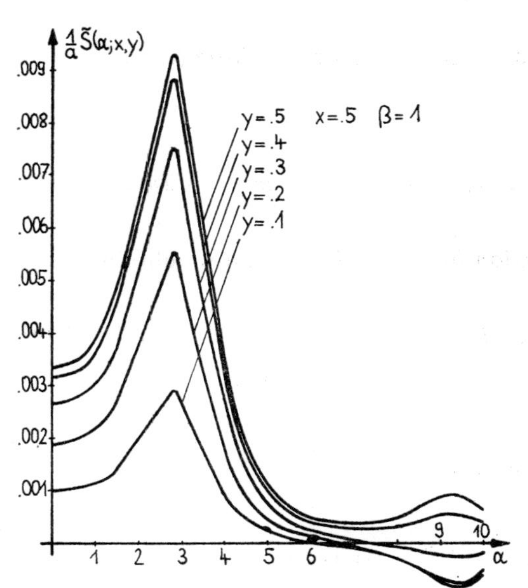

Fig. 3.9. Spectral density of the stationary limit function for different values of (x, y)

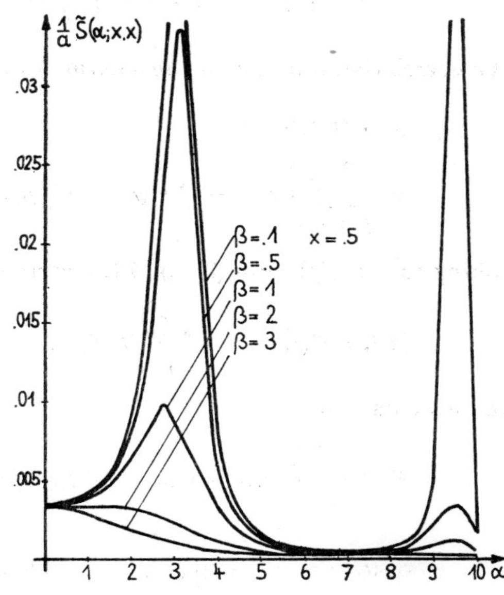

Fig. 3.10. Spectral density of the stationary limit function for different values of β

3.1.3. Comparing results

In this section the results determined by the limit distribution of weakly correlated functions are to be compared with results obtained from the vibration problem with respect to a realistic random function. The comparison has to be restricted to the correlation relations and the spectral densities since further results are only difficult to get in general without the assumption of a Gaussian random function.

A correlation function involving in the vibration problem is assumed to have the form

$$\langle \overline{p}(t, x)\, \overline{p}(r, y)\rangle = \sigma^2 \exp\left(-\gamma\, |t - r| - \varkappa\, |x - y|\right) \tag{3.22}$$

in order to secure the simplification of the following calculations. Furthermore, let the operator L have the simple form

$$Lu = -Au_{xx} \quad \text{on} \quad [0, l]$$

with the boundary conditions

$$u(t, 0) = u(t, l) = 0$$

(see also the example in the previous section). Then it follows

$$\lambda_k = A\left(\frac{k\pi}{l}\right)^2; \quad f_k(x) = \sqrt{\frac{2}{l}}\sin\left(\frac{k\pi x}{l}\right) \quad \text{for} \quad k = 1, 2, \ldots$$

The correlation function of the solution of (3.1) can be determined from (3.4):

$$\langle \bar{u}(t, x)\, \bar{u}(r, y)\rangle$$
$$= \sum_{i,j=1}^{\infty} \int_0^t \int_0^r h_i(t - s)\, h_j(r - \bar{s}) \left\langle \left(\overline{p}(s, .), f_i\right) \left(\overline{p}(\bar{s}, .), f_j\right)\right\rangle \mathrm{d}s\, \mathrm{d}\bar{s}\; f_i(x)\, f_j(y).$$

Taking the special form (3.22) of the correlation function of $\overline{p}(t, x, \omega)$ we obtain

$$\langle \bar{u}(t, x)\, \bar{u}(r, y)\rangle = \sigma^2 \sum_{i,j=1}^{\infty} R_{ij} S_{ij}(t, r)\, f_i(x)\, f_j(y) \tag{3.23}$$

where we have put

$$R_{ij} = \int_0^l \int_0^l \exp\left(-\varkappa\, |v - \bar{v}|\right) f_i(v)\, f_j(\bar{v})\, \mathrm{d}v\, \mathrm{d}\bar{v},$$

$$S_{ij}(t, r) = \int_0^t \int_0^r \exp\left(-\gamma\, |s - \bar{s}|\right) h_i(t - s)\, h_j(r - \bar{s})\, \mathrm{d}s\, \mathrm{d}\bar{s}.$$

Now, the quantities R_{ij}, $S_{ij}(t, r)$ are calculated. Setting

$$w_i \doteq \frac{i\pi}{l}$$

the quantity R_{ij} can be written as

$$R_{ij} = \int_0^l f_i(v) \left[e^{-\varkappa v} \int_0^v e^{\varkappa \bar{v}} f_j(\bar{v}) \, d\bar{v} + e^{\varkappa v} \int_v^l e^{-\varkappa \bar{v}} f_j(\bar{v}) \, d\bar{v} \right] dv$$

$$= \int_0^l f_i(v) \left[e^{-\varkappa v} \int_0^v e^{\varkappa \bar{v}} f_j(\bar{v}) \, d\bar{v} - (-1)^j e^{-\varkappa(l-v)} \int_0^{l-v} e^{\varkappa \bar{v}} f_j(\bar{v}) \, d\bar{v} \right] dv$$

$$= \frac{1}{w_j^2 + \varkappa^2} \int_0^l f_i(v) \left[2\varkappa f_j(v) + \sqrt{\frac{2}{l}} \, w_j \left(e^{-\varkappa v} - (-1)^j \, e^{-\varkappa(l-v)} \right) \right] dv$$

and we obtain

$$R_{ij} = \begin{cases} \dfrac{2}{l} \dfrac{1}{\varkappa^2 + w_i^2} \left[l\varkappa + \dfrac{2w_i^2}{\varkappa^2 + w_i^2} \left(1 - (-1)^i \, e^{-\varkappa l}\right) \right] & \text{for} \quad i = j, \\ \dfrac{2w_i w_j}{l(\varkappa^2 + w_i^2)\,(\varkappa^2 + w_j^2)} \left[1 + (-1)^{i+j} - \left((-1)^i + (-1)^j\right) e^{-\varkappa l} \right] & \text{for} \quad i \neq j. \end{cases}$$

For $t \leqq r$ the above expression for S_{ij} can be reduced to

$$S_{ij}(t, r) = \frac{\exp\left(-\beta(t + r)\right)}{\beta_i \beta_j} \int_0^t e^{\beta s} \sin\left(\beta_i(t - s)\right)$$

$$\times \left[e^{-\gamma s} \int_0^s e^{(\beta+\gamma)\bar{s}} \sin\left(\beta_j(r - \bar{s})\right) d\bar{s} + e^{\gamma s} \int_s^r e^{(\beta-\gamma)\bar{s}} \sin\left(\beta_j(r - \bar{s})\right) d\bar{s} \right] ds$$

$$= \frac{\exp\left(-\beta(t + r)\right)}{\beta_i \beta_j} \left[I_{ij}(t, r; \gamma) - I_{ij}(t, r; -\gamma) \right.$$

$$\left. - \{(\beta + \gamma) \sin(\beta_j r) + \beta_j \cos(\beta_j r)\} \, J_{ij}(t; -\gamma) + \beta_j \, e^{(\beta-\gamma)r} \, J_{ij}(t; \gamma) \right]$$

where I_{ij} and J_{ij} are defined by

$$I_{ij}(t, r; \gamma) \doteq \frac{1}{(\beta + \gamma)^2 + \beta_j^2} \int_0^t e^{2\beta s} \sin\left(\beta_i(t - s)\right)$$

$$\times \left[(\beta + \gamma) \sin\left(\beta_j(r - s)\right) + \beta_j \cos\left(\beta_j(r - s)\right) \right] ds,$$

$$J_{ij}(t; \gamma) \doteq \frac{1}{(\beta - \gamma)^2 + \beta_j^2} \int_0^t e^{(\beta+\gamma)s} \sin\left(\beta_i(t - s)\right) ds.$$

I_{ij} can be calculated by

$$I_{ij}(t, r; \gamma) = \frac{1}{(\beta + \gamma)^2 + \beta_j^2} \left[\frac{1}{2} (\beta + \gamma) \{a(t, r; \beta_i, -\beta_j) - a(t, r; \beta_i, \beta_j)\} \right.$$

$$\left. + \frac{1}{2} \beta_j \{b(t, r; \beta_i, -\beta_j) + b(t, r; \beta_i, \beta_j)\} \right]$$

if we use

$$a(t, r; \varepsilon, \delta) \doteq \int_0^t e^{2\beta s} \cos\left(\varepsilon t + \delta r - (\varepsilon + \delta) s\right) ds$$

$$= \frac{1}{4\beta^2 + (\varepsilon + \delta)^2} \left[e^{2\beta t}\left\{2\beta \cos\left(\delta(r - t)\right) - (\varepsilon - \delta) \sin\left(\delta(r - t)\right)\right\}\right.$$

$$\left. - 2\beta \cos(\varepsilon t + \delta r) + (\varepsilon + \delta) \cos(\varepsilon t + \delta r)\right]$$

and

$$b(t, r; \varepsilon, \delta) \doteq \int_0^t e^{2\beta s} \sin\left(\varepsilon t + \delta r - (\varepsilon + \delta) s\right) ds$$

$$= \frac{1}{4\beta^2 + (\varepsilon + \delta)^2} \left[e^{2\beta t}\left\{2\beta \sin\left(\delta(r - t)\right) + (\varepsilon + \delta) \cos\left(\delta(r - t)\right)\right\}\right.$$

$$\left. - 2\beta \sin(\varepsilon t + \delta r) - (\varepsilon + \delta) \cos(\varepsilon t + \delta r)\right].$$

The quantity J_{ij} can be obtained as

$$J_{ij}(t; \gamma)$$

$$= \frac{1}{\left((\beta - \gamma)^2 + \beta_j^2\right)\left((\beta + \gamma)^2 + \beta_i^2\right)} \left[\beta_i e^{(\beta+\gamma)t} - (\beta + \gamma) \sin(\beta_i t) - \beta_i \cos(\beta_i t)\right].$$

With the help of these considerations $S_{ij}(t, r)$ can be calculated for $t \leqq r$.

For the investigation of the stationary part of the correlation function we consider

$$\lim_{\substack{t,r \to \infty \\ r-t=s=\text{const} \geqq 0}} S_{ij}(t, r) = \frac{1}{\beta_i \beta_j} [\mathring{I}_{ij}(s; \gamma) - \mathring{I}_{ij}(s; -\gamma)] + \frac{\exp(-\gamma s)}{c_{ij}(\gamma)} \doteq \mathring{S}_{ij}(s) \tag{3.24}$$

where

$$\mathring{I}_{ij}(s; \gamma) \doteq \frac{1}{(\beta + \gamma)^2 + \beta_j^2} \left[\frac{1}{2} (\beta + \gamma) \{\mathring{a}(s; \beta_i, -\beta_j) - \mathring{a}(s; \beta_i, \beta_j)\}\right.$$

$$\left. + \frac{1}{2} \beta_j \{\mathring{b}(s; \beta_i, -\beta_j) + \mathring{b}(s; \beta_i, \beta_j)\}\right],$$

$$\mathring{a}(s; \varepsilon, \delta) \doteq \lim_{\substack{t,r \to \infty \\ r-t=s=\text{const} \geqq 0}} e^{-\beta(t+r)} a(t, r; \varepsilon, \delta)$$

$$= \frac{1}{4\beta^2 + (\varepsilon + \delta)^2} e^{-\beta s} \{2\beta \cos(\delta s) - (\varepsilon + \delta) \sin(\delta s)\},$$

$$\mathring{b}(s; \varepsilon, \delta) \doteq \lim_{\substack{t,r \to \infty \\ r-t=s=\text{const} \geqq 0}} e^{-\beta(t+r)} b(t, r; \varepsilon, \delta)$$

$$= \frac{1}{4\beta^2 + (\varepsilon + \delta)^2} e^{-\beta s} \{2\beta \sin(\delta s) + (\varepsilon + \delta) \cos(\delta s)\},$$

and

$$c_{ij}(\gamma) \doteq \left((\beta - \gamma)^2 + \beta_j^2\right)\left((\beta + \gamma)^2 + \beta_i^2\right).$$

Hence, we can deduce

$$\mathring{I}_{ij}(s;\gamma) = \frac{e^{-\beta s}}{(\beta+\gamma)^2+\beta_i^2}\left[\frac{1}{4\beta^2+(\beta_i+\beta_j)^2}\left\{\left(-\beta(\beta+\gamma)+\frac{1}{2}\beta_j(\beta_i+\beta_j)\right)\cos(\beta_j s)\right.\right.$$
$$\left.+\left(\frac{1}{2}(\beta+\gamma)(\beta_i+\beta_j)+\beta\beta_j\right)\sin(\beta_j s)\right\}$$
$$+\frac{1}{4\beta^2+(\beta_i-\beta_j)^2}\left\{\left(\beta(\beta+\gamma)+\frac{1}{2}\beta_j(\beta_i-\beta_j)\right)\cos(\beta_j s)\right.$$
$$\left.\left.+\left(\frac{1}{2}(\beta+\gamma)(\beta_i-\beta_j)-\beta\beta_j\right)\sin(\beta_j s)\right\}\right]$$

and from this

$$\mathring{I}_{ii}(s;\gamma) = \frac{1}{4}\frac{e^{-\beta s}}{(\beta+\gamma)^2+\beta_i^2}\left[\left(2+\frac{\gamma}{\beta}-\frac{\beta(2\beta+\gamma)}{\lambda_i}\right)\cos(\beta_i s)\right.$$
$$\left.+\left(\frac{(2\beta+\gamma)\beta_i}{\lambda_i}-\frac{\lambda_i}{\beta}\right)\sin(\beta_i s)\right].$$

Using (3.23) for $s \geqq 0$ it follows

$$\mathring{R}(s,x,y) \doteq \lim_{\substack{t,r\to\infty \\ r-t=s=\text{const}\geqq 0}} \langle\bar{u}(t,x)\,\bar{u}(r,y)\rangle = \sigma^2\sum_{i,j=1}^{\infty} R_{ij}\mathring{S}_{ij}(s)\,f_i(x)\,f_j(y) \tag{3.25}$$

and for $s < 0$

$$\mathring{R}(s,x,y) = \sigma^2\sum_{i,j=1}^{\infty} R_{ij}\mathring{S}_{ij}(-s)\,f_i(x)\,f_j(y)$$
$$= \sigma^2\sum_{i,j=1}^{\infty} R_{ij}\mathring{S}_{ij}(-s)\,f_i(y)\,f_j(x) = \mathring{R}(-s,y,x)$$

since

$$\lim_{\substack{t,r\to\infty \\ r-t=s=\text{const}\leqq 0}} S_{ij}(t,r) = \lim_{\substack{t,r\to\infty \\ r-t=s=\text{const}\leqq 0}} S_{ji}(r,t) = \mathring{S}_{ji}(t-r) = \mathring{S}_{ji}(-s).$$

In particular, from (3.25) we obtain

$$\mathring{R}(0,x,y) = \sigma^2\sum_{i,j=1}^{\infty} R_{ij}\mathring{S}_{ij}(0)\,f_i(x)\,f_j(y)$$
$$= \sigma^2\left[\sum_{i=1}^{\infty} R_{ii}\mathring{S}_{ii}(0)\,f_i(x)\,f_i(y) + \sum_{\substack{i,j=1 \\ j<i}}^{\infty} R_{ij}\mathring{S}_{ij}(0)\left(f_i(x)\,f_j(y)+f_i(y)\,f_j(x)\right)\right]$$

using

$$R_{ij} = R_{ji} \quad\text{and}\quad \mathring{S}_{ij}(0) = \mathring{S}_{ji}(0).$$

In order to compare the results from the theory of the weakly correlated random functions with the results following from the correlation function (3.22) we consider the relations

$$\lim_{\gamma\to\infty} \gamma \mathring{S}_{ij}(s) = \frac{1}{\beta_i\beta_j} \{\mathring{a}(s; \beta_i, -\beta_j) - \mathring{a}(s; \beta_i, \beta_j)\},$$

and

$$\lim_{\varkappa\to\infty} \varkappa R_{ii} = 2,$$

$$\lim_{\varkappa\to\infty} \varkappa R_{ij} = 0 \quad \text{where} \quad R_{ij} = O(\varkappa^{-4}) \quad \text{for} \quad i \neq j.$$

Then we have

$$\lim_{\gamma\to\infty} \gamma \mathring{S}_{ii}(s) = \frac{1}{2\lambda_i} e^{-\beta s} \left(\frac{1}{\beta} \cos(\beta_i s) + \frac{1}{\beta_i} \sin(\beta_i s)\right)$$

and

$$\lim_{\gamma\to\infty} \gamma \mathring{S}_{ij}(s) = \beta_i e^{-\beta s} \frac{4\beta\beta_j \cos(\beta_j s) + (4\beta^2 + \lambda_i - \lambda_j) \sin(\beta_j s)}{[4\beta^2 + (\beta_i + \beta_j)^2] [4\beta^2 + (\beta_i - \beta_j)^2]}.$$

Thus, (3.25) leads to

$$\lim_{\substack{\varkappa\to\infty \\ \gamma\to\infty}} \varkappa\gamma \mathring{R}(s, x, y) = 4\sigma^2 \sum_{i=1}^{\infty} \tilde{T}_i(s) f_i(x) f_i(y). \tag{3.26}$$

The intensity of a weakly correlated random function having an approximative correlation function of the form (3.22) can be determined from

$$\frac{1}{\varepsilon^2} \iint_{\mathcal{K}_\varepsilon(0)} \exp(-\gamma |s| - \varkappa |z|) \, ds \, dz \approx \frac{1}{\varepsilon^2} \int_{-\varepsilon}^{\varepsilon}\int_{-\varepsilon}^{\varepsilon} \exp(-\gamma |s| - \varkappa |z|) \, ds \, dz$$

$$= \frac{4}{\gamma\varkappa\varepsilon^2} (1 - e^{-\varepsilon\gamma}) (1 - e^{-\varepsilon\varkappa})$$

and we obtain

$$\varepsilon^2 a \approx \frac{4\sigma^2}{\gamma\varkappa}$$

since ε is chosen that

$$\exp(-\gamma\varepsilon) \approx 0, \qquad \exp(-\varkappa\varepsilon) \approx 0;$$

i.e. for large values γ, $\varkappa$ the correlation length ε can then also be taken small. Consequently, (3.26) leads to

$$\mathring{R}(s, x, y) \approx \varepsilon^2 \tilde{R}(s, x, y) = \varepsilon^2 a \sum_{i=1}^{\infty} \tilde{T}_i(s) f_i(x) f_i(y)$$

for large values of γ, $\varkappa$. Thus, for the correlation function, we have obtained the connection between the result from the weakly correlated theory and the result following from (3.22).

In Fig. 3.11 the correlation function

$$R_\gamma(s) = \sigma^2 \exp(-\gamma |s|)$$

is plotted for different values of γ. This function $R_\gamma(s)$ shows the characteristic figure of the correlation function

$$\langle \overline{p}(t, x)\, \overline{p}(r, y) \rangle$$

from (3.22). A comparison with the correlation function of the weakly correlated theory considered in the limit case gives an information on the connection between $\tilde{R}(s, x, y)$ and $\mathring{R}(s, x, y)$.

Setting

$$\tilde{K}(s, x, y) \doteq \frac{1}{a}\, \tilde{R}(s, x, y), \qquad \mathring{K}(s, x, y) \doteq \frac{\gamma \varkappa}{4\sigma^2}\, \mathring{R}(s, x, y)$$

we obtain

$$\lim_{\substack{\varkappa \to \infty \\ \gamma \to \infty}} \mathring{K}(s, x, y) = \tilde{K}(s, x, y).$$

In Fig. 3.11, in the case of $s = 0$, the modified correlation function $\tilde{K}(s, x, y)$ obtained

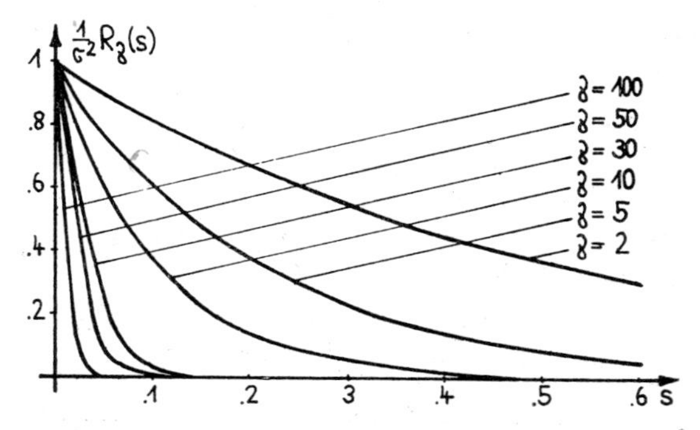

Fig. 3.11. Correlation function $R_\gamma(s) = \sigma^2 \exp(-\gamma\,|s|)$ for different values of γ

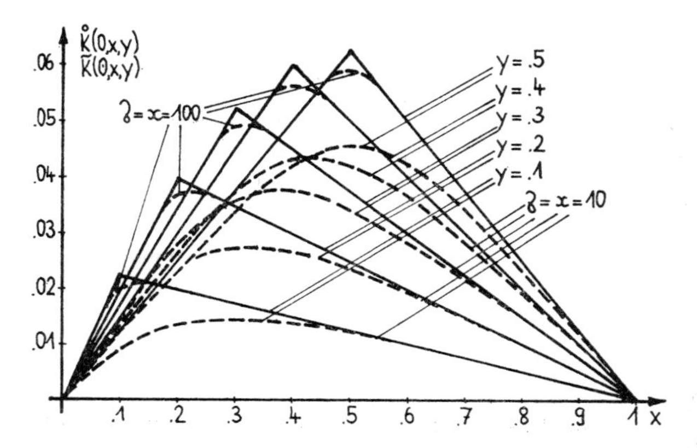

Fig. 3.12. Comparison between $\mathring{K}(0, x, y)$ and $\tilde{K}(0, x, y)$ for $\gamma = \varkappa = 10, 100$

by the weakly correlated theory as

$$\tilde{K}(0, x, y) = \frac{l}{4\beta A}\left[\min\left\{\frac{x}{l}, \frac{y}{l}\right\} - \frac{x}{l}\frac{y}{l}\right]$$

is compared with the correlation function $\mathring{K}(s, x, y)$ calculated on the basis of the correlation function (3.22) of $\overline{p}(t, x, \omega)$. We have put $\beta = 1$, $l = 1$, $A = 1$, and $\gamma = \varkappa$ in this figure and we will also set these values in the figures of this section. The function following from the weakly correlated theory is plotted by a bold line and the results from the correlation function (3.22) by a hatched line.

Fig. 3.13 shows $\tilde{K}(0, x, y)$ and $\mathring{K}(0, x, y)$ in dependence on x for different values of y. It can be seen that the inequality

$$\mathring{K}(0, x, y) \leqq \tilde{K}(0, x, y)$$

is fulfilled in the considered example; i.e. the results from the weakly correlated theory are upper estimations. Naturally, the same property can be observed for $\mathring{K}(0, x, x)$

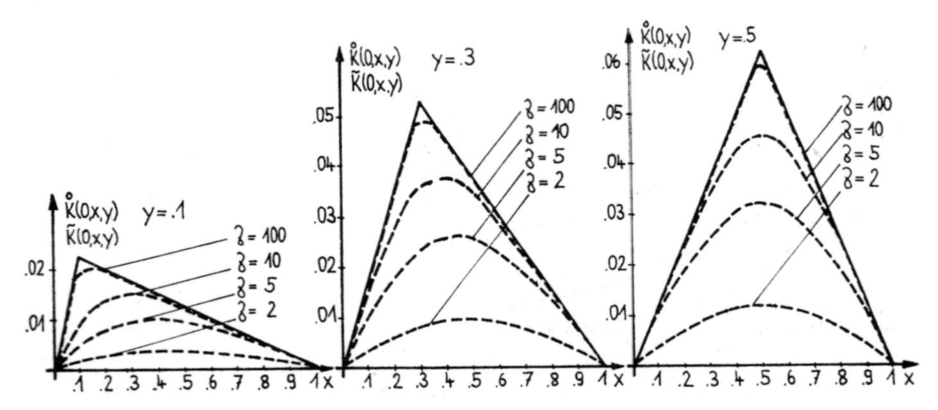

Fig. 3.13. Comparison between $\mathring{K}(0, x, y)$ and $\tilde{K}(0, x, y)$ in dependence on x for different values of y

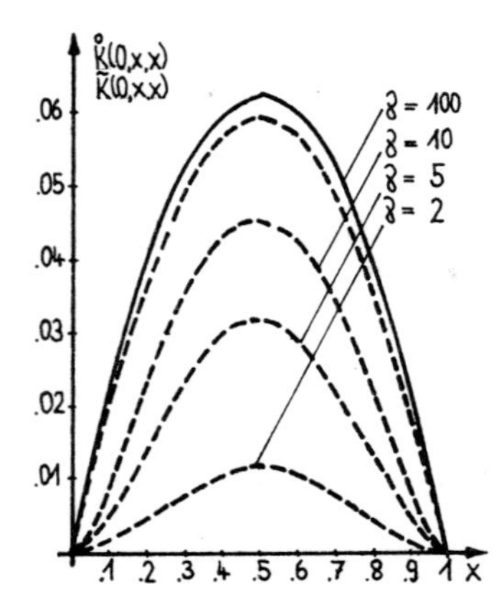

Fig. 3.14. Comparison between $\mathring{K}(0, x, x)$ and $\tilde{K}(0, x, x)$ in dependence on x

and $\tilde{K}(0, x, x)$ shown in Fig. 3.14. Finally, in Fig. 3.15 the correlation function $\tilde{K}(s, x, x)$ is compared with the correlation function $\mathring{K}(s, x, x)$ in dependence on s. We have put $x = 0.5$ and $\beta = 0.1$; 1. The function $\mathring{K}(s, x, x)$ was calculated for $\gamma = \varkappa = 5$; 10. The largest deviations of the compared functions can be found at the relative maxima or relative minima. It is observed that $\tilde{K}(s, x, x)$ approximates very well the correlation function for relatively small values of $\gamma = \varkappa$, too.

The great efforts for calculating $\mathring{K}(s, x, y)$ must be taken into account if these results are considered. In order to reduce the numerical calculations we have already chosen a very simple correlation function (3.22).

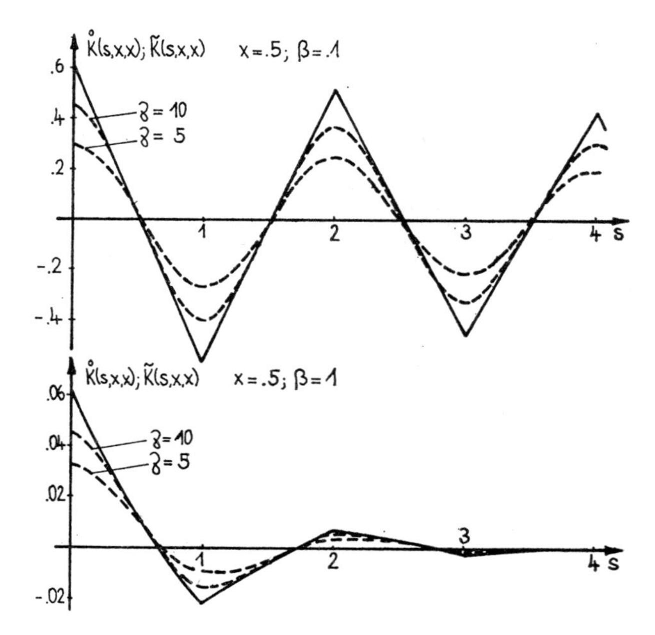

Fig. 3.15. Comparison between $\mathring{K}(s, x, x)$ and $\tilde{K}(s, x, x)$ in dependence on s

Let us consider the determination of the spectral density of the stationary solution of the vibration problem having the correlation function (3.22) for the inhomogeneous term. In the previous considerations we have obtained (3.25). It is

$$\mathring{K}(s, x, y) = \mathring{K}(-s, y, x)$$

since

$$\mathring{S}_{ij}(s) = \mathring{S}_{ji}(-s) \quad \text{for} \quad s < 0.$$

The spectral density $\mathring{S}(\alpha; x, y)$ is defined by

$$\mathring{S}(\alpha; x, y) \doteq \frac{1}{2\pi} \int_{-\infty}^{\infty} \mathrm{e}^{-\mathrm{i}s\alpha} \mathring{K}(s, x, y) \, \mathrm{d}s$$

and we obtain the relation

$$\mathring{S}(\alpha; x, y) = \sigma^2 \sum_{i,j=1}^{\infty} R_{ij}\mathring{s}_{ij}(\alpha) f_i(x) f_j(y)$$
$$= \sigma^2 \left[\sum_{i=1}^{\infty} R_{ii}\mathring{s}_{ii}(\alpha) f_i(x) f_i(y) + \sum_{\substack{i,j=1\\ i\neq j}}^{\infty} R_{ij}\mathring{s}_{ij}(\alpha) f_i(x) f_j(y) \right] \tag{3.27}$$

setting

$$\mathring{s}_{pq}(\alpha) \doteq \frac{1}{2\pi} \int_{-\infty}^{\infty} e^{-i\alpha s} \mathring{S}_{pq}(s)\, ds.$$

Since $\mathring{S}_{ij}(-s) = \mathring{S}_{ji}(s)$ we obtain

$$\mathring{s}_{pq}(\alpha) = \frac{1}{2\pi} \left[\int_0^{\infty} e^{-i\alpha s} \mathring{S}_{pq}(s)\, ds + \int_0^{\infty} e^{i\alpha s} \mathring{S}_{qp}(s)\, ds \right]$$

and then

$$\mathring{s}_{pq}(\alpha) = \frac{1}{2\pi\beta_p\beta_q} \left[\int_0^{\infty} e^{-i\alpha s} \{\mathring{I}_{pq}(s;\gamma) - \mathring{I}_{pq}(s;-\gamma)\}\, ds \right.$$
$$\left. + \int_0^{\infty} e^{i\alpha s} \{\mathring{I}_{qp}(s;\gamma) - \mathring{I}_{qp}(s;-\gamma)\}\, ds \right]$$
$$+ \frac{1}{2\pi} \left[\frac{1}{c_{pq}(\gamma)} \int_0^{\infty} e^{-i\alpha s - \gamma s}\, ds + \frac{1}{c_{qp}(\gamma)} \int_0^{\infty} e^{i\alpha s - \gamma s}\, ds \right].$$

In particular, the relations

$$\mathring{s}_{pp}(\alpha) = \frac{1}{\pi\beta_p^2} \int_0^{\infty} \cos(\alpha s) \{\mathring{I}_{pp}(s;\gamma) - \mathring{I}_{pp}(s;-\gamma)\}\, ds + \frac{1}{\pi c_{pp}(\gamma)} \int_0^{\infty} \cos(\alpha s)\, e^{-\gamma s}\, ds$$

and

$$\mathring{s}_{pq}(\alpha) + \mathring{s}_{qp}(\alpha) = \frac{1}{\pi} \left\{ \frac{1}{c_{pq}(\gamma)} + \frac{1}{c_{qp}(\gamma)} \right\} \int_0^{\infty} \cos(\alpha s)\, e^{-\gamma s}\, ds + \frac{1}{\pi\beta_p\beta_q} \int_0^{\infty} \cos(\alpha s)$$
$$\times \{\mathring{I}_{pq}(s;\gamma) - \mathring{I}_{pq}(s;-\gamma) + \mathring{I}_{qp}(s;\gamma) - \mathring{I}_{qp}(s;-\gamma)\}\, ds$$

can be deduced. Hence, for $x = y$ the spectral density

$$\mathring{S}(\alpha; x, x) = \sigma^2 \left[\sum_{i=1}^{\infty} R_{ii}\mathring{s}_{ii}(\alpha) f_i^2(x) + \sum_{\substack{i,j=1\\ i>j}}^{\infty} R_{ij}\{\mathring{s}_{ij}(\alpha) + \mathring{s}_{ji}(\alpha)\} f_i(x) f_j(y) \right]$$

is a real function. Now, the function $\mathring{L}_{ij}(\alpha;\gamma)$ is introduced by

$$\mathring{L}_{ij}(\alpha;\gamma) \doteq \frac{1}{\pi} \int_0^{\infty} \cos(\alpha s)\, \mathring{I}_{ij}(s;\gamma)\, ds.$$

Then, we have

$$\mathring{s}_{ii}(\alpha) = \frac{1}{\beta_i^2} [\mathring{L}_{ii}(\alpha; \gamma) - \mathring{L}_{ii}(\alpha; -\gamma)] + \frac{1}{\pi c_{ii}(\gamma)} \frac{\gamma}{\gamma^2 + \alpha^2}$$

and

$$\mathring{s}_{ij}(\alpha) + \mathring{s}_{ji}(\alpha) = \frac{1}{\beta_i \beta_j} [\mathring{L}_{ij}(\alpha; \gamma) - \mathring{L}_{ij}(\alpha; -\gamma) + \mathring{L}_{ji}(\alpha; \gamma) - \mathring{L}_{ji}(\alpha; -\gamma)]$$

$$+ \frac{1}{\pi} \left\{ \frac{1}{c_{ij}(\gamma)} + \frac{1}{c_{ji}(\gamma)} \right\} \frac{\gamma}{\gamma^2 + \alpha^2}.$$

For $\mathring{L}_{ij}(\alpha; \gamma)$ it follows

$$\mathring{L}_{ij}(\alpha; \gamma) = \frac{1}{(\beta + \gamma)^2 + \beta_j^2} \left[\frac{1}{2} (\beta + \gamma) \{\mathring{A}(\alpha; \beta_i, -\beta_j) - \mathring{A}(\alpha; \beta_i, \beta_j)\} \right.$$

$$\left. + \frac{1}{2} \beta_j \{\mathring{B}(\alpha; \beta_i, -\beta_j) + \mathring{B}(\alpha; \beta_i, \beta_j)\} \right]$$

where

$$\mathring{A}(\alpha; \varepsilon, \delta) = \frac{1}{\pi} \int_0^\infty \cos(\alpha s)\, \mathring{a}(s; \varepsilon, \delta)\, ds$$

$$= \frac{1}{4\beta^2 + (\varepsilon + \delta)^2} \frac{1}{2\pi} \left[\frac{2\beta^2 - (\varepsilon + \delta)(\delta - \alpha)}{\beta^2 + (\alpha - \delta)^2} + \frac{2\beta^2 - (\varepsilon + \delta)(\delta + \alpha)}{\beta^2 + (\delta + \alpha)^2} \right]$$

and

$$\mathring{B}(\alpha; \varepsilon, \delta) = \frac{1}{\pi} \int_0^\infty \cos(\alpha s)\, \mathring{b}(s; \varepsilon, \delta)\, ds$$

$$= \frac{1}{4\beta^2 + (\varepsilon + \delta)^2} \frac{\beta}{2\pi} \left[\frac{2(\delta - \alpha) + \varepsilon + \delta}{\beta^2 + (\delta - \alpha)^2} + \frac{2(\delta + \alpha) + \varepsilon + \delta}{\beta^2 + (\delta + \alpha)^2} \right].$$

It is easy to see that

$$\lim_{\gamma \to \infty} \gamma \mathring{L}_{ij}(\alpha; \pm\gamma) = \pm \frac{1}{2} \{\mathring{A}(\alpha; \beta_i, -\beta_j) - \mathring{A}(\alpha; \beta_i, \beta_j)\},$$

and

$$\lim_{\gamma \to \infty} \gamma \mathring{s}_{ii}(\alpha) = \frac{1}{\beta_i^2} \{\mathring{A}(\alpha; \beta_i, -\beta_i) - \mathring{A}(\alpha; \beta_i, \beta_i)\} = \frac{1}{\pi} \frac{1}{(\lambda_i + \alpha^2)^2 - 4\alpha^2\beta_i^2},$$

$$\lim_{\gamma \to \infty} \gamma\big(\mathring{s}_{ij}(\alpha) + \mathring{s}_{ji}(\alpha)\big) = \frac{1}{\beta_i \beta_j} \{\mathring{A}(\alpha; \beta_i, -\beta_j) - \mathring{A}(\alpha; \beta_i, \beta_j) + \mathring{A}(\alpha; \beta_j, -\beta_i)$$

$$- \mathring{A}(\alpha; \beta_j, \beta_i)\}.$$

Using (3.27) we obtain

$$\lim_{\substack{\varkappa \to \infty \\ \gamma \to \infty}} \gamma \varkappa \mathring{S}(\alpha; x, y) = \frac{2\sigma^2}{\pi} \sum_{i=1}^{\infty} \frac{f_i(x)\, f_i(y)}{(\lambda_i + \alpha^2)^2 - 4\alpha^2\beta_i^2}$$

and then

$$\mathring{S}(\alpha; x, y) \approx \frac{2\sigma^2}{\pi\gamma\varkappa} \sum_{i=1}^{\infty} \frac{f_i(x)\, f_i(y)}{(\lambda_i + \alpha^2)^2 - 4\alpha^2\beta_i^2} = \tilde{S}(\alpha; x, y)$$

for large numbers γ, $\varkappa$ since

$$a \approx \frac{4\sigma^2}{\gamma\varkappa}.$$

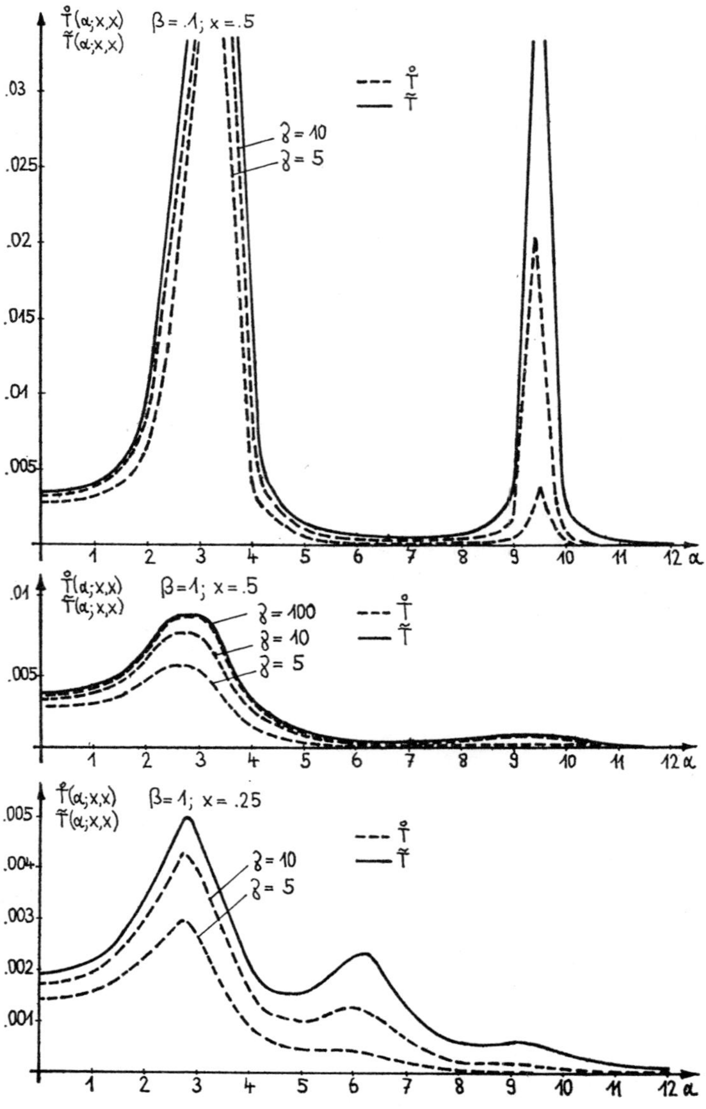

Fig. 3.16. Comparison between the spectral density $\mathring{T}(\alpha; x, x)$ and $\tilde{T}(\alpha; x, x)$ for $\gamma = \varkappa = 5, 10, 100$

$\tilde{S}(\alpha; x, y)$ is the spectral density resulting from the application of the weakly correlated functions (see (3.21)).

Fig. 3.16 shows some numerical results of a comparison between

$$\tilde{T}(\alpha; x, x) \doteq \frac{1}{a} \tilde{S}(\alpha; x, x)$$

and

$$\mathring{T}(\alpha; x, x) \doteq \frac{\gamma \varkappa}{4\sigma^2} \mathring{S}(\alpha; x, x)$$

for different values of γ ($\gamma = \varkappa$). A good agreement between the results from the weakly correlated theory and the results following from the correlation function (3.22) can be observed. We also have in this case that the spectral density $\tilde{S}(\alpha; x, x)$ is an upper bound.

The frequencies α around

$$\alpha_k \doteq \sqrt{\lambda_k - 2\beta^2}$$

in dependence on the damping measure β are critical since the spectral density has maximum values in the neighbourhood of these frequencies for $x = 0.5$. From this consideration of $\tilde{S}(\alpha; x, x)$ as to critical frequencies we also obtain information about realistic inhomogeneous terms.

3.1.4. Partially weakly correlated excitations as to the spatial variables

Consider the boundary-initial-value problem (3.8) with given boundary and initial conditions. It is assumed that the excitation $p(t, x, \omega)$ leads to a partially weakly correlated function with respect to x,

$$\overline{p}_\varepsilon(t, x, \omega) \doteq p(t, x, \omega) - \langle p(t, x) \rangle .$$

The properties of the definition of a partially weakly correlated function refer to a group of variables (e.g. to x from (t, x)) in distinction to a weakly correlated function. The averaged solution $U_\varepsilon(t, \omega)$ can be represented as

$$U_\varepsilon(t, \omega) = \int_0^t \int_{\mathcal{D}} F(t - s, y)\, \overline{p}_\varepsilon(s, y, \omega)\, \mathrm{d}y\, \mathrm{d}s$$

where

$$F(t, y) = \sum_{k=1}^{\infty} h_k(t)\, f_k(y)\, (f_k, g)$$

and the same notations were selected as in Section 3.1.1. Let the weakly correlated function $\overline{p}_\varepsilon(t, x, \omega)$ as to t possess smooth sample functions a.s. Then, with the help of an integration by parts we have

$$U_\varepsilon(t, \omega) = \int_{\mathcal{D}} \left[\hat{F}(t, y)\, \overline{p}_\varepsilon(0, y, \omega) + \int_0^t \hat{F}(t - s, y)\, \overline{p}_{\varepsilon, s}(s, y, \omega)\, \mathrm{d}s \right] \mathrm{d}y \tag{3.28}$$

where

$$\hat{F}(t, y) \doteq \sum_{k=1}^{\infty} H_k(t)\, f_k(y)\, (f_k, g),$$

$$H_k(t) \doteq \int_0^t h_i(t - s)\, ds = \int_0^t h_i(s)\, ds.$$

Using

$$\hat{F}(t, x, y) \doteq \sum_{k=1}^{\infty} H_k(t)\, f_k(x)\, f_k(y) \in \mathbf{L}_2([0, T] \times \mathcal{D} \times \mathcal{D})$$

it follows

$$U_\varepsilon(t, \omega) = \left(\int_{\mathcal{D}} \hat{F}(t, y, .)\, \overline{p}_\varepsilon(0, y, \omega)\, dy + \int_{\mathcal{D}} \int_0^t \hat{F}(t - s, y, .)\, \overline{p}_{\varepsilon, s}(s, y, \omega)\, ds\, dy, g \right).$$

Assuming the uniform convergence of the series

$$\sum_{k=1}^{\infty} H_k^2(t)\, f_k^2(x) \quad \text{as to} \quad x \in \mathcal{D} \quad \text{for all} \quad t \in [0, T],$$

the first factor in the above scalar product is a continuous function with respect to these variables for which the scalar is taken:

$$\Bigg| \int_{\mathcal{D}} \left(\hat{F}(t, y, x) - \hat{F}(t, y, \tilde{x}) \right) \overline{p}_\varepsilon(0, y, \omega)\, dy$$

$$+ \int_{\mathcal{D}} \int_0^t \left(\hat{F}(t - s, y, x) - \hat{F}(t - s, y, \tilde{x}) \right) \overline{p}_{\varepsilon, s}(s, y, \omega)\, ds\, dy \Bigg|$$

$$\leqq \sqrt{\int_{\mathcal{D}} \left(\hat{F}(t, y, x) - \hat{F}(t, y, \tilde{x}) \right)^2 dy}\; \|\overline{p}_\varepsilon(0, ., \omega)\|_{\mathcal{D}}$$

$$+ \sqrt{\int_{\mathcal{D}} \int_0^t \left(\hat{F}(t - s, y, x) - \hat{F}(t - s, y, \tilde{x}) \right)^2 ds\, dy}\; \|\overline{p}_{\varepsilon, s}(., ., \omega)\|_{[0,t] \times \mathcal{D}}$$

$$= C_1(\omega) \sqrt{\sum_{k=1}^{\infty} H_k^2(t) \left(f_k(x) - f_k(\tilde{x}) \right)^2} + C_2(\omega) \sqrt{\int_0^t \sum_{k=1}^{\infty} H_k^2(s) \left(f_k(x) - f_k(\tilde{x}) \right)^2 ds}.$$

Hence, we obtain for the solution of (3.8)

$$\bar{u}_\varepsilon(t, x, \omega) = \int_{\mathcal{D}} \hat{F}(t, y, x)\, \overline{p}_\varepsilon(0, y, \omega)\, dy$$

$$+ \int_{\mathcal{D}} \int_0^t \hat{F}(t - s, y, x)\, \overline{p}_{\varepsilon, s}(s, y, \omega)\, ds\, dy \tag{3.29}$$

since

$$\overline{U}_\varepsilon(t, \omega) = \left(\bar{u}_\varepsilon(t, ., \omega), g \right).$$

For the derivatives

$$\mathrm{D}_t^p \mathrm{D}_x^q \bar{u}_\varepsilon \doteq \frac{\partial^{p+q} \bar{u}_\varepsilon}{\partial t^p \, \partial x_1^{q_1} \ldots \partial x_n^{q_n}} \qquad \left(\sum_{s=1}^{n} q_s = q; \quad p = 0, 1 \right)$$

it follows

$$\mathrm{D}_t^p \mathrm{D}_x^q \bar{u}_\varepsilon(t, x, \omega) = \int_{\mathcal{D}} \mathrm{D}_t^p \mathrm{D}_x^q \hat{F}(t, y, x) \, \bar{p}_\varepsilon(0, y, \omega) \, \mathrm{d}y + \int_{\mathcal{D}} \int_0^t \mathrm{D}_t^p \mathrm{D}_x^q \hat{F}(t - s, y, x) \, \bar{p}_{\varepsilon,s}(s, y, \omega) \, \mathrm{d}s \, \mathrm{d}y \tag{3.30}$$

where

$$\mathrm{D}_t^p \mathrm{D}_x^q \hat{F}(t, y, x) \doteq \sum_{k=1}^{\infty} \mathrm{D}_t^p H_k(t) \, \mathrm{D}_x^q f_k(x) \, f_k(y)$$

using the condition

$$\mathrm{D}_t^p \mathrm{D}_x^q \hat{F}(t, y, x) \in \mathbf{L}_2([0, T] \times \mathcal{D} \times \mathcal{D})$$

and the continuity of the right hand side of (3.30) as to x. In particular, the continuity is fulfilled if the series

$$\sum_{k=1}^{\infty} \left(\mathrm{D}_t^p H_k(t)\right)^2 \left(\mathrm{D}_x^q f_k(x)\right)^2$$

converges uniformly in x.

It is advantageous for the following applications that a second integration by parts is done in the second summand of (3.28). Thus

$$\int_0^t \hat{F}(t - s, y) \, \bar{p}_{\varepsilon,s}(s, y, \omega) \, \mathrm{d}s = \hat{\hat{F}}(t, y) \, \bar{p}_{\varepsilon,t}(0, y, \omega) + \int_0^t \hat{\hat{F}}(t - s, y) \, \bar{p}_{\varepsilon,ss}(s, y, \omega) \, \mathrm{d}s$$

where $\hat{\hat{F}}(t, y)$ is defined by

$$\hat{\hat{F}}(t, y) \doteq \sum_{k=1}^{\infty} \int_0^t H_k(s) \, \mathrm{d}s f_k(y) \, (f_k, g) \, .$$

It can be calculated

$$H_k(t) = \frac{1}{\lambda_k \beta_k} \left(\beta_k - \mathrm{e}^{-\beta t} (\beta \sin(\beta_k t) + \beta_k \cos(\beta_k t))\right),$$

$$\int_0^t H_k(s) \, \mathrm{d}s = \frac{t}{\lambda_k} - \frac{2\beta}{\lambda_k^2} + \frac{\mathrm{e}^{-\beta t}}{\lambda_k^2} \left[\frac{2\beta^2 - \lambda_k}{\beta_k} \sin(\beta_k t) + 2\beta \cos(\beta_k t) \right].$$

Using

$$\hat{\hat{F}}(t, x, y) \doteq \sum_{k=1}^{\infty} \int_0^t H_i(s) \, \mathrm{d}s f_k(x) \, f_k(y) \in \mathbf{L}_2([0, T] \times \mathcal{D} \times \mathcal{D})$$

it follows

$$U_\varepsilon(t,\omega) = \left(\int_{\mathcal{D}} \hat{F}(t,y,.)\,\overline{p}_\varepsilon(0,y,\omega)\,\mathrm{d}y + \int_{\mathcal{D}} \hat{\hat{F}}(t,y,.)\,\overline{p}_{\varepsilon,t}(0,y,\omega)\,\mathrm{d}y \right.$$
$$\left. + \int_{\mathcal{D}} \int_0^t \hat{\hat{F}}(t-s,y,.)\,\overline{p}_{\varepsilon,ss}(s,y,\omega)\,\mathrm{d}s\,\mathrm{d}y, g \right).$$

From this the relation

$$\bar{u}_\varepsilon(t,x,\omega) = \int_{\mathcal{D}} \hat{F}(t,y,x)\,\overline{p}_\varepsilon(0,y,\omega)\,\mathrm{d}y + \int_{\mathcal{D}} \hat{\hat{F}}(t,y,x)\,\overline{p}_{\varepsilon,t}(0,y,\omega)\,\mathrm{d}y$$
$$+ \int_{\mathcal{D}} \int_0^t \hat{\hat{F}}(t-s,y,x)\,\overline{p}_{\varepsilon,ss}(s,y,\omega)\,\mathrm{d}s\,\mathrm{d}y \tag{3.31}$$

can be deduced if, for example, the series $\sum_{k=1}^{\infty} \lambda_k^{-2}$ converges since the right hand side of (3.31) has to be continuous.

In comparison with (3.30) for the derivatives it follows

$$\mathrm{D}_t^p \mathrm{D}_x^q \bar{u}_\varepsilon(t,x,\omega) = \int_{\mathcal{D}} \mathrm{D}_t^p \mathrm{D}_x^q \hat{F}(t,y,x)\,\overline{p}_\varepsilon(0,y,\omega)\,\mathrm{d}y$$
$$+ \int_{\mathcal{D}} \mathrm{D}_t^p \mathrm{D}_x^q \hat{\hat{F}}(t,y,x)\,\overline{p}_{\varepsilon,t}(0,y,\omega)\,\mathrm{d}y$$
$$+ \int_{\mathcal{D}} \int_0^t \mathrm{D}_t^p \mathrm{D}_x^q \hat{\hat{F}}(t-s,y,x)\,\overline{p}_{\varepsilon,ss}(s,y,\omega)\,\mathrm{d}y \tag{3.32}$$

for $p = 0, 1$; $q = 0, 1, 2, \ldots$ We have put

$$\mathrm{D}_t^p \mathrm{D}_x^q \hat{\hat{F}}(t,y,x) \doteq \sum_{k=1}^{\infty} D_t^p \int_0^t H_k(s)\,\mathrm{d}s\, \mathrm{D}_x^q f_k(x)\, f_k(y)$$

and it is convenient to assume that the series

$$\sum_{k=1}^{\infty} (\mathrm{D}_t^p H_k(t))^2\,(\mathrm{D}_x^q f_k(x))^2 \quad \text{and} \quad \sum_{k=1}^{\infty} \left(\mathrm{D}_t^p \int_0^t H_k(s)\,\mathrm{d}s\right)^2 (\mathrm{D}_x^q f_k(x))^2$$

converge uniformly as to $x \in \mathcal{D}$.

Supposing $\overline{p}_\varepsilon(0,y,\omega) \equiv 0$ a.s. the first term of the right hand side of (3.32) is zero and the uniform convergence of the series

$$\sum_{k=1}^{\infty} \left(\mathrm{D}_t^p \int_0^t H_k(s)\,\mathrm{d}s\right)^2 (\mathrm{D}_x^q f_k(x))^2$$

as to x is sufficient. The condition $\overline{p}_\varepsilon(0,y,\omega) = 0$ a.s. is fulfilled if $p(0,y,\omega)$ is a non-random function or we have $p(0,y,\omega) = 0$, respectively. The last condition means that the external load vanishes for the time $t = 0$. This assumption will be important for the calculation of the stresses.

Before the limit distribution of

$$\mathrm{D}_t^p \mathrm{D}_x^q \bar{u}_\varepsilon(t,x,\omega) \quad \text{as} \quad \varepsilon \downarrow 0$$

is treated we deal with the special case

$$\overline{p}_\varepsilon(t, x, \omega) \doteq p_0(t)\, \overline{p}_{1\varepsilon}(x, \omega).$$

Supposing this special case from (3.32) it follows

$$\mathrm{D}_t^p \mathrm{D}_x^q \bar{u}_\varepsilon(t, x, \omega) = \int_{\mathscr{D}} \mathrm{D}_t^p \mathrm{D}_x^q T(t, y, x)\, \overline{p}_{1\varepsilon}(y, \omega)\, \mathrm{d}y \tag{3.33}$$

where

$$T(t, y, x) \doteq \sum_{k=1}^{\infty} T_k(t)\, f_k(x)\, f_k(y),$$

$$T_k(t) \doteq \int_0^t h_k(t - s)\, p_0(s)\, \mathrm{d}s,$$

and

$$\mathrm{D}_t^p \mathrm{D}_x^q T(t, y, x) \doteq \sum_{k=1}^{\infty} \mathrm{D}_t^p T_k(t)\, \mathrm{D}_x^q f_k(x)\, f_k(y)$$

assuming that

$$\sum_{k=1}^{\infty} \left(\mathrm{D}_t^p T_k(t)\right)^2 \left(\mathrm{D}_x^q f_k(x)\right)^2$$

converges uniformly as to $x \in \mathscr{D}$ for all $t \in [0, T]$. It can be seen that

$$\begin{aligned} T_k(t) &= H_k(t)\, p_0(0) + \int_0^t H_k(s)\, \mathrm{d}s \cdot p_0'(0) + \int_0^t \int_0^{t-s} H_k(r)\, \mathrm{d}r \cdot p_0''(s)\, \mathrm{d}s \\ &= \int_0^t h_k(t - s)\, p_0(s)\, \mathrm{d}s \end{aligned}$$

and hence, in general,

$$T_k(t) \sim \frac{1}{\lambda_k}, \qquad T_k'(t) \sim \frac{1}{\beta_k},$$

but for $p_0(0) = 0$ then

$$T_k(t) \sim \frac{1}{\lambda_k}, \qquad T_k'(t) \sim \frac{1}{\lambda_k}.$$

Furthermore, we note that the order of $T_k(t)$, $T_k'(t)$ in the case of $p_0(0) = 0$, $p_0'(0) = 0$ is equal to the case of $p_0(0) = 0$.

A limit theorem can also be proved for partially weakly correlated connected vector functions so that for $\mathrm{D}_t^p \mathrm{D}_x^q \bar{u}_\varepsilon(t, x, \omega)$ given by (3.32) a limit relation

$$\lim_{\varepsilon \downarrow 0} \frac{1}{\sqrt{\varepsilon}^{\,n}}\, \mathrm{D}_t^p \mathrm{D}_x^q \bar{u}_\varepsilon(t, x, \omega) = \eta_{pq}(t, x, \omega) \quad \text{in distribution} \tag{3.34}$$

is true where η_{pq} denotes a Gaussian function having the moments

$$\langle \eta_{pq}(t, x) \rangle = 0,$$

$$\langle \eta_{pq}(t, x)\, \eta_{pq}(r, y) \rangle \tag{3.35}$$

$$= \sum_{i,j=1}^{3} \int_0^t \int_0^r \int_{\mathcal{D}} F_{ipq}(t, s_1, x, z)\, F_{jpq}(r, s_2, y, z)\, a_{ij}(s_1, s_2, z)\, dz\, ds_2\, ds_1$$

for $t, r > 0$ (see Section 1.6.1). We have put

$$F_{1pq}(t, s, x, y) \doteq \frac{1}{t}\, D_t^p D_x^q \hat{F}(t, y, x),$$

$$F_{2pq}(t, s, x, y) \doteq \frac{1}{t}\, D_t^p D_x^q \hat{\hat{F}}(t, y, x),$$

$$F_{3pq}(t, s, x, y) \doteq D_t^p D_x^q \hat{\hat{F}}(t - s, y, x)$$

and

$$a_{ij}(t, r, y) = \lim_{\varepsilon \downarrow 0} \frac{1}{\varepsilon^n} \int_{\mathcal{K}_\varepsilon(0)} \langle \overline{p}_{i\varepsilon}(t, y)\, \overline{p}_{j\varepsilon}(r, y + z) \rangle\, dz, \qquad i, j = 1, 2, 3,$$

with

$$\overline{p}_{1\varepsilon}(t, y, \omega) \doteq \overline{p}_\varepsilon(0, y, \omega),$$

$$\overline{p}_{2\varepsilon}(t, y, \omega) \doteq \overline{p}_{\varepsilon,t}(0, y, \omega),$$

$$\overline{p}_{3\varepsilon}(t, y, \omega) \doteq \overline{p}_{\varepsilon,tt}(t, y, \omega)$$

have been introduced. The relation (3.34) is true if the uniform convergence as to x of

$$\sum_{k=1}^{\infty} (D_t^p H_k(t))^2\, (D_x^q f_k(x))^2 \quad \text{and} \quad \sum_{k=1}^{\infty} \left(D_t^p \int_0^t H_k(s)\, ds \right)^2 (D_x^q f_k(x))^2$$

is fulfilled and these functions shall be contained in $\mathbf{L}_1([0, T])$. In the case of $\overline{p}_\varepsilon(0, y, \omega) = 0$ a.s. the condition for the first series may be omitted.

If the uniform convergence of

$$\lim_{\varepsilon \downarrow 0} \frac{1}{\varepsilon^n} \int_{\mathcal{K}_\varepsilon(0)} \langle \overline{p}_{\varepsilon,tt}(t, y)\, \overline{p}_{\varepsilon,rr}(r, y + z) \rangle\, dz$$

is assumed then, with the help of

$$a(t, r, y) = \lim_{\varepsilon \downarrow 0} \frac{1}{\varepsilon^n} \int_{\mathcal{K}_\varepsilon(0)} \langle \overline{p}_\varepsilon(t, y)\, \overline{p}_\varepsilon(r, y + z) \rangle\, dz,$$

we obtain

$$a_{11}(t, r, y) = a(0, 0, y),$$

$$a_{21}(t, r, y) = a_t(0, 0, y),$$

$$a_{31}(t, r, y) = a_{tt}(t, 0, y),$$
$$a_{12}(t, r, y) = a_r(0, 0, y),$$
$$a_{22}(t, r, y) = a_{tr}(0, 0, y),$$
$$a_{32}(t, r, y) = a_{ttr}(t, 0, y),$$
$$a_{13}(t, r, y) = a_{rr}(0, r, y),$$
$$a_{23}(t, r, y) = a_{trr}(0, r, y),$$
$$a_{33}(t, r, y) = a_{ttrr}(t, r, y)$$

for use of (3.34).

For example, in the case of

$$p_\varepsilon(t, x, \omega) = p_0(t)\, p_{1\varepsilon}(x, \omega)$$

the given relations can be applied and it is

$$a(t, r, y) = p_0(t)\, p_0(r)\, a(y),$$
$$a(y) = \lim_{\varepsilon \downarrow 0} \frac{1}{\varepsilon^n} \int\limits_{\mathcal{K}_\varepsilon(0)} \langle \overline{p}_{1\varepsilon}(y)\, \overline{p}_{1\varepsilon}(y + z)\rangle\, \mathrm{d}z.$$

A similar limit relation like that in (3.34) can be proved for vector functions of the form

$$\left(\mathrm{D}_t^{p_1}\mathrm{D}_x^{q_1}\bar{u}_\varepsilon(t, x), \ldots, \mathrm{D}_t^{p_k}\mathrm{D}_x^{q_k}\bar{u}_\varepsilon(t, x)\right), \quad p_i = 0, 1;\ q_i = 0, 1, 2, \ldots;\ i = 1, 2, \ldots, k,\text{i}$$

particularly for

$$\left(\bar{u}_\varepsilon(t, x),\ \bar{u}_{\varepsilon,t}(t, x)\right)$$

and

$$\left(\bar{u}_{\varepsilon,x_ix_i}(t, x),\ \bar{u}_{\varepsilon,x_ix_it}(t, x)\right)$$

so that the expected number of threshold crossings for displacements and stresses can be calculated approximately.

3.1.5. Applications

First we are concerned with the problem

$$u_{tt} + 2\beta u_t + Lu = p_\varepsilon(t, x, \omega), \qquad x \in \mathcal{D},$$
zero boundary conditions,
initial conditions: $u(0, x) = 0,\ u_t(0, x) = 0$.

The eigenvalues of L are denoted by λ_i and the eigenfunctions by $f_i(x)$. Examples of these elements are summarized in Table 3.1 for different operators L and domains $\mathcal{D}$ and the order of the terms

$$A_{i(j)} \doteq \mathrm{D}_t^p H_{i(j)}(t)\, \mathrm{D}_x^q f_{i(j)}(x),$$

$$B_{i(j)} \doteq \mathrm{D}_t^p \int\limits_0^t H_{i(j)}(s)\, \mathrm{d}s\, \mathrm{D}_x^q f_{i(j)}(x)$$

Table 3.1

L	$\mathcal{D}$	$i(j)$	$f_{i(j)}$
$L_1u = -Au_{x_1x_1}$	$[0, l]$	$A\left(\frac{i\pi}{l}\right)^2$	$\sqrt{\frac{2}{l}}\sin\left(\frac{i}{l}\pi x_1\right)$
$L_2u = Au_{x_1x_1x_1x_1}$	$[0, l]$	$A\left(\frac{i\pi}{l}\right)^4$	$\sqrt{\frac{2}{l}}\sin\left(\frac{i}{l}\pi x_1\right)$
$L_3u = -A\Delta u$	$[0, l] \times [0, m]$	$A\pi^2\left(\left(\frac{i}{l}\right)^2 + \left(\frac{j}{m}\right)^2\right)$	$\frac{2}{\sqrt{lm}}\sin\left(\frac{i}{l}\pi x_1\right)\sin\left(\frac{j}{m}\pi x_2\right)$
$L_4u = A\Delta\Delta u$	$[0, l] \times [0, m]$	$A\pi^4\left(\left(\frac{i}{l}\right)^2 + \left(\frac{j}{m}\right)^2\right)^2$	$\frac{2}{\sqrt{lm}}\sin\left(\frac{i}{l}\pi x_1\right)\sin\left(\frac{j}{m}\pi x_2\right)$

is given in Table 3.2 for these cases contained in Table 3.1. Hence, in the case of a partially weakly correlated function $\overline{p}_\varepsilon(t, x, \omega)$ as to x the limit relation (3.34) for $\bar{u}_\varepsilon(t, x, \omega)$ can be applied for all cases considered in Table 3.1 since

$$\sum_{i(j)=1}^{\infty} \lambda_{i(j)}^{-2} < \infty .$$

Furthermore, we have the limit relation for $\bar{u}_{\varepsilon,t}(t, x, \omega)$ in the case of L_2 and also in the cases of L_1, L_3, L_4 assumed that $\overline{p}_\varepsilon(0, x, \omega) = 0$. Limit relations for $\bar{u}_{\varepsilon,x_rx_s}$ ($r, s = 1$ or $r, s = 1, 2$, respectively) can be obtained for L_2, L_4 but cannot be obtained for L_1, L_3.

Table 3.2

p	q	$A_{i(j)}$	$B_{i(j)}$
0	0	$1/\lambda_{i(j)}$	$1/\lambda_{i(j)}$
1	0	$1/\beta_{i(j)}$	$1/\lambda_{i(j)}$
0	2	$i^2/\lambda_{i(j)}$	$i^2/\lambda_{i(j)}$
1	2	$i^2/\beta_{i(j)}$	$i^2/\lambda_{i(j)}$

Finally, limit relations for $\bar{u}_{\varepsilon,x_rx_st}$ follow in the case of L_2, L_4 assumed $\overline{p}_\varepsilon(0, x, \omega) = 0$. For example, the expected number of threshold crossings with respect to a fixed point x can be determined for $\bar{u}_\varepsilon(t, x, \omega)$ in the case of L_2 and also for L_1, L_3, L_4 supposed that $\overline{p}_\varepsilon(0, x, \omega) = 0$. Assuming again $\overline{p}_\varepsilon(0, x, \omega) = 0$ we also obtain the expected number of threshold crossings of $\bar{u}_{\varepsilon,x_rx_s}(t, x, \omega)$ in the case of L_2 and L_4.

Let us consider a simply supported beam. Hence, we have to investigate the boundary-initial-value problem

$$u_{tt} + 2\beta u_t + Au_{xxxx} = p_\varepsilon(t, x, \omega); \qquad \mathcal{D} = [0, l],$$

boundary conditions: $u(t, 0) = u(t, l) = u_{xx}(t, 0) = u_{xx}(t, l) = 0,$

initial conditions: $u(0, x) = u_t(0, x) = 0.$

The random function $p_\varepsilon(t, x, \omega)$ represents an external random load. Let $\overline{p}_\varepsilon(t, x, \omega)$ be partially weakly correlated. We put

$$p_\varepsilon(t, x, \omega) \doteq p_0(t)\, p_{1\varepsilon}(x, \omega)\,; \qquad p_0(0) = 0\,.$$

Furthermore, let $\overline{p}_{1\varepsilon}(x, \omega)$ be a wide-sense stationary process. In this case we obtain

$$a(t, r, x) = p_0(t)\, p_0(r)\, a$$

where

$$a = \lim_{\varepsilon \downarrow 0} \frac{1}{\varepsilon} \int_{-\varepsilon}^{\varepsilon} \langle \overline{p}_{1\varepsilon}(x)\, \overline{p}_{1\varepsilon}(x + z)\rangle\, \mathrm{d}z\,.$$

This example describes the vibration of a beam loaded by a random force. The randomness is generated by the random choice of the place of the external load because $p_\varepsilon(t, x, \omega)$ is a non-random function as to the time.

The expectation $\langle u(t, x)\rangle$ or other statistical characteristics can be taken as mean values of a number of such trials in which the external load randomly varies the points of application. The eigenvalues and eigenfunctions as to the operator $Lu = Au_{xxxx}$ and the given boundary conditions can be determined as

$$\lambda_i = A \left(\frac{i\pi}{l}\right)^4 \quad \text{and} \quad f_i(x) = \sqrt{\frac{2}{l}}\, \sin\left(\frac{i\pi x}{l}\right),$$

respectively. Using (3.33) we obtain for the solution

$$\bar{u}_\varepsilon(t, x, \omega) = u_\varepsilon(t, x, \omega) - w(t, x)$$

the equation

$$\mathrm{D}_t^p \mathrm{D}_x^q \bar{u}_\varepsilon(t, x, \omega) = \int_{\mathscr{D}} \sum_{i=1}^{\infty} \mathrm{D}_t^p T_i(t)\, \mathrm{D}_x^q f_i(x)\, f_i(z)\, \overline{p}_{1\varepsilon}(z, \omega)\, \mathrm{d}z$$

and then the limit relation

$$\lim_{\varepsilon \downarrow 0} \frac{1}{\sqrt{\varepsilon}} \big(\bar{u}_\varepsilon(t, x, \omega), \bar{u}_{\varepsilon,t}(t, x, \omega)\big) = \big(\eta_1(t, x, \omega), \eta_2(t, x, \omega)\big).$$

The random vector function (η_1, η_2) is Gaussian and the first two moments can be determined by

$$\langle \eta_i(t, x)\rangle = 0$$

and

$$\langle \eta_i(t, x)\, \eta_j(r, y)\rangle = a \sum_{s=1}^{\infty} T_{si}(t)\, T_{sj}(r)\, f_s(x)\, f_s(y)$$

where

$$T_{p1}(t) \doteq T_p(t)\,, \qquad T_{p2}(t) \doteq T_p'(t)$$

with

$$T_p(t) = \int_0^t h_p(t - s)\, p_0(s)\, \mathrm{d}s\,.$$

Furthermore, we can deduce

$$\lim_{\varepsilon \downarrow 0} \frac{1}{\sqrt{\varepsilon}} \big(\bar{u}_{\varepsilon,xx}(t, x, \omega), \bar{u}_{\varepsilon,xxt}(t, x, \omega)\big) = \big(\zeta_1(t, x, \omega), \zeta_2(t, x, \omega)\big),$$

where

$$\langle \zeta_i(t, x) \rangle = 0$$

and

$$\langle \zeta_i(t, x)\, \zeta_j(r, y) \rangle = a \sum_{s=1}^{\infty} T_{si}(t)\, T_{sj}(r)\, f_s''(x)\, f_s''(y)$$

are the first two moments of the Gaussian vector function (ζ_1, ζ_2). In this way, the distribution of the bending stress of the beam

$$\sigma(t, x, \omega) = Ehu_{xx}(t, x, \omega)$$

(E — modulus of elasticity; h — thickness of beam) and its derivative as to t can be calculated approximately. Applying the relation

$$\lim_{\varepsilon \downarrow 0} \frac{1}{\sqrt{\varepsilon}} \big(\bar{u}_{\varepsilon}(t, x, \omega), \bar{u}_{\varepsilon,xx}(t, x, \omega)\big) = \big(\eta_1(t, x, \omega), \zeta_1(t, x, \omega)\big)$$

the stochastic dependence between the displacement $u(t, x, \omega)$ and the bending stress $\sigma(t, x, \omega)$ can be determined. To do this, we take into consideration the Gaussian distribution of (η_1, ζ_1) having the moments

$$\langle \eta_1(t, x) \rangle = \langle \zeta_1(t, x) \rangle = 0$$

and

$$\langle \eta_1(t, x)\, \zeta_1(r, y) \rangle = a \sum_{s=1}^{\infty} T_s(t)\, T_s(r)\, f_s(x)\, f_s''(y).$$

From the limit distributions of $(\bar{u}_\varepsilon, \bar{u}_{\varepsilon,t})$ and $(\bar{u}_{\varepsilon,xx}, \bar{u}_{\varepsilon.xxt})$ it is possible to calculate the expected number of threshold crossings $\langle N(n, t, r, x) \rangle$ of the displacement u or the bending stress σ at n with respect to x.

We calculate the expected number of threshold crossings for $u(t, x, \omega)$. Using the moments

$$\langle u(t, x) \rangle \approx w(t, x),$$

$$\langle u_t(t, x) \rangle \approx w_t(t, x),$$

$$m_{00}(t, x) \doteq \big\langle \big(u(t, x) - w(t, x)\big)^2 \big\rangle \approx \varepsilon \langle \eta_1^2(t, x) \rangle = \varepsilon a \sum_{i=1}^{\infty} T_i^2(t)\, f_i^2(x),$$

$$m_{01}(t, x) \doteq \big\langle \big(u(t, x) - w(t, x)\big) \big(u_t(t, x) - w_t(t, x)\big) \big\rangle$$

$$\approx \varepsilon \langle \eta_1(t, x)\, \eta_2(t, x) \rangle = \varepsilon a \sum_{i=1}^{\infty} T_i(t)\, T_i'(t)\, f_i^2(x),$$

$$m_{11}(t, x) \doteq \big\langle \big(u_t(t, x) - w_t(t, x)\big)^2 \big\rangle \approx \varepsilon \langle \eta_2^2(t, x) \rangle = \varepsilon a \sum_{i=1}^{\infty} T_i'^2(t)\, f_i^2(x)$$

we obtain

$$\langle N(n, t, r, x)\rangle = \int_t^r r(n, s, x)\,\mathrm{d}s$$

where $r(n, s, x)$ denotes the expected rate of threshold crossings at n,

$$r(n, s, x) = \sqrt{\frac{m_{11}}{m_{00}}}\frac{1}{\sqrt{2\pi}}\exp\left(-\frac{(n-w)^2}{2m_{00}}\right)$$
$$\times\left\{2\sqrt{1-\varrho^2}\frac{1}{\sqrt{2\pi}}\exp\left(-\frac{H^2(n)}{2(1-\varrho^2)}\right)+H(n)\left[2\Phi\left(\frac{H(n)}{\sqrt{1-\varrho^2}}\right)-1\right]\right\} \tag{3.36}$$

(see SOONG [1]; HEINRICH, HENNIG [1]). In (3.36) it is put

$$\varrho \doteq \frac{m_{01}}{\sqrt{m_{00}m_{11}}},$$

$$H(n) \doteq \frac{w_t}{\sqrt{m_{11}}} + \varrho\frac{n-w}{\sqrt{m_{00}}},$$

$$\Phi(r) \doteq \frac{1}{\sqrt{2\pi}}\int_{-\infty}^{r}\exp\left(-\frac{1}{2}p^2\right)\mathrm{d}p.$$

A similar formula for the expected number of crossings is obtained in the case of bending stresses. The corresponding moments only have to be substituted by the moments $m_{00}^b(t, x)$, $m_{01}^b(t, x)$, and $m_{11}^b(t, x)$, respectively.

Now we investigate the example of a force p with

$$p_0(t) \doteq 1 - \mathrm{e}^{-\delta t}.$$

This case characterizes a load that possesses the property to be independent of time for large time values. This property follows from the expressions

$$T_i(t) = \int_0^t h_i(t-s)\,p_0(s)\,\mathrm{d}s$$
$$= \frac{1}{\lambda_i} - \frac{\mathrm{e}^{-\beta t}}{\lambda_i\beta_i}\left(\beta\sin(\beta_i t) + \beta_i\cos(\beta_i t)\right) - \frac{\mathrm{e}^{-\delta t}}{\lambda_i - 2\beta\delta + \delta^2}$$
$$+ \frac{\mathrm{e}^{-\beta t}}{\beta_i(\lambda_i - 2\beta\delta + \delta^2)}\left((\beta-\delta)\sin(\beta_i t) + \beta_i\cos(\beta_i t)\right)$$

and

$$T_i'(t) = \frac{\delta}{\lambda_i - 2\delta\beta + \delta^2}\left[\mathrm{e}^{-\delta t} - \mathrm{e}^{-\beta t}\left(\frac{\beta-\delta}{\beta_i}\sin(\beta_i t) + \cos(\beta_i t)\right)\right].$$

Furthermore, the second moments show the convergence properties

$$\begin{aligned}
m_{00}(t, x) &\approx \varepsilon a \sum_{i=1}^{\infty} T_i^2(t)\, f_i^2(x) \xrightarrow[t\to\infty]{} \varepsilon a \sum_{i=1}^{\infty} \frac{1}{\lambda_i^2}\, f_i^2(x) \\
&= \frac{\varepsilon a l^7}{A} \frac{4x^2}{45} \left(\frac{1}{42} - \frac{x^2}{12} + \frac{x^4}{6} - \frac{x^5}{7} + \frac{x^6}{28} \right), \\
m_{01}(t, x) &\approx \varepsilon a \sum_{i=1}^{\infty} T_i(t)\, T_i'(t)\, f_i^2(x) \xrightarrow[t\to\infty]{} 0, \\
m_{11}(t, x) &\approx \varepsilon a \sum_{i=1}^{\infty} T_i'^2(t)\, f_i^2(x) \xrightarrow[t\to\infty]{} 0.
\end{aligned} \tag{3.37}$$

We obtain

$$r(n, s, x) \xrightarrow[t\to\infty]{} 0$$

and

$$\langle N(n, 0, T, x) \rangle \approx \int_0^{t_0} r(n, s, x)\, \mathrm{d}s$$

if $r(n, s, x) \approx 0$ can be put for $s \geqq t_0$. If

$$\lim_{t\to\infty} m_{11}(t, x) = 0$$

then the velocity of the beam is zero after a time t_0 and the displacement is determined by the force being only dependent on the coordinate x (we assumed that the force has acted on the beam since the time $t = 0$).

In particular, in the case of $\delta \to \infty$ we easily find that

$$\lim_{\delta\to\infty} T_i(t) = T_{i0}(t) = \frac{1}{\lambda_i} \left(1 - \mathrm{e}^{-\beta t} \left(\frac{\beta}{\beta_i} \sin(\beta_i t) + \cos(\beta_i t) \right) \right),$$

$$\lim_{\delta\to\infty} T_i'(t) = T_{i0}'(t) = \frac{1}{\beta_i}\, \mathrm{e}^{-\beta t} \sin(\beta_i t).$$

Let $A = 1$, $l = 1$ be for numerical calculations. Putting $\delta = 10$ and $\beta = 1;\ 5$, the moments $m_{ij}(t, x)$, $i, j = 0, 1$, of the displacement $u(t, x)$ and whose velocity $u_t(t, x)$ are illustrated in Fig. 3.17 in dependence on the time t with $x = 0.5$. It can be observed that the magnitude of the damping coefficient β has a great influence on the vibration of the beam. The relations given by (3.37) are satisfied very well after a relatively short time by the functions plotted in this figure.

In Fig. 3.18 some correlation coefficients are plotted as functions of time for $x = 0.5$ where we define

$$\varrho(t, x) \doteq \frac{m_{01}(t, x)}{\sqrt{m_{00}(t, x)\, m_{11}(t, x)}}, \qquad \varrho^b(t, x) \doteq \frac{m_{01}^b(t, x)}{\sqrt{m_{00}^b(t, x)\, m_{11}^b(t, x)}},$$

$$\tilde{\varrho}(t, x) \doteq \frac{\langle u(t, x)\, \sigma(t, x) \rangle}{\sqrt{m_{00}(t, x)\, m_{00}^b(t, x)}}.$$

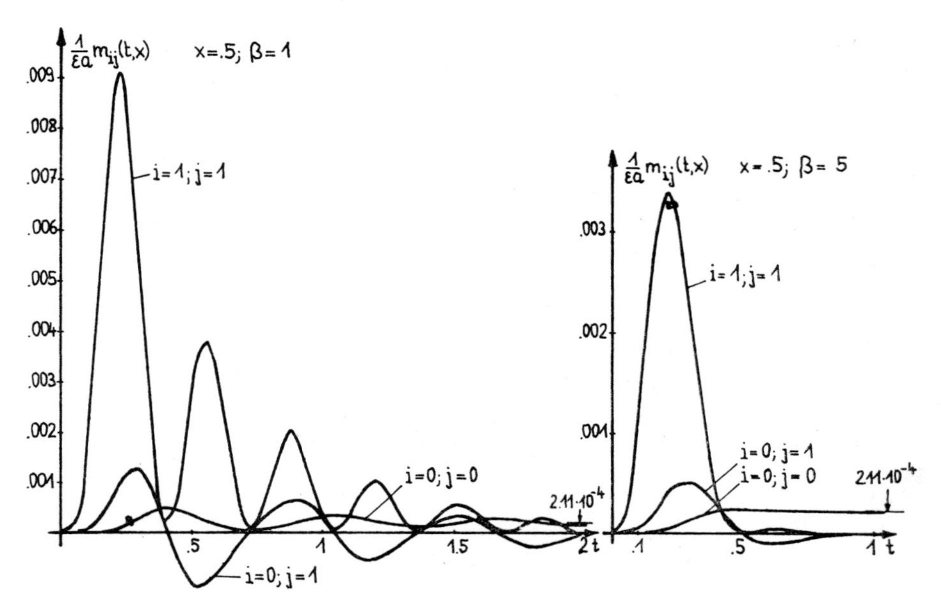

Fig. 3.17. Moments m_{00}, m_{01}, m_{11} of the displacement $u(t, x, \omega)$ and whose velocity $u_t(t, x, \omega)$ as a function of time for $x = 0.5$ and $\beta = 1; 5, \delta = 10$

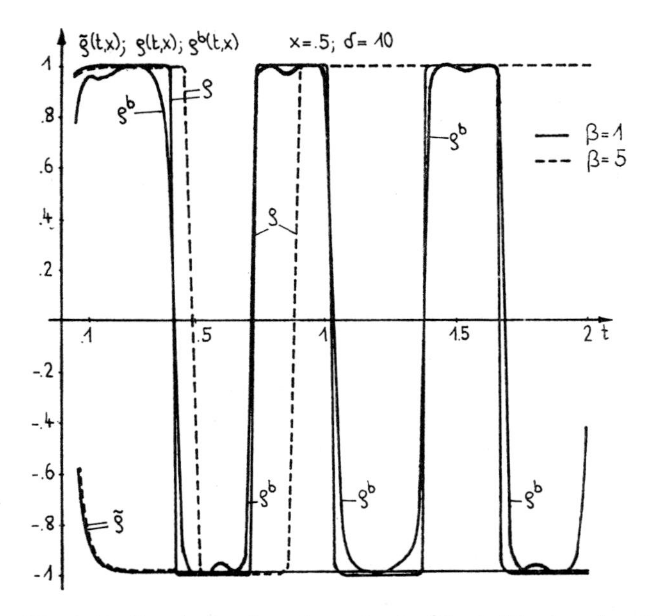

Fig. 3.18. Correlation functions $\varrho(t, x)$, $\varrho^b(t, x)$, $\tilde{\varrho}(t, x)$ as functions of time

$m_{ij}^b(t, x)$ denotes the moment $m_{ij}(t, x)$ taken for the bending stress. From $|\tilde{\varrho}| \approx 1$ for all $t \geqq 0.2$ it follows a nearly linear connection between the displacement $u(t, x)$ and the bending stress $\sigma(t, x)$,

$$\sigma(t, x) = \tilde{A}(t, x)\, u(t, x).$$

The correlation coefficients $\varrho(t, x)$ and $\varrho^b(t, x)$ demonstrate a similar behaviour. The linear relations between u and u_t, and σ, σ_t are described by $A(t, x)$ and $A^b(t, x)$, respectively. The functions $A(t, x)$, $\tilde{A}(t, x)$ are plotted in Fig. 3.19. We note that the bending stress and the displacement satisfy the relation

$$\sigma(t, x, \omega) \approx -9.88 Ehu(t, x, \omega)$$

for $x = 0.5$. Furthermore, we have

$$u_t(t, x, \omega) = A(t, x)\, u(t, x, \omega)$$

where $A(t, x)$ is given by Fig. 3.19 for $x = 0.5$. Hence, it is

$$m_{11}(t, x) = A^2(t, x)\, m_{00}(t, x),$$

$$m_{01}(t, x) = A(t, x)\, m_{00}(t, x).$$

These relations are satisfied by the results plotted in Fig. 3.17. Further regression equations can be given. By this, the moments $m_{ij}^b(t, x)$ can be deduced from the moments $m_{ij}(t, x)$. The reason why we have such a linear connection is that the first summand in the series contained in the solution for the displacement $u(t, x, \omega)$ and the bending stress $\sigma(t, x, \omega)$ determines essentially the value of the series; i.e. the series converge very well.

Using (3.36) the expected rate of crossings $r(n, t, x)$ of $u(t, x, \omega)$ can be calculated where $w(t, x) \equiv 0$ because $\langle p_1(x) \rangle = 0$. The function $r(n, t, x)$ is plotted in Fig. 3.20 as a function of t for $x = 0.5$, $\beta = 1; 5$. The level n is given by $\sqrt{\varepsilon a}$ units. The larger damping constant effects a faster decay of the vibrations. Fig. 3.22 shows the expected rate of crossings $r^b(n, t, x)$ of the bending stress $\sigma(t, x)$ for $x = 0.5$, $\beta = 1; 5$. It is clear that the expected rate of crossings has a minimum for those times for which the moment m_{11} has a minimum. For these times the beam is nearly quiescent and an increase as to the expected number of crossings cannot be obtained. By means of these figures the vibration behaviour of the beam for $x = 0.5$ can be investigated very well.

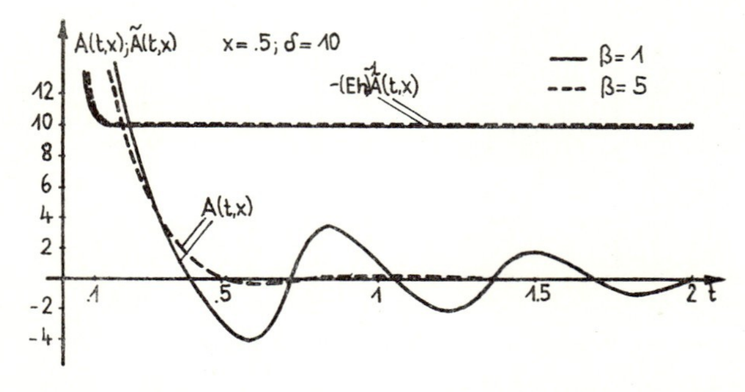

Fig. 3.19. Factors of the linear connection $A(t, x)$ and $\tilde{A}(t, x)$ as functions of time

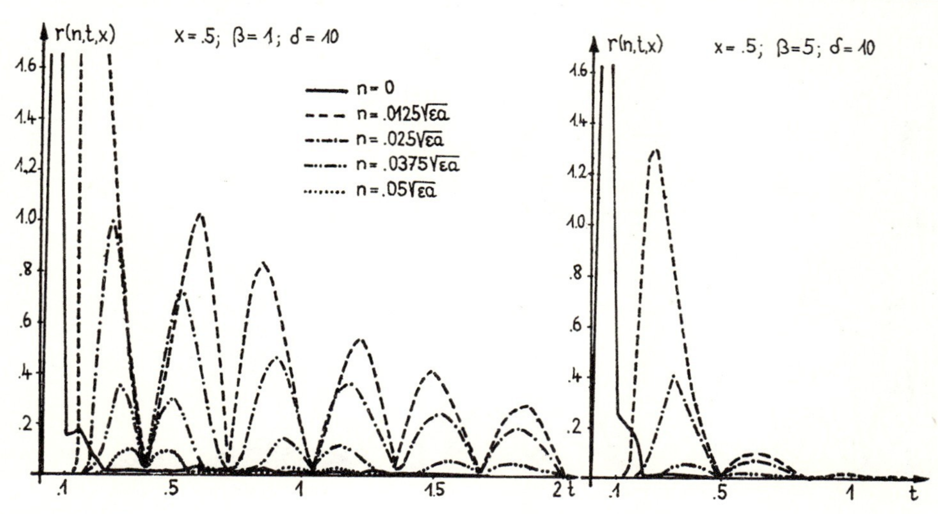

Fig. 3.20. Expected rate of crossings $r(n, t, x)$ of $u(t, x, \omega)$ in dependence on t

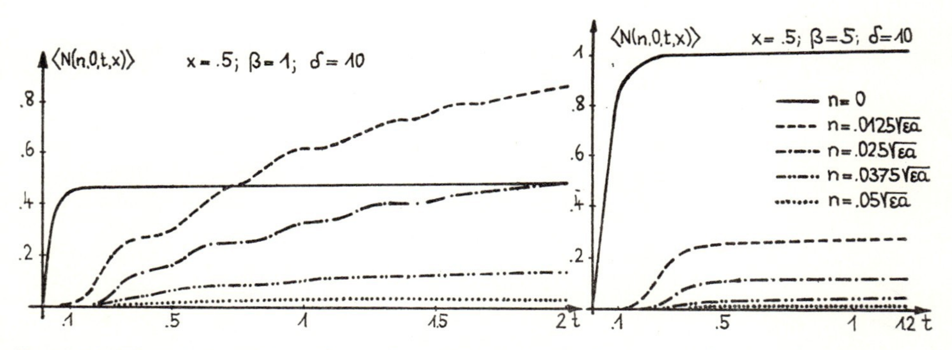

Fig. 3.21. Expected number of crossings $\langle N(n, 0, t, x)\rangle$ of $u(t, x, \omega)$ in dependence on t

In Fig. 3.21 the expected number of crossings $\langle N(n, 0, t, x)\rangle$ of the displacement u is illustrated for $x = 0.5$, $\delta = 10$, and $\beta = 1; 5$. In Fig. 3.23 we have drawn the corresponding value $\langle N^b(n, 0, t, x)\rangle$ for the bending stress. The essential increase of the mean values in the case of $n = 0$ is produced a short time after the vibration has begun, and the increase in the case of higher levels is produced later. The impact of β is clearly determined by a comparison of the two parts of Fig. 3.21, on the one hand, and of Fig. 3.23, on the other hand.

3.1.6. Partially weakly correlated excitations as to the time variable

This section concerns with the case that the random function $\overline{p}_\varepsilon(t, x, \omega)$ which corresponds to the external force is partially weakly correlated with respect to the time. Similar limit distributions of the displacement and the stresses can be obtained as in the previous section.

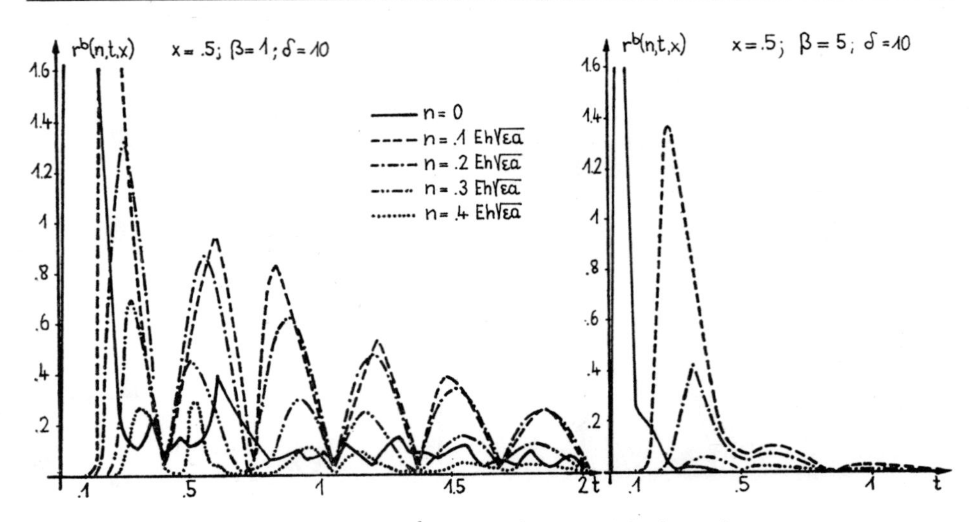

Fig. 3.22. Expected rate of crossings $r^b(n, t, x)$ of $\sigma(t, x, \omega)$ in dependence on t

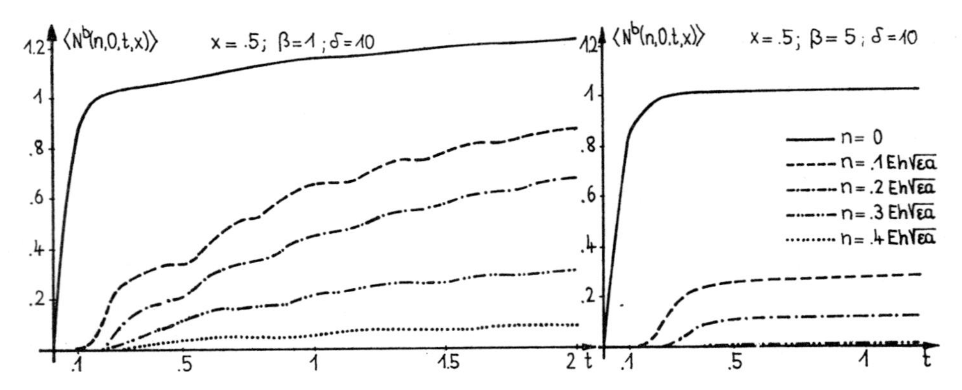

Fig. 3.23. Expected number of crossings $\langle N^b(n, 0, t, x)\rangle$ of $\sigma(t, x, \omega)$ in dependence on t

The averaged solution of the boundary-initial-value problem (3.8) is given by

$$U_\varepsilon(t, \omega) = \int_0^t \int_{\mathcal{D}} F(t - s, y)\, \overline{p}_\varepsilon(s, y, \omega)\, \mathrm{d}y\, \mathrm{d}s$$

where

$$F(t, y) = \sum_{i=1}^{\infty} h_i(t)\, f_i(y)\, (f_i, g)$$

and we have put

$$U_\varepsilon \doteq (\bar{u}_\varepsilon, g) \quad \text{with} \quad g \in \mathrm{C}_0^\infty(\mathcal{D}).$$

Let the partially weakly correlated function $\overline{p}_\varepsilon(t, x, \omega)$ as to t have sufficiently smooth sample functions. Applying $g \in \mathbf{C}_0^\infty(\mathcal{D})$ we obtain

$$\left(\mathrm{D}_t^p \mathrm{D}_x^q \bar{u}_\varepsilon(t, .), g\right) = \int_0^t \int_{\mathcal{D}} \sum_{i=1}^\infty \mathrm{D}_t^p h_i(t - s) f_i(y) (\mathrm{D}_x^q f_i, g) \bar{p}_\varepsilon(s, y, \omega) \, \mathrm{d}y \, \mathrm{d}s$$

$$= \int_0^t \sum_{i=1}^\infty \mathrm{D}_t^p h_i(t - s) \left(f_i, \bar{p}_\varepsilon(s, .)\right) (\mathrm{D}_x^q f_i, g) \, \mathrm{d}s$$

for $p = 0, 1; q = 0, 1, 2, \dots$ From this it follows

$$\left(\mathrm{D}_t^p \mathrm{D}_x^q \bar{u}_\varepsilon(t, .), g\right) = \left(\int_0^t \sum_{i=1}^\infty \mathrm{D}_t^p h_i(t - s) \mathrm{D}_x^q f_i(.) \left(f_i, \bar{p}_\varepsilon(s, .)\right) \mathrm{d}s, g\right)$$

if the condition

$$\sum_{i=1}^\infty \mathrm{D}^p h_i(t - s) \mathrm{D}_x^q f_i(x) \left(f_i, \bar{p}_\varepsilon(s, .)\right) \in \mathbf{L}_2([0, t]_s \times \mathcal{D})$$

is assumed. Finally, it is

$$\mathrm{D}_t^p \mathrm{D}_x^q \bar{u}_\varepsilon(t, x, \omega) = \int_0^t \sum_{i=1}^\infty \mathrm{D}_t^p h_i(t - s) \mathrm{D}_x^q f_i(x) \left(f_i, \bar{p}_\varepsilon(s, .)\right) \mathrm{d}s \tag{3.38}$$

under the assumption of the continuity as to x of the right-hand side. In particular, this assumption is satisfied if the series (3.38) is uniformly convergent as to x. If $\bar{p}_\varepsilon(t, x, \omega)$ fulfils certain boundary conditions then the uniform convergence follows from

$$\left(f_i, \bar{p}_\varepsilon(t, .)\right) = \frac{1}{\lambda_i^r} \left(f_i, L^r \bar{p}_\varepsilon(t, .)\right) \quad \text{for} \quad r = 1, 2, \dots$$

Certainly, weaker conditions as to $\bar{p}_\varepsilon$ are also sufficient. For a suitable r we have

$$\mathrm{D}_t^p \mathrm{D}_x^q \bar{u}_\varepsilon(t, x, \omega) = \int_{\mathcal{D}} \int_0^t F_{pq}^r(t - s, x, y) L^r \bar{p}_\varepsilon(s, y, \omega) \, \mathrm{d}s \, \mathrm{d}y \tag{3.39}$$

where

$$F_{pq}^r(t, x, y) \doteq \sum_{i=1}^\infty \frac{1}{\lambda_i^r} \mathrm{D}_t^p h_i(t) \mathrm{D}_x^q f_i(x) f_i(y).$$

Let $\bar{p}_\varepsilon(t, x, \omega)$ be a partially weakly correlated function with the relations

$$a(t, x, y) = \lim_{\varepsilon \downarrow 0} \frac{1}{\varepsilon} \int_{-\varepsilon}^{\varepsilon} \langle \bar{p}_\varepsilon(t, x) \bar{p}_\varepsilon(t + s, y) \rangle \, \mathrm{d}s,$$

$$L_x^r L_y^r a(t, x, y) = \lim_{\varepsilon \downarrow 0} \frac{1}{\varepsilon} \int_{-\varepsilon}^{\varepsilon} \langle L_x^r \bar{p}_\varepsilon(t, x) L_y^r \bar{p}_\varepsilon(t + s, y) \rangle \, \mathrm{d}s.$$

Then we deduce from a limit theorem (see Section 1.6.1) the convergence in distribution

$$\lim_{\varepsilon \downarrow 0} \frac{1}{\sqrt{\varepsilon}} \mathrm{D}_t^p \mathrm{D}_x^q \bar{u}_\varepsilon(t, x, \omega) = \zeta_{pq}(t, x, \omega)$$

where $\zeta_{pq}(t, x, \omega)$ denotes a Gaussian function having zero-mean and the second moment

$$\langle \zeta_{pq}(t, x)\, \zeta_{pq}(r, y) \rangle = \int_{\mathcal{D}} \int_{\mathcal{D}} \int_0^{\bar{t}} F_{pq}^r(t - s, x, u)\, F_{pq}^r(r - s, y, v)\, L_u^r L_v^r a(s, u, v)\, \mathrm{d}s\, \mathrm{d}u\, \mathrm{d}v, \tag{3.40}$$

$\bar{t} \doteq \min\{t, r\}$. The necessary assumptions with regard to $F_{pq}^r(t, x, y)$ are satisfied by a suitable r.

In the special case

$$\bar{p}_\varepsilon(t, x, \omega) \doteq \bar{p}_{0\varepsilon}(t, \omega)\, p_1(x)$$

from (3.38) it follows

$$\mathrm{D}_t^p \mathrm{D}_x^q \bar{u}_\varepsilon(t, x, \omega) = \int_0^t G_{pq}(t - s, x)\, \bar{p}_{0\varepsilon}(s, \omega)\, \mathrm{d}s$$

where

$$G_{pq}(t, x) \doteq \sum_{i=1}^{\infty} \mathrm{D}_t^p h_i(t)\, \mathrm{D}_x^q f_i(x)\, (f_i, p_1)$$

if

$$G_{pq}(s, x) \in \mathbf{L}_2([0, t]_s \times \mathcal{D})$$

and the integral

$$\int_0^t G_{pq}(t - s, x)\, \bar{p}_{0\varepsilon}(s, \omega)\, \mathrm{d}s$$

is continuous as to x. From the limit theorem we obtain

$$\lim_{\varepsilon \downarrow 0} \frac{1}{\sqrt{\varepsilon}} \mathrm{D}_t^p \mathrm{D}_x^q \bar{u}_\varepsilon(t, x, \omega) = \zeta_{pq}(t, x, \omega) \quad \text{in distribution.}$$

The Gaussian function $\zeta_{pq}(t, x, \omega)$ possesses the moments $\langle \zeta_{pq}(t, x) \rangle = 0$ and

$$\langle \zeta_{pq}(t, x)\, \zeta_{pq}(r, y) \rangle = \int_0^{\bar{t}} G_{pq}(t - s, x)\, G_{pq}(r - s, y)\, a(s)\, \mathrm{d}s, \tag{3.41}$$

$\bar{t} \doteq \min\{t, r\}$, where

$$a(t) = \lim_{\varepsilon \downarrow 0} \frac{1}{\varepsilon} \int_{-\varepsilon}^{\varepsilon} \langle \bar{p}_{0\varepsilon}(t)\, \bar{p}_{0\varepsilon}(t + s) \rangle\, \mathrm{d}s.$$

The relation (3.41) can also be deduced from (3.40) taking into consideration

$$a(t, x, y) = a(t)\, p_1(x)\, p_1(y).$$

It is also possible to transfer such limit statements to random vector functions

$$\frac{1}{\sqrt{\varepsilon}}\left(\mathrm{D}_t^{p_1}\mathrm{D}_x^{q_1}\bar{u}_\varepsilon(t,x,\omega),\ldots,\mathrm{D}_t^{p_k}\mathrm{D}_x^{q_k}\bar{u}_\varepsilon(t,x,\omega)\right),$$

for example, to

$$\frac{1}{\sqrt{\varepsilon}}\left(\bar{u}_\varepsilon(t,x,\omega),\bar{u}_{\varepsilon,t}(t,x,\omega)\right)$$

and

$$\frac{1}{\sqrt{\varepsilon}}\left(\bar{u}_{\varepsilon,xx}(t,x,\omega),\bar{u}_{\varepsilon,xxt}(t,x,\omega)\right).$$

For the further investigations the random function $\bar{p}_\varepsilon(t,x,\omega)$ is assumed to be wide-sense stationary, i.e. we assume

$$\langle\bar{p}_\varepsilon(t,x)\,\bar{p}_\varepsilon(r,y)\rangle = R_\varepsilon(|t-r|,x,y).$$

Thus

$$a(t,x,y) = a(x,y) \quad \text{and} \quad a(x,y) = a(y,x).$$

We define

$$T_{ij}^{p_1p_2}(t,r) \doteq \int_0^{\bar{t}} h_i^{(p_1)}(t-s)\,h_j^{(p_2)}(r-s)\,\mathrm{d}s$$

for $p_1, p_2 = 0, 1$. In a similar form as deduced in (3.40), we obtain

$$\begin{aligned}&\langle\zeta_{p_1q_1}(t,x)\,\zeta_{p_2q_2}(r,y)\rangle\\&=\int_{\mathcal{D}}\int_{\mathcal{D}}\sum_{i,j=1}^{\infty}\frac{1}{\lambda_i^r\lambda_j^r}T_{ij}^{p_1p_2}(t,r)\,\mathrm{D}_x^{q_1}f_i(x)\,\mathrm{D}_y^{q_2}f_j(y)\,f_i(u)\,f_j(v)\,L_u^rL_v^ra(u,v)\,\mathrm{d}u\,\mathrm{d}v.\end{aligned}$$

In it, $(\zeta_{p_1q_1}(t,x),\zeta_{p_2q_2}(t,x))$ denotes the random function to which

$$\frac{1}{\sqrt{\varepsilon}}\left(\mathrm{D}_t^{p_1}\mathrm{D}_x^{q_1}\bar{u}_\varepsilon(t,x,\omega),\mathrm{D}_t^{p_2}\mathrm{D}_x^{q_2}\bar{u}_\varepsilon(t,x,\omega)\right)$$

converges as to $\varepsilon\downarrow 0$. The variance of $\mathrm{D}_t^p\mathrm{D}_x^q\bar{u}_\varepsilon(t,x,\omega)$ is approximately given by

$$\begin{aligned}\left\langle\left(\mathrm{D}_t^p\mathrm{D}_x^q\bar{u}_\varepsilon(t,x)\right)^2\right\rangle &\approx \varepsilon\langle\zeta_{pq}^2(t,x)\rangle\\&=\varepsilon\int_{\mathcal{D}}\int_{\mathcal{D}}\sum_{i,j=1}^{\infty}\frac{1}{\lambda_i^r\lambda_j^r}T_{ij}^{pp}(t,t)\,\mathrm{D}_x^qf_i(x)\,\mathrm{D}_x^qf_j(x)\,f_i(u)\,f_j(v)\\&\quad\times L_u^rL_v^ra(u,v)\,\mathrm{d}u\,\mathrm{d}v\end{aligned}$$

and in a similar way we can also calculate other moments. In a straightforward manner it can be obtained

$$\begin{aligned}&\lim_{\substack{t,r\to\infty\\ r-t=s=\text{const}\geqq 0}} T_{ij}^{00}(t,r)\\&=\frac{\mathrm{e}^{-\beta s}}{8\beta^2(\lambda_i+\lambda_j)+(\lambda_j-\lambda_i)^2}\left[4\beta\cos(\beta_js)+\frac{4\beta^2+\lambda_i-\lambda_j}{\beta_j}\sin(\beta_js)\right]\doteq\tilde{T}_{ij}^{00}(s),\end{aligned}$$

$$\lim_{\substack{t,r\to\infty \\ r-t=s=\text{const}\geqq 0}} T_{ij}^{10}(t,r)$$

$$= \frac{e^{-\beta s}}{8\beta^2(\lambda_i+\lambda_j)+(\lambda_j-\lambda_i)^2}\left[(\lambda_j-\lambda_i)\cos(\beta_j s)+\frac{\beta}{\beta_j}(\lambda_i+3\lambda_j)\sin(\beta_j s)\right]$$

$$\doteq \tilde{T}_{ij}^{10}(s) = -\frac{d}{ds}\tilde{T}_{ij}^{00}(s), \tag{3.42}$$

$$\lim_{\substack{t,r\to\infty \\ r-t=s=\text{const}\geqq 0}} T_{ij}^{11}(t,r) = \frac{e^{-\beta s}}{8\beta^2(\lambda_i+\lambda_j)+(\lambda_j-\lambda_i)^2}\left[2\beta(\lambda_i+\lambda_j)\cos(\beta_j s)\right.$$

$$\left.-\frac{1}{\beta_j}\big(2\beta^2(\lambda_i+\lambda_j)+\lambda_j(\lambda_j-\lambda_i)\big)\sin(\beta_j s)\right]$$

$$\doteq \tilde{T}_{ij}^{11}(s) = -\frac{d^2}{ds^2}\tilde{T}_{ij}^{00}(s).$$

From this, it follows that

$$\big(\zeta_{0q}(t,x,\omega), \zeta_{1q}(t,x,\omega)\big)$$

is approximately a stationary Gaussian vector function as to t for large times which is denoted by

$$\big(\xi_{0q}(t,x,\omega), \xi_{1q}(t,x,\omega)\big).$$

Hence, we have

$$\langle \xi_{p_1q_1}(t,x)\,\xi_{p_2q_2}(r,y)\rangle$$

$$= \int_{\mathscr{D}}\int_{\mathscr{D}} \sum_{i,j=1}^{\infty} \frac{1}{\lambda_i^r\lambda_j^r}\tilde{T}_{ij}^{p_1p_2}(r-t)\,D_x^{q_1}f_i(x)\,D_y^{q_2}f_j(y)\,f_i(u)\,f_j(v)\,L_u^rL_v^ra(u,v)\,du\,dv$$

for $r - t \geqq 0$. Furthermore, for $s > 0$ it can be shown

$$\tilde{T}_{ij}^{00}(-s) = \tilde{T}_{ji}^{00}(s); \qquad \tilde{T}_{ij}^{11}(-s) = \tilde{T}_{ji}^{11}(s); \qquad \tilde{T}_{ij}^{10}(-s) = -\tilde{T}_{ji}^{10}(s)$$

since

$$T_{ij}^{10}(t,r) = h_i(t)\,h_j(r) - T_{ji}^{10}(r,t) \quad \text{for} \quad t - r \geqq 0.$$

In particular, if $L_u^rL_v^ra(u,v)$ is a symmetric function of u, v then it is

$$\langle \xi_{0q}(t,x)\,\xi_{1q}(t,x)\rangle = 0,$$

i.e. $\xi_{0q}(t,x,\omega)$ and $\xi_{1q}(t,x,\omega)$ are independent random variables. Hence, on the same point (t,x) the displacement and the velocity are independent.

The spectral density $S_{p_1qp_2q}$ as to $\langle \xi_{p_1q}(t,x)\,\xi_{p_2q}(r,y)\rangle$ is defined by

$$S_{p_1qp_2q}(\alpha; x, y) \doteq \frac{1}{2\pi}\int_{-\infty}^{\infty} e^{-is\alpha}\langle \xi_{p_1q}(t,x)\,\xi_{p_2q}(t+s,y)\rangle\,ds.$$

We obtain

$$\frac{1}{2\pi} \int_{-\infty}^{\infty} e^{-is\alpha} \tilde{T}_{vw}^{pq}(s)\, ds = U_{vw}^{pq}(\alpha) + i V_{vw}^{pq}(\alpha)$$

where

$$U_{vw}^{pq}(\alpha) = \frac{1}{2\pi} \int_{0}^{\infty} \cos(\alpha s) \left(\tilde{T}_{vw}^{pq}(s) + \tilde{T}_{vw}^{pq}(-s)\right) ds,$$

$$V_{vp}^{pq}(\alpha) = \frac{1}{2\pi} \int_{0}^{\infty} \sin(\alpha s) \left(-\tilde{T}_{vw}^{pq}(s) + \tilde{T}_{vw}^{pq}(-s)\right) ds$$

for $p, q = 0, 1$; $v, w = 1, 2, \ldots$ The functions $U_{vw}^{pq}(\alpha)$, $V_{vw}^{pq}(\alpha)$ can be calculated with the help of the explicit terms of $\tilde{T}_{vw}^{pq}(s)$ in (3.42). Thus

$$W_{vw}^{00}(\alpha) \doteq \frac{1}{2\pi} \int_{0}^{\infty} \cos(\alpha s)\, \tilde{T}_{vw}^{00}(s)\, ds = \frac{8\beta^2 \lambda_w + (\lambda_v - \lambda_w)(\lambda_w - \alpha^2)}{2\pi A_{vw}(\alpha, \beta)},$$

$$\frac{1}{2\pi} \int_{0}^{\infty} \sin(\alpha s)\, \tilde{T}_{vw}^{00}(s)\, ds = \frac{2\alpha\beta(8\beta^2 + 2\alpha^2 + \lambda_v - 3\lambda_w)}{2\pi A_{vw}(\alpha, \beta)},$$

$$W_{vw}^{11}(\alpha) \doteq \frac{1}{2\pi} \int_{0}^{\infty} \cos(\alpha s)\, \tilde{T}_{vw}^{11}(s)\, ds = \frac{4\beta^2\alpha^2(\lambda_v + \lambda_w) + (\alpha^2 - \lambda_w)\, \lambda_w(\lambda_w - \lambda_v)}{2\pi A_{vw}(\alpha, \beta)}$$

$$= \frac{1}{2\pi} \frac{d}{ds} \tilde{T}_{vw}^{00}(0) + \frac{\alpha^2}{2\pi} \int_{0}^{\infty} \cos(\alpha s)\, \tilde{T}_{vw}^{00}(s)\, ds,$$

$$\frac{1}{2\pi} \int_{0}^{\infty} \sin(\alpha s)\, \tilde{T}_{vw}^{11}(s)\, ds = \frac{2\alpha\beta\left((\lambda_v + \lambda_w)\, \alpha^2 - 2\lambda_w^2\right)}{2\pi A_{vw}(\alpha, \beta)}$$

$$= -\frac{\alpha}{2\pi} \tilde{T}_{vw}^{00}(0) + \frac{\alpha^2}{2\pi} \int_{0}^{\infty} \sin(\alpha s)\, \tilde{T}_{vw}^{00}(s)\, ds$$

where A_{vw} is defined by

$$A_{vw}(\alpha, \beta) \doteq [8\beta^2(\lambda_v + \lambda_w) + (\lambda_v - \lambda_w)^2]\, [(\lambda_w + \alpha^2)^2 - 4\beta_w^2\alpha^2].$$

For example, the spectral density of $\xi_{0q}(t, x, \omega)$ for $x = y$ can be determined by

$$S_{0q0q}(\alpha; x, x) = \int_{\mathcal{D}} \int_{\mathcal{D}} \sum_{v,w=1}^{\infty} \frac{1}{\lambda_v^r \lambda_w^r} U_{vw}^{00}(\alpha)\, D_x^q f_v(x)\, D_x^q f_w(x)\, f_v(y)\, f_w(z)\, L_y^r L_z^r a(y, z)\, dy\, dz$$

if the relations

$$U^{00}_{vw} = U^{00}_{wv}, \qquad V^{00}_{vw} = -V^{00}_{wv}$$

are taken into consideration. From the given considerations the equation

$$S_{1q1q}(\alpha; x, x) = \alpha^2 S_{0q0q}(\alpha; x, x)$$

can easily be deduced. In particular, in the case of

$$\overline{p}_\varepsilon(t, x, \omega) \doteq \overline{p}_{0\varepsilon}(t, \omega)\, p_1(x)$$

we obtain

$$S_{0q0q}(\alpha; x, x) = a \sum_{v,w=1}^{\infty} U^{00}_{vw}(\alpha)\, \mathrm{D}^q_x f_v(x)\, \mathrm{D}^q_x f_w(x)\, (f_v, p_1)\, (f_w, p_1)$$

$$= 2a \sum_{v,w=1}^{\infty} W^{00}_{vw}(\alpha)\, \mathrm{D}^q_x f_v(x)\, \mathrm{D}^q_x f_w(x)\, (f_v, p_1)\, (f_w, p_1)$$

by means of the above relations. The intensity $a(t)$ is a constant if the process $\overline{p}_{0\varepsilon}$ has the property of the stationarity in the wide sense.

As in Section 3.1.5 we consider vibrations of a simple supported beam which are described by the boundary-initial-value problem

$$u_{tt} + 2\beta u_t + A u_{xxxx} = p_\varepsilon(t, x, \omega), \qquad x \in [0, l],$$

boundary conditions: $u(t, 0) = u(t, l) = u_{xx}(t, 0) = u_{xx}(t, l) = 0$,

initial conditions: $u(0, x) = u_t(0, x) = 0$

where we put

$$p_\varepsilon(t, x, \omega) = p_{0\varepsilon}(t, \omega)\, p_1(x).$$

Let $p_{0\varepsilon}(t, \omega)$ be a weakly correlated process being wide-sense stationary. The intensity of the process $p_{0\varepsilon}$ is denoted by a. In the given case we obtain

$$\lambda_i = A \left(\frac{i\pi}{l}\right)^4, \qquad f_i(x) = \sqrt{\frac{2}{l}} \sin\left(\frac{i\pi x}{l}\right).$$

If we assume that

$$G_{pq}(s, x) = \sum_{i=1}^{\infty} \mathrm{D}^p_s h_i(s)\, \mathrm{D}^q_x f_i(x)\, (f_i, p_1)$$

is from $\mathbf{L}_2([0, t] \times \mathcal{D})$ and is continuous with respect to x in $\mathbf{L}_2([0, t])$ then the statements for $\mathrm{D}^p_t \mathrm{D}^q_x \bar{u}_\varepsilon(t, x, \omega)$ carried out above are true. In particular, the before-mentioned conditions as to G_{pq} are satisfied if the series defining the function G_{pq} is uniformly convergent as to $s \in [0, t]$ and $x \in \mathcal{D}$. This property can be fulfilled by the choice of sufficiently smooth functions $p_1(x)$ which have to satisfy certain boundary conditions. In order to investigate $u_\varepsilon(t, x, \omega)$ we do not demand conditions for $p_1(x)$ since in this case the series defining $G_{pq}(s, x)$ is uniformly convergent. If we demand $p_1(0) = p_1(l) = 0$ and the existence of the second derivative then the results for $u_{\varepsilon,t}(t, x, \omega)$ and $u_{\varepsilon,xx}(t, x, \omega)$ follow from the uniform convergence of the corresponding series. Finally, the limit theorems for $u_{\varepsilon,xxt}(t, x, \omega)$ are obtained from the conditions

$$p_1(0) = p_1(l) = p_1''(0) = p_1''(l) = 0, \qquad p_1 \in \mathbf{C}^4.$$

For simplification we choose

$$p_1(x) = M \sin\left(\frac{m\pi x}{l}\right).$$

The case of $m = 1$ is particularly interesting since it describes an important load of the beam. The function $p_1(x)$ satisfies all above assumptions. Hence, the function $\frac{1}{\sqrt{\varepsilon}} \mathrm{D}_t^p \mathrm{D}_x^q u_\varepsilon(t, x, \omega)$ converges in distribution to $\zeta_{pq}(t, x, \omega)$ and $\xi_{pq}(t, x, \omega)$ denotes the corresponding stationary function.

Applying

$$(f_i, p_1) = M \sqrt{\frac{l}{2}}\, \delta_{im}$$

the correlation function of the limit function ξ_{pq} can be determined by

$$\langle \xi_{pq}(t, x)\, \xi_{pq}(r, y)\rangle = \frac{aM^2l}{2}\, \tilde{T}_{mm}^{pp}(s)\, \mathrm{D}_x^q f_m(x)\, \mathrm{D}_x^q f_m(y)$$

for $s = r - t \geqq 0$. In particular, we have

$$\langle \xi_{0q}(t, x)\, \xi_{0q}(r, y)\rangle = \frac{aM^2l}{2}\, \frac{\mathrm{e}^{-\beta s}}{4\lambda_m} \left[\frac{1}{\beta} \cos(\beta_m s) + \frac{1}{\beta_m} \sin(\beta_m s)\right] \mathrm{D}_x^q f_m(x)\, \mathrm{D}_y^q f_m(y),$$

$$\langle \xi_{1q}(t, x)\, \xi_{1q}(r, y)\rangle = \frac{aM^2l}{2}\, \frac{\mathrm{e}^{-\beta s}}{4} \left[\frac{1}{\beta} \cos(\beta_m s) - \frac{1}{\beta_m} \sin(\beta_m s)\right] \mathrm{D}_x^q f_m(x)\, \mathrm{D}_y^q f_m(y).$$

It is clear that the correlation function

$$\langle \xi_{00}(t, x)\, \xi_{00}(r, y)\rangle$$

converges to zero as $m \to \infty$. This property is not true for the other interesting moments.

The spectral density $S_{0q0q}(\alpha; x, y)$ belonging to $\langle \xi_{0q}(t, x)\, \xi_{0q}(r, y)\rangle$ follows from

$$\frac{1}{2\pi} \int_{-\infty}^{\infty} \mathrm{e}^{-\mathrm{i}\alpha s} \tilde{T}_{mm}^{00}(s)\, \mathrm{d}s = U_{mm}^{00}(\alpha) = 2W_{mm}^{00}(\alpha)$$

as

$$S_{0q0q}(\alpha; x, y) = \frac{aM^2l}{4\pi}\, \frac{1}{(\lambda_m + \alpha^2)^2 - 4\beta_m^2\alpha^2}\, \mathrm{D}_x^q f_m(x)\, \mathrm{D}_y^q f_m(y)$$

$$= \frac{aM^2l}{4\pi}\, \frac{1}{(\lambda_m - \alpha^2)^2 + 4\beta^2\alpha^2}\, \mathrm{D}_x^q f_m(x)\, \mathrm{D}_y^q f_m(y).$$

In Fig. 3.24 the function

$$S_m(\alpha, \beta) = \frac{1}{4\pi}\, \frac{1}{(\lambda_m - \alpha^2)^2 + 4\beta^2\alpha^2}$$

is plotted in dependence on α which essentially determines the spectral density $S_{0q0q}(\alpha; x, y)$. This function has a maximum for

$$\alpha_0 = \sqrt{\lambda_m - 2\beta^2}$$

and it is

$$S_m(\alpha_0, \beta) = \frac{1}{16\pi\beta^2\beta_m^2}.$$

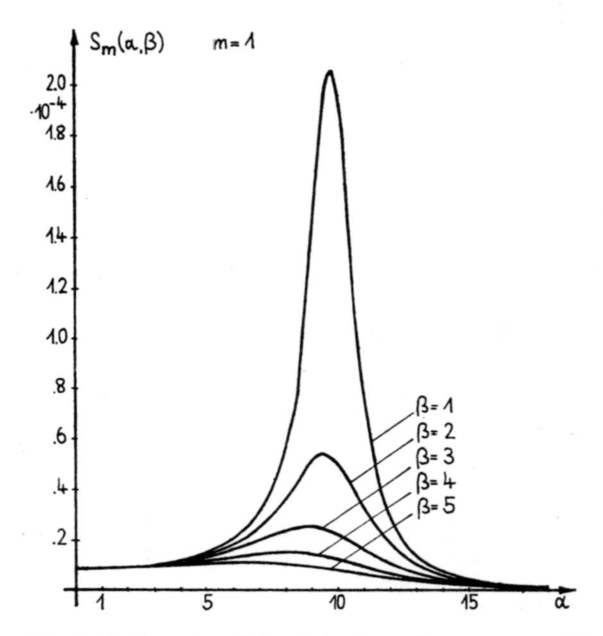

Fig. 3.24. Function $S_m(\alpha, \beta)$ in dependence on α and β

This function $S_1(\alpha, \beta)$ is the predominant term of the spectral density, also in the case if we choose other functions $p_1(x)$ than

$$p_1(x) = M \sin\left(\frac{\pi x}{l}\right).$$

This statement can be deduced from the special form of the spectral density. In our considered case the displacement $u(t, x, \omega)$ and the bending stress

$$\sigma(t, x, \omega) = Ehu_{xx}(t, x, \omega)$$

have proportional spectral densities.

It is also possible to calculate the cross-correlation function

$$\langle \xi_{p_1q_1}(t, x)\, \xi_{p_2q_2}(r, y) \rangle = \frac{aM^2l}{2}\, \tilde{T}_{mm}^{p_1p_2}(s)\, \mathrm{D}_x^{q_1} f_m(x)\, \mathrm{D}_y^{q_2} f_m(y)$$

for $s = r - t \geqq 0$. In particular, it follows

$$\langle \xi_{00}(t, x)\, \xi_{02}(t, x) \rangle = \frac{aM^2l}{2}\, \tilde{T}^{00}_{mm}(0)\, f_m(x)\, f''_m(x) = \frac{aM^2l}{8}\, \frac{1}{\beta\lambda_m}\, f_m(x)\, f''_m(x)$$

and then the correlation coefficient $\tilde{\varrho}$ between the random variables $\xi_{00}(t, x, \omega)$ and $\xi_{02}(t, x, \omega)$ as

$$\tilde{\varrho} = \frac{\langle \xi_{00}(t, x)\, \xi_{02}(t, x) \rangle}{\sqrt{\langle \xi_{00}^2(t, x) \rangle \langle \xi_{02}^2(t, x) \rangle}} = \frac{f_m(x)\, f''_m(x)}{|f_m(x)\, f''_m(x)|} = -1.$$

The relation

$$u(t, x, \omega) = M \sqrt{\frac{l}{2}} \int_0^t h_m(t - s)\, p_{0\varepsilon}(s, \omega)\, \mathrm{d}s \cdot f_m(x)$$

explains this correlation coefficient $\tilde{\varrho}$.

For the second moments of u and u_t we obtain

$$\langle u^2(t, x) \rangle \approx \varepsilon \langle \xi_{00}^2(t, x) \rangle = \frac{\varepsilon a M^2 l}{8}\, \frac{f_m^2(x)}{\beta\lambda_m},$$

$$\langle u_t^2(t, x) \rangle \approx \varepsilon \langle \xi_{10}^2(t, x) \rangle = \frac{\varepsilon a M^2 l}{8}\, \frac{f_m^2(x)}{\beta},$$

and of σ and σ_t then

$$\langle \sigma^2(t, x) \rangle \approx \varepsilon (Eh)^2 \langle \xi_{02}^2(t, x) \rangle = \frac{\varepsilon a l (MEh)^2}{8A}\, \frac{f_m^2(x)}{\beta},$$

$$\langle \sigma_t^2(t, x) \rangle \approx \varepsilon (Eh)^2 \langle \xi_{12}^2(t, x) \rangle = \frac{\varepsilon a l (MEh)^2}{8A}\, \frac{\lambda_m f_m^2(x)}{\beta}.$$

From the stationarity it follows that the expected rates of crossings are independent of the time. We obtain

$$r(n, x) = \frac{1}{\pi} \sqrt{\lambda_m} \exp\left(-\frac{\lambda_m c_1}{f_m^2(x)}\, n^2\right) \quad \text{where} \quad c_1 \doteq \frac{4\beta}{\varepsilon a M^2 l},$$

$$r^b(n, x) = \frac{1}{\pi} \sqrt{\lambda_m} \exp\left(-\frac{c_2}{f_m^2(x)}\, n^2\right) \quad \text{where} \quad c_2 \doteq \frac{4A\beta}{\varepsilon a l (MEh)^2}.$$

These expected rates of crossings are illustrated for $m = 1, 2, 3$ and $x = 0.5$, 0.25 in Fig. 3.25 in the case of $A = 1$, $l = 1$, $\beta = 1$. Considering $r(\hat{n}, x)$ higher levels $\hat{n}$ are not exceeded if m increases. This property is not true for $r^b(\bar{n}, x)$ and this can be explained from mechanical considerations. For $m = 2$ and $x = 0.5$ we obtain

$$\xi_{00}(t, 0.5, \omega) = \xi_{02}(t, 0.5, \omega) = 0 \quad \text{a.s.}$$

since $f_2(0.5) = 0$; i.e. both the displacement and the bending stress are zero at this point for all times t. In the case of r the level $\hat{n}$ is given by

$$\hat{n} = \sqrt{\frac{4\beta}{\varepsilon a M^2 l}}\, n\,.$$

For example, putting

$$r\left(n_0 \sqrt{\frac{4\beta}{\varepsilon a M^2 l}}, x\right) = r_0$$

then r_0 gives the expected rate of crossings for a higher level if M, l, or εa (this quantity includes the variance of the force acting on the beam) increase or β decreases. These

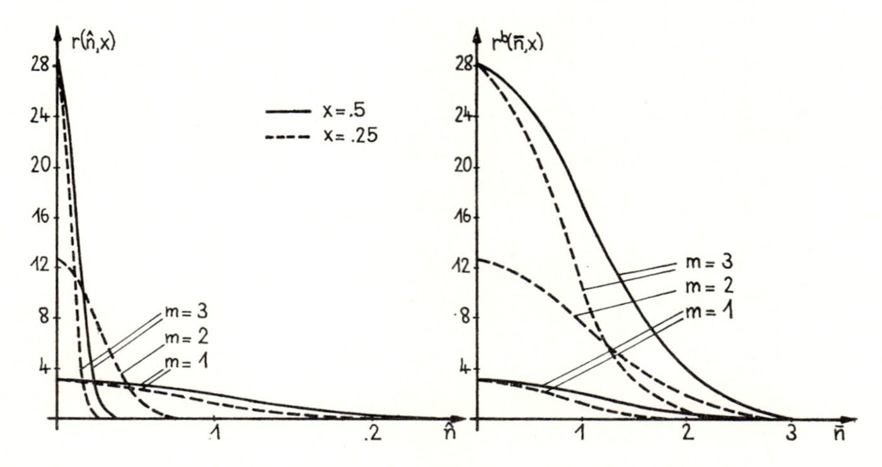

Fig. 3.25. Expected rate of crossings $r(\hat{n}, x)$, $r^b(\bar{n}, x)$ in dependence on the level $\hat{n}$, $\bar{n}$

properties are clear from the mechanical point of view. Similar considerations can be deduced for the expected rate of crossings r^b of bending stress. In this case we have put

$$\bar{n} = \sqrt{\frac{4A\beta}{\varepsilon a l (MEh)^2}}\, n\,.$$

The function

$$K_m(s) = \frac{e^{-\beta s}}{8\lambda_m}\left[\frac{1}{\beta}\cos(\beta_m s) + \frac{1}{\beta_m}\sin(\beta_m s)\right]$$

essentially determines the correlation functions

$$\langle \xi_{0q}(t, x)\, \xi_{0q}(t + s, y)\rangle$$

for $q = 0, 2$. In Fig. 3.26 we have drawn the function $K_m(s)$ for $m = 1, 2$ and $\beta = 1, 2, 3$. $K_m(s)$ is already for $m = 1$ a strong oscillating function. The function represented in Fig. 3.26 describes the correlation between $u(t, x)$ and $u(t + s, y)$ except the factor

$\varepsilon a M^2 l f_m(x) f_m(y)$ and the correlation between the bending stresses $\sigma(t, x)$ and $\sigma(t + s, y)$ except the factor

$$\varepsilon a l (MEh)^2 \left(\frac{m\pi}{l}\right)^4 f_m(x) f_m(y) .$$

Also in the case of

$$p_1(x) = \delta(x - x_0)$$

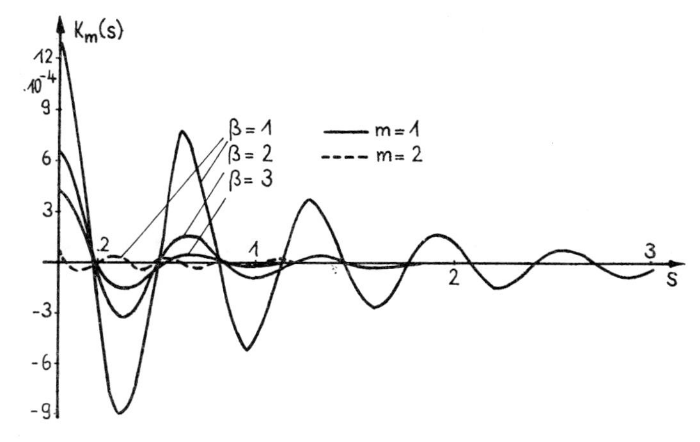

Fig. 3.26. Function $K_m(s)$ in dependence on s

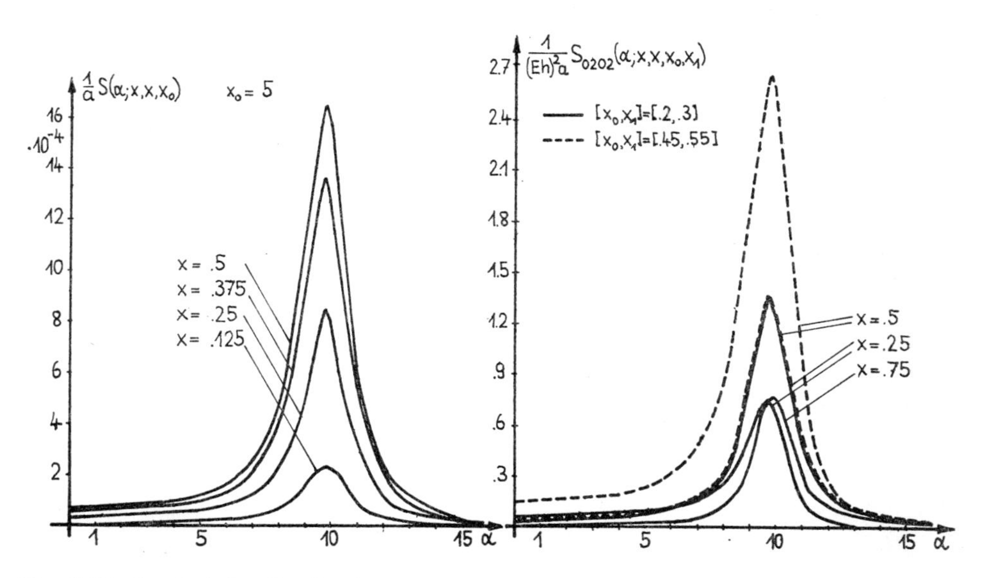

Fig. 3.27. Spectral density $S_{0000}(\alpha; x, x; x_0)$ in dependence on α for $x_0 = 0.5$ and different values of x

Fig. 3.28. Spectral density $S_{0202}(\alpha; x, x; x_0, x_1)$ for the bending stress generated by a force acting on $[x_0, x_1]$ in dependence on α

we obtain results as to the displacement $u(t, x, \omega)$ of the beam. Using this force which acts in x_0 the approximate correlation function of u can be given by

$$\langle \xi_{00}(t, x)\, \xi_{00}(r, y)\rangle = a \sum_{i,j=1}^{\infty} \tilde{T}_{ij}^{00}(r - t)\, f_i(x)\, f_j(y)\, f_i(x_0)\, f_j(x_0)$$

for $t \leqq r$ and the spectral density for $x = y$ by

$$S_{0000}(\alpha; x, x; x_0) = 2a \sum_{i,j=1}^{\infty} W_{ij}^{00}(\alpha)\, f_i(x)\, f_j(x)\, f_i(x_0)\, f_j(x_0).$$

In Fig. 3.27, putting $x_0 = 0.5$, the spectral density is plotted for different values of x.
Similar results can be calculated for a function $p_1(x)$ of the form

$$p_1(x) = \begin{cases} \left[\dfrac{4}{(x_1 - x_0)^2}(x - x_0)(x - x_1)\right]^4 & \text{for} \quad x \in [x_0, x_1], \\ 0 & \text{for} \quad x \in [0, 1] \setminus [x_0, x_1]. \end{cases}$$

This function also satisfies all conditions in order to calculate the characteristics of the bending stress. The coefficients

$$r_k \doteq (f_k, p_1), \qquad k = 1, 2, 3, \ldots,$$

are given by

$$r_k = \frac{256}{d(k\pi)^2}\left\{\left[-\frac{24}{(dk\pi)^3} + \frac{4320}{(dk\pi)^5} - \frac{40320}{(dk\pi)^7}\right][\cos(k\pi x_0) - \cos(k\pi x_1)]\right.$$
$$\left. + \left[\frac{480}{(dk\pi)^4} - \frac{20160}{(dk\pi)^6}\right][\sin(k\pi x_0) + \sin(k\pi x_1)]\right\}$$

where $d \doteq x_1 - x_0$. In Fig. 3.28 the spectral density for the bending stress is plotted. The bold lines are drawn in the case of $x_0 = 0.2$, $x_1 = 0.3$ and the hatched lines in the case of $x_0 = 0.45$, $x_1 = 0.55$. For example, the spectral density taken for $x = 0.5$ is essentially dependent on the place on which the force acts.

In a similar way, we can investigate a whole series of concrete vibration problems. These problems can be both one-dimensional and multi-dimensional with regard to x. By means of suitable force functions a good survey can be given without extensive numerical calculations.

3.1.7. Comparing results

As done in the sections concerning weakly correlated loads as to the time variable and spatial ones we will now deal with some comparing results. The non-weakly correlated function $p(t, x, \omega)$ is assumed to be separated:

$$p(t, x, \omega) \doteq p_0(t, \omega)\, p_1(x).$$

Let $p_0(t, \omega)$ have the correlation function

$$\langle \overline{p}_0(t)\, \overline{p}_0(r)\rangle = \sigma^2 \exp(-\gamma\, |r - t|) \tag{3.43}$$

where $\gamma > 0$. The Gaussian distribution of $p_0(t, \omega)$ is assumed for all those cases of this comparison in which the distribution of the solution $u(t, x, \omega)$ plays a role. Using the given conditions the solution of problem (3.8) has the form

$$\bar{u}(t, x, \omega) = \int_0^t \sum_{i=1}^{\infty} h_i(t-s)\, f_i(x)\, (f_i, p_1)\, \overline{p}_0(s, \omega)\, ds\,.$$

By means of this solution it is possible to determine the correlation function

$$\langle D_t^{p_1} D_x^{q_1} \bar{u}(t, x)\, D_r^{p_2} D_y^{q_2} \bar{u}(r, y) \rangle = \sum_{i,j=1}^{\infty} S_{ij}^{p_1p_2}(t, r)\, D_x^{q_1} f_i(x)\, D_y^{q_2} f_j(y)\, (f_i, p_1)\, (f_j, p_1) \quad (3.44)$$

where

$$S_{ij}^{p_1p_2}(t, r) \doteq \int_0^t \int_0^r D_t^{p_1} h_i(t-s)\, D_r^{p_2} h_j(r-\bar{s})\, \langle \overline{p}_0(s)\, \overline{p}_0(\bar{s}) \rangle\, ds\, d\bar{s}$$

for $p_1, p_2 = 0, 1$; $q_1, q_2 = 0, 1, \ldots$ Applying the functions $S_{ij}(t, r)$ of Section 3.1.3 it follows

$$S_{ij}^{00}(t, r) = S_{ij}(t, r)\,.$$

In the following considerations we use the notations and results of Section 3.1.3. Furthermore, we assume that the present series converge without making remarks.

The case

$$Lu = Au_{xxxx} \quad \text{on} \quad [0, l]$$

with

$$u(0, t) = u(l, t) = u_{xx}(0, t) = u_{xx}(l, t) = 0$$

is investigated and leads to

$$\lambda_i = A \left(\frac{i\pi}{l}\right)^4, \quad f_i(x) = \sqrt{\frac{2}{l}} \sin\left(\frac{i\pi x}{l}\right) \quad \text{for} \quad i = 1, 2, \ldots$$

It is easy to see that

$$S_{ij}^{p_1p_2}(t, r) = D_t^{p_1} D_r^{p_2} S_{ij}(t, r) \quad \text{for} \quad p_1, p_2 = 0, 1$$

and furthermore

$$\lim_{\substack{t,r\to\infty \\ r-t=s\geqq 0}} S_{ij}^{p_1p_2}(t, r) = \mathring{S}_{ij}^{p_1p_2}(s) = \begin{cases} \mathring{S}_{ij}(s) & \text{for} \quad p_1 = p_2 = 0, \\ \dfrac{d}{ds} \mathring{S}_{ij}(s) & \text{for} \quad p_1 = 0,\ p_2 = 1, \\ -\dfrac{d}{ds} \mathring{S}_{ij}(s) & \text{for} \quad p_1 = 1,\ p_2 = 0, \\ -\dfrac{d^2}{ds^2} \mathring{S}_{ij}(s) & \text{for} \quad p_1 = p_2 = 1. \end{cases}$$

The function $\mathring{S}_{ij}(s)$ is given by (3.24). The stationary part of the correlation function

(3.44) can be written as

$$\mathring{R}^{p_1p_2q_1q_2}(s, x, y) \doteq \lim_{\substack{t,r\to\infty \\ r-t=s\geqq 0}} \langle \mathrm{D}_t^{p_1}\mathrm{D}_x^{q_1}\bar{u}(t, x)\, \mathrm{D}_r^{p_2}\mathrm{D}_y^{q_2}\bar{u}(r, y)\rangle$$

$$= \sigma^2 \sum_{i,j=1}^{\infty} \mathring{S}_{ij}^{p_1p_2}(s)\, \mathrm{D}_x^{q_1}f_i(x)\, \mathrm{D}_y^{q_2}f_j(y)\, (f_i, p_1)\, (f_j, p_1)$$

for $s \geqq 0$. In the case of $s < 0$, $\mathring{R}^{p_1p_2q_1q_2}(s, x, y)$ follows from

$$\mathring{S}_{ij}^{p_1p_2}(s) = \mathring{S}_{ji}^{p_2p_1}(-s).$$

In order to calculate the derivatives of $\mathring{S}_{ij}(s)$ we deduce from (3.24)

$$\frac{\mathrm{d}^p}{\mathrm{d}s^p}\, \mathring{S}_{ij}(s) = \frac{1}{\beta_i\beta_j}\left[\frac{\mathrm{d}^p}{\mathrm{d}s^p}\, \mathring{I}_{ij}(s;\gamma) - \frac{\mathrm{d}^p}{\mathrm{d}s^p}\, \mathring{I}_{ij}(s;-\gamma)\right] + (-\gamma)^p\, \frac{\mathrm{e}^{-\gamma s}}{c_{ij}(\gamma)}$$

for $p = 0, 1, 2$ and furthermore

$$\frac{\mathrm{d}^p}{\mathrm{d}s^p}\, \mathring{I}_{ij}(s;\gamma) = \frac{1}{(\beta+\gamma)^2 + \beta_j^2}\left[\frac{1}{2}\,(\beta+\gamma)\, \frac{\mathrm{d}^p}{\mathrm{d}s^p}\, \{\mathring{a}(s;\beta_i,-\beta_j) - \mathring{a}(s;\beta_i,\beta_j)\}\right.$$

$$\left. + \frac{1}{2}\,\beta_j\, \frac{\mathrm{d}^p}{\mathrm{d}s^p}\, \{\mathring{b}(s;\beta_i,-\beta_j) - \mathring{b}(s;\beta_i,\beta_j)\}\right].$$

The derivatives of $\mathring{a}$ and $\mathring{b}$ are given by

$$\frac{\mathrm{d}}{\mathrm{d}s}\, \mathring{a}(s;\varepsilon,\delta) = \frac{\mathrm{e}^{-\beta s}}{4\beta^2 + (\varepsilon+\delta)^2}\, \left[-\left(2\beta^2 + \delta(\varepsilon+\delta)\right)\cos(\delta s) + \beta(\varepsilon-\delta)\sin(\delta s)\right],$$

$$\frac{\mathrm{d}^2}{\mathrm{d}s^2}\, \mathring{a}(s;\varepsilon,\delta) = \frac{\mathrm{e}^{-\beta s}}{4\beta^2 + (\varepsilon+\delta)^2}\, \left[2\beta(\beta^2 + \delta\varepsilon)\cos(\delta s)\right.$$

$$\left. + \left(\beta^2(3\delta-\varepsilon) + \delta^2(\varepsilon+\delta)\right)\sin(\delta s)\right],$$

$$\frac{\mathrm{d}}{\mathrm{d}s}\, \mathring{b}(s;\varepsilon,\delta) = \frac{\mathrm{e}^{-\beta s}}{4\beta^2 + (\varepsilon+\delta)^2}\, \left[\beta(\delta-\varepsilon)\cos(\delta s) - \left(2\beta^2 + \delta(\varepsilon+\delta)\right)\sin(\delta s)\right],$$

$$\frac{\mathrm{d}^2}{\mathrm{d}s^2}\, \mathring{b}(s;\varepsilon,\delta) = \frac{\mathrm{e}^{-\beta s}}{4\beta^2 + (\varepsilon+\delta)^2}\, \left[-\left(\beta^2(3\delta+\varepsilon) + \delta^2(\varepsilon+\delta)\right)\cos(\delta s)\right.$$

$$\left. + 2\beta(\beta^2 + \delta\varepsilon)\sin(\delta s)\right].$$

From the theory of weakly correlated functions we obtain for the stationary part of the correlation function

$$\langle \mathrm{D}_t^{p_1}\mathrm{D}_x^{q_1}\bar{u}(t, x)\, \mathrm{D}_r^{p_2}\mathrm{D}_y^{q_2}\bar{u}(r, y)\rangle \approx \varepsilon\langle \zeta_{p_1q_1}(t, x)\, \zeta_{p_2q_2}(r, y)\rangle$$

$$= \varepsilon a \sum_{i,j=1}^{\infty} \tilde{T}_{ij}^{p_1p_2}(r-t)\, \mathrm{D}_x^{q_1}f_i(x)\, \mathrm{D}_y^{q_2}f_j(y)\, (f_i, p_1)\, (f_j, p_1)$$

$$= \varepsilon\tilde{R}^{p_1p_2q_1q_2}(r-t, x, y)$$

if the process $\overline{p}_{0\varepsilon}(t, \omega)$ is assumed to be wide-sense stationary. The quantities $\tilde{T}_{ij}^{p_1p_2}$ are given by (3.42). The intensity of a weakly correlated process having a correlation function corresponding to (3.43) can be determined by

$$\frac{1}{\varepsilon}\int\limits_{-\varepsilon}^{\varepsilon} \mathrm{e}^{-\gamma|s|}\,\mathrm{d}s \approx \frac{1}{\varepsilon}\int\limits_{-\infty}^{\infty} \mathrm{e}^{-\gamma|s|}\,\mathrm{d}s = \frac{2}{\varepsilon\gamma}.$$

We obtain

$$\varepsilon a \approx \frac{2\sigma^2}{\gamma}.$$

Taking $\gamma \to \infty$ from the given relations it follows

$$\lim_{\gamma\to\infty} \gamma \mathring{S}_{ij}^{p_1p_2}(s) = 2\tilde{T}_{ij}^{p_1p_2}(s) \quad \text{for} \quad p_1, p_2 = 0, 1 \quad \text{and} \quad i, j = 1, 2, \dots$$

and then

$$\mathring{R}^{p_1p_2q_1q_2}(s, x, y) \approx \varepsilon\tilde{R}^{p_1p_2q_1q_2}(s, x, y)$$

for large values γ.

The accuracy of these relations shall be investigated by some numerical calculations. Therefore, we turn to the case of

$$p_1(x) = M \sin\left(\frac{m\pi x}{l}\right).$$

Thus

$$\mathring{R}_m^{p_1p_2q_1q_2}(s, x, y) = \frac{M^2\sigma^2 l}{2}\,\mathring{S}_{mm}^{p_1p_2}(s)\,\mathrm{D}_x^{q_1}f_m(x)\,\mathrm{D}_y^{q_2}f_m(y)$$

and

$$\tilde{R}_m^{p_1p_2q_1q_2}(s, x, y) = \frac{M^2 a l}{2}\,\tilde{T}_{mm}^{p_1p_2}(s)\,\mathrm{D}_x^{q_1}f_m(x)\,\mathrm{D}_y^{q_2}f_m(y).$$

In Fig. 3.29 the functions

$$\mathring{K}_m^{p_1p_2}(s, \gamma) = \frac{\gamma}{2}\,\frac{2}{M^2\sigma^2 l}\,\frac{1}{\mathrm{D}_x^{q_1}f_m(x)\,\mathrm{D}_y^{q_2}f_m(y)}\,\mathring{R}_m^{p_1p_2q_1q_2}(s, x, y),$$

$$\tilde{K}_m^{p_1p_2}(s) = \frac{2}{M^2 a l}\,\frac{1}{\mathrm{D}_x^{q_1}f_m(x)\,\mathrm{D}_y^{q_2}f_m(y)}\,\tilde{R}_m^{p_1p_2q_1q_2}(s, x, y)$$

are illustrated; i.e. we have plotted the functions

$$\frac{\gamma}{2}\,\mathring{S}_{mm}^{p_1p_2}(s) \quad \text{and} \quad \tilde{T}_{mm}^{p_1p_2}(s).$$

In this figure and in the following ones it was put $A = 1$, $l = 1$, and also $\beta = 1$, in general. The bold lines show the weakly correlated case and the hatched lines belong to $\mathring{K}_m^{p_1p_2}(s)$.

Fig. 3.30 shows $\tilde{K}_m^{pp}(0)$, $\mathring{K}_m^{pp}(0, \gamma)$ and the values of γ can be read off for which the agreement of $\mathring{K}$ and $\tilde{K}$ is sufficient. We have used the relations

$$\mathring{K}_m^{00}(0, \gamma) = \frac{\gamma}{2} \mathring{S}_{mm}^{00}(0) = \frac{\gamma}{2} \frac{1}{(\lambda_m + \gamma^2)^2 - 4\beta^2\gamma^2} \left[1 + \left(\frac{1}{2\beta} - \frac{2\beta}{\lambda_m}\right) + \frac{\gamma^3}{2\beta\lambda_m}\right],$$

$$\mathring{K}_m^{11}(0, \gamma) = \frac{\gamma}{2} \mathring{S}_{mm}^{11}(0) = \frac{\gamma^2}{2} \frac{1}{(\lambda_m + \gamma^2)^2 - 4\beta^2\gamma^2} \left[\frac{1}{2\beta} (\lambda_m + \gamma^2) - \gamma\right].$$

Because of the wide-sense stationarity of $\bar{u}$ the expected rate of crossings is independent of the time t. Hence, assuming a Gaussian process $p_0(t, \omega)$ with $\langle p_0(t) \rangle = 0$ and the corre-

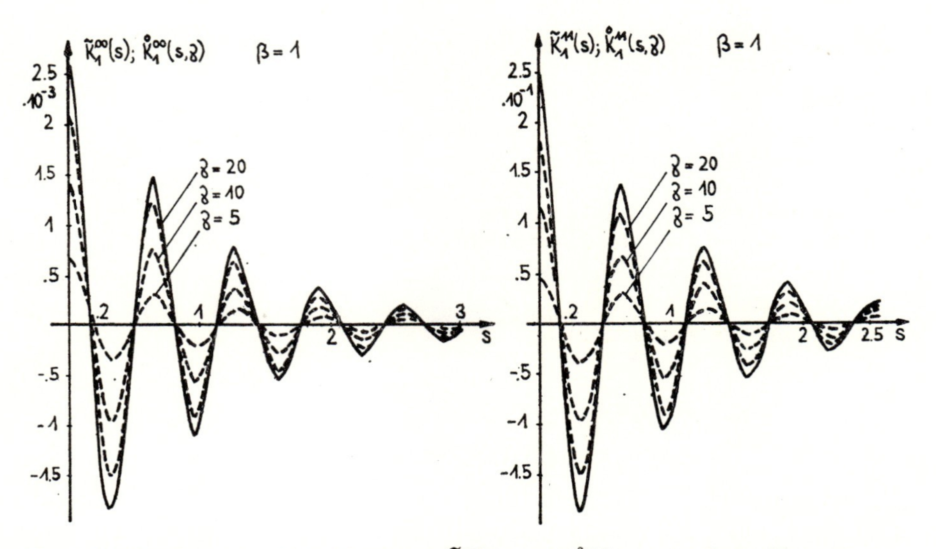

Fig. 3.29. Comparison of the functions $\tilde{K}_1^{pp}(s)$ and $\mathring{K}_1^{pp}(s, \gamma)$ in dependence on s for $p = 0, 1$ and different values γ

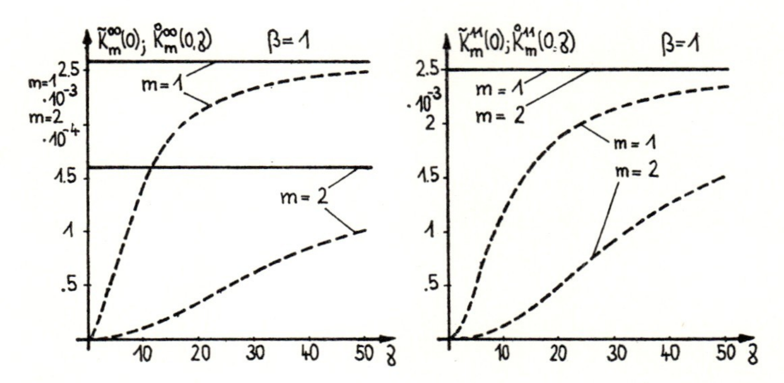

Fig. 3.30. Comparison of the functions $\tilde{K}_m^{pp}(0)$ and $\mathring{K}_m^{pp}(0, \gamma)$ in dependence on γ for $p = 0, 1$; $m = 1, 2$

lation function given by (3.43), the expected rate of crossings of u can be obtained by

$$\mathring{r}_{u,m}(n, x) = \frac{1}{\pi} \sqrt{\frac{\mathring{S}^{11}_{mm}(0)}{\mathring{S}_{mm}(0)}} \exp\left(-\frac{n^2}{M^2\sigma^2 l \mathring{S}_{mm}(0) f_m^2(x)}\right)$$

(with $p_1(x) = M \sin(m\pi x/l)$). The expected rate of crossings of the bending stress $\sigma(t, x) = Ehu_{xx}(t, x)$ has the form

$$\mathring{r}_{\sigma,m}(n, x) = \frac{1}{\pi} \sqrt{\frac{\mathring{S}^{11}_{mm}(0)}{\mathring{S}_{mm}(0)}} \exp\left(-\frac{n^2}{M^2\sigma^2 l \mathring{S}_{mm}(0) \left(Ehf''_m(x)\right)^2}\right).$$

The terms corresponding to r for the weakly correlated theory follow from $\mathring{r}_{u,m}(n, x)$ and $\mathring{r}_{\sigma,m}(n, x)$ if the quantities $\mathring{S}^{pp}_{mm}(0)$ are substituted by $\frac{2}{\gamma} \tilde{T}^{pp}_{mm}(0)$. Fig. 3.31 shows the comparison between $\mathring{r}_{u,m}(n, x)$ and $\tilde{r}_{u.m}(n, x)$ if $\bar{n}$ is defined by

$$\bar{n} = \frac{n}{M\sigma \sqrt{l}\, |f_m(x)|}$$

and the comparison between $\mathring{r}_{\sigma,m}(n, x)$ and $\tilde{r}_{\sigma,m}(n, x)$ if

$$\bar{n} = \frac{n}{M\sigma \sqrt{l}\, Eh\, |f''_m(x)|}$$

is put. All results in Fig. 3.31 are given in the case of $m = 1$. The values of $\tilde{r}_m(\bar{n})$ are drawn by a bold line and those of $\mathring{r}_m(\bar{n})$ by a hatched line. We observe that

$$\mathring{r}_1(\bar{n}) \leqq \tilde{r}_1(\bar{n});$$

i.e. the results from the weakly correlated theory are upper bounds compared to the Gaussian input function with the correlation function given by (3.43). Similar results

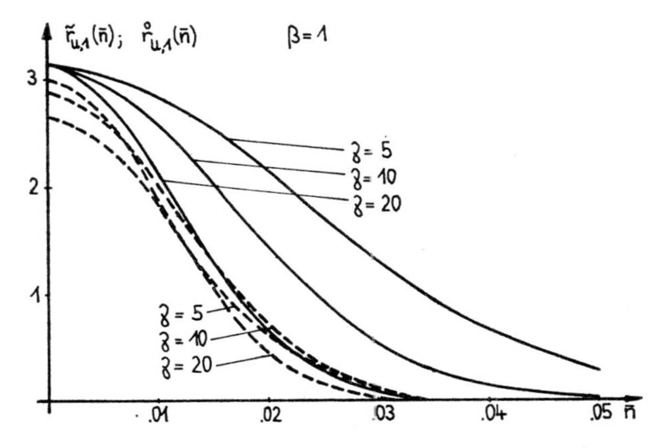

Fig. 3.31. Comparison of expected rates of crossings of u and σ as function of the level $\bar{n}$ for $m = 1$

were obtained in Section 3.1.3. The maximum of the difference

$$|\tilde{r}_1(\bar{n}) - \mathring{r}_1(\bar{n})|$$

decreases if γ increases (see Table 3.3). The moments needed in the above calculations are summarized in Table 3.4. A fixed value $\bar{n}$ corresponds to a level

$$n = \bar{n} M \sigma \sqrt{l}\, |f_m(x)|$$

in the case of displacements, and to a level

$$n = \bar{n} M \sigma \sqrt{l}\, Eh \left(\frac{m\pi}{l}\right)^2 |f_m(x)|$$

in the case of bending stresses. Fig. 3.32 shows the expected rates of crossings $\mathring{r}_1$ and $\tilde{r}_1$ in dependence on γ for different values of $\bar{n}$. In this figure the convergence of $\mathring{r}_1$ to $\tilde{r}_1$

Table 3.3

	$\gamma = 5$	$\gamma = 10$	$\gamma = 20$
$\sup_{\bar{n}} |\tilde{r}_1(\bar{n}) - \mathring{r}_1(\bar{n})|$	1.52	0.76	0.29

Table 3.4

	$\gamma = 5$	$\gamma = 10$	$\gamma = 20$
$\mathring{S}_{11}(0)$	2.71 E-4	2.83 E-4	2.10 E-4
$\mathring{S}_{11}^{11}(0)$	1.888 E-2	2.300 E-2	1.861 E-2
$\frac{2}{\gamma}\tilde{T}_{11}^{00}(0)$	1.028 E-3	0.514 E-3	0.257 E-3
$\frac{2}{\gamma}\tilde{T}_{11}^{11}(0)$	1 E-1	0.5 E-1	0.25 E-2

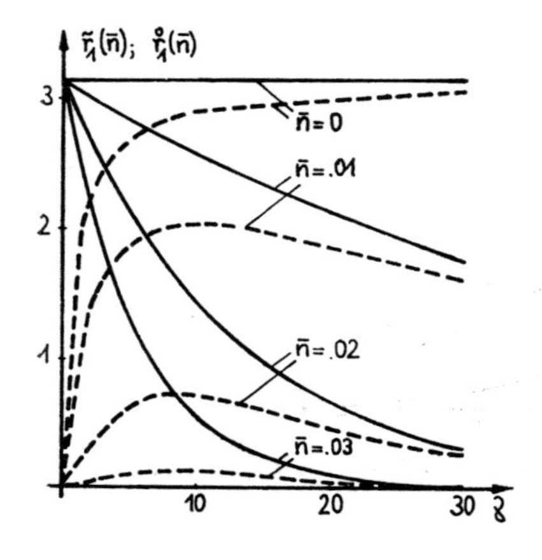

Fig. 3.32. Comparison of expected rates of crossings in dependence on γ for different values of $\bar{n}$

as $\gamma \to \infty$ can be observed very well. As before-mentioned it is

$$\mathring{r}_1(\bar{n}) \leqq \tilde{r}_1(\bar{n});$$

i.e. the expected rates of crossings from the weakly correlated theory are upper bounds of the $\mathring{r}_1$-values. The lines according to $\mathring{r}_1$ were hatchedly drawn and the lines as to $\tilde{r}_1$ boldly.

As a force function which differs from $p_1(x) = M \sin (m\pi x/l)$ we consider

$$p_1(x) = \delta(x - x_0).$$

In this way, we obtain a force function which acts at x_0 and is random in time. By means of the weakly correlated theory the correlation function as to $\bar{\xi}_{00}(t, x, \omega)$ can be determined as

$$\tilde{R}(s, x, y) = \langle \bar{\xi}_{00}(t, x)\, \bar{\xi}_{00}(t + s, y)\rangle = a \sum_{i,j=1}^{\infty} \tilde{T}_{ij}^{00}(s)\, f_i(x)\, f_j(y)\, f_i(x_0)\, f_j(x_0).$$

Using (3.43) as correlation function of $p_0(t, \omega)$ then we obtain the stationary part

$$\mathring{R}(s, x, y) = \lim_{\substack{t,r\to\infty \\ r-t=s\geqq 0}} \langle \bar{u}(t, x)\, \bar{u}(r, y)\rangle = \sigma^2 \sum_{i,j=1}^{\infty} \mathring{S}_{ij}(s)\, f_i(x)\, f_j(y)\, f_i(x_0)\, f_j(x_0).$$

In the following considerations we turn to the spectral densities. From the weakly correlated theory the stationary part of the correlation function leads to the spectral density

$$\tilde{S}_{0q0q}(\alpha; x, x) = 2a\varepsilon \sum_{i,j=1}^{\infty} \tilde{W}_{ij}(\alpha)\, \mathrm{D}_x^q f_i(x)\, \mathrm{D}_x^q f_j(x)\, (f_i, p_1)\, (f_j, p_1)$$

with the function

$$\tilde{W}_{ij}(\alpha) = \frac{8\beta^2\lambda_j + (\lambda_i - \lambda_j)(\lambda_j - \alpha^2)}{2\pi A_{ij}(\alpha, \beta)}$$

given in Section 3.1.5 where

$$A_{ij}(\alpha, \beta) = [8\beta^2(\lambda_i + \lambda_j) + (\lambda_i - \lambda_j)^2]\,[(\lambda_j + \alpha^2)^2 - 4\beta_j^2\alpha^2].$$

By means of the correlation function (3.43) as to the input function the spectral density of $\mathrm{D}_x^q \bar{u}(t, x)$ can be calculated as

$$\mathring{S}_{0q0q}(\alpha; x, x) = \sigma^2 \sum_{i,j=1}^{\infty} U_{ij}(\alpha)\, \mathrm{D}_x^q f_i(x)\, \mathrm{D}_x^q f_j(x)\, (f_i, p_1)\, (f_j, p_1)$$

where $U_{ij}(\alpha)$ is given by

$$U_{ij}(\alpha) = \frac{1}{2\pi} \int_{-\infty}^{\infty} \cos (\alpha s)\, \mathring{S}_{ij}(s)\, \mathrm{d}s = \frac{1}{2\pi} \int_{0}^{\infty} \cos (\alpha s)\, [\mathring{S}_{ij}(s) + \mathring{S}_{ji}(s)]\, \mathrm{d}s$$

using

$$\mathring{S}_{ij}(s) = \mathring{S}_{ji}(-s).$$

Hence we have

$$\mathring{S}_{0q0q}(\alpha; x, x) = 2\sigma^2 \sum_{i,j=1}^{\infty} \mathring{W}_{ij}(\alpha)\, \mathrm{D}_x^q f_i(x)\, \mathrm{D}_x^q f_j(x)\, (f_i, p_1)\, (f_j, p_1)$$

and $\mathring{W}_{ij}(\alpha)$ is defined by

$$\mathring{W}_{ij}(\alpha) \doteq \frac{1}{2\pi} \int_0^{\infty} \cos(\alpha s)\, \mathring{S}_{ij}(s)\, \mathrm{d}s .$$

It follows

$$\mathring{W}_{ij}(\alpha) = \frac{1}{2\beta_i\beta_j} \big(L_{ij}(\alpha; \gamma) - L_{ij}(\alpha; -\gamma)\big) + \frac{1}{2\pi c_{ij}(\gamma)} \frac{\gamma}{\gamma^2 + \alpha^2}$$

where

$$\mathring{L}_{ij}(\alpha; \gamma) = \frac{1}{\pi} \int_0^{\infty} \cos(\alpha s)\, \mathring{I}_{ij}(s; \gamma)\, \mathrm{d}s$$

has been calculated in Section 3.1.3. From

$$\lim_{\gamma\to\infty} \gamma \mathring{W}_{ij}(\alpha) = \frac{1}{2\beta_i\beta_j} \{\mathring{A}(\alpha; \beta_i, -\beta_j) - \mathring{A}(\alpha; \beta_i, \beta_j)\}$$

$$= \frac{8\beta^2\lambda_j + (\lambda_i - \lambda_j)(\lambda_j - \alpha^2)}{\pi A_{ij}(\alpha, \beta)} = 2\widetilde{W}_{ij}(\alpha)$$

it can be deduced

$$\lim_{\gamma\to\infty} \gamma \mathring{S}_{0q0q}(\alpha; x, x) = \frac{2\sigma^2}{a\varepsilon} \widetilde{S}_{0q0q}(\alpha; x, x)$$

and by means of $a\varepsilon \approx \frac{2}{\gamma}\sigma^2$ then for large γ

$$\mathring{S}_{0q0q}(\alpha; x, x) \approx \widetilde{S}_{0q0q}(\alpha; x, x) .$$

Now, the accuracy of this relation shall be treated by some numerical calculations. If

$$p_1(x) = M \sin\left(\frac{m\pi x}{l}\right)$$

is put then we obtain

$$\widetilde{S}_{0q0q}(\alpha; x, x) = a\varepsilon M^2 l \widetilde{W}_{mm}(\alpha)\, \big(\mathrm{D}_x^q f_m(x)\big)^2$$

$$= \frac{2\sigma^2}{\gamma} M^2 l \widetilde{W}_{mm}(\alpha)\, \big(\mathrm{D}_x^q f_m(x)\big)^2$$

and

$$\mathring{S}_{0q0q}(\alpha; x, x) = \sigma^2 M^2 l \mathring{W}_{mm}(\alpha)\, \big(\mathrm{D}_x^q f_m(x)\big)^2 .$$

In Fig. 3.33 the functions

$$\tilde{S}_m(\alpha) \doteq \frac{\gamma}{2} \frac{1}{\sigma^2 M^2 l (\mathrm{D}_x^q f_m(x))^2} \tilde{S}_{0q0q}(\alpha; x, x) = \widetilde{W}_{mm}(\alpha),$$

$$\mathring{S}_m(\alpha; \gamma) \doteq \frac{\gamma}{2} \frac{1}{\sigma^2 M^2 l (\mathrm{D}_x^q f_m(x))^2} \mathring{S}_{0q0q}(\alpha; x, x) = \frac{\gamma}{2} \mathring{W}_{mm}(\alpha)$$

are plotted. The lines as to $\tilde{S}_1(\alpha)$ are drawn by a bold line and the lines with respect to $\mathring{S}_1(\alpha; \gamma)$ by a hatched line. In the first part of Fig. 3.33 it was taken $\beta = 1$ and in the second part $\beta = 5$. The influence of β on the spectral density is clear from this figure. We have again the inequality

$$\mathring{S}_{0q0q}(\alpha; x, x) \leqq \tilde{S}_{0q0q}(\alpha; x, x).$$

Assuming

$$p_1(x) = \delta(x - x_0)$$

it follows for $q = 0$

$$\tilde{S}_{0000}(\alpha; x, x) = 2a\varepsilon \sum_{i,j=1}^{\infty} \widetilde{W}_{ij}(\alpha) f_i(x) f_j(x) f_i(x_0) f_j(x_0),$$

$$\mathring{S}_{0000}(\alpha; x, x) = 2\sigma^2 \sum_{i,j=1}^{\infty} \mathring{W}_{ij}(\alpha) f_i(x) f_j(x) f_i(x_0) f_j(x_0)$$

Now we still investigate the functions

$$\tilde{S}(\alpha; x, x_0) \doteq \frac{1}{a} \tilde{S}_{0000}(\alpha; x, x),$$

$$\mathring{S}(\alpha; x; x_0) \doteq \frac{\gamma}{2\sigma^2} \mathring{S}_{0000}(\alpha; x, x).$$

Fig. 3.34 illustrates the functions $\tilde{S}$ and $\mathring{S}$. The same behaviour as in the case of

$$p_1(x) = M \sin\left(\frac{m\pi x}{l}\right)$$

can be established. Fig. 3.34 shows these functions of α in dependence on γ in the first part for $\beta = 1$, $x_0 = 0.5$, $x = 0.5$ and in a second part for $\beta = 1$, $x_0 = 0.5$, $x = 0.25$. It is clear that $\tilde{S}$ and $\mathring{S}$ are symmetric as to x and x_0. In order to compare the spectral densities with different β-values we additionally draw the spectral densities for $\beta = 5$ in a third part of Fig. 3.34. The maximum of the spectral densities is determined by the maximum of $\widetilde{W}_{11}(\alpha)$. This maximum is taken for

$$\alpha_M = \sqrt{\lambda_1 - 2\beta^2},$$

i.e. for

$$\alpha_M = 9.7678 \quad \text{in the case of } \beta = 1,$$

$$\alpha_M = 6.8854 \quad \text{in the case of } \beta = 5.$$

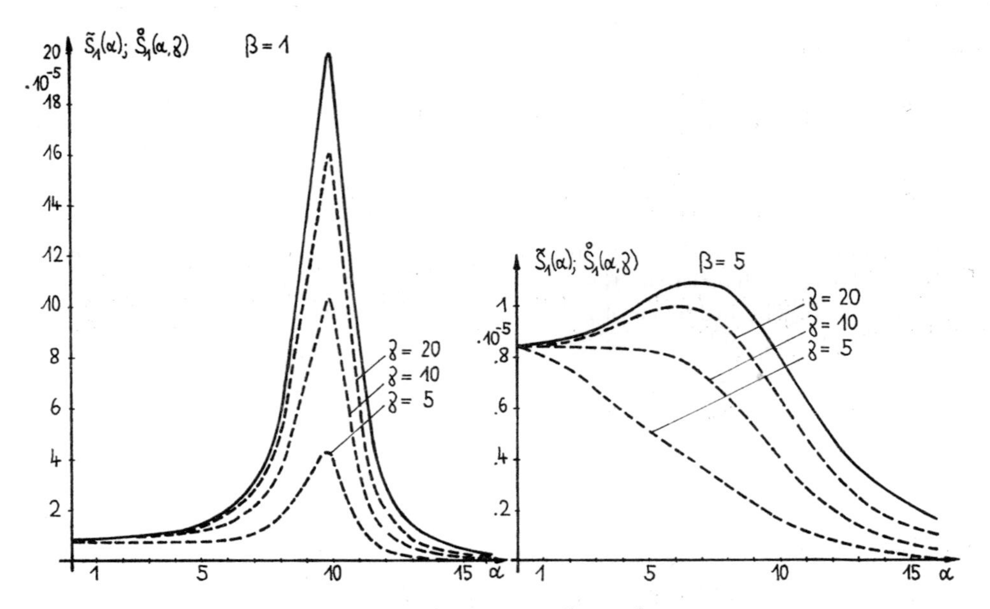

Fig. 3.33. Comparison between the spectral densities $\tilde{S}_m(\alpha)$, $\mathring{S}_m(\alpha; \gamma)$ in dependence on α for $m = 1$

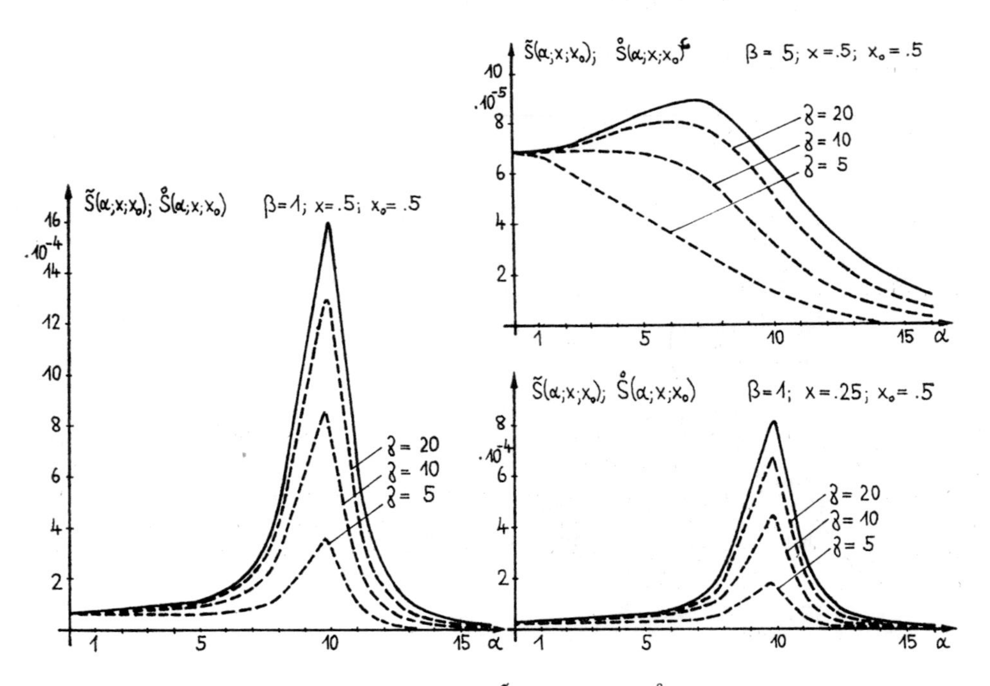

Fig. 3.34. Comparison of the spectral densities $\tilde{S}(\alpha; x, x_0)$ and $\mathring{S}(\alpha; x, x_0)$ as functions of α for different values of γ, $x_0 = 0.5$ and $x = 0.5; 0.25$

Thus

$$\widetilde{W}_{11}(\alpha_M) = \frac{1}{16\pi\beta^2\beta_1^2} = \begin{cases} 2.06\text{ E-4} & \text{for} \quad \beta = 1, \\ 0.11\text{ E-4} & \text{for} \quad \beta = 5. \end{cases}$$

In the case of smaller β-values we establish a sharp maximum whereas this property is not observed for larger β-values.

3.2. Random vibrations of discrete mechanical systems

3.2.1. Linear vibration systems with non-weakly correlated input

A linear vibration system of n masses excited by random loads can be described by the linear differential equation system of second order

$$Ax'' + Bx' + Cx = g(t, \omega); \qquad x(t_0) = x_0, \qquad x'(t_0) = x_1 \tag{3.45}$$

where A, B, and C denote real quadratic $n \times n$-matrices and A, C are assumed to be regular, i.e. A^{-1}, C^{-1} exist. The random excitation is denoted by

$$g(t, \omega) = \big(g_1(t, \omega), \ldots, g_r(t, \omega), 0, \ldots, 0\big)^\mathsf{T}, \qquad 0 < r \leqq n,$$

and the random solution vector process by

$$x(t, \omega) = \big(x_i(t, \omega)\big)^\mathsf{T}_{1\leqq i\leqq n}.$$

The $x_i(t, \omega)$ describe the deviation of the masses. Setting $y = x'$ from (3.45) the initial value problem of first order

$$Mz' + Nz = g(t, \omega), \qquad z(t_0) = z_0 \tag{3.46}$$

can be obtained which is equivalent to (3.45). We have defined the $2n \times 2n$-matrices

$$M = \begin{pmatrix} A & 0 \\ 0 & I \end{pmatrix}, \quad N = \begin{pmatrix} B & C \\ -I & 0 \end{pmatrix} \quad \text{with} \quad I = (\delta_{ij})_{1\leqq i,j\leqq n},$$

the random excitation

$$g(t, \omega) = \big(g_1(t, \omega), \ldots, g_r(t, \omega), 0, \ldots, 0\big)^\mathsf{T},$$

the random solution vector process

$$z(t, \omega) = \big(y_1(t, \omega), \ldots, y_n(t, \omega), x_1(t, \omega), \ldots, x_n(t, \omega)\big)^\mathsf{T},$$

and the initial vector

$$z_0 = ({}^2x_1, \ldots, {}^2x_n, {}^1x_1, \ldots, {}^1x_n)^\mathsf{T}.$$

For this investigations the random vector process $g(t, \omega)$ is assumed to be wide-sense stationary, i.e.

$$\langle g(t)\rangle = \text{const},$$

$$\big\langle \big(g(t_1) - \langle g\rangle\big) \big(g(t_2) - \langle g\rangle\big)^\mathsf{T}\big\rangle = C_{gg}(t_2 - t_1).$$

Then we have the spectral density matrix $S_{gg}(\alpha)$ which is defined by the Fourier transform of the correlation function,

$$S_{gg}(\alpha) \doteq \frac{1}{2\pi} \int_{-\infty}^{\infty} e^{-i\alpha t} C_{gg}(t)\, dt ,$$

$$C_{gg}(t) = \int_{-\infty}^{\infty} e^{i\alpha t} S_{gg}(\alpha)\, d\alpha .$$

The solution of the initial value problem can be written as

$$z(t, \omega) = \exp\big(R(t - t_0)\big)\, z_0 + \int_{t_0}^{t} \exp\big(R(t - s)\big)\, h(s, \omega)\, ds$$

where R and $h(s, \omega)$ are given by

$$R = -M^{-1}N , \qquad h(s, \omega) = M^{-1}g(s, \omega) .$$

In the following considerations we only consider vibration systems in which the matrix R possesses only eigenvalues with negative real parts. From this assumption it follows that the solution of the homogeneous differential equation system $\exp\big(R(t - t_0)\big)\, z_0$ is approximately zero after a certain transient time (see e.g. BUNKE [2]). Since the difference of the solutions

$$\int_{t_0}^{t} \exp\big(R(t - s)\big)\, h(s, \omega)\, ds \quad \text{and} \quad \int_{-\infty}^{t} \exp\big(R(t - s)\big)\, h(s, \omega)\, ds$$

of (3.46) is a solution of the homogeneous system belonging to (3.46) we can consider

$$z(t, \omega) = \int_{-\infty}^{t} \exp\big(R(t - s)\big)\, h(s, \omega)\, ds = \int_{0}^{\infty} \exp(Rs)\, h(t - s, \omega)\, ds \tag{3.47}$$

as a solution of the initial value problem (3.46) after a certain transient time. This solution vector process (3.47) is wide-sense stationary and with $g(t, \omega)$ wide-sense stationary connected since we obtain immediately from (3.47)

$$\langle z(t) \rangle = \int_{0}^{\infty} \exp\big(R(t - s)\big)\, ds \langle h \rangle = \text{const} ,$$

$$K_{zz}(t_1, t_2) = \Big\langle \big(z(t_1) - \langle z \rangle\big) \big(z(t_2) - \langle z \rangle\big)^{\mathsf{T}} \Big\rangle$$

$$= \int_{0}^{\infty} \int_{0}^{\infty} \exp(Rs_1)\, M^{-1} C_{gg}(t_2 - s_2 - t_1 + s_1)\, M^{-\mathsf{T}} \exp(R^{\mathsf{T}} s_2)\, ds_1\, ds_2$$

$$\doteq C_{zz}(t_2 - t_1) ,$$

$$K_{zg}(t_1, t_2) = \Big\langle \big(z(t_1) - \langle z \rangle\big) \big(g(t_2) - \langle g \rangle\big)^{\mathsf{T}} \Big\rangle$$

$$= \int_{0}^{\infty} \exp(Rs)\, M^{-1} C_{gg}(t_2 - t_1 + s)\, ds \doteq C_{zg}(t_2 - t_1) .$$

Now, the relations between the spectral density matrices of the input process and the output ones are deduced. The centred process

$$w(t, \omega) = z(t, \omega) - \langle z \rangle$$

is the solution of

$$Mw' + Nw = g(t, \omega) - \langle g \rangle, \qquad w(t_0) = 0, \tag{3.48}$$

and is wide-sense stationary and wide-sense stationary connected as to g. Furthermore, we have

$$C_{ww}(t) = C_{zz}(t), \qquad C_{gw}(t) = C_{gz}(t)$$

and hence

$$S_{ww}(\alpha) = S_{zz}(\alpha), \qquad S_{gw}(\alpha) = S_{gz}(\alpha).$$

Using (3.48) it is

$$\begin{aligned} C_{gg}(t_2 - t_1) &= \langle [Mw'(t_1) + Nw(t_1)] [Mw'(t_2) + Nw(t_2)]^\mathsf{T} \rangle \\ &= -MC''_{ww}(t_2 - t_1) M^\mathsf{T} - MC'_{ww}(t_2 - t_1) N^\mathsf{T} + NC'_{ww}(t_2 - t_1) M^\mathsf{T} \\ &\quad + NC_{ww}(t_2 - t_1) N^\mathsf{T} \end{aligned}$$

and then

$$S_{ww}(\alpha) = H(-\mathrm{i}\alpha) S_{gg}(\alpha) H(\mathrm{i}\alpha)^\mathsf{T} \tag{3.49}$$

where the transfer matrix $H(\mathrm{i}\alpha)$ is defined by

$$H(\mathrm{i}\alpha) = (\mathrm{i}\alpha M + N)^{-1}. \tag{3.50}$$

By similar considerations we obtain from

$$\begin{aligned} C_{gg}(t_2 - t_1) &= \langle [g(t_1) - \langle g \rangle] [Mw'(t_2) + Nw(t_2)]^\mathsf{T} \rangle \\ &= C'_{gw}(t_2 - t_1) M^\mathsf{T} + C_{gw}(t_2 - t_1) N^\mathsf{T} \end{aligned}$$

the relation

$$S_{gw}(\alpha) = S_{gg}(\alpha) H(\mathrm{i}\alpha)^\mathsf{T}. \tag{3.51}$$

The transfer matrix $H(\mathrm{i}\alpha)$ exists for all real α, i.e.

$$\det(\mathrm{i}\alpha M + N) \neq 0$$

for all real α. This property follows from

$$\det(\lambda M + N) = \det(M) \det(\lambda I - R)$$

and $\mathrm{Re}(\lambda) < 0$ for all λ which satisfy the equation

$$\det(\lambda I - R) = 0.$$

Now we consider, for example, a car as a n-mass system which goes with constant velocity v on a random road profile $f(t, \omega)$. This car is excited by the stationary processes

$$p_1(t, \omega), \ldots, p_r(t, \omega)$$

on the r wheels. The numeration of the wheels takes place from in front to behind and s_k denotes the distance between the k-th wheel and the r-th wheel, $k = 1, 2, \ldots, r$. The behaviour of this model is described by a differential equation system of the form (3.45). $x = (x_1, \ldots, x_n)^\mathsf{T}$ is the vector of the deviations of the masses. The inhomogeneous term of (3.45) can be written as

$$g(t, \omega) = Dp(t, \omega) + Ep'(t, \omega) \tag{3.52}$$

where

$$D_{kj} = \begin{cases} d_k & \text{for} \quad k = j;\ k = 1, 2, \ldots, r, \\ 0 & \text{otherwise}, \end{cases}$$

$$E_{kj} = \begin{cases} e_k & \text{for} \quad k = j;\ k = 1, 2, \ldots, r, \\ 0 & \text{otherwise} \end{cases}$$

and

$$p(t, \omega) = \big(p_1(t, \omega), \ldots, p_r(t, \omega), 0, \ldots, 0\big)^\mathsf{T}$$

Let $d_k \neq 0$ for $k = 1, 2, \ldots, r$.

Using (3.49) and (3.51) we obtain

$$\begin{aligned} S_{x_k x_j}(\alpha) &= S_{z_{n+k} z_{n+j}}(\alpha) \\ &= \sum_{u,v=1}^{r} H_{n+ku}(-\mathrm{i}\alpha)\, H_{n+jv}(\mathrm{i}\alpha)\, S_{g_u g_v}(\alpha) \quad \text{for} \quad k, j = 1, 2, \ldots, n, \end{aligned} \tag{3.53}$$

$$\begin{aligned} S_{g_k x_j}(\alpha) &= S_{g_k z_{n+j}}(\alpha) \\ &= \sum_{u=1}^{r} H_{n+ju}(\mathrm{i}\alpha)\, S_{g_k g_u}(\alpha) \quad \text{for} \quad k = 1, 2, \ldots, r;\ j = 1, 2, \ldots, n. \end{aligned} \tag{3.54}$$

With the help of (3.52) it follows

$$S_{g_k x_j}(\alpha) = (d_k - \mathrm{i}\alpha e_k)\, S_{p_k x_j}(\alpha)$$

or

$$S_{p_k x_j}(\alpha) = \frac{d_k + \mathrm{i}\alpha e_k}{d_k^2 + \alpha^2 e_k^2}\, S_{g_k x_j}(\alpha) \quad \text{for} \quad k = 1, 2, \ldots, r;\ j = 1, 2, \ldots, n,$$

$$S_{g_k g_j}(\alpha) = (d_k - \mathrm{i}\alpha e_k)\,(d_j + \mathrm{i}\alpha e_j)\, S_{p_k p_j}(\alpha) \quad \text{for} \quad k, j = 1, 2, \ldots, r$$

and then

$$\begin{aligned} S_{x_k x_j}(\alpha) &= \sum_{u,v=1}^{r} H_{n+ku}(-\mathrm{i}\alpha)\, H_{n+jv}(\mathrm{i}\alpha)\, (d_u - \mathrm{i}\alpha e_u)\, (d_v + \mathrm{i}\alpha e_v)\, S_{p_u p_v}(\alpha) \\ &\text{for} \quad k, j = 1, 2, \ldots, n, \end{aligned} \tag{3.55}$$

$$\begin{aligned} S_{p_k x_j}(\alpha) &= \frac{d_k + \mathrm{i}\alpha e_k}{d_k^2 + \alpha^2 e_k^2} \sum_{u=1}^{r} H_{n+ju}(\mathrm{i}\alpha)\, (d_k - \mathrm{i}\alpha e_k)\, (d_u + \mathrm{i}\alpha e_u)\, S_{p_k p_u}(\alpha) \\ &\text{for} \quad k = 1, 2, \ldots, r;\ j = 1, 2, \ldots, n. \end{aligned} \tag{3.56}$$

Furthermore, we take into consideration

$$p_k(t, \omega) = f\left(t + \frac{1}{v} s_k, \omega\right) \quad \text{for} \quad k = 1, 2, \dots, r$$

and this leads to

$$\langle p_k(t) \rangle = \text{const} \quad \text{for} \quad k = 1, 2, \dots, r,$$

$$C_{p_k p_j}(t) = C_{ff}\left(t + \frac{1}{v}(s_j - s_k)\right) \quad \text{for} \quad k, j = 1, 2, \dots, r,$$

$$S_{p_k p_j}(\alpha) = \exp\left(\mathrm{i}\,\frac{\alpha}{v}(s_j - s_k)\right) S_{ff}(\alpha) \quad \text{for} \quad k, j = 1, 2, \dots, r. \tag{3.57}$$

Replacing $S_{p_k p_j}$ in (3.55), (3.56) by (3.57) the following results for the spectral densities of the output process in dependence on the spectral density of the road profile can be given:

$$S_{x_k x_j}(\alpha) = S_{ff}(\alpha) \sum_{u,q=1}^{r} H_{n+ku}(-\mathrm{i}\alpha)\, H_{n+jq}(\mathrm{i}\alpha)\, [P_{uq}(\alpha) + \mathrm{i}Q_{uq}(\alpha)] \tag{3.58}$$

for $k, j = 1, 2, \dots, n$,

$$S_{p_k x_j}(\alpha) = S_{ff}(\alpha) \frac{d_k + \mathrm{i}\alpha e_k}{d_k^2 + \alpha^2 e_k^2} \sum_{u=1}^{r} H_{n+ju}(\mathrm{i}\alpha)\, [P_{ku}(\alpha) + \mathrm{i}Q_{ku}(\alpha)] \tag{3.59}$$

for $k = 1, 2, \dots, r$; $j = 1, 2, \dots, n$

where

$$P_{uq}(\alpha) + \mathrm{i}Q_{uq}(\alpha) = (d_u - \mathrm{i}\alpha e_u)(d_q + \mathrm{i}\alpha e_q) \exp\left(\mathrm{i}\,\frac{\alpha}{v}(s_q - s_u)\right),$$

$$P_{uq}(\alpha) = (d_u d_q + \alpha^2 e_u e_q) \cos\left(\frac{\alpha}{v}(s_q - s_u)\right) + \alpha(e_u d_q - e_q d_u) \sin\left(\frac{\alpha}{v}(s_q - s_u)\right),$$

$$Q_{uq}(\alpha) = (d_u d_q + \alpha^2 e_u e_q) \sin\left(\frac{\alpha}{v}(s_q - s_u)\right) + \alpha(e_q d_u - e_u d_q) \cos\left(\frac{\alpha}{v}(s_q - s_u)\right).$$

For numerical calculations the transfer matrix $H(\mathrm{i}\alpha)$ has to be determined for α_u, $u = 1, 2, \dots$ The usual methods of the inversion of $(\mathrm{i}\alpha M + N)$ are not suitable since the inversion has to be carried out for every value α_u. Furthermore, (3.58) and (3.59) show that for fixed k, j it is sufficient to calculate the first r elements of the $(n + k)$-th and $(n + j)$-th row. We obtain

$$H_{uv}(\beta) = (-1)^{u+v} \frac{\det(\beta M^{uv} + N^{uv})}{\det(\beta M + N)}, \qquad u, v = 1, 2, \dots, n, \tag{3.60}$$

where A^{uv} is a matrix which arises from A by deleting the row and the column to which the element a_{uv} belongs. Now the determinants contained in (3.60) are calculated by the method of Hessenberg and Wilkinson (see e.g. ZURMÜHL [1, 2]) where these determinants are expanded as to β. Now it is possible to make available the coefficients of the polynomials of numerator and denumerator of (3.60). The calculation of $H_{uv}(\beta)$ at $\beta = \mathrm{i}\alpha$ can be realized by the determination of polynomials for these values.

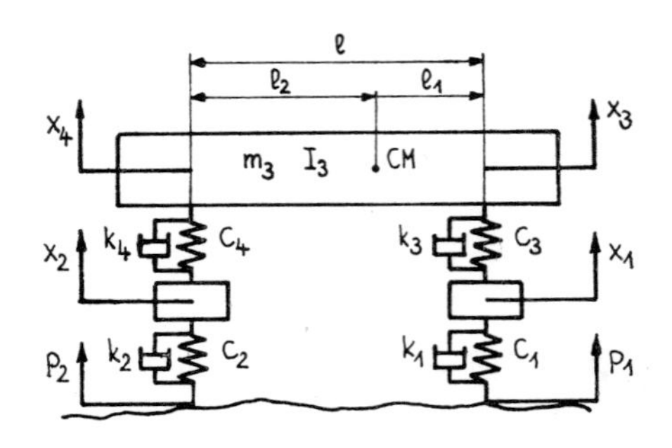

Fig. 3.35. Simple model of a car with two wheels

Consider a simple example of a vehicle with two wheels (see Fig. 3.35). I_3 denotes the moment of inertia as to the center of mass CM. This vibration system can be described by the differential equation system:

$$m_1x_1'' + k_1(x_1' - p_1') + c_1(x_1 - p_1) - k_3(x_3' - x_1') - c_3(x_3 - x_1) = 0,$$

$$m_2x_2'' + k_2(x_2' - p_2') + c_2(x_2 - p_2) - k_4(x_4' - x_2') - c_4(x_4 - x_2) = 0,$$

$$m_4x_3'' + k_3(x_3' - x_1') + c_3(x_3 - x_1) + m_6x_4'' = 0,$$

$$m_5x_4'' + k_4(x_4' - x_2') + c_4(x_4 - x_2) + m_6x_3'' = 0$$

where

$$m_4 = \frac{1}{l^2}(m_3l_2^2 + I_3), \qquad m_5 = \frac{1}{l^2}(m_3l_1^2 + I_3), \qquad m_6 = \frac{1}{l^2}(m_3l_1l_2 - I_3)$$

(see MITSCHKE [1]). The matrix form can be written as

$$Ax'' + Bx' + Cx = Dp + Ep'$$

where the matrices are given by

$$A = \begin{pmatrix} m_1 & 0 & 0 & 0 \\ 0 & m_2 & 0 & 0 \\ 0 & 0 & m_4 & m_6 \\ 0 & 0 & m_6 & m_5 \end{pmatrix}, \qquad B = \begin{pmatrix} k_1 + k_3 & 0 & -k_3 & 0 \\ 0 & k_2 + k_4 & 0 & -k_4 \\ -k_3 & 0 & k_3 & 0 \\ 0 & -k_4 & 0 & k_4 \end{pmatrix},$$

$$C = \begin{pmatrix} c_2 + c_3 & 0 & -c_3 & 0 \\ 0 & c_2 + c_4 & 0 & -c_4 \\ -c_3 & 0 & c_3 & 0 \\ 0 & -c_4 & 0 & c_4 \end{pmatrix},$$

$$D = \begin{pmatrix} c_1 & 0 & 0 & 0 \\ 0 & c_2 & 0 & 0 \\ 0 & 0 & 0 & 0 \\ 0 & 0 & 0 & 0 \end{pmatrix}, \qquad E = \begin{pmatrix} k_1 & 0 & 0 & 0 \\ 0 & k_2 & 0 & 0 \\ 0 & 0 & 0 & 0 \\ 0 & 0 & 0 & 0 \end{pmatrix}.$$

For a statistical investigation of $x_3(t, \omega)$ we assume

$$\begin{pmatrix} m_1 & l_1 & c_1 & k_1 \\ m_2 & l_2 & c_2 & k_2 \\ m_3 & v & c_3 & k_3 \\ I_3 & 0 & c_4 & k_4 \end{pmatrix} = \begin{pmatrix} 537\text{ kg} & 1.435\text{ m} & 1.92\text{E}6\text{ N/m} & 0 \\ 992\text{ kg} & 1.765\text{ m} & 2.94\text{E}6\text{ N/m} & 0 \\ 2810\text{ kg} & 80\text{ km/h} & 2.30\text{E}5\text{ N/m} & 1.13\text{E}4\text{ Ns/m} \\ 6200\text{ kg} & 0 & 7.00\text{E}5\text{ N/m} & 1.13\text{E}4\text{ Ns/m} \end{pmatrix}$$

as a variant I and a variant II

$$\begin{pmatrix} \bar{m}_1 & \bar{l}_1 & \bar{c}_1 & \bar{k}_1 \\ \bar{m}_2 & \bar{l}_2 & \bar{c}_2 & \bar{k}_2 \\ \bar{m}_3 & \bar{v} & \bar{c}_3 & \bar{k}_3 \\ \bar{I}_3 & 0 & \bar{c}_4 & \bar{k}_4 \end{pmatrix} = \begin{pmatrix} m_1 & l_1 & c_1 & k_1 \\ m_2 & l_2 & c_2 & k_2 \\ m_3 & v & 2c_3 & 2k_3 \\ I_3 & 0 & c_4 & k_4 \end{pmatrix}.$$

In order to study the influence of the vibration system to the output process $x_3(t, \omega)$ we put $S_{ff}(\alpha) \equiv 1$ where

$$p_1(t, \omega) = f\left(t + \frac{l}{v}, \omega\right), \qquad p_2(t, \omega) = f(t, \omega).$$

Fig. 3.36 shows the correlation function $C_{x_3x_3}(t)$ for the variants I and II and in Fig. 3.37 the spectral densities are illustrated. In the case of I we have a variance

$$\sigma_{x_3} = 10.33$$

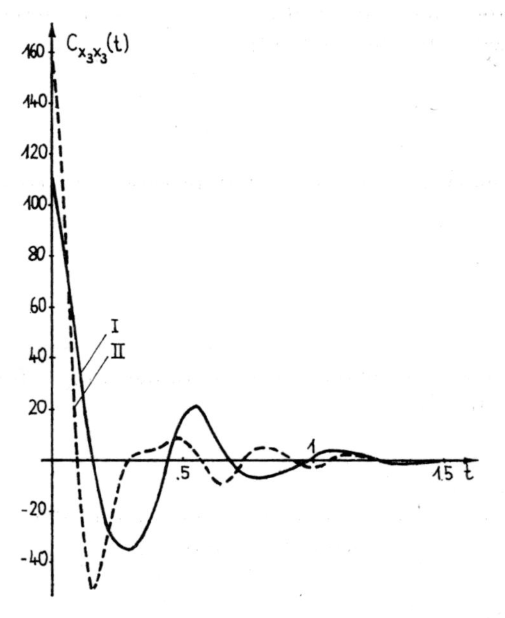

Fig. 3.36. Correlation function $C_{x_3x_3}(t)$

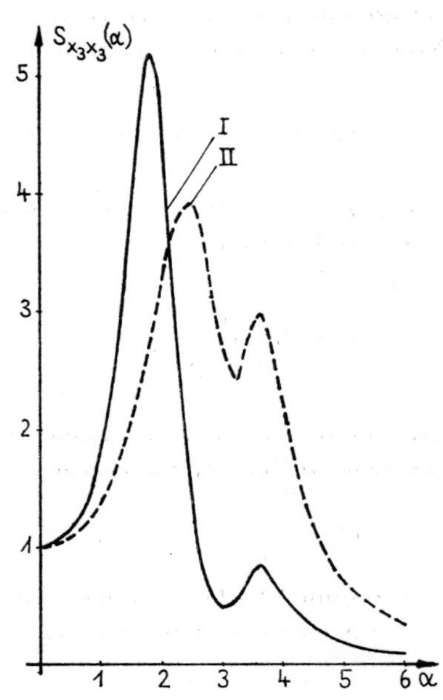

Fig. 3.37. Spectral density $S_{x_3x_3}(\alpha)$

and for II

$$\sigma_{x_3} = 12.45.$$

Considering the variant II we see that the essential frequencies are displaced to higher values.

3.2.2. Linear vibration systems with weakly correlated input

3.2.2.1. Approximation of random road surfaces

The excitation of a vibration system is based on a random road surface. The differential equations also contain derivatives of the road surface. On the other hand, derivatives of the road surface are needed if derivatives of solutions are investigated. Hence, random excitations with sufficiently many differentiable sample functions are needed for the analytic theory of the sample function solutions. In investigations of vibration problems the approximations of spectral densities of random road surfaces

$$S_1(\alpha) = c\alpha^{-w}; \qquad S_2(\alpha) = c(\alpha_0^2 + \alpha^2)^{-1}$$

are often used which were found by BRAUN [1, 2]. We note that $S_1(\alpha)$ possesses a singularity at $\alpha = 0$ and the correlation function

$$C_2(t) = d\ \mathrm{e}^{-b|t|}$$

belonging to $S_2(\alpha)$ is not differentiable at $t = 0$. Hence, these spectral densities are not suitable for input processes of vibration problems for which differentiable sample function solutions are investigated and then statistical characteristics are determined.

Now, the road surface is approximated by the random process

$$f(t, \omega) = \int_{-\infty}^{t} Q(t - s)\, f_\varepsilon(s, \omega)\, \mathrm{d}s \tag{3.61}$$

where $f_\varepsilon(t, \omega)$ is a wide-sense stationary weakly correlated process with correlation length ε and the intensity

$$a = \lim_{\varepsilon \downarrow 0} \frac{1}{\varepsilon} \int_{-\varepsilon}^{\varepsilon} \langle f_\varepsilon(t)\, f_\varepsilon(t + s)\rangle\, \mathrm{d}s.$$

Furthermore, $f_\varepsilon(t, \omega)$ is assumed to have continuously differentiable sample functions and a twice differentiable correlation function

$$C_\varepsilon(t_2 - t_1) = \langle f_\varepsilon(t_1)\, f_\varepsilon(t_2)\rangle.$$

$Q(t)$ is assumed to be a smooth function for $t \geqq 0$ for which the integral (3.61) exists. Using expansions of moments with respect to ε (see (1.23)) we obtain

$$C_{ff}(t) = a\varepsilon \int_0^\infty Q(u)\, Q(|t| + u)\, \mathrm{d}u + O(\varepsilon^2).$$

The derivatives of $f(t, \omega)$ are given by

$$f'(t, \omega) = \int_{-\infty}^{t} Q'(t - s) f_\varepsilon(s, \omega) \, ds,$$

$$f''(t, \omega) = \int_{-\infty}^{t} Q''(t - s) f_\varepsilon(s, \omega) \, ds$$

assuming

$$Q(0) = 0; \qquad Q'(0) = 0; \qquad Q^{(k)}(t) \in \mathbf{L}_2(0, \infty) \quad \text{for} \quad k = 0, 1, 2.$$

Expansions as to ε lead to

$$C_{f^{(k)}f^{(k)}}(t) = C^1_{f^{(k)}f^{(k)}}(t) + O(\varepsilon^2) \quad \text{for} \quad k = 0, 1, 2. \tag{3.62}$$

$$S_{ff}(\alpha) = S^1_{ff}(\alpha) + O(\varepsilon^2) \tag{3.63}$$

where

$$C^1_{f^{(k)}f^{(k)}}(t) \doteq a\varepsilon \int_0^\infty Q^{(k)}(u) \, Q^{(k)}(|t| + u) \, du,$$

$$S^1_{ff}(\alpha) \doteq \frac{a\varepsilon}{\pi} \int_0^\infty \cos(\alpha t) \int_0^\infty Q(u) \, Q(t + u) \, du \, dt.$$

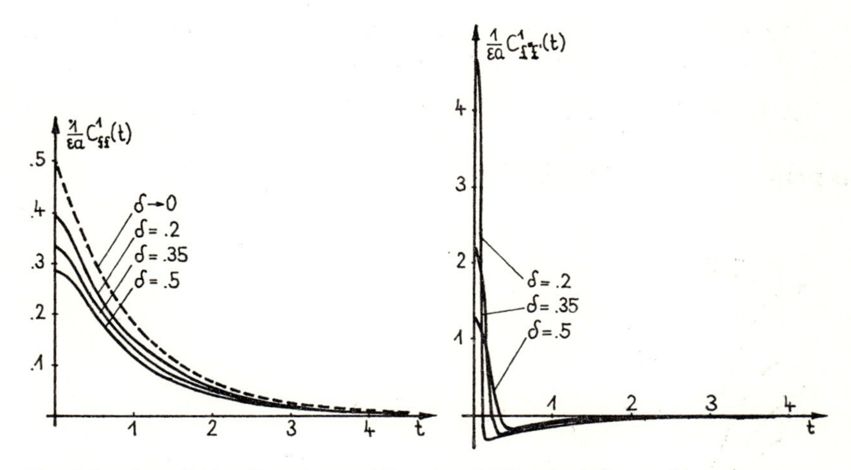

Fig. 3.38. Correlation functions of $f(t, \omega)$ and $f'(t, \omega)$ of first order as to ε

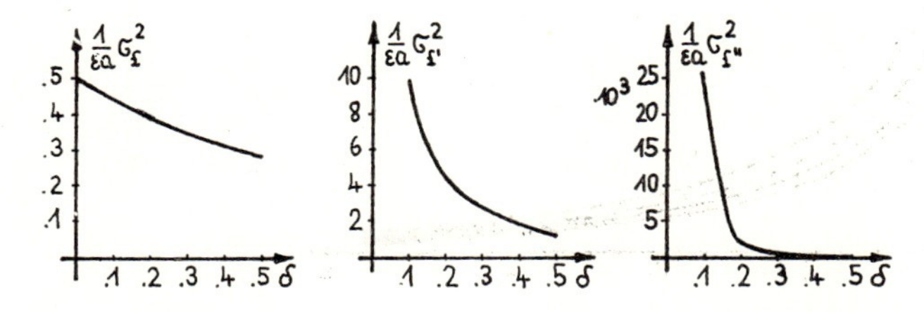

Fig. 3.39. Variances of $f(t, \omega)$, $f'(t, \omega)$, and $f''(t, \omega)$ in dependence on δ

It can be put

$$Q(s) = Q_0(s)\,\mathrm{e}^{-\gamma s}, \qquad Q_0(s) = \begin{cases} \left(\dfrac{s}{\delta}\right)^2 \left(\dfrac{s}{\delta} - 2\right)^2 & \text{for } 0 \leqq s \leqq \delta, \\ 1 & \text{for } s > \delta. \end{cases}$$

Fig. 3.38 shows the correlation functions of $f(t, \omega)$ and $f'(t, \omega)$ of first order for different values of δ and $\gamma = 1$. It is

$$\lim_{\delta \to 0} C^1_{ff}(t) = a\varepsilon \int_0^\infty \mathrm{e}^{-\gamma(2u + |t|)}\,\mathrm{d}u = \frac{a\varepsilon}{2\gamma}\,\mathrm{e}^{-\gamma|t|}.$$

The variances of $f^{(k)}(t, \omega)$ for $k = 0, 1, 2$ are illustrated in Fig. 3.39. The spectral density $S^1_{ff}(\alpha)$ of first order fulfils the limit relation

$$\lim_{\delta \to 0} S^1_{ff}(\alpha) = \frac{a\varepsilon}{2\pi}\,\frac{1}{\gamma^2 + \alpha^2}$$

and $S_{ff}(\alpha)$ can be taken as a spectral density of a road surface of the form $S_2(\alpha)$. Furthermore, we have

$$S^1_{f'f'}(\alpha) = \alpha^2 S^1_{ff}(\alpha) \quad \text{and} \quad S^1_{f''f''}(\alpha) = \alpha^4 S^1_{ff}(\alpha).$$

The first approximation of $S_{ff}(\alpha)$ as to ε is plotted in Fig. 3.40 and of $S_{f^{(k)}f^{(k)}}(\alpha)$ for $k = 1, 2$ in Fig. 3.41. For $\delta > 0$ we have the relation

$$\lim_{\alpha \to 0} S_{f^{(k)}f^{(k)}}(\alpha) = 0 \quad \text{for} \quad k = 0, 1, 2$$

but this property is not fulfilled for $\delta = 0$ and $k = 1, 2$.

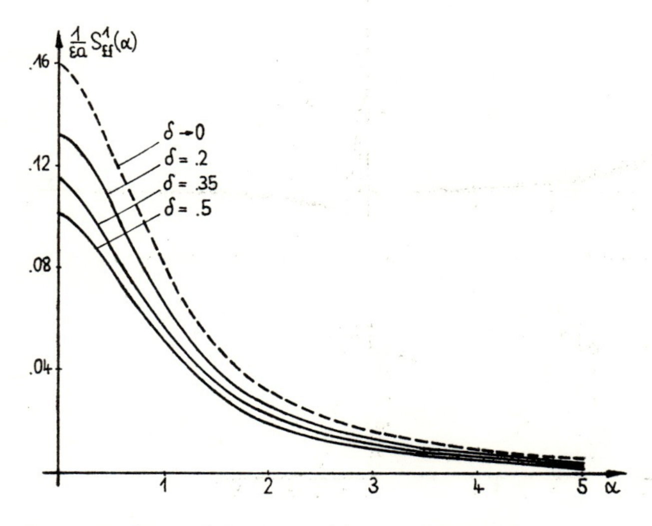

Fig. 3.40. Spectral densities of $f(t, \omega)$ of first order as to ε

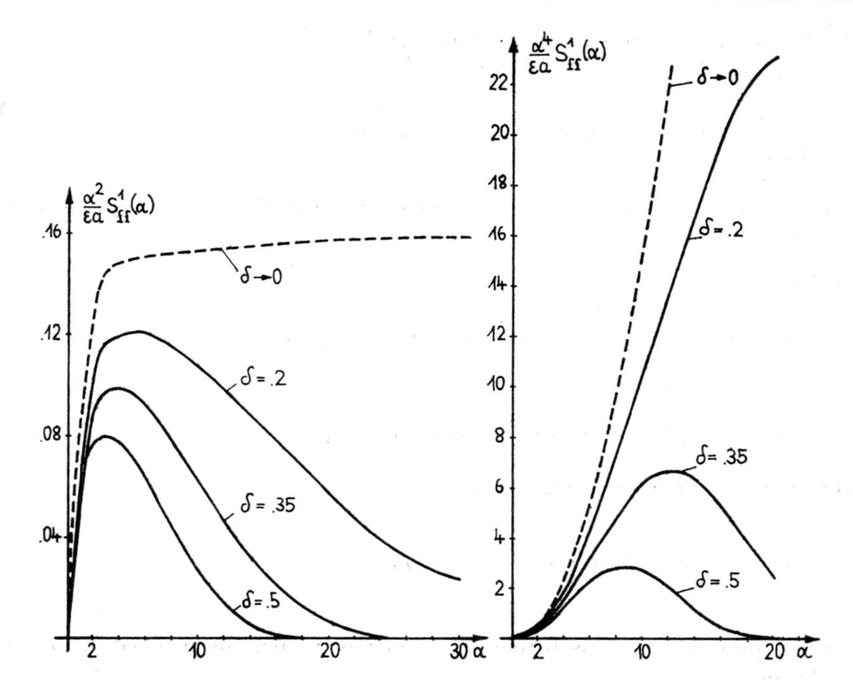

Fig. 3.41. Spectral densities of $f'(t, \omega)$ and $f''(t, \omega)$ of first order as to ε

3.2.2.2. Direct weakly correlated input

Linear systems are considered which can be described by a system of ordinary differential equations

$$Mz' + Nz = g(t, \omega), \qquad z(t_0) = z_0. \tag{3.64}$$

In our considerations we will deal with vibration systems in which the matrix $R = -M^{-1}N$ possesses only eigenvalues with negative real parts. Then the solution

$$z(t, \omega) = \int_{-\infty}^{t} G(t-s)\, g(s, \omega)\, \mathrm{d}s; \qquad G(t) = \exp\left(-M^{-1}Nt\right) M^{-1},$$

is studied (see (3.47)) which can be interpreted as solution of the initial value problem (3.64) after a certain transient time. The difference between the solution of the random problem and the solution of the averaged problem belonging to (3.64) is given by

$$w(t, \omega) = z(t, \omega) - \langle z(t) \rangle = \int_{-\infty}^{t} G(t-s)\left[g(s, \omega) - \langle g(s) \rangle\right] \mathrm{d}s. \tag{3.65}$$

Now, first of all, we put

$$g(t, \omega) = h(t) + Pf_\varepsilon(t, \omega) \tag{3.66}$$

where $f_\varepsilon(t, \omega)$ denotes a weakly correlated connected vector, $h(t)$ a non-random vector function, and P a real $n \times n$-matrix.

First, the solution (3.65) is investigated with regard to the lowest order as to ε. Using the results of Section 1.6.1 it follows

$$\lim_{\varepsilon \downarrow 0} \frac{1}{\sqrt{\varepsilon}} w(t, \omega) = \xi(t, \omega) \qquad \text{in distribution}$$

and the convergence of the corresponding moments. The Gaussian vector process $\xi(t, \omega)$ has the moments

$$\langle \xi(t) \rangle = 0,$$

$$\langle \xi(t_1)\, \xi(t_2)^\mathsf{T} \rangle = \int_{-\infty}^{t_{12}} G(t_1 - s)\, Pa(s)\, P^\mathsf{T} G(t_2 - s)^\mathsf{T} \, \mathrm{d}s \tag{3.67}$$

where

$$t_{12} \doteq \min \{t_1, t_2\} \quad \text{and} \quad a(s) = \big(a_{ij}(s)\big)_{1 \leq i, j \leq n}$$

denotes the matrix of intensities,

$$a_{ij}(s) = \lim_{\varepsilon \downarrow 0} \frac{1}{\varepsilon} \int_{-\varepsilon}^{\varepsilon} \langle f_{i\varepsilon}(s)\, f_{j\varepsilon}(s + u) \rangle \, \mathrm{d}u .$$

If the vector process $f_\varepsilon(t, \omega)$ is wide-sense stationary then the intensities are constant and we obtain

$$\langle \xi(t_1)\, \xi(t_2)^\mathsf{T} \rangle = C_{\xi\xi}(t_2 - t_1)$$

where

$$C_{\xi\xi}(t) = \begin{cases} \int_0^\infty G(s)\, PaP^\mathsf{T} G(t + s)^\mathsf{T} \, \mathrm{d}s & \text{for } t \geqq 0, \\ \int_0^\infty G(s - t)\, PaP^\mathsf{T} G(s)^\mathsf{T} \, \mathrm{d}s & \text{for } t \leqq 0. \end{cases} \tag{3.68}$$

In order to determine the spectral density $S_{\xi\xi}(\alpha)$ the transfer matrix

$$H(t) = (tM + N)^{-1}$$

is introduced. Then it follows the relation

$$\int_0^\infty \mathrm{e}^{\mathrm{i}\alpha t} G(s + t) \, \mathrm{d}t = H(-\mathrm{i}\alpha)\, MG(s) .$$

Using this relation we can calculate the spectral density

$$S_{\xi\xi}(\alpha) = \frac{1}{2\pi} \int_{-\infty}^{\infty} \mathrm{e}^{-\mathrm{i}\alpha t} C_{\xi\xi}(t) \, \mathrm{d}t = \frac{1}{2\pi} [H(-\mathrm{i}\alpha)\, MC_{\xi\xi}(0) + C_{\xi\xi}(0)\, M^\mathsf{T} H^\mathsf{T}(\mathrm{i}\alpha)] . \tag{3.69}$$

We note that $H(t)$ exists for all imaginary numbers because of

$$tM + N = M(tI + M^{-1}N) .$$

$C_{\xi\xi}(0)$ can be determined on the one hand by

$$C_{\xi\xi}(0) = \int_0^\infty G(s)\, PaP^\mathsf{T} G(s)^\mathsf{T}\, \mathrm{d}s \tag{3.70}$$

and on the other hand as the unique solution of the linear equation system

$$MC_{\xi\xi}(0)\, N^\mathsf{T} + NC_{\xi\xi}(0)\, M^\mathsf{T} = PaP^\mathsf{T}. \tag{3.71}$$

The existence of the solution $C_{\xi\xi}(0)$ from (3.71) follows from the existence of the integral of (3.70) and integration by parts. The uniqueness of the solution of (3.71) is given with the existence of the solution for any inhomogeneous terms of (3.71).

The random excitation (3.66) is assumed to have the form

$$f_\varepsilon(t, \omega) = \big(f_\varepsilon(t + s_1, \omega), f_\varepsilon(t + s_2, \omega), \dots, f_\varepsilon(t + s_n, \omega)\big)^\mathsf{T} \tag{3.72}$$

where $f_\varepsilon(t, \omega)$ denotes a weakly correlated process. Such input functions have to be considered if vibrations of cars are to be investigated where the vibrations are caused by a random surface. Using (3.65) we obtain

$$\begin{aligned} w_i(t, \omega) &= \sum_{p,q=1}^{n} \int_{-\infty}^{t} G_{ip}(t - s)\, P_{pq} f_\varepsilon(s + s_q)\, \mathrm{d}s \\ &= \sum_{p,q=1}^{n} \int_{-\infty}^{t+s_q} G_{ip}(t + s_q - s)\, P_{pq} f_\varepsilon(s)\, \mathrm{d}s \quad \text{for} \quad i = 1, 2, \dots, n. \end{aligned} \tag{3.73}$$

The Gaussian limit process $\xi(t, \omega)$ has the moments

$$\langle \xi_i(t) \rangle = 0,$$

$$\langle \xi_i(t_1)\, \xi_j(t_2) \rangle = \sum_{p_1,p_2,q_1,q_2=1}^{n} \int_{-\infty}^{t_{q_1q_2}} G_{ip_1}(t_1 + s_{q_1} - s)\, G_{jp_2}(t_2 + s_{q_2} - s)\, P_{p_1q_1} P_{p_2q_2} a(s)\, \mathrm{d}s$$

where

$$t_{pq} \doteq \min \{t_1 + s_p, t_2 + s_q\}$$

and $a(t)$ denotes the intensity of $f_\varepsilon(t, \omega)$. If we assume that $f_\varepsilon(t, \omega)$ is wide-sense stationary then it follows

$$\langle \xi_i(t_1)\, \xi_j(t_2) \rangle = C_{\xi_i\xi_j}(t_2 - t_1)$$

where

$$C_{\xi_i\xi_j}(t) = \sum_{p_1,p_2,q_1,q_2=1}^{n} T_{ijp_1p_2}(t + s_{q_2} - s_{q_1})\, P_{p_1q_1} P_{p_2q_2} a \tag{3.74}$$

and

$$T_{ijpq}(u) = \begin{cases} \displaystyle\int_0^\infty G_{ip}(s)\, G_{jq}(u + s)\, \mathrm{d}s & \text{for} \quad u \geqq 0, \\ \displaystyle\int_0^\infty G_{ip}(s - u)\, G_{jq}(s)\, \mathrm{d}s & \text{for} \quad u \leqq 0. \end{cases}$$

Hence, $\xi(t, \omega)$ is stationary connected. The spectral density $S_{\xi\xi}(\alpha)$ is given by

$$S_{\xi\xi}(\alpha) = \frac{a}{2\pi} [H(-i\alpha) MR(\alpha) + R(\alpha) M^{\mathsf{T}}H^{\mathsf{T}}(i\alpha)] \tag{3.75}$$

where $R(\alpha)$ is defined by

$$R(\alpha) = \int_0^\infty G(s) PV(\alpha) P^{\mathsf{T}}G(s)^{\mathsf{T}} ds$$

with

$$V(\alpha) = \big(V_{pq}(\alpha)\big)_{1\leq p,q\leq n}; \qquad V_{pq}(\alpha) = \exp\big(i\alpha(s_q - s_p)\big).$$

$R(\alpha)$ can be determined again as unique solution of the linear equation system

$$MR(\alpha) N^{\mathsf{T}} + NR(\alpha) M^{\mathsf{T}} = PV(\alpha) P^{\mathsf{T}}.$$

In order to determine higher approximations of moments or density functions as to ε the statistical characteristics ${}^pA_q$ have to be calculated with respect to

$$w(t, \omega) = \int_{-\infty}^t G(t - s) Pf_\varepsilon(s, \omega) ds.$$

We put

$$\begin{aligned} w_i(t, \omega) &= \sum_{p=1}^n r_{ip}(t, \omega); \\ r_{ip}(t, \omega) &= \int_{-\infty}^t \tilde{G}_{ip}(t - s) f_{p\varepsilon}(s, \omega) ds; \qquad \tilde{G}_{ip}(t) = \big(G(t) P\big)_{ip} \end{aligned} \tag{3.76}$$

and using Section 1.6.1 we obtain

$$^2A_1\big(\tilde{G}_{ip}(t_1 - .), \tilde{G}_{jq}(t_2 - .)\big) = \int_{-\infty}^{t_{12}} \tilde{G}_{ip}(t_1 - s) \tilde{G}_{jq}(t_2 - s) a_{pq}(s) ds \tag{3.77}$$

and using Section 1.6.2

$$^2A_2\big(\tilde{G}_{ip}(t_1 - .), \tilde{G}_{jq}(t_2 - .)\big)$$
$$= \begin{cases} \int_{-\infty}^{t_1} \tilde{G}_{ip}(t_1 - s) \tilde{G}_{jq}(t_2 - s) b_{pq}(s) ds - \int_{-\infty}^{t_1} \tilde{G}_{ip}(t_1 - s) \tilde{G}'_{jq}(t_2 - s) \alpha_{pq}(s) ds & \text{for } t_1 < t_2, \\ \int_{-\infty}^{t} \tilde{G}_{ip}(t - s) \tilde{G}_{jq}(t - s) b_{pq}(s) ds - \int_{-\infty}^{t} \tilde{G}_{ip}(t - s) G'_{jq}(t - s) \alpha_{pq}(s) ds & \\ \quad + \tilde{G}_{ip}(0) \tilde{G}_{jq}(0) \bar{S}_{pq}(t; -1, 0) & \text{for } t_1 = t_2 = t, \\ \int_{-\infty}^{t_2} \tilde{G}_{ip}(t_1 - s) \tilde{G}_{jq}(t_2 - s) b_{qp}(s) ds - \int_{-\infty}^{t_2} \tilde{G}'_{ip}(t_1 - s) \tilde{G}_{jq}(t_2 - s) \alpha_{qp}(s) ds & \text{for } t_1 > t_2 \end{cases} \tag{3.78}$$

where

$$\int_{\mathcal{K}_\varepsilon(t)} \langle f_{p\varepsilon}(t)\, f_{q\varepsilon}(s)\rangle \, \mathrm{d}s = a_{pq}(t)\, \varepsilon + b_{pq}(t)\, \varepsilon^2 + o(\varepsilon^2),$$

$$\lim_{\varepsilon \downarrow 0} \frac{1}{\varepsilon^2} \int_{\mathcal{K}_\varepsilon(t)} (s - t) \langle f_{p\varepsilon}(t)\, f_{q\varepsilon}(s)\rangle \, \mathrm{d}s = \alpha_{pq}(t),$$

$$\bar{S}_{pq}(t; y_1, y_2) = \int_{y_1}^{y_2} \bar{a}_{pq}(u; t)\, \mathrm{d}u - a_{pq}(t),$$

$$\lim_{\varepsilon \downarrow 0} \left[\frac{1}{\varepsilon} \int_{\mathcal{K}_\varepsilon(r) \cap [t_0, t]} \langle f_{p\varepsilon}(r)\, f_{q\varepsilon}(v)\rangle \, \mathrm{d}v \right]_{r = t + \varepsilon u} = \bar{a}_{pq}(u; t).$$

Under the assumption that $f_\varepsilon(t, \omega)$ is wide-sense stationary we obtain

- a_{pq}, b_{pq}, α_{pq} are constants and $\bar{a}_{pq}$ is only dependent on u,
- it is $a_{pq} = a_{qp}$, $b_{pq} = b_{qp}$, and $\alpha_{pq} = -\alpha_{qp}$,
- ${}^2A_1$ and ${}^2A_2$ are functions of t_1, t_2 only depending on $|t_2 - t_1|$.

With the help of (1.36) it follows

$$\langle w_i(t_1)\, w_j(t_2)\rangle = \sum_{p,q=1}^{n} \langle r_{ip} r_{jq}\rangle = \sum_{p,q=1}^{n} {}^2A_1\big(\tilde{G}_{ip}(t_1 - .), \tilde{G}_{jq}(t_2 - .)\big)\, \varepsilon$$
$$+ \sum_{p,q=1}^{n} {}^2A_2\big(\tilde{G}_{ip}(t_1 - .), \tilde{G}_{jq}(t_2 - .)\big)\, \varepsilon^2 + O(\varepsilon^3). \tag{3.79}$$

The characteristics

$${}^3A_2\big(\tilde{G}_{i_1p_1}(t_1 - .), \tilde{G}_{i_2p_2}(t_2 - .), \tilde{G}_{i_3p_3}(t_3 - .)\big),$$

$${}^4\underline{A}_3\big(\tilde{G}_{i_1p_1}(t_1 - .), \tilde{G}_{i_2p_2}(t_2 - .), \tilde{G}_{i_3p_3}(t_3 - .), \tilde{G}_{i_4p_4}(t_4 - .)\big)$$

can be determined from (1.139) and (1.140), respectively. Then the density function of $w_i(t, \omega)$ can be obtained from (1.87) and of

$$\big(w_i(t, \omega), w_j(t, \omega)\big)^\top$$

from (1.89). If we assume that

$$\big(f_1(t, \omega), \ldots, f_n(t, \omega)\big)^\top$$

is symmetrically distributed then it is

$${}^3A_2(\ldots) = 0$$

and in the case of a Gaussian vector process $f(t, \omega)$ we have

$${}^3A_2(\ldots) = 0, \qquad {}^4\underline{A}_3(\ldots) = 0.$$

Under the assumption of a Gaussian vector process $f(t, \omega)$ the density function of

$w_i(t, \omega)$ follows explicitly from (1.87) as

$$p_i(u; t) = \frac{1}{\sqrt{2\pi}} \frac{1}{\sqrt{\varepsilon\, {}^2A_{1,ii}(t,t)}} \exp\left(-\frac{1}{2} \frac{u^2}{\varepsilon\, {}^2A_{1,ii}(t,t)}\right)$$
$$\times \left\{1 + \frac{1}{2} \frac{{}^2A_{2,ii}(t,t)}{{}^2A_{1,ii}(t,t)} H_2\left(\frac{u}{\sqrt{\varepsilon\, {}^2A_{1,ii}(t,t)}}\right) \varepsilon\right\} + o(\varepsilon) \tag{3.80}$$

where

$${}^2A_{k,ij}(t_1, t_2) \doteq \sum_{p,q=1}^{n} {}^2A_k\big(\tilde{G}_{ip}(t_1 - \cdot), \tilde{G}_{jq}(t_2 - \cdot)\big).$$

Consider the simple example of an oscillator described by the differential equation

$$mx'' + kx' + cx = -m[h(t) + f_\varepsilon(t, \omega)]; \qquad m, k, c > 0, \qquad 4mc \geqq k^2.$$

Using

$$z(t) \doteq \big(x'(t), x(t)\big)^{\mathsf{T}}$$

it follows

$$Mz' + Nz = h(t) + Pf_\varepsilon(t, \omega)$$

where

$$M = \begin{pmatrix} m & 0 \\ 0 & 1 \end{pmatrix}, \qquad N = \begin{pmatrix} k & c \\ -1 & 0 \end{pmatrix}, \qquad h(t) = -mh(t)\begin{pmatrix} 1 \\ 0 \end{pmatrix},$$

$$P = \begin{pmatrix} -m & 0 \\ 0 & 0 \end{pmatrix}$$

and the transfer matrix

$$H(t) = \frac{1}{mt^2 + tk + c} \begin{pmatrix} t & -c \\ 1 & tm + k \end{pmatrix}.$$

Since $\exp(-M^{-1}Nt)\, y_0$ is the solution of

$$y' = -M^{-1}Ny; \qquad y(0) = y_0;$$

the matrix $\exp(-M^{-1}Nt)$ can be calculated as

$$\exp(-M^{-1}Nt) = e^{-\varkappa t} \begin{pmatrix} \cos(\omega t) - \dfrac{\varkappa}{\omega} \sin(\omega t) & -\dfrac{1}{\omega} (\varkappa^2 + \omega^2) \sin(\omega t) \\ \dfrac{1}{\omega} \sin(\omega t) & \cos(\omega t) + \dfrac{\varkappa}{\omega} \sin(\omega t) \end{pmatrix}$$

where

$$\varkappa = \frac{k}{2m}, \qquad \omega = \frac{1}{2m}\sqrt{4mc - k^2}.$$

Furthermore, we obtain

$$\tilde{G}(t) = e^{-\varkappa t} \begin{pmatrix} \frac{\varkappa}{\omega} \sin(\omega t) - \cos(\omega t) & 0 \\ -\frac{1}{\omega} \sin(\omega t) & 0 \end{pmatrix}.$$

In order to reduce the calculations $f_\varepsilon(t, \omega)$ is supposed to be wide-sense stationary. Then the correlation function $C^1_{w_i w_j}(t_2 - t_1)$ with respect to the first order as to ε can be determined by (3.68) or by the first summand of (3.79). Thus

$$\begin{aligned} C^1_{w_1 w_1}(t) &= a_{11}\varepsilon \int_0^\infty \tilde{G}_{11}(u)\, \tilde{G}_{11}(|t| + u)\, du \\ &= \frac{1}{4} a_{11}\varepsilon\, e^{-\varkappa |t|} \left[\frac{1}{\varkappa} \cos(\omega t) - \frac{1}{\omega} \sin(\omega |t|)\right], \\ C^1_{w_2 w_2}(t) &= a_{11}\varepsilon \int_0^\infty \tilde{G}_{21}(u)\, \tilde{G}_{21}(|t| + u)\, du \\ &= \frac{a_{11}\varepsilon}{4(\varkappa^2 + \omega^2)} e^{-\varkappa |t|} \left[\frac{1}{\varkappa} \cos(\omega t) + \frac{1}{\omega} \sin(\omega |t|)\right], \\ C^1_{w_1 w_2}(t) &= a_{11}\varepsilon \begin{cases} \int_0^\infty \tilde{G}_{11}(u)\, \tilde{G}_{21}(t + s)\, ds & \text{for } t \geqq 0, \\ \int_0^\infty \tilde{G}_{11}(-t + s)\, \tilde{G}_{21}(s)\, ds & \text{for } t \leqq 0 \end{cases} \\ &= \frac{1}{4\omega\varkappa} a_{11}\varepsilon\, e^{-\varkappa |t|} \sin(\omega |t|). \end{aligned} \tag{3.81}$$

For $m = 1$, $k = 4$, and $c = 12$ the normalized correlation functions

$$\bar{C}^1_{ii}(t) = \frac{C^1_{w_i w_i}(t)}{C^1_{w_i w_i}(0)} \quad \text{for } i = 1, 2,$$

$$\bar{C}^1_{12}(t) = \frac{C^1_{w_1 w_2}(t)}{\sqrt{C^1_{w_1 w_1}(0)\, C^1_{w_2 w_2}(0)}}$$

are illustrated in Fig. 3.42.

Now we assume a correlation function of $f_{1\varepsilon}(t, \omega)$ of the form

$$R_{1\varepsilon}(t_1, t_2) = \begin{cases} \sigma^2 \left(1 - \frac{1}{\varepsilon} |t_1 - t_2|\right) & \text{for } |t_1 - t_2| < \varepsilon, \\ 0 & \text{otherwise.} \end{cases}$$

Section 1.6.3 contains the terms which are used in the following considerations:

$$a_{11} = \sigma^2; \qquad b_{11} = 0; \qquad \alpha_{11} = 0;$$

$$\bar{S}_{11}(t; -1, 0) = -\frac{1}{6} \sigma^2; \qquad \underline{S}_{11}(t; 0, 1) = -\frac{1}{6} \sigma^2.$$

Hence, we obtain

$$^2A_{2,ij}(t_1, t_2) = {}^2A_2(\tilde{G}_{i1}(t_1 - .), \tilde{G}_{j1}(t_2 - .)) = -\frac{1}{6}\sigma^2\tilde{G}_{i1}(0)\,\tilde{G}_{j1}(0)$$

$$= \begin{cases} -\frac{1}{6}\sigma^2 & \text{for } i = j = 1 \text{ and } t_1 = t_2, \\ 0 & \text{otherwise}. \end{cases}$$

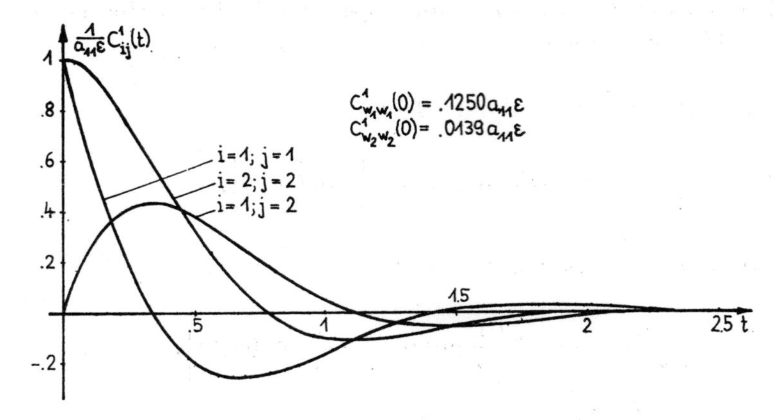

Fig. 3.42. Normalized correlation function

The cause that $^2A_{2,11}(t_1, t_2)$ is not zero only for $t_2 - t_1 = 0$ can be found by the derivation of this formula in Section 1.6.2 in which is assumed that ε is small in comparison to $t_2 - t_1$. If the solution of the initial value problem (3.64) is considered then $w(t, \omega)$ has the form

$$w(t, \omega) = \int_{t_0}^{t} G(t - s)\, Pf_\varepsilon(s, \omega)\, \mathrm{d}s$$

and the additional term

$$\tilde{G}_{ip}(t_1 - t_0)\,\tilde{G}_{jq}(t_2 - t_0)\,\underline{S}_{pq}(t_0; 0, 1)$$

is contained as a summand in the formula of

$$^2A_2(\tilde{G}_{ip}(t_1 - .), \tilde{G}_{jq}(t_2 - .)).$$

Then it is

$$^2A_{2,ij}(t_1, t_2) = {}^2A_2(\tilde{G}_{i1}(t_1 - .), \tilde{G}_{j1}(t_2 - .))$$

$$= -\frac{1}{6}\sigma^2\tilde{G}_{i1}(t_1 - t_0)\,\tilde{G}_{j1}(t_2 - t_0)$$

$$+ \begin{cases} 0 & \text{for } t_1 \neq t_2, \\ -\frac{1}{6}\sigma^2 G_{i1}(0)\, G_{j1}(0) & \text{for } t_1 = t_2. \end{cases}$$

The second approximations of the variances $C^2_{w_iw_i}(0)$ lead to

$$C^2_{w_iw_i}(0) = {}^2A_{1,ii}(t,t)\,\varepsilon + {}^2A_{2,ii}(t,t)\,\varepsilon^2$$

$$= \begin{cases} \sigma^2\left(\dfrac{\varepsilon}{4\varkappa} - \dfrac{1}{6}\,\varepsilon^2\right) & \text{for } i = 1, \\[2ex] \dfrac{\sigma^2\varepsilon}{4\varkappa(\varkappa^2+\omega^2)} & \text{for } i = 2. \end{cases}$$

Using (3.71) it follows

$$\begin{pmatrix} 2kmC^1_{w_1w_1} + 2mcC^1_{w_2w_2} & -mC^1_{w_1w_1} + kC^1_{w_1w_2} + cC^1_{w_2w_2} \\ -mC^1_{w_1w_1} + kC^1_{w_2w_1} + cC^1_{w_2w_2} & -C^1_{w_2w_1} - C^1_{w_1w_2} \end{pmatrix} = \begin{pmatrix} m^2a_{11}\varepsilon & 0 \\ 0 & 0 \end{pmatrix}$$

and then also

$$C^1_{w_iw_j}(0) = \begin{cases} \dfrac{m}{2k}\,a_{11}\varepsilon & \text{for } i = j = 1, \\[2ex] \dfrac{m^2}{2kc}\,a_{11}\varepsilon & \text{for } i = j = 2, \\[2ex] 0 & \text{otherwise}. \end{cases}$$

The approximations of the variances $C_{w_iw_i}(0)$ are illustrated in Fig. 3.43 as a function of ε. Consider the spectral density. With the help of (3.69) we can deduce that

$$S^1_{w_1w_1}(\alpha) = \frac{m}{\pi}\,C^1_{w_1w_1}(0)\,\mathrm{Re}\,(H_{11}) = \frac{m\alpha^2a_{11}\varepsilon}{2\pi[(c-m\alpha^2)^2+\alpha^2k^2]},$$

$$S^1_{w_2w_2}(\alpha) = \frac{1}{\pi}\,C^1_{w_2w_2}(0)\,\mathrm{Re}\,(H_{22}) = \frac{m^2a_{11}\varepsilon}{2\pi[(c-m\alpha^2)^2+\alpha^2k^2]},$$

$$S^1_{w_1w_2}(\alpha) = \frac{1}{2\pi}\left(C^1_{w_1w_2}(0)\,\overline{H}_{12} + C^1_{w_2w_1}(0)\,H_{21}\right) = \frac{m(-\mathrm{i}\alpha)\,a_{11}\varepsilon}{2\pi[(c-m\alpha^2)^2+\alpha^2k^2]}$$

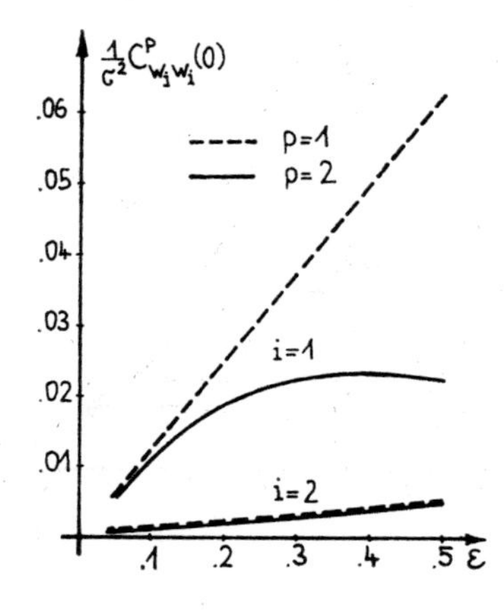

Fig. 3.43. Variances of $w_i(t, \omega)$ in first and second order

where

$$(H_{pq})_{1\le p,q\le 2} = H(\mathrm{i}\alpha)$$

and $\overline{H}_{pq}$ denotes the conjugate complex number as to H_{pq}. The relations

$$S^1_{w_1w_1}(\alpha) = \alpha^2 S^1_{w_2w_2}(\alpha); \qquad S^1_{w_1w_2}(\alpha) = (-\mathrm{i}\alpha)\, S^1_{w_2w_2}(\alpha)$$

are obvious. Using the explicite term for $C^1_{w_2w_2}(t)$ we obtain

$$S^1_{w_2w_2}(\alpha) = \frac{1}{2\pi}\int\limits_{-\infty}^{\infty} \mathrm{e}^{-\mathrm{i}\alpha t} C^1_{w_2w_2}(t)\,\mathrm{d}t = \frac{1}{\pi}\int\limits_{0}^{\infty} \cos(\alpha t)\, C^1_{w_2w_2}(t)\,\mathrm{d}t$$

$$= \frac{m^2 a_{11}\varepsilon}{2\pi[(c - m\alpha^2)^2 + \alpha^2 k^2]}.$$

This spectral density is plotted in Fig. 3.44 in dependence on m, k, and c.

In order to discuss the density function $\varphi_i(u; t)$ of $w_i(t, \omega)$ we assume the simplest properties of $f_\varepsilon(t, \omega)$; $f_{1\varepsilon}(t, \omega)$ is supposed to be Gaussian. In this case ${}^3A_2$, ${}^4\underline{A}_3$ vanish and we have (3.80). Applying the given correlation function $R_{1\varepsilon}(t_1, t_2)$ it is

$$p_1^2(u; t) = \varphi_1^1(u) = \sqrt{\frac{k}{\pi m\sigma^2\varepsilon}} \exp\left(-\frac{ku^2}{m\sigma^2\varepsilon}\right)\left\{1 - \frac{k}{6m} H_2\left(\frac{u\sqrt{2k}}{\sqrt{m\sigma^2\varepsilon}}\right)\varepsilon\right\},$$

$$p_2^2(u; t) = \varphi_2^2(u) = \sqrt{\frac{kc}{\pi m^2\sigma^2\varepsilon}} \exp\left(-\frac{kcu^2}{m^2\sigma^2\varepsilon}\right).$$

Fig. 3.45 shows $p_2^1(u) = p_2^2(u)$ for different values of $p \doteq kc/m^2$ and for $\sigma^2 = 1$, $\varepsilon = 0.1$. In the case of $i = 1$ Fig. 3.46 illustrates $p_1^1(u)$ and $p_1^2(u)$ for $q \doteq k/m = 4$, $\sigma^2 = 1$. This

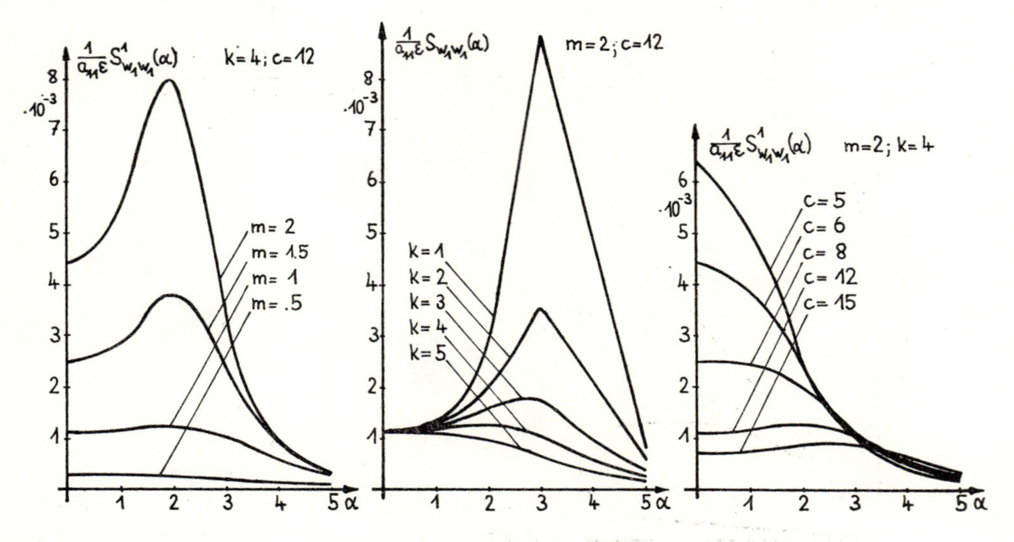

Fig. 3.44. Spectral density of first order for different values of m, k, and c

density function is independent of c and only depends on q as to the system parameters m, k, c.

Consider the correlation function $\overline{R}_3(t)$ of Section 1.6.3. Then we have

$$a_{11}(t) = \frac{3}{4} d^2; \qquad b_{11}(t) = -\frac{1}{2} d^2; \qquad \alpha_{11}(t) = 0;$$

$$\overline{S}_{11}(t; -1, 0) = -\frac{7}{80} d^2; \qquad \underline{S}_{11}(t; 0, 1) = -\frac{53}{80} d^2.$$

${}^2A_{1,ij}$ is given by

$${}^2A_{1,ij}(t_1, t_2) = \frac{1}{\varepsilon} C^1_{w_iw_j}(|t_2 - t_1|)$$

where $C^1_{w_iw_j}(t)$ can be taken from (3.81) with $a_{11} = \frac{3}{4} d^2$. Furthermore, it is

$$\begin{aligned} {}^2A_{2,ij}(t_1, t_2) &= {}^2A_2\big(\tilde{G}_{i1}(t_1 - .), \tilde{G}_{j1}(t_2 - .)\big) \\ &= -\frac{1}{2} d^2 \int\limits_{-\infty}^{t_{12}} \tilde{G}_{i1}(t_1 - s)\, \tilde{G}_{j1}(t_2 - s)\, \mathrm{d}s \\ &\quad - \begin{cases} \frac{7}{80} d^2\, \tilde{G}_{i1}(0)\, \tilde{G}_{j1}(0) & \text{for} \quad t_1 = t_2, \\ 0 & \text{otherwise}. \end{cases} \end{aligned}$$

Under this input assumptions it is also possible to obtain similar results as in the previous case. But in this case the term ${}^2A_{2,ij}(t_1, t_2)$ does not vanish for $t_1 \neq t_2$ and hence the approximations $S^1_{w_iw_j}(\alpha)$ and $S^2_{w_iw_j}(\alpha)$ of the spectral density $S_{w_iw_j}(\alpha)$ differ.

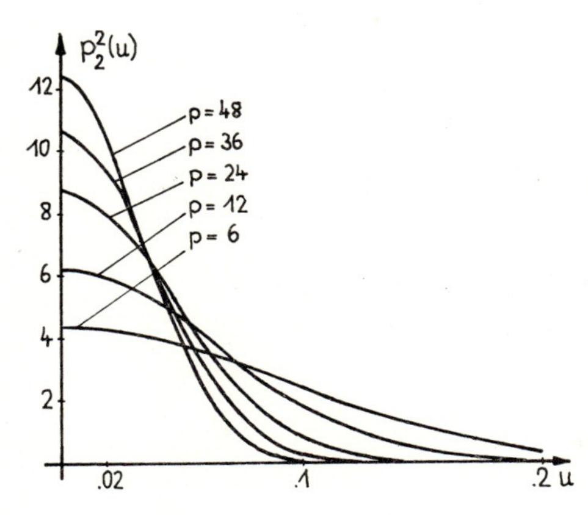

Fig. 3.45. $p_2^2(u)$ for different system parameters p and $\sigma^2 = 1$, $\varepsilon = 0.1$

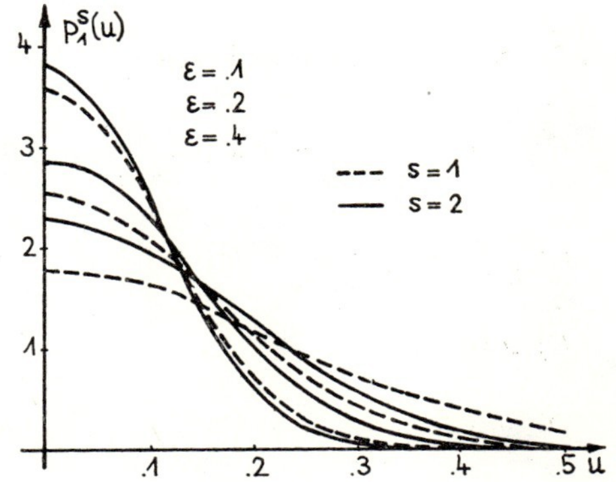

Fig. 3.46. $p_1^s(u)$, $s = 1, 2$, for different correlation lengths and $\sigma^2 = 1$, $q = 4$

We deal with a *two-mass vibration system* as a second example illustrated in Fig. 3.47. This system is described by the differential equation system

$$m_2x_2'' + k_2(x_2' - x_1') + c_2(x_2 - x_1) = 0,$$

$$m_1x_1'' + c_1(x_1 - f_\varepsilon) - k_2(x_2' - x_1') - c_2(x_2 - x_1) = 0$$

where $f_\varepsilon(t, \omega)$ denotes a weakly correlated process. $f_\varepsilon(t, \omega)$ is assumed to be wide-sense stationary. It is $h(t) = 0$ and the matrices which are contained in (3.64), (3.66) have the form

$$M = \begin{pmatrix} m_1 & 0 & 0 & 0 \\ 0 & m_2 & 0 & 0 \\ 0 & 0 & 1 & 0 \\ 0 & 0 & 0 & 1 \end{pmatrix},$$

$$N = \begin{pmatrix} k_2 & -k_2 & c_1 + c_2 & -c_2 \\ -k_2 & k_2 & -c_2 & c_2 \\ -1 & 0 & 0 & 0 \\ 0 & -1 & 0 & 0 \end{pmatrix} \qquad P = \begin{pmatrix} c_1 & 0 & 0 & 0 \\ 0 & 0 & 0 & 0 \\ 0 & 0 & 0 & 0 \\ 0 & 0 & 0 & 0 \end{pmatrix}.$$

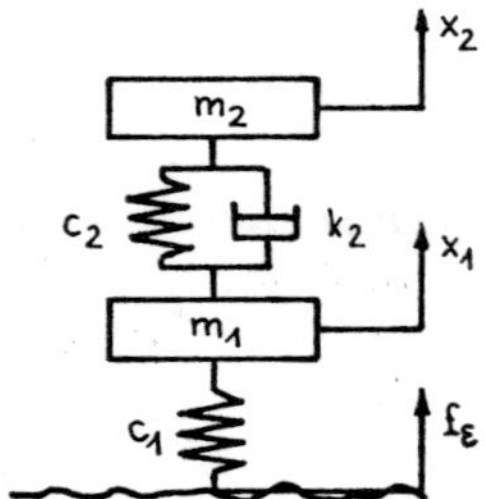

Fig. 3.47. Two-mass vibration system

Using (3.71) some calculations lead to

$$C^1_{x_1x_1}(0) = \frac{a\varepsilon}{2m_2^2k_2c_1}\left[c_1k_2^2(m_1 + m_2) + \left(c_2(m_1 + m_2) - m_2c_1\right)^2 + m_2^2c_1c_2\right],$$

$$C^1_{x_1x_2}(0) = \frac{a\varepsilon}{2m_2^2k_2c_1}\left[c_1k_2^2(m_1 + m_2) + c_2^2(m_1 + m_2)^2 - c_1c_2m_1m_2\right],$$

$$C^1_{x_1x_1'}(0) = 0,$$

$$C^1_{x_1x_2'}(0) = \frac{c_1a\varepsilon}{2m_2},$$

$$C^1_{x_2x_2}(0) = \frac{a\varepsilon}{2m_2^2k_2c_1}\left[c_1k_2^2(m_1 + m_2) + c_2^2(m_1 + m_2)^2 + m_2^2c_1c_2\right],$$

$$C^1_{x_2x_1'}(0) = -\frac{c_1 a\varepsilon}{2m_2},$$

$$C^1_{x_2x_2'}(0) = 0,$$

$$C^1_{x_1'x_1'}(0) = \frac{a\varepsilon}{2m_1m_2^2k_2}\,[k_2^2c_1m_1 + (m_1c_2 - m_2c_1)^2 + c_2^2m_1m_2],$$

$$C^1_{x_1'x_2'}(0) = \frac{a\varepsilon}{2k_2m_2^2}\,[m_1c_2^2 + k_2^2c_1 - m_2c_2(c_1 - c_2)],$$

$$C^1_{x_2'x_2'}(0) = \frac{a\varepsilon}{2k_2m_2^2}\,[c_2^2(m_1 + m_2) + k_2^2c_1]$$

and now, for example, it is possible to discuss the variances of $x_1(t, \omega)$ and $x_2(t, \omega)$ in dependence on the parameters m_1, m_2, c_1, c_2, k_2 of the system.

Using (3.69) we obtain

$$S^1_{w_pw_p}(\alpha) = \frac{1}{2\pi}\left(\overline{H}_{pu}M_{uv}C_{w_vw_p}(0) + C_{w_pw_u}(0)\,M_{vu}H_{pv}\right)$$

$$= \frac{1}{\pi}\,\mathrm{Re}\,(H_{pu})\,M_{uv}C_{w_vw_p}(0)$$

where $H_{pq} = H_{pq}(\mathrm{i}\alpha)$ and $\overline{H}_{pq}$ denotes the conjugate complex function. We define

$$R(t) \doteq (tM + N)$$

and R^{pq} denotes the minor determinant of R if we delete the row and the column to which the element R_{pq} belongs. Then it is

$$H_{pq} = \frac{1}{\det\left(R(\mathrm{i}\alpha)\right)}\,(-1)^{p+q}\,R^{qp}(\mathrm{i}\alpha)$$

and

$$\det\left(R(\mathrm{i}\alpha)\right) = \left(\alpha^4m_1m_2 - \alpha^2(m_2(c_1 + c_2) + m_1c_2) + c_1c_2\right) + \mathrm{i}\alpha k_2\left(-\alpha^2(m_1 + m_2) + c_1\right).$$

Now it follows

$$S^1_{w_pw_p}(\alpha) = \frac{1}{\pi}\,\mathrm{Re}\left[\frac{1}{\det\left(R(\mathrm{i}\alpha)\right)}\left(-m_1R^{1p}C_{w_1w_p}(0) + m_2R^{2p}C_{w_2w_p}(0) - R^{3p}C_{w_3w_p}(0) + R^{4p}C_{w_4w_p}(0)\right)\right]$$

and we can determine

$$S^1_{x_1x_1}(\alpha) = \frac{a\varepsilon c_1^2}{2\pi N(\alpha)}\,[m_2^2\alpha^4 + (k_2^2 - 2m_2c_2)\,\alpha^2 + c_2^2],$$

$$S^1_{x_2x_2}(\alpha) = \frac{a\varepsilon c_1^2}{2\pi N(\alpha)}\,[k_2^2\alpha^2 + c_2^2]$$

where $N(\alpha)$ is given by

$$N(\alpha) = \det\left(R(i\alpha)\right)\overline{\det\left(R(i\alpha)\right)}$$
$$= \left(\alpha^4 m_1 m_2 - \alpha^2(m_2(c_1 + c_2) + m_1 c_2) + c_1 c_2\right)^2 + \alpha^2 k_2^2\left(\alpha^2(m_1 + m_2) - c_1\right)^2.$$

These results are to be compared with the classical theory (see Section 3.2.1). For this we put

$$m_1 = 1; \qquad m_2 = 10; \qquad c_1 = 10000; \qquad c_2 = 2000; \qquad k_2 = 200$$

and $f_\varepsilon(t, \omega)$ is assumed to have a correlation function of the form

$$C_\varepsilon(t) = \begin{cases} \sigma^2\left(1 - \dfrac{|t|}{\varepsilon}\right) & \text{for} \quad |t| < \varepsilon, \\ 0 & \text{otherwise} \end{cases}$$

and hence the spectral density

$$S_\varepsilon(\alpha) = \frac{\sigma^2}{\pi\varepsilon\alpha^2}\left(1 - \cos\left(\alpha\varepsilon\right)\right). \tag{3.82}$$

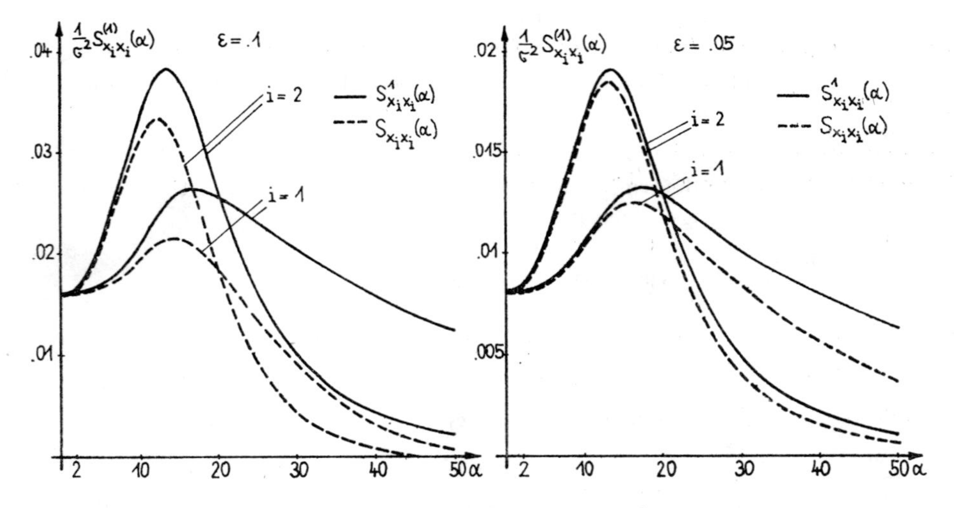

Fig. 3.48. Spectral densities $S_{x_ix_i}(\alpha)$ and approximations $S^1_{x_ix_i}(\alpha)$ for $i = 1, 2$ and $\varepsilon = 0{,}1;\ 0.05$

The intensity can be calculated as $a = \sigma^2$. Fig. 3.48 shows the spectral density $S_{x_px_p}(\alpha)$ for $p = 1, 2$ and the approximation $S^1_{x_px_p}(\alpha)$ determined by the weakly correlated theory. The function $S_{x_px_p}(\alpha)$ can be calculated by (3.49):

$$S_{w_pw_p}(\alpha) = \overline{H}_{pu}H_{pv}S_{g_ug_v}(\alpha) = c_1^2\,|H_{p1}|^2\,S_\varepsilon(\alpha)$$

where

$$g(t, \omega) = \left(c_1 f_\varepsilon(t, \omega), 0, 0, 0\right)^\mathsf{T}$$

was applied. Now it is

$$S_{x_px_p}(\alpha) = S_{w_{p+2}w_{p+2}}(\alpha) \quad \text{for} \quad p = 1, 2$$

and

$$S_{x_1x_1}(\alpha) = \frac{c_1^2}{N(\alpha)} \left[m_2^2\alpha^4 + (k_2^2 - 2c_2m_2)\,\alpha^2 + c_2^2\right] S_\varepsilon(\alpha),$$

$$S_{x_2x_2}(\alpha) = \frac{c_1^2}{N(\alpha)} \left[k_2^2\alpha^2 + c_2^2\right] S_\varepsilon(\alpha).$$

In this case we obtain $S^1_{x_ix_i}(\alpha)$ from $S_{x_ix_i}(\alpha)$ by replacing $S_\varepsilon(\alpha)$ by $S_\varepsilon(0)$. The difference between $S_\varepsilon(\alpha)$ and $S_\varepsilon(0)$ is illustrated in Fig. 3.49 for the chosen spectral density (3.82).

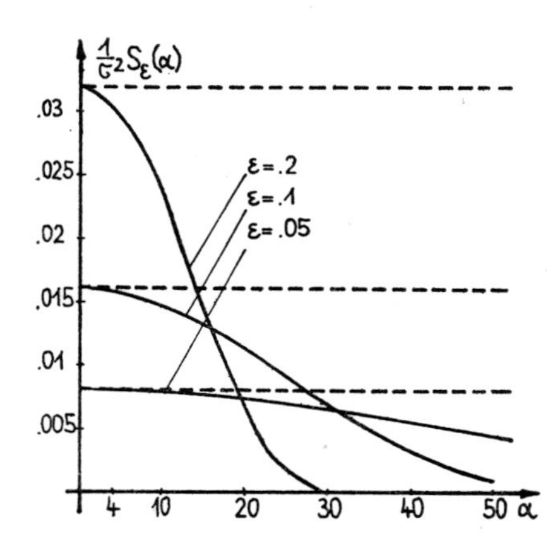

Fig. 3.49. Spectral density of the input process for different ε

Now we want to calculate statistical characteristics of the relative movements

$$y_1 = x_1 - f_\varepsilon; \qquad y_2 = x_2 - x_1.$$

These statistical characteristics are determined from those of x_1, x_2. It is

$$C_{y_1y_1}(t) = C_{x_1x_1}(t) + C_\varepsilon(t) - C_{x_1f_\varepsilon}(t) - C_{x_1f_\varepsilon}(-t),$$

$$C_{y_2y_2}(t) = C_{x_2x_2}(t) + C_{x_1x_1}(t) - C_{x_1x_2}(t) - C_{x_1x_2}(-t),$$

$$C_{y_1y_2}(t) = C_{x_1x_2}(t) + C_{f_\varepsilon x_1}(t) - C_{x_1x_1}(t) - C_{f_\varepsilon x_2}(t)$$

and the corresponding formulas for the spectral densities where we use

$$S_{xy}(\alpha) = \overline{S_{yx}(\alpha)}.$$

Considering $C_{x_1f_\varepsilon}(t)$ we obtain

$$C_{x_1f_\varepsilon}(t) = \langle x_1(s)\, f_\varepsilon(t+s)\rangle$$

$$= c_1 \int\limits_{-\infty}^{s} G_{31}(s-u)\, C_\varepsilon(t+s-u)\,\mathrm{d}u = c_1 \int\limits_{0}^{\infty} G_{31}(u)\, C_\varepsilon(t+u)\,\mathrm{d}u$$

and hence for $t = 0$

$$C_{x_1f_\varepsilon}(0) = c_1 \int\limits_0^\varepsilon G_{31}(u)\, C_\varepsilon(u)\, \mathrm{d}u = \frac{1}{2}\, \alpha\varepsilon c_1 G_{31}(0) + O(\varepsilon^2),$$

for $t < 0$

$$C_{x_1f_\varepsilon}(t) = c_1 \int\limits_{-\varepsilon}^{\varepsilon} G_{31}(u - t)\, C_\varepsilon(u)\, \mathrm{d}u = \alpha\varepsilon c_1 G_{31}(-t) + O(\varepsilon^2).$$

Summarizing it is

$$C^1_{x_1f_\varepsilon}(t) = \begin{cases} 0 & \text{for} \quad t > 0, \\ \dfrac{1}{2}\, \alpha\varepsilon c_1 G_{31}(0) & \text{for} \quad t = 0, \\ \alpha\varepsilon c_1 G_{31}(-t) & \text{for} \quad t < 0. \end{cases}$$

If we use the results for $C^1_{x_ix_j}(0)$ it follows

$$C^1_{y_1y_1}(0) = \frac{\alpha\varepsilon}{2m_2^2k_2c_1} \left(c_1k_2^2(m_1 + m_2) + (c_2(m_1 + m_2) - m_2c_1)^2 + m_2^2c_1c_2\right) + \sigma^2,$$

$$C^1_{y_2y_2}(0) = \frac{\alpha\varepsilon c_1}{2k_2},$$

$$C^1_{y_1y_2}(0) = \frac{\alpha\varepsilon}{2m_2k_2} \left(m_1c_2 + m_2(c_2 - c_1)\right).$$

The spectral density $S^1_{x_1f_\varepsilon}(\alpha)$ can be obtained from

$$S^1_{x_1f_\varepsilon}(\alpha) = \frac{1}{2\pi} \int\limits_{-\infty}^{\infty} \mathrm{e}^{-\mathrm{i}\alpha t} C^1_{x_1f_\varepsilon}(t)\, \mathrm{d}t = \frac{\alpha\varepsilon c_1}{2\pi}\, \overline{H}_{31}$$

and $S^1_{x_1x_2}(\alpha)$ from

$$S^1_{x_1x_2}(\alpha) = \frac{1}{2\pi}\, [m_1\overline{H}_{31}C^1_{x_1'x_2}(0) + (\overline{H}_{33} + H_{44})\, C^1_{x_1x_2}(0) + \overline{H}_{34}C^1_{x_2x_2}(0)$$
$$+ m_2H_{42}C^1_{x_1x_2'}(0) + H_{43}C^1_{x_1x_1}(0)].$$

Thus

$$\operatorname{Re}\left(S^1_{x_1f_\varepsilon}(\alpha)\right) = \frac{\alpha\varepsilon c_1}{2\pi N(\alpha)}\, [-m_1m_2^2\alpha^6 + \{2m_1m_2c_2 + m_2^2(c_1 + c_2)$$
$$- k_2^2(m_1 + m_2)\}\, \alpha^4 - \{2c_1c_2m_2 + m_2c_2^2 + m_1c_2^2 - k_2^2c_1\}\, \alpha^2 + c_1c_2^2],$$

$$\operatorname{Re}\left(S^1_{x_1x_2}(\alpha)\right) = \frac{\alpha\varepsilon c_1^2}{2\pi N(\alpha)}\, [(k_2^2 - c_2m_2)\, \alpha^2 + c_2^2].$$

Now the approximation of the spectral densities can be written as

$$S^1_{y_1y_1}(\alpha) = \frac{a\varepsilon}{2\pi} + \frac{a\varepsilon c_1}{2\pi N(\alpha)} [2m_1m_2^2\alpha^6 + \{2k_2^2(m_1 + m_2) - m_2^2c_1 - 2m_1m_2c_2$$

$$- 2m_2c_2(m_1 + m_2)\} \alpha^4 + \{2c_1c_2m_2 - c_1k_2^2 + 2c_2^2(m_1 + m_2)\} \alpha^2 - c_1c_2^2],$$

$$S^1_{y_2y_2}(\alpha) = \frac{a\varepsilon c_1^2}{2\pi N(\alpha)} m_2^2\alpha^6$$

where

$$S_\varepsilon(\alpha) = \frac{a\varepsilon}{2\pi} + O(\varepsilon^2)$$

was taken into consideration.

The exact spectral densities are given by

$$S_{y_1y_1}(\alpha) = \Bigg[1 + \frac{c_1}{N(\alpha)} [2m_1m_2^2\alpha_6$$

$$+ \{2k_2^2(m_1 + m_2) - m_2^2c_1 - 2m_1m_2c_2 - 2m_2c_2(m_1 + m_2)\} \alpha^4$$

$$+ \{2c_1c_2m_2 - c_1k_2^2 + 2c_2^2(m_1 + m_2)\} \alpha^2 - c_1c_2^2]\Bigg] S_\varepsilon(\alpha),$$

$$S_{y_2y_2}(\alpha) = \frac{c_1^2}{N(\alpha)} m_2^2\alpha^4S_\varepsilon(\alpha).$$

Figs. 3.50, 3.51, 3.52 illustrate the correspondence between the spectral density $S_{y_iy_i}(\alpha)$ and the approximation $S^1_{y_iy_i}(\alpha)$ for different small ε. For all ε and $i = 1$ a great difference is to observe between the values of $S_{y_iy_i}(\alpha)$ and $S^1_{y_iy_i}(\alpha)$ for great values α. The reason is founded by the term $a\varepsilon/2\pi$ which follows from the fact that the spectral density $S_\varepsilon(\alpha)$

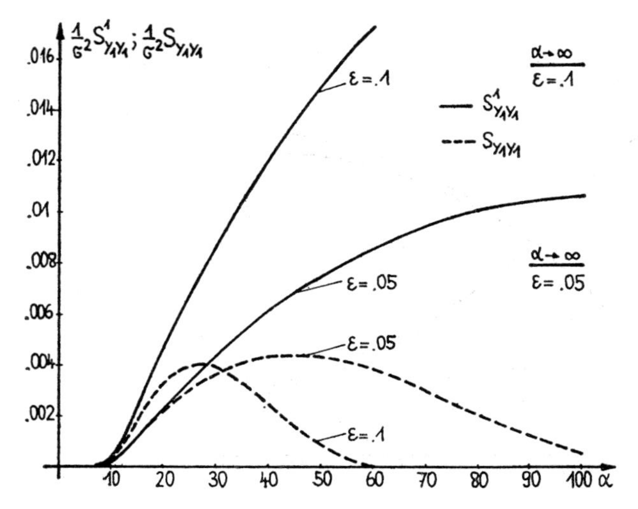

Fig. 3.50. Spectral density S and its approximation S^1 for the relative movement y_1 and $\varepsilon = 0.1$; 0.05

of the weakly correlated process $f_\varepsilon(t, \omega)$ is contained in this result. We also have

$$\lim_{\alpha \to \infty} S^1_{y_1 y_1}(\alpha) = \frac{a\varepsilon}{2\pi}$$

and

$$\lim_{\alpha \to \infty} S_{y_1 y_1}(\alpha) = 0 .$$

In the case of $i = 2$ we observe a better coincidence between $S^1_{y_i y_i}(\alpha)$ and $S_{y_i y_i}(\alpha)$ for the considered ε.

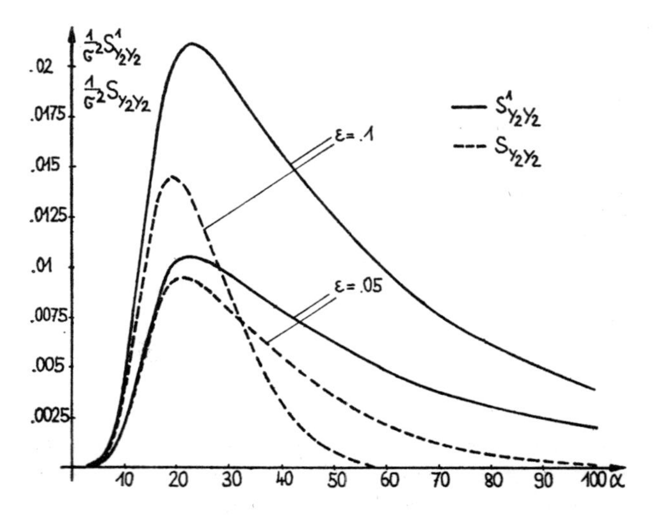

Fig. 3.51. Spectral density S and its approximation S^1 for the relative movement y_2 and $\varepsilon = 0.1; 0.05$

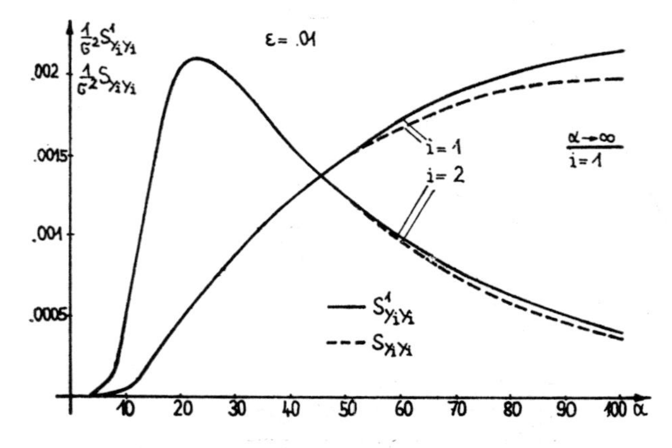

Fig. 3.52. Spectral density S and its approximation S^1 for the relative movements y_1, y_2 and $\varepsilon = 0.01$

3.2.2.3. Indirect weakly correlated input

In this section an input function of the form

$$g(t, \omega) = h(t) + \sum_{l=0}^{2} P_l f^{(l)}(t, \omega)$$
$$= h(t) + \int_{-\infty}^{t} \sum_{l=0}^{2} P_l Q^{(l)}(t - s) f_\varepsilon(s, \omega) \, ds \tag{3.83}$$

is supposed where $Q(t)$ denotes an $n \times n$-matrix with elements discussed in Section 3.2.2.1. Using this input function (3.83) and (3.65) the stationary solution of (3.64) is obtained as

$$w(t, \omega) = \int_{-\infty}^{t} G(t - s) \int_{-\infty}^{s} \sum_{l=0}^{2} P_l Q^{(l)}(s - u) f_\varepsilon(u, \omega) \, du \, ds .$$

A commutation of the integrals leads to

$$w(t, \omega) = \int_{-\infty}^{t} \left[\int_{u}^{t} G(t - s) \sum_{l=0}^{2} P_l Q^{(l)}(s - u) \, ds \right] f_\varepsilon(u, \omega) \, du$$
$$= \int_{-\infty}^{t} \hat{G}(t - s) f_\varepsilon(s, \omega) \, ds \tag{3.84}$$

where $\hat{G}(t)$ is defined by

$$\hat{G}(t) \doteq \sum_{l=0}^{2} \int_{0}^{t} G(t - u) P_l Q^{(l)}(u) \, du$$
$$= \sum_{l=0}^{2} \int_{0}^{t} \exp\left(-M^{-1}N(t - u)\right) M^{-1} P_l Q^{(l)}(u) \, du . \tag{3.85}$$

In the case of $P_2 = 0$ we obtain

$$w'(t, \omega) = \int_{-\infty}^{t} \hat{G}'(t - s) f_\varepsilon(s, \omega) \, ds$$

and $\hat{G}'(t)$ is given by

$$\hat{G}'(t) = \sum_{l=0}^{1} \left[G(0) P_l Q^{(l)}(t) + \int_{0}^{t} G'(t - u) P_l Q^{(l)}(u) \, du \right]$$
$$= \sum_{l=0}^{1} \int_{0}^{t} G(t - u) P_l Q^{(l+1)}(u) \, du$$

if the second summand is integrated by parts. A similar formula can be deduced for $w''(t, \omega)$ assuming $P_1 = P_2 = 0$; it is

$$w''(t, \omega) = \int_{-\infty}^{t} \hat{G}''(t - s) f_\varepsilon(s, \omega) \, ds$$

where

$$\hat{G}''(t) = \int_0^t G(t-u)\, P_0 Q''(u)\, \mathrm{d}u .$$

In order to determine the statistical characteristics ${}^2A_1(F_1, F_2)$ we define the matrices

$$\hat{G}_{lm}(t) = \int_0^t G(t-u)\, P_l Q^{(m)}(u)\, \mathrm{d}u \tag{3.86}$$

and consider

$$w_{lm}(t, \omega) = \int_{-\infty}^t \hat{G}_{lm}(t-s)\, f_\varepsilon(s, \omega)\, \mathrm{d}s . \tag{3.87}$$

Under the corresponding assumptions we now can write

$$w(t, \omega) = w_{00}(t, \omega) + w_{11}(t, \omega) + w_{22}(t, \omega),$$

$$w'(t, \omega) = w_{01}(t, \omega) + w_{12}(t, \omega),$$

$$w''(t, \omega) = w_{02}(t, \omega).$$

We deal with the moments

$$\langle w_{l_1m_1}(t_1)\, w_{l_2m_2}^{\mathsf{T}}(t_2)\rangle$$

and consequently we have to investigate the term

$${}^2A_1\big(\hat{G}_{l_1m_1}(t_1 - .), \hat{G}_{l_2m_2}(t_2 - .)\big) = \int_{-\infty}^{t_{12}} \hat{G}_{l_1m_1}(t_1 - s)\, a(s)\, \hat{G}_{l_2m_2}^{\mathsf{T}}(t_2 - s)\, \mathrm{d}s$$

where the intensity $a(t)$ of the vector process $f_\varepsilon(t, \omega)$ is defined by

$$a(t) \doteq \lim_{\varepsilon \downarrow 0} \frac{1}{\varepsilon} \int_{-\varepsilon}^{\varepsilon} \langle f_\varepsilon(t)\, f_\varepsilon^{\mathsf{T}}(t+s)\rangle\, \mathrm{d}s$$

and $t_{12} \doteq \min\{t_1, t_2\}$. Let $f_\varepsilon(t, \omega)$ be wide-sense stationary. Then $a(t)$ is a constant matrix, $a(t) \equiv a$.

Defining

$${}^2A_1^{l_1m_1l_2m_2}(t_1, t_2) \doteq {}^2A_1\big(\hat{G}_{l_1m_1}(t_1 - .), \hat{G}_{l_2m_2}(t_2 - .)\big)$$

we obtain

$${}^2A_1^{l_1m_1l_2m_2}(t_1, t_2) = \begin{cases} \displaystyle\int_0^\infty \hat{G}_{l_1m_1}(u)\, a\hat{G}_{l_2m_2}^{\mathsf{T}}(t+u)\, \mathrm{d}u & \text{for} \quad t = t_2 - t_1 \geqq 0, \\ \displaystyle\int_0^\infty \hat{G}_{l_1m_1}(t+u)\, a\hat{G}_{l_2m_2}^{\mathsf{T}}(u)\, \mathrm{d}u & \text{for} \quad t = t_1 - t_2 \geqq 0. \end{cases} \tag{3.88}$$

The first approximation of the correlation function

$$\langle w_{l_1m_1}(t_1)\, w^{\mathsf{T}}_{l_2m_2}(t_2)\rangle$$

has the form

$$C^{1,l_1m_1l_2m_2}(t) = {}^2A_1^{l_1m_1l_2m_2}(0, t)\, \varepsilon$$

and for $t \geqq 0$

$$C^{1,l_1m_1l_2m_2}(t) = \varepsilon \int_0^\infty \hat{G}_{l_1m_1}(u)\, a\hat{G}^{\mathsf{T}}_{l_2m_2}(t+u)\, \mathrm{d}u$$

where

$$\hat{G}_{lm}(t) = G(t)\, A_{lm}(t)\,, \qquad A_{lm}(t) \doteq M \int_0^t G(-s)\, P_l Q^{(m)}(s)\, \mathrm{d}s\,. \tag{3.89}$$

In particular, the first approximation for $t_1 = t_2$ can be written as

$$C^{1,l_1m_1l_2m_2}(0) = \varepsilon \int_0^\infty \hat{G}_{l_1m_1}(u)\, a\hat{G}^{\mathsf{T}}_{l_2m_2}(u)\, \mathrm{d}u$$

$$= \varepsilon \int_0^\infty G(u)\, A_{l_1m_1}(u)\, aA^{\mathsf{T}}_{l_2m_2}(u)\, G^{\mathsf{T}}(u)\, \mathrm{d}u\,. \tag{3.90}$$

The first approximation $S^{1,l_1m_1l_2m_2}(\alpha)$ of the spectral density $S^{l_1m_1l_2m_2}(\alpha)$ is the Fourier transform of $C^{1,l_1m_1l_2m_2}(t)$. We have

$$S^{1,l_1m_1l_2m_2}(\alpha) = \frac{1}{2\pi} \int_{-\infty}^{\infty} \mathrm{e}^{-\mathrm{i}\alpha t} C^{1,l_1m_1l_2m_2}(t)\, \mathrm{d}t$$

$$= \frac{\varepsilon}{2\pi} \left[\int_0^\infty \mathrm{e}^{-\mathrm{i}\alpha t} \int_0^\infty \hat{G}_{l_1m_1}(u)\, a\hat{G}^{\mathsf{T}}_{l_2m_2}(t+u)\, \mathrm{d}u\, \mathrm{d}t \right.$$

$$\left. + \int_0^\infty \mathrm{e}^{\mathrm{i}\alpha t} \int_0^\infty \hat{G}_{l_1m_1}(t+u)\, a\hat{G}^{\mathsf{T}}_{l_2m_2}(u)\, \mathrm{d}u\, \mathrm{d}t \right]. \tag{3.91}$$

An integration by parts leads to

$$\int_0^\infty \mathrm{e}^{\mathrm{i}\alpha t} \hat{G}_{lm}(t+u)\, \mathrm{d}t = H(-\mathrm{i}\alpha)\, [M\hat{G}_{lm}(u) + P_l U_m(\alpha; u)]$$

where $U_m(\alpha; u)$ is defined by

$$U_m(\alpha; u) \doteq \int_0^\infty \mathrm{e}^{\mathrm{i}\alpha t} Q^{(m)}(t+u)\, \mathrm{d}t\,. \tag{3.92}$$

Using this relation and (3.91) we obtain

$$S^{1,l_1m_1l_2m_2}(\alpha) = \frac{1}{2\pi}\left[H(-\mathrm{i}\alpha)\left\{MC^{1,l_1m_1l_2m_2}(0) + \varepsilon P_{l_1}\int_0^\infty U_{m_1}(\alpha;u)\, a\hat{G}^{\mathsf{T}}_{l_2m_2}(u)\,\mathrm{d}u\right\}\right.$$
$$\left. + \left\{C^{1,l_1m_1l_2m_2}(0)\,M^{\mathsf{T}} + \varepsilon\int_0^\infty \hat{G}_{l_1m_1}(u)\,aU^{\mathsf{T}}_{m_2}(-\alpha;u)\,\mathrm{d}u\cdot P^{\mathsf{T}}_{l_2}\right\}H^{\mathsf{T}}(\mathrm{i}\alpha)\right]. \tag{3.93}$$

We consider the example of an oscillator as in Section 3.2.2. with a changed inhomogeneous term,

$$mx'' + kx' + cx = -m\int_{-\infty}^{t} Q(t-s)\, f_\varepsilon(s,\omega)\,\mathrm{d}s\,.$$

M, N are given above and P_0, Q by

$$P_0 = \begin{pmatrix} -m & 0 \\ 0 & 0 \end{pmatrix}, \qquad Q(t) = \begin{pmatrix} Q_{11}(t) & 0 \\ 0 & 0 \end{pmatrix}$$

with

$$Q_{11}(t) = Q(t) = Q_0(t)\,\mathrm{e}^{-\gamma t}$$

(see 3.2.2.1). Then we find

$$\hat{G}_{0m}(t) = \int_0^t G(t-u)\,P_0Q^{(m)}(u)\,\mathrm{d}u = G(t)\,A_{0m}(t)$$

with

$$A_{0m}(t) = M\int_0^t G(-s)\,P_0Q^{(m)}(s)\,\mathrm{d}s$$

or

$$\hat{G}_{0m,hk}(t) = \sum_{p=1}^{2} G_{hp}(t)\,A_{0m,pk}(t)$$

with

$$A_{0m,hk}(t) = -mM_{hh}\int_0^t G_{h1}(-s)\,Q_{11}^{(m)}(s)\,\mathrm{d}s\cdot\delta_{1k}\,.$$

Now, we can give the approximation of $\langle x(s)\,x(s+t)\rangle$ by

$$C_{22}^{1,0000}(t) = \varepsilon a\int_0^\infty \hat{G}_{00,21}(u)\,\hat{G}_{00,21}(t+u)\,\mathrm{d}u \tag{3.94}$$

and of $\langle x'(s)\,x'(s+t)\rangle$ by

$$C_{11}^{1,0000}(t) = \varepsilon a\int_0^\infty \hat{G}_{00,11}(u)\,\hat{G}_{00,11}(t+u)\,\mathrm{d}u \tag{3.95}$$

for $t \geqq 0$. It is

$$w'(t, \omega) = \int_{-\infty}^{t} \hat{G}'(t - s) f_\varepsilon(s, \omega) \, ds$$

or

$$w_{01}(t, \omega) = \int_{-\infty}^{t} \hat{G}_{01}(t - s) f_\varepsilon(s, \omega) \, ds$$

where

$$\hat{G}'(t) = \hat{G}_{01}(t) = \int_{0}^{t} G(t - u) P_0 Q'(u) \, du$$

and the first approximation of the correlation function of $x''(t, \omega)$ is given by

$$C_{11}^{1,0101}(t) = \varepsilon a \int_{0}^{\infty} \hat{G}_{01,11}(u) \, \hat{G}_{01,11}(t + u) \, du .$$

It is easy to see that the relation

$$C_{22}^{1,0101}(t) = C_{11}^{1,0000}(t) \tag{3.96}$$

is fulfilled. The spectral density $S^{1,0000}(\alpha)$ is determined by direct Fourier transformation.

The matrix $G(s)$ is calculated in 3.2.2.2 and we have

$$A_{00,11}(t) = -m^2 \int_{0}^{t} G_{11}(-s) \, Q_{11}(s) \, ds,$$

$$A_{00,21}(t) = -m \int_{0}^{t} G_{21}(-s) \, Q_{11}(s) \, ds,$$

$$A_{01,11}(t) = -m^2 \int_{0}^{t} G_{11}(-s) \, Q'_{11}(s) \, ds,$$

$$A_{01,21}(t) = -m \int_{0}^{t} G_{21}(-s) \, Q'_{11}(s) \, ds$$

and then

$$\hat{G}_{00,pq}(t) = G_{p1}(t) \, A_{00,1q}(t) + G_{p2}(t) \, A_{00,2q}(t),$$

$$\hat{G}_{01,pq}(t) = G_{p1}(t) \, A_{01,1q}(t) + G_{p2}(t) \, A_{01,2q}(t).$$

Finally,

$$C_{22}^{1,0000}(t), \qquad C_{11}^{1,0000}(t), \quad \text{and} \quad C_{11}^{1,0101}(t)$$

can be obtained. The spectral density $S_{22}^{1,0000}(\alpha)$ of $x(t, \omega)$ follows from

$$S_{22}^{1,0000}(\alpha) = \frac{1}{\pi} \int_{0}^{\infty} \cos(\alpha t) \, C_{22}^{1,0000}(t) \, dt$$

and it is

$$S_{11}^{1,0000}(\alpha) = \alpha^2 S_{22}^{1,0000}(\alpha)\,; \qquad S_{11}^{1,0101}(\alpha) = \alpha^4 S_{22}^{1,0000}(\alpha)\,.$$

Figs. 3.53a,b show the approximation of the correlation function of $x(t, \omega)$, $x'(t, \omega)$, and $x''(t, \omega)$ for different values of (m, k, c). We refer to the essentially different vari-

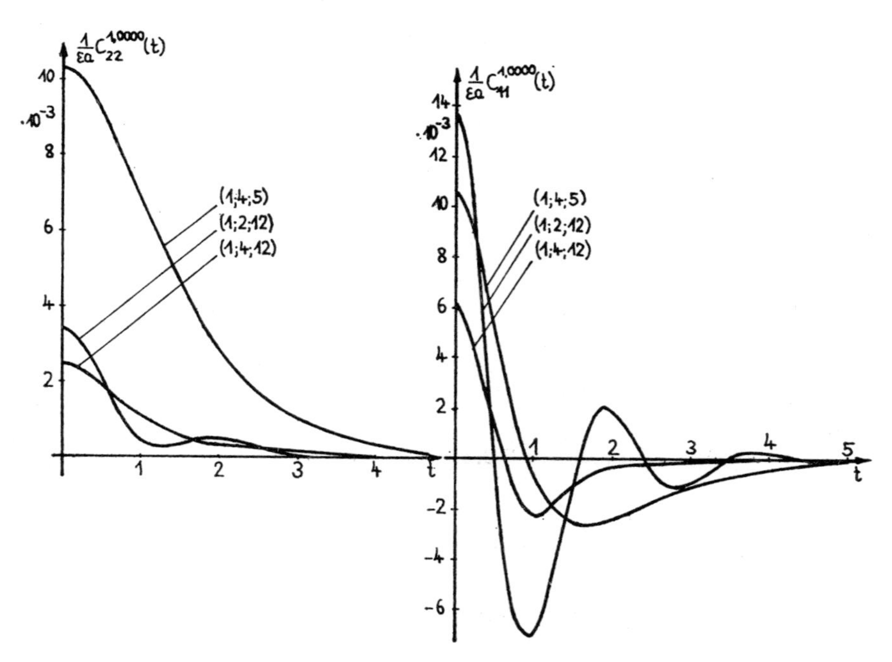

Fig. 3.53a. Correlation functions of $x(t, \omega)$, $x'(t, \omega)$ for different values of (m, k, c)

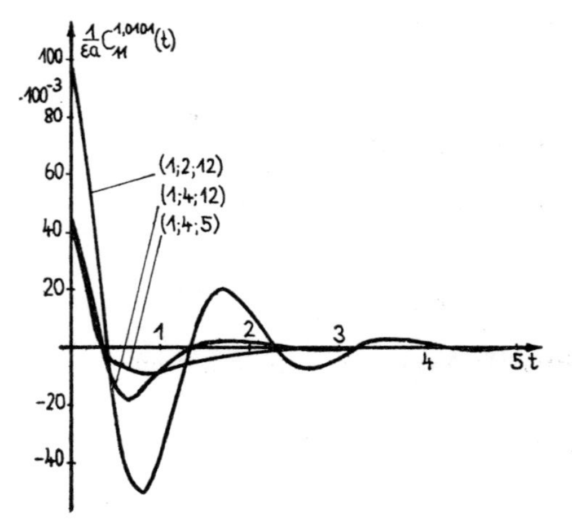

Fig. 3.53b. Correlation function of $x''(t, \omega)$ for different values of (m, k, c)

ances σ_x^2, $\sigma_{x'}^2$, $\sigma_{x''}^2$ for the different values (m, k, c). Furthermore, the approximation of the spectral density of $x^{(k)}(t, \omega)$ for $k = 0, 1, 2$ is plotted in Fig. 3.54. In order to obtain these numerical results it was put $\delta = 0.2$, $\gamma = 1$.

Now we deal with a one-mass system in which the process

$$f(t, \omega) = \int_{-\infty}^{t} Q(t - s) f_\varepsilon(s, \omega) \, ds$$

is the road surface. The corresponding differential equation can be written as

$$mx'' + kx' + cx = cf(t, \omega) + kf'(t, \omega)$$

for the absolute movement $x(t, \omega)$ and as

$$my'' + ky' + cy = -mf''(t, \omega)$$

for the relative movement

$$y(t, \omega) = x(t, \omega) - f(t, \omega).$$

The stationary solutions have the form

$$\begin{aligned} w_x(t, \omega) &= \int_{-\infty}^{t} \hat{G}_{00}(t - s) f_\varepsilon(s, \omega) \, ds + \int_{-\infty}^{t} \hat{G}_{11}(t - s) f_\varepsilon(s, \omega) \, ds \\ &= w_{x,00}(t, \omega) + w_{x,11}(t, \omega), \end{aligned}$$

$$w_y(t, \omega) = \int_{-\infty}^{t} \hat{G}_{22}(t - s) f_\varepsilon(s, \omega) \, ds = w_{y,22}(t, \omega)$$

where P_i, $i = 0, 1, 2$, are given by

$$P_0 = \begin{pmatrix} c & 0 \\ 0 & 0 \end{pmatrix}; \quad P_1 = \begin{pmatrix} k & 0 \\ 0 & 0 \end{pmatrix}; \quad P_2 = \begin{pmatrix} -m & 0 \\ 0 & 0 \end{pmatrix}.$$

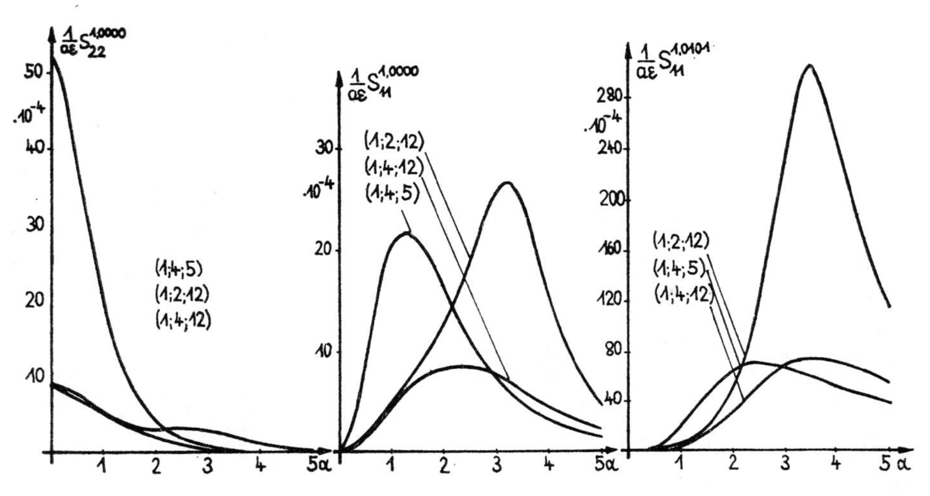

Fig. 3.54. Spectral densities of $x(t, \omega)$, $x'(t, \omega)$, and $x''(t, \omega)$ for different values of (m, k, c)

We determine the approximations of the correlation function and spectral density of $x(t, \omega)$, $x'(t, \omega)$, $y(t, \omega)$, and $y'(t, \omega)$. $C_x^1(t)$, $C_y^1(t)$ denote the first approximations of

$$\langle w_x(t_1)\, w_x^{\mathsf{T}}(t_2)\rangle, \qquad \langle w_y(t_1)\, w_y^{\mathsf{T}}(t_2)\rangle,$$

respectively. We obtain

$$C_x^1(t) = C^{1,0000}(t) + C^{1,1111}(t) + C^{1,0011}(t) + C^{1,1100}(t),$$

$$C_y^1(t) = C^{1,2222}(t).$$

Using (3.88) and (3.89) it follows

$$C_{11}^{1,l_1m_1l_2m_2}(t) = \varepsilon a \int_0^\infty \hat{G}_{l_1m_1,11}(u)\, \hat{G}_{l_2m_2,11}(u+t)\, \mathrm{d}u,$$

$$C_{22}^{1,l_1m_1l_2m_2}(t) = \varepsilon a \int_0^\infty \hat{G}_{l_1m_1,21}(u)\, \hat{G}_{l_2m_2,21}(u+t)\, \mathrm{d}u$$

for $t \geqq 0$ where

$$\hat{G}_{lm,11} = G_{11}A_{lm,11} + G_{12}A_{lm,21},$$

$$\hat{G}_{lm,21} = G_{21}A_{lm,11} + G_{22}A_{lm,21};$$

$$A_{lm,11} = mP_{l,11} \int_0^t G_{11}(-s)\, Q_{11}^{(m)}(s)\, \mathrm{d}s,$$

$$A_{lm,21} = P_{l,11} \int_0^t G_{21}(-s)\, Q_{11}^{(m)}(s)\, \mathrm{d}s.$$

Hence, we can establish that

$$C_{ii}^{1,l_1m_1l_2m_2}(t) = \frac{1}{(P_{2,11})^2}\, P_{l_1,11}P_{l_2,11}C_{ii}^{1,2m_12m_2} \quad \text{for} \quad i = 1, 2.$$

The relation

$$C_{22}^{1,l2l2} = C_{11}^{1,l1l1}$$

leads to the statement that the approximative correlation function of $y(t, \omega)$ coincides with $C_{11}^{1,0101}(t)$ from (3.96) and some numerical results can be found in Figs. 3.53a,b, 3.54. The approximative spectral density as to $y'(t, \omega)$ is illustrated in Fig. 3.55 for $m = 1$, $k = 2, 4$, $c = 12$.

Now we deal with $C_x^1(t)$. It is

$$C_{11}^{1,0011}(t) = \varepsilon a \int_0^\infty \hat{G}_{00,11}(u)\, \hat{G}_{11,11}(u+t)\, \mathrm{d}u$$

$$= \varepsilon a \int_0^\infty \hat{G}_{00,11}(u)\, \hat{G}'_{10,11}(u+t)\, \mathrm{d}u$$

and an integration by parts leads to

$$C_{11}^{1,0011}(t) = -\varepsilon a \int_0^\infty \hat{G}_{01,11}(u)\, \hat{G}_{10,11}(u + t)\, \mathrm{d}u$$

$$= -\varepsilon a \int_0^\infty \hat{G}_{11,11}(u)\, \hat{G}_{00,11}(u + t)\, \mathrm{d}u = -C_{11}^{1,1100}(t)$$

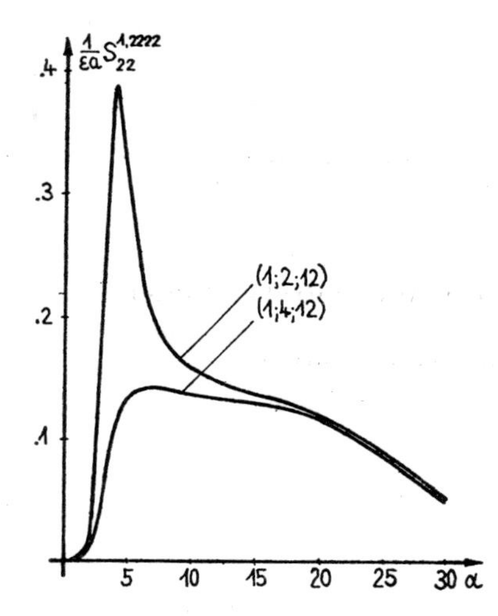

Fig. 3.55. Spectral density of $y'(t, \omega)$ for $m = 1$; $k = 2, 4$; $c = 12$

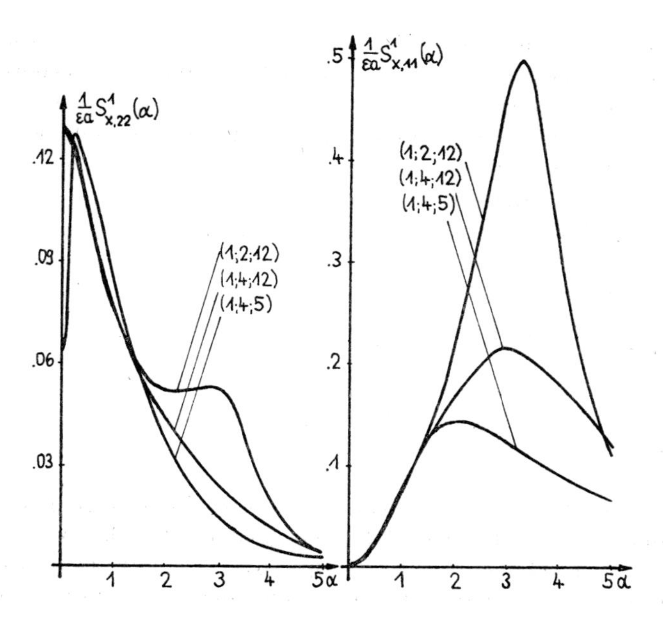

Fig. 3.56. Spectral densities of $x(t, \omega)$ and $x'(t, \omega)$ for different values (m, k, c)

so that we obtain

$$C^1_{x,ii}(t) = C^{1,0000}_{ii}(t) + C^{1,1111}_{ii}(t)$$

and

$$S^1_{x,ii}(\alpha) = S^{1,0000}_{ii}(\alpha) + S^{1,1111}_{ii}(\alpha).$$

In connection with the foregoing results we have

$$S^1_{x,11}(\alpha) = \left(\frac{\alpha}{m}\right)^2 (c^2 + \alpha^2 k^2)\, S^{1,0000}_{22}(\alpha),$$

$$S^1_{x,22}(\alpha) = \frac{1}{m^2} (c^2 + \alpha^2 k^2)\, S^{1,0000}_{22}(\alpha)$$

and these functions are plotted in Fig. 3.56 for different values (m, k, c).

3.2.3. Non-linear vibration systems with weakly correlated input

3.2.3.1. Expansions of first and second moments of solutions

Non-linear systems of the form

$$Ax'' + Bx' + Cx + \eta \sum_{k=2}^{m} B_k(x, x') = D\bar{f}_\varepsilon(t, \omega); \quad x(t_0) = x_0, \quad x'(t_0) = x_1, \quad (3.97)$$

are considered where

$$x(t, \omega) = \big(x_i(t, \omega)\big)^{\mathsf{T}}_{1 \le i \le n},$$

A, B, C, D denote real $n \times n$-matrices, and $\bar{f}_\varepsilon$, x_0, x_1, B_k, $k = 2, \ldots, m$, vectors. Let η be a small parameter. The components of B_k are assumed to be homogeneous polynomials of the k-th degree as to the components of x and x'.

For the further investigations it is advantageous to consider an equivalent system of differential equations of first order

$$Mz' + Nz + \eta \sum_{k=2}^{m} B_k(z) = Pf_\varepsilon(t, \omega); \qquad z(t_0) = z_0, \qquad (3.98)$$

where M, N, P are $2n \times 2n$-matrices,

$$z^{\mathsf{T}}(t, \omega) = \big(x'^{\mathsf{T}}(t, \omega), x^{\mathsf{T}}(t, \omega)\big),$$

and

$$B_{k,l}(z) = \sum_{i_1, i_2, \ldots, i_k = 1}^{2n} b_{l i_1 \ldots i_k} z_{i_1} z_{i_2} \ldots z_{i_k}.$$

We choose the same notations as in 3.2.1. The linear problem belonging to (3.98) is obtained for $\eta = 0$ and was considered in 3.2.2. The matrix $R = M^{-1}N$ is supposed to have eigenvalues with positive real parts, only. Finally, $f_\varepsilon(t, \omega)$ denotes a wide-sense stationary weakly correlated connected vector process having the intensity matrix a.

Now, the solution $z(t, \omega)$ of (3.98) is expanded as to powers of η,

$$z(t) = \sum_{p=0}^{\infty} {}^{p}z(t)\, \eta^{p}. \tag{3.99}$$

To obtain the differential equations for ${}^{p}z(t)$, $p = 0, 1, 2, \ldots$, we substitute this series (3.99) into the differential equations (3.98) and pick out the terms with η^{p}. Then we obtain the equations

$$M\,{}^{0}z' + N\,{}^{0}z = Pf_{\varepsilon}(t, \omega), \tag{3.100}$$

$$M\,{}^{p}z' + N\,{}^{p}z = -\sum_{k=2}^{m} {}^{p-1}B_k(z) \quad \text{for} \quad p = 1, 2, \ldots \tag{3.101}$$

In this, ${}^{p-1}B_k(z)$ denotes the homogeneous terms of $(p-1)$-th order as to η of the non-linear term $B_k(z)$,

$$^{q}B_{k,l}(z) = \sum_{i_1,\ldots,i_k=1}^{2n} b_{li_1\ldots i_k} \sum_{\substack{p_1,\ldots,p_k=0\\ \sum_{s=1}^{k} p_s = q}} {}^{p_1}z_{i_1} \ldots {}^{p_k}z_{i_k}. \tag{3.102}$$

The term on the right-hand side of (3.101) is only dependent on ${}^{q}z(t, \omega)$ with $q < p$ and hence, this term is known if ${}^{p}z(t, \omega)$ is determined from (3.101). The stationary solution of (3.100) can be written as

$$^{0}z(t, \omega) = \int_{-\infty}^{t} G(t-s)\, Pf_{\varepsilon}(s, \omega)\, ds \tag{3.103}$$

(see (3.65)) and of (3.101) as

$$\begin{aligned} ^{p}z(t, \omega) &= -\int_{-\infty}^{t} G(t-s) \sum_{k=2}^{m} {}^{p-1}B_k\big(z(s)\big)\, ds \\ &= -\sum_{k=2}^{m} \int_{-\infty}^{t} G(t-s)\, {}^{p-1}B_k\big(z(s)\big)\, ds. \end{aligned} \tag{3.104}$$

Consider the second moments of the solution $z(t, \omega)$,

$$\begin{aligned} \langle z(t_1)\, z^{\mathsf T}(t_2)\rangle &= \langle {}^{0}z(t_1)\, {}^{0}z^{\mathsf T}(t_2)\rangle + [\langle {}^{0}z(t_1)\, {}^{1}z^{\mathsf T}(t_2)\rangle + \langle {}^{1}z(t_1)\, {}^{0}z^{\mathsf T}(t_2)\rangle]\, \eta \\ &\quad + [\langle {}^{0}z(t_1)\, {}^{2}z^{\mathsf T}(t_2)\rangle + \langle {}^{1}z(t_1)\, {}^{1}z^{\mathsf T}(t_2)\rangle + \langle {}^{2}z(t_1)\, {}^{0}z^{\mathsf T}(t_2)\rangle]\, \eta^2 + \cdots \end{aligned} \tag{3.105}$$

We now have the goal to expand the moments contained in (3.105). First, it follows

$$\begin{aligned} \langle {}^{0}z_i(t_1)\, {}^{0}z_j(t_2)\rangle &= \sum_{u,v=1}^{2n} {}^{2}A_1\big(\tilde G_{iu}(t_1 - .), \tilde G_{jv}(t_2 - .)\big)\, \varepsilon \\ &\quad + \sum_{u,v=1}^{2n} {}^{2}A_2\big(\tilde G_{iu}(t_1 - .), \tilde G_{jv}(t_2 - .)\big)\, \varepsilon^2 + O(\varepsilon^3) \end{aligned} \tag{3.106}$$

(see (3.79)). Furthermore, the linear terms as to η demand the study of

$$\langle {}^0z_i(t_1)\, {}^1z_j(t_2)\rangle = -\sum_{k=2}^{m} \sum_{u,v=1}^{2n} \int_{-\infty}^{t_1} \int_{-\infty}^{t_2} \tilde{G}_{iu}(t_1 - s_1)\, G_{jv}(t_2 - s_2) \times \langle f_{\varepsilon,u}(s_1)\, {}^0B_{k,v}(z(s_2))\rangle\, ds_1\, ds_2 \tag{3.107}$$

where

$${}^0B_{k,v}(z) = \sum_{i_1,\ldots,i_k=1}^{2n} b_{v i_1\ldots i_k}\, {}^0z_{i_1}\, {}^0z_{i_2} \ldots {}^0z_{i_k}$$

and, in particular,

$${}^0B_{2,v}(z) = \sum_{i_1,i_2=1}^{2n} b_{v i_1 i_2}\, {}^0z_{i_1}\, {}^0z_{i_2},$$

$${}^0B_{3,v}(z) = \sum_{i_1,i_2,i_3=1}^{2n} b_{v i_1 i_2 i_3}\, {}^0z_{i_1}\, {}^0z_{i_2}\, {}^0z_{i_3}.$$

For abbreviation, we define

$$B_{k,ijv}(t_1, t_2) \doteq \int_{-\infty}^{t_2} G_{jv}(t_2 - s)\, \langle {}^0z_i(t_1)\, {}^0B_{k,v}(z(s))\rangle\, ds$$

and obtain immediately

$$B_{k,ijv}(t_1, t_2) = O(\varepsilon^3) \quad \text{for} \quad k \geqq 4$$

using (1.24), (1.26). Setting $k = 2, 3$ it is

$$B_{2,ijv}(t_1, t_2) = \sum_{i_1,i_2=1}^{2n} b_{v i_1 i_2} \int_{-\infty}^{t_2} G_{jv}(t_2 - s)\, \langle {}^0z_i(t_1)\, {}^0z_{i_1}(s)\, {}^0z_{i_2}(s)\rangle\, ds\,,$$

$$B_{3,ijv}(t_1, t_2) = \sum_{i_1,i_2,i_3=1}^{2n} b_{v i_1 i_2 i_3} \int_{-\infty}^{t_2} G_{jv}(t_2 - s)\, \langle {}^0z_i(t_1)\, {}^0z_{i_1}(s)\, {}^0z_{i_2}(s)\, {}^0z_{i_3}(s)\rangle\, ds\,.$$

The equations (1.24), (1.26) lead to

$$\langle {}^0z_i(t)\, {}^0z_{i_1}(s)\, {}^0z_{i_2}(s)\rangle = {}^3A_{2,ii_1i_2}(t, s, s)\, \varepsilon^2 + O(\varepsilon^3)\,,$$

$$\langle {}^0z_i(t)\, {}^0z_{i_1}(s)\, {}^0z_{i_2}(s)\, {}^0z_{i_3}(s)\rangle = {}^2A_{1,ii_1i_2i_3}(t, s, s, s)\, \varepsilon^2 + O(\varepsilon^3)$$

with

$${}^2A_{1,i_1i_2i_3i_4}(t_1, t_2, t_3, t_4) \doteq {}^2A_{1,i_1i_2}(t_1, t_2)\, {}^2A_{1,i_3i_4}(t_3, t_4) + {}^2A_{1,i_1i_3}(t_1, t_3)\, {}^2A_{1,i_2i_4}(t_2, t_4) + {}^2A_{1,i_1i_4}(t_1, t_4)\, {}^2A_{1,i_2i_3}(t_2, t_3) \tag{3.108}$$

where we define

$${}^2A_{1,ij}(t, r) \doteq \sum_{p,q=1}^{2n} \int_{-\infty}^{\min\{t,r\}} \tilde{G}_{ip}(t - u)\, \tilde{G}_{jq}(r - u)\, a_{pq}(u)\, du\,,$$

$${}^3A_{2,ijk}(t, r, s) \doteq \sum_{p,q,v=1}^{2n} \int_{-\infty}^{\min\{t,r,s\}} \tilde{G}_{ip}(t - u)\, \tilde{G}_{jq}(r - u)\, \tilde{G}_{kv}(s - u)\, a_{pqv}(u)\, du$$

and the results of Section 1.6.1 and 1.6.6 are taken into consideration. This shows that the moment

$$\langle {}^0z_i(t_1)\, {}^1z_j(t_2)\rangle$$

has the form

$$\langle {}^0z_i(t_1)\, {}^1z_j(t_2)\rangle = -\sum_{v=1}^{2n} \left[\sum_{i_1,i_2=1}^{2n} b_{vi_1i_2} \int_{-\infty}^{t_2} G_{jv}(t_2 - s)\, {}^3A_{2,ii_1i_2}(t_1, s, s)\, \mathrm{d}s \right.$$
$$\left. + \sum_{i_1,i_2,i_3=1}^{2n} b_{vi_1i_2i_3} \int_{-\infty}^{t_2} G_{jv}(t_2 - s)\, {}^2A_{1,ii_1i_2i_3}(t_1, s, s, s)\, \mathrm{d}s \right] \varepsilon^2 + O(\varepsilon^3). \tag{3.109}$$

To obtain all terms of the order 2 as to ε we have to investigate the third summand on the right-hand side of (3.105). Thus,

$$\langle {}^1z_i(t_1)\, {}^1z_j(t_2)\rangle = \sum_{k,h=2}^{m} \sum_{u,v=1}^{2n} \int_{-\infty}^{t_1} \int_{-\infty}^{t_2} G_{iu}(t_1 - s_1)\, G_{jv}(t_2 - s_2)$$
$$\times \left\langle {}^0B_{k,u}\big(z(s_1)\big)\, {}^0B_{h,v}\big(z(s_2)\big)\right\rangle \mathrm{d}s_1\, \mathrm{d}s_2$$
$$= \sum_{u,v=1}^{2n} \int_{-\infty}^{t_1} \int_{-\infty}^{t_2} G_{iu}(t_1 - s_1)\, G_{jv}(t_2 - s_2)$$
$$\times \left\langle {}^0B_{2,u}\big(z(s_1)\big)\, {}^0B_{2,v}\big(z(s_2)\big)\right\rangle \mathrm{d}s_1\, \mathrm{d}s_2 + O(\varepsilon^3)$$
$$= \sum_{u,v=1}^{2n} \sum_{i_1,i_2,i_3,i_4=1}^{2n} b_{ui_1i_2} b_{vi_3i_4} \int_{-\infty}^{t_1} \int_{-\infty}^{t_2} G_{iu}(t_1 - s_1)\, G_{jv}(t_2 - s_2)$$
$$\times \langle {}^0z_{i_1}(s_1)\, {}^0z_{i_2}(s_1)\, {}^0z_{i_3}(s_2)\, {}^0z_{i_4}(s_2)\rangle\, \mathrm{d}s_1\, \mathrm{d}s_2 + O(\varepsilon^3)$$

and using the definition (3.108) the moment $\langle {}^1z_i(t_1)\, {}^1z_j(t_2)\rangle$ has the expansion

$$\langle {}^1z_i(t_1)\, {}^1z_j(t_2)\rangle = \sum_{u,v=1}^{2n} \sum_{i_1,i_2,i_3,i_4=1}^{2n} b_{ui_1i_2} b_{vi_3i_4}$$
$$\times \int_{-\infty}^{t_1} \int_{-\infty}^{t_2} G_{iu}(t_1 - s)\, G_{jv}(t_2 - r)\, {}^2A_{1,i_1i_2i_3i_4}(s, s, r, r)\, \mathrm{d}s\, \mathrm{d}r \cdot \varepsilon^2 + O(\varepsilon^3). \tag{3.110}$$

The moment $\langle {}^0z_i(t_1)\, {}^2z_j(t_2)\rangle$ can be written as

$$\langle {}^0z_i(t_1)\, {}^2z_j(t_2)\rangle = -\sum_{k=2}^{m} \sum_{v=1}^{2n} \int_{-\infty}^{t_2} G_{jv}(t_2 - s) \left\langle {}^0z_i(t_1)\, {}^1B_{k,v}\big(z(s)\big)\right\rangle \mathrm{d}s$$

where ${}^1B_{k,v}$ is given by

$$ {}^1B_{k,v}\big(z(s)\big) = \sum_{i_1,\dots,i_k=1}^{2n} b_{vi_1\dots i_k} \left\{ \sum_{p=1}^{k} {}^1z_{i_p}(s) \prod_{\substack{q=1\\ q\neq p}}^{k} {}^0z_{i_q}(s) \right\}$$

and, in particular, ${}^1B_{2,v}$ by

$$ {}^1B_{2,v}\big(z(s)\big) = 2 \sum_{i_1,i_2=1}^{2n} b_{vi_1i_2}\, {}^1z_{i_1}(s)\, {}^0z_{i_2}(s) \,. $$

We consider

$$ \langle {}^1z_{i_1}(t_1)\, {}^0z_{i_2}(t_2)\, {}^0z_{i_3}(t_3)\rangle $$

$$ = -\sum_{k=2}^{m} \sum_{v=1}^{2n} \int_{-\infty}^{t_1} G_{i_1v}(t_1 - s) \left\langle {}^0B_{k,v}\big(z(s)\big)\, {}^0z_{i_2}(t_2)\, {}^0z_{i_3}(t_3)\right\rangle \mathrm{d}s $$

$$ = -\sum_{v=1}^{2n} \sum_{j_1,j_2=1}^{2n} b_{vj_1j_2} \int_{-\infty}^{t_1} G_{i_1v}(t_1 - s) \langle {}^0z_{j_1}(s)\, {}^0z_{j_2}(s)\, {}^0z_{i_2}(t_2)\, {}^0z_{i_3}(t_3)\rangle \,\mathrm{d}s + O(\varepsilon^3) $$

$$ = -\sum_{v=1}^{2n} \sum_{j_1,j_2=1}^{2n} b_{vj_2j_2} \int_{-\infty}^{t_1} G_{i_1v}(t_1 - s)\, {}^2A_{1,j_1j_2i_2i_3}(s, s, t_2, t_3) \,\mathrm{d}s \cdot \varepsilon^2 + O(\varepsilon^3) $$

and with this

$$ \langle {}^0z_i(t_1)\, {}^2z_j(t_2)\rangle = -\sum_{u=1}^{2n} \sum_{i_1,i_2=1}^{2n} b_{ui_1i_2} \int_{-\infty}^{t_2} G_{ju}(t_2 - s) $$

$$ \times [\langle {}^0z_i(t_1)\, {}^1z_{i_1}(s)\, {}^0z_{i_2}(s)\rangle + \langle {}^0z_i(t_1)\, {}^0z_{i_1}(s)\, {}^1z_{i_2}(s)\rangle] \,\mathrm{d}s + O(\varepsilon^3) $$

$$ = \sum_{u,v=1}^{2n} \sum_{i_1,i_2=1}^{2n} \sum_{j_1,j_2=1}^{2n} b_{ui_1i_2} b_{vj_1j_2} \int_{-\infty}^{t_2} G_{ju}(t_2 - s) $$

$$ \times \left[\int_{-\infty}^{s} G_{i_1v}(s - r)\, {}^2A_{1,j_1j_2i_2i}(r, r, s, t_1) \,\mathrm{d}r \right. $$

$$ \left. + \int_{-\infty}^{s} G_{i_2v}(s - r)\, {}^2A_{1,j_1j_2ii_1}(r, r, t_1, s) \,\mathrm{d}r \right] \mathrm{d}s \cdot \varepsilon^2 + O(\varepsilon^3) $$

$$ = 2 \sum_{u,v=1}^{2n} \sum_{i_1,i_2=1}^{2n} \sum_{j_1,j_2=1}^{2n} b_{ui_1i_2} b_{vj_1j_2} \int_{-\infty}^{t_2} G_{ju}(t_2 - s) \int_{-\infty}^{s} G_{i_1v}(s - r) $$

$$ \times\, {}^2A_{1,j_1j_2i_2i}(r, r, s, t_1) \,\mathrm{d}r \,\mathrm{d}s \cdot \varepsilon^2 + O(\varepsilon^3) \tag{3.111} $$

where we used the relations

$$ {}^2A_{1,ij}(t, r) = {}^2A_{1,ji}(r, t) \,, $$

$$ {}^2A_{1,i_1i_2ij}(t_1, t_2, r, t) = {}^2A_{1,i_1i_2ji}(t_1, t_2, t, r) \,, $$

$$ b_{vij} = b_{vji} \,. $$

We deal with terms of the form

$$ \langle {}^pz_i(t_1)\, {}^qz_j(t_2)\rangle \quad \text{with} \quad p + q \geqq 3. $$

These terms lead to expressions which are characterized by

$$\langle {}^{p-1}B_k \, {}^{q-1}B_h \rangle \quad \text{for} \quad k, h \geqq 2 .$$

The lowest order as to ε is obtained for $k = h = 2$ and $p = 1, q = 2$, for example. Hence, terms of the form

$$\langle {}^1z_{i_1} \, {}^0z_{i_2} \, {}^0z_{i_3} \, {}^0z_{i_4} \rangle$$

have to be considered. But these expressions have the order $O(\varepsilon^3)$ and must not be taken into consideration.

Finally, replacing (3.106), (3.109), and (3.111) in (3.105) we get

$$\begin{aligned}
\langle z_i(t_1)\, z_j(t_2) \rangle &= {}^2A_{1,ij}(t_1, t_2)\, \varepsilon \\
&+ \Bigg[{}^2A_{2,ij}(t_1, t_2) - \Bigg\{ \sum_{i_1,i_2,u=1}^{2n} b_{ui_1i_2} \Bigg(\int_{-\infty}^{t_2} G_{ju}(t_2 - s)\, {}^3A_{2,ii_1i_2}(t_1, s, s)\, \mathrm{d}s \\
&+ \int_{-\infty}^{t_1} G_{iu}(t_1 - s)\, {}^3A_{2,ji_1i_2}(t_2, s, s)\, \mathrm{d}s \Bigg) \\
&+ \sum_{i_1,i_2,i_3,u=1}^{2n} b_{ui_1i_2i_3} \Bigg(\int_{-\infty}^{t_2} G_{ju}(t_2 - s)\, {}^2A_{1,ii_1i_2i_3}(t_1, s, s, s)\, \mathrm{d}s \\
&+ \int_{-\infty}^{t_1} G_{iu}(t_1 - s)\, {}^2A_{1,ji_1i_2i_3}(t_2, s, s, s)\, \mathrm{d}s \Bigg) \Bigg\} \eta + 2 \sum_{i_1,i_2,i_3,i_4,u,v=1}^{2n} b_{ui_1i_2} b_{vi_3i_4} \\
&\times \Bigg(\int_{-\infty}^{t_2} G_{ju}(t_2 - s) \int_{-\infty}^{s} G_{i_1v}(s - r)\, {}^2A_{1,ii_2i_3i_4}(t_1, s, r, r)\, \mathrm{d}r\, \mathrm{d}s \\
&+ \int_{-\infty}^{t_1} G_{iu}(t_1 - s) \int_{-\infty}^{s} G_{i_1v}(s - r)\, {}^2A_{1,ji_2i_3i_4}(t_2, s, r, r)\, \mathrm{d}r\, \mathrm{d}s \\
&+ \frac{1}{2} \int_{-\infty}^{t_1} \int_{-\infty}^{t_2} G_{iu}(t_1 - s)\, G_{jv}(t_2 - r)\, {}^2A_{1,i_1i_2i_3i_4}(s, s, r, r)\, \mathrm{d}r\, \mathrm{d}s \Bigg) \eta^2 \Bigg] \\
&\times \varepsilon^2 + O(\varepsilon^3).
\end{aligned} \tag{3.112}$$

Assuming

$$a_{pq}(u) = a_{pq} = \text{const}; \qquad a_{pqv}(u) = a_{pqv} = \text{const} \tag{3.113}$$

we can prove that $\langle z_i(t_1)\, z_j(t_2) \rangle$ is only a function of $|t_2 - t_1|$. For this purpose, the term

$$\int_{-\infty}^{t_2} G_{ju}(t_2 - s)\, {}^3A_{2,ii_1i_2}(t_1, s, s)\, \mathrm{d}s$$

leads to a sum of expressions of the form

$$\begin{cases} a_{pqv}\left[\int\limits_0^\infty G_{ju}(t_2 - t_1 + s) \int\limits_0^\infty \tilde{G}_{ip}(s + r)\, \tilde{G}_{i_1q}(r)\, \tilde{G}_{i_2v}(r)\, \mathrm{d}r\, \mathrm{d}s \right. \\ \left. + \int\limits_0^{t_2-t_1} G_{ju}(t_2 - t_1 - s) \int\limits_0^\infty \tilde{G}_{ip}(r)\, \tilde{G}_{i_1q}(r + s)\, \tilde{G}_{i_2v}(r + s)\, \mathrm{d}r\, \mathrm{d}s\right] & \text{for} \quad t_1 \leqq t_2, \\ a_{pqv} \int\limits_0^\infty G_{ju}(s) \int\limits_0^\infty \tilde{G}_{ip}(t_1 - t_2 + s + r)\, \tilde{G}_{i_1q}(r)\, \tilde{G}_{i_2v}(r)\, \mathrm{d}r\, \mathrm{d}s & \text{for} \quad t_2 \leqq t_1 \end{cases}$$

and, hence, this term is only dependent on $|t_2 - t_1|$. Now, it is

$$\int\limits_{-\infty}^{t_2} G_{ju}(t_2 - s)\, {}^2A_{1,ii_1i_2i_3}(t_1, s, s, s)\, \mathrm{d}s$$

$$= \int\limits_0^\infty G_{ju}(s)\, {}^2A_{1,ii_1i_2i_3}(t_1, t_2 - s, t_2 - s, t_2 - s)\, \mathrm{d}s \tag{3.114}$$

and

$${}^2A_{1,ijkl}(t_1, t_2 - s, t_2 - s, t_2 - s)$$

$$= {}^2A_{1,ij}(t_1, t_2 - s)\, {}^2A_{1,kl}(0, 0) + {}^2A_{1,ik}(t_1, t_2 - s)\, {}^2A_{1,jl}(0, 0)$$

$$+ {}^2A_{1,il}(t_1, t_2 - s)\, {}^2A_{1,jk}(0, 0).$$

For ${}^2A_{1,ij}(t, r)$ we obtain

$${}^2A_{1,ij}(t, r) = \sum_{p,q=1}^{2n} a_{pq} \begin{cases} \int\limits_0^\infty \tilde{G}_{ip}(s)\, \tilde{G}_{jq}(r - t + s)\, \mathrm{d}s & \text{for} \quad t \leqq r, \\ \int\limits_0^\infty \tilde{G}_{ip}(t - r + s)\, G_{jq}(s)\, \mathrm{d}s & \text{for} \quad r \leqq t \end{cases} \tag{3.115}$$

and (3.114) is a function of $|t_2 - t_1|$. We deal with

$$\int\limits_{-\infty}^{t_2} G_{ju}(t_2 - s) \int\limits_{-\infty}^{s} G_{i_1v}(s - r)\, {}^2A_{1,ii_2i_3i_4}(t_1, s, r, r)\, \mathrm{d}r\, \mathrm{d}s$$

$$= \int\limits_0^\infty G_{ju}(s) \int\limits_0^\infty G_{i_1v}(r)\, {}^2A_{1,ii_2i_3i_4}(t_1, t_2 - s, t_2 - s - r, t_2 - s - r)\, \mathrm{d}r\, \mathrm{d}s.$$

(3.108) shows

$${}^2A_{1,ijkl}(t_1, t_2 - s, t_2 - s - r, t_2 - s - r)$$

$$= {}^2A_{1,ij}(t_1, t_2 - s)\, {}^2A_{1,kl}(0, 0) + {}^2A_{1,ik}(t_1, t_2 - s - r)\, {}^2A_{1,jl}(0, -r)$$

$$+ {}^2A_{1,il}(t_1, t_2 - s - r)\, {}^2A_{1,jk}(0, -r)$$

and the above expression is again dependent on $|t_2 - t_1|$. Finally, the term

$$\int_{-\infty}^{t_1} \int_{-\infty}^{t_2} G_{iu}(t_1 - s)\, G_{jv}(t_2 - r)\, {}^2A_{1,i_1i_2i_3i_4}(s, s, r, r)\, \mathrm{d}r\, \mathrm{d}s$$

$$= \int_0^\infty \int_0^\infty G_{iu}(s)\, G_{jv}(r)\, {}^2A_{1,i_1i_2i_3i_4}(t_1 - s, t_1 - s, t_2 - r, t_2 - r)\, \mathrm{d}s\, \mathrm{d}r \tag{3.116}$$

is considered. From

$${}^2A_{1,ijkl}(t_1 - s, t_1 - s, t_2 - r, t_2 - r)$$
$$= {}^2A_{1,ij}(0, 0)\, {}^2A_{1,kl}(0, 0) + {}^2A_{1,ik}(t_1 - s, t_2 - r)\, {}^2A_{1,jl}(t_1 - s, t_2 - r)$$
$$+ {}^2A_{1,il}(t_1 - s, t_2 - r)\, {}^2A_{1,jk}(t_1 - s, t_2 - r)$$

we also have the same dependence of (3.116). With the help of (3.78) and the assumption of the wide-sense stationarity of $f_\varepsilon(t, \omega)$ the term ${}^2A_{2,ij}(t_1, t_2)$ is only dependent on $|t_2 - t_1|$.

The expectation $\langle z(t) \rangle$ of the solution of (3.98) can also be determined by the same method. It is

$$\langle z(t) \rangle = \langle {}^1z(t) \rangle\, \eta + \langle {}^2z(t) \rangle\, \eta^2 + \langle {}^3z(t) \rangle\, \eta^3 + \cdots$$

where $\langle {}^0z(t) \rangle = 0$ was applied. Using (3.104) we obtain

$$\langle {}^pz(t) \rangle = -\sum_{k=2}^{m} \int_{-\infty}^{t} G(t - s) \left\langle {}^{p-1}B_k\big(z(s)\big)\right\rangle \mathrm{d}s \quad \text{for} \quad p = 1, 2, \ldots$$

First, we start with $p = 1$,

$$\langle {}^1z_i(t) \rangle = -\sum_{q=1}^{2n} \int_{-\infty}^{t} G_{iq}(t - s) \left[\left\langle {}^0B_{2,q}\big(z(s)\big)\right\rangle + \left\langle {}^0B_{3,q}\big(z(s)\big)\right\rangle\right.$$
$$\left. + \left\langle {}^0B_{4,q}\big(z(s)\big)\right\rangle\right] \mathrm{d}s + O(\varepsilon^3)$$
$$= -\sum_{q=1}^{2n} \int_{-\infty}^{t} G_{iq}(t - s) \left[\sum_{i_1,i_2=1}^{2n} b_{qi_1i_2} \langle {}^0z_{i_1}(s)\, {}^0z_{i_2}(s) \rangle\right.$$
$$+ \sum_{i_1,i_2,i_3=1}^{2n} b_{qi_1i_2i_3} \langle {}^0z_{i_1}(s)\, {}^0z_{i_2}(s)\, {}^0z_{i_3}(s) \rangle$$
$$\left. + \sum_{i_1,i_2,i_3,i_4=1}^{2n} b_{qi_1i_2i_3i_4} \langle {}^0z_{i_1}(s)\, {}^0z_{i_2}(s)\, {}^0z_{i_3}(s)\, {}^0z_{i_4}(s) \rangle\right] \mathrm{d}s + O(\varepsilon^3)$$
$$= -\sum_{q=1}^{2n} \sum_{i_1,i_2=1}^{2n} b_{qi_1i_2} \int_{-\infty}^{t} G_{iq}(t - s)\, {}^2A_{1,i_1i_2}(s, s)\, \mathrm{d}s \cdot \varepsilon$$
$$- \sum_{p=1}^{2n} \left[\sum_{i_1,i_2=1}^{2n} b_{qi_1i_2} \int_{-\infty}^{t} G_{iq}(t - s)\, {}^2A_{2,i_1i_2}(s, s)\, \mathrm{d}s\right.$$

$$+ \sum_{i_1,i_2,i_3=1}^{2n} b_{qi_1i_2i_3} \int_{-\infty}^{t} G_{iq}(t-s)\, {}^3A_{2,i_1i_2i_3}(s,s,s)\, \mathrm{d}s$$

$$\left. + \sum_{i_1,i_2,i_3,i_4=1}^{2n} b_{qi_1i_2i_3i_4} \int_{-\infty}^{t} G_{iq}(t-s)\, {}^2A_{1,i_1i_2i_3i_4}(s,s,s,s)\, \mathrm{d}s \right] \varepsilon^2 + O(\varepsilon^3). \tag{3.117}$$

For $p = 2$ it follows

$$\langle {}^2z_i(t)\rangle = -\sum_{q=1}^{2n} \int_{-\infty}^{t} G_{iq}(t-s) \left[\langle {}^1B_{2,q}(z(s))\rangle + \langle {}^1B_{3,q}(z(s))\rangle\right] \mathrm{d}s + O(\varepsilon^3)$$

$$= -\sum_{q=1}^{2n} \int_{-\infty}^{t} G_{iq}(t-s) \left[2 \sum_{i_1,i_2=1}^{2n} b_{qi_1i_2} \langle {}^1z_{i_1}(s)\, {}^0z_{i_2}(s)\rangle\right.$$

$$\left. + 3 \sum_{i_1,i_2,i_3=1}^{2n} b_{qi_1i_2i_3} \langle {}^1z_{i_1}(s)\, {}^0z_{i_2}(s)\, {}^0z_{i_3}(s)\rangle \right] \mathrm{d}s + O(\varepsilon^3)$$

$$= \sum_{p,q=1}^{2n} \left[2 \sum_{i_1,i_2,i_3,i_4=1}^{2n} b_{pi_1i_2} b_{qi_3i_4} \int_{-\infty}^{t} \int_{-\infty}^{s} G_{ip}(t-s)\, G_{i_1q}(s-r)\, {}^3A_{2,i_2i_3i_4}(s,r,r)\right.$$

$$+ 2 \sum_{i_1,i_2,i_3,i_4,i_5=1}^{2n} b_{pi_1i_2} b_{qi_3i_4i_5}$$

$$\times \int_{-\infty}^{t} \int_{-\infty}^{s} G_{ip}(t-s)\, G_{i_1q}(s-r)\, {}^2A_{1,i_2i_3i_4i_5}(s,r,r,r)\, \mathrm{d}r\, \mathrm{d}s$$

$$+ 3 \sum_{i_1,i_2,i_3,i_4,i_5=1}^{2n} b_{pi_1i_2i_3} b_{qi_4i_5}$$

$$\left. \times \int_{-\infty}^{t} \int_{-\infty}^{s} G_{ip}(t-s)\, G_{i_1q}(s-r)\, {}^2A_{1,i_2i_3i_4i_5}(s,s,r,r)\, \mathrm{d}r\, \mathrm{d}s \right] \varepsilon^2 + O(\varepsilon^3), \tag{3.118}$$

and for $p = 3$ the expression

$$\langle {}^3z_i(t)\rangle = -\sum_{q=1}^{2n} \int_{-\infty}^{t} G_{iq}(t-s) \langle {}^2B_{2,q}(z(s))\rangle\, \mathrm{d}s + O(\varepsilon^3)$$

$$= -\sum_{q=1}^{2n} \int_{-\infty}^{t} G_{iq}(t-s) \sum_{i_1,i_2=1}^{2n} b_{qi_1i_2}$$

$$\times [2\langle {}^2z_{i_1}(s)\, {}^0z_{i_2}(s)\rangle + \langle {}^1z_{i_1}(s)\, {}^1z_{i_2}(s)\rangle\, \mathrm{d}s.$$

We use (3.110) and (3.111) and get

$$\langle {}^3z_i(t)\rangle = -\sum_{q,u,v=1}^{2n} \sum_{i_1,i_2,i_3,i_4,i_5,i_6=1}^{2n} b_{qi_1i_2} b_{ui_3i_4} b_{vi_5i_6} \int_{-\infty}^{t} G_{ip}(t-s)$$

$$\times \left[4 \int_{-\infty}^{s} \int_{-\infty}^{r} G_{i_1u}(s-r)\, G_{i_2v}(r-x)\, {}^2A_{1,i_2i_4i_5i_6}(s,r,x,x)\, \mathrm{d}x\, \mathrm{d}r \right.$$

$$+ \int_{-\infty}^{s} \int_{-\infty}^{s} G_{i_1 u}(s-r)\, G_{i_2 v}(s-x)\, {}^2A_{1,\, i_3 i_4 i_5 i_6}(r, r, x, x)\, \mathrm{d}x\, \mathrm{d}r \Bigg]$$

$$\times \mathrm{d}s \cdot \varepsilon^2 + O(\varepsilon^3). \tag{3.119}$$

Summarizing we can write

$$\langle z_i(t) \rangle^1 = -\sum_{p=1}^{2n} \sum_{i_1, i_2 = 1}^{2n} b_{p i_1 i_2} \int_{-\infty}^{t} G_{ip}(t-s)\, {}^2A_{1,\, i_1 i_2}(s, s)\, \mathrm{d}s \cdot \varepsilon\eta, \tag{3.120}$$

$$\langle z_i(t) \rangle^2 = \langle {}^1z_i(t) \rangle^2\, \eta + \langle {}^2z_i(t) \rangle^2\, \eta^2 + \langle {}^3z_i(t) \rangle^2\, \eta^3 \tag{3.121}$$

where

$$\langle z_i(t) \rangle^p, \qquad \langle {}^kz_i(t) \rangle^p, \qquad p = 1, 2; \quad k = 1, 2, 3,$$

denote the expectation of $z_i(t, \omega)$ of the order p as to ε and $\langle {}^1z_i(t) \rangle^2$ is given by (3.117), $\langle {}^2z_i(t) \rangle^2$ by (3.118), and $\langle {}^3z_i(t) \rangle^2$ by (3.119).

The approximative solution of the averaging problem of (3.98) is given by $\langle z(t) \rangle^p$, $p = 1, 2$, since the stationary solution of the averaged problem belonging to (3.98) vanishes. The first approximation can be written as

$$\langle z_i(t) \rangle^1 = -\sum_{p=1}^{2n} \sum_{i_1, i_2 = 1}^{2n} b_{p i_1 i_2}\, {}^2A_{1,\, i_1 i_2}(0, 0) \int_{0}^{\infty} G_{ip}(s)\, \mathrm{d}s \cdot \varepsilon\eta$$

if $a_{pq}(u) = a_{pq} = \text{const}$ is assumed; i.e. $\langle z_i(t) \rangle^1$ is constant. By means of

$${}^2A_{1,\, ij}(0, 0) = \sum_{u,v=1}^{2n} a_{uv} \int_{0}^{\infty} G_{iu}(s)\, G_{jv}(s)\, \mathrm{d}s,$$

$$\int_{0}^{\infty} G(s)\, \mathrm{d}s = \int_{0}^{\infty} \exp\left(-M^{-1}Ns\right) M^{-1}\, \mathrm{d}s = N^{-1}$$

it follows

$$\langle z_i(t) \rangle^1 = -\sum_{p=1}^{2n} (N^{-1})_{ip} \sum_{i_1, i_2, i_3, i_4 = 1}^{2n} b_{p i_1 i_2} a_{i_3 i_4} \int_{0}^{\infty} G_{i_1 i_3}(s)\, G_{i_2 i_4}(s)\, \mathrm{d}s \cdot \varepsilon\eta.$$

Now, it is possible to determine the approximations of the correlation function of $z(t, \omega)$. We have

$$\Big\langle \big(z_i(t_1) - \langle z_i(t_1) \rangle\big) \big(z_j(t_2) - \langle z_j(t_2) \rangle\big) \Big\rangle = \langle z_i(t_1)\, z_j(t_2) \rangle - \langle z_i(t_1) \rangle \langle z_j(t_2) \rangle$$

and consequently

$$C_{ij}^1(t_1, t_2) = \langle z_i(t_1)\, z_j(t_2) \rangle^1,$$

$$C_{ij}^2(t_1, t_2) = \langle z_i(t_1)\, z_j(t_2) \rangle^2 - \langle z_i(t_1) \rangle^1 \langle z_j(t_2) \rangle^1$$

where $\langle z_i(t_1)\, z_j(t_2) \rangle^p$, $p = 1, 2$, is given by (3.112) and $\langle z_i(t_k) \rangle^1$ by (3.120). The results of the first order as to $C_{ij}(t_1, t_2)$ do not differ to the results of the linear case ($\eta = 0$). This is easy to see from

$$C_{ij}^1(t_1, t_2) = {}^2A_{1,\, ij}(t_1, t_2)\, \varepsilon.$$

By means of (3.113) it follows

$$C_{ij}^p(t_1, t_2) = C_{ij}^p(|t_2 - t_1|) \quad \text{for} \quad p = 1, 2.$$

In this case the approximations of the spectral densities are given by

$$S_{ij}^p(\alpha) = \frac{1}{\pi} \int_0^\infty \cos(\alpha t)\, C_{ij}^p(t)\, dt. \tag{3.122}$$

If we replace the input function of (3.98) by

$$\sum_{l=0}^{2} P_l \int_{-\infty}^{t} Q^{(l)}(t-s)\, f_\varepsilon(s, \omega)\, ds$$

(see 3.2.2.3) then we obtain

$$M\,{}^0z' + N\,{}^0z = \sum_{l=0}^{2} P_l \int_{-\infty}^{t} Q^{(l)}(t-s)\, f_\varepsilon(s, \omega)\, ds,$$

$$M\,{}^pz' + N\,{}^pz = -\sum_{k=2}^{m} {}^{p-1}B_k(z) \quad \text{for} \quad p = 1, 2, \ldots$$

in comparison with (3.100), (3.101). To determine the solution of the first equation we use (3.84) and can write

$${}^0z(t, \omega) = \int_{-\infty}^{t} \hat{G}(t-s)\, f_\varepsilon(s, \omega)\, ds$$

where $\hat{G}(t)$ is given by (3.85). For ${}^pz(t, \omega)$ we get the same relation as (3.104). Now it is possible to deduce similar results as to $\langle z(t) \rangle$ and $\langle z(t_1)\, z^\mathsf{T}(t_2) \rangle$ by replacing the matrix $G(t)$ in connection with ${}^0z(t, \omega)$ by $\hat{G}(t)$. In this case we can deal with non-linear problems in which the inhomogeneous term contains derivatives.

3.2.3.2. Examples

Consider a vibration equation with a cubic non-linearity,

$$mx'' + kx' + cx + \eta bx^3 = -mf_\varepsilon(t, \omega) \tag{3.123}$$

and assume that $f_\varepsilon(t, \omega)$ is symmetrically distributed. Then we get

$${}^3A_{2,ijk} = 0.$$

Setting

$$z(t, \omega) = \big(x'(t, \omega), x(t, \omega)\big)^\mathsf{T}$$

it is

$$Mz' + Nz + \eta B_3(z) = Pf_\varepsilon(t, \omega)$$

where

$$M = \begin{pmatrix} m & 0 \\ 0 & 1 \end{pmatrix}; \quad N = \begin{pmatrix} k & c \\ -1 & 0 \end{pmatrix}; \quad P = \begin{pmatrix} -m & 0 \\ 0 & 0 \end{pmatrix}; \quad B_3(z) = \begin{pmatrix} bz_2^3 \\ 0 \end{pmatrix}.$$

For the further investigations we can use the notations of Section 3.2.2.2.

Using (3.121) it is

$$\langle z_i(t)\rangle^p = 0 \quad \text{for} \quad p = 1, 2$$

and the approximations of the correlation function $\langle z_i(t_1)\, z_j(t_2)\rangle$ can be obtained from (3.112),

$$\langle z_i(t_1)\, z_j(t_2)\rangle^1 = {}^2A_{1,ij}(t_1, t_2)\, \varepsilon,$$

$$\langle z_i(t_1)\, z_j(t_2)\rangle^2 = {}^2A_{1,ij}(t_1, t_2)\, \varepsilon + \left[{}^2A_{2,ij}(t_1, t_2) - b\eta \left(\int_{-\infty}^{t_2} G_{j1}(t_2 - s)\, {}^2A_{1,i222}(t_1, s, s, s)\, ds + \int_{-\infty}^{t_1} G_{i1}(t_1 - s)\, {}^2A_{1,j222}(t_2, s, s, s)\, ds\right)\right] \varepsilon^2 .$$

Assume that $a_{11}(u) = a = \text{const}$. Then we have

$$ {}^2A_{1,ij}(t_1, t_2) = a \int_{-\infty}^{t_{12}} \tilde{G}_{i1}(t_1 - u)\, \tilde{G}_{j1}(t_2 - u)\, du = {}^2A_{1,ij}(|t_2 - t_1|)$$

where $\tilde{G}_{i2}(u) = 0$ for $i = 1, 2$ were used. This functions ${}^2A_{1,ij}(t)$ are given by (3.81). The first approximation of the correlation function as to the non-linear system is the same as for the adequate linear system. A distinction can be established with respect to the second approximation. The term ${}^2A_{1,i222}(t, s, s, s)$ can be calculated by

$$ {}^2A_{1,i222}(t, s, s, s) = 3\, {}^2A_{1,i2}(t, s)\, {}^2A_{1,22}(0) = \frac{3a}{4\varkappa(\varkappa^2 + \omega^2)}\, {}^2A_{1,i2}(t, s)$$

and it follows

$$\langle z_i(t_1)\, z_j(t_2)\rangle^2 = {}^2A_{1,ij}(t_1, t_2)\, \varepsilon + \left[{}^2A_{2,ij}(t_1, t_2) - \frac{3ab\eta}{4\varkappa(\varkappa^2 + \omega^2)} \int_0^\infty \{G_{j1}(u)\, {}^2A_{1,i2}(t_1, t_2 - u) + G_{i1}(u)\, {}^2A_{1,j2}(t_2, t_1 - u)\}\, du\right] \varepsilon^2$$

and this expression is a function of $|t_2 - t_1|$ (see the considerations after (3.112)). We put $t_2 = t_1 + t$ with $t \geqq 0$ and obtain

$$\int_0^\infty [G_{j1}(u)\, {}^2A_{1,i2}(t_1, t_1 + t - u) + G_{i1}(u)\, {}^2A_{1,j2}(t_1 + t, t_1 - u)]\, du$$

$$= \int_0^t G_{j1}(u)\, {}^2A_{1,i2}(t_1, t_1 + t - u)\, du + \int_t^\infty G_{j1}(u)\, {}^2A_{1,i2}(t_1, t_1 + t - u)\, du + \int_0^\infty G_{i1}(u)\, {}^2A_{1,j2}(t_1 + t, t_1 - u)\, du$$

where

$$^2A_{1,i2}(t_1, t_1 + t - u) = a \int_0^\infty \tilde{G}_{i1}(v)\, \tilde{G}_{21}(t - u + v)\, \mathrm{d}v \quad \text{for} \quad t - u \geqq 0,$$

$$^2A_{1,i2}(t_1, t_1 + t - u) = a \int_0^\infty \tilde{G}_{i1}(u - t + v)\, \tilde{G}_{21}(v)\, \mathrm{d}v \quad \text{for} \quad u - t \geqq 0,$$

$$^2A_{1,i2}(t_1 + t, t_1 - u) = a \int_0^\infty \tilde{G}_{i1}(u + t + v)\, \tilde{G}_{21}(v)\, \mathrm{d}v \quad \text{for} \quad u \geqq 0.$$

Some calculations lead to

$$\int_0^\infty G_{i1}(u) \left({}^2A_{1,i2}(t_1, t_1 + t - u) + {}^2A_{1,i2}(t_1 + t, t_1 - u)\right) \mathrm{d}u$$

$$= \begin{cases} \dfrac{a\,\mathrm{e}^{-\varkappa t}}{8\omega m} \left(\left(\dfrac{t}{\varkappa} - \dfrac{1}{\omega^2}\right) \sin(\omega t) + \dfrac{t}{\omega} \cos(\omega t)\right) & \text{for} \quad i = 1, \\ \dfrac{a\,\mathrm{e}^{-\varkappa t}}{8m\omega(\varkappa^2 + \omega^2)} \left(\left(\dfrac{1}{\omega^2} + \dfrac{2}{\varkappa^2 + \omega^2} + \dfrac{t}{\varkappa}\right) \sin(\omega t) + \left(\dfrac{2\omega}{\varkappa(\varkappa^2 + \omega^2)} - \dfrac{t}{\omega}\right) \cos(\omega t)\right) & \text{for} \quad i = 2 \end{cases}$$

and to the second approximation of the correlation functions for $t = t_2 - t_1 \geqq 0$,

$$\begin{aligned} \langle z_1(t_1)\, z_1(t_2)\rangle^2 = {} & \frac{a\varepsilon}{4}\, \mathrm{e}^{-\varkappa t} \left[\frac{1}{\varkappa} \cos(\omega t) - \frac{1}{\omega} \sin(\omega t)\right] \\ & + \varepsilon^2 \left[{}^2A_{2,11}(t_1, t_2) - \frac{3a^2 b\eta\, \mathrm{e}^{-\varkappa t}}{32\varkappa\omega m(\varkappa^2 + \omega^2)} \right. \\ & \left. \times \left(\left(\frac{t}{\varkappa} - \frac{1}{\omega^2}\right) \sin(\omega t) + \frac{t}{\omega} \cos(\omega t)\right)\right], \end{aligned} \tag{3.124}$$

$$\begin{aligned} \langle z_2(t_1)\, z_2(t_2)\rangle^2 = {} & \frac{a\varepsilon\, \mathrm{e}^{-\varkappa t}}{4(\varkappa^2 + \omega^2)} \left[\frac{1}{\varkappa} \cos(\omega t) + \frac{1}{\omega} \sin(\omega t)\right] \\ & + \varepsilon^2 \left[{}^2A_{2,22}(t_1, t_2) - \frac{3a^2 b\eta\, \mathrm{e}^{-\varkappa t}}{32m\omega\varkappa(\varkappa^2 + \omega^2)^2} \right. \\ & \times \left(\left(\frac{1}{\omega^2} + \frac{2}{\varkappa^2 + \omega^2} + \frac{t}{\varkappa}\right) \sin(\omega t) \right. \\ & \left.\left. + \left(\frac{2\omega}{\varkappa(\varkappa^2 + \omega^2)} - \frac{t}{\omega}\right) \cos(\omega t)\right)\right]. \end{aligned} \tag{3.125}$$

In particular, it is

$$\langle z_1(t)^2\rangle^2 = \frac{a\varepsilon}{4\varkappa} + {}^2A_{2,11}(t, t)\, \varepsilon^2, \tag{3.126}$$

$$\langle z_2(t)^2\rangle^2 = \frac{a\varepsilon}{4\varkappa(\varkappa^2 + \omega^2)} + \left[{}^2A_{2,22}(t, t) - \frac{3a^2 b\eta}{16m\varkappa^2(\varkappa^2 + \omega^2)^3}\right] \varepsilon^2. \tag{3.127}$$

The function ${}^2A_{2,ij}(t_1, t_2)$ was discussed in Section 3.2.2.2 and we have there obtained

$$ {}^2A_{2,ij}(t_1, t_2) = \begin{cases} -\dfrac{1}{6}\sigma^2 & \text{for} \quad i = j = 1 \quad \text{and} \quad t_1 = t_2, \\ 0 & \text{otherwise;} \end{cases} $$

$a = \sigma^2$ for an example of a simple correlation function as to $f_\varepsilon(t, \omega)$. This ${}^2A_{2,ij}(t_1, t_2)$ is kept in the following considerations. The approximations of the spectral densities $S_{ij}^p(\alpha)$ are defined by

$$ S_{ij}^p(\alpha) = \frac{1}{\pi} \int\limits_0^\infty \cos(\alpha t)\, C_{ij}^p(t)\, \mathrm{d}t $$

and it follows

$$ S_{22}^1(\alpha) = \frac{a\varepsilon m^2}{2\pi N(\alpha)} \tag{3.128} $$

and

$$ S_{22}^2(\alpha) = \frac{a\varepsilon m^2}{2\pi N(\alpha)} \left(1 - \frac{3a\varepsilon b\eta m^2(c - m\alpha^2)}{kcN(\alpha)}\right) \tag{3.129} $$

where $N(\alpha)$ is given by

$$ N(\alpha) = (c - m\alpha^2)^2 + k^2\alpha^2. $$

We remark that $S_{11}^p(\alpha)$ can be determined by

$$ S_{11}^p(\alpha) = \alpha^2 S_{22}^p(\alpha) \quad \text{for} \quad p = 1, 2. $$

In order to compare these results we deal with (3.123) assuming a Gaussian input function $g(t, \omega)$ with mean zero and the correlation function $C(t)$. The solution $z(t, \omega)$ is chosen to have the form

$$ z(t, \omega) = \sum_{p=0}^\infty {}^pz(t, \omega)\, \eta^p. $$

Then the vectors ${}^pz(t, \omega)$, $p = 0, 1, 2$, are solutions of the equations

$$ M\,{}^0z' + N\,{}^0z = -mg(t, \omega), $$

$$ M\,{}^1z' + N\,{}^1z = -\begin{pmatrix} b({}^0z_2)^3 \\ 0 \end{pmatrix}, \qquad M\,{}^2z' + N\,{}^2z = -\begin{pmatrix} 3b({}^0z_2)^2\,{}^1z_2 \\ 0 \end{pmatrix} $$

which are solved by

$$ {}^0z(t, \omega) = -m \int\limits_{-\infty}^t G(t - s)\, g(s, \omega)\, \mathrm{d}s, $$

$$ {}^1z(t, \omega) = -b \int\limits_{-\infty}^t G(t - s) \begin{pmatrix} ({}^0z_2)^3 \\ 0 \end{pmatrix} (s)\, \mathrm{d}s, $$

$$ {}^2z(t, \omega) = -3b \int\limits_{-\infty}^t G(t - s) \begin{pmatrix} ({}^0z_2)^2\,{}^1z_2 \\ 0 \end{pmatrix} (s)\, \mathrm{d}s $$

The expectation of $z(t, \omega)$ is given by

$$\langle z(t)\rangle = \langle {}^0z(t)\rangle + \langle {}^1z(t)\rangle\, \eta + \langle {}^2z(t)\rangle\, \eta^2 + O(\eta^3) = O(\eta^3)$$

since ${}^0z(t, \omega)$ is a Gaussian vector process with mean zero and hence the expectation of the processes

$${}^1z_j(t, \omega) = -b \int_{-\infty}^{t} G_{j1}(t - s) \left({}^0z_2(s, \omega)\right)^3 \mathrm{d}s$$

and

$${}^2z_j(t, \omega) = 3b^2 \int_{-\infty}^{t} \int_{-\infty}^{s} G_{j1}(t - s)\, G_{21}(s - u) \left({}^0z_2(s)\right)^2 \left({}^0z_2(u)\right)^3 \mathrm{d}u\, \mathrm{d}s$$

vanishes. Now we turn to the correlation function,

$$\begin{aligned}\langle z_k(t_1)\, z_j(t_2)\rangle = {}& \langle {}^0z_k(t_1)\, {}^0z_j(t_2)\rangle + [\langle {}^0z_k(t_1)\, {}^1z_j(t_2)\rangle + \langle {}^1z_k(t_1)\, {}^0z_j(t_2)\rangle]\, \eta \\ &+ [\langle {}^0z_k(t_1)\, {}^2z_j(t_2)\rangle + \langle {}^1z_k(t_1)\, {}^1z_j(t_2)\rangle + \langle {}^2z_k(t_1)\, {}^0z_j(t_2)\rangle]\, \eta^2 + O(\eta^3).\end{aligned}$$

Thus,

$$\begin{aligned}\langle {}^0z_k(t_1)\, {}^0z_j(t_2)\rangle &= m^2 \int_{-\infty}^{t_1} \int_{-\infty}^{t_2} G_{k1}(t_1 - s_1)\, G_{j1}(t_2 - s_2)\, C(s_2 - s_1)\, \mathrm{d}s_2\, \mathrm{d}s_1 \\ &= m^2 \int_0^{\infty} \int_0^{\infty} G_{k1}(s_1)\, G_{j1}(s_2)\, C(t_2 - t_1 + s_1 - s_2)\, \mathrm{d}s_2\, \mathrm{d}s_1 = C^0_{kj}(t_2 - t_1).\end{aligned}$$

The spectral density of $g(t, \omega)$ is denoted by $S(\alpha)$ and the relation

$$\int_0^{\infty} \mathrm{e}^{\mathrm{i}\alpha t} G(t + s)\, \mathrm{d}t = H(-\mathrm{i}\alpha)\, MG(s) \tag{3.130}$$

leads to

$$\int_0^{\infty} \mathrm{e}^{\mathrm{i}\alpha t} G_{j1}(t)\, \mathrm{d}t = H_{j1}(-\mathrm{i}\alpha) \quad \text{where} \quad H = H(\mathrm{i}\alpha) = (\mathrm{i}\alpha M + N)^{-1}.$$

With this we can easy see that

$$C^0_{kj}(t) = m^2 \int_{-\infty}^{\infty} \mathrm{e}^{\mathrm{i}\alpha t} H_{k1}(-\mathrm{i}\alpha)\, H_{j1}(\mathrm{i}\alpha)\, S(\alpha)\, \mathrm{d}\alpha \tag{3.131}$$

and then

$$S^0_{kj}(\alpha) = m^2 H_{k1}(-\mathrm{i}\alpha)\, H_{j1}(\mathrm{i}\alpha)\, S(\alpha)$$

and

$$S^0_{jj}(\alpha) = m^2\, |H_{j1}(\mathrm{i}\alpha)|^2\, S(\alpha) \quad \text{for} \quad j, k = 1, 2.$$

In particular, we have found

$$S^0_{22}(\alpha) = \frac{m^2}{N(\alpha)}\, S(\alpha) \quad \text{and} \quad S^0_{11}(\alpha) = \alpha^2 S^0_{22}(\alpha). \tag{3.132}$$

We deal with the terms of the order 1 as to η. The Gaussian character of ${}^0z_i(t, \omega)$ implies

$$\langle {}^0z_k(t_1)\, {}^1z_j(t_2)\rangle = -b \int\limits_{-\infty}^{t_2} G_{j1}(t_2 - s) \left\langle {}^0z_k(t_1) \left({}^0z_2(s)\right)^3\right\rangle \mathrm{d}s$$

$$= -3bC^0_{22}(0) \int\limits_0^{\infty} G_{j1}(s)\, C^0_{k2}(t_2 - t_1 - s)\, \mathrm{d}s$$

$$= -3bC^0_{22}(0) \int\limits_{-\infty}^{\infty} \mathrm{e}^{\mathrm{i}\alpha(t_2-t_1)} H_{j1}(\mathrm{i}\alpha)\, S^0_{k2}(\alpha)\, \mathrm{d}\alpha$$

and the first approximations are

$$C^1_{kj}(t) = C^0_{kj}(t) - 3b\eta C^0_{22}(0) \left[\int\limits_0^{\infty} G_{j1}(s)\, C^0_{k2}(t - s)\, \mathrm{d}s + \int\limits_0^{\infty} G_{k1}(s)\, C^0_{j2}(t + s)\, \mathrm{d}s\right], \tag{3.133}$$

$$S^1_{kj}(\alpha) = S^0_{kj}(\alpha) - 3b\eta C^0_{22}(0) \left[H_{j1}(\mathrm{i}\alpha)\, S^0_{k2}(\alpha) + H_{k1}(-\mathrm{i}\alpha)\, S^0_{j2}(-\alpha)\right]. \tag{3.134}$$

For $k = j$ it follows

$$S^1_{22}(\alpha) = S^0_{22}(\alpha) \left(1 - 6b\eta C^0_{22}(0)\, \mathrm{Re}\,(H_{21}(\mathrm{i}\alpha))\right)$$

$$= \frac{m^2}{N(\alpha)} \left(1 - 6b\eta C^0_{22}(0)\, \frac{c - m\alpha^2}{N(\alpha)}\right) S(\alpha),$$

$$S^1_{11}(\alpha) = \alpha^2 S^0_{22}(\alpha).$$

Considering the weakly correlated theory we have also taken into consideration terms of the second order as to η and hence we have to investigate these terms in the Gaussian input case, too. But, we remember that for our special case the terms of second order as to η lead to $O(\varepsilon^3)$ if we deal with second moments of the solution, for example. First, we get

$$\langle {}^0z_k(t_1)\, {}^2z_j(t_2)\rangle = 3b^2 \int\limits_{-\infty}^{t_2} \int\limits_{-\infty}^{s} G_{j1}(t_2 - s)\, G_{21}(s - u) \left\langle {}^0z_k(t_1) \left({}^0z_2(s)\right)^2 \left({}^0z_2(u)\right)^3\right\rangle \mathrm{d}u\, \mathrm{d}s$$

$$= 3b^2 \int\limits_{-\infty}^{t_2} \int\limits_{-\infty}^{s} G_{j1}(t_2 - s)\, G_{21}(s - u) \left\{6C^0_{k2}(s - t_1)\, C^0_{22}(u - s)\, C_{22}(0)\right.$$

$$\left. + 3C^0_{k2}(u - t_1) \left[C^0_{22}(0)^2 + 2C^0_{22}(u - s)^2\right]\right\} \mathrm{d}u\, \mathrm{d}s$$

using the Gaussian character of ${}^0z_k(t, \omega)$. We replace $C^0_{kj}(t)$ by

$$C^0_{kj}(t) = \int\limits_{-\infty}^{\infty} \mathrm{e}^{\mathrm{i}\alpha t} S^0_{kj}(\alpha)\, \mathrm{d}\alpha$$

and apply (3.130):

$$\langle {}^0z_k(t_1)\, {}^2z_j(t_2)\rangle = 3b^2 \Bigg[3C_{22}^0(0)^2 \int_{-\infty}^{\infty} e^{i\alpha(t_2-t_1)} H_{j1}(i\alpha)\, H_{21}(i\alpha)\, S_{k2}^0(\alpha)\, d\alpha$$

$$+ 6C_{22}^0(0) \int_{-\infty}^{\infty} \int_{-\infty}^{\infty} e^{i\alpha_1(t_2-t_1)} H_{j1}(i\alpha_1)\, H_{21}(i\alpha_2)\, S_{k2}^0(\alpha_1)\, S_{22}^0(\alpha_2)\, d\alpha_1\, d\alpha_2$$

$$+ 6 \int_{-\infty}^{\infty} \int_{-\infty}^{\infty} \int_{-\infty}^{\infty} e^{i\alpha_1(t_2-t_1)} H_{j1}(i\alpha_1)\, H_{21}\big(i(\alpha_1 + \alpha_2 + \alpha_3)\big)$$

$$\times S_{k2}^0(\alpha_1)\, S_{22}^0(\alpha_2)\, S_{22}^0(\alpha_3)\, d\alpha_1\, d\alpha_2\, d\alpha_3 \Bigg]. \qquad (3.135)$$

With the help of the same method we determine $\langle {}^1z_k(t_1)\, {}^1z_j(t_2)\rangle$ and obtain

$$\langle {}^1z_k(t_1)\, {}^1z_j(t_2)\rangle = b^2 \int_{-\infty}^{t_1} \int_{-\infty}^{t_2} G_{k1}(t_1 - s)\, G_{j1}(t_2 - u)\, \big\langle \big({}^0z_2(s)\big)^3 \big({}^0z_2(u)\big)^3 \big\rangle\, du\, ds$$

$$= 3b^2 \int_0^{\infty} \int_0^{\infty} G_{k1}(s)\, G_{j1}(u)\, \{3C_{22}^0(0)^2\, C_{22}^0(t_2 - t_1 + s - u)$$

$$+ 2C_{22}^0(t_2 - t_1 + s - u)^3\}\, du\, ds$$

and then

$$\langle {}^1z_k(t_1)\, {}^1z_j(t_2)\rangle = 3b^2 \Bigg[3C_{22}^0(0)^2 \int_{-\infty}^{\infty} e^{i\alpha(t_2-t_1)} H_{k1}(-i\alpha)\, H_{j1}(i\alpha)\, S_{22}^0(\alpha)\, d\alpha$$

$$+ 2 \int_{-\infty}^{\infty} \int_{-\infty}^{\infty} \int_{-\infty}^{\infty} e^{i\alpha_1(t_2-t_1)} H_{k1}(-i\alpha_1)\, H_{j1}(i\alpha_1)\, S_{22}^0(\alpha_1 - \alpha_2 - \alpha_3)$$

$$\times S_{22}^0(\alpha_2)\, S_{22}^0(\alpha_3)\, d\alpha_3\, d\alpha_2\, d\alpha_1 \Bigg]. \qquad (3.136)$$

The formula for the correlation function $\langle z_k(t_1)\, z_j(t_2)\rangle$ up to terms of the second order as to η can be summarized easy. Using (3.135) and (3.136) $S_{kj}^2(\alpha)$ is given by

$$S_{kj}^2(\alpha) = S_{kj}^1(\alpha) + 3b^2\eta^2 \Bigg[3C_{22}^0(0)^2 \big(H_{j1}(i\alpha)\, H_{21}(i\alpha)\, S_{k2}^0(\alpha)$$

$$+ H_{k1}(-i\alpha)\, H_{21}(-i\alpha)\, S_{k2}^0(-i\alpha)\big)$$

$$+ 6C_{22}^0(0) \left(H_{j1}(i\alpha)\, S_{k2}^0(\alpha) + H_{k1}(-i\alpha)\, S_{j2}^0(-\alpha) \int_{-\infty}^{\infty} H_{21}(i\beta)\, S_{22}^0(\beta)\, d\beta \right)$$

$$+ 6 \int_{-\infty}^{\infty} \int_{-\infty}^{\infty} \big(H_{j1}(i\alpha)\, H_{21}(i(\alpha + \beta + \gamma))\, S_{k2}^0(\alpha)$$

$$+ H_{k1}(-i\alpha)\, H_{21}(i(-\alpha + \beta + \gamma))\, S_{j2}^0(-\alpha)\big)\, S_{22}^0(\beta)\, S_{22}^0(\gamma)\, d\beta\, d\gamma$$

$$+ 3C_{22}^0(0)^2\, H_{k1}(-i\alpha)\, H_{j1}(i\alpha)\, S_{22}^0(\alpha)$$

$$+ 2H_{k1}(-i\alpha)\, H_{j1}(i\alpha) \int_{-\infty}^{\infty} \int_{-\infty}^{\infty} S_{22}^0(\alpha - \beta - \gamma)\, S_{22}^0(\beta)\, S_{22}^0(\gamma)\, d\beta\, d\gamma \Bigg].$$

Now we apply our results of the weakly correlated theory to examples and compare these with the corresponding results under the assumption of a Gaussian input. The statistical characteristics which follow from the Gaussian input are denoted by a "o" over the corresponding sign. For example, $\mathring{S}_{ii}^{p}(\alpha)$ stands for the approximative spectral density of $z_i(t, \omega)$ with Gaussian input and up to terms of the order η^p. On the other hand, $S_{ii}^{p}(\alpha)$ denotes the approximative spectral density of $z_i(t, \omega)$ up to terms of the order ε^p using the weakly correlated theory.

The results from the weakly correlated theory can be discussed easy in dependence on the parameters of the vibration system. For example, we deal with the relative extrema of the approximative spectral density $S_{22}^{p}(\alpha)$ of $z_2(t, \omega) = x(t, \omega)$.

We can easy show that in the case of

$$2mc - k^2 \leqq 0 : S_{22}^{1}(\alpha) \quad \text{possesses for } \alpha_1 = 0 \text{ a relative maximum}$$

and

$$2mc - k^2 > 0 : S_{22}^{1}(\alpha) \quad \text{possesses for } \alpha_1 = 0 \text{ a relative minimum and for}$$

$$\alpha_2 = \frac{1}{2m} \sqrt{4mc - 2k^2} \quad \text{a relative maximum.}$$

Furthermore, the relative extrema of $S_{22}^{2}(\alpha)$ are investigated. $\alpha_1 = 0$ is also the place of a relative extrema; $S_{22}^{2}(\alpha)$ possesses for $\alpha_1 = 0$ a relative maximum if $2mc - k^2 \leqq 0$ and a relative minimum if $2mc - k^2 > 0$. This statement is true for small values of η and follows from

$$(S_{22}^{2})''(0) = \frac{2mc - k^2}{c^4} + \frac{3a\varepsilon m^2}{kc^5} \left(m + \frac{2}{c} (k^2 - 2mc) \right) \eta b .$$

In the case of $2mc - k^2 > 0$ the function $S_{22}^{2}(\alpha)$ has for α_2 an additional relative maximum and α_2 can be deduced to have the expansion

$$\alpha_2(\eta) = \frac{1}{2m} \sqrt{4mc - 2k^2} + \frac{3a\varepsilon m^2 b\eta}{2kc\sqrt{4mc - 2k^2}} + O(\eta^2)$$

as to η. We can see that in the nonlinear case the place of a relative maximum $\alpha_2(\eta)$ is displaced in direction of higher frequencies in comparison with the linear case ($\eta = 0$).

We deal with the variances $C_{jj}^{p}(0)$ for $j = 1, 2$; $p = 1, 2$ and $\mathring{C}_{jj}^{q}(0)$ for $j = 1, 2$; $q = 0, 1$. $f_\varepsilon(t, \omega)$ and $g_\varepsilon(t, \omega)$ are assumed to have a correlation function

$$C_\varepsilon(t) = \begin{cases} \sigma^2 \left(1 - \dfrac{1}{\varepsilon} |t|\right) & \text{for } |t| < \varepsilon, \\ 0 & \text{otherwise} \end{cases}$$

and hence a spectral density

$$S_\varepsilon(\alpha) = \frac{\sigma^2}{\pi \varepsilon \alpha^2} \left(1 - \cos(\alpha\varepsilon)\right).$$

Then we can show that $a = \sigma^2$ and ${}^2A_{2,ij}(t_1, t_2)$ can be found in Section **3.2.2.2** (see example of an oscillator). The variances $C^p_{jj}(0)$ are given by (3.126), (3.127) as

$$C^1_{11}(0) = \frac{1}{2k}\, m\sigma^2\varepsilon\,,$$

$$C^2_{11}(0) = \frac{1}{2k}\, m\sigma^2\varepsilon - \frac{1}{6}\,\sigma^2\varepsilon^2\,,$$

$$C^1_{22}(0) = \frac{1}{2kc}\, m^2\sigma^2\varepsilon,$$

$$C^2_{22}(0) = \frac{1}{2kc}\, m^2\sigma^2\varepsilon - \frac{3m^4\sigma^4 b\eta\varepsilon^2}{4k^2c^3}.$$

In the case of a Gaussian input the calculation of $\mathring{C}^q_{jj}(0)$ is exacter but more difficult, too. $\mathring{C}^q_{jj}(0)$ can be determined with the help of the spectral densities,

$$\mathring{C}^0_{jj}(0) = 2\int\limits_0^\infty \mathring{S}^0_{jj}(\alpha)\,\mathrm{d}\alpha = 2m^2\int\limits_0^\infty |H_{j1}(\mathrm{i}\alpha)|^2\, S(\alpha)\,\mathrm{d}\alpha,$$

$$\mathring{C}^1_{jj}(0) = \mathring{C}^0_{jj}(0) - 6b\eta m^2\mathring{C}^0_{22}(0)\int\limits_{-\infty}^\infty |H_{j1}(\mathrm{i}\alpha)|^2\,\mathrm{Re}\,\big(H_{21}(\mathrm{i}\alpha)\big)\, S(\alpha)\,\mathrm{d}\alpha$$

using (3.134). It is

$$H_{11}(\mathrm{i}\alpha) = \frac{\alpha}{N(\alpha)}\big(k\alpha + \mathrm{i}(c - m\alpha^2)\big), \qquad |H_{11}(\mathrm{i}\alpha)|^2 = \frac{\alpha^2}{N(\alpha)},$$

$$H_{21}(\mathrm{i}\alpha) = \frac{1}{N(\alpha)}\big((c - m\alpha^2) - \mathrm{i}k\alpha\big), \qquad |H_{21}(\mathrm{i}\alpha)|^2 = \frac{1}{N(\alpha)}$$

and $\mathring{C}^0_{jj}(0)$ can be calculated by numerical integration. The results are illustrated in Fig. 3.57 setting $m = 1$, $k = 4$, $c = 12$. The nonlinear term in the differential equation effects a reduction of the variances. This fact is clear if $\mathring{C}^0_{ii}(0)$ and $\mathring{C}^1_{ii}(0)$ or $C^1_{ii}(0)$ and $C^2_{ii}(0)$ are compared. Furthermore, we remark that $C^p_{ii}(0)$ denotes the approximative

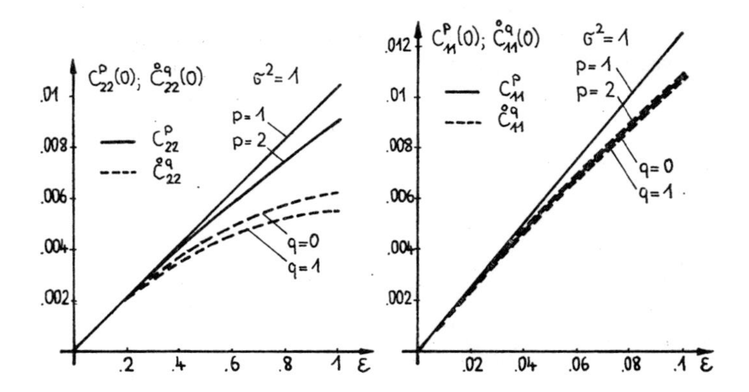

Fig. 3.57. Approximative variances of $x(t, \omega)$ and $x'(t, \omega)$ for $\eta b = 50$

variances of $x(t, \omega)$ or $x'(t, \omega)$, respectively, if all terms are considered up to the order ε^p. $\mathring{C}^q_{ii}(0)$ takes into consideration all terms up to the order η^q. Hence, a direct comparison is not possible; we have determined approximately the variances on two different methods. Fig. 3.58 shows these approximative variances in dependence on the variance of the wide-sense stationary input process where we put $m = 1, k = 4, c = 12, \eta b = 50, \varepsilon = 0.1$. Furthermore, it is possible to discuss these variances in dependence on the system parameters m, k, c.

The approximative spectral densities $S^p_{22}(\alpha)$, $p = 1, 2$, are illustrated in Fig. 3.59. We remark essential differences between $S^1_{22}(\alpha)$ and $S^2_{22}(\alpha)$ for $\varepsilon = 0.5$. This difference can be explained by the nonlinear term of (3.123). In the case of $\varepsilon = 0.1$ the spectral

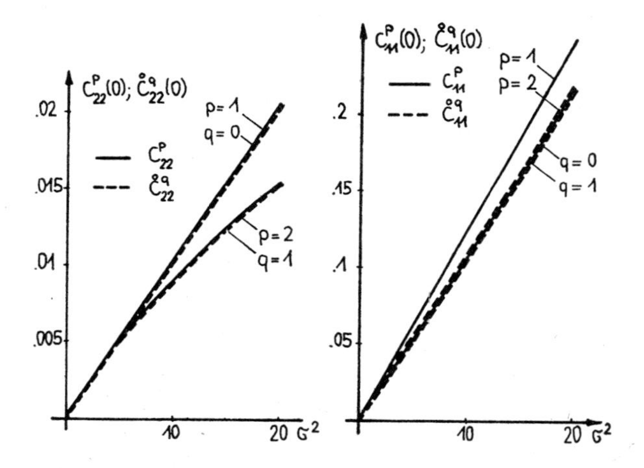

Fig. 3.58. Approximative variances of $x(t, \omega)$, $x'(t, \omega)$ in dependence on σ^2 for $\eta b = 50$

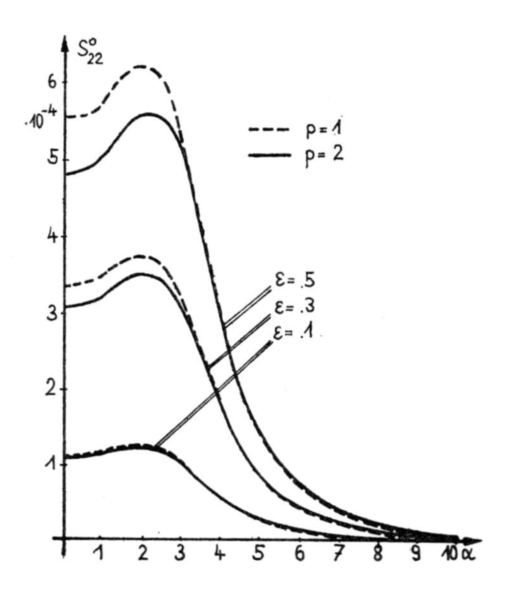

Fig. 3.59. Approximative spectral densities $S^p_{22}(\alpha)$ of $x(t, \omega)$ for $p = 1, 2$

densities of the linear equation and the nonlinear equation coincide very well and it is sufficient to investigate the linear equation. Finally, Fig. 3.60 illustrates a comparison between $S^1_{22}(\alpha)$ and $\mathring{S}^0_{22}(\alpha)$ in the first part of this figure and between $S^2_{22}(\alpha)$ and $\mathring{S}^1_{22}(\alpha)$ in the second part. For small ε we remark a very good coincidence. The differences between $S^1_{22}(\alpha)$ and $\mathring{S}^0_{22}(\alpha)$ for greater ε can be explained by the expansion of the moments of $g(t, \omega)$ as to powers of ε. On the other hand, the differences between $S^2_{22}(\alpha)$ and $\mathring{S}^1_{22}(\alpha)$ follow from the above mentioned expansion as to ε and from the fact that $\mathring{S}^1_{22}(\alpha)$ only takes into consideration linear terms of the solution with respect to η and $S^2_{22}(\alpha)$ also takes into consideration terms of higher order than first order.

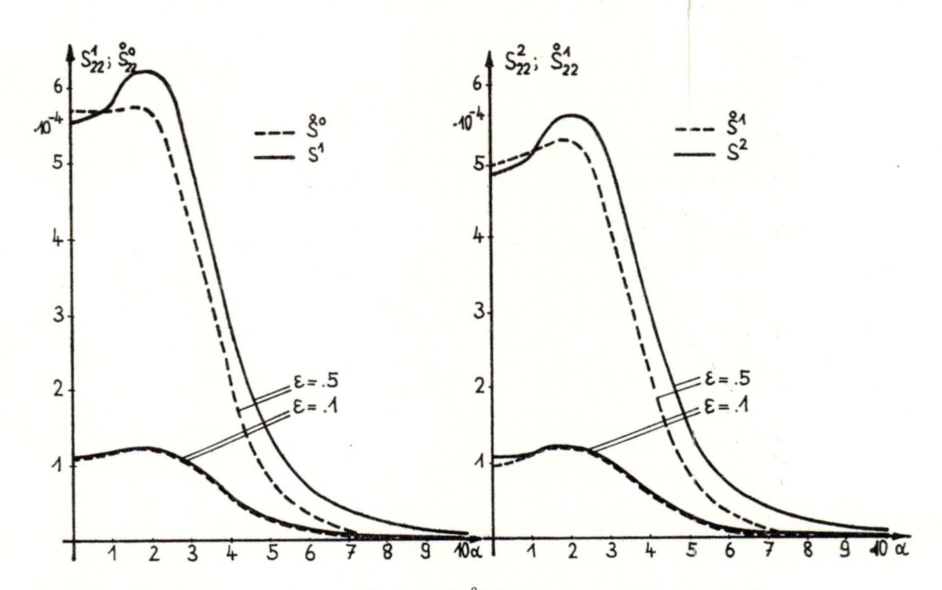

Fig. 3.60. Comparison between $S^p_{22}(\alpha)$ and $\mathring{S}^q_{22}(\alpha)$ for $p = 1, 2$; $q = 0, 1$

4. Random temperature propagation

4.1. Deterministic solutions

First, we will consider the simple random boundary-initial value problem

$$\frac{\partial u}{\partial t} = \alpha \Delta u \quad \text{with} \quad \Delta u \doteq \frac{\partial^2 u}{\partial x^2} + \frac{\partial^2 u}{\partial y^2},$$

$$\text{initial condition:} \quad u(0, x, y) = u_0(x, y), \tag{4.1}$$

$$\text{boundary condition:} \quad \frac{\partial u}{\partial y}(t, x, 0) = P(t, \omega)$$

for a half-plane (see Fig. 4.1). The compatibility condition

$$\frac{\partial u_0}{\partial y}(x, y)\Big|_{y=0} = P(0, \omega)$$

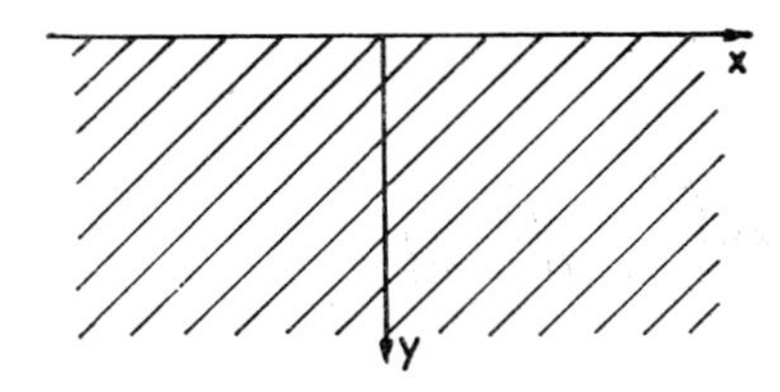

Fig. 4.1. Half-plane with coordinate axis

is assumed to be fulfilled. We deal with the function

$$\tilde{u}(t, x, y, \omega) \doteq u(t, x, y, \omega) - w(t, x, y)$$

where w is the solution of the averaged problem

$$w_t = \alpha \Delta w,$$

$$w(0, x, y) = u_0(x, y); \qquad \frac{\partial w}{\partial y}(t, x, 0) = \langle P(t) \rangle. \tag{4.2}$$

Hence, $\bar{u}$ is obtained from

$$\bar{u}_t = \alpha \Delta \bar{u},$$
$$\bar{u}(0, x, y) = 0; \qquad \frac{\partial \bar{u}}{\partial y}(t, x, 0) = \bar{P}(t, \omega)$$

with

$$\bar{P}(t, \omega) = P(t, \omega) - \langle P(t) \rangle.$$

The solution $\bar{u}(t, x, y, \omega)$ is independent of x because of the simple boundary condition and hence we have

$$\begin{aligned} &\bar{u}_t = \alpha \bar{u}_{yy}, \\ &\bar{u}(0, y) = 0; \qquad \frac{\partial \bar{u}}{\partial y}(t, 0) = \bar{P}(t, \omega). \end{aligned} \tag{4.3}$$

Lemma 4.1. *The solution of* (4.3) *can be determined as*

$$\bar{u}(t, y, \omega) = \int_0^t F(t - s, y)\, \bar{P}(s, \omega)\, \mathrm{d}s \tag{4.4}$$

with

$$F(t, y) = -\sqrt{\frac{\alpha}{\pi t}} \exp\left(-\frac{y^2}{4\alpha t}\right)$$

if $\bar{P}(t, \omega)$ *is a process with a.s. continuous sample functions. The compatibility condition can be written as*

$$\bar{P}(0, \omega) = 0 \qquad a.s.$$

Proof. First $\bar{u}(t, y, \omega)$ from (4.4) fulfils the differential equation

$$\bar{u}_t = \alpha \bar{u}_{yy} \quad \text{for} \quad t > 0,\ y > 0$$

since

$$\bar{u}_t - \alpha \bar{u}_{yy} = -\sqrt{\frac{\alpha}{\pi}} \left\{ \lim_{s \to t-0} \frac{\bar{P}(s, \omega)}{\sqrt{t - s}} \exp\left(-\frac{y^2}{4\alpha(t - s)}\right) \right.$$
$$\left. + \int_0^t \bar{P}(s, \omega) \left[\frac{\partial}{\partial t} - \alpha \frac{\partial^2}{\partial y^2}\right] \frac{1}{\sqrt{t - s}} \exp\left(-\frac{y^2}{4\alpha(t - s)}\right) \mathrm{d}s \right\} = 0$$

because of

$$\lim_{s \to t-0} \frac{\bar{P}(s, \omega)}{\sqrt{t - s}} \exp\left(-\frac{y^2}{4\alpha(t - s)}\right) = 0$$

and

$$\left(\frac{\partial}{\partial t} - \alpha \frac{\partial^2}{\partial y^2}\right) \frac{1}{\sqrt{t - s}} \exp\left(-\frac{y^2}{4\alpha(t - s)}\right) = 0.$$

From the relation

$$\lim_{t\to 0+0} \int_0^t \frac{\overline{P}(s,\omega)}{\sqrt{t-s}} \exp\left(-\frac{y^2}{4\alpha(t-s)}\right) ds = 0$$

using

$$\left|\frac{1}{\sqrt{s}} \exp\left(-\frac{y^2}{4\alpha s}\right)\right| \leqq c$$

for small values s we see that $\bar{u}(0, x, \omega) = 0$ a.s. Furthermore, for $y > 0$ we obtain

$$\begin{aligned}\frac{\partial \bar{u}}{\partial y}(t, y) &= \frac{y}{2\sqrt{\alpha\pi}} \int_0^t \frac{\overline{P}(t-s)}{\sqrt{s}^3} \exp\left(-\frac{y^2}{4\alpha s}\right) ds \\ &= \frac{y}{2\sqrt{\alpha\pi}} \left[\int_{s_0}^t \frac{\overline{P}(t-s)}{\sqrt{s}^3} \exp\left(-\frac{y^2}{4\alpha s}\right) ds + \int_0^{s_0} \frac{\overline{P}(t-s)-\overline{P}(t)}{\sqrt{s}^3} \exp\left(-\frac{y^2}{4\alpha s}\right) ds \right. \\ &\qquad \left. + \overline{P}(t) \int_0^{s_0} \frac{1}{\sqrt{s}^3} \exp\left(-\frac{y^2}{4\alpha s}\right) ds\right]\end{aligned}$$

and from

$$\lim_{y\to 0} y \int_{s_0}^t \frac{\overline{P}(t-s)}{\sqrt{s}^3} \exp\left(-\frac{y^2}{4\alpha s}\right) ds = 0 \quad \text{for all } s_0 > 0,$$

$$\lim_{y\to 0+0} y \int_0^{s_0} \frac{1}{\sqrt{s}^3} \exp\left(-\frac{y^2}{4\alpha s}\right) ds = \lim_{y\to 0+0} 4\sqrt{\alpha} \int_{\frac{y}{2\sqrt{\alpha s_0}}}^{\infty} \exp(-s^2)\, ds = 2\sqrt{\alpha\pi} \quad \text{for all } s_0 > 0,$$

and

$$\left|\lim_{y\to 0+0} y \int_0^{s_0} \frac{\overline{P}(t-s)-\overline{P}(t)}{\sqrt{s}^3} \exp\left(-\frac{y^2}{4\alpha s}\right) ds\right| \leqq 2\eta\sqrt{\alpha\pi}$$

for an s_0 with

$$|\overline{P}(t-s) - \overline{P}(t)| \leqq \eta \quad \text{for} \quad 0 \leqq s \leqq s_0$$

then the relation

$$\lim_{y\to 0+0} \frac{\partial \bar{u}}{\partial y}(t, y, \omega) = \overline{P}(t, \omega) \qquad \text{a.s.}$$

The function $F(t, y)$ has the property

$$F(., y) \in \mathbf{L}_2(0, t) \quad \text{for every } y > 0.$$

This is proved by

$$\int_0^t F^2(s, y)\,\mathrm{d}s = \frac{\alpha}{\pi}\int_0^t \frac{1}{s}\exp\left(-\frac{y^2}{2\alpha s}\right)\mathrm{d}s = \frac{2\alpha}{\pi}\int_{\frac{y}{\sqrt{2\alpha t}}}^{\infty}\frac{1}{s}\exp(-s^2)\,\mathrm{d}s < \infty$$

for $y > 0$. ◀

Now we turn to the boundary-initial value problem for a bounded domain

$$\mathcal{G} = \{(x, y)\colon |x| \leqq R;\ 0 \leqq y \leqq L\}\colon$$

$$\frac{\partial u}{\partial t} = \alpha \Delta u \quad \text{where} \quad \Delta u \doteq \frac{\partial^2 u}{\partial x^2} + \frac{\partial^2 u}{\partial y^2},$$

$$\text{initial condition:} \qquad u(0, x, y) = u_0(x, y), \tag{4.5}$$

$$\text{boundary conditions:}\ \frac{\partial u}{\partial y}(t, x, y)\big|_{(\partial\mathcal{G})_1} = P(t, x, \omega),$$

$$\left[\lambda \frac{\partial u}{\partial n}(t, x, y) + \alpha_i\{u(t, x, y) - u_R(t, x, y)\}\right]_{(\partial\mathcal{G})_i} = 0 \quad \text{for} \quad i = 2, 3, 4.$$

$(\partial\mathcal{G})_i$, $i = 1, 2, 3, 4$, denote the parts of the boundary of $\mathcal{G}$ illustrated in Fig. 4.2 and $\partial/\partial n$ the normal derivative where the normal vector n shows in the exterior of $\mathcal{G}$. If we explain this problem as a temperature propagation problem with free heat transition as to $(\partial\mathcal{G})_i$, $i = 2, 3, 4$, then α represents the temperature conductivity and λ the thermal conductivity number, and α_i, $i = 2, 3, 4$, the heat transition number. Furthermore, $u_0(x, y)$ denotes a non-random initial temperature, u_R a non-random boundary temperature; and $P(t, x, \omega)$ a random function on $(\partial\mathcal{G})_1$ with respect to a given surface power. The special case $\alpha_i = 0$ corresponds to an adiabatic part of the boundary and $\alpha_i \to \infty$ to the case of a given surface temperature u_R. A technological application of this model consists in the investigation on the temperature propagation of brakes of vehicles.

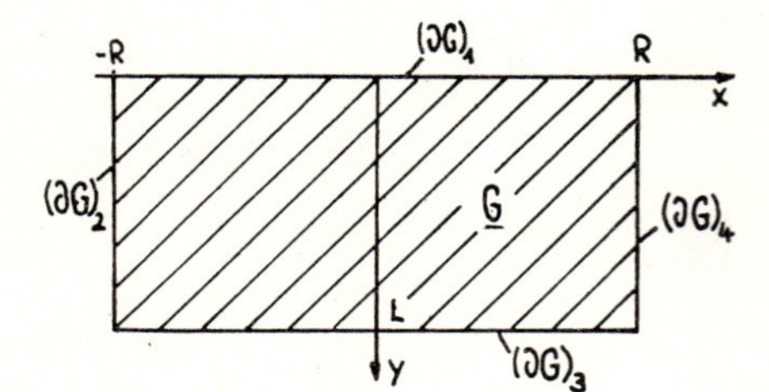

Fig. 4.2. Domain $\mathcal{G}$ for the investigated problem

The compatibility conditions are

$$P(0, x, \omega) = \frac{\partial u_0}{\partial y}(x, 0), \tag{4.6}$$

$$\lambda \frac{\partial P}{\partial n}(t, \pm R, \omega) + \alpha_i\left\{P(t, \pm R, \omega) - \frac{\partial u_R}{\partial y}(t, \pm R, 0)\right\} = 0 \quad \text{for} \quad i = 2, 4. \tag{4.7}$$

We assume that $P(t, x, \omega)$ possesses a.s. continuously differentiable sample functions.

The averaged problem concerning the above problem (4.5) has the form

$$\begin{aligned}
&\frac{\partial w}{\partial t} = \alpha \Delta w, \\
&w(0, x, y) = u_0(x, y), \\
&\frac{\partial w}{\partial y}(t, x, y)\Big|_{(\partial \mathcal{G})_1} = \langle P(t, x) \rangle, \\
&\left[\lambda \frac{\partial w}{\partial n}(t, x, y) + \alpha_i \{w(t, x, y) - u_R(t, x, y)\}\right]_{(\partial \mathcal{G})_i} = 0 \quad \text{for} \quad i = 2, 3, 4.
\end{aligned} \tag{4.8}$$

Hence, the random function

$$\bar{u}(t, x, y, \omega) \doteq u(t, x, y, \omega) - w(t, x, y)$$

denotes the solution of the boundary-initial value problem

$$\begin{aligned}
&\frac{\partial \bar{u}}{\partial t} = \alpha \Delta \bar{u}, \\
&\bar{u}(0, x, y) = 0, \\
&\frac{\partial \bar{u}}{\partial y}(t, x, y)\Big|_{(\partial \mathcal{G})_1} = \bar{P}(t, x, \omega), \\
&\left[\lambda \frac{\partial \bar{u}}{\partial n}(t, x, y) + \alpha_i \bar{u}(t, x, y)\right]_{(\partial \mathcal{G})_i} = 0 \quad \text{for} \quad i = 2, 3, 4
\end{aligned} \tag{4.9}$$

where $\bar{P}(t, x, \omega)$ is the centred function

$$\bar{P}(t, x, \omega) \doteq P(t, x, \omega) - \langle P(t, x) \rangle$$

with the compatibility conditions

$$\begin{aligned}
&\bar{P}(0, x, \omega) = 0, \\
&\left[\lambda \frac{\partial \bar{P}}{\partial n}(t, x, \omega) + \alpha_i \bar{P}(t, x, \omega)\right]_{(\partial \mathcal{G})_i} = 0 \quad \text{for} \quad i = 2, 4.
\end{aligned} \tag{4.10}$$

Now a weak solution of problem (4.9) is determined. In order to apply the Fourier method the function

$$\tilde{u}(t, x, y, \omega) = 2L\bar{u}(t, x, y, \omega) + (L - y)^2 \, \bar{P}(t, x, \omega)$$

is introduced to obtain a boundary-initial value problem for $\tilde{u}$ having homogeneous boundary conditions but an inhomogeneous differential equation. The function $\tilde{u}$ is determined from the problem

$$\begin{aligned}
&\tilde{u}_t = \alpha \Delta \tilde{u} - 2\alpha \bar{P} + (L - y)^2 \, [\bar{P}_t - \alpha \bar{P}_{xx}], \\
&\tilde{u}(0, x, y) = 0, \\
&\frac{\partial \tilde{u}}{\partial y}(t, x, y)\Big|_{(\partial \mathcal{G})_1} = 0, \\
&\left[\lambda \frac{\partial \tilde{u}}{\partial n} + \alpha_i \tilde{u}\right]_{(\partial \mathcal{G})_i} = 0 \quad \text{for} \quad i = 2, 3, 4
\end{aligned} \tag{4.11}$$

if the compatibility conditions (4.10) were taken into consideration. Using the Fourier method a formal solution $\tilde{u}$ of the problem (4.11) will be calculated. Starting from the series

$$\tilde{u}(t, x, y) = \sum_{k,l=0}^{\infty} c_{kl}(t) f_{kl}(x, y) \tag{4.12}$$

we obtain

$$\sum_{k,l=0}^{\infty} \left(c'_{kl}(t) + \alpha\lambda_{kl}c_{kl}(t)\right) f_{kl}(x, y) = \sum_{k,l=0}^{\infty} (g, f_{kl}) f_{kl}(x, y) \tag{4.13}$$

from the differential equation of (4.11) if the right-hand side of the differential equation is denoted by $g(t, x, y)$,

$$g(t, x, y) \doteq (L - y)^2 [\overline{P}_t(t, x) - \alpha\overline{P}_{xx}(t, x)] - 2\alpha\overline{P}(t, x). \tag{4.14}$$

Furthermore, $f_{kl}(x, y)$ are taken as the eigenfunctions with respect to the differential operator $-\Delta$ and the given boundary conditions,

$$\begin{gathered} -\Delta f = \lambda f, \\ -\frac{\partial f}{\partial n}\bigg|_{(\partial\mathcal{G})_1} = 0, \qquad \left[\lambda \frac{\partial f}{\partial n} + \alpha_i f\right]_{(\partial\mathcal{G})_i} = 0 \quad \text{for} \quad i = 2, 3, 4. \end{gathered} \tag{4.15}$$

(g, f) is defined as the scalar product on $\mathbf{L}_2(\mathcal{G})$,

$$(g, f) \doteq \int_{\mathcal{G}} g(x, y) f(x, y) \,\mathrm{d}x \,\mathrm{d}y.$$

Using (4.13) and the initial condition in (4.11) the function $c_{kl}(t)$ can be determined from the initial value problem

$$c'_{kl}(t) + \alpha\lambda_{kl}c_{kl}(t) = \left(g(t, ., .), f_{kl}\right); \qquad c_{kl}(0) = 0.$$

Thus

$$\begin{aligned} c_{kl}(t) &= \int_0^t (g, f_{kl}) \exp\left(-\alpha\lambda_{kl}(t - s)\right) \mathrm{d}s \\ &= \int_0^t \int_{-R}^{R} \int_0^L \{(L - y)^2 [\overline{P}_s(s, x) - \alpha\overline{P}_{xx}(s, x)] - 2\alpha\overline{P}(s, x)\} \\ &\qquad \times f_{kl}(x, y) \,\mathrm{d}x \,\mathrm{d}y \exp\left(-\alpha\lambda_{kl}(t - s)\right) \mathrm{d}s. \end{aligned} \tag{4.16}$$

Now the eigenvalues λ_{kl} and the eigenfunctions f_{kl} of (4.15) can be calculated by separation. It is convenient to write $f(x, y)$ as a product,

$$f(x, y) = \varphi_1(x) \varphi_2(y).$$

This assumption leads to the both problems

$$\begin{aligned} &\varphi_1'' + c\varphi_1 = 0; \qquad -\lambda\varphi_1'(-R) + \alpha_2\varphi_1(-R) = 0, \qquad \lambda\varphi_1'(R) + \alpha_4\varphi_1(R) = 0, \\ &\varphi_2'' + d\varphi_2 = 0; \qquad \varphi_2'(0) = 0, \qquad \lambda\varphi_2'(L) + \alpha_3\varphi_2(L) = 0. \end{aligned}$$

The eigenvalues

$$\lambda_{kl} = c_k + d_l, \qquad k, l = 0, 1, 2, \ldots,$$

are obtained as solutions from the equations

$$(\lambda^2 c_k - \alpha_2\alpha_4) \sin\left(2R\sqrt{c_k}\right) = \lambda\sqrt{c_k}(\alpha_2 + \alpha_4) \cos\left(2R\sqrt{c_k}\right),$$

$$\lambda d_l \sin\left(L\sqrt{d_l}\right) = \alpha_3\sqrt{d_l} \cos\left(L\sqrt{d_l}\right)$$

which can be solved by a suitable numerical method. The eigenfunctions

$$f_{kl}(x, y) = \varphi_{1k}(x)\, \varphi_{2l}(y)$$

follow from

$$\varphi_{1k}(x) = g_1 \sin\left(\sqrt{c_k}x\right) + g_2 \cos\left(\sqrt{c_k}x\right),$$

$$\varphi_{2l}(y) = h_1 \sin\left(\sqrt{d_l}y\right) + h_2 \cos\left(\sqrt{d_l}y\right)$$

where g_1, g_2 and h_1, h_2 are non-trivial solutions of

$$-\left(\lambda\sqrt{c_k} \cos\left(\sqrt{c_k}R\right) + \alpha_2 \sin\left(\sqrt{c_k}R\right)\right) g_1 + \left(\alpha_2 \cos\left(\sqrt{c_k}R\right) - \lambda\sqrt{c_k} \sin\left(\sqrt{c_k}R\right)\right) g_2 = 0,$$

$$\left(\lambda\sqrt{c_k} \cos\left(\sqrt{c_k}R\right) + \alpha_4 \sin\left(\sqrt{c_k}R\right)\right) g_1 + \left(\alpha_4 \cos\left(\sqrt{c_k}R\right) - \lambda\sqrt{c_k} \sin\left(\sqrt{c_k}R\right)\right) g_2 = 0$$

and

$$\sqrt{d_l}h_1 = 0,$$

$$\left(\lambda\sqrt{d_l} \cos\left(\sqrt{d_l}L\right) + \alpha_3 \sin\left(\sqrt{d_l}L\right)\right) h_1 + \left(\alpha_3 \cos\left(\sqrt{d_l}L\right) - \lambda\sqrt{d_l} \sin\left(\sqrt{d_l}L\right)\right) h_2 = 0,$$

respectively. For example, assuming

$$-\lambda\sqrt{c_k} \sin\left(\sqrt{c_k}R\right) + \alpha_2 \cos\left(\sqrt{c_k}R\right) \neq 0, \qquad d_2 \neq 0$$

we obtain the eigenfunctions

$$\varphi_{1k}(x) = g(c_k)\left[\sin\left(\sqrt{c_k}x\right) + \frac{\lambda\sqrt{c_k} \cos\left(\sqrt{c_k}R\right) + \alpha_2 \sin\left(\sqrt{c_k}R\right)}{\alpha_2 \cos\left(\sqrt{c_k}R\right) - \lambda\sqrt{c_k} \sin\left(\sqrt{c_k}R\right)} \cos\left(\sqrt{c_k}x\right)\right],$$

$$\varphi_{2l}(y) = h(d_l) \cos\left(\sqrt{d_l}y\right)$$

where $g(c_k)$ and $h(d_l)$ can be calculated from

$$\int_{-R}^{R} \varphi_{1k}^2(x)\, \mathrm{d}x = \int_{0}^{L} \varphi_{2l}^2(y)\, \mathrm{d}y = 1.$$

A simple calculation leads to

$$g^{-2}(c_k) = \left(R + \frac{\sin\left(2\sqrt{c_k}R\right)}{2\sqrt{c_k}}\right)\left(\frac{\lambda\sqrt{c_k} \cos\left(\sqrt{c_k}R\right) + \alpha_2 \sin\left(\sqrt{c_k}R\right)}{\alpha_2 \cos\left(\sqrt{c_k}R\right) + \lambda\sqrt{c_k} \sin\left(\sqrt{c_k}R\right)}\right)^2 + R - \frac{\sin\left(2\sqrt{c_k}R\right)}{2\sqrt{c_k}},$$

$$h^{-2}(d_l) = \frac{1}{2}\left(L + \frac{\sin\left(2\sqrt{d_l}L\right)}{2\sqrt{d_l}}\right).$$

In the adiabatic special case $\alpha_2 = \alpha_3 = \alpha_4 = 0$ we have

$$\lambda_{kl} = c_k + d_l \quad \text{with} \quad c_k = \left(\frac{k\pi}{2R}\right)^2 \quad \text{for} \quad k = 0, 1, 2, \dots,$$

$$d_l = \left(\frac{l\pi}{L}\right)^2 \quad \text{for} \quad l = 0, 1, 2, \dots,$$

$$f_{kl}(x, y) = \varphi_{1k}(x)\, \varphi_{2l}(y)$$

with

$$\varphi_{1k}(x) = \sqrt{\frac{1}{R}} \begin{cases} 1 & \text{for } k = 0, \\ \sin\left(\dfrac{k\pi}{2R}\, x\right) & \text{for } k = 1, 3, 5, \dots, \\ \cos\left(\dfrac{k\pi}{2R}\, x\right) & \text{for } k = 2, 4, 6, \dots, \end{cases}$$

$$\varphi_{2l}(y) = \sqrt{\frac{2}{L}} \begin{cases} \dfrac{1}{\sqrt{2}} & \text{for } l = 0, \\ \cos\left(\dfrac{l\pi}{L}\, y\right) & \text{for } l = 1, 2, 3, \dots \end{cases}$$

and for $\alpha_i \to \infty$, $i = 2, 3, 4$, we obtain the results

$$\lambda_{kl} = c_k + d_l \quad \text{with} \quad c_k = \left(\frac{k\pi}{2R}\right)^2 \quad \text{for} \quad k = 1, 2, 3, \dots,$$

$$d_l = \left(\frac{(2l-1)\,\pi}{2L}\right)^2 \quad \text{for} \quad l = 1, 2, 3, \dots,$$

$$f_{kl}(x, y) = \varphi_{1k}(x)\, \varphi_{2l}(y)$$

with

$$\varphi_{1k}(x) = \sqrt{\frac{1}{R}} \begin{cases} \cos\left(\dfrac{k\pi}{2R}\, x\right) & \text{for } k = 1, 3, 5, \dots, \\ \sin\left(\dfrac{k\pi}{2R}\, x\right) & \text{for } k = 2, 4, 6, \dots, \end{cases}$$

$$\varphi_{2l}(y) = \sqrt{\frac{2}{L}} \cos\left(\frac{(2l-1)\,\pi}{2L}\, y\right) \quad \text{for } l = 1, 2, 3, \dots$$

With the help of these eigenvalues λ_{kl} and eigenfunctions it is possible to determine the functions $c_{kl}(t)$. Using

$$p_l \doteq \int_0^L (L - y)^2\, \varphi_{2l}(y)\, \mathrm{d}y$$

$$= \begin{cases} \dfrac{1}{3} \sqrt{L^5} & \text{if } d_0 = 0 \text{ is eigenvalue}, \\ \dfrac{2Lh(d_l)}{d_l} \left(1 - \dfrac{\sin\left(\sqrt{d_l} L\right)}{\sqrt{d_l} L}\right) & \text{otherwise}, \end{cases}$$

$$q_l \doteq \int\limits_0^L \varphi_{2l}(y)\,\mathrm{d}y$$

$$= \begin{cases} \sqrt{L} & \text{if } d_0 = 0 \text{ is eigenvalue,} \\ h(d_l)\,\dfrac{\sin\left(\sqrt{d_l}L\right)}{\sqrt{d_l}} & \text{otherwise} \end{cases}$$

we have

$$(g, f_{kl}) = p_l \int\limits_{-R}^{R} \left(\overline{P}_t(t, x) - \alpha \overline{P}_{xx}(t, x)\right) \varphi_{1k}(x)\,\mathrm{d}x - 2\alpha q_l \int\limits_{-R}^{R} \overline{P}(t, x)\,\varphi_{1k}(x)\,\mathrm{d}x .$$

Now, the compatibility conditions (4.10) and the boundary conditions as to $\varphi_{1k}(x)$ are applied and lead to

$$\int\limits_{-R}^{R} \overline{P}_{xx}(t, x)\,\varphi_{1k}(x)\,\mathrm{d}x = \overline{P}_x\varphi_{1k}\big|_{-R}^{R} - \int\limits_{-R}^{R} \overline{P}_x\varphi_{1k}'\,\mathrm{d}x$$

$$= (\overline{P}_x\varphi_{1k} - \overline{P}\varphi_{1k}')\big|_{-R}^{R} + \int\limits_{-R}^{R} \overline{P}_x\varphi_{1k}''\,\mathrm{d}x$$

$$= -c_k \int\limits_{-R}^{R} \overline{P}(t, x)\,\varphi_{1k}(x)\,\mathrm{d}x$$

and furthermore to

$$(g, f_{kl}) = p_l \int\limits_{-R}^{R} \overline{P}_t(t, x)\,\varphi_{1k}(x)\,\mathrm{d}x + \alpha(p_l c_k - 2q_l) \int\limits_{-R}^{R} \overline{P}(t, x)\,\varphi_{1k}(x)\,\mathrm{d}x .$$

We find that

$$c_{kl}(t) = \int\limits_0^t \left(g(t, ., .), f_{kl}\right) \exp\left(-\alpha\lambda_{kl}(t - s)\right) \mathrm{d}s$$

$$= p_l \int\limits_{-R}^{R} \overline{P}(t, x)\,\varphi_{1k}(x)\,\mathrm{d}x$$

$$- \alpha(p_l d_l + 2q_l) \int\limits_0^t \int\limits_{-R}^{R} \overline{P}(s, x)\,\varphi_{1k}(x) \exp\left(-\alpha\lambda_{kl}(t - s)\right) \mathrm{d}s\,\mathrm{d}x \tag{4.16}$$

where

$$\int\limits_0^t \overline{P}_s(s, x) \exp\left(-\alpha\lambda_{kl}(t - s)\right) \mathrm{d}s$$

$$= \overline{P}(t, x) - \alpha\lambda_{kl} \int\limits_0^t \overline{P}(s, x) \exp\left(-\alpha\lambda_{kl}(t - s)\right) \mathrm{d}s .$$

Then the formal solution $\tilde{u}(t, x, y, \omega)$ follows from (4.12).

We will deal with the "averaged" solution

$$U(t) \doteq \left(\bar{u}(t, ., .), \psi\right) \tag{4.17}$$

to guarantee the convergence of the series. It is possible to prove that $\bar{u}(t, x, y)$ is a weak solution of the given problem. Using the definition of $\tilde{u}$ we obtain

$$U(t) = \frac{1}{2L} \left[(\tilde{u}, \psi) - \left((L - y)^2 \overline{P}, \psi \right) \right]$$

where

$$(\tilde{u}, \psi) = \sum_{k,l=0}^{\infty} c_{kl}(t) \, (f_{kl}, \psi)$$

(with $c_{kl}(t)$ from (4.16)) and

$$\left((L - y)^2 \overline{P}, \psi \right) = \sum_{k,l=0}^{\infty} \left((L - y)^2 \overline{P}, f_{kl} \right) (f_{kl}, \psi)$$
$$= \sum_{k,l=1}^{\infty} p_l \int_{-R}^{R} \overline{P}(t, x) \, \varphi_{1k}(x) \, \mathrm{d}x (f_{kl}, \psi) .$$

Furthermore we have

$$U(t) = -\frac{\alpha}{2L} \sum_{k,l=0}^{\infty} (p_l d_l + 2q_l) \int_0^t \int_{-R}^{R} \overline{P}(s, x) \, \varphi_{1k}(x) \exp \left(-\alpha \lambda_{kl}(t - s) \right) \mathrm{d}s \, \mathrm{d}x (f_{kl}, \psi)$$

or

$$U(t) = \int_0^t \int_{-R}^{R} F(t - s, x) \, \overline{P}(s, x) \, \mathrm{d}s \, \mathrm{d}x \tag{4.18}$$

where

$$\begin{aligned} F(t, x) &\doteq \sum_{k,l=0}^{\infty} C_l \exp(-\alpha \lambda_{kl} t) \, \varphi_{1k}(x) \, (f_{kl}, \psi) , \\ C_l &= -\frac{\alpha}{2L} (p_l d_l + 2q_l) \end{aligned} \tag{4.19}$$

since the series contained in (4.19) possesses strong properties as to the convergence.

Considering the both special cases it is

$$C_l = -\frac{\alpha}{2L} (p_l d_l + 2q_l) = -\alpha \sqrt{\frac{2}{L}} .$$

4.2. Random temperatures of a half-plane

First, the random boundary-initial value problem (4.1) is considered and the solution $\bar{u}$ is given by (4.4). Using the condition

$$F(., y) \in \mathbf{L}_2(0, t)$$

and assuming the weak correlation of $\overline{P}(t, \omega) = \overline{P}_\varepsilon(t, \omega)$ the limit theorem

$$\lim_{\varepsilon \downarrow 0} \frac{1}{\sqrt{\varepsilon}} \bar{u}_\varepsilon(t, y, \omega) = \xi(t, y, \omega) \qquad \text{in distribution} \tag{4.20}$$

can be derived from Section 1.4. This is the result of Section 1.4 taking the smallest order as to ε. The random function $\xi(t, y, \omega)$ is a Gaussian function having the mean

$$\langle \xi(t, y) \rangle = 0$$

and the correlation function

$$\langle \xi(t_1, y_1)\, \xi(t_2, y_2) \rangle = \int_0^{t_{12}} F(t_1 - s, y_1)\, F(t_2 - s, y_2)\, a(s)\, \mathrm{d}s \tag{4.21}$$

where

$$t_{12} \doteq \min\{t_1, t_2\}.$$

The intensity $a(t)$ of the process $\overline{\boldsymbol{P}}_\varepsilon(t, \omega)$ is defined by

$$a(t) = \lim_{\varepsilon \downarrow 0} \frac{1}{\varepsilon} \int_{-\varepsilon}^{\varepsilon} \langle \overline{\boldsymbol{P}}_\varepsilon(t)\, \overline{\boldsymbol{P}}_\varepsilon(t + s) \rangle\, \mathrm{d}s.$$

In order to fulfil the compatibility condition $\overline{\boldsymbol{P}}(0, \omega) = 0$ a.s. we put

$$\overline{\boldsymbol{P}}(t, \omega) = \psi(t)\, P_1(t, \omega)$$

where $\psi(t)$ is a function with the properties

$$0 \leqq \psi(t) \leqq 1 \quad \text{for} \quad 0 \leqq t \leqq \eta,$$

$$\psi(0) = 0, \qquad \psi(t) = 1 \quad \text{for} \quad t \geqq \eta$$

and $P_1(t, \omega)$ is a given weakly correlated process. For example, $\psi(t)$ can be taken as a continuous function:

$$\psi(t) = \begin{cases} \dfrac{t}{\eta} & \text{for} \quad 0 \leqq t \leqq \eta, \\ 1 & \text{for} \quad \eta < t \end{cases}$$

or a continuously differentiable function:

$$\psi(t) = \begin{cases} \left(\dfrac{t}{\eta}\right)^2 \left(-2\,\dfrac{t}{\eta} + 3\right) & \text{for} \quad 0 \leqq t \leqq \eta, \\ 1 & \text{for} \quad \eta < t. \end{cases}$$

Then we obtain

$$a(t) = \psi(t) \lim_{\varepsilon \downarrow 0} \frac{1}{\varepsilon} \int_{-\varepsilon}^{\varepsilon} \psi(t + s)\, \langle P_1(t)\, P_1(t + s) \rangle\, \mathrm{d}s$$

and assuming the positivity of the correlation function of $P_1(t, \omega)$ then

$$a(t) = \psi^2(t)\, a_1(t)$$

where $a_1(t)$ denotes the intensity of $P_1(t, \omega)$. Now we have

$$K(t_1, y_1; t_2, y_2) \doteq \langle \xi(t_1, y_1)\, \xi(t_2, y_2) \rangle$$

$$= \frac{\alpha}{\pi} \int\limits_0^{t_{12}} \exp\left(-\frac{y_1^2}{4\alpha(t_1 - s)} - \frac{y_2^2}{4\alpha(t_2 - s)} \right) \frac{a(s)}{\sqrt{(t_1 - s)\,(t_2 - s)}}\, \mathrm{d}s. \tag{4.22}$$

Supposing that $P_1(t, \omega)$ is a wide-sense stationary process with the intensity $a_1(t) \equiv a_1$ then the correlation function $K(t_1, y_1; t_2, y_2)$ can be calculated approximately by

$$\tilde{K}(t_1, y_1; t_2, y_2) = \frac{\alpha a_1}{\pi} \int\limits_0^{t_{12}} \exp\left(-\frac{y_1^2}{4\alpha(t_1 - s)} - \frac{y_2^2}{4\alpha(t_2 - s)} \right) \frac{\mathrm{d}s}{\sqrt{(t_1 - s)\,(t_2 - s)}}. \tag{4.23}$$

Especially, the variance can be determined from

$$\langle \xi^2(t, y) \rangle \approx \tilde{K}(t, y; t, y)$$

$$= \frac{\alpha a_1}{\pi} \int\limits_0^t \frac{1}{s} \exp\left(-\frac{y^2}{2\alpha s} \right) \mathrm{d}s = -\frac{\alpha a_1}{\pi} \int\limits_{-\infty}^{-\frac{y^2}{2\alpha t}} \frac{1}{s} \exp(s)\, \mathrm{d}s$$

$$= -\frac{\alpha a_1}{\pi} \operatorname{Ei}\left(-\frac{y^2}{2\alpha t} \right) \tag{4.24}$$

and $\operatorname{Ei}(-x)$ is defined by

$$\operatorname{Ei}(-x) \doteq \int\limits_{-\infty}^{-x} \frac{1}{s}\, \mathrm{e}^s\, \mathrm{d}s.$$

For small positive numbers x we can calculate $\operatorname{Ei}(-x)$ by $\operatorname{Ei}(-x) \approx \ln(\gamma x)$ with the Euler constant $\gamma = 1.781072418$. From the relation above we see that

$$\lim_{t \to \infty} \tilde{K}(t, y; t, y) = \infty \quad \text{for} \quad y > 0.$$

Taking the linear function ψ we have

$$K(t, y; t, y)$$

$$= \begin{cases} \dfrac{\alpha a_1}{\pi \eta^2} \displaystyle\int\limits_0^t \exp\left(-\frac{y^2}{2\alpha p} \right) \frac{(t - p)^2}{p}\, \mathrm{d}p & \text{for } t \leqq \eta, \\[2ex] \dfrac{\alpha a_1}{\pi \eta^2} \displaystyle\int\limits_{t-\eta}^t \exp\left(-\frac{y^2}{2\alpha p} \right) \frac{(t - p)^2}{p}\, \mathrm{d}p - \frac{\alpha a_1}{\pi} \int\limits_{-\infty}^{-\frac{y^2}{2\alpha(t-\eta)}} \frac{1}{p} \exp(p)\, \mathrm{d}p & \text{for } t > \eta. \end{cases}$$

Fig. 4.3 shows $K(t, y; t, y)$ as a function of t for $y = 10^{-6}$, $y = 10^{-4}$ and $\eta = 0.5$, $\eta = 1$, $\eta = 2$. The relation

$$\tilde{K}(t, y; t, y) \approx K(t, y; t, y)$$

for great values of t in comparison with η can be observed very well. The temperature conductivity α was assumed to be

$$\alpha = 1.278 \cdot 10^{-5} \frac{\mathrm{m}^2}{\mathrm{s}}.$$

The function $\tilde{K}(t, y; t, y)$ is plotted in Fig. 4.4 as a function of t and as a function of y. The variance increases if t goes to greater values. But the variance decreases with in-

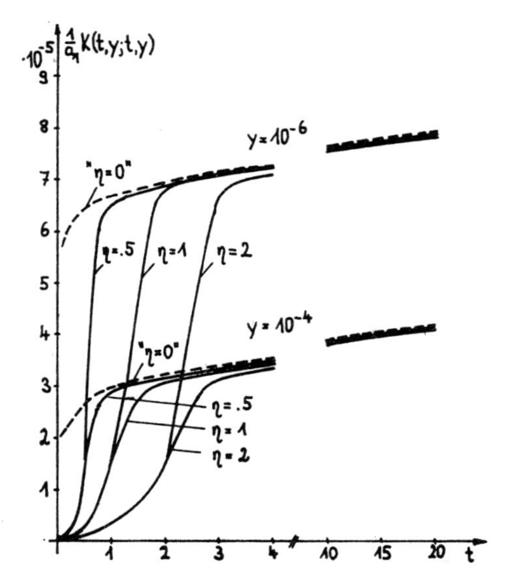

Fig. 4.3. $K(t, y; t, y)$ for different values of η and y as a function of t

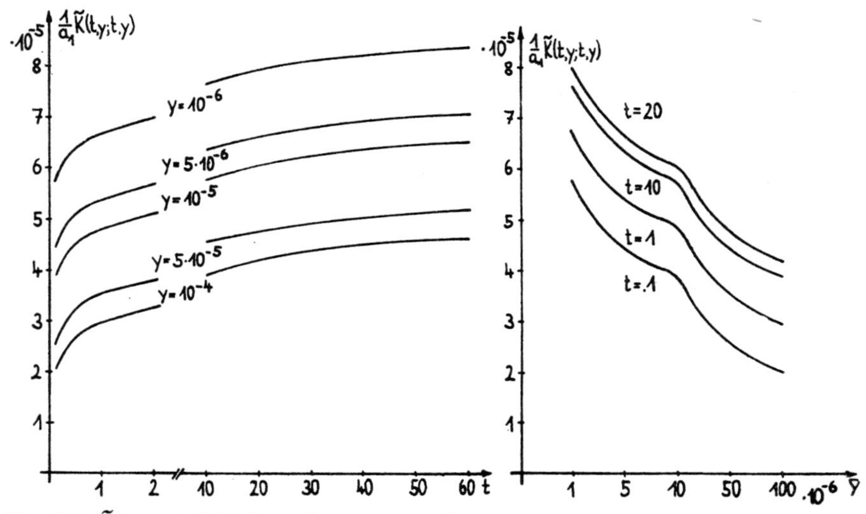

Fig. 4.4. $\tilde{K}(t, y; t, y)$ in dependence on t and y

creasing depth y. Setting

$$P(t, \omega) = \begin{cases} \frac{1}{\lambda} P_m & \text{with probability } p, \\ 0 & \text{with probability } 1 - p \end{cases}$$

it follows

$$\langle P(t) \rangle = \frac{1}{\lambda} P_m p \quad \text{and} \quad \langle \bar{P}(t)^2 \rangle = \left(\frac{1}{\lambda} P_m\right)^2 (1 - p)\, p .$$

In the following a weakly correlated process $\bar{P}(t, \omega)$ is assumed the variance of which is

$$\langle \bar{P}(t)^2 \rangle = \left(\frac{1}{\lambda} P_m\right)^2 (1 - p)\, p$$

and the correlation function is denoted by $R(s)$ or $R_\varepsilon(s)$, respectively. An example of a correlation function of P_1 is

$$R_\varepsilon(s) = \begin{cases} \left(\frac{1}{\lambda} P_m\right)^2 p(1 - p) \left(1 - \frac{1}{\varepsilon} |s|\right) & \text{for } |s| \leqq \varepsilon, \\ 0 & \text{otherwise}. \end{cases} \tag{4.25}$$

Realistic values are the following:

$$\alpha = 1.278 \cdot 10^{-5} \frac{\mathrm{m}^2}{\mathrm{s}}; \qquad P_m = 1.26 \cdot 10^7 \frac{\mathrm{W}}{\mathrm{m}^2}; \qquad \lambda = 45.3 \frac{\mathrm{W}}{\mathrm{m} \cdot \mathrm{K}};$$

$$t_{\mathrm{cor}} = 65 \cdot 10^{-6}\ \mathrm{s}.$$

Taking $\varepsilon = t_{\mathrm{cor}}$ we obtain

$$a_1 = \lim_{\varepsilon \downarrow 0} \frac{1}{\varepsilon} \int_{-\varepsilon}^{\varepsilon} R_\varepsilon(s)\, \mathrm{d}s = \left(\frac{1}{\lambda} P_m\right)^2 p(1 - p)$$

and for $p = 1/2$ then

$$\varepsilon a_1 = 1.2571817 \cdot 10^6\ \mathrm{K}^2 .$$

In Fig. 4.5 the dispersion $\sqrt{\langle \bar{u}^2(t, y) \rangle}$ is illustrated where the above concrete values were used. The dispersion increases considerably for small values y and a fixed t.

For $t_1 = t_2$ (4.23) leads to

$$\tilde{K}(t, y_1; t, y_2) = \frac{\alpha a_1}{\pi} \int_0^t \exp\left(-\frac{y_1^2 + y_2^2}{4\alpha(t - s)}\right) \frac{\mathrm{d}s}{t - s}$$

$$= -\frac{\alpha a_1}{\pi} \operatorname{Ei}\left(-\frac{1}{4\alpha t} (y_1^2 + y_2^2)\right).$$

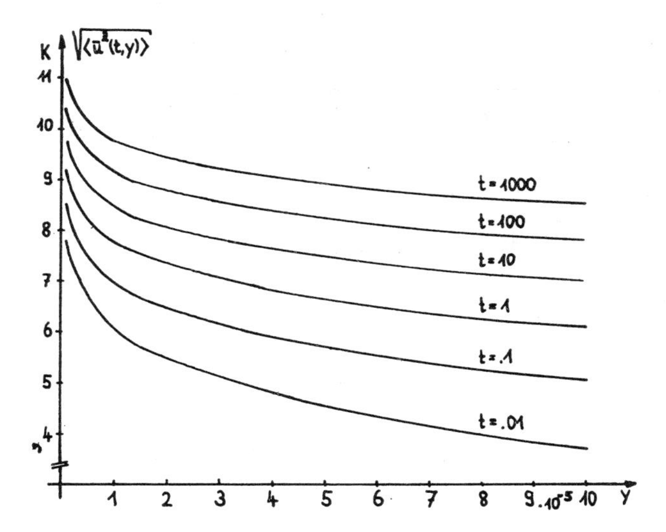

Fig. 4.5. Dispersion of $\bar{u}(t, y, \omega)$

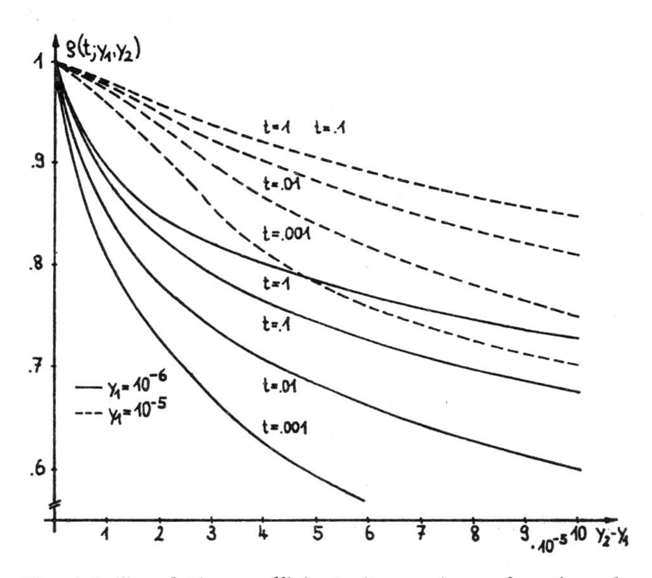

Fig. 4.6. Correlation coefficient $\varrho(t, y_1, y_2)$ as a function of $y_2 - y_1$

Hence, the correlation coefficient $\varrho(t, y_1, y_2)$ of $\bar{u}(t, y_1, \omega)$ and $\bar{u}(t, y_2, \omega)$ can be determined as

$$\varrho(t, y_1, y_2) \approx \frac{-\mathrm{Ei}\left(-\frac{1}{4\alpha t}(y_1^2 + y_2^2)\right)}{\sqrt{\mathrm{Ei}\left(-\frac{1}{2\alpha t} y_1^2\right) \mathrm{Ei}\left(-\frac{1}{2\alpha t} y_2^2\right)}}.$$

The correlation coefficient $\varrho(t, y_1, y_2)$ is plotted in Fig. 4.6 as a function of $y_2 - y_1 \geqq 0$ where we have put $y_1 = 10^{-6}$ m (10^{-5} m). For smaller y_1 the correlation coefficient decreases faster in dependence on $y_2 - y_1$ as for greater values y_1. Fig. 4.6 shows clearly the influence of the time t on the correlation coefficient.

If we want to calculate the second approximation, for example, as to the variance of $\bar{u}(t, y, \omega)$ we apply (1.35) and obtain

$$\langle \bar{u}^2(t, y) \rangle = {}^2A_1(t, y)\, \varepsilon + {}^2A_2(t, y)\, \varepsilon^2 + o(\varepsilon^2).$$

The function ${}^2A_1(t, y)$ has the form

$${}^2A_1(t, y) = \int_0^t F^2(t - s, y)\, a(s)\, ds$$

where $a(s)$ is given by

$$a(s) = \psi^2(s)\, a_1(s).$$

Theorem 1.9 leads to

$${}^2A_2(t, y) = \int_0^t F^2(t - s, y)\, b(s)\, ds + F^2(0, y)\, \bar{S}(t; -1, 0) + F^2(t, y)\, \underline{S}(0; 0, 1)$$

with

$$F(0, y) = 0$$

and

$$\underline{S}(0; 0, 1) = \int_0^1 \underline{a}(u; 0)\, du - a(0),$$

$$\underline{a}(u; 0) = \lim_{\varepsilon \downarrow 0} \frac{1}{\varepsilon} \int_0^{\varepsilon(u+1)} \langle \bar{P}(\varepsilon u)\, \bar{P}(s) \rangle\, ds.$$

Using a correlation function of the form (4.25) we have

$$b(s) = 0,$$

$$\underline{a}(u; 0) = \lim_{\varepsilon \downarrow 0} \psi(\varepsilon u) \frac{1}{\varepsilon} \int_0^{\varepsilon(u+1)} \psi(s)\, R_\varepsilon(s - \varepsilon u)\, ds$$

$$= \psi(0) \lim_{\varepsilon \downarrow 0} \frac{1}{\varepsilon} \int_0^{\varepsilon(u+1)} \psi(s)\, R_\varepsilon(s - \varepsilon u)\, ds = 0$$

and then

$${}^2A_2(t, y) = 0.$$

Hence, the expression ${}^2A_1(t, y)\, \varepsilon$ describes the variance $\langle \bar{u}^2(t, y) \rangle$ up to terms of the order $o(\varepsilon^2)$.

The temporal derivative of the temperature $\bar{u}(t, y, \omega)$ can be obtained from (4.4):

$$\bar{u}_t(t, y, \omega) = \frac{1}{2}\sqrt{\frac{\alpha}{\pi}} \int_0^t \frac{1}{\sqrt{t-s}^3} \exp\left(-\frac{y^2}{4\alpha(t-s)}\right)\left\{1 - \frac{y^2}{2\alpha(t-s)}\right\} \bar{P}(s, \omega)\, ds$$

for $t > 0$ and $y > 0$. Thus

$$\bar{u}_t(t, y, \omega) = \int_0^t F_1(t-s, y)\, \bar{P}(s, \omega)\, ds$$

where

$$F_1(t, y) = \frac{1}{2}\sqrt{\frac{\alpha}{\pi}} \frac{1}{\sqrt{t}^3} \exp\left(-\frac{y^2}{4\alpha t}\right)\left\{1 - \frac{y^2}{2\alpha t}\right\}.$$

By application of a limit theorem for vector functions we obtain

$$\lim_{\varepsilon \downarrow 0} \frac{1}{\sqrt{\varepsilon}} \big(\bar{u}(t, y, \omega), \bar{u}_t(t, y, \omega)\big) = \big(\xi(t, y, \omega), \xi_1(t, y, \omega)\big)$$

in distribution and

$$\big(\xi(t, y, \omega), \xi_1(t, y, \omega)\big)$$

is a Gaussian vector function. The conditions for the application of the limit theorem are fulfilled since

$$F_1(\cdot, y) \in \mathbf{L}_2(0, t) \quad \text{for} \quad y > 0.$$

In addition to the correlation relation (4.22) we still have

$$\begin{aligned}\langle \xi(t_1, y_1)\, \xi_1(t_2, y_2)\rangle &= \int_0^{t_{12}} F(t_1 - s, y_1)\, F_1(t_2 - s, y_2)\, a(s)\, ds \\ &= -\frac{1}{2}\frac{\alpha}{\pi} \int_0^{t_{12}} \frac{1}{\sqrt{t_1 - s}\,\sqrt{t_2 - s}^3} \left(1 - \frac{y_2^2}{2\alpha(t_2 - s)}\right) \\ &\qquad \times \exp\left(-\frac{y_1^2}{4\alpha(t_1 - s)} - \frac{y_2^2}{4\alpha(t_2 - s)}\right) a(s)\, ds\end{aligned}$$

and

$$\begin{aligned}&\langle \xi_1(t_1, y_1)\, \xi_1(t_2, y_2)\rangle \\ &= \int_0^{t_{21}} F_1(t_1 - s, y_1)\, F_1(t_2 - s, y_2)\, a(s)\, ds \\ &= \frac{\alpha}{4\pi} \int_0^{t_{12}} \frac{1}{\sqrt{t_1 - s}^3\,\sqrt{t_2 - s}^3} \left(1 - \frac{y_1^2}{2\alpha(t_1 - s)}\right)\left(1 - \frac{y_2^2}{2\alpha(t_2 - s)}\right) \\ &\qquad \times \exp\left(-\frac{y_1^2}{4\alpha(t_1 - s)} - \frac{y_2^2}{4\alpha(t_2 - s)}\right) a(s)\, ds.\end{aligned}$$

We want to investigate the expected rate of crossings and so we calculate the correlation relations for $t_1 = t_2 = t$ and $y_1 = y_2 = y$. Assuming a stationary process $\bar{P}(t, \omega)$

with the intensity a we obtain

$$\langle \xi(t, y)\, \xi_1(t, y) \rangle = -\frac{a\alpha}{2\pi} \int\limits_0^t \frac{1}{(t-s)^2} \left(1 - \frac{y^2}{2\alpha(t-s)}\right) \exp\left(-\frac{y^2}{2\alpha(t-s)}\right) \mathrm{d}s$$

$$= -\frac{a\alpha^2}{\pi y^2} \int\limits_{\frac{y^2}{2\alpha t}}^{\infty} (1-s)\, \mathrm{e}^{-s}\, \mathrm{d}s = \frac{a\alpha}{2\pi t} \exp\left(-\frac{y^2}{2\alpha t}\right)$$

and

$$\langle \xi_1^2(t, y) \rangle = \frac{a\alpha}{4\pi} \int\limits_0^t \frac{1}{(t-s)^3} \left(1 - \frac{y^2}{2\alpha(t-s)}\right)^2 \exp\left(-\frac{y^2}{2\alpha(t-s)}\right) \mathrm{d}s$$

$$= \frac{a\alpha^3}{\pi y^4} \int\limits_{\frac{y^2}{2\alpha t}}^{\infty} s(1-s)^2\, \mathrm{e}^{-s}\, \mathrm{d}s$$

$$= \frac{a\alpha^3}{\pi y^4} \left(1 + \frac{y^2}{2\alpha t}\right) \left(3 + \left(\frac{y^2}{2\alpha t}\right)^2\right) \exp\left(-\frac{y^2}{2\alpha t}\right).$$

Considering (4.24) we can see that the variance of the temperature increases if the time t increases or y decreases. The variance of $\bar{u}_t(t, y, \omega)$ shows a similar behaviour since the function

$$f(x) = (1 + x)\,(3 + x^2)\, \mathrm{e}^{-x}$$

is a monotonously decreasing function for $x > 0$. We have

$$\lim_{t \to \infty} \langle \xi_1^2(t, y) \rangle = \frac{3a\alpha^3}{\pi y^4} \quad \text{for} \quad y > 0$$

and

$$\lim_{y \to 0+0} \langle \xi_1^2(t, y) \rangle = \infty.$$

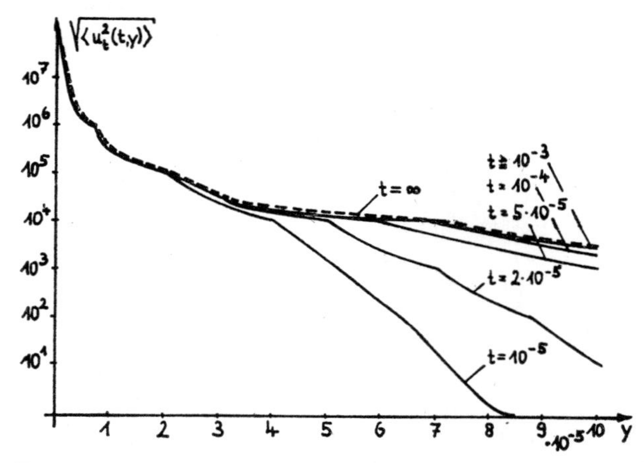

Fig. 4.7. Dispersion of $\bar{u}_t(t, y, \omega)$

The dispersion of $\bar{u}_t(t, y, \omega)$ is plotted in Fig. 4.7. The figure shows very well the properties discussed above.

The correlation coefficient between $\xi(t, y, \omega)$ and $\xi_1(t, y, \omega)$ is given by

$$\varrho_{01}(t, y) = \frac{\langle \xi(t, y)\, \xi_1(t, y)\rangle}{\sqrt{\langle \xi^2(t, y)\rangle \langle \xi_1^2(t, y)\rangle}}$$

and we find that

$$\varrho_{01}(t, y) = \frac{\dfrac{y^2}{2\alpha t} \exp\left(-\dfrac{y^2}{4\alpha t}\right)}{\sqrt{-\left(1 + \dfrac{y^2}{2\alpha t}\right)\left(3 + \left(\dfrac{y^2}{2\alpha t}\right)^2\right) \operatorname{Ei}\left(-\dfrac{y^2}{2\alpha t}\right)}}$$

and

$$\lim_{t\to\infty} \varrho_{01}(t, y) = 0 \quad \text{for} \quad y > 0,$$

$$\lim_{y\to 0+0} \varrho_{01}(t, y) = 0 \quad \text{for} \quad t > 0.$$

Setting

$$z = \frac{y^2}{2\alpha t}$$

it follows

$$\varrho_{01}(t, y) = \frac{z\, \mathrm{e}^{-(1/2)z}}{\sqrt{-\operatorname{Ei}(-z)\,(1 + z)\,(3 + z^2)}}$$

and

$$\lim_{z\to 0+0} \varrho_{01}(t, y) = 0, \qquad \lim_{z\to\infty} \varrho_{01}(t, y) = 1.$$

The averaged problem belonging to the problem (4.1) has the form

$$w_t = \alpha \Delta w; \qquad w(0, y) = c, \qquad \frac{\partial w}{\partial y}(t, 0) = \langle P(t)\rangle$$

if the initial temperature $u_0(y)$ is assumed to be constant,

$$u_0(y) \equiv c.$$

Using (4.4) the solution can be written as

$$w(t, y) = -\sqrt{\frac{\alpha}{\pi}} \int_0^t \frac{\langle P(s)\rangle}{\sqrt{t - s}} \exp\left(-\frac{y^2}{4\alpha(t - s)}\right) \mathrm{d}s + c.$$

By means of

$$\langle P(t)\rangle = \psi(t)\, P_m \frac{p}{\lambda} \approx \frac{p}{\lambda} P_m \quad \text{for} \quad t \geqq 0$$

we obtain

$$w(t, y) = -\sqrt{\frac{\alpha}{\pi}}\,\frac{p}{\lambda}\,P_m \int_0^t \frac{1}{\sqrt{t-s}} \exp\left(-\frac{y^2}{4\alpha(t-s)}\right) \mathrm{d}s + c$$

$$= \frac{2p}{\lambda}\,P_m \left\{-y\left(1 - \Phi\left(\frac{y}{\sqrt{2\alpha t}}\right)\right) + \sqrt{\frac{\alpha t}{\pi}} \exp\left(-\frac{y^2}{4\alpha t}\right)\right\} + c \tag{4.26}$$

and

$$w_t(t, y) = \frac{p}{\lambda}\,P_m \sqrt{\frac{\alpha}{\pi t}} \exp\left(-\frac{y^2}{4\alpha t}\right).$$

We have

$$\lim_{y \to 0+0} \big(w(t, y) - c\big) = \frac{2p}{\lambda}\,P_m \sqrt{\frac{\alpha t}{\pi}} \quad \text{for} \quad t > 0.$$

Fig. 4.8 shows the behaviour of the averaged solution $w(t, y)$ where we have taken the given physical constants and $p = 1/2$. Fig. 4.9 illustrates the temperature domains in which $u(t, y, \omega) - c$ is contained with probability 0.9987. In this figure we have put $p = 0.5$ and 0.1. It follows that the temperature oscillations can be considerable. In general, the mean $w(t, y) - c$ is multiplied by p (see (4.26)) and the dispersion of $u(t, y, \omega)$ by $\sqrt{p(1-p)}$ since

$$a_1 = \left(\frac{1}{\lambda}\,P_m\right)^2 p(1-p).$$

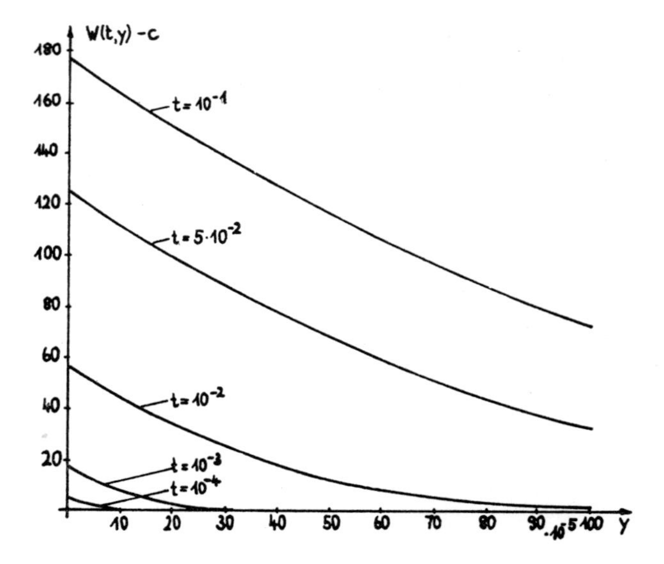

Fig. 4.8. Solution $w(t, y)$ of the averaged problem for $p = 1/2$

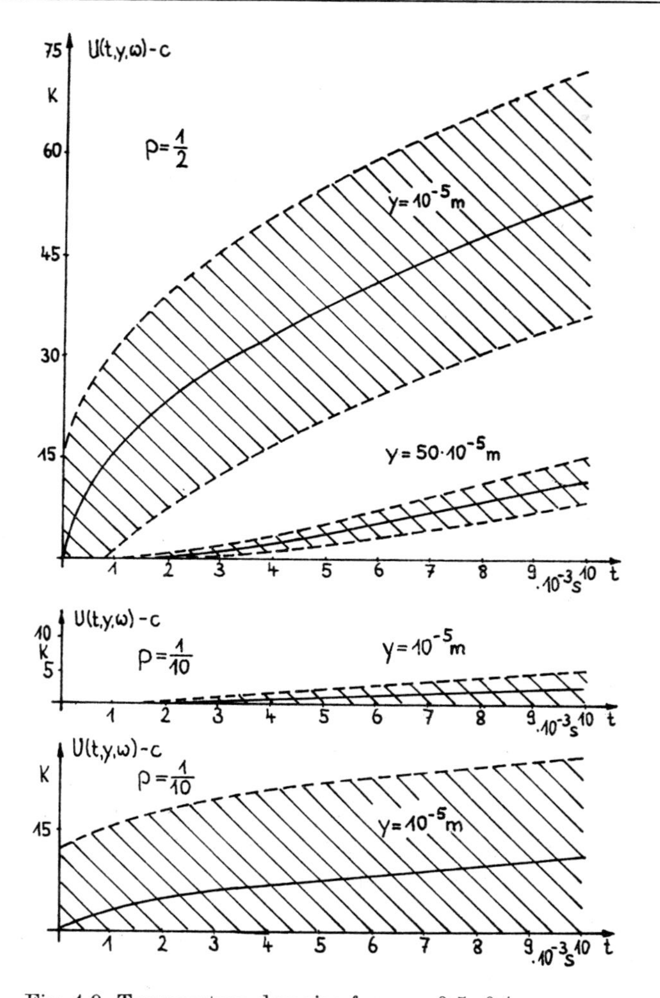

Fig. 4.9. Temperature domains for $p = 0.5$; 0.1

For smaller p the mean value decreases faster in dependence on p than the dispersion because

$$\frac{p}{\sqrt{p(1-p)}} \leqq 1 .$$

The expected number of threshold crossings of the temperature $u(t, y, \omega)$ for the value y and the time $t_1 \leqq t \leqq t_2$ is given by

$$\langle N(n; t_1, t_2, y)\rangle = \int_{t_1}^{t_2} r(n; t, y) \, dt$$

where n denotes the assumed level. The expected rate of crossings $r(n; t, y)$ can be written as

$$r(n; t, y) = \sqrt{\frac{m_{11}}{m_{00}}} \frac{1}{\sqrt{2\pi}} \exp\left(-\frac{(n-w)^2}{2m_{00}}\right)$$
$$\times \left\{2\sqrt{1-\varrho_{01}^2} \frac{1}{\sqrt{2\pi}} \exp\left(-\frac{H^2(n)}{2(1-\varrho_{01}^2)}\right) + H(n)\left[2\Phi\left(\frac{H(n)}{\sqrt{1-\varrho_{01}^2}}\right) - 1\right]\right\}$$

where we have put

$$m_{00} = m_{00}(t, y) = \langle \bar{u}^2(t, y)\rangle,$$

$$m_{01} = m_{01}(t, y) = \langle \bar{u}(t, y)\, \bar{u}_t(t, y)\rangle,$$

$$m_{11} = m_{11}(t, y) = \langle \bar{u}_t^2(t, y)\rangle,$$

$$\varrho_{01} = \varrho_{01}(t, y) = \frac{m_{01}(t, y)}{\sqrt{m_{00}(t, y)\, m_{11}(t, y)}},$$

$$H(n) = H(n; t, y) = \frac{w_t(t, y)}{\sqrt{m_{11}(t, y)}} + \varrho_{01}(t, y) \frac{n - w(t, y)}{\sqrt{m_{00}(t, y)}},$$

$$\Phi(r) = \frac{1}{\sqrt{2\pi}} \int_{-\infty}^{r} \exp\left(-\frac{1}{2}s^2\right) ds.$$

The functions which are necessary for the calculation of $r(n; t, y)$ were made available in this section. In Fig. 4.10 $r(n; t, y)$ is plotted as a function of t for different values y and for $n = 30 + c$, $n = 500 + c$. The maximum of r can be found to the time for which we have $n = w(t, y)$. This figure shows the strong dependence of the expected rate on the depth y. In the case of $n = 30 + c$ the function $r(n; t, y)$ is different of zero for values of t with

$$0.0005\ \text{s} \leqq t \leqq 0.008\ \text{s}$$

and in the case of $n = 500 + c$ for values of t with

$$0.72\ \text{s} \leqq t \leqq 0.9\ \text{s}.$$

This shows that the system needs a certain time up to the temperatures e.g. around $(500 + c)$ K. For $t \geqq 0.9$ s the temperature $u(t, y, \omega)$ has values which are greater than $(500 + c)$ K. For a fixed level n the function $\langle N\rangle$ depends on y very strongly (see Fig. 4.11). The function $\langle N(n; 0, \infty, y)\rangle$ increases essentially if n increases. This is a sequence of the slow growing for great times t. For example, $\langle N(500 + c; 0, t, 10^{-5})\rangle$ increases essentially for $0.74\ \text{s} \leqq t \leqq 0.86$ s; i.e. $\langle N\rangle$ increases during a time $T = 0.12$ s. On the other hand $\langle N(30 + c; 0, t, 10^{-5})\rangle$ increases for $0.001\ \text{s} \leqq t \leqq 0.007$ s and hence during a time $T = 0.006$ s. The function $\langle N\rangle$ leads to hints on times for which given temperatures are reached.

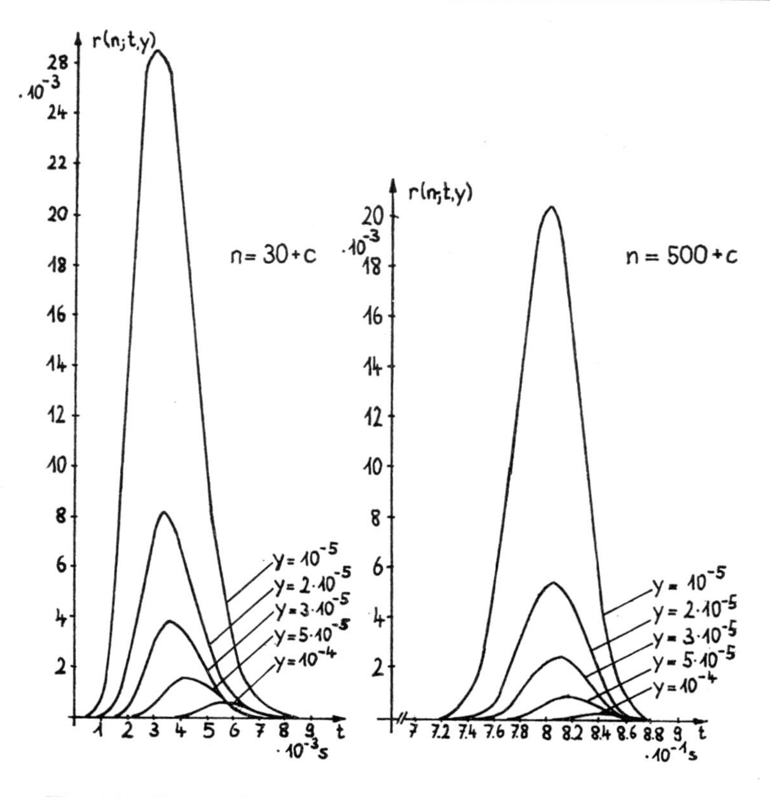

Fig. 4.10. Expected rates of crossings as function of t for different y

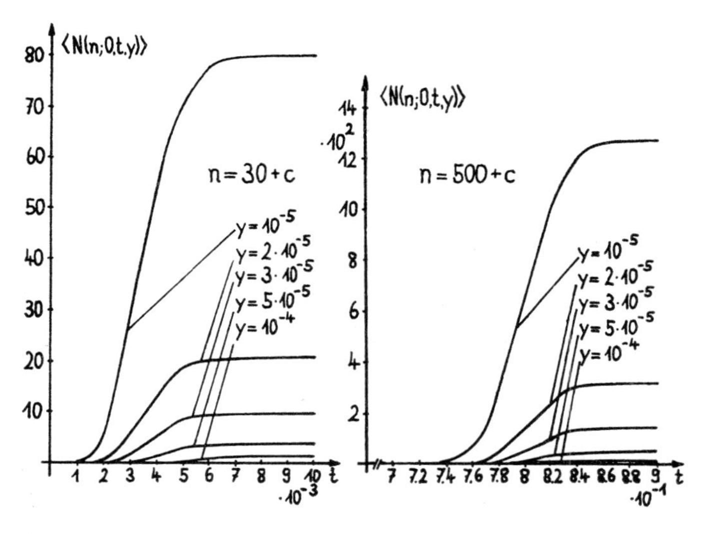

Fig. 4.11. Expected number of threshold crossings as function of t for different y

4.3. Random temperatures of a bounded domain

4.3.1. Weakly correlated temperature gradient

In this chapter the boundary-initial value problem (4.9) is considered where the random function $\overline{P}(t, x, \omega)$ is assumed to be a weakly correlated function as to (t, x). Limit theorems (see Section 1.6.1) can be applied to $U(t, \omega)$ from (4.18) and it is

$$\lim_{\varepsilon \downarrow 0} \frac{1}{\varepsilon} U(t, \omega) = \xi(t, \omega) \qquad \text{in distribution}$$

where $\xi(t, \omega)$ denotes a Gaussian process having the moments

$$\begin{aligned} &\langle \xi(t) \rangle = 0, \\ &\langle \xi(t_1)\, \xi(t_2) \rangle = \int_0^{t_{12}} \int_{-R}^{R} F(t_1 - s, x)\, F(t_2 - s, x)\, a(s, x)\, \mathrm{d}s\, \mathrm{d}x \end{aligned} \tag{4.27}$$

with $t_{12} \doteq \min\{t_1, t_2\}$. The function $a(t, x)$ is the intensity of the weakly correlated function $\overline{P}(t, x, \omega)$ defined by

$$a(t, x) = \lim_{\varepsilon \downarrow 0} \frac{1}{\varepsilon^2} \iint_{\mathcal{K}_\varepsilon(t,x)} \langle \overline{P}(t, x)\, \overline{P}(s, y) \rangle\, \mathrm{d}s\, \mathrm{d}y. \tag{4.28}$$

In this term it is integrated as to a ball $\mathcal{K}_\varepsilon(t, x)$ with the centre (t, x) and the radius ε,

$$\mathcal{K}_\varepsilon(t, x) = \{(s, y) : |(s, y) - (t, x)| \leqq \varepsilon\}.$$

Supposing the property

$$\langle \overline{P}(t, x)\, \overline{P}(s, y) \rangle = R(s - t, y - x) \tag{4.29}$$

then

$$\begin{aligned} a(t, x) &= \lim_{\varepsilon \downarrow 0} \frac{1}{\varepsilon^2} \iint_{\mathcal{K}_\varepsilon(t,x)} R(s - t, y - x)\, \mathrm{d}s\, \mathrm{d}y \\ &= \lim_{\varepsilon \downarrow 0} \frac{1}{\varepsilon^2} \iint_{\mathcal{K}_\varepsilon(0,0)} R(s, y)\, \mathrm{d}s\, \mathrm{d}y = a = \text{const}. \end{aligned}$$

If the compatibility conditions (4.10) are taken into consideration then an assumption of the form (4.29) is impossible, but we can assume

$$a(t, x) = a = \text{const} \quad \text{for} \quad t \geqq \delta, \qquad |x| \leqq R - \delta.$$

Since the results for this special case with respect to the correlation function (4.27) slightly differ from the results applying $a(t, x) = a$ for $t \geqq 0$, $|x| \leqq R$ the further considerations are made with the assumption $a(t, x) = a = \text{const}$.

If the correlation function of $\bar{P}(t, x, \omega)$ is selected to be

$$R_1(s, y) = \langle \bar{P}(t, x)\, \bar{P}(t + s, x + y)\rangle$$
$$= \begin{cases} \sigma^2 \left(1 - \frac{1}{\tilde{\varepsilon}}\, |s|\right)\left(1 - \frac{1}{\tilde{\varepsilon} b}\, |y|\right) & \text{for} \quad |(s, y)| \geqq \tilde{\varepsilon}, \\ 0 & \text{otherwise} \end{cases} \tag{4.30}$$

where $|(t, x)|$ is defined by

$$|(t, x)| \doteq \max \left\{|t|, \frac{1}{b}\, |x|\right\}$$

and the correlation length ε is $\varepsilon = \tilde{\varepsilon}\sqrt{1 + b^2}$, then

$$\frac{1}{\varepsilon^2} \iint\limits_{\mathcal{K}_\varepsilon(0,0)} R_1(s, y)\, \mathrm{d}s\, \mathrm{d}y = \frac{1}{\varepsilon^2} \int\limits_{-\tilde{\varepsilon}}^{\tilde{\varepsilon}} \int\limits_{-\tilde{\varepsilon} b}^{\tilde{\varepsilon} b} R_1(s, y)\, \mathrm{d}s\, \mathrm{d}y = \frac{\sigma^2 b}{1 + b^2}$$

follows. In this case the intensity

$$a_1 = \frac{\sigma^2 b}{1 + b^2}$$

is obtained.

Using the correlation function

$$R_2(s, y) = \sigma^2 \exp\left(-\delta_1\, |s| - \delta_2\, |y|\right) \tag{4.31}$$

with $\delta_1, \delta_2 > 0$ and setting $R_2(s, y) \approx 0$ for

$$|(s, y)| = \max \left\{|s|, \frac{1}{b}\, |y|\right\} \geqq \tilde{\varepsilon}\,.$$

it can be calculated

$$\frac{1}{\varepsilon^2} \iint\limits_{\mathcal{K}_\varepsilon(0,0)} R_2(s, y)\, \mathrm{d}s\, \mathrm{d}y = \left(\frac{\sigma}{\varepsilon}\right)^2 \int\limits_{-\tilde{\varepsilon}}^{\tilde{\varepsilon}} \exp\left(-\delta_1\, |s|\right) \mathrm{d}s \int\limits_{-\tilde{\varepsilon} b}^{\tilde{\varepsilon} b} \exp\left(-\delta_2\, |y|\right) \mathrm{d}y$$
$$= \frac{4\sigma^2}{\varepsilon^2 \delta_1 \delta_2} \left(1 - \exp\left(-\delta_1 \tilde{\varepsilon}\right)\right)\left(1 - \exp\left(-\delta_2 \tilde{\varepsilon} b\right)\right) \approx \frac{4\sigma^2}{\varepsilon^2 \delta_1 \delta_2}.$$

It follows as intensity

$$a_2 = \frac{4\sigma^2}{\varepsilon^2 \delta_1(\tilde{\varepsilon})\, \delta_2(\tilde{\varepsilon})} \quad \text{with} \quad \varepsilon = \tilde{\varepsilon}\sqrt{1 + b^2}.$$

For example the condition $R_2(s, y) \approx 0$ for $|(s, y)| \geqq \tilde{\varepsilon}$ is fulfilled in the case of $\delta_1 = 3/\tilde{\varepsilon}$, $\delta_2 = 3/\tilde{\varepsilon} b$ and it is

$$a_2 = \frac{4}{9}\, \frac{b\sigma^2}{1 + b^2}.$$

Finally we obtain from

$$R_3(s, y) = \sigma^2 \exp(-\delta_1 s^2 - \delta_2 y^2) \tag{4.32}$$

with $\delta_1, \delta_2 > 0$ the relation

$$\frac{1}{\varepsilon^2} \iint_{\mathcal{K}_\varepsilon(0,0)} R_3(s, y) \, ds \, dy = \frac{\sigma^2 \pi}{\varepsilon^2 \sqrt{\delta_1 \delta_2}} \left(2\Phi\left(\sqrt{2\delta_1}\tilde{\varepsilon}\right) - 1\right) \left(2\Phi\left(\sqrt{2\delta_2}\tilde{\varepsilon}b\right) - 1\right)$$

if we put $R_3(s, y) \approx 0$ for $|(s, y)| \geqq \tilde{\varepsilon}$. For $\delta_1 = 3\tilde{\varepsilon}^{-2}$, $\delta_2 = 3(b\tilde{\varepsilon})^{-2}$ the intensity is calculated to be

$$a_3 \approx \frac{\sigma^2 \pi b}{3(1 + b^2)} \left(2\Phi\left(\sqrt{6}\right) - 1\right)^2 \approx \frac{\sigma^2 \pi b}{3(1 + b^2)}.$$

Fig. 4.12 shows the above discussed correlation functions for $y = 0$ and the adequate intensities in dependence on b.

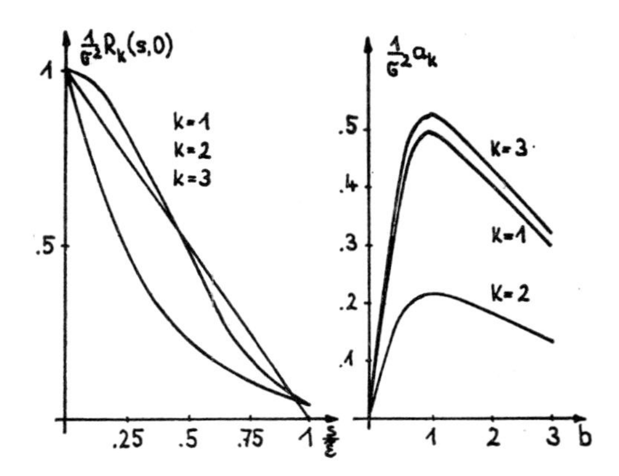

Fig. 4.12. Different correlation functions and intensities

After these considerations of the intensity we again turn to the correlation function of the limit process $\xi(t, \omega)$. Using (4.19) it follows

$$\langle \xi(t_1) \, \xi(t_2) \rangle$$
$$= \sum_{k_1, k_2 = 0}^{\infty} \sum_{l_1, l_2 = 0}^{\infty} C_{l_1} C_{l_2} \int_0^{t_{12}} \int_{-R}^{R} \varphi_{1k_1}(x) \, \varphi_{1k_2}(x)$$
$$\times \exp\left(-\varkappa \lambda_{k_1 l_1}(t_1 - s) - \alpha \lambda_{k_2 l_2}(t_2 - s)\right) (f_{k_1 l_1}, \psi) (f_{k_2 l_2}, \psi) \, a(s, x) \, ds \, dx . \tag{4.33}$$

With the help of the assumption $a(s, x) = a = \text{const}$ the correlation function has the form

$$\langle \xi(t_1) \, \xi(t_2) \rangle$$
$$= a \sum_{k=0}^{\infty} \sum_{l_1, l_2 = 0}^{\infty} C_{l_1} C_{l_2} \int_0^{t_{12}} \exp\left(-\alpha \lambda_{kl_1}(t_1 - s) - \alpha \lambda_{kl_2}(t_2 - s)\right) (f_{kl_1}, \psi) (f_{kl_2}, \psi) \, ds$$

because of

$$\int_{-R}^{R} \varphi_{1k_1}(x)\, \varphi_{1k_2}(x)\, \mathrm{d}x = \delta_{k_1k_2}.$$

Applying

$$\int_0^{t_{12}} \exp\left(-\alpha\lambda_{kl_1}(t_1 - s) - \alpha\lambda_{kl_2}(t_2 - s)\right) \mathrm{d}s$$
$$= \frac{\exp\left(-\alpha\lambda_{kl_1}(t_1 - t_{12}) - \alpha\lambda_{kl_2}(t_2 - t_{12})\right) - \exp\left(-\alpha\lambda_{kl_1}t_1 - \alpha\lambda_{kl_2}t_2\right)}{\alpha(\lambda_{kl_1} + \lambda_{kl_2})}$$

it follows

$$R(t_1, t_2) \doteq \langle \xi(t_1)\, \xi(t_2) \rangle$$
$$= \frac{a}{\alpha} \sum_{k=0}^{\infty} \sum_{l_1,l_2=0}^{\infty} C_{l_1} C_{l_2} \frac{(f_{kl_1}, \psi)\,(f_{kl_2}, \psi)}{\lambda_{kl_1} + \lambda_{kl_2}}$$
$$\times \{\exp(-\alpha\lambda_{kl_1}s) - \exp(-\alpha\lambda_{kl_1}t_1 - \alpha\lambda_{kl_2}t_2)\} \tag{4.34}$$

where $s \doteq |t_2 - t_1|$. If the correlation function $R(t_1, t_2)$ is only considered for large times t_1, t_2 then we obtain

$$\lim_{\substack{t_1,t_2\to\infty \\ |t_2-t_1|=s=\text{const}}} \langle \xi(t_1)\, \xi(t_2) \rangle \doteq \tilde{R}(s)$$
$$= \frac{a}{\alpha} \sum_{k=0}^{\infty} \sum_{l_1,l_2=0}^{\infty} C_{l_1} C_{l_2} \frac{(f_{kl_1}, \psi)\,(f_{kl_2}, \psi)}{\lambda_{kl_1} + \lambda_{kl_2}} \exp(-\alpha\lambda_{kl_1}s) \tag{4.35}$$

and $\tilde{\xi}(t, \omega)$ denotes the stationary Gaussian process belonging to $\xi(t, \omega)$ the correlation function of which is given by (4.35). For the following considerations the function $\psi(x, y)$ is selected as

$$\psi(x, y) = \chi_1(x)\, \chi_2(y).$$

Then from

$$f_{kl}(x, y) = \varphi_{1k}(x)\, \varphi_{2l}(y)$$

we deduce

$$(f_{kl}, \psi) = \iint_{\mathscr{G}} f_{kl}(x, y)\, \psi(x, y)\, \mathrm{d}x\, \mathrm{d}y$$
$$= \int_{-R}^{R} \varphi_{1k}(x)\, \chi_1(x)\, \mathrm{d}x \int_0^L \varphi_{2l}(y)\, \chi_2(y)\, \mathrm{d}y = r_{1k} r_{2l}$$

where

$$r_{1k} \doteq \int_{-R}^{R} \varphi_{1k}(x)\, \chi_1(x)\, \mathrm{d}x, \qquad r_{2l} \doteq \int_0^L \varphi_{2l}(y)\, \chi_2(y)\, \mathrm{d}y.$$

By means of (4.34) we have

$$R(t_1, t_2) = \frac{a}{\alpha} \sum_{k=0}^{\infty} \sum_{l_1,l_2=0}^{\infty} C_{l_1} C_{l_2} \frac{r_{1k}^2 r_{2l_1} r_{2l_2}}{\lambda_{kl_1} + \lambda_{kl_2}}$$
$$\times \{\exp(-\alpha\lambda_{kl_1}s) - \exp(-\alpha\lambda_{kl_1}t_1 - \alpha\lambda_{kl_2}t_2)\}. \tag{4.36}$$

It is $C_l = O(1)$ and the convergence conditions are to fulfil by

$$\frac{r_{1k}^2 r_{2l_1} r_{2l_2}}{\lambda_{kl_1} + \lambda_{kl_2}}.$$

For example, in the cases of $\alpha_i = 0$ for $i = 2, 3, 4$ and $\alpha_i \to \infty$ for $i = 2, 3, 4$ we have

$$\left|\frac{r_{1k}^2 r_{2l_1} r_{2l_2}}{\lambda_{kl_1} + \lambda_{kl_2}}\right| \leq \frac{\bar{C}}{k^2 l_1 l_2 (k^2 + l_1^2 + l_2^2)} \leq \frac{C}{k^2 l_1^2 l_2^2}$$

for

$$\chi_1(x) = 1_{[x_1, x_2]}(x), \qquad \chi_2(y) = 1_{[y_1, y_2]}(y)$$

and the convergence of the series in (4.36) is given. Taking into consideration the convergence of $\frac{1}{\varepsilon} U(t, \omega)$ to $\xi(t, \omega)$ or $\bar{\xi}(t, \omega)$, respectively, we obtain

$$\langle U(t_1) \, U(t_2) \rangle \approx \varepsilon^2 R(t_1, t_2).$$

Because of the continuity a.s. of the function $\bar{u}(t, x, y, \omega)$ the relation

$$U(t) = \left(\bar{u}(t, \cdot, \cdot), \chi_1 \chi_2\right) \approx \bar{u}(t, x_0, y_0) \int\limits_{\mathcal{D}_1(x_0)} \chi_1(x) \, dx \int\limits_{\mathcal{D}_2(y_0)} \chi_2(y) \, dy$$

is obtained if χ_1 is zero outside of a small neighbourhood $\mathcal{D}_1(x_0)$ of x_0 and χ_2 is zero outside of a small neighbourhood $\mathcal{D}_2(y_0)$ of y_0. Thus

$$\langle \bar{u}(t_1, x, y) \, \bar{u}(t_2, x, y) \rangle \approx \frac{\varepsilon^2 R(t_1, t_2)}{\left(\int\limits_{\mathcal{D}_1(x)} \chi_1(t) \, dt \int\limits_{\mathcal{D}_2(y)} \chi_2(t) \, dt\right)^2}. \tag{4.37}$$

For the following considerations, first, we deal with the case $\alpha_i \to \infty$, $i = 2, 3, 4$, in order to obtain some numerical results. First, we put

$$\chi_1(t) = 1_{[x-\delta_1, x+\delta_1]}(t), \qquad \chi_2(t) = 1_{[y-\delta_2, y+\delta_2]}(t) \tag{4.38}$$

and get

$$r_{1k} = \int\limits_{x-\delta_1}^{x+\delta_1} \varphi_{1k}(t) \, dt = \begin{cases} \dfrac{4\sqrt{R}}{k\pi} \cos\left(\dfrac{k\pi x}{2R}\right) \sin\left(\dfrac{1}{2R} k\pi\delta_1\right) & \text{for} \quad k = 1, 3, \ldots, \\ \dfrac{4\sqrt{R}}{k\pi} \sin\left(\dfrac{k\pi x}{2R}\right) \sin\left(\dfrac{1}{2R} k\pi\delta_1\right) & \text{for} \quad k = 2, 4, \ldots, \end{cases}$$

$$r_{2l} = \int\limits_{y-\delta_2}^{y+\delta_2} \varphi_{2l}(t) \, dt = \frac{4\sqrt{2L}}{(2l-1)\pi} \cos\left(\frac{(2l-1) \, y\pi}{2L}\right) \sin\left(\frac{1}{2L} (2l-1) \, \pi\delta_2\right)$$

$$\text{for} \quad l = 1, 2, \ldots$$

In order to determine the correlation function $\tilde{R}(s)$ we have

$$\tilde{R}(s) = \frac{2a\alpha}{L} \sum_{k=1}^{\infty} \sum_{p=1}^{\infty} r_{1k}^2 r_{2p} \exp\left(-\alpha s(c_k + d_p)\right) \sum_{q=1}^{\infty} \frac{r_{2q}}{2c_k + d_p + d_g}$$

from (4.35) and we can explicitly calculate the series over q. By means of

$$\sum_{q=1}^{\infty} \frac{\cos(qx)}{p^2+q^2} = \frac{\pi}{2p} \frac{\operatorname{ch}\left(p(\pi-x)\right)}{\operatorname{sh}(p\pi)} - \frac{1}{2p^2} \quad \text{for} \quad 0 < x < 2\pi$$

we obtain

$$\sum_{q=1}^{\infty} \frac{\cos\left((2q-1)x\right)}{p^2+(2q-1)^2} = \sum_{q=1}^{\infty} \frac{\cos(qx)}{p^2+q^2} - \frac{1}{4} \sum_{q=1}^{\infty} \frac{\cos\left(q(2x)\right)}{(p/2)^2+q^2}$$
$$= \frac{\pi}{2p}\left(\frac{\operatorname{ch}\left(p(\pi-x)\right)}{\operatorname{sh}(p\pi)} - \frac{1}{2}\frac{\operatorname{ch}\left(p(\pi/2-x)\right)}{\operatorname{sh}(p\pi/2)}\right) \quad \text{for} \quad 0 \leqq x < \pi$$

since this relation is also fulfilled for $x = 0$ because of

$$\tanh\left(\frac{p\pi}{2}\right) = \frac{4p}{\pi} \sum_{q=1}^{\infty} \frac{1}{p^2+(2q-1)^2}.$$

With

$$m_{kp}^2 \doteq 4L^2(2c_k + d_p)$$

it follows that

$$\sum_{q=1}^{\infty} \frac{r_{2q}}{2c_k+d_p+d_q} = \sum_{q=1}^{\infty} \frac{1}{2c_k+d_p+d_q} \int_{y-\delta_2}^{y+\delta_2} \varphi_{2q}(s)\,ds$$
$$= \sqrt{\frac{2}{L}} \int_{y-\delta_2}^{y+\delta_2} \sum_{q=1}^{\infty} \frac{\cos\left(\sqrt{d_q}\,s\right)}{2c_k+d_p+d_q}\,ds$$
$$= \sqrt{\frac{2}{L}} \left(\frac{2L}{\pi}\right)^2 \int_{y-\delta_2}^{y+\delta_2} \sum_{q=1}^{\infty} \frac{\cos\left(\frac{(2q-1)\pi}{2L}s\right)}{\left(\frac{m_{kp}}{\pi}\right)^2+(2q-1)^2}\,ds$$

since

$$0 \leqq \frac{\pi}{2L}(y-\delta_2) \leqq t = \frac{\pi s}{2L} \leqq \frac{\pi}{2L}(y+\delta_2) \leqq \frac{\pi}{2}.$$

Furthermore, we obtain

$$\sum_{q=1}^{\infty} \frac{r_{2q}}{2c_k+d_p+d_q} = \sqrt{\frac{L}{2}} \left(\frac{2L}{m_{kp}}\right)^2 \operatorname{sh}\left(\frac{1}{2L}\delta_2 m_{kp}\right)$$
$$\times \left\{\left(\tanh\left(\frac{1}{2}m_{kp}\right)-1\right)X_{kp} + \left(\tanh\left(\frac{1}{2}m_{kp}\right)+1\right)X_{kp}^{-1}\right\} \quad (4.39)$$

where

$$X_{kp} \doteq \exp\left(\frac{y}{2L}m_{kp}\right)$$

and finally

$$\tilde{R}(s) = 2a\alpha \sqrt{2L}^3 \sum_{k=1}^{\infty} \sum_{p=1}^{\infty} r_{1k}^2 r_{2p} \frac{\operatorname{sh}\left(\frac{1}{2L} \delta_2 m_{kp}\right)}{m_{kp}^2}$$

$$\times \left\{\left(\tanh\left(\frac{1}{2} m_{kp}\right) - 1\right) X_{kp} + \left(\tanh\left(\frac{1}{2} m_{kp}\right) + 1\right) X_{kp}^{-1}\right\}$$

$$\times \exp\left(-\alpha s(c_k + d_p)\right).$$

We can use this relation for the numerical calculations of the correlation function as to the wide-sense stationary process $\xi(t, \omega)$. For $L = 2$, $R = 1$ we have

$$m_{kp} \geqq m_{11} = 3\pi \quad \text{and} \quad 1 \geqq \tanh\left(\frac{1}{2} m_{kp}\right) \geqq \tanh\left(\frac{1}{2} m_{11}\right) = 0.999838.$$

Hence, it is put

$$\tanh\left(\frac{1}{2} m_{kp}\right) = 1$$

and we obtain

$$\tilde{R}(s) = 8a\alpha\delta_2 \sum_{k=1}^{\infty} \sum_{p=1}^{\infty} r_{1k}^2 r_{2p} \frac{\operatorname{sh}\left(\frac{1}{4} \delta_2 m_{kp}\right)}{\frac{1}{4} \delta_2 m_{kp}} \frac{1}{m_{kp}} \exp\left(-\frac{1}{4} y m_{kp}\right)$$

$$\times \exp\left(-\alpha s(c_k + d_p)\right).$$

In order to determine correlation values of $\bar{u}(t, x, y, \omega)$ the function $\tilde{V}(s, x, y)$ is defined by

$$\tilde{V}(s, x, y; \delta_1, \delta_2) \doteq \frac{\varepsilon^2}{(2\delta_1)^2 (2\delta_2)^2} \tilde{R}(s) \approx \langle \bar{u}(t, x, y)\, \bar{u}(t + s, x, y) \rangle$$

and it is

$$\tilde{V}(s, x, y; \delta_1, \delta_2)$$

$$= 4a\alpha\varepsilon^2 \sum_{k=1}^{\infty} \begin{Bmatrix} \cos^2\left(\frac{k\pi x}{2}\right) \\ \sin^2\left(\frac{k\pi x}{2}\right) \end{Bmatrix} \frac{\sin^2\left(\frac{k\pi}{2} \delta_1\right)}{\left(\frac{k\pi}{2} \delta_1\right)^2} \sum_{l=1}^{\infty} \cos\left(\frac{1}{4} (2l - 1) \pi y\right)$$

$$\times \frac{\sin\left(\frac{1}{4} (2l - 1) \pi\delta_2\right)}{\frac{1}{4} (2l - 1) \pi\delta_2} \frac{\operatorname{sh}\left(\frac{1}{4} \delta_2 m_{kl}\right)}{\frac{1}{4} \delta_2 m_{kl}} \frac{1}{m_{kl}} \exp\left(-\frac{1}{4} y m_{kl}\right)$$

$$\times \exp\left(-\alpha s(c_k + d_l)\right).$$

Using $\delta_1 \ll 1$, $\delta_2 \ll 1$ it follows

$$\tilde{V}(s, x, y; \delta_1, \delta_2) \approx \tilde{V}(s, x, y)$$

$$= 4a\alpha\varepsilon^2 \sum_{k=1}^{\infty} \begin{Bmatrix} \cos^2\left(\frac{k\pi x}{2}\right) \\ \sin^2\left(\frac{k\pi x}{2}\right) \end{Bmatrix} \sum_{l=1}^{\infty} \cos\left(\frac{1}{4}(2l-1)\pi y\right) \frac{1}{m_{kl}} \exp\left(-\frac{1}{4} y m_{kl}\right)$$

$$\times \exp\left(-\alpha s(c_k + d_l)\right)$$

and the moment

$$\langle \bar{u}(t, x, y)\, \bar{u}(t + s, x, y)\rangle$$

is given for large times t by $\tilde{V}(s, x, y)$.

In the following a weakly correlated function $\bar{P}(t, x, \omega)$ is assumed the variance of which is

$$\langle \bar{P}(t, x)^2\rangle = \left(\frac{1}{\lambda} P_{\mathrm{m}}\right)^2 p(1 - p)$$

(see Section 4.2) and the correlation function of which is denoted by $R(s, y)$. An example of a correlation function is $R_1(s, y)$ and other approximative examples are $R_2(s, y)$, $R_3(s, y)$. Realistic values are

$$\alpha = 1.278 \cdot 10^{-5} \frac{\mathrm{m}^2}{\mathrm{s}}, \qquad P_{\mathrm{m}} = 1.26 \cdot 10^7 \frac{\mathrm{W}}{\mathrm{m}^2}, \qquad \lambda = 45.3 \frac{\mathrm{W}}{\mathrm{m} \cdot \mathrm{K}},$$

$$t_{\mathrm{cor}} = 65\ \mu\mathrm{s}, \qquad x_{\mathrm{cor}} = 5\ \mu\mathrm{m}.$$

We have denoted by

- α the temperature conductivity,
- P_{m} the maximum surface power,
- λ the thermal conductivity,
- t_{cor} the temporal correlation period,
- x_{cor} the spatial correlation distance.

In the case of R_1 and $p = 1/2$ we obtain

$$b = 7.6923 \cdot 10^{-2} \frac{\mathrm{m}}{\mathrm{s}} \quad \text{and} \quad a_1 = 1.4790358 \cdot 10^9 \frac{\mathrm{K}^2}{\mathrm{ms}}$$

with $\tilde{\varepsilon} = 65\ \mu\mathrm{s}$. Fig. 4.13 shows

$$\sqrt{\frac{\tilde{V}(s, x, y)}{a\varepsilon^2}}$$

for $s = 0$, $x = 0$ and $10^{-5} \leqq y \leqq 10^{-4}$ in the first part of this figure and for $0.1 \leqq y \leqq 0.5$ in the second part. It is easy to see that the dispersion strongly decreases if y increases. The correlation values $\langle \bar{u}(t, 0, y)\, \bar{u}(t + s, 0, y)\rangle$ for large times t in dependence

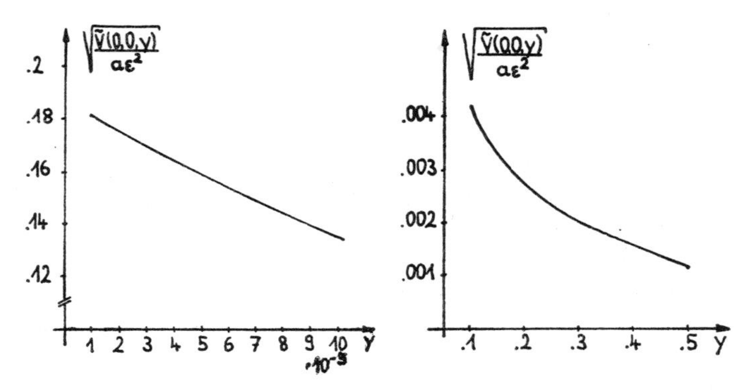

Fig. 4.13. Normalized dispersions in dependence on y for different domains as to y

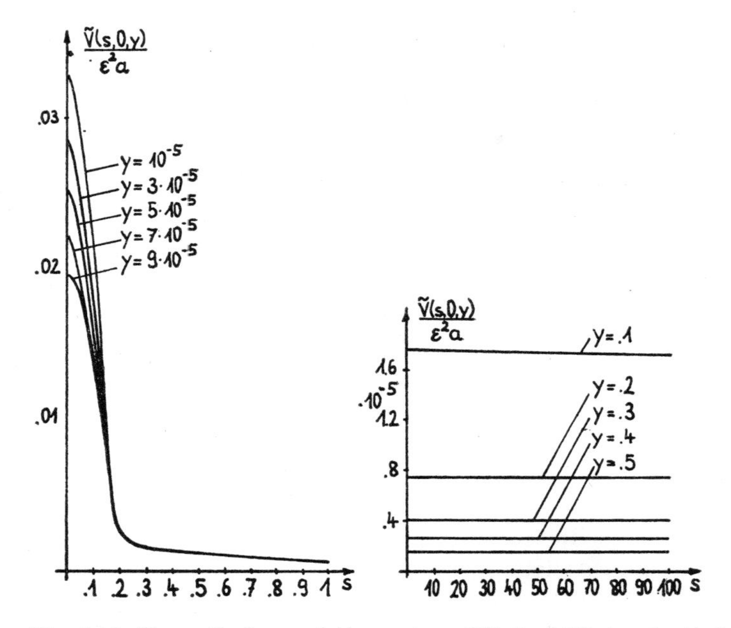

Fig. 4.14. Normalized correlation values $\langle \bar{u}(t, 0, y)\, \bar{u}(t + s, 0, y) \rangle$ for large times t in dependence on s

on s are plotted in Fig. 4.14. For small values y $(10^{-5} \leqq y \leqq 10^{-4})$ we can observe a strong decrease of the correlation values in dependence on small times s. For greater values y $(0.1 \leqq y \leqq 0.5)$ the correlation values stay constant for large times s. These values allow to deduce some properties of the temperature $\bar{u}(t, x, y, \omega)$ near by the surface and also for deeper layers. These properties are much different.

Now, we turn to the case of $0 < \tilde{\alpha} = \alpha_2 = \alpha_3 = \alpha_4$. Then the eigenvalues $\lambda_{kl} = c_k + d_l$ can be calculated from

$$(\lambda^2 c - \tilde{\alpha}^2) \sin\left(2R\sqrt{c}\right) = 2\tilde{\alpha}\lambda\sqrt{c}\cos\left(2R\sqrt{c}\right)$$

and

$$\lambda \sqrt{d} \sin \left(L \sqrt{d}\right) = \tilde{\alpha} \cos \left(L \sqrt{d}\right)$$

where $c = 0$, $d = 0$ are not eigenvalues. The eigenfunctions can be calculated as given in Section 4.1. For numerical calculations we take the same physical values given in connection with the case $\alpha_2, \alpha_3, \alpha_4 \to \infty$. We put $1\ \mathrm{W/m^2K} \leqq \tilde{\alpha} < \infty$. Fig. 4.15 shows the stationary dispersion in dependence on great values y for selected values of $\tilde{\alpha}$. The line for $\tilde{\alpha} \approx \infty$ corresponds with the results contained in Fig. 4.13. We can see that these values decrease with increasing values of $\tilde{\alpha}$ and also of y. In order to determine these values we have used (4.35) with the functions $\chi_1(t)$, $\chi_2(t)$ from (4.38) where $\delta_1 = 0.05$, $\delta_2 = 0.01$ are put.

The dependence of the stationary dispersion $\sqrt{\tilde{V}(0, x, y)}$ on x and y shows Fig. 4.16. We have chosen $\tilde{\alpha} = 10\ \mathrm{W/m^2K}$ and $\delta_1 = \delta_2 = 0.01$. The variance of $\bar{u}(t, x, y, \omega)$ is denoted by $V(t, x, y)$ and we have

$$\lim_{t \to \infty} V(t, x, y) = \tilde{V}(0, x, y).$$

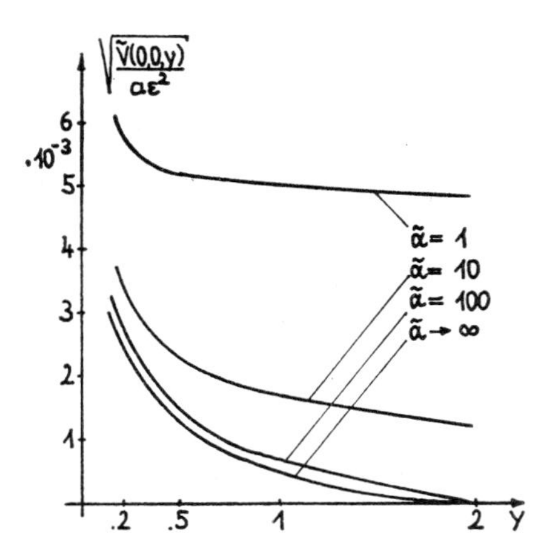

Fig. 4.15. Normalized dispersions as function of y in dependence on $\tilde{\alpha}$

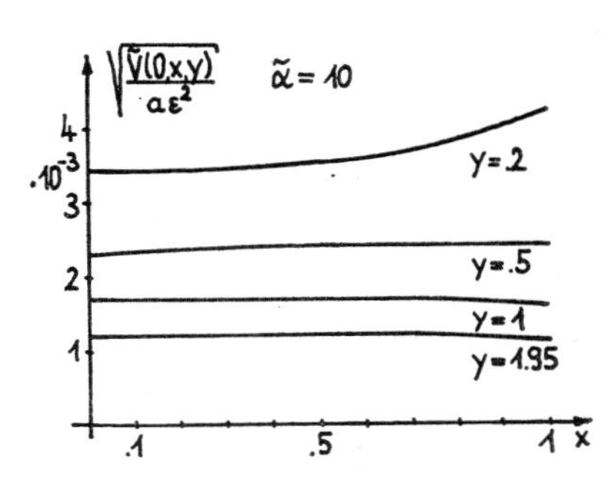

Fig. 4.16. Normalized dispersions as function of x in dependence on y

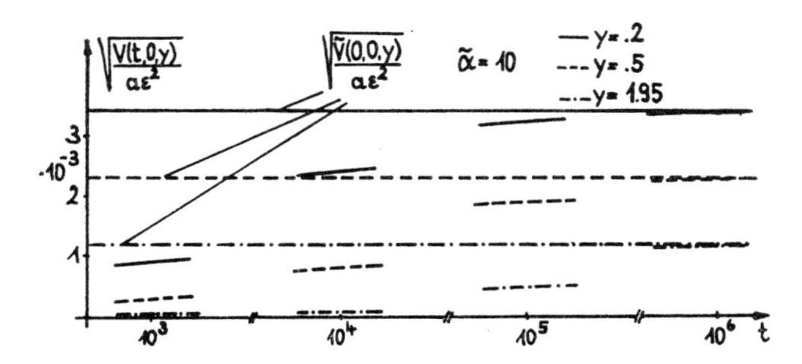

Fig. 4.17. Normalized dispersions as function of t in dependence on y

The behaviour of the dispersion $\sqrt{V(t, x, y)}$ with respect to $\sqrt{\tilde{V}(0, x, y)}$ is illustrated in Fig. 4.17 thereby $\tilde{\alpha} = 10$ W/m²K is chosen. We have put $\delta_1 = 0.05$, $\delta_2 = 0.01$. This figure shows the velocity of the convergence $\lim_{t\to\infty} V = \tilde{V}$. For $y \geqq 0.2$ m we have

$$V(t, 0, y) \approx \tilde{V}(0, 0, y) \quad \text{for} \quad t \geqq 10^6 \text{ s}.$$

We give a further example of the calculation of the variance in comparison with electroanalogical investigations in physics. Let be

$$R = 48\ \mu\text{m}, \qquad L = 25.2\ \mu\text{m}, \qquad x_{\text{cor}} = 5\ \mu\text{m}, \qquad t_{\text{cor}} = 5\ \mu\text{s},$$

$$\alpha = 1.278 \cdot 10^{-5}\ \frac{\text{m}^2}{\text{s}}, \qquad \lambda = 45.3\ \frac{\text{W}}{\text{m} \cdot \text{K}}, \qquad P_{\text{m}} = 6.91 \cdot 10^7\ \frac{\text{W}}{\text{m}^2}.$$

Furthermore, we consider the case of $\alpha_2 = \alpha_4 = 0$ and $\alpha_3 = 10$ W/m²K. The probability p is chosen as 0.5. The eigenvalues $\lambda_{kl} = c_k + d_l$ are given by

$$c_k = \left(\frac{k\pi}{2R}\right)^2 \quad \text{for} \quad k = 0, 1, 2, \ldots$$

and

$$d_l: \ d \sin\left(L\sqrt{d}\right) = \alpha_3 \sqrt{d} \cos\left(L\sqrt{d}\right)$$

where $d = 0$ is not an eigenvalue. The eigenfunctions $f_{kl}(x, y) = \varphi_{1k}(x)\, \varphi_{2l}(y)$ have the form

$$\varphi_{1k}(x) = \sqrt{\frac{1}{R}} \begin{cases} 1 & \text{for} \quad k = 0, \\ \sin\left(\dfrac{k\pi x}{2R}\right) & \text{for} \quad k \text{ odd}, \\ \cos\left(\dfrac{k\pi x}{2R}\right) & \text{for} \quad k \text{ even}, \end{cases}$$

$$\varphi_{2l}(y) = \frac{\sqrt{2}}{\sqrt{L + \dfrac{1}{2\sqrt{d_l}} \sin\left(2L\sqrt{d_l}\right)}} \cos\left(\sqrt{d_l}\, y\right) \quad \text{for} \quad l = 1, 2, \ldots$$

Considering the correlation function $R_1(t, x)$ we obtain the dispersion presented in Fig. 4.18. For the value $t = 34.1\ \mu$s we have comparison values of the dispersion measured

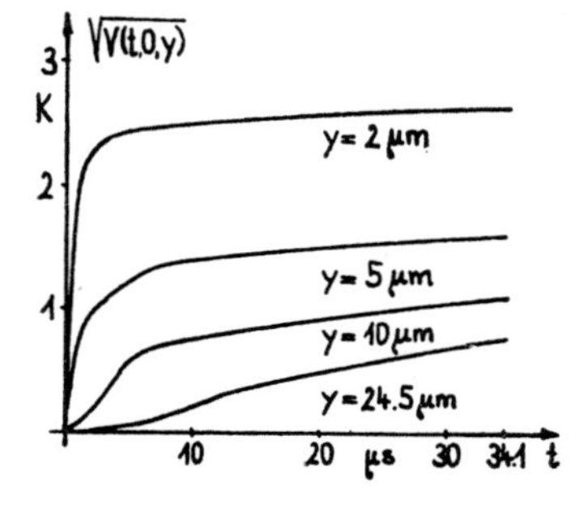

Fig. 4.18. Dispersion as function of t for different values of y

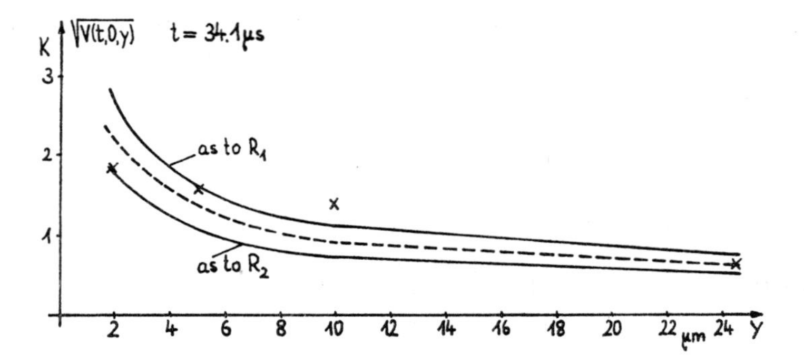

Fig. 4.19. Dispersion as function of y for different intensities and measured values

with physical methods mentioned above. In Fig. 4.19 these values ("$\times$") and our calculated values for the two correlation functions R_1 and R_2 are plotted. Thereby we can see that the values with the function R_1 possess a good agreement with the measured values. We have applied the intensity

$$a_1 = 2.9085043 \cdot 10^{11} \frac{\mathrm{K}^2}{\mathrm{ms}} \quad \text{for} \quad R_1$$

and

$$a_2 = 1.2926686 \cdot 10^{11} \frac{\mathrm{K}^2}{\mathrm{ms}} \quad \text{for} \quad R_2 .$$

With respect to the measured values the best intensity would be

$$a_{\text{best}} = 2.0039801 \cdot 10^{11} \frac{\mathrm{K}^2}{\mathrm{ms}}$$

where for this intensity the sum over the squares of the difference between the measured and calculated values is a minimum in dependence on the intensity. The values with a_{best} are plotted by a hatched line in Fig. 4.19.

In order to determine the second approximation of the variance of $\bar{u}(t, x, y, \omega)$ we use (1.35) and obtain

$$\langle U^2(t) \rangle = {}^2A_1(F, F)\,\varepsilon^2 + {}^2A_2(F, F)\,\varepsilon^3 + \cdots$$

where $F(t, x)$ is defined by (4.19). The term ${}^2A_1(F, F)$ is calculated by (4.34):

$$\begin{aligned} {}^2A_1(F, F) &= R(t, t) \\ &= \frac{a}{\alpha} \sum_{k=0}^{\infty} \sum_{l_1,l_2=0}^{\infty} C_{l_1} C_{l_2} \frac{(f_{kl_1}, \psi)\,(f_{kl_2}, \psi)}{\lambda_{kl_1} + \lambda_{kl_2}} \left\{1 - \exp\left(-\alpha t(\lambda_{kl_1} + \lambda_{kl_2})\right)\right\}. \end{aligned}$$

${}^2A_2(F, F)$ is calculated by Theorem 1.10. If the function ψ is from $\mathbf{C}^2$ then the function

$$F(t, x) = \sum_{k,l=0}^{\infty} C_l \varphi_{1k}(x)\,(f_{kl}, \psi) \exp\left(-\alpha \lambda_{kl} t\right)$$

is continuously differentiable on $\mathcal{D} = (0, t) \times (-R, R)$. Then it follows $\mathcal{D}_1 = \mathcal{D}_2 = \mathcal{D}$ and

$$(\partial\mathcal{D}_2)_1^0 = \partial\mathcal{D}, \qquad (\partial\mathcal{D}_2)_1^i = \emptyset, \qquad (\partial\mathcal{D}_2)_1^b = \partial\mathcal{D},$$

$$I_{12}(x) = (0, 1) \quad \text{for} \quad x \in \partial\mathcal{D}.$$

We use

$$\bar{P}(t, x, \omega) = \varkappa_1(t)\, \varkappa_2(x)\, \tilde{P}(t, x, \omega)$$

with

$$\varkappa_1(0) = 0, \qquad 0 \leqq \varkappa_1(t) \leqq 1, \qquad \varkappa_1(t) = 1 \quad \text{for} \quad t \geqq \eta, \quad \text{monotonically increasing;}$$

$$\varkappa_2(\pm R) = \varkappa_2'(\pm R) = 0, \qquad 0 \leqq \varkappa_2(x) \leqq 1 \quad \text{for} \quad -R \leqq x \leqq -R + \eta, \quad R - \eta \leqq x \leqq R,$$

$$\varkappa_2(x) = 1 \quad \text{for} \quad -R + \eta \leqq x \leqq R - \eta,$$

monotonically increasing for $-R \leqq x \leqq -R + \eta$,
monotonically decreasing for $R - \eta \leqq x \leqq R$

where η is small but arbitrary. Then the compatibility conditions (4.10) are fulfilled in the case of $\alpha_2 = \alpha_4 = 0$. Furthermore, we consider the correlation function

$$\langle \tilde{P}(t, x)\, \tilde{P}(s, y) \rangle = \sigma^2 \begin{cases} \left(1 - \dfrac{1}{\tilde{\varepsilon}}\, |t - s|\right)\left(1 - \dfrac{1}{\tilde{\varepsilon} b}\, |x - y|\right) & \text{for} \quad \max\left\{|t - s|, \dfrac{1}{b}\, |x - y|\right\} \leqq \tilde{\varepsilon}, \\ 0 & \text{otherwise} \end{cases}$$

and we have the real correlation length $\varepsilon = \sqrt{1 + b^2}\, \tilde{\varepsilon}$ according to the Euclidean metric. Using this special correlation function the further computations are carried out similarly to Section 1.6.3. First we obtain

$$\int_{\mathcal{K}_\varepsilon(0,0)} \langle \bar{P}(t, x)\, \bar{P}(t + s, x + y) \rangle \,\mathrm{d}s\, \mathrm{d}y$$

$$= \sigma^2 \varkappa_1(t) \int_{-\tilde{\varepsilon}}^{\tilde{\varepsilon}} \varkappa_1(t + s) \left(1 - \frac{1}{\tilde{\varepsilon}}\, |s|\right) \mathrm{d}s \cdot \varkappa_2(x) \int_{-b\tilde{\varepsilon}}^{b\tilde{\varepsilon}} \varkappa_2(x + y) \left(1 - \frac{1}{b\tilde{\varepsilon}}\, |y|\right) \mathrm{d}y$$

and take

$$\varkappa_1(t) = -\frac{2}{\eta^3}\left[t^3 - \frac{3}{2}\, \eta t^2\right] \quad \text{for } 0 \leqq t \leqq \eta, \text{ and } \varkappa_1(t) = 1 \text{ for } t \geqq \eta;$$

$$\varkappa_2(x) = \begin{cases} -\dfrac{2}{\eta^3}\left[x^3 - \dfrac{3}{2}\, (\eta - 2R)\, x^2 + 3R(R - \eta)\, x - \dfrac{1}{2}\, R^2(3\eta - 2R)\right] & \text{for} \quad -R \leqq x \leqq -R + \eta, \\ 1 & \text{for} \quad -R + \eta \leqq x \leqq R - \eta, \\ -\dfrac{2}{\eta^3}\left[x^3 + \dfrac{3}{2}\, (\eta - 2R)\, x^2 + 3R(R - \eta)\, x + \dfrac{1}{2}\, R^2(3\eta - 2R)\right] & \text{for} \quad R - \eta \leqq x \leqq R. \end{cases}$$

Because of

$$\int_{-\bar{\varepsilon}}^{\bar{\varepsilon}} (p+q)^k \left(1 - \frac{1}{\bar{\varepsilon}} |q|\right) \mathrm{d}q = \begin{cases} \bar{\varepsilon} & \text{for } k = 0, \\ p\bar{\varepsilon} & \text{for } k = 1, \\ p^2\bar{\varepsilon} + \dfrac{1}{6} \bar{\varepsilon}^3 & \text{for } k = 2, \\ p^3\bar{\varepsilon} + \dfrac{1}{2} p\bar{\varepsilon}^3 & \text{for } k = 3 \end{cases}$$

we have for $(s, x) \in [0, t] \times [-R, R]$; $s \neq \eta$; $x \neq -R + \eta, R - \eta$

$$a(s, x) = \frac{1}{1+b^2} \left(\sigma^2 \varkappa_1^2(s)\, \varkappa_2^2(x)\, b\right)$$

$$b(s, x) = 0$$

and furthermore

$$a(u; s, x) = 0 \quad \text{for} \quad s = 0; \qquad -R < x < R \quad \text{and} \quad 0 < s < t; \quad x = \pm R.$$

For $s = t$ and $\bar{z} \doteq z + (s, y)$ it follows

$$\left[\int_{\mathcal{D} \cap \mathcal{K}_\varepsilon(z)} \langle \bar{P}(z)\, \bar{P}(\bar{z}) \rangle \,\mathrm{d}\bar{z}\right]_{z=(t-\varepsilon u, x)}$$

$$= \sigma^2 \varkappa_2(x) \int_{-\bar{\varepsilon}b}^{\bar{\varepsilon}b} \varkappa_2(x+y) \left(1 - \frac{1}{\bar{\varepsilon}b} |y|\right) \mathrm{d}y$$

$$\times \begin{cases} \varkappa_1(t - \varepsilon u) \displaystyle\int_{-\bar{\varepsilon}}^{\varepsilon u} \varkappa_1(t - \varepsilon u + s) \left(1 - \frac{1}{\bar{\varepsilon}} |s|\right) \mathrm{d}s & \text{for } 0 \leqq u \leqq \dfrac{1}{\sqrt{1+b^2}}, \\ \varkappa_1(t - \varepsilon u) \displaystyle\int_{-\bar{\varepsilon}}^{\bar{\varepsilon}} \varkappa_1(t - \varepsilon u + s) \left(1 - \frac{1}{\bar{\varepsilon}} |s|\right) \mathrm{d}s & \text{for } \dfrac{1}{\sqrt{1+b^2}} \leqq u \leqq 1, \end{cases}$$

with $t > \eta$

$$a(u; t, x) = \frac{\varkappa_2^2(x)\, b\sigma^2}{1+b^2} \begin{cases} \dfrac{1}{2} \left(1 + 2\sqrt{1+b^2}\, u - (1+b^2)\, u^2\right) & \text{for } 0 \leqq u \leqq \dfrac{1}{\sqrt{1+b^2}}, \\ 1 & \text{for } \dfrac{1}{\sqrt{1+b^2}} \leqq u \leqq 1, \end{cases}$$

and finally

$$\int_0^1 a(u; t, x) \,\mathrm{d}u = \frac{\varkappa_2^2(x)\, b\sigma^2}{1+b^2} \left(1 - \left(\frac{1}{6\sqrt{1+b^2}}\right)\right) \quad \text{for} \quad t > \eta,\ -R < x < R.$$

Now we obtain

$$ {}^2A_2(F, F) = \int_{\partial\mathcal{D}} F^2(t - s, x) \left[\int_0^1 a(u; s, x) \, du - a(s, x) \right] dS $$

$$ = -\frac{b\sigma^2}{6\sqrt{1 + b^2}^3} \int_{-R}^{R} F^2(0, x) \, \varkappa_2^2(x) dx \approx -\frac{b\sigma^2}{6\sqrt{1 + b^2}^3} \int_{-R}^{R} F^2(0, x) \, dx $$

and

$$ \int_{-R}^{R} F^2(0, x) \, dx = \sum_{k=0}^{\infty} \sum_{p,q=0}^{\infty} C_p C_q (f_{kp}, \psi) \, (f_{kq}, \psi) . $$

In order to determine this sum we take up the special case $\alpha_i \to \infty$ for $i = 2, 3, 4$. Then we have $C_p = -\alpha \sqrt{2/L}$ and

$$ \int_{-R}^{R} F^2(0, x) \, dx = \frac{2\alpha^2}{L} \sum_{k=1}^{\infty} r_{1k}^2 \left(\sum_{p=1}^{\infty} r_{2p} \right)^2 $$

with

$$ r_{2p} = \int_0^L \chi_2(s) \, \varphi_{2p}(s) \, ds , $$

$$ \varphi_{2p}(s) = \sqrt{\frac{2}{L}} \cos\left(\sqrt{d_p} s\right), \qquad d_p = \left(\frac{(2p - 1)\,\pi}{2L} \right)^2 . $$

The convergence of the series follows from the choice of $\chi_2(s)$,

$$ \chi_2(s) = \begin{cases} \dfrac{1}{\delta} (s + \delta - y) & \text{for} \quad y - \delta < s \leqq y , \\ \dfrac{1}{\delta} (-s + \delta + y) & \text{for} \quad y \leqq s < y + \delta ; \quad 0 < y < L , \\ 0 & \text{otherwise} . \end{cases} $$

After some calculations we obtain

$$ r_{2p} = \int_{y-\delta}^{y+\delta} \chi_2(s) \, \varphi_{2p}(s) \, ds = \frac{2}{\delta d_p} \sqrt{\frac{2}{L}} \cos\left(\sqrt{d_p} y\right) \left(1 - \cos\left(\sqrt{d_p} \delta\right)\right) $$

and then

$$ \sum_{p=1}^{\infty} r_{2p} = \frac{2}{\delta} \sqrt{\frac{2}{L}} \left(\sum_{p=1}^{\infty} \frac{1}{d_p} \cos\left(\sqrt{d_p} y\right) - \frac{1}{2} \sum_{p=1}^{\infty} \frac{1}{d_p} \cos\left(\sqrt{d_p}(y - \delta)\right) \right. $$
$$ \left. - \frac{1}{2} \sum_{p=1}^{\infty} \frac{1}{d_p} \cos\left(\sqrt{d_p}(y + \delta)\right) \right) . $$

We have

$$ \sum_{p=1}^{\infty} \frac{1}{p^2} \cos(px) = \frac{\pi^2}{6} - \frac{\pi x}{2} + \frac{x^2}{4} \quad \text{for} \quad 0 \leqq x \leqq 2\pi $$

and

$$\sum_{p=1}^{\infty} \frac{1}{(2p-1)^2} \cos\left((2p-1)\,x\right) = \sum_{p=1}^{\infty} \frac{1}{p^2} \cos(px) - \sum_{p=1}^{\infty} \frac{1}{(2p)^2} \cos(2px)$$

$$= \frac{\pi}{8}(\pi - 2x) \quad \text{for} \quad 0 \leqq x \leqq \pi.$$

These considerations result in

$$\sum_{p=1}^{\infty} r_{2p} = 0$$

and then ${}^2A_2(F, F) = 0$.

In a similar way numerical investigations also lead to ${}^2A_2(F, F) \approx 0$ in the case of $\alpha_2 = \alpha_4 = 0$, $\alpha_3 \neq 0$. Hence, the results plotted in Fig. 4.19 are correct up to terms of the order $o(\varepsilon^3)$.

It is also possible to use limit theorems for random vectors. Thus

$$\lim_{\varepsilon \downarrow 0} \frac{1}{\varepsilon}\left(U_1(t, \omega), U_2(t, \omega)\right) = \left(\xi_1(t, \omega), \xi_2(t, \omega)\right) \quad \text{in distribution}$$

where

$$U_i(t, \omega) = \int_0^t \int_{-R}^{R} F_i(t - s, x)\, \overline{P}(s, x)\, ds\, dx \quad \text{for} \quad i = 1, 2$$

and $\left(\xi_1(t, \omega), \xi_2(t, \omega)\right)$ denotes a Gaussian vector process having the moments

$$\langle \xi_i(t) \rangle = 0 \quad \text{for} \quad i = 1, 2,$$

$$\langle \xi_i(t_1)\, \xi_j(t_2) \rangle = \int_0^{t_{12}} \int_{-R}^{R} F_i(t_1 - s, x)\, F_j(t_2 - s, x)\, a(s, x)\, ds\, dx \quad \text{for} \quad i, j \in \{1, 2\}$$

(see also (4.27)). For some applications we investigate processes $U_i(t, \omega)$, $i = 1, 2$, with

$$F_i(t, x) = \sum_{k,l=0}^{\infty} C_l \exp(-\alpha\lambda_{kl} t)\, \varphi_{1k}(x)\, (f_{kl}, \psi_i) \quad \text{for} \quad i = 1, 2$$

and the intensity $a(t, x)$ of $\overline{P}(t, x, \omega)$ is assumed to be constant. Then it follows

$$\begin{aligned} R_{ij}(t_1, t_2) &\doteq \langle \xi_i(t_1)\, \xi_j(t_2) \rangle \\ &= \frac{a}{\alpha} \sum_{k=0}^{\infty} \sum_{l_1, l_2 = 0}^{\infty} C_{l_1} C_{l_2} \frac{1}{\lambda_{kl_1} + \lambda_{kl_2}} (f_{kl_1}, \psi_i)\, (f_{kl_2}, \psi_j) \\ &\quad \times \left\{\exp\left(-\alpha\lambda_{kl_1}(t_1 - t_{12}) - \alpha\lambda_{kl_2}(t_2 - t_{12})\right) - \exp\left(-\alpha\lambda_{kl_1} t_1 - \alpha\lambda_{kl_2} t_2\right)\right\} \end{aligned} \tag{4.40}$$

and for large times we obtain the correlation function

$$\begin{aligned} \tilde{R}_{ij}(s) &= \lim_{\substack{t_1, t_2 \to \infty \\ |t_2 - t_1| = s = \text{const}}} \langle \xi_i(t_1)\, \xi_j(t_2) \rangle \\ &= \frac{a}{\alpha} \sum_{k=0}^{\infty} \sum_{l_1, l_2 = 0}^{\infty} C_{l_1} C_{l_2} \frac{(f_{kl_1}, \psi_i)\, (f_{kl_2}, \psi_j)}{\lambda_{kl_1} + \lambda_{kl_2}} \begin{cases} \exp(-\alpha\lambda_{kl_2} s) & \text{for} \quad t_{12} = t_1, \\ \exp(-\alpha\lambda_{kl_1} s) & \text{for} \quad t_{12} = t_2 \end{cases} \end{aligned} \tag{4.41}$$

where $s \doteq |t_1 - t_2|$. $\psi_i(x, y)$ is assumed to have the form

$$\psi_i(x, y) = \chi_{i1}(x)\, \chi_{i2}(y)$$

where $\chi_{i1}(x)$ is zero outside of a small neighbourhood $\mathcal{D}_{i1}(x_i)$ of x_i and $\chi_{i2}(y)$ is zero outside of $\mathcal{D}_{i2}(y_i)$, $i = 1, 2$.

For example, it is possible to use the functions

$$\chi_{i1}(x) = \mathbf{1}_{[x_i - \delta_{i1}, x_i + \delta_{i1}]}(x),$$

$$\chi_{i2}(y) = \mathbf{1}_{[y_i - \delta_{i2}, y_i + \delta_{i2}]}(y).$$

Then, it follows

$$\langle \bar{u}(t_1, x_i, y_i)\, \bar{u}(t_2, x_j, y_j)\rangle$$

$$\approx \frac{\varepsilon^2 R_{ij}(t_1, t_2)}{\int\limits_{\mathcal{D}_{i1}(x_i)} \chi_{i1}(t)\, \mathrm{d}t \int\limits_{\mathcal{D}_{j1}(x_j)} \chi_{j1}(t)\, \mathrm{d}t \int\limits_{\mathcal{D}_{i2}(y_i)} \chi_{i2}(t)\, \mathrm{d}t \int\limits_{\mathcal{D}_{j2}(y_j)} \chi_{j2}(t)\, \mathrm{d}t}. \tag{4.42}$$

The notations for variances are given by $V(t, x, y)$, $\tilde{V}(s, x, y)$ as in the above considerations. Now, the vector

$$\big(u(t_1, x_i, y_i), u(t_2, x_j, y_j)\big)$$

is a Gaussian vector with mean

$$\big(w(t_1, x_i, y_i), w(t_2, x_j, y_j)\big)$$

and the correlation values

$$\frac{\varepsilon^2}{16\delta_1^2\delta_2^2} R_{ij}(t_1, t_2) \tag{4.43}$$

where is put $\delta_1 = \delta_{i1} = \delta_{j1}$, $\delta_2 = \delta_{i2} = \delta_{j2}$.

Now we are interested in some probabilities concerning special temperature values being decisive for the considered technical component.

Firstly, we will determine the probability that the temperature exceeds a critical value $Z(t)$ at a fixed time t and a fixed point (x, y). Then we have from (4.43)

$$\mathsf{P}\big(u(t, x, y) \geqq Z(t)\big) \approx 1 - \Phi\left(\frac{Z(t) - w(t, x, y)}{\sqrt{V(t, x, y)}}\right) \tag{4.44}$$

where

$$\Phi(u) \doteq \frac{1}{\sqrt{2\pi}} \int\limits_{-\infty}^{u} \exp\left(-\frac{1}{2}s^2\right) \mathrm{d}s.$$

It follows in the case of $Z(t) = w(t, x, y) + c$, $c = \text{const} > 0$,

$$\mathsf{P}\big(u(t, x, y) - w(t, x, y) \geqq c\big) \approx 1 - \Phi\left(\frac{c}{\sqrt{V(t, x, y)}}\right) \tag{4.45}$$

and for large times t

$$\mathsf{P}\big(u(t, x, y) - w(t, x, y) \geqq c\big) \approx 1 - \Phi\left(\frac{c}{\sqrt{\tilde{V}(0, x, y)}}\right). \tag{4.46}$$

These probabilities can be interpreted in the following manner. (4.44) corresponds to the consideration of time-depending deviations between $u(t, x, y)$ and $w(t, x, y)$. (4.45) means that the considered deviation is independent of time and it is equal to c. Formula (4.46) can be used in the case of (4.45) for large times.

Secondly, we consider the temperature differences concerning the times t_1, t_2 in a fixed point (x, y). Thereby, we restrict us to the case of large times and the consideration of $\bar{u}(t, x, y)$. Using the Gaussian distribution of $(\bar{u}(t_1, x, y), \bar{u}(t_2, x, y))$ we can calculate approximately the probability

$$\mathsf{P}\big(\bar{u}(t_1, x, y) \leqq M;\, \bar{u}(t_2, x, y) \geqq N\big) \tag{4.47}$$

which can be interpreted as probability of temperature changes during the time $t_2 - t_1$. For short times we have to substitute u for $\bar{u}$ and 0 for w.

Thirdly, we consider temperature differences in two fixed neighbouring points (x_1, y_1), (x_2, y_2) at the same time t, i.e.

$$\mathsf{P}\big(u(t, x_1, y_1) \leqq M;\, u(t, x_2, y_2) \geqq N\big). \tag{4.48}$$

Furthermore it can be noted that the general case of (t_1, x_1, y_1) and (t_2, x_2, y_2) can be considered analogously.

Finally, some examples of probabilities are given. We use the values

$$R = 1, \qquad L = 2; \qquad \alpha = 1.3 \cdot 10^{-5}, \qquad \lambda = 45$$

as well as

$$x_{\text{cor}} = 10^{-4}, \qquad y_{\text{cor}} = 10^{-3} \quad \text{and} \quad \alpha_2 = \alpha_3 = \alpha_4 = 10.$$

The intensity is chosen as $a = 0.31 \cdot 10^{13}$. Furthermore, we set $\delta_1 = 0.05$, $\delta_2 = 0.01$. Using

$$\sqrt{V(10^4, 0, 0.2)} = 4.28 \quad \text{and} \quad \sqrt{\tilde{V}(0, 0, 0.2)} = 6.01$$

we obtain the probabilities $\mathsf{P}\big(\bar{u}(t, x, y) \geqq c\big)$ plotted in Fig. 4.20. Because of the monotony of $V(t, x, y)$ as to t these probabilities increase in t.

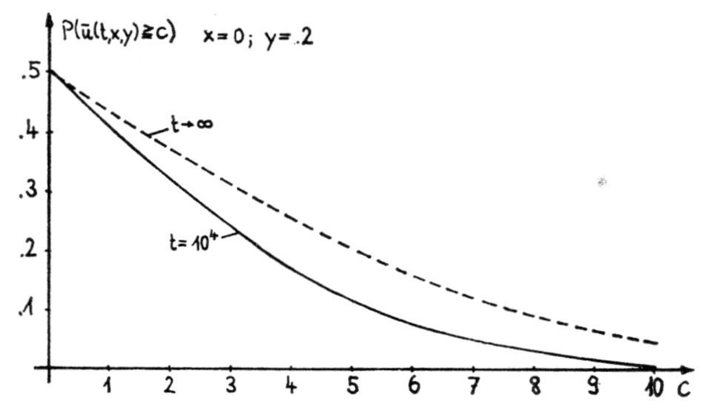

Fig. 4.20. Probabilities $\mathsf{P}\big(\bar{u}(t, x, y) \geqq c\big)$ in dependence on c

Now we will approximately calculate the probability (4.47) for $M = -1$, $N = 1$, large times and different values of $s = t_2 - t_1$ using numerical methods. Thereby we consider again the point $x = 0$, $y = 0.2$ with

$$\sqrt{\tilde{V}(0, 0, 0.2)} = 6.01 .$$

It can be seen that the values of P increase with increasing values of s and it tends to the probability $\mathsf{P} = 0.19$ which is also the result of the case of independence of $\bar{u}(t_1, x, y)$ and $\bar{u}(t_2, x, y)$ with $\varrho = 0$.

Table 4.1

x_2	y_2	$\tilde{V}(0, x_2, y_2)$	ϱ	P
0	0.21	6.01	0.98	0.010
0	0.3	5.02	0.97	0.012
0	0.4	4.45	0.91	0.036
0.05	0.2	6.01	0.99	0.001
0.1	0.2	6.02	0.98	0.012
0.2	0.2	6.03	0.91	0.040

The probabilities (4.48) are also determined for large times, $x_1 = 0$, $y_1 = 0.2$ and different values of x_2, y_2. Thereby $M = -0.5, N = 0.5$ is set. Some results are summarized in Table 4.1 where we have written

$$\varrho = \varrho(\bar{u}(t, x_1, y_1), \bar{u}(t, x_2, y_2))$$

and

$$P = \mathsf{P}(\bar{u}(t, x_1, y_1) \leqq M ; \quad \bar{u}(t, x_2, y_2) \geqq N) .$$

Finally, we consider the correlation between two points (x_1, y_1), (x_2, y_2) for different times t_1, t_2. Let be $(x_1, y_1) = (0, 0.2)$, $(x_2, y_2) = (0, 0.4)$ and $t_1 = 10^4$, $t_2 = 10^5$. Then we have firstly for $\bar{u}(t_1, x_1, y_1)$ and $\bar{u}(t_2, x_2, y_2)$ the correlation coefficient $\varrho = 0.127$. Secondly, for $\bar{u}(t_2, x_1, y_1)$ and $\bar{u}(t_1, x_2, y_2)$ this value is $\varrho = 0.098$. These results are in accordance with empirical considerations. Fig. 4.21 shows some further correlation values

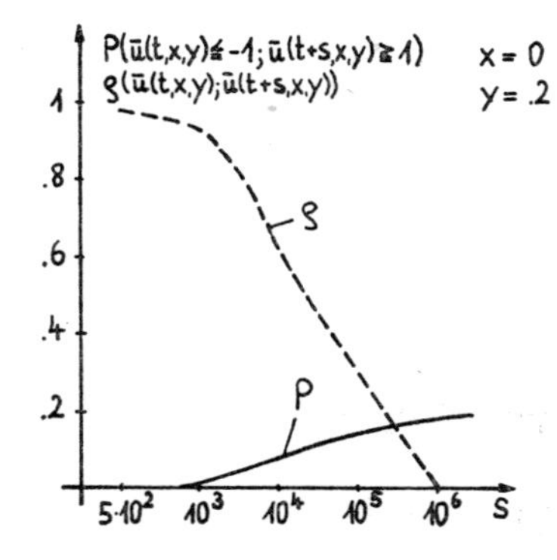

Fig. 4.21. Correlation coefficients $\varrho(\bar{u}(t, x, y), \bar{u}(t + s, x, y))$ and probabilities $\mathsf{P}(\bar{u}(t,x,y) \leqq -1; \bar{u}(t + s, x, y) \geqq 1)$ in dependence on s

4.3.2. Partially weakly correlated temperature gradient

In this section $\bar{P}$ is supposed to have the form

$$\bar{P}(t, x, \omega) = P_0(x) P_1(t, \omega) \tag{4.49}$$

and let $P_1(t, \omega)$ be a weakly correlated process. The "averaged" solution of problem (4.9) is obtained as

$$\begin{aligned} U_i(t, \omega) &= \int_0^t \sum_{k=0}^{\infty} \sum_{p=0}^{\infty} C_p \int_{-R}^{R} P_0(x) \varphi_{1k}(x) \, dx \exp\left(-\alpha\lambda_{kp}(t-s)\right) P_1(s, \omega) \, ds (f_{kp}, \psi_i) \\ &= \int_0^t F_{1i}(t-s) P_1(s, \omega) \, ds \end{aligned}$$

where

$$F_{1i}(t) \doteq \sum_{k=0}^{\infty} \sum_{p=0}^{\infty} C_p W_k \exp(-\alpha\lambda_{kp} t) (f_{kp}, \psi_i) \tag{4.50}$$

and

$$W_k = \int_{-R}^{R} \varphi_{1k}(x) P_0(x) \, dx = (\varphi_{1k}, P_0).$$

In this case it follows

$$\lim_{\varepsilon \downarrow 0} \frac{1}{\sqrt{\varepsilon}} \left(U_1(t, \omega), U_2(t, \omega)\right) = \left(\eta_1(t, \omega), \eta_2(t, \omega)\right) \quad \text{in distribution} \tag{4.51}$$

where $\left(\eta_1(t, \omega), \eta_2(t, \omega)\right)$ is a Gaussian vector process with mean $\langle \eta_i(t) \rangle = 0$ and correlation relations

$$\langle \eta_i(t_1) \eta_j(t_2) \rangle = \int_0^{t_{12}} F_{1i}(t_1 - s) F_{1j}(t_2 - s) a(s) \, ds. \tag{4.52}$$

$a(t)$ is the intensity of the weakly correlated process $P_1(t, \omega)$ defined by

$$a(t) \doteq \lim_{\varepsilon \downarrow 0} \frac{1}{\varepsilon} \int_{-\varepsilon}^{\varepsilon} \langle P_1(t) P_1(t+s) \rangle \, ds$$

and we set $a(t) = a = \text{const}$, i.e. $P_1(t, \omega)$ is assumed to be a wide-sense stationary process. Setting

$$R_1(s) = \langle P_1(t) P_1(t+s) \rangle = \begin{cases} \sigma^2 \left(1 - \frac{1}{\varepsilon} |s|\right) & \text{for} \quad |s| \leqq \varepsilon, \\ 0 & \text{otherwise} \end{cases}$$

then $a_1 = \sigma^2$. If the correlation function

$$R_2(s) = \langle P_1(t) P_1(t+s) \rangle = \sigma^2 \exp(-\delta |s|) \qquad \text{with} \;\; \delta = \frac{3}{\varepsilon}$$

is taken it follows the intensity $a_2 = 2\sigma^2/3$. Similarly, we obtain for

$$R_3(s) = \langle P_1(t)\, P_1(t+s)\rangle = \sigma^2 \exp(-\delta s^2) \qquad \text{with} \quad \delta = \frac{3}{\varepsilon^2}$$

the intensity

$$a_3 = 2\sigma^2 \sqrt{\frac{\pi}{3}} \left(\Phi\left(\sqrt{6}\right) - \frac{1}{2}\right) = 1.00867\sigma^2 .$$

In the following a is assumed to be constant although there are some difficulties in connection with the compatibility condition $P_1(0, \omega) = 0$ a.s. We refer to the considerations as to this problem in the previous section.

Assuming that at least one $\alpha_i \neq 0$, $i \in \{2, 3, 4\}$, from (4.52) and (4.50) we obtain

$$\begin{aligned} &\langle \eta_i(t_1)\, \eta_j(t_2)\rangle \\ &= \frac{a}{\alpha} \sum_{k,h=0}^{\infty} \sum_{p,q=0}^{\infty} C_p C_q W_k W_h \frac{(f_{kp}, \psi_i)\,(f_{hq}, \psi_j)}{\lambda_{kp} + \lambda_{hq}} \\ &\quad \times \left[\exp\left(-\alpha\lambda_{kp}(t_1 - t_{12}) - \alpha\lambda_{hq}(t_2 - t_{12})\right) - \exp\left(-\alpha\lambda_{kp}t_1 - \alpha\lambda_{hq}t_2\right)\right] \\ &\doteq R_{1ij}(t_1, t_2) \end{aligned} \tag{4.53}$$

and for large times

$$\begin{aligned} \tilde{R}_{1ij}(s) &\doteq \lim_{\substack{t_1, t_2 \to \infty \\ |t_2 - t_1| = s = \text{const}}} R_{1ij}(t_1, t_2) \\ &= \frac{a}{\alpha} \sum_{k,h=0}^{\infty} \sum_{p,q=0}^{\infty} C_p C_q W_k W_h \frac{(f_{kp}, \psi_i)\,(f_{hq}, \psi_j)}{\lambda_{kp} + \lambda_{hq}} \begin{cases} \exp(-\alpha\lambda_{kp}s) & \text{for} \quad t_{12} = t_2, \\ \exp(-\alpha\lambda_{hq}s) & \text{for} \quad t_{12} = t_1. \end{cases} \end{aligned} \tag{4.54}$$

As in the previous section we choose

$$\psi_i(x, y) = \chi_{i1}(x)\, \chi_{i2}(y)$$

and obtain

$$(f_{kp}, \psi_i) = r_{1k}^i r_{2p}^i \quad \text{with} \quad r_{1k}^i = \int_{-R}^{R} \varphi_{1k}(x)\, \chi_{i1}(x)\, \mathrm{d}x ,$$

$$r_{2p}^i = \int_0^L \varphi_{2p}(y)\, \chi_{i2}(y)\, \mathrm{d}y \quad \text{for} \quad i = 1, 2 .$$

For numerical calculations an important example is

$$\begin{aligned} \chi_{i1}(x) &= \mathbf{1}_{[x_i - \delta_{i1}, x_i + \delta_{i1}]}(x) , \\ \chi_{i2}(x) &= \mathbf{1}_{[y_i - \delta_{i2}, y_i + \delta_{i2}]}(x) . \end{aligned} \tag{4.55}$$

Now, we have

$$\begin{aligned} &\langle \bar{u}(t_1, x_i, y_i)\, \bar{u}(t_2, x_j, y_j)\rangle \\ &\approx \frac{R_{1ij}(t_1, t_2)}{\int_{\mathcal{D}_{i1}(x_i)} \chi_{i1}(t)\, \mathrm{d}t \int_{\mathcal{D}_{j1}(x_j)} \chi_{j1}(t)\, \mathrm{d}t \int_{\mathcal{D}_{i2}(y_i)} \chi_{i2}(t)\, \mathrm{d}t \int_{\mathcal{D}_{j2}(y_j)} \chi_{j2}(t)\, \mathrm{d}t} . \end{aligned} \tag{4.56}$$

Now, we investigate the case of $\alpha_i \to \infty$ for $i = 2, 3, 4$ with $x_i = x_j = x$, $y_i = y_j = y$. Using (4.39) the stationary correlation function is given by

$$\tilde{R}_{111}(s) = \frac{2a\alpha}{L} \sum_{k,h=1}^{\infty} \sum_{p=1}^{\infty} W_k W_h r_{1k} r_{1h} r_{2p} \exp(-\alpha \lambda_{kp} s) \sum_{q=1}^{\infty} \frac{r_{2q}}{c_k + c_h + d_p + d_q}$$

$$= 2\sqrt{2L}^3\, a\alpha \sum_{k,h=1}^{\infty} W_k W_h r_{1k} r_{1h}$$

$$\times \sum_{p=1}^{\infty} r_{2p} \operatorname{sh}\left(\frac{1}{2L}\, \delta_2 m_{khp}\right) \frac{1}{m_{khp}^2} \left\{\left(\tanh\left(\frac{1}{2}\, m_{khp}\right) - 1\right) X_{khp}\right.$$

$$\left. + \left(\tanh\left(\frac{1}{2}\, m_{khp}\right) + 1\right) X_{khp}^{-1}\right\} \exp(-\alpha \lambda_{kp} s)$$

where

$$m_{khp}^2 \doteq 4L^2(c_k + c_h + d_p),$$

$$X_{khp} \doteq \exp\left(\frac{1}{2L}\, m_{khp} y\right)$$

and we have taken the χ-functions in the form of (4.55). Since $p_{111} = 9.424778$ for $R = 1$, $L = 2$ we obtain

$$\tanh\left(\frac{1}{2}\, p_{111}\right) = 0.999839$$

and then

$$\tanh\left(\frac{1}{2}\, m_{khp}\right) \approx 1.$$

Hence

$$\tilde{V}(s, x, y; \delta_1, \delta_2) \doteq \frac{1}{16\delta_1^2\delta_2^3}\, \varepsilon \tilde{R}_{111}(s)$$

has the form

$$\tilde{V}(s, x, y; \delta_1, \delta_2)$$

$$= \frac{4a\alpha\varepsilon}{R} \sum_{k,h=1}^{\infty} W_k W_h \begin{Bmatrix} \cos\left(\frac{k\pi x}{2R}\right) \\ \sin\left(\frac{k\pi x}{2R}\right) \end{Bmatrix} \begin{Bmatrix} \cos\left(\frac{h\pi x}{2R}\right) \\ \sin\left(\frac{h\pi x}{2R}\right) \end{Bmatrix} \frac{\sin\left(\frac{k\pi}{2R}\, \delta_1\right)}{\frac{k\pi}{2R}\, \delta_1} \frac{\sin\left(\frac{h\pi}{2R}\, \delta_1\right)}{\frac{h\pi}{2R}\, \delta_1}$$

$$\times \sum_{p=1}^{\infty} \cos\left(\frac{(2p-1)\,\pi y}{2L}\right) \frac{\sin\left(\frac{(2p-1)\,\pi}{2L}\, \delta_2\right)}{\frac{(2p-1)\,\pi}{2L}\, \delta_2} \frac{\operatorname{sh}\left(\frac{1}{2L}\, \delta_2 m_{khp}\right)}{\frac{1}{2L}\, \delta_2 m_{khp}} \frac{1}{m_{khp}}$$

$$\times \exp\left(-\frac{1}{2L}\, m_{khp} y\right) \exp(-\alpha \lambda_{kp} s)$$

and using $\delta_1 \ll 1$, $\delta_2 \ll 1$ we obtain

$$\tilde{V}(s, x, y; \delta_1, \delta_2) \approx \tilde{V}(s, x, y)$$

$$\doteq \frac{4a\alpha\varepsilon}{R} \sum_{k,h=1}^{\infty} W_k W_h \begin{Bmatrix} \cos\left(\dfrac{k\pi x}{2R}\right) \\ \sin\left(\dfrac{k\pi x}{2R}\right) \end{Bmatrix} \begin{Bmatrix} \cos\left(\dfrac{h\pi x}{2R}\right) \\ \sin\left(\dfrac{h\pi x}{2R}\right) \end{Bmatrix}$$

$$\times \sum_{p=1}^{\infty} \cos\left(\frac{(2p-1)\pi y}{2L}\right) \frac{1}{m_{khp}} \exp\left(-\frac{1}{2L} m_{khp} y\right) \exp\left(-\alpha\lambda_{kp} s\right) \tag{4.57}$$

where the upper terms are taken for k, h odd and the lower terms for k, h even. Now, for large times t, it follows

$$\langle \bar{u}(t, x, y)\, \bar{u}(t+s, x, y)\rangle = \tilde{V}(s, x, y).$$

First, we choose

$$P_0(x) = c\varphi_{11}(x) = \frac{c}{\sqrt{R}} \cos\left(\frac{\pi x}{2R}\right).$$

Thus

$$W_k = (P_0, \varphi_{1k}) = c(\varphi_{11}, \varphi_{1k}) = c\delta_{1k}$$

and the correlation function $R_{111}(t_1, t_2)$ is reduced essentially. We denote

$$V(t_1, t_2, x, y; \delta_1, \delta_2) \approx \langle \bar{u}(t_1, x, y)\, \bar{u}(t_2, x, y)\rangle$$

and obtain for $\delta_1 \ll 1$, $\delta_2 \ll 1$

$$V(t_1, t_2, x, y; \delta_1, \delta_2) \approx V(t_1, t_2, x, y)$$

$$= \frac{4\alpha^2 a\varepsilon}{RL^2} \cos^2\left(\frac{\pi x}{2R}\right) \sum_{p,q=1}^{\infty} \cos\left(\frac{(2p-1)\pi y}{2L}\right) \cos\left(\frac{(2q-1)\pi y}{2L}\right)$$

$$\times \frac{\exp\left(-\alpha\lambda_{1p}\,|t_2 - t_1|\right) - \exp\left(-\alpha\lambda_{1p} t_1 - \alpha\lambda_{1q} t_2\right)}{\lambda_{1p} + \lambda_{1q}}. \tag{4.58}$$

For the stationary correlation function (4.57) leads to

$$\tilde{V}(s, x, y) = \frac{4}{R}\, a\alpha c^2 \varepsilon \cos^2\left(\frac{\pi x}{2R}\right) A(s, y) \tag{4.59}$$

where

$$A(s, y) \doteq \sum_{p=1}^{\infty} \cos\left(\frac{(2p-1)\pi y}{2L}\right) \frac{1}{m_{11p}} \exp\left(-\frac{1}{2L} m_{11p} y\right) \exp\left(-\alpha\lambda_{1p} s\right).$$

In this formula $A(s, y)$ can be written in the form

$$A(s, y) \doteq \sum_{p=0}^{\infty} A(2p+1, s, y).$$

Since the terms of this series change slowly for small y we can calculate $A(s, y)$ with the help of

$$A(s, y) \approx \sum_{p=0}^{P} \sum_{q=pm}^{(p+1)m-1} A(2q+1, s, y) \approx m \sum_{p=0}^{P} A\big((2p+1)\, m, s, y\big).$$

A more favourable method of calculating $A(s, y)$ is the approximation of $A(2p+1, s, y)$ by a quadratic function as to p for $p_1 \leqq p \leqq p_2$. We have

$$A(s, y) = \sum_{p=0}^{\infty} A(2p+1, s, y) = \sum_{p=0}^{\infty} \sum_{q=pm}^{m(p+1)-1} A(2q+1, s, y)$$

and put

$$\begin{aligned} a(p) &\doteq A(2q+1, s, y)|_{q=pm} &&= A(2mp+1, s, y), \\ b(p) &\doteq A(2q+1, s, y)|_{q=mp+(m-1)/2} &&= A\big((2p+1)\, m, s, y\big), \\ c(p) &\doteq A(2q+1, s, y)|_{q=m(p+1)-1} &&= A\big(2(p+1)\, m - 1, s, y\big). \end{aligned}$$

Now we approximate $A(2q+1, s, y)$ on the interval $pm \leqq l \leqq (p+1)\, m - 1$ by

$$B(q) = a_2 q^2 + a_1 q + a_0$$

where we demand

$$B(pm) = a(p), \qquad B\left(pm + \frac{m-1}{2}\right) = b(p), \qquad B\big((p+1)\, m - 1\big) = c(p).$$

It follows

$$B(q) = a(p) + z(q)\,\big(b(p) - a(p)\big) + \frac{1}{2}\, z(q)\,\big(z(q) - 1\big)\,\big(c(p) - 2b(p) + a(p)\big)$$

with

$$z(q) = \frac{q - pm}{\frac{m-1}{2}} = \frac{2}{m-1}\,(q - pm).$$

Thus

$$\begin{aligned} \sum_{q=pm}^{(p+1)m-1} A(2p+1, s, y) &\approx \sum_{q=pm}^{(p+1)m-1} B(q) \\ &= \sum_{t=0}^{m-1} \left(a(p) + \frac{2t}{m-1}\,\big(b(p) - a(p)\big) + \frac{t}{m-1}\left(\frac{2t}{m-1} - 1\right)\right. \\ &\qquad \times \left.\big(c(p) - 2b(p) + a(p)\big)\right) \\ &= m\left[b(p) + \frac{1}{6}\,\frac{m+1}{m-1}\,\big(c(p) - 2b(p) + a(p)\big)\right] \end{aligned}$$

and then

$$A(s, y) \approx m \sum_{p=0}^{\infty} \left[b(p) + \frac{1}{6}\,\frac{m+1}{m-1}\,\big(c(p) - 2b(p) + a(p)\big)\right].$$

For $m = 3$ the exact sum is obtained by this method. Applying this method it is favourable to choose $m = 3$ for the first terms and greater m for the further terms.

Fig. 4.22 shows

$$\sqrt{\frac{1}{\varepsilon a} V(s, x, y)}$$

in dependence on y for $s = 0$ and different values of x. It is put $c = 1$, $R = 1$, $L = 2$. In Fig. 4.23 the function $\sqrt{V(0, x, y)/\varepsilon a}$ is plotted in dependence on x for different values of y.

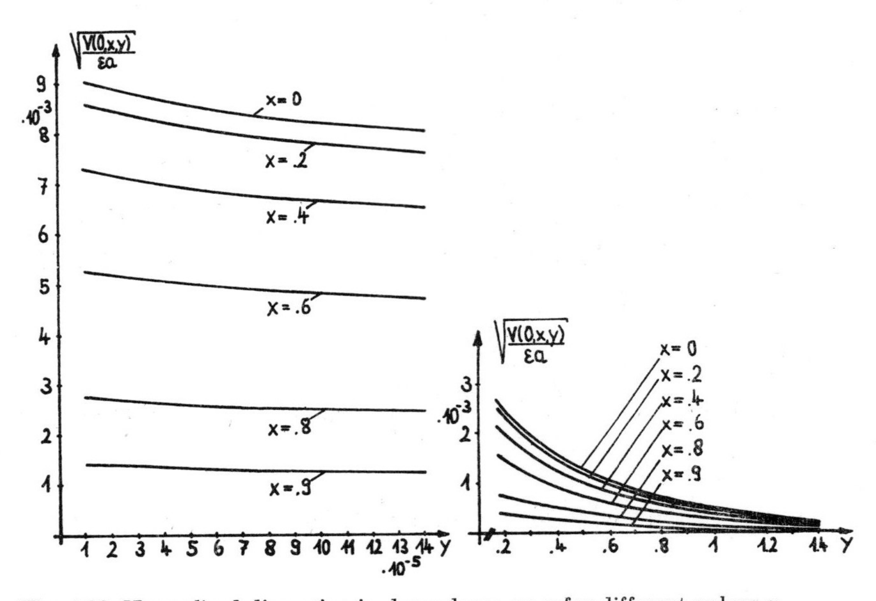

Fig. 4.22. Normalized dispersion in dependence on y for different values x

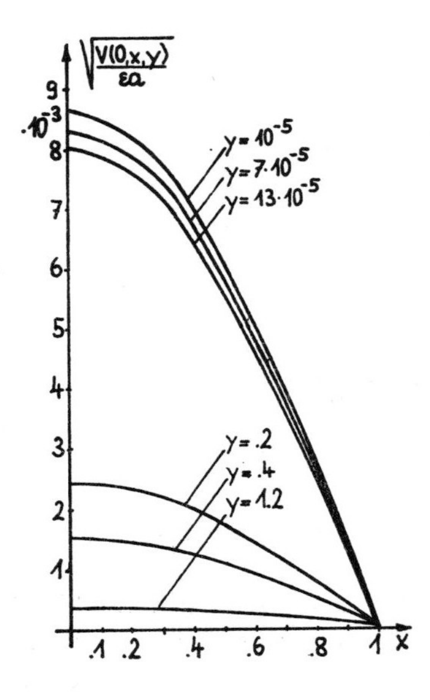

Fig. 4.23. Normalized dispersion in dependence on x for different values y

If the correlation function of $P_1(t, \omega)$ has the form $R_1(s)$ then it can be found that

$$\varepsilon a = t_{\text{cor}}\sigma^2 = t_{\text{cor}} \left(\frac{1}{\lambda} P_m\right)^2 p(1-p)$$

where

$$P_m = 1.26 \cdot 10^7 \frac{\text{W}}{\text{m}^2}, \qquad \lambda = 45.3 \frac{\text{W}}{\text{m} \cdot \text{K}}, \qquad t_{\text{cor}} = 65 \cdot 10^{-6}\ \text{s}$$

are put. Setting $p = 1/2$ we have

$$\varepsilon a = 1.2571817 \cdot 10^6 \frac{\text{sK}^2}{\text{m}^2}.$$

It follows

$$\sqrt{V(0, x, y)} = 9.70\ \text{K} \quad \text{for} \quad x = 0,\ y = 10^{-5} \quad \text{and} \quad p = \frac{1}{2},$$

$$\sqrt{V(0, x, y)} = 8.40\ \text{K} \quad \text{for} \quad x = 0,\ y = 10^{-5} \quad \text{and} \quad p = \frac{1}{4} \left(p = \frac{3}{4}\right)$$

and in connection with Fig. 4.22 we have a good survey of the dispersions with the given physical constants.

The normalized correlation function $\varrho(t, x, y; \bar{t}, \bar{x}, \bar{y})$ is defined by

$$\varrho(t, x, y; \bar{t}, \bar{x}, \bar{y}) = \frac{\langle \tilde{u}(t, x, y)\, \tilde{u}(\bar{t}, \bar{x}, \bar{y})\rangle}{\sqrt{\langle \tilde{u}^2(t, x, y)\rangle \langle \tilde{u}^2(\bar{t}, \bar{x}, \bar{y})\rangle}}. \tag{4.60}$$

Using (4.54), $s = t - \bar{t} \geqq 0$ and the form (4.55) of χ_{ij} we obtain

$$\tilde{V}(s; x, y, \bar{x}, \bar{y}) = \frac{4}{R} a\alpha \sum_{k,h=1}^{\infty} W_k W_h \begin{Bmatrix} \cos\left(\frac{k\pi x}{2R}\right) \\ \sin\left(\frac{k\pi x}{2R}\right) \end{Bmatrix} \begin{Bmatrix} \cos\left(\frac{h\pi \bar{x}}{2R}\right) \\ \sin\left(\frac{h\pi \bar{x}}{2R}\right) \end{Bmatrix}$$

$$\times \sum_{p=1}^{\infty} \cos\left(\frac{(2p-1)\,\pi y}{2L}\right) \frac{1}{m_{khp}} \exp\left(-\frac{\bar{y}}{2L} m_{khp}\right) \exp\left(-\alpha\lambda_{kp} s\right)$$

and, in particular, for $P_0(x) = c\varphi_{11}(x)$

$$\tilde{V}(s, x, y, \bar{x}, \bar{y}) = \frac{4}{R} a\alpha c^2 \cos\left(\frac{\pi x}{2R}\right) \cos\left(\frac{\pi \bar{x}}{2R}\right) A(s, y, \bar{y})$$

where $A(s, y, \bar{y})$ is defined by

$$A(s, y, \bar{y}) \doteq \sum_{p=1}^{\infty} \cos\left(\frac{(2p-1)\,\pi y}{2L}\right) \frac{1}{m_{11p}} \exp\left(-\frac{1}{2L} m_{11p}\bar{y}\right) \exp\left(-\alpha\lambda_{1p} s\right).$$

The normalized stationary correlation function can be written as

$$\tilde{\varrho}(t, x, y; \bar{t}, \bar{x}, \bar{y}) = \frac{A(s, y, \bar{y})}{\sqrt{A(0, y)\, A(0, \bar{y})}}.$$

It is easy to see that

$$\tilde{\varrho}(t, x, y; t, \bar{x}, y) = \frac{A(0, y, y)}{A(0, y)} = 1$$

and then it is

$$\bar{u}(t, \bar{x}, y, \omega) = \beta_1 \bar{u}(t, x, y, \omega) + \beta_2$$

where β_1, β_2 are non-random. From $\langle \bar{u} \rangle = 0$ it follows $\beta_2 = 0$ and furthermore we obtain

$$\beta_1 = \frac{\langle \bar{u}(t, \bar{x}, y)\, \bar{u}(t, x, y) \rangle}{\langle \bar{u}^2(t, x, y) \rangle} = \frac{\cos\left(\dfrac{\pi \bar{x}}{2R}\right)}{\cos\left(\dfrac{\pi x}{2R}\right)}.$$

Hence, we have

$$\bar{u}(t, \bar{x}, y, \omega) = \frac{\cos\left(\dfrac{\pi \bar{x}}{2R}\right)}{\cos\left(\dfrac{\pi x}{2R}\right)} \bar{u}(t, x, y, \omega) \quad \text{a.s.}$$

from which

$$\langle \bar{u}^2(t, \bar{x}, y, \omega) \rangle = \cos^2\left(\frac{\pi \bar{x}}{2R}\right) \langle \bar{u}^2(t, 0, y, \omega) \rangle$$

follows (see also Fig. 4.23).

Now we consider

$$\tilde{\varrho}(t, x, y; \bar{t}, x, y) = \frac{A(s, y, y)}{A(0, y)} = \frac{A(s, y)}{A(0, y)}$$

with $s = t - \bar{t} \geqq 0$. $\tilde{\varrho}$ is illustrated in dependence on s for different values y in Fig. 4.24. In the case of $y = 0.2 \cdot k$, $k = 1, 3, 5, 7$, the temperatures change very slowly as to the time on (x, y) since $\tilde{\varrho}(t, x, y; \bar{t}, x, y) \approx 1$ for times $\bar{t}$ with

$$t \leqq \bar{t} \leqq t + 1000.$$

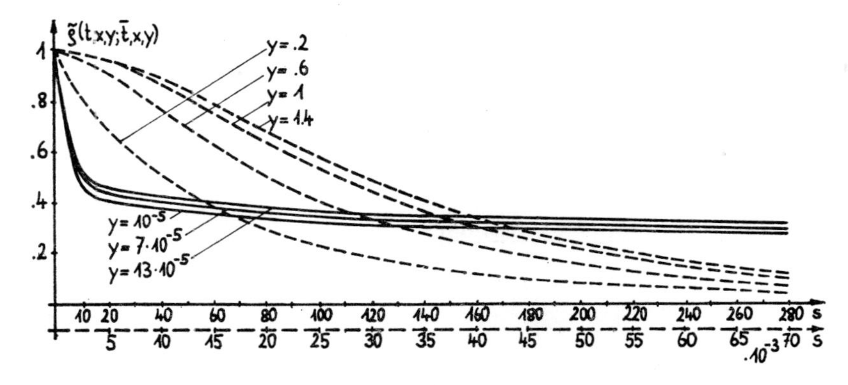

Fig. 4.24. Correlation function $\tilde{\varrho}$ in dependence on s for different values of y

Finally, we consider the interaction of $\bar{u}(t, x, y, \omega)$ and $\bar{u}(t, x, \bar{y}, \omega)$ by means of the correlation coefficient $\tilde{\varrho}$,

$$\tilde{\varrho}(t, x, y; t, x, \bar{y}) = \frac{A(0, y, \bar{y})}{\sqrt{A(0, y)\, A(0, \bar{y})}}.$$

This correlation coefficient is plotted in Fig. 4.25 in dependence on $\bar{y}$ for $y = 10^{-5}$, $7 \cdot 10^{-5}$, $15 \cdot 10^{-5}$. Because of $\tilde{\varrho} \approx 1$ it is possible to put

$$\bar{u}(t, x, \bar{y}, \omega) = \beta_1 \bar{u}(t, x, y, \omega) + \beta_2$$

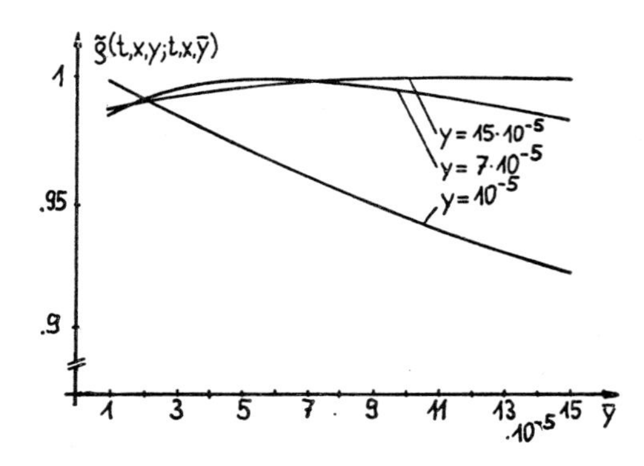

Fig. 4.25. Correlation function in dependence on $\bar{y}$ for different small values of y

and we obtain $\beta_2 = 0$ since $\langle \bar{u} \rangle = 0$ and

$$\beta_1 = \frac{\langle \bar{u}(t, x, \bar{y})\, \bar{u}(t, x, y) \rangle}{\langle \bar{u}^2(t, x, y) \rangle} = \sqrt{\frac{A(0, \bar{y})}{A(0, y)}}.$$

Hence, it follows approximately

$$\bar{u}(t, x, \bar{y}, \omega) = \sqrt{\frac{A(0, \bar{y})}{A(0, y)}}\, \bar{u}(t, x, y, \omega)$$

for the considered values $y, \bar{y}$.

We now investigate variances and correlation relations for small times, i.e. for times which are not considered by the investigation of the stationary behaviour. Using (4.53) it follows

$$V(t, x, y; \bar{t}, \bar{x}, \bar{y})$$
$$= \frac{4a\alpha c^2 \varepsilon}{RL^2} \cos\left(\frac{\pi x}{2R}\right) \cos\left(\frac{\pi \bar{x}}{2R}\right) \sum_{p,q=0}^{\infty} \cos\left(\frac{(2p+1)\,\pi y}{2L}\right) \cos\left(\frac{(2q+1)\,\pi \bar{y}}{2L}\right)$$
$$\times \frac{\exp\left(-\alpha\lambda_{1p}(t - \bar{\bar{t}}) - \alpha\lambda_{1q}(\bar{t} - \bar{\bar{t}})\right) - \exp\left(-\alpha\lambda_{1p} t - \alpha\lambda_{1q} \bar{t}\right)}{\lambda_{1p} + \lambda_{1q}} \tag{4.61}$$

where $\bar{\bar{t}} = \min\{t, \bar{t}\}$. We define by $B(2p + 1, 2q + 1; t, \bar{t}, y, \bar{y})$ the general term in the series of (4.61) and

$$B(t, \bar{t}, y, \bar{y}) = \sum_{p,q=0}^{\infty} B(2p + 1, 2q + 1; t, \bar{t}, y, \bar{y}).$$

By specialization we have

$$B(2p+1, 2q+1; t, t, y, y)$$
$$= \cos\left(\frac{(2p+1)\pi y}{2L}\right)\cos\left(\frac{(2q+1)\pi y}{2L}\right)\frac{1-\exp\left(-\alpha(\lambda_{1p}+\lambda_{1q})\,t\right)}{\lambda_{1p}+\lambda_{1q}}$$

and

$$B(t, t, y, y) = \sum_{p=0}^{\infty}\sum_{q=0}^{\infty}\sum_{r=mp}^{m(p+1)-1}\sum_{s=mq}^{m(q+1)-1} B(2r+1, 2s+1; t, t, y, y).$$

An approximation by quadratic functions (see above) leads to

$$\begin{aligned}
&B(t, t, y, y)\\
&\approx m^2 \sum_{p=0}^{\infty}\sum_{q=0}^{\infty}\Big[\bar{c}^2 B\big((2p+1)\,m, (2q+1)\,m\big)\\
&\quad + c\bar{c}\Big\{B\big((2p+1)\,m, 2m(q+1)-1\big) + B\big((2p+1)\,m, 2mq+1\big)\\
&\quad + B\big(2mp+1, (2q+1)\,m\big) + B\big(2m(p+1)-1, (2q+1)\,m\big)\Big\}\\
&\quad + c^2\Big\{B\big(2mp+1, 2m(q+1)-1\big) + B(2mp+1, 2mq+1)\\
&\quad + B\big(2m(p+1)-1, 2m(q+1)-1\big) + B\big(2m(p+1)-1, 2mq+1\big)\Big\}\Big]
\end{aligned} \tag{4.62}$$

where c, $\bar{c}$ are defined by

$$c \doteq \frac{1}{6}\cdot\frac{m+1}{m-1}, \qquad \bar{c} \doteq 1-2c.$$

Applying the symmetry of $B(2p+1, 2q+1; t, t, y, y)$ as to p, q we obtain

$$B(t, t, y, y) \approx 2m^2\left[\frac{1}{2}\sum_{p=0}^{\infty}\bar{B}(p, p) + \sum_{p=1}^{\infty}\sum_{q=0}^{p-1}\bar{B}(p, q)\right] \tag{4.63}$$

where $\bar{B}(p, q)$ denotes the general term of the series in (4.62). If we consider the function $B(t, t, y, y)$ as a function of y for a fixed t then the decreasing behaviour is obvious from

$$\langle \bar{u}^2(t, x, y)\rangle \sim B(t, t, y, y).$$

For $t = 10^{-3}$ s the dispersion is approximately zero for $y = 14\cdot 10^{-5}$ m and it increases with the time t. These properties are illustrated in Fig. 4.26. The variances for great times (stationary behaviour) can be deduced from (4.59) for $s = 0$. Comparing the stationary variance with (4.59) we have

$$\lim_{t\to\infty} B(t, t, y, y) = L^2 A(0, y).$$

Fig. 4.27 shows the convergence of $B(t, t, y, y)$ to $L^2A(0, y)$ as $t \to \infty$ for $y = 10^{-5}$ m, $y = 14\cdot 10^{-5}$ m, and $y = 0.2$ m. We have used

$$L^2A(0, y)|_{y=10^{-5}} = 6.3934, \qquad L^2A(0, y)|_{y=14\cdot 10^{-5}} = 5.0276,$$

$$L^2A(0, y)|_{y=0.2} = 0.1180$$

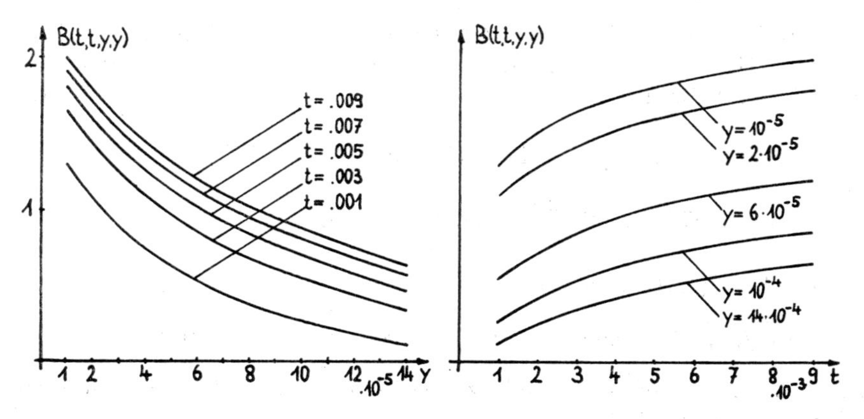

Fig. 4.26. $B(t, t, y, y)$ in dependence on small depths y and small times t

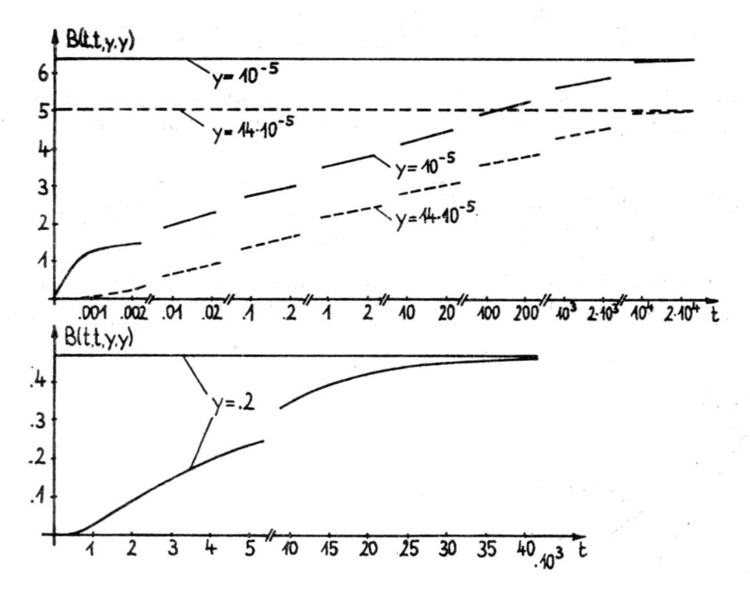

Fig. 4.27. Convergence of $B(t, t, y, y)$ to $L^2\,A(0, y)$ as $t \to \infty$ for different values of y

(see also Fig. 4.22). The time t_0 for which

$$B(t, t, y, y) \approx L^2 A(0, y) \quad \text{for} \quad t \geqq t_0$$

is based on the temperature conductivity assumed to be

$$\alpha = 1.278 \cdot 10^{-5} \frac{\text{m}^2}{\text{s}}.$$

For greater values α the time t_0 will essentially decrease. We can note that assuming the given properties of the material the function $B(t, t, y, y)|_{y=0.2}$ (or also the variance) is approximately zero for a long time ($t \leq 1000$ s) and then it increases to the maximum value up to a time $t_0 \approx 30000$ s. This behaviour for both small values and greater values of y coincides very well in the treated example with the imagination on the heat propagation.

Assuming the correlation function $R_1(s)$ for $\overline{P}_1(t, \omega)$ it follows

$$\varepsilon a = 1.2571817 \cdot 10^6 \, \frac{\mathrm{sK}^2}{\mathrm{m}^2} \qquad \text{for} \quad p = \frac{1}{2}.$$

With the help of (4.61) we obtain

$$\sqrt{\langle \bar{u}^2(t, x, y) \rangle} = 5.66 \text{ K} \quad \text{for} \quad t = 0.009,\ x = 0,\ y = 10^{-5}$$

with $c = 1$. Because of the boundary conditions for $x = \pm R$ the dispersion goes to zero if x converges to $\pm R$.

Now we deal with correlation relations for small times. From (4.61) we have

$$\langle \bar{u}(t, x, y)\, \bar{u}(\bar{t}, \bar{x}, \bar{y}) \rangle \approx \frac{4a\varepsilon\alpha c^2}{RL^2} \cos\left(\frac{\pi x}{2R}\right) \cos\left(\frac{\pi \bar{x}}{2R}\right) B(t, \bar{t}, y, \bar{y}).$$

Let $\bar{t} \leq t$. We will only investigate the temporal correlation, i.e. we put $x = \bar{x}$, $y = \bar{y}$. For the calculation of $B(t, \bar{t}, y, y)$ (4.63) is applied. The correlation coefficient ϱ is given by

$$\varrho(t, \bar{t}, y, \bar{y}) = \frac{\langle \bar{u}(t, x, y)\, \bar{u}(\bar{t}, \bar{x}, \bar{y}) \rangle}{\sqrt{\langle \bar{u}^2(t, x, y) \rangle \langle \bar{u}^2(\bar{t}, \bar{x}, \bar{y}) \rangle}} = \frac{B(t, \bar{t}, y, \bar{y})}{\sqrt{B(t, t, y, y)\, B(\bar{t}, \bar{t}, \bar{y}, \bar{y})}}.$$

In Fig. 4.28 these correlation coefficients are plotted. For $y = 2 \cdot 10^{-5}$ and $\bar{t} = 0.001$ the values $\varrho(t, \bar{t}, y, y)$ are small even for small $t - \bar{t}$. Furthermore, these values are

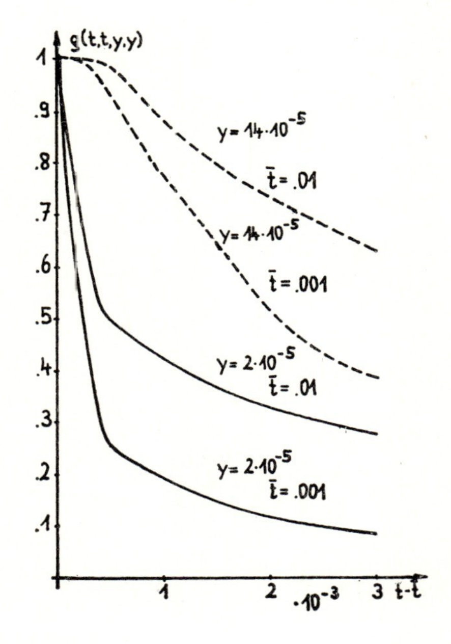

Fig. 4.28. Correlation coefficients for small times

greater for the same values $t - \bar{t}$ but for $\bar{t} = 0.01$. If the correlation values ϱ are considered for the same values $t, \bar{t}$ but for $y = 14 \cdot 10^{-5}$ then these are much greater, i.e. the temporal fluctuations of u are more "arranged" in deeper layers than in higher layers. The discussion as to $\bar{t} = 0.001$ and $\bar{t} = 0.01$ is similar as in the case of $y = 2 \cdot 10^{-5}$.

The previous considerations were taken for

$$P_0(x) = \frac{c}{\sqrt{R}} \cos\left(\frac{\pi x}{2R}\right).$$

Other functions $P_0(x)$ will lead to similar results.

Neglecting the compatibility conditions we put

$$P_0(x) \equiv c.$$

Thus

$$W_k = \int_{-R}^{R} P_0(x)\, \varphi_{1k}(x)\, \mathrm{d}x$$

$$= \frac{c}{\sqrt{R}} \begin{cases} \displaystyle\int_{-R}^{R} \cos\left(\frac{k\pi x}{2R}\right) \mathrm{d}x & \text{for } k \text{ odd}, \\ \displaystyle\int_{-R}^{R} \sin\left(\frac{k\pi x}{2R}\right) \mathrm{d}x & \text{for } k \text{ even} \end{cases} = \begin{cases} \dfrac{4\sqrt{R}c}{k\pi} (-1)^{\frac{k-1}{2}} & \text{for } k \text{ odd}, \\ 0 & \text{for } k \text{ even}. \end{cases}$$

Using (4.57) we obtain

$$\tilde{V}(s, x, y) = \frac{64\varkappa a\varepsilon c^2}{\pi^2} \sum_{\substack{k,h=1\\ \text{odd}}}^{\infty} \frac{1}{kh} (-1)^{\frac{k+h}{2}-1} \cos\left(\frac{k\pi x}{2R}\right) \cos\left(\frac{h\pi x}{2R}\right)$$
$$\times \sum_{p=1}^{\infty} \cos\left(\frac{(2p-1)\pi y}{2L}\right) \frac{1}{m_{khp}} \exp\left(-\frac{1}{2L} m_{khp} y\right) \exp(-\varkappa\lambda_{kp} s).$$

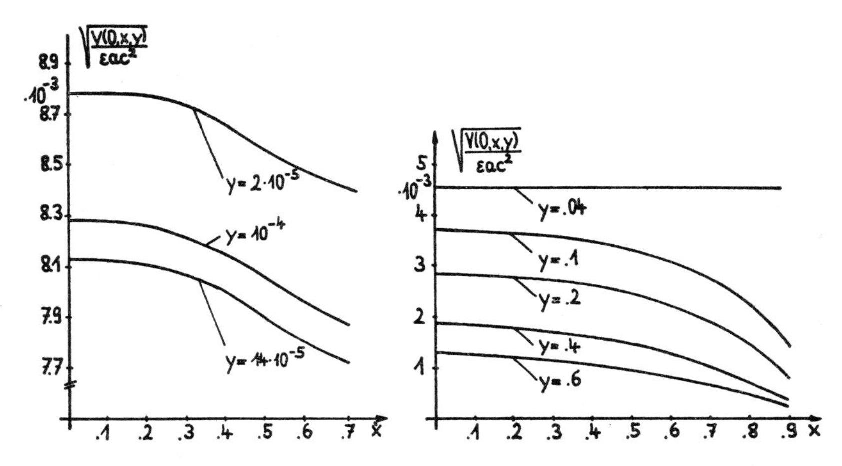

Fig. 4.29. Normalized dispersion of u in dependence on x in the case of $P_0 \equiv c$

Fig. 4.29 shows the normalized dispersion of u as function of x for some values y. In comparison with the case

$$P_0(x) = \frac{c}{\sqrt{R}} \cos\left(\frac{\pi x}{2R}\right)$$

the dispersion of $\bar{u}$ has nearly for $x = 0$ the same values (see also Fig. 4.22).

Now we investigate the case

$$P_0(x) = c\delta(x - x_0)$$

and obtain

$$W_k = \int_{-R}^{R} P_0(x)\, \varphi_{1k}(x)\, dx = \frac{c}{\sqrt{R}} \begin{cases} \cos\left(\dfrac{k\pi}{2R}\, x_0\right) & \text{for } k \text{ odd}, \\ \sin\left(\dfrac{k\pi}{2R}\, x_0\right) & \text{for } k \text{ even}. \end{cases}$$

Assuming the stationary case (4.57) leads to

$$\tilde{V}(s, x, y) = \frac{4\alpha a\varepsilon c^2}{R^2} \sum_{k,h=1}^{\infty} \begin{Bmatrix} \cos\left(\dfrac{k\pi}{2R}\, x_0\right) \cos\left(\dfrac{k\pi x}{2R}\right) \\ \sin\left(\dfrac{k\pi}{2R}\, x_0\right) \sin\left(\dfrac{k\pi x}{2R}\right) \end{Bmatrix} \begin{Bmatrix} \cos\left(\dfrac{h\pi}{2R}\, x_0\right) \cos\left(\dfrac{h\pi x}{2R}\right) \\ \sin\left(\dfrac{h\pi}{2R}\, x_0\right) \sin\left(\dfrac{h\pi x}{2R}\right) \end{Bmatrix}$$

$$\times \sum_{p=1}^{\infty} \cos\left(\frac{(2p-1)\,\pi y}{2L}\right) \frac{1}{m_{khp}} \exp\left(-\frac{1}{2L}\, m_{khp} y\right) \exp\left(-\alpha\lambda_{kp} s\right).$$

Fig. 4.30 shows

$$\sqrt{\frac{\tilde{V}(0, x, y)}{\varepsilon a c^2}}\Bigg|_{x_0}$$

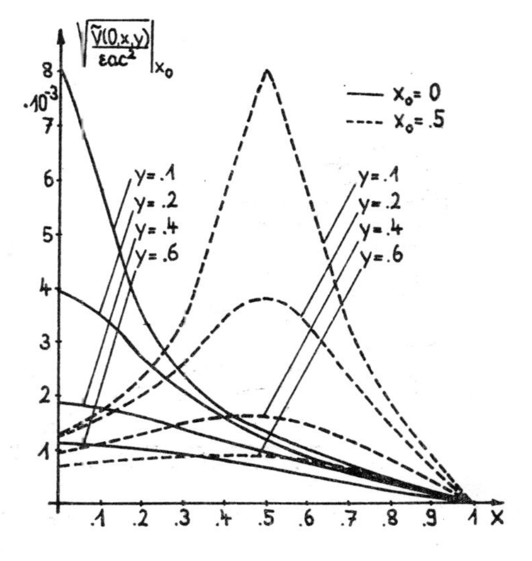

Fig. 4.30. Normalized dispersion of u for $P_0(x) = c\delta(x - x_0)$

for $x_0 = 0$ (bold lines) and for $x_0 = 0.5$ (hatched lines) as a function of x for different values y.

$$\left.\sqrt{\frac{\tilde{V}(0, x, y)}{\varepsilon a c^2}}\right|_{x_0=0.5} = 0.008$$

corresponds to a dispersion

$$\sqrt{\tilde{V}(0, x, y)} = 8.97 \text{ K}$$

using $c = 1$ and

$$\varepsilon a = 1.2571817 \cdot 10^6 .$$

It seems clearly that a maximum of the dispersion can be found for $x = x_0$. The dispersions decrease if one goes away from the point $x = x_0$.

For more investigations it is put

$$\overline{P}(t, x, \omega) = \sum_{i=1}^{g} P_{0i}(x) \, P_{1i}(t, \omega)$$

where $\left(P_{11}(t, \omega), P_{12}(t, \omega), \ldots, P_{1g}(t, \omega)\right)$ denotes a weakly correlated connected vector process. Then, (4.18) leads to

$$U(t, \omega) = \sum_{i=1}^{g} \int_0^t G_i(t - s) \, P_{1i}(s, \omega) \, \mathrm{d}s \tag{4.64}$$

with

$$\begin{aligned} G_i(t) &\doteq \sum_{k,l=0}^{\infty} C_l W_{ki}(f_{kl}, \psi) \exp\left(-\alpha \lambda_{kl} t\right), \\ W_{ki} &\doteq \int_{-R}^{R} \varphi_{1k}(x) \, P_{0i}(x) \, \mathrm{d}x . \end{aligned} \tag{4.65}$$

By application of limit theorems we obtain

$$\langle \eta(t_1) \, \eta(t_2) \rangle = \sum_{i,j=1}^{g} \int_0^{t_{12}} G_i(t_1 - s) \, G_j(t_2 - s) \, a_{ij}(s) \, \mathrm{d}s$$

with respect to the limit process $\eta(t, \omega)$ where

$$a_{ij}(s) = \lim_{\varepsilon \downarrow 0} \frac{1}{\varepsilon} \int_{-\varepsilon}^{\varepsilon} \langle P_{1i}(s) \, P_{1j}(s + r) \rangle \, \mathrm{d}r .$$

Let $\left(P_{11}(t, \omega), \ldots, P_{1g}(t, \omega)\right)$ be stationary connected then

$$a_{ij}(s) = a_{ij} = \text{const}$$

and

$$\begin{aligned} \langle \eta(t_1) \, \eta(t_2) \rangle = \sum_{i,j=1}^{g} a_{ij} \sum_{k,h=0}^{\infty} \sum_{p,q=0}^{\infty} C_p C_q W_{ki} W_{hj} (f_{kp}, \psi) \, (f_{hq}, \psi) \\ \times \frac{\exp\left(-\alpha(t_1 - t_{12}) \, \lambda_{kp} - \alpha(t_2 - t_{12}) \, \lambda_{hq}\right) - \exp\left(-\alpha t_1 \lambda_{kp} - \alpha t_2 \lambda_{hq}\right)}{\lambda_{kp} + \lambda_{hq}} . \end{aligned}$$

The stationary variance can be written as

$$\langle \tilde{\eta}(t)^2 \rangle = \sum_{i,j=1}^{g} a_{ij} \sum_{k,h=0}^{\infty} \sum_{p,q=0}^{\infty} C_p C_q W_{ki} W_{hj} (f_{kp}, \psi)\,(f_{hq}, \psi)\, \frac{1}{\lambda_{kp} + \lambda_{hq}}.$$

Now, we deal with the case of $\alpha_i \to \infty$ for $i = 2, 3, 4$. We obtain

$$\tilde{V}(0, x, y) = \sum_{i,j=1}^{g} \tilde{V}_{ij}(0, x, y)$$

with

$$\tilde{V}_{ij}(0, x, y) = \frac{4\alpha a_{ij} \varepsilon}{R} \sum_{k,h=1}^{\infty} W_{ki} W_{hj} \begin{Bmatrix} \cos\left(\frac{k\pi x}{2R}\right) \\ \sin\left(\frac{k\pi x}{2R}\right) \end{Bmatrix} \begin{Bmatrix} \cos\left(\frac{h\pi x}{2R}\right) \\ \sin\left(\frac{h\pi x}{2R}\right) \end{Bmatrix}$$
$$\times \sum_{p=1}^{\infty} \cos\left(\frac{(2p-1)\,\pi y}{2L}\right) \frac{1}{m_{khp}} \exp\left(-\frac{1}{2L} m_{khp} y\right).$$

The typical behaviour is studied for $g = 2$ and

$$P_{01}(x) = c_1 \delta(x - x_{01}), \qquad P_{02}(x) = c_2 \delta(x - x_{02}).$$

In this case we obtain

$$\bar{P}(t, x, \omega) = c_1 P_{11}(t, \omega)\, \delta(x - x_{01}) + c_2 P_{12}(t, \omega)\, \delta(x - x_{02}).$$

Let $P_{11}(t, \omega)$ be independent of $P_{12}(t, \omega)$. Thus

$$\tilde{V}(0, x, y) = \frac{4a\alpha\varepsilon}{R} [\tilde{V}_{11}(0, x, y) + \tilde{V}_{22}(0, x, y)], \tag{4.66}$$

i.e. the variances obtained for

$$\bar{P} = c_1 P_{11} \delta(x - x_{01}) \quad \text{and} \quad \bar{P} = c_2 P_{12} \delta(x - x_{02})$$

are added.

Let $P_{12}(t, \omega) = P_{11}(t, \omega)$ a.s. Hence, $\tilde{V}(0, x, y)$ has the form

$$\tilde{V}(0, x, y) = \tilde{V}_{11}(0, x, y) + 2\tilde{V}_{12}(0, x, y) + \tilde{V}_{22}(0, x, y)$$
$$= \frac{4\alpha a_{11} \varepsilon}{R} \sum_{k,h=1}^{\infty} [W_{k1} W_{h1} + 2W_{k1} W_{h2} + W_{k2} W_{h2}]$$
$$\times \begin{Bmatrix} \cos\left(\frac{k\pi x}{2R}\right) \\ \sin\left(\frac{k\pi x}{2R}\right) \end{Bmatrix} \begin{Bmatrix} \cos\left(\frac{h\pi x}{2R}\right) \\ \sin\left(\frac{h\pi x}{2R}\right) \end{Bmatrix}$$
$$\times \sum_{p=1}^{\infty} \cos\left(\frac{(2p-1)\,\pi y}{2L}\right) \frac{1}{m_{khp}} \exp\left(-\frac{1}{2L} m_{khp} y\right). \tag{4.67}$$

Fig. 4.31 shows the case of independent processes $P_{1i}(t, \omega)$, $i = 1, 2$, with $a_{11} = a_{22}$ (hatched lines) and the case of strongly dependent processes $P_{1i}(t, \omega)$, $P_{11}(t, \omega) = P_{12}(t, \omega)$ a.s. (bold lines). It was chosen $x_{01} = 0$, $x_{02} = 1/2$, and $R = 1$, $L = 2$. The averaging

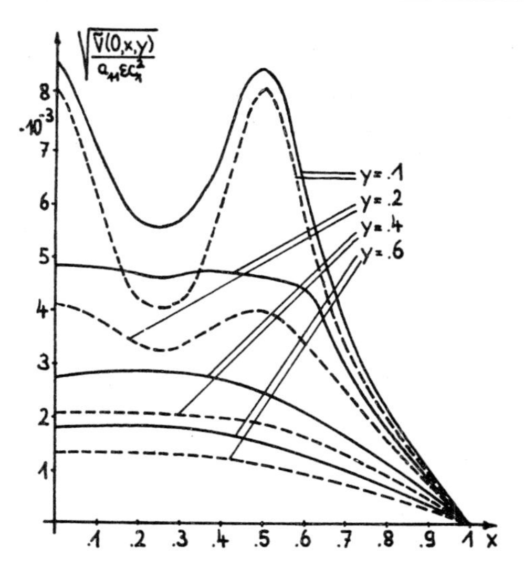

Fig. 4.31. Normalized dispersion of u for $\overline{P} = c\left(P_{11}\delta(x) + P_{12}\delta\left(x - \frac{R}{2}\right)\right)$

character of the temperatures for increasing values y is obvious. For smaller values y the dispersion has a maximum at that places where the boundary condition is different of zero. This property disappears for increasing values y. Figs. 4.22, 4.23, and 4.29 in connection with Figs. 4.30 and 4.31 show a comparison in which way the boundary conditions take an influence on the variance of the temperatures.

Now we still deal with the case of

$$P_{12}(t, \omega) = P_{11}(t + \beta, \omega) \quad \text{a.s.}$$

Then $P_{11}(t, \omega)$ and $P_{12}(t, \omega)$ are not weakly correlated connected. From (4.64) it is possible to show that $U(t, \omega)$ has the form

$$U(t, \omega) = \int_0^t G_1(t - s)\, P_{11}(s, \omega)\, \mathrm{d}s + \int_\beta^{t+\beta} G_2(t + \beta - s)\, P_{11}(s, \omega)\, \mathrm{d}s$$

and we have

$$\lim_{\varepsilon \downarrow 0} \frac{1}{\sqrt{\varepsilon}}\, U(t, \omega) = \eta(t, \omega) \quad \text{in distribution}.$$

The correlation function of $\eta(t, \omega)$ can be written as

$$\begin{aligned}
\langle \eta(t_1)\, \eta(t_2) \rangle = & \int_0^{\min\{t_1, t_2\}} G_1(t_1 - s)\, G_1(t_2 - s)\, a(s)\, \mathrm{d}s \\
& + \int_{\max\{0,\beta\}}^{\min\{t_1, t_2+\beta\}} G_1(t_1 - s)\, G_2(t_2 + \beta - s)\, a(s)\, \mathrm{d}s \\
& + \int_{\max\{0,\beta\}}^{\min\{t_2, t_1+\beta\}} G_1(t_2 - s)\, G_2(t_1 + \beta - s)\, a(s)\, \mathrm{d}s \\
& + \int_\beta^{\min\{t_1+\beta, t_2+\beta\}} G_2(t_1 + \beta - s)\, G_2(t_2 + \beta - s)\, a(s)\, \mathrm{d}s.
\end{aligned}$$

For simplification we will only treat the case of the variance and $\beta > 0$. Then it follows

$$\langle \eta^2(t) \rangle = a \left\{ \int_0^t G_1^2(t-s)\,\mathrm{d}s + 2 \int_\beta^t G_1(t_1 - s)\, G_2(t+\beta-s)\,\mathrm{d}s + \int_\beta^{t+\beta} G_2^2(t+\beta-s)\,\mathrm{d}s \right\}$$

for $t > \beta$ and a denotes the intensity of the weakly correlated process $P_{11}(t, \omega)$ assumed to be wide-sense stationary. Using (4.65) we obtain

$$\langle \eta^2(t) \rangle = \sum_{k,h=0}^{\infty} \sum_{p,q=0}^{\infty} C_p C_q r_{1k} r_{1h} r_{2p} r_{2q}$$
$$\times \left\{ [W_{k1} W_{h1} + W_{k2} W_{h2}] \int_0^t \exp\left(-\alpha(t-s)\,(\lambda_{kp} + \lambda_{hq})\right) \mathrm{d}s \right.$$
$$\left. + \, 2 W_{k1} W_{h2} \int_\beta^t \exp\left(-\alpha(t-s)\,\lambda_{kp} - \alpha(t+\beta-s)\,\lambda_{hq}\right) \mathrm{d}s \right\}$$

and for the stationary variance

$$\langle \tilde{\eta}^2(t) \rangle = \frac{1}{\alpha} \sum_{k,h=0}^{\infty} \sum_{p,q=0}^{\infty} C_p C_q r_{1k} r_{1h} r_{2p} r_{2q} \frac{1}{\lambda_{kp} + \lambda_{hq}}$$
$$\times \{W_{k1} W_{h1} + W_{k2} W_{h2} + 2 W_{k1} W_{h2} \exp(-\alpha\beta\lambda_{hq})\}\,.$$

Now we still consider the case of $\alpha_i \to \infty$ for $i = 2, 3, 4$ and find that

$$\tilde{V}_\beta(0, x, y)$$
$$= \frac{4\alpha a \varepsilon}{R} \sum_{k,h=1}^{\infty} \sum_{p=1}^{\infty} \{W_{k1} W_{h1} + W_{k2} W_{h2} + 2 W_{k1} W_{h2} \exp(-\alpha\beta\lambda_{hp}\}$$
$$\times \begin{Bmatrix} \cos\left(\dfrac{k\pi x}{2R}\right) \\ \sin\left(\dfrac{k\pi x}{2R}\right) \end{Bmatrix} \cdot \begin{Bmatrix} \cos\left(\dfrac{h\pi x}{2R}\right) \\ \sin\left(\dfrac{h\pi x}{2R}\right) \end{Bmatrix} \cdot \cos\left(\frac{(2p-1)\,\pi y}{2L}\right) \frac{1}{m_{khp}} \exp\left(-\frac{y}{2L}\, m_{khp}\right).$$

For $\beta = 0$ the result of (4.67) is obtained and for $\beta \to \infty$ the result of (4.66).

Finally, we give some results in the case of

$$0 < \alpha_2, \alpha_3, \alpha_4 < \infty\,.$$

Fig. 4.32 shows

$$\sqrt{\frac{\tilde{V}(0, x, y)}{\varepsilon a c^2}}$$

for $\tilde{\alpha} = \alpha_2 = \alpha_3 = \alpha_4 = 10$ and $P_0(x) = c$ with the physical constants given above. For the numerical calculations we have chosen $\delta_1 = \delta_2 = 0.01$. If we put the values $\alpha_2 = \alpha_4 = 0.1$ and $\alpha_3 = 10$, then, in this case we obtain

$$\left.\sqrt{\frac{\tilde{V}(0, x, y)}{\varepsilon a c^2}}\right|_{y=0.2} = 5.6 \cdot 10^{-3} \quad \text{for} \quad 0 \leq x \leq 0.95\,.$$

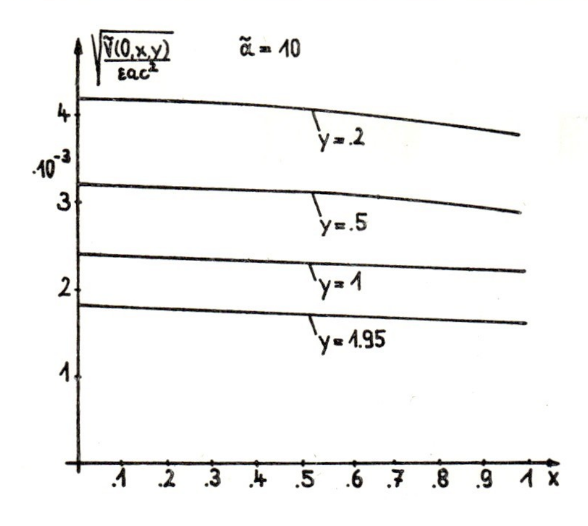

Fig. 4.32. Normalized dispersion in the case of $P_0(x) = c$

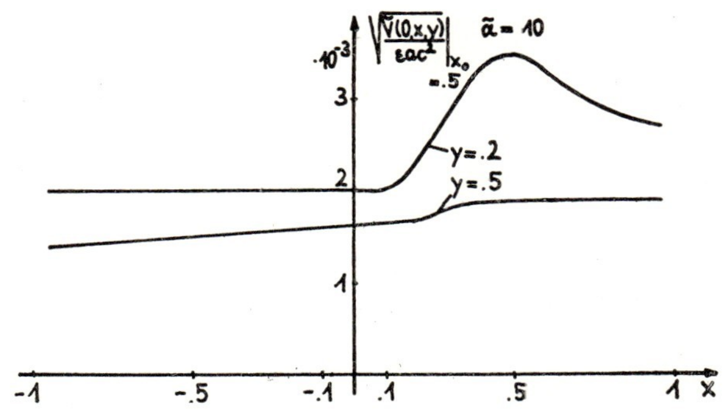

Fig. 4.33. Normalized dispersion in the case of $P_0(x) = c\delta\left(x - \frac{1}{2}\right)$

Fig. 4.33 illustrates the case of

$$P_0(x) = c\delta(x - x_0)$$

with $x_0 = 0.5$ and $\tilde\alpha = 10$ where the same numerical conditions have been selected.

4.3.3. Sliding temperature gradient

In this section the case of a sliding temperature gradient

$$\overline{P}(t, x, \omega) = P_0(x - vt, \omega) \tag{4.68}$$

is considered where these investigations are justified by practical considerations. The velocity is denoted by v and the process $P_0(r, \omega)$ is assumed to be weakly correlated with correlation length ε. Fig. 4.34 illustrates this sliding procedure. For example, if the roughness of a surface is the initial point of the random influence then the correlation length can be obtained by considerations as to the distances of the contact points.

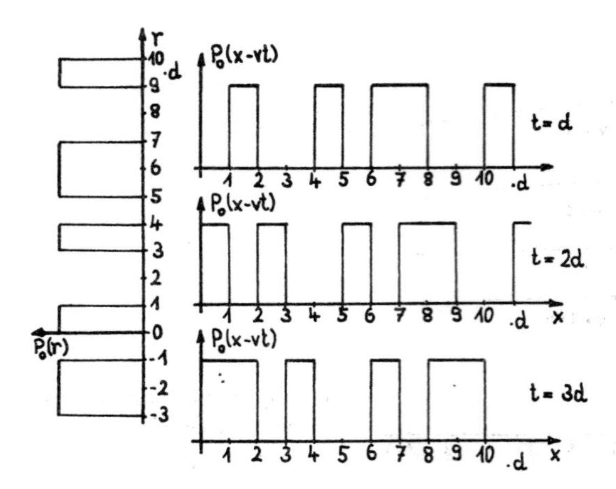

Fig. 4.34. Sliding procedure with $v = 1$

As in the above investigations the solution of problem (4.9) is given by

$$U_i(t, \omega) = \int_0^t \int_{-R}^{R} F_i(t - s, x)\, P_0(x - vs, \omega)\, \mathrm{d}s\, \mathrm{d}x$$

where

$$F_i(t, x) = -\alpha \sqrt{\frac{2}{L}} \sum_{k,p=1}^{\infty} \exp\,(-\alpha\lambda_{kp}t)\, \varphi_{1k}(x)\, (f_{kp}, \psi_i)$$

and we only consider the case of $\alpha_i \to \infty$ for $i = 2, 3, 4$. If we consider the two cases,

$$\text{case } 1\colon R - vt \geqq -R \quad \text{or} \quad t \leqq \frac{2R}{v},$$

$$\text{case } 2\colon R - vt < -R \quad \text{or} \quad t > \frac{2R}{v},$$

then we can obtain as solution

$$U_i(t, \omega) = \int_{-R-vt}^{R} {}^kF_{0i}(t, r)\, P_0(r)\, \mathrm{d}r$$

where k refers to the case k. The functions ${}^kF_{0i}(t, r)$ are given by

$${}^kF_{0i}(t, r) = {}^kF_{1i}(t, r) + {}^kF_{2i}(t, r) + {}^kF_{3i}(t, r) \quad \text{for} \quad k = 1, 2,$$

and

$${}^1F_{1i}(t, r) = \int_{-\frac{1}{v}(R+r)}^{t} F_i(t - s, r + vs)\, \mathrm{d}s \cdot 1_{[-R-vt,-R]}(r),$$

$${}^1F_{2i}(t, r) = \int_{0}^{t} F_i(t - s, r + vs)\, \mathrm{d}s \cdot 1_{[-R,R-vt]}(r),$$

$$
\begin{aligned}
{}^{1}F_{3i}(t, r) &= \int\limits_{0}^{\frac{1}{v}(R-r)} F_i(t - s, r + vs)\, \mathrm{d}s \cdot \mathbf{1}_{[R-vt,R]}(r)\,,\\
{}^{2}F_{1i}(t, r) &= \int\limits_{-\frac{1}{v}(R+r)}^{t} F_i(t - s, r + vs)\, \mathrm{d}s \cdot \mathbf{1}_{[-R-vt,R-vt]}(r)\,,\\
{}^{2}F_{2i}(t, r) &= \int\limits_{-\frac{1}{v}(R+r)}^{\frac{1}{v}(R-r)} F_i(t - s, r + vs)\, \mathrm{d}s \cdot \mathbf{1}_{[R-vt,-R]}(r)\,,\\
{}^{2}F_{3i}(t, r) &= \int\limits_{0}^{\frac{1}{v}(R-r)} F_i(t - s, r + vs)\, \mathrm{d}s \cdot \mathbf{1}_{[-R,R]}(r)\,.
\end{aligned}
$$

Since the functions ${}^{k}F_{0i}(t, r)$ are quadratically integrable as to r the correlation relation of the limit vector process of

$$
\lim_{\varepsilon \downarrow 0} \frac{1}{\sqrt{\varepsilon}} \big(U_1(t, \omega), U_2(t, \omega)\big) = \big(\xi_1(t, \omega), \xi_2(t, \omega)\big) \qquad \text{in distribution}
$$

can be determined by

$$
R_{ij}(t_1, t_2) \doteq \langle \xi_i(t_1)\, \xi_j(t_2) \rangle = \int\limits_{-R-vt_{12}}^{R} {}^{k_1}F_{0i}(t_1, r)\, {}^{k_2}F_{0j}(t_2, r)\, a(r)\, \mathrm{d}r
$$

where k_i is given by t_i.

Firstly, we deal with great times, $t_2 \geqq t_1 > 2R/v$. In this case we have $k_1 = k_2 = 2$. s is introduced to be $s \doteq t_2 - t_1$. Then we obtain

$$
R_{ij}(t_1, t_2) = \begin{cases}
\displaystyle\int\limits_{-R-vt_1}^{R-vt_2} {}^{2}F_{1i}(t_1, r)\, {}^{2}F_{1j}(t_2, r)\, a(r)\, \mathrm{d}r \\
\displaystyle + \int\limits_{R-vt_2}^{R-vt_1} {}^{2}F_{1i}(t_1, r)\, {}^{2}F_{2j}(t_2, r)\, a(r)\, \mathrm{d}r \\
\displaystyle + \int\limits_{R-vt_1}^{-R} {}^{2}F_{2i}(t_1, r)\, {}^{2}F_{2j}(t_2, r)\, a(r)\, \mathrm{d}r \\
\displaystyle + \int\limits_{-R}^{R} {}^{2}F_{3i}(t_1, r)\, {}^{2}F_{3j}(t_2, r)\, a(r)\, \mathrm{d}r \quad \text{for} \quad s < \frac{2R}{v}, \\[2ex]
\displaystyle\int\limits_{-R-vt_1}^{R-vt_1} {}^{2}F_{1i}(t_1, r)\, {}^{2}F_{2j}(t_2, r)\, a(r)\, \mathrm{d}r \\
\displaystyle + \int\limits_{R-vt_1}^{-R} {}^{2}F_{2i}(t_1, r)\, {}^{2}F_{2j}(t_2, r)\, a(r)\, \mathrm{d}r \\
\displaystyle + \int\limits_{-R}^{R} {}^{2}F_{3i}(t_1, r)\, {}^{2}F_{3j}(t_2, r)\, a(r)\, \mathrm{d}r \quad \text{for} \quad s \geqq \frac{2R}{v}.
\end{cases} \tag{4.69}
$$

Secondly, in the case of $t_1 \leqq 2R/v < t_2$ we have

$$R_{ij}(t_1, t_2) = \begin{cases} \int\limits_{-R-vt_1}^{R-vt_2} {}^1F_{1i}(t_1, r)\, {}^2F_{1j}(t_2, r)\, a(r)\, dr \\ + \int\limits_{R-vt_2}^{-R} {}^1F_{1i}(t_1, r)\, {}^2F_{2j}(t_2, r)\, a(r)\, dr \\ + \int\limits_{-R}^{R-vt_1} {}^1F_{2i}(t_1, r)\, {}^2F_{3j}(t_2, r)\, a(r)\, dr \\ + \int\limits_{R-vt_1}^{R} {}^1F_{3i}(t_1, r)\, {}^2F_{3j}(t_2, r)\, a(r)\, dr \quad \text{for} \quad s < \dfrac{2R}{v}, \\ \\ \int\limits_{-R-vt_1}^{-R} {}^1F_{1i}(t_1, r)\, {}^2F_{2j}(t_2, r)\, a(r)\, dr \\ + \int\limits_{-R}^{R-vt_1} {}^1F_{2i}(t_1, r)\, {}^2F_{3j}(t_2, r)\, a(r)\, dr \\ + \int\limits_{R-vt_1}^{R} {}^1F_{3i}(t_1, r)\, {}^2F_{3j}(t_2, r)\, a(r)\, dr \quad \text{for} \quad s \geqq \dfrac{2R}{v}. \end{cases}$$

Finally, we give $R_{ij}(t_1, t_2)$ in the case of $t_1 \leqq t_2 < 2R/v$:

$$\begin{aligned} R_{ij}(t_1, t_2) = {} & \int\limits_{-R-vt_1}^{-R} {}^1F_{1i}(t_1, r)\, {}^1F_{1j}(t_2, r)\, a(r)\, dr \\ & - \int\limits_{-R}^{R-vt_2} {}^1F_{2i}(t_1, r)\, {}^1F_{2j}(t_2, r)\, a(r)\, dr \\ & + \int\limits_{R-vt_2}^{R-vt_1} {}^1F_{2i}(t_1, r)\, {}^1F_{3j}(t_2, r)\, a(r)\, dr \\ & + \int\limits_{R-vt_1}^{R} {}^1F_{3i}(t_1, r)\, {}^1F_{3j}(t_2, r)\, a(r)\, dr. \end{aligned}$$

We want only to investigate nearlier the case $t_2 \geqq t_1 > 2R/v$. The indicator functions contained in the definition of the functions ${}^kF_{ui}(t, r)$, $u = 1, 2, 3$, can be neglected since these are taken into consideration by the limits of integration. Thus

$$\begin{aligned} {}^2F_{1i}(t, r) = {} & -\alpha \sqrt{\frac{2}{LR}} \sum_{k,p=1}^{\infty} \frac{(f_{kp}, \psi_i)}{\alpha^2\lambda_{kp}^2 + v^2 c_k} \\ & \times \left[\alpha\lambda_{kp} \cdot \begin{Bmatrix} \cos\left(\sqrt{c_k}\,(r + vt)\right) \\ \sin\left(\sqrt{c_k}\,(r + vt)\right) \end{Bmatrix} + v\sqrt{c_k} \cdot \begin{Bmatrix} \sin\left(\sqrt{c_k}\,(r + vt)\right) \\ -\cos\left(\sqrt{c_k}\,(r + vt)\right) \end{Bmatrix} \right. \\ & \left. + v\sqrt{c_k}\, \exp\left(-\alpha\lambda_{kp}\left(t + \frac{R + r}{v}\right)\right) \cdot \begin{Bmatrix} (-1)^{\frac{k-1}{2}} \\ (-1)^{k/2} \end{Bmatrix} \right], \end{aligned}$$

$$
\begin{aligned}
&{}^2F_{2i}(t, r) \\
&= -\alpha \sqrt{\frac{2}{LR}} \sum_{k,p=1}^{\infty} \frac{(f_{kp}, \psi_i)}{\alpha^2\lambda_{kp}^2 + v^2 c_k} v\sqrt{c_k} \left[\exp\left(-\alpha\lambda_{kp}\left(t - \frac{R-r}{v}\right)\right) \cdot \begin{Bmatrix} (-1)^{\frac{k-1}{2}} \\ (-1)^{\frac{k+2}{2}} \end{Bmatrix}\right. \\
&\quad \left. + \exp\left(-\alpha\lambda_{kp}\left(t + \frac{R+r}{v}\right)\right) \cdot \begin{Bmatrix} (-1)^{\frac{k-1}{2}} \\ (-1)^{k/2} \end{Bmatrix}\right],
\end{aligned}
$$

$$
\begin{aligned}
&{}^2F_{3i}(t, r) \\
&= -\alpha \sqrt{\frac{2}{LR}} \sum_{k,p=1}^{\infty} \frac{(f_{kp}, \psi_i)}{\alpha^2\lambda_{kp}^2 + v^2 c_k} \left[v\sqrt{c_k} \exp\left(-\alpha\lambda_{kp}\left(t - \frac{R-r}{v}\right)\right) \cdot \begin{Bmatrix} (-1)^{\frac{k-1}{2}} \\ (-1)^{\frac{k+2}{2}} \end{Bmatrix}\right. \\
&\quad \left. - \exp\left(-\alpha\lambda_{kp}t\right)\left(\alpha\lambda_{kp} \cdot \begin{Bmatrix} \cos\left(\sqrt{c_k}r\right) \\ \sin\left(\sqrt{c_k}r\right) \end{Bmatrix} + v\sqrt{c_k} \cdot \begin{Bmatrix} \sin\left(\sqrt{c_k}r\right) \\ -\cos\left(\sqrt{c_k}r\right) \end{Bmatrix}\right)\right]
\end{aligned}
$$

where in every formula the upper line has to be taken for k odd and the lower line for k even.

For simplification the process $P_0(r, \omega)$ is assumed to be wide-sense stationary and hence, the intensity is constant, $a(r) = a = \text{const}$.

Now, the integrals of (4.69) are to be determined for great times t_1, t_2. It is

$$\lim_{t \to \infty} {}^2F_{3,i}(t, r) = 0 \quad \text{for} \quad -R \leqq r \leqq R$$

because of

$$0 < t - \frac{2R}{v} \leqq t - \frac{R-r}{v}$$

and

$$\lim_{\substack{t_1, t_2 \to \infty \\ t_2 - t_1 = s = \text{const}}} \int_{-R}^{R} {}^2F_{3i}(t_1, r)\, {}^2F_{3j}(t_2, r)\, dr = 0.$$

Furthermore, we can obtain

$$
\begin{aligned}
&\lim_{\substack{t_1, t_2 \to \infty \\ t_2 - t_1 = s = \text{const}}} \int_{R - vt_1}^{-R} {}^2F_{2i}(t_1, r)\, {}^2F_{2j}(t_2, r)\, dr \\
&= \frac{2\alpha v^3}{LR} \sum_{k,h=1}^{\infty} \sum_{p,q=1}^{\infty} \frac{(f_{kp}, \psi_i)(f_{hq}, \psi_j)}{(\alpha^2\lambda_{kp}^2 + v^2 c_k)(\alpha^2\lambda_{hq}^2 + v^2 c_h)} \sqrt{c_k c_h} \frac{\exp(-\alpha\lambda_{hq}s)}{\lambda_{kp} + \lambda_{hq}} \\
&\quad \times \left(1 - (-1)^k \exp\left(-\frac{2R\alpha}{v}\lambda_{kp}\right)\right)\left(1 - (-1)^h \exp\left(-\frac{2R\alpha}{v}\lambda_{hq}\right)\right) \\
&\quad \times \begin{cases} (-1)^{\frac{k+h+2}{2}} & \text{for } k, h \text{ odd}, \\ (-1)^{\frac{k+h+1}{2}} & \text{for } k + h \text{ odd}, \\ (-1)^{\frac{k+h}{2}} & \text{for } k, h \text{ even}. \end{cases}
\end{aligned}
\tag{4.70}
$$

It is possible to determine the other integrals in (4.69) but the calculations and the results are to extensive in order to represent here.

Restricting we deal with the case of the variance, i.e. $t_1 = t_2 = t$. Then we obtain that the second integral in (4.69) is zero and it follows

$$\int\limits_{-R-vt}^{R-vt} {}^2F_{1i}(t, r)\, {}^2F_{1j}(t, r)\, \mathrm{d}r = \int\limits_{-R}^{R} {}^2F_{1i}(t, r - vt)\, {}^2F_{1j}(t, r - vt)\, \mathrm{d}r$$

where

$$\begin{aligned} {}^2F_{1i}(t, r - vt) &= \int\limits_0^{\frac{R+r}{v}} F_i(u, r - vu)\, \mathrm{d}u \\ &= -\sqrt{\frac{2}{LR}} \sum_{k=1}^{\infty} \sum_{p=1}^{\infty} \frac{(f_{kp}, \psi_i)}{\alpha^2\lambda_{kp}^2 + v^2 c_k} \\ &\quad \times \left[v\sqrt{c_k} \exp\left(-\alpha\lambda_{kp} \frac{R+r}{v}\right) \cdot \begin{Bmatrix} (-1)^{\frac{k-1}{2}} \\ (-1)^{k/2} \end{Bmatrix} \right. \\ &\quad \left. + \alpha\lambda_{kp} \cdot \begin{Bmatrix} \cos\left(\sqrt{c_k} r\right) \\ \sin\left(\sqrt{c_k} r\right) \end{Bmatrix} + v\sqrt{c_k} \cdot \begin{Bmatrix} \sin\left(\sqrt{c_k} r\right) \\ -\cos\left(\sqrt{c_k} r\right) \end{Bmatrix} \right]. \end{aligned}$$

Some calculations lead to

$$\begin{aligned} &\int\limits_{-R-vt}^{R-vt} {}^2F_{1i}(t, r)\, {}^2F_{1j}(t, r)\, \mathrm{d}r \\ &= \frac{2\alpha^2 a}{LR} \sum_{k,h=1}^{\infty} \sum_{p,q=1}^{\infty} \frac{(f_{kp}, \psi_i)(f_{hq}, \psi_j)}{(\alpha^2\lambda_{kp}^2 + v^2 c_k)(\alpha^2\lambda_{hq}^2 + v^2 c_h)} \\ &\quad \times \left\{ v^3\alpha\sqrt{c_k c_h} \left[\frac{1 - e_{kp} e_{hq}}{\alpha^2(\lambda_{kp} + \lambda_{hq})} + \frac{(\lambda_{hq} - \lambda_{kp})\, e_{hkp}}{\alpha^2\lambda_{kp}^2 + v^2 c_h} + \frac{(\lambda_{kp} - \lambda_{hq})\, e_{khq}}{\alpha^2\lambda_{hq}^2 + v^2 c_k} \right] \right. \\ &\quad \times \begin{cases} (-1)^{\frac{k+h-2}{2}} & \text{for } k, h \text{ odd}, \\ (-1)^{\frac{k+h-1}{2}} & \text{for } k + h \text{ odd}, \\ (-1)^{\frac{k+h}{2}} & \text{for } k, h \text{ even} \end{cases} \\ &\quad + R\left(\alpha^2\lambda_{kp}\lambda_{hq} + v^2\sqrt{c_k c_h}\right) \delta_{kh} \\ &\quad \left. + \frac{2\alpha v\sqrt{c_k c_h}}{c_k - c_h} (\lambda_{hq} - \lambda_{kp})(-1)^{\frac{k+h-1}{2}} (k + h) \bmod (2) \right\} \end{aligned} \tag{4.71}$$

where e_{kp} and e_{hkp} are defined by

$$e_{kp} \doteq \exp\left(-\frac{2R\alpha}{v} \lambda_{kp}\right), \qquad e_{hkp} \doteq 1 - (-1)^h e_{kp}.$$

Substituting (4.70) and (4.71) in (4.69) if follows

$$\lim_{t\to\infty} R_{ij}(t, t)$$

$$= \frac{2\alpha^2 a}{LR} \sum_{k,h=1}^{\infty} \sum_{p,q=1}^{\infty} \frac{(f_{kp}, \psi_i)(f_{hq}, \psi_j)}{(\alpha^2\lambda_{kp}^2 + v^2 c_k)(\alpha^2\lambda_{hq}^2 + v^2 c_h)}$$

$$\times \left\{ v^3\alpha \sqrt{c_k c_h} \left[\frac{e_{kkp}e_{hhq}}{\alpha^2(\lambda_{kp} + \lambda_{hq})} + \frac{1 - e_{kp}e_{hq}}{\alpha^2(\lambda_{kp} + \lambda_{hq})} + \frac{(\lambda_{hq} - \lambda_{kp})\, e_{hkp}}{\alpha^2\lambda_{kp}^2 + v^2 c_h} \right.\right.$$

$$\left. + \frac{(\lambda_{kp} - \lambda_{hq})\, e_{khq}}{\alpha^2\lambda_{hq}^2 + v^2 c_k} \right] \cdot \begin{cases} (-1)^{\frac{k+h-2}{2}} & \text{for } k, h \text{ odd}, \\ (-1)^{\frac{k+h-1}{2}} & \text{for } k + h \text{ odd}, \\ (-1)^{\frac{k+h}{2}} & \text{for } k, h \text{ even} \end{cases}$$

$$+ R\left(\alpha^2\lambda_{kp}\lambda_{hq} + v^2\sqrt{c_k c_h}\right) \delta_{kh}$$

$$\left. + \frac{2\alpha v \sqrt{c_k c_h}}{c_k c_h} (\lambda_{hq} - \lambda_{kp})\, (-1)^{\frac{k+h-1}{2}} (k + h) \bmod (2) \right\}.$$

As to the solution $u(t, x, y, \omega)$ we have

$$\langle \bar{u}(t_1, x_1, y_1)\, \bar{u}(t_2, x_2, y_2) \rangle \approx \frac{\varepsilon R_{12}(t_1, t_2)}{16\delta_{11}\delta_{21}\delta_{12}\delta_{22}}$$

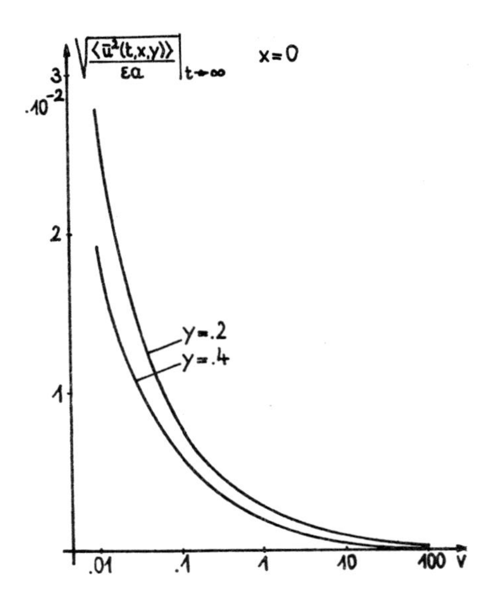

Fig. 4.35. Normalized dispersion in dependence on the velocity v

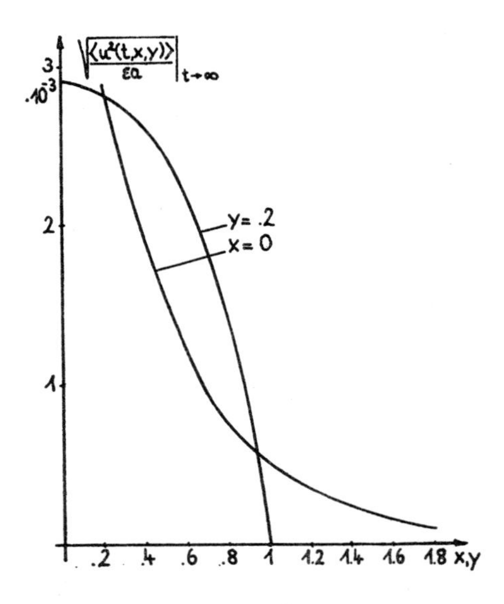

Fig. 4.36. Normalized dispersion in dependence on x and y for $v = 1$

with

$$\chi_{i1}(x) = 1_{[x_i - \delta_{i1}, x_i + \delta_{i1}]}(x),$$

$$\chi_{i2}(y) = 1_{[y_i - \delta_{i2}, y_i + \delta_{i2}]}(y).$$

Fig. 4.35 shows the behaviour of the normalized dispersion in dependence on the velocity v. For increasing velocities the normalized dispersion decreases. For $v = 1$ Fig. 4.36 illustrates the function

$$\left. \sqrt{\frac{\langle \bar{u}^2(t, x, y) \rangle}{\varepsilon a}} \right|_{t \to \infty}$$

as function of x and y.

5. Random boundary value problems

5.1. Ordinary differential operators

In this section we consider the stochastic boundary value problem

$$L(\omega)\, u = g(x, \omega),$$
$$U_k[u] = 0, \qquad k = 1, 2, \dots, 2n, \qquad 0 \leqq x \leqq 1, \tag{5.1}$$

where the operator $L(\omega)$ is given by

$$L(\omega)\, u \doteq (-1)^n \, [f_n(x)\, u^{(n)}]^{(n)} + \sum_{i=0}^{n-1} (-1)^i \, [f_i(x, \omega)\, u^{(i)}]^{(i)}$$

and U_k by

$$U_k[u] = \sum_{i=0}^{2n-1} \big(a_{ki} u^{(k)}(0) + b_{ki} u^{(k)}(1)\big) = 0, \qquad k = 1, 2, \dots, 2n.$$

The boundary conditions are constituted so that

$$\big(L(\omega)\, u, v\big) = \sum_{i=0}^{n} \int_0^1 (-1)^i \, [f_i(x)\, u^{(i)}]^{(i)} \, v \, dx$$
$$= \sum_{i=0}^{n} \int_0^1 f_i(x)\, u^{(i)} v^{(i)} \, dx \tag{5.2}$$

is fulfilled for all permissible functions u, v. A function u is said to be permissible if it possesses $2n$ continuous derivatives and satisfies the boundary conditions. The scalar product of $\mathbf{L}_2(0, 1)$ is denoted by $(.,.)$. Then the random operator $L(\omega)$ is symmetric relative to all permissible functions,

$$\big(L(\omega)\, u, v\big) = \big(u, L(\omega)\, v\big).$$

The vector process

$$\big(\bar{f}_{0\varepsilon}(x, \omega), \bar{f}_{1\varepsilon}(x, \omega), \dots, \bar{f}_{n-1\varepsilon}(x, \omega), \bar{g}_\varepsilon(x, \omega)\big)^\mathsf{T}$$

with

$$\bar{h}(x, \omega) \doteq h(x, \omega) - \langle h(x) \rangle$$

is assumed to be weakly correlated connected with correlation length ε. The non-random function $f_n(x)$ is supposed to be not equal to zero. Furthermore, let

$$|\bar{f}_{i\varepsilon}^{(s)}(x, \omega)| \leqq \eta \quad \text{for} \quad i = 0, 1, \ldots, n-1; \qquad s = 0, 1, \ldots, i$$

with a small η where $\bar{f}_{i\varepsilon}(x, \omega)$ are to be a.s. differentiable up to the order i.

Then the boundary value problem (5.1) can be written by the following form:

$$\begin{aligned} L(\omega)\, u &= \langle L\rangle\, u + L_1(\omega)\, u = \langle g(x)\rangle + \bar{g}(x, \omega), \\ U_k[u] &= 0, \qquad k = 1, 2, \ldots, 2n, \end{aligned} \tag{5.3}$$

where $\langle L\rangle$ and $L_1(\omega)$ are given by

$$\langle L\rangle\, u = \sum_{i=0}^{n} (-1)^i\, [\langle f_i(x)\rangle\, u^{(i)}]^{(i)},$$

$$L_1(\omega)\, u = \sum_{i=0}^{n-1} (-1)^i\, [\bar{f}_i(x, \omega)\, u^{(i)}]^{(i)}.$$

We assume that

$$\langle L\rangle\, w = 0; \qquad U_i[w] = 0, \qquad i = 1, 2, \ldots, n,$$

only possesses the trivial solution $w \equiv 0$.

Now we assume an expansion of the form

$$u(x, \omega) = \sum_{j=0}^{\infty} u_j(x, \omega) \tag{5.4}$$

where $u_j(x, \omega)$ denotes the terms of the solution $u(x, \omega)$ which are homogeneous of j-th order as to $\bar{g}_\varepsilon(x, \omega), \bar{f}_{i\varepsilon}(x, \omega)$, $i = 0, 1, \ldots, n-1$. We determine formally the expressions $u_j(x, \omega)$ by substituting the expansion (5.4) in the boundary value problem (5.3). Then the functions $u_j(x, \omega)$, $j = 0, 1, 2, \ldots$, follow successively from the equations

$$\begin{aligned} &\langle L\rangle\, u_0 = \langle g\rangle; && U_i[u_0] = 0, && i = 1, 2, \ldots, 2n, \\ &\langle L\rangle\, u_1 = \bar{g} - L_1(\omega)\, u_0; && U_i[u_1] = 0, && i = 1, 2, \ldots, 2n, \\ &\langle L\rangle\, u_j = -L_1(\omega)\, u_{j-1}; && U_i[u_j] = 0, && i = 1, 2, \ldots, 2n; \quad j = 2, 3, \ldots \end{aligned} \tag{5.5}$$

Let $G(x, y)$ be the Green's function with respect to $\langle L\rangle$ and the boundary conditions $U_i[.] = 0$, $i = 1, 2, \ldots, 2n$. $G(x, y)$ is defined on

$$\mathcal{D} = \{(x, y): 0 \leqq x \leqq 1,\ 0 \leqq y \leqq 1\}$$

having the following properties:

- $G(x, y)$ is $2n$ times continuously differentiable as to x on both domains

  $$\mathcal{D}_1 = \{(x, y) \in \mathcal{D}: y \leqq x\}, \qquad \mathcal{D}_2 = \{(x, y) \in \mathcal{D}: x \leqq y\}$$

  and satisfies the homogeneous differential equation $\langle L\rangle\, u = 0$,
- $G(x, y)$ is continuous on $\mathcal{D}$,
- for $0 < y < 1$ the Green's function $G(x, y)$ fulfils the relation

  $$\frac{\partial^{2n-1}}{\partial x^{2n-1}}\, G(y+0, y) - \frac{\partial^{2n-1}}{\partial x^{2n-1}}\, G(y-0, y) = \frac{(-1)^n}{f_n(y)},$$
- for $0 < y < 1$ the Green's function $G(x, y)$ satisfies the boundary value conditions $U_i[G] = 0$, $i = 1, 2, \ldots, 2n$, as function of x.

Then we obtain

$$u_0(x) = \int_0^1 G(x, y) \langle g(y) \rangle \, dy,$$

i.e. $u_0(x)$ is a deterministic function. With the help of (5.2) the term $u_1(x, \omega)$ is given by

$$\begin{aligned} u_1(x, \omega) &= \int_0^1 G(x, y) \{\bar{g}_\varepsilon(y, \omega) - L_1(\omega)\, u_0(y)\} \, dy \\ &= \int_0^1 G(x, y)\, \bar{g}_\varepsilon(y, \omega) \, dy - \sum_{i=0}^{n-1} \int_0^1 u_0^{(i)}(y) \frac{\partial^i G(x, y)}{\partial y^i} \bar{f}_{i\varepsilon}(y, \omega) \, dy \\ &= \int_0^1 \overset{1}{H}(x; y)\, \bar{g}_\varepsilon(y, \omega) \, dy + \sum_{i=0}^{n-1} \int_0^1 \overset{2}{H}_i(x; y)\, \bar{f}_{i\varepsilon}(y, \omega) \, dy \end{aligned}$$

with

$$\overset{1}{H}(x; y) = G(x, y), \qquad \overset{2}{H}_i(x; y) = -u_0^{(i)}(y) \frac{\partial^i G(x, y)}{\partial y^i}. \tag{5.6}$$

Now we assume

$$\begin{aligned} u_{j-1}(x) = & \sum_{i_1, \ldots, i_{j-2}=0}^{n-1} \int_0^1 \ldots \int_0^1 \overset{1}{H}_{i_1 \ldots i_{j-2}}(x; y_1, \ldots, y_{j-1}) \\ & \times \bar{f}_{i_1\varepsilon}(y_1) \ldots \bar{f}_{i_{j-2}\varepsilon}(y_{j-2})\, \bar{g}_\varepsilon(y_{j-1}) \, dy_1 \ldots dy_{j-1} \\ & + \sum_{i_1, \ldots, i_{j-1}=0}^{n-1} \int_0^1 \ldots \int_0^1 \overset{2}{H}_{i_1 \ldots i_{j-1}}(x; y_1, \ldots, y_{j-1})\, \bar{f}_{i_1\varepsilon}(y_1) \ldots \bar{f}_{i_{j-1}\varepsilon}(y_{j-1}) \, dy_1 \ldots dy_{j-1} \end{aligned}$$

and with this $u_j(x)$ is calculated. (5.5) leads to

$$\begin{aligned} u_j(x) &= -\int_0^1 G(x, y)\, L_1 u_{j-1}(y) \, dy \\ &= -\sum_{i=0}^{n-1} \int_0^1 u_{j-1}^{(i)}(y) \frac{\partial^i G(x, y)}{\partial y^i} \bar{f}_{i\varepsilon}(y) \, dy \\ &= -\sum_{i=0}^{n-1} \Bigg[\sum_{i_1, \ldots, i_{j-2}=0}^{n-1} \int_0^1 \ldots \int_0^1 \overset{1}{H}{}^{(i)}_{i_1 \ldots i_{j-2}}(y; y_1, \ldots, y_{j-1}) \\ &\quad \times \frac{\partial^i G(x, y)}{\partial y^i} \bar{f}_{i_1\varepsilon}(y_1) \ldots \bar{f}_{i_{j-2}\varepsilon}(y_{j-2})\, \bar{f}_{i\varepsilon}(y)\, \bar{g}_\varepsilon(y_{j-1}) \, dy \, dy_1 \ldots dy_{j-1} \\ &\quad + \sum_{i_1, \ldots, i_{j-1}=0}^{n-1} \int_0^1 \ldots \int_0^1 \overset{2}{H}{}^{(i)}_{i_1, \ldots, i_{j-1}}(y; y_1, \ldots, y_{j-1}) \\ &\quad \times \frac{\partial^i G(x, y)}{\partial y^i} \bar{f}_{i_1\varepsilon}(y_1) \ldots \bar{f}_{i_{j-1}\varepsilon}(y_{j-1})\, \bar{f}_{i\varepsilon}(y) \, dy \, dy_1 \ldots dy_{j-1} \Bigg]. \end{aligned} \tag{5.7}$$

Then $u_j(x)$ has the form of $u_{j-1}(x)$ with

$$\begin{aligned} \overset{1}{H}_{i_1 \ldots i_{j-1}}(x; y_1, \ldots, y_j) &= -\overset{1}{H}{}^{(i_{j-1})}_{i_1 \ldots i_{j-2}}(y_{j-1}; y_1, \ldots, y_{j-2}, y_j)\, G^{(i_{j-1})}_{y_{j-1}}(x, y_{j-1}), \\ \overset{2}{H}_{i_1 \ldots i_j}(x; y_1, \ldots, y_j) &= -\overset{2}{H}{}^{(i_j)}_{i_1 \ldots i_{j-1}}(y_j; y_1, \ldots, y_{j-1})\, G^{(i_j)}_{y_j}(x, y_j) \end{aligned} \tag{5.8}$$

where

$$G_{xy}^{(ij)}(x, y) \doteq \frac{\partial^{i+j} G(x, y)}{\partial x^i \, \partial y^j}.$$

The i-th derivatives

$$\frac{\partial^i}{\partial x^i} \overset{1}{H}_{i_1 \dots i_{j-2}}(x; y_1, \dots, y_{j-1}) \doteq H^{(i)}_{i_1 \dots i_{j-2}}(x; y_1, \dots, y_{j-1}),$$

$$\frac{\partial^i}{\partial x^i} \overset{2}{H}_{i_1 \dots i_{j-1}}(x; y_1, \dots, y_{j-1}) \doteq H^{(i)}_{i_1 \dots i_{j-1}}(x; y_1, \dots, y_{j-1})$$

exists for $i = 1, 2, \dots, n - 1$ since the derivatives

$$G_{xy}^{(ij)}(x, y) \quad \text{for} \quad i = 0, 1, \dots, n - 1; \qquad j = 0, 1, \dots, n - 1,$$

exists. Hence, we get

$$\begin{aligned}
&\overset{1}{H}(x; y_1) = G(x, y_1), \\
&\overset{1}{H}_{i_1}(x; y_1, y_2) = -G_{y_1}^{(i_1)}(x, y_1)\, G_{y_1}^{(i_1)}(y_1, y_2), \\
&\overset{1}{H}_{i_1 i_2}(x; y_1, y_2, y_3) = G_{y_2}^{(i_2)}(x, y_2)\, G_{y_1 y_2}^{(i_1 i_2)}(y_2, y_1)\, G_{y_1}^{(i_1)}(y_1, y_3), \\
&\overset{1}{H}_{i_1 i_2 i_3}(x; y_1, y_2, y_3, y_4) = -G_{y_3}^{(i_3)}(x, y_3)\, G_{y_1 y_2}^{(i_1 i_2)}(y_2, y_1)\, G_{y_2}^{(i_2)}(y_3, y_2)\, G_{y_1}^{(i_1)}(y_1, y_4)
\end{aligned} \tag{5.9}$$

as to $\overset{1}{H}$ and

$$\begin{aligned}
&\overset{2}{H}_{i_1}(x; y_1) = -G_{y_1}^{(i_1)}(x; y_1)\, u_0^{(i_1)}(y_1), \\
&\overset{2}{H}_{i_1 i_2}(x; y_1, y_2) = G_{y_2}^{(i_2)}(x, y_2)\, G_{y_1 y_2}^{(i_1 i_2)}(y_2, y_1)\, u_0^{(i_1)}(y_1), \\
&\overset{2}{H}_{i_1 i_2 i_3}(x; y_1, y_2, y_3) = -G_{y_3}^{(i_3)}(x, y_3)\, G_{y_1 y_2}^{(i_1 i_2)}(y_2, y_1)\, G_{y_2}^{(i_2)}(y_3, y_2)\, u_0^{(i_1)}(y_1), \\
&\overset{2}{H}_{i_1 i_2 i_3 i_4}(x; y_1, y_2, y_3, y_4) = G_{y_4}^{(i_4)}(x, y_4)\, G_{y_1 y_2}^{(i_1 i_2)}(y_2, y_1)\, G_{y_3}^{(i_3)}(y_4, y_3)\, G_{y_2}^{(i_2)}(y_3, y_2)\, u_0^{(i_1)}(y_1).
\end{aligned} \tag{5.10}$$

By these functions the first four terms u_j, $j = 1, 2, 3, 4$, can be determined and these u_j are needed in order to calculate the moment $\langle u(x) \rangle$ up to terms of the order ε^2.

Now we deal with questions of the convergence of the expansion (5.4). Since $G_{xy}^{(rs)}(x, y)$ for $r, s = 0, 1, \dots, 2n$ with $0 \leqq r + s \leqq 2n$ is continuous for $0 \leqq x < y \leqq 1$ or $0 \leqq y < x \leqq 1$ it follows for $(x, y) \in [0, 1] \times [0, 1]$ that

$$|G_{xy}^{(rs)}(x, y)| \leqq c$$

is fulfilled for $r, s = 0, 1, \dots, 2n$ with $0 \leqq r + s \leqq 2n$. Furthermore, we have

$$|u_0^{(r)}(x)| \leqq c_0 \quad \text{for} \quad 0 \leqq r \leqq 2n \quad \text{and} \quad x \in [0, 1],$$

$$\left| \frac{1}{f_n(x)} \right| \leqq b \qquad \text{for} \quad x \in [0, 1],$$

and

$$|\bar{f}_{i\varepsilon}^{(r)}(x, \omega)| \leqq \eta \tag{5.11}$$

almost surely for $0 \leqq i \leqq n-1$, $0 \leqq r \leqq i$, $x \in [0, 1]$. c, c_0 and b are constants. Using properties of the Green's function it is

$$u_k^{(r)}(x, \omega) = -\int_0^1 G_x^{(r)}(x, y)\, L_1 u_{k-1}(y)\, dy \quad \text{for} \quad r = 0, 1, \ldots, 2n-1,$$

$$u_k^{(2n)}(x, \omega) = -\int_0^1 G_x^{(2n)}(x, y)\, L_1 u_{k-1}(y)\, dy - \frac{(-1)^n}{f_n(x)}\, L_1 u_{k-1}(x)$$

and therefore

$$|u_k^{(r)}(x, \omega)| \leqq (c + b) \max_{x \in [0,1]} |L_1 u_{k-1}(x)| \quad \text{for} \quad r = 0, 1, \ldots, 2n.$$

For $|L_1 u_0(x)|$ we have the inequalities

$$|L_1 u_0(x)| \leqq \sum_{i=0}^{n-1} \left| \sum_{r=0}^{i} \binom{i}{r} \bar{f}_{i\varepsilon}^{(r)} u_0^{(2i-r)} \right| \leqq \eta c_0 \sum_{i=0}^{n-1} \sum_{r=0}^{i} \binom{i}{r} \leqq 2^n \eta c_0$$

and then

$$|u_1^{(r)}(x, \omega)| \leqq 2^n \eta c_0 (c + b).$$

Finally, in a similar way we get

$$|u_j^{(r)}(x, \omega)| \leqq \big(2^n \eta (c + b)\big)^j c_0 \quad \text{for} \quad j = 0, 1, \ldots, \quad \text{and} \quad x \in [0, 1],$$

$$\text{for} \quad r = 0, 1, \ldots, 2n.$$

Hence

$$\sum_{j=0}^{\infty} |u_j^{(r)}(x, \omega)|$$

converges almost surely and uniformly in $0 \leqq x \leqq 1$ supposed

$$|\eta| < \frac{1}{2^n (c + b)}.$$

In this case

$$\sum_{j=0}^{\infty} \langle |u_j^{(r)}(x)| \rangle$$

also converges uniformly as to $0 \leqq x \leqq 1$. Now we have proved that

$$u(x, \omega) = \sum_{j=0}^{\infty} u_j(x, \omega)$$

is the solution of the random boundary value problem (5.1). These results are summarized in Theorem 5.1.

Theorem 5.1. *Assuming the conditions $|\bar{f}_{i\varepsilon}^{(r)}(x, \omega)| \leqq \eta$ a.s. for $0 \leqq i \leqq n-1$ then the solution $u(x, \omega)$ of the boundary value problem* (5.1) *can be written as*

$$u(x, \omega) = \sum_{j=0}^{\infty} u_j(x, \omega).$$

$G(x, y)$ denotes the Green's function as to $\langle L \rangle$ and the boundary conditions $U_i[.] = 0$, $i = 1, 2, \ldots, 2n$. Then it is

$$u_0(x) = \int_0^1 G(x, y) \langle g(y) \rangle \, dy$$

and

$$u_j(x, \omega) = \sum_{i_1, \ldots, i_{j-1}=0}^{n-1} \int_0^1 \cdots \int_0^1 \overset{1}{H}_{i_1 \ldots i_{j-1}}(x; y_1, \ldots, y_j) \bar{f}_{i_1\varepsilon}(y_1) \ldots \bar{f}_{i_{j-1}\varepsilon}(y_{j-1}) \bar{g}_\varepsilon(y_j) \, dy_1 \ldots dy_j$$
$$+ \sum_{i_1, \ldots, i_j=0}^{n-1} \int_0^1 \cdots \int_0^1 \overset{2}{H}_{i_1 \ldots i_j}(x; y_1, \ldots, y_j) \bar{f}_{i_1\varepsilon}(y_1) \ldots \bar{f}_{i_j\varepsilon}(y_j) \, dy_1 \ldots dy_j$$

for $j = 1, 2, \ldots$ where

$$\overset{1}{H}_{i_1 \ldots i_{j-1}}(x; y_1, \ldots, y_j), \qquad \overset{2}{H}_{i_1 \ldots i_j}(x; y_1, \ldots, y_j)$$

is given by (5.6) for $j = 1$ and by (5.8) for $j > 1$.

Using the results of Chapter 1 we can formulate Theorem 5.2.

Theorem 5.2. *The vector process*

$$\left(\bar{f}_{0\varepsilon}(x, \omega), \bar{f}_{1\varepsilon}(x, \omega), \ldots, \bar{f}_{n-1\varepsilon}(x, \omega), \bar{g}_\varepsilon(x, \omega)\right)^{\mathrm{T}}$$

is assumed to be weakly correlated connected. Then the process

$$\frac{1}{\sqrt{\varepsilon}} \left(u(x, \omega) - u_0(x)\right)$$

converges in distribution to a Gaussian process $\xi(x, \omega)$. The process $\xi(x, \omega)$ possesses the mean $\langle \xi(x) \rangle = 0$ and the correlation function

$$\langle \xi(x) \, \xi(y) \rangle = \int_0^1 G(x, z) \, G(y, z) \, a_{gg}(z) \, dz$$
$$- \sum_{i=0}^{n-1} \int_0^1 \{G(x, z) \, G_z^{(i)}(y, z) \, a_{gi}(z) + G(y, z) \, G_z^{(i)}(x, z) \, a_{ig}(z)\} \, u_0^{(i)}(z) \, dz$$
$$+ \sum_{i,j=0}^{n-1} \int_0^1 G_z^{(i)}(x, z) \, G_z^{(j)}(y, z) \, u_0^{(i)}(z) \, u_0^{(j)}(z) \, a_{ij}(z) \, dz$$

where $a_{gg}(x)$ denotes the intensity of the process $\bar{g}_\varepsilon(x, \omega)$; $a_{gi}(x)$ the intensity between $\bar{g}_\varepsilon(x, \omega)$ and $\bar{f}_{i\varepsilon}(x, \omega)$; and $a_{ij}(x)$ between $\bar{f}_{i\varepsilon}(x, \omega)$ and $\bar{f}_{j\varepsilon}(x, \omega)$.

Proof. We put

$$k_{u(x,\omega)} \doteq \sum_{j=0}^{k} u_j(x, \omega).$$

Using similar considerations as applied in Section 1.3 as to the expansion of moments of linear functionals we see that

$$\lim_{\varepsilon \downarrow 0} \frac{1}{\sqrt{\varepsilon}^p} \left\langle \prod_{q=1}^{p} \left({}^k u(x_{i_q}) - u_0(x_{i_q})\right) \right\rangle = \begin{cases} \lim_{\varepsilon \downarrow 0} \frac{1}{\sqrt{\varepsilon}^p} \left\langle \prod_{q=1}^{p} u_1(x_{i_q}) \right\rangle & \text{for } p \text{ even}, \\ 0 & \text{for } p \text{ odd} \end{cases}$$

where $i_q \in \{1, \ldots, s\}$ and $\{x_1, \ldots, x_s\}$ denotes a set of s points from $[0, 1]$. Now, the processes $b_s(x, \omega)$, $s = 0, 1, \ldots, n$, are introduced by

$$b_n(x, \omega) \doteq \int_0^1 G(x, y)\, \bar{g}_\varepsilon(y, \omega)\, \mathrm{d}y,$$

$$b_s(x, \omega) \doteq -\int_0^1 G_y^{(s)}(x, y)\, u_0^{(s)}(y)\, \bar{f}_{s\varepsilon}(y, \omega)\, \mathrm{d}y \quad \text{for} \quad s = 0, 1, \ldots, n-1$$

and hence $u_1(x, \omega)$ can be written as

$$u_1(x, \omega) = \sum_{s=0}^{n} b_i(x, \omega).$$

With the help of (1.23), (1.24), and (1.32) it follows

$$\lim_{\varepsilon \downarrow 0} \frac{1}{\sqrt{\varepsilon}^p} \left\langle \prod_{q=1}^{p} u_1(x_{i_q}) \right\rangle = \begin{cases} \sum\limits_{s_1, \ldots, s_p = 0}^{n} \sum\limits_{\{u_1, v_1\}, \ldots, \{u_{p/2}, v_{p/2}\}} \prod\limits_{q=1}^{p/2} {}^2A_{1, u_q v_q} & \text{for } p \text{ even}, \\ 0 & \text{for } p \text{ odd} \end{cases}$$

where the sum

$$\sum_{\{u_1, v_1\}, \ldots, \{u_{p/2}, v_{p/2}\}}$$

is taken over all nonequivalent separations

$$\bigl\{\{u_1, v_1\}, \ldots, \{u_{p/2}, v_{p/2}\}\bigr\} \quad \text{of} \quad \{1, \ldots, p\}$$

and ${}^2A_{1, u_q v_q}$ is defined by

$${}^2A_{1, u_q v_q} \doteq \lim_{\varepsilon \downarrow 0} \frac{1}{\varepsilon} \left\langle b_{s_{u_q}}(x_{i_{u_q}})\, b_{s_{v_q}}(x_{i_{v_q}}) \right\rangle.$$

Now, we define the Gaussian process $\xi(x, \omega)$ by $\langle \xi(x) \rangle = 0$ and

$$\langle \xi(x)\, \xi(y) \rangle = \sum_{s_1, s_2 = 0}^{n} \lim_{\varepsilon \downarrow 0} \frac{1}{\varepsilon} \langle b_{s_1}(x)\, b_{s_2}(y) \rangle. \tag{5.12}$$

Then, it follows

$$\sum_{s_1, \ldots, s_p = 0}^{n} \sum_{\{u_1, v_1\}, \ldots, \{u_{p/2}, v_{p/2}\}} \prod_{q=1}^{p/2} {}^2A_{1, u_q v_q} = \sum_{\{u_1, v_1\}, \ldots, \{u_{p/2}, v_{p/2}\}} \prod_{q=1}^{p/2} \langle \xi(x_{i_{u_q}})\, \xi(x_{i_{v_q}}) \rangle.$$

Because of the proved convergence of moments of ${}^k u(x, \omega) - u_0(x)$ to the adequate moments of $\xi(x, \omega)$ the statement of Theorem 5.2 is showed for

$$\frac{1}{\sqrt{\varepsilon}} \left({}^k u(x, \omega) - u_0(x)\right).$$

This will be clear completely if the limit value contained in (5.12) is determined. It is

$$\lim_{\varepsilon \downarrow 0} \frac{1}{\varepsilon} \sum_{s_1, s_2 = 0}^{n} \langle b_{s_1}(x)\, b_{s_2}(y) \rangle$$

$$= \lim_{\varepsilon \downarrow 0} \frac{1}{\varepsilon} \left[\langle b_n(x)\, b_n(y) \rangle + \sum_{s=0}^{n-1} \{ \langle b_n(x)\, b_s(y) \rangle + \langle b_s(x)\, b_n(y) \rangle \} + \sum_{s,t=0}^{n-1} \langle b_s(x)\, b_t(y) \rangle \right]$$

where

$$\lim_{\varepsilon\downarrow 0} \frac{1}{\varepsilon} \langle b_n(x)\, b_n(y)\rangle = \int_0^1 G(x, z)\, G(y, z)\, a_{gg}(z)\, dz,$$

$$\lim_{\varepsilon\downarrow 0} \frac{1}{\varepsilon} \langle b_n(x)\, b_s(y)\rangle = -\int_0^1 G(x, z)\, G_z^{(s)}(y, z)\, u_0^{(s)}(z)\, a_{gs}(z)\, dz,$$

$$\lim_{\varepsilon\downarrow 0} \frac{1}{\varepsilon} \langle b_s(x)\, b_n(y)\rangle = -\int_0^1 G(y, z)\, G_z^{(s)}(x, z)\, u_0^{(s)}(z)\, a_{sg}(z)\, dz,$$

$$\lim_{\varepsilon\downarrow 0} \frac{1}{\varepsilon} \langle b_s(x)\, b_t(y)\rangle = \int_0^1 G_z^{(s)}(x, z)\, G_z^{(t)}(y, z)\, u_0^{(s)}(z)\, u_0^{(t)}(z)\, a_{st}(z)\, dz.$$

The complete proof of this theorem follows from the uniform convergence as to ε,

$$\lim_{k\to\infty} \frac{1}{\varepsilon} \left\langle \left({}^k u(x) - u(x)\right)^2 \right\rangle = 0. \tag{5.13}$$

First, we assume the relation (5.13) and prove this theorem. From the convergence in $\mathbf{L}_2$-mean it follows the convergence in probability

$$\mathsf{P} - \lim_{k\to\infty} \frac{1}{\sqrt{\varepsilon}} \left({}^k u(x, \omega) - u_0(x)\right) = \frac{1}{\sqrt{\varepsilon}} \left(u(x, \omega) - u_0(x)\right) \quad \text{uniformly as to } \varepsilon$$

and also

$$\lim_{k\to\infty} \mathsf{P}\left(\frac{1}{\sqrt{\varepsilon}} \left({}^k u(x, \omega) - u_0(x)\right) < t\right)$$

$$= \mathsf{P}\left(\frac{1}{\sqrt{\varepsilon}} \left(u(x, \omega) - u_0(x)\right) < t\right) \qquad \text{uniformly as to } \varepsilon. \tag{5.14}$$

We have shown in the first part of this proof

$$\lim_{\varepsilon\downarrow 0} {}^k F_\varepsilon(t_1, \ldots, t_s) = \Phi(t_1, \ldots, t_s) \qquad \text{for all} \quad k \geqq 1 \tag{5.15}$$

defining

$${}^k F_\varepsilon(t_1, \ldots, t_s) = \mathsf{P}\left(\frac{1}{\sqrt{\varepsilon}} \left({}^k u(x_1, \omega) - u_0(x_1)\right) < t_1, \ldots, \frac{1}{\sqrt{\varepsilon}} \left({}^k u(x_s, \omega) - u_0(x_s)\right) < t_s\right)$$

and

$$\Phi(t_1, \ldots, t_s) = \mathsf{P}\left(\xi(x_1, \omega) < t_1, \ldots, \xi(x_s, \omega) < t_s\right).$$

Let $F_\varepsilon(t_1, \ldots, t_s)$ be the distribution function of

$$\left(\frac{1}{\sqrt{\varepsilon}} \left(u(x_1, \omega) - u_0(x_1)\right), \ldots, \frac{1}{\sqrt{\varepsilon}} \left(u(x_s, \omega) - u_0(x_s)\right)\right).$$

Hence, from relations as in (5.14) we can obtain

$$|{}^k F_\varepsilon(t_1, \ldots, t_s) - F_\varepsilon(t_1, \ldots, t_s)| < \eta$$

for all $k \geqq k_0(\eta)$ uniformly as to ε and with (5.15)

$$\Phi(t_1, \ldots, t_s) - \eta \leqq \overline{\lim_{\varepsilon \downarrow 0}} F_\varepsilon(t_1, \ldots, t_s) \leqq \Phi(t_1, \ldots, t_s) + \eta$$

for arbitrary $\eta > 0$. Then it follows

$$\lim_{\varepsilon \downarrow 0} F_\varepsilon(t_1, \ldots, t_s) = \Phi(t_1, \ldots, t_s)$$

and the proof of this theorem.

Now, we have to prove the $\mathbf{L}_2$-mean convergence of (5.13). Consider

$$\frac{1}{\varepsilon} \langle u_p(x)\, u_q(x) \rangle \quad \text{with} \quad p, q \geqq N_0$$

in order to prove the relation (5.13). Using the results of Theorem 5.1 we obtain the inequality

$$\begin{aligned} &\left| \frac{1}{\varepsilon} \langle u_p(x)\, u_q(x) \rangle \right| \\ &\leqq \frac{1}{\varepsilon} C^{p+q} \sum_{i_1, \ldots, i_{p+q-2}=0}^{n-1} \int_0^1 \cdots \int_0^1 |\langle \bar{f}_{i_1\varepsilon}(y_1) \ldots \bar{f}_{i_{p+q-2}\varepsilon}(y_{p+q-2})\, \bar{g}_\varepsilon(y_{p+q-1})\, \bar{g}_\varepsilon(y_{p+q}) \rangle| \, \mathrm{d}y_1 \ldots \mathrm{d}y_{p+q} \\ &+ \frac{2}{\varepsilon} C_0 C^{p+q} \sum_{i_1, \ldots, i_{p+q-1}=0}^{n-1} \int_0^1 \cdots \int_0^1 |\langle \bar{f}_{i_1\varepsilon}(y_1) \ldots \bar{f}_{i_{p+q-1}\varepsilon}(y_{p+q-1})\, \bar{g}_\varepsilon(y_{p+q}) \rangle| \, \mathrm{d}y_1 \ldots \mathrm{d}y_{p+q} \\ &+ \frac{1}{\varepsilon} C_0^2 C^{p+q} \sum_{i_1, \ldots, i_{p+q}=0}^{n-1} \int_0^1 \cdots \int_0^1 |\langle \bar{f}_{i_1\varepsilon}(y_1) \ldots \bar{f}_{i_{p+q}\varepsilon}(y_{p+q}) \rangle| \, \mathrm{d}y_1 \ldots \mathrm{d}y_{p+q}. \end{aligned} \tag{5.16}$$

Let $h_\varepsilon(y, \omega)$ be a weakly correlated process with correlation length ε and $|h_\varepsilon(y, \omega)| < \eta$ a.s. Then we deal with an integral of the form

$$\int_0^1 \cdots \int_0^1 |\langle h_\varepsilon(y_1) \ldots h_\varepsilon(y_l) \rangle| \, \mathrm{d}y_1 \ldots \mathrm{d}y_l \quad \text{with} \quad l \geqq 2.$$

For $l^2 \geqq 1/2\varepsilon$ we get

$$\int_0^1 \cdots \int_0^1 |\langle h_\varepsilon(y_1) \ldots h_\varepsilon(y_l) \rangle| \, \mathrm{d}y_1 \ldots \mathrm{d}y_l \leqq \eta^l \leqq 2\varepsilon l^2 \eta^l$$

and for $l^2 < 1/2\varepsilon$ the relation

$$\int_0^1 \cdots \int_0^1 |\langle h_\varepsilon(y_1) \ldots h_\varepsilon(y_l) \rangle| \, \mathrm{d}y_1 \ldots \mathrm{d}y_l \leqq \eta^l (1 - V_l)$$

where V_l denotes the volume of points $(y_1, \ldots, y_l) \in [0, 1]^l$ for which the maximally ε-neighbouring separation is

$$\{(y_1), (y_2), \ldots, (y_l)\}.$$

It follows

$$V_l = \int_0^1 \mathrm{d}y_1 \int_{|y_1 - y_2| > \varepsilon} \mathrm{d}y_2 \ldots \int_{\substack{|y_1 - y_l| > \varepsilon \\ \vdots \\ |y_{l-1} - y_l| > \varepsilon}} \mathrm{d}y_l \geqq \prod_{t=1}^{l-1} (1 - 2\varepsilon t) \geqq 1 - 2\varepsilon l^2$$

and for all $l \geqq 2$ we obtain

$$\int_0^1 \dots \int_0^1 |\langle h_\varepsilon(y_1) \dots h_\varepsilon(y_l)\rangle| \, \mathrm{d}y_1 \dots \mathrm{d}y_l \leqq 2\varepsilon l^2 \eta^l .$$

This relation is also true for a weakly correlated connected vector process

$$\big(h_{1\varepsilon}(y, \omega), h_{2\varepsilon}(y, \omega), \dots, h_{s\varepsilon}(y, \omega)\big);$$

i.e. we have

$$\int_0^1 \dots \int_0^1 |\langle h_{i_1\varepsilon}(y_1) \dots h_{i_l\varepsilon}(y_l)\rangle| \, \mathrm{d}y_1 \dots \mathrm{d}y_l \leqq 2\varepsilon l^2 \eta^l$$

if $|h_{i\varepsilon}(y, \omega)| < \eta$ a.s. for $i = 1, 2, \dots, s$ and $i_q \in \{1, 2, \dots, s\}$ for $q = 1, 2, \dots, l$. This inequality is applied to (5.16) in a modified form and with

$$|\bar{g}(x, \omega)| \leqq g \qquad \text{a.s.}$$

it is

$$\left| \frac{1}{\varepsilon} \langle u_p(x) \, u_q(x)\rangle \right| \leqq 2(g + C_0 n\eta)^2 \, C^2 (nC\eta)^{p+q-2} \, (p + q)^2 .$$

The uniform convergence of (5.13) is obtained by the convergence of the series

$$\sum_{p,q=1}^{\infty} (p + q)^2 \, (\eta C n)^{p+q-2} \quad \text{for} \quad |\eta C n| < 1 .$$

Theorem 5.2 is proved. ◀

Corollary 5.1. *Let $\bar{f}_{i\varepsilon}(x, \omega) \equiv 0$ for $i = 0, 1, \dots, n - 1$, i.e. let L be a deterministic operator, then Theorem 5.2 is true without someone restriction on the weakly correlated process $\bar{g}_\varepsilon(x, \omega)$ as to the smallness.*

This follows from

$$u(x, \omega) = u_0(x) + \int_0^1 G(x, y) \, \bar{g}_\varepsilon(y, \omega) \, \mathrm{d}y$$

with

$$u_0(x) = \int_0^1 G(x, y) \, \langle g(y)\rangle \, \mathrm{d}y .$$

Hence, we obtain in this case that

$$\frac{1}{\sqrt{\varepsilon}} \big(u(x, \omega) - u_0(x)\big) \tag{5.17}$$

converges in distribution to a Gaussian process $\xi(x, \omega)$ having the moments

$$\langle \xi(x)\rangle = 0 ,$$

$$\langle \xi(x) \, \xi(y)\rangle = \int_0^1 G(x, z) \, G(y, z) \, a_{gg}(z) \, \mathrm{d}z .$$

Boyce [2] dealt with this case of a stochastic boundary value problem and he showed that (5.17) is a Gaussian random variable for all $x \in [0, 1]$ if ε goes to zero. The general problem (5.1) in the case of $n = 1$ is investigated by Boyce, Xia [1]. The consideration of a random coefficient with respect to the highest derivative in the differential operator L is incompatible with the result of Theorem 5.1 on which this paper of Boyce and Xia is based.

Corollary 5.2. *Let $w_i(x)$, λ_i, $i = 1, 2, \ldots$, be the eigenfunctions and eigenvalues, respectively, as to $\langle L \rangle$ and the given boundary conditions where*

$$\int_0^1 w_i(x)\, w_j(x)\, \mathrm{d}x = \delta_{ij} \quad \text{for} \quad i, j = 1, 2, \ldots$$

is assumed. The correlation function of $\xi(x, \omega)$ can be determined by

$$\langle \xi(x)\, \xi(y) \rangle = \sum_{p,q=1}^{\infty} \left\{ b_{pq} - \sum_{i=0}^{n-1} (c_{pq}^{gi} + c_{qp}^{ig}) + \sum_{i,j=0}^{n-1} d_{pq}^{ij} \right\} \frac{1}{\lambda_p \lambda_q}\, w_p(x)\, w_q(y)$$

where we define

$$b_{pq} \doteq \int_0^1 w_p(z)\, w_q(z)\, a_{gg}(z)\, \mathrm{d}z,$$

$$c_{pq}^{gi} \doteq \int_0^1 w_p(z)\, w_q^{(i)}(z)\, u_0^{(i)}(z)\, a_{gi}(z)\, \mathrm{d}z,$$

$$d_{pq}^{ij} \doteq \int_0^1 w_p^{(i)}(z)\, w_q^{(j)}(z)\, u_0^{(i)}(z)\, u_0^{(j)}(z)\, a_{ij}(z)\, \mathrm{d}z.$$

Proof. We use Mercer's theorem

$$G(x, y) = \sum_{p=1}^{\infty} \frac{1}{\lambda_p}\, w_p(x)\, w_p(y)$$

where this series converges absolutely and uniformly on $[0, 1] \times [0, 1]$. Furthermore, the series

$$\sum_{p=1}^{\infty} \frac{1}{\lambda_p}\, w_p(x)\, w_p^{(i)}(y)$$

also converges absolutely and uniformly on $[0, 1] \times [0, 1]$ for $i = 0, 1, \ldots, n - 1$. This follows from the inequalities

$$\sum_{p=s}^{t} \left| \frac{1}{\lambda_p}\, w_p^{(i)}(x)\, w_p^{(j)}(y) \right| \leq \left[\sum_{p=s}^{t} \frac{1}{\lambda_p} \left(w_p^{(i)}(x)\right)^2 \sum_{p=s}^{t} \frac{1}{\lambda_p} \left(w_p^{(j)}(y)\right)^2 \right]^{1/2}$$

and

$$\sum_{p=s}^{t} \frac{1}{\lambda_p} \left(w_p^{(i)}(y)\right)^2 \leq \left(\frac{\partial^{2i} G(x, y)}{\partial x^i\, \partial y^i} \right)\Bigg|_{x=y} \leq C \quad \text{for} \quad i = 0, 1, \ldots, n - 1$$

(see Collatz [1]). Now, Corollary 5.2 can be obtained from Theorem 5.2. ◀

We consider the boundary value problem (5.1) using the conditions given above. $\{\varphi_p\}_{1 \leq p < \infty}$ denotes a system of functions of the energetic space $\mathscr{H}_{\langle L \rangle}$. This system is assumed to be complete. The solution $u(x, \omega)$ of the boundary value problem (5.1) is the minimum of the energetic functional

$$(Lu, u) - 2(g, u)$$

in the energetic space. Using the Ritz-method with the coordinate functions $\{\varphi_p\}_{1 \leq p < \infty}$ we put

$$u_m(x, \omega) = \sum_{p=1}^{m} x_p^{(m)}(\omega)\, \varphi_p(x) \tag{5.18}$$

and get

$$(Lu_m, u_m) - 2(g, u_m) = \sum_{p,q=1}^{m} (L\varphi_p, \varphi_q)\, x_p^{(m)} x_q^{(m)} - 2 \sum_{p=1}^{m} (g, \varphi_p)\, x_p^{(m)}.$$

We choose $\{x_p^{(m)}\}_{1 \leq p \leq m}$ so that this functional has a minimum and this leads to the system of linear equations

$$\sum_{p=1}^{m} \big(L(\omega)\, \varphi_p, \varphi_s\big)\, x_s^{(m)} = (g, \varphi_s), \qquad s = 1, 2, \ldots, n. \tag{5.19}$$

By means of the solution of (5.19) we get from (5.18) the m-th Ritz-approximation of the solution $u(x, \omega)$ of (5.1). Now, (5.19) can be written in the form

$$\big(A_0 + B(\omega)\big)\, X^{(m)} = b_0 + c(\omega) \tag{5.20}$$

where the matrices A_0, $B(\omega)$, b_0, $c(\omega)$ are given by

$$A_0 = \big((\langle L \rangle\, \varphi_p, \varphi_q)\big)_{1 \leq p,q \leq m}; \qquad b_0 = \big((\langle g \rangle, \varphi_p)\big)^{\mathsf{T}}_{1 \leq p \leq m};$$
$$B(\omega) = \big((L_1(\omega)\, \varphi_p, \varphi_q)\big)_{1 \leq p,q \leq m}; \quad c(\omega) = \big((\bar{g}, \varphi_p)\big)^{\mathsf{T}}_{1 \leq p \leq m}.$$

Applying the conditions as to $\{\varphi_p\}_{1 \leq p < \infty}$ the matrix A_0 is regular and it is

$$A_0^{-1} \doteq (\alpha_{pq})_{1 \leq p,q \leq m}.$$

It is possible to expand the solution of (5.19) with respect to the elements of the matrices $B(\omega)$ and $c(\omega)$,

$$X^{(m)}(\omega) = \sum_{r=0}^{\infty} (-1)^r\, X_r^{(m)}(\omega). \tag{5.21}$$

For the matrices $X_r^{(m)}(\omega)$ we obtain

$$X_0^{(m)} = A_0^{-1} b_0; \qquad X_1^{(m)} = A_0^{-1}(BX_0^{(m)} - c);$$

$$X_r^{(m)} = A_0^{-1} B X_{r-1}^{(m)} \quad \text{for} \quad r = 2, 3, \ldots$$

and furthermore

$$X_0^{(m)} = A_0^{-1} b_0,$$
$$X_r^{(m)} = (A_0^{-1}B)^r A_0^{-1} b_0 - (A_0^{-1}B)^{r-1} A_0^{-1} c \quad \text{for} \quad r = 1, 2, 3, \ldots$$

Assuming

$$|B(\omega)| \leq |A_0^{-1}|^{-1}\, \eta \qquad \text{a.s. with} \quad 0 < \eta < 1$$

we have the inequalities

$$\left|\sum_{r=0}^{\infty} (-1)^r X_r^{(m)}(\omega)\right| \leq \sum_{r=0}^{\infty} |A_0^{-1}B|^r\, |A_0^{-1}b_0| + \sum_{r=1}^{\infty} |A_0^{-1}B|^{r-1}\, |A_0^{-1}c|$$

$$\leq |A_0^{-1}b_0| \sum_{r=0}^{\infty} \eta^r + |A_0^{-1}c| \sum_{r=1}^{\infty} \eta^{r-1} = \frac{1}{1-\eta} (|A_0^{-1}b_0| + |A_0^{-1}c|)$$

and the series contained in (5.21) converges a.s.

Considering the m-th Ritz-approximation $u_m(x, \omega)$ from (5.18) it follows

$$u_m(x, \omega) = \sum_{p=1}^{m} x_{0p}^{(m)} \varphi_p(x) - \sum_{p=1}^{m} x_{1p}^{(m)} \varphi_p(x) + O_2$$

where $X_r^{(m)} = (x_{rp}^{(m)})_{1 \leq p \leq m}$ and O_2 denotes the terms which contains elements of $B(\omega)$ and $c(\omega)$ of a higher order than 1. Further, we get

$$u_m(x, \omega) - \sum_{p=1}^{m} x_{0p}^{(m)} \varphi_p(x) = \sum_{p=1}^{m} \left(A_0^{-1}(c - BX_0^{(m)})\right)_p \varphi_p(x) + O_2$$

$$= \sum_{p=1}^{m} c_p \bar{\varphi}_p(x) - \sum_{p,q=1}^{m} b_{pq} x_{q0}^{(m)} \bar{\varphi}_p(x) + O_2 \tag{5.22}$$

where $\bar{\varphi}_p(x)$ is defined by

$$\bar{\varphi}_p(x) \doteq (A_0^{-\mathsf{T}} \varphi)_p = \sum_{q=1}^{m} \alpha_{qp} \varphi_q(x).$$

Furthermore, we put

$$h^{(m)}(x, z) \doteq \sum_{p=1}^{m} \bar{\varphi}_p(x)\, \varphi_p(z),$$

$$h_i^{(m)}(x, z) \doteq \sum_{p=1}^{m} \bar{\varphi}_p(x)\, \varphi_p^{(i)}(z) \sum_{p=1}^{m} x_{p0}^{(m)} \varphi_p^{(i)}(z)$$

and (5.22) leads to

$$u_m(x, \omega) - \sum_{p=1}^{m} x_{0p}^{(m)} \varphi_p(x) = \int_0^1 h^{(m)}(x, z) \bar{g}_\varepsilon(z, \omega)\, \mathrm{d}z - \sum_{i=0}^{m-1} \int_0^1 h_i^{(m)}(x, z) \bar{f}_{i\varepsilon}(z, \omega)\, \mathrm{d}z + O_2.$$

Hence, we have

$$\lim_{\varepsilon \downarrow 0} \frac{1}{\sqrt{\varepsilon}} \left(u_m(x, \omega) - u_0(x)\right) = \xi_m(x, \omega) \qquad \text{in distribution} \tag{5.23}$$

where $\xi_m(x, \omega)$ is Gaussian distributed with the moments

$$\langle \xi_m(x) \rangle = 0,$$

$$\langle \xi_m(x)\, \xi_m(y) \rangle = \int_0^1 h^{(m)}(x, z)\, h^{(m)}(y, z)\, a_{gg}(z)\, \mathrm{d}z$$

$$- \sum_{i=0}^{n-1} \int_0^1 \{h^{(m)}(x, z)\, h_i^{(m)}(y, z)\, a_{gi}(z) + h^{(m)}(y, z)\, h_i^{(m)}(x, z)\, a_{ig}(z)\}\, \mathrm{d}z$$

$$+ \sum_{i,j=0}^{n-1} \int_0^1 h_i^{(m)}(x, z)\, h_j^{(m)}(y, z)\, a_{ij}(z)\, \mathrm{d}z. \tag{5.24}$$

The correlation function $\langle \xi_m(x)\, \xi_m(y) \rangle$ can be written as

$$\begin{aligned}
&\langle \xi_m(x)\, \xi_m(y) \rangle \\
&= \sum_{p,q,u,v=1}^{m} \alpha_{up}\alpha_{vq}\varphi_u(x)\, \varphi_v(y) \int_0^1 \varphi_p(z)\, \varphi_q(z)\, a_{gg}(z)\, \mathrm{d}z \\
&\quad - \sum_{i=0}^{n-1} \sum_{p,q,u,v,w=1}^{m} \alpha_{vp}\alpha_{wq} x_{u0}^{(m)} \varphi_v(x)\, \varphi_w(y) \\
&\quad \times \int_0^1 \{\varphi_p(z)\, \varphi_q^{(i)}(z)\, \varphi_u^{(i)}(z)\, a_{gi}(z) + \varphi_q(z)\, \varphi_p^{(i)}(z)\, \varphi_u^{(i)}(z)\, a_{ig}(z)\}\, \mathrm{d}z \\
&\quad + \sum_{i,j=0}^{n-1} \sum_{p,q,u,v,w,s=1}^{m} \alpha_{wp}\alpha_{sq} x_{u0}^{(m)} x_{v0}^{(m)} \varphi_w(x)\, \varphi_s(y) \int_0^1 \varphi_p^{(i)}(z)\, \varphi_q^{(j)}(z)\, \varphi_u^{(i)}(z)\, \varphi_v^{(j)}(z)\, a_{ij}(z)\, \mathrm{d}z
\end{aligned} \tag{5.25}$$

if this expression is represented in dependence on the original terms α_{pq}, $x_{p0}^{(m)}$, and $\varphi_p(x)$.

Now, we choose the eigenfunctions $w_i(x)$ as to $\langle L \rangle$ and the boundary conditions from (5.1) as coordinate functions. We obtain

$$\alpha_{ij} = \frac{1}{\lambda_i}\, \delta_{ij}, \qquad i, j = 1, 2, \ldots, m,$$

where λ_i, $i = 1, 2, \ldots, m$, denote the eigenvalues with respect to $w_i(x)$ and with this

$$\begin{aligned}
&\langle \xi_m(x)\, \xi_m(y) \rangle \\
&= \sum_{p,q=1}^{m} \frac{1}{\lambda_p\lambda_q}\, \varphi_p(x)\, \varphi_q(y) \int_0^1 \varphi_p(z)\, \varphi_q(z)\, a_{gg}(z)\, \mathrm{d}z - \sum_{i=0}^{n-1} \sum_{p,q,u=1}^{m} \frac{1}{\lambda_p\lambda_q}\, x_{u0}^{(m)} \varphi_p(x)\, \varphi_q(y) \\
&\quad \times \int_0^1 \{\varphi_p(z)\, \varphi_q^{(i)}(z)\, \varphi_u^{(i)}(z)\, a_{gi}(z) + \varphi_q(z)\, \varphi_p^{(i)}(z)\, \varphi_u^{(i)}(z)\, a_{ig}(z)\}\, \mathrm{d}z \\
&\quad + \sum_{i,j=0}^{n-1} \sum_{p,q,u,v=1}^{m} \frac{1}{\lambda_p\lambda_q}\, x_{u0}^{(m)} x_{v0}^{(m)} \varphi_p(x)\, \varphi_q(y) \int_0^1 \varphi_p^{(i)}(z)\, \varphi_q^{(j)}(z)\, \varphi_u^{(i)}(z)\, \varphi_v^{(j)}(z)\, a_{ij}(z)\, \mathrm{d}z.
\end{aligned}$$

We define

$$G_{m,z}^{(i)}(x, z) \doteq \sum_{p=1}^{m} \frac{1}{\lambda_p}\, w_p(x)\, w_p^{(i)}(z) \quad \text{for} \quad i = 0, 1, \ldots, n-1;$$

$$u_{0,m}(z) \doteq \sum_{p=1}^{m} x_{p0}^{(m)} w_p(z)$$

and get

$$\begin{aligned}
&\langle \xi_m(x)\, \xi_m(y) \rangle \\
&= \int_0^1 G_m(x, z)\, G_m(y, z)\, a_{gg}(z)\, \mathrm{d}z \\
&\quad - \sum_{i=0}^{n-1} \int_0^1 \{G_m(x, z)\, G_{m,z}^{(i)}(y, z)\, a_{gi}(z) + G_m(y, z)\, G_{m,z}^{(i)}(x, z)\, a_{ig}(z)\}\, u_{0,m}^{(i)}(z)\, \mathrm{d}z \\
&\quad + \sum_{i,j=0}^{n-1} \int_0^1 G_{m,z}^{(i)}(x, z)\, G_{m,z}^{(j)}(y, z)\, u_{0,m}^{(i)}(z)\, u_{0,m}^{(j)}(z)\, a_{ij}(z)\, \mathrm{d}z.
\end{aligned}$$

Using the relations

$$\lim_{m\to\infty} G^{(i)}_{m,z}(x, z) = G^{(i)}_z(x, z) \quad \text{for} \quad i = 0, 1, \ldots, n-1,$$

$$\lim_{m\to\infty} u^{(i)}_{0,m}(z) = u^{(i)}_0(z) \quad \text{for} \quad i = 0, 1, \ldots, n-1$$

uniformly as to z we have

$$\begin{aligned}
&\lim_{m\to\infty} \langle \xi_m(x)\, \xi_m(y)\rangle \\
&= \int_0^1 G(x, z)\, G(y, z)\, a_{gg}(z)\, \mathrm{d}z \\
&\quad - \sum_{i=0}^{n-1} \int_0^1 \{G(x, z)\, G^{(i)}_z(y, z)\, a_{gi}(z) + G(y, z)\, G^{(i)}_z(x, z)\, a_{ig}(z)\}\, u^{(i)}_0(z)\, \mathrm{d}z \\
&\quad + \sum_{i,j=0}^{n-1} \int_0^1 G^{(i)}_z(x, z)\, G^{(j)}_z(y, z)\, u^{(i)}_0(z)\, u^{(j)}_0(z)\, a_{ij}(z)\, \mathrm{d}z
\end{aligned}$$

and this result coincides with the correlation function written in Theorem 5.2.

Example 5.1. We consider the simple random boundary value problem

$$-u'' = \bar{g}_\varepsilon(x, \omega); \qquad u(0) = u(1) = 0$$

where $\bar{g}_\varepsilon(x, \omega)$ denotes a wide-sense stationary weakly correlated process. Then it is

$$a_{gg}(z) = a_{gg} = \text{const}.$$

The Green's function $G(x, y)$ as to the averaged operator

$$\langle L\rangle\, u = -u''$$

and the boundary conditions

$$u(0) = u(1) = 0$$

can be given by

$$G(x, z) = \begin{cases} (1 - z)\, x & \text{for} \quad 0 \leqq x < z \leqq 1, \\ (1 - x)\, z & \text{for} \quad 0 \leqq z < x \leqq 1. \end{cases}$$

It follows from the previous considerations that $\frac{1}{\sqrt{\varepsilon}}\, u(x, \omega)$ converges to $\xi(x, \omega)$ in distribution as to $\varepsilon \downarrow 0$ where $\xi(x, \omega)$ is Gaussian distributed with

$$\langle \xi(x)\rangle = 0,$$

$$\langle \xi(x)\, \xi(y)\rangle = a_{gg} \int_0^1 G(x, z)\, G(y, z)\, \mathrm{d}z$$

$$= a_{gg} \begin{cases} \frac{1}{6}\, x^3(y - 1) + \frac{1}{6}\, xy(y^2 - 3y + 2) & \text{for} \quad 0 \leqq x \leqq y \leqq 1, \\ \frac{1}{6}\, y^3(x - 1) + \frac{1}{6}\, xy(x^2 - 3x + 2) & \text{for} \quad 0 \leqq y \leqq x \leqq 1. \end{cases}$$

The variance is given by

$$\langle \xi^2(x) \rangle = \frac{1}{3} a_{gg} x^2 (1 - x)^2 .$$

Considering this simple example the calculation of the variance is produced with the help of the Ritz-method. We use the functions

$$\varphi_p(x) = \sqrt{2p + 1} \int_0^x P_p(2t - 1) \, dt$$

where $P_p(t)$, $p = 1, 2, \ldots$, denote the Legendre's polynomials. By a simple calculation we see that

$$\alpha_{pq} = (\langle L \rangle \varphi_p, \varphi_q) = \sqrt{(2p + 1)(2q + 1)} \int_0^1 P_p(2t - 1) P_q(2t - 1) \, dt = \delta_{pq} .$$

Applying (5.25) we have

$$\langle \xi_m^2(x) \rangle = a_{gg} \sum_{p,q=1}^{m} (\varphi_p, \varphi_q) \varphi_p(x) \varphi_q(y)$$

and for $m = 3$

$$\langle \xi_3^2(x) \rangle = a_{gg}[1.944444x^8 - 7.777778x^7 + 12.420634x^6 - 10.039682x^5 + 4.603174x^4 - 1.547619x^3 + 0.396820x^2] .$$

Table 5.1 shows the comparison between the exact values $\langle \xi^2(x) \rangle$ and the approximative values $\langle \xi_m^2(x) \rangle$ for $m = 3$.

Table 5.1

x	$\frac{1}{a_{gg}} \langle \xi_3^2(x) \rangle$	$\frac{1}{a_{gg}} \langle \xi^2(x) \rangle$
0	0	0
0.1	0.0028	0.0027
0.2	0.0083	0.0085
0.3	0.0143	0.0147
0.4	0.0189	0.0192
0.5	0.0206	0.0208
0.6	0.0189	0.0192
0.7	0.0143	0.0147
0.8	0.0083	0.0085
0.9	0.0028	0.0027
1	0	0

Example 5.2. Let $\bar{g}_\varepsilon(x, \omega)$ be a wide-sense stationary, weakly correlated process. The boundary value problem

$$-u'' + bu = \bar{g}_\varepsilon(x, \omega), \qquad u(0) = u(1) = 0$$

with $b =$ const possesses the averaged problem

$$-w'' + bw = 0, \qquad w(0) = w(1) = 0 .$$

This problem only has the trivial solution in the case of

$$b \neq -(k\pi)^2, \qquad k = 1, 2, \ldots$$

Assuming $b > 0$ the Green's function $G_b(x, y)$ can be written as

$$G_b(x, z) = \begin{cases} \dfrac{\operatorname{sh}(\beta x) \operatorname{sh}\big(\beta(1-z)\big)}{\beta \operatorname{sh}(\beta)} & \text{for} \quad 0 \leqq x < z \leqq 1, \\ \dfrac{\operatorname{sh}(\beta z) \operatorname{sh}\big(\beta(1-x)\big)}{\beta \operatorname{sh}(\beta)} & \text{for} \quad 0 \leqq z < x \leqq 1 \end{cases} \tag{5.26}$$

where $\beta \doteq \sqrt{b}$ and for $b < 0$ with $\bar{\beta} \doteq \sqrt{-b}$ we get

$$G_b(x, z) = \begin{cases} \dfrac{\sin(\bar{\beta} x) \sin\big(\bar{\beta}(1-z)\big)}{\bar{\beta} \sin(\bar{\beta})} & \text{for} \quad 0 \leqq x < z \leqq 1, \\ \dfrac{\sin(\bar{\beta} z) \sin\big(\bar{\beta}(1-x)\big)}{\bar{\beta} \sin(\bar{\beta})} & \text{for} \quad 0 \leqq z < x \leqq 1. \end{cases} \tag{5.26'}$$

Hence the correlation function for the limit process $\xi_b(x, \omega)$ from Theorem 5.2 has the form

$$\langle \xi_b(x)\, \xi_b(y) \rangle = a_{gg} \int_0^1 G_b(x, z)\, G_b(y, z)\, \mathrm{d}z$$

and can easy be determined in a straightforward manner. The variance is given by

$$\langle \xi_b^2(x) \rangle = a_{gg} \int_0^1 G_b(x, z)^2\, \mathrm{d}z = a_{gg} \left[\int_0^x G_b(x, z)^2\, \mathrm{d}z + \int_x^1 G_b(x, z)^2\, \mathrm{d}z \right]$$

and then by

$$\langle \xi_b^2(x) \rangle = \begin{cases} \dfrac{a_{gg}}{\beta^2 \operatorname{sh}^2(\beta)} \big(h_\beta(x) + h_\beta(1-x)\big) & \text{for} \quad b > 0, \\ \dfrac{a_{gg}}{\bar{\beta}^2 \sin^2(\bar{\beta})} \big(\bar{h}_{\bar{\beta}}(x) + \bar{h}_{\bar{\beta}}(1-x)\big) & \text{for} \quad b < 0 \quad \text{and} \quad b \neq -(k\pi)^2, \\ & \qquad k = 1, 2, \ldots, \end{cases}$$

where $h_\beta(x)$, $\bar{h}_\beta(x)$ are defined by

$$h_\beta(x) \doteq \operatorname{sh}^2\big(\beta(1-x)\big) \left[\frac{1}{4\beta} \operatorname{sh}(2\beta x) - \frac{1}{2} x \right],$$

$$\bar{h}_\beta(x) \doteq -\sin^2\big(\beta(1-x)\big) \left[\frac{1}{4\beta} \sin(2\beta x) - \frac{1}{2} x \right].$$

In particular, it is

$$\lim_{b \to 0} \langle \xi_b^2(x) \rangle = \frac{1}{3} a_{gg} x^2 (1-x)^2,$$

$$\lim_{b \to -(k\pi)^2} \langle \xi_b^2(x) \rangle = \infty \quad \text{for} \quad 0 < x < 1.$$

Fig. 5.1 shows the variance of the limit process $\xi_b(x, \omega)$ for some values of the parameter b. The relation

$$\langle \xi_b^2(x) \rangle = \langle \xi_b^2(1 - x) \rangle$$

is clear. The variance $\langle \xi_b^2(x) \rangle|_{x=0,5}$ is plotted in Fig. 5.2 in dependence on $-b$.

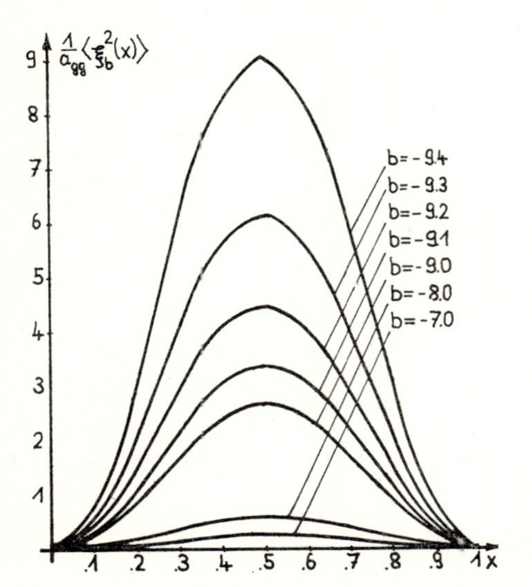

Fig. 5.1. Variance of the limit process $\xi_b(x, \omega)$ for different values b in the case of random $\bar{g}_\varepsilon$

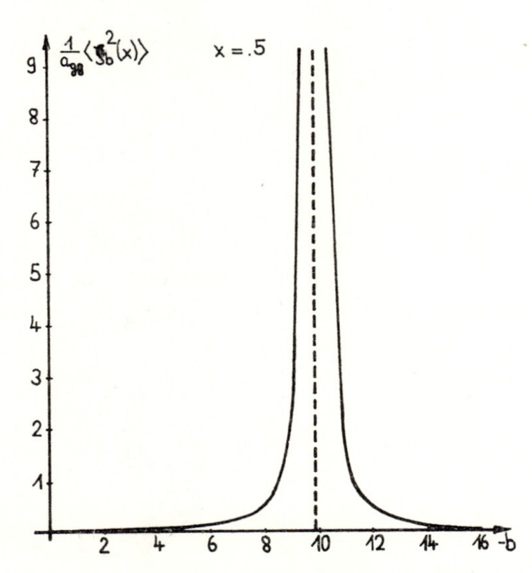

Fig. 5.2. Variance of the limit process $\xi_b(x, \omega)$ in dependence on $-b$ for $x = 0.5$ in the case of random $\bar{g}_\varepsilon$

Example 5.3. Now we consider an example in which a coefficient of the differential operator is random,

$$-u'' + \left(b + \bar{f}_{0\varepsilon}(x, \omega)\right) u = c + \bar{g}_\varepsilon(x, \omega); \qquad u(0) = u(1) = 0$$

where $b, c = \text{const}$. In order to reduce this problem the processes $\bar{f}_{0\varepsilon}(x, \omega)$ and $\bar{g}_\varepsilon(x, \omega)$ are assumed to be independent.

The Green's function belonging to $\langle L \rangle$ and the given boundary conditions is written in (5.26), (5.26'). Hence the correlation function of the limit process $\xi_b(x, \omega)$ can be determined by Theorem 5.2,

$$\langle \xi_b(x)\, \xi_b(y) \rangle = a_{gg} \int_0^1 G_b(x, z)\, G_b(y, z)\, \mathrm{d}z + a_{00} \int_0^1 G_b(x, z)\, G_b(y, z)\, u_0^2(z)\, \mathrm{d}z$$

where the processes $\bar{g}_\varepsilon(x, \omega)$, $\bar{f}_{0\varepsilon}(x, \omega)$ are supposed to be wide-sense stationary. The function $u_0(x)$ is given by

$$u_0(x) = c \int_0^1 G_b(x, z)\, \mathrm{d}z = \frac{c}{\bar{\beta}^2 \sin(\bar{\beta})} \left(\sin(\bar{\beta}x) - \sin(\bar{\beta}) + \sin\left(\bar{\beta}(1 - x)\right)\right)$$

for $0 > b = -\bar{\beta}^2$ (see Theorem 5.1). In order to get some results as to the variance we deal with

$$V(x) = \int_0^1 \big(G_b(x, z)\, u_0(z)\big)^2\, dz.$$

Thus,

$$V(x) = \int_0^x \big(G_b(x, z)\, u_0(z)\big)^2\, dz + \int_x^1 \big(G_b(x, z)\, u_0(z)\big)^2\, dz$$
$$= \left(\frac{c \sin\big(\bar\beta(1-x)\big)}{\bar\beta^3 \sin^2(\bar\beta)}\right)^2 V_1(x; \bar\beta) + \left(\frac{c \sin(\bar\beta x)}{\bar\beta^3 \sin^2(\bar\beta)}\right)^2 V_1(1-x; \bar\beta)$$

where $V_1(x; \bar\beta)$ is defined by

$$V_1(x; \bar\beta) = \int_0^x \left[\sin(\bar\beta z)\left\{\sin(\bar\beta z) - \sin(\bar\beta) + \sin\big(\bar\beta(1-z)\big)\right\}\right]^2 dz.$$

We put $a_{gg} = a_{00} = a$, $c = 1$. Fig. 5.3 shows the variance $V(x)$ of the limit process $\xi_b(x, \omega)$ in the cases of $\bar f_{0\varepsilon}(x, \omega) \neq 0$, $\bar g_\varepsilon(x, \omega) = 0$ and $\bar f_{0\varepsilon}(x, \omega) = 0$, $\bar g_\varepsilon(x, \omega) \neq 0$ and in the case of $\bar f_{0\varepsilon}(x, \omega) \neq 0$, $\bar g_\varepsilon(x, \omega) \neq 0$ but independent. In this figure it is chosen $b = -9$

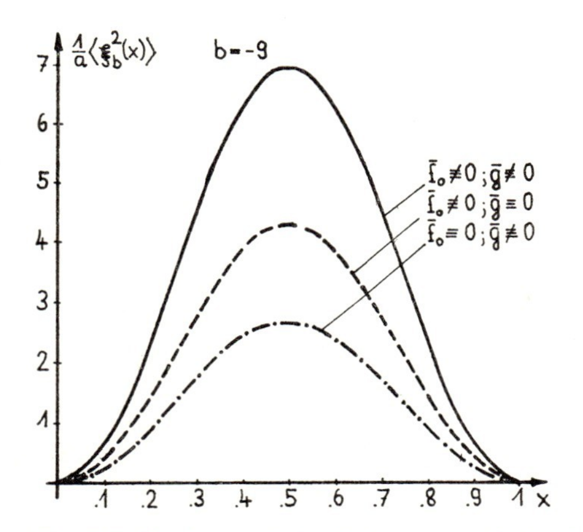

Fig. 5.3. Variance of the limit process $\xi_b(x, \omega)$ in the case of random $\bar g_\varepsilon$, $\bar f_{0\varepsilon}$ for $b = -9$

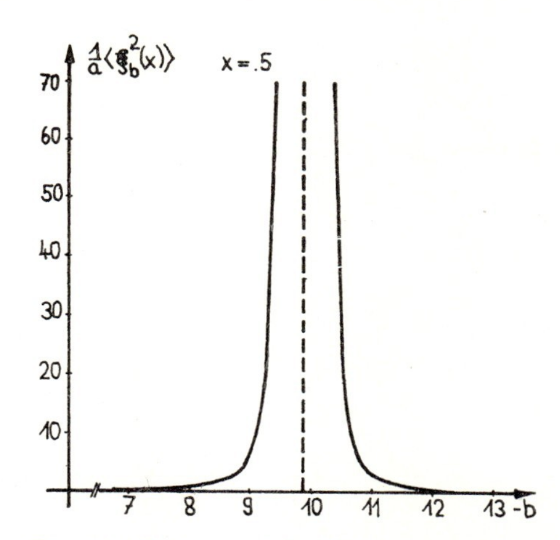

Fig. 5.4. Variance of the limit process $\xi_b(x, \omega)$ in the case of random $\bar g_\varepsilon$, $\bar f_{0\varepsilon}$ for $x = 0.5$

and the variances are plotted in dependence on x. Fig. 5.4 illustrates the variance of $\xi_b(x, \omega)$ in dependence on b for $x = 0.5$ in the case of independent processes $\bar f_{0\varepsilon}(x, \omega)$, $\bar g_\varepsilon(x, \omega)$. In comparison to Fig. 5.2 a faster increasing to infinity can be noted.

Investigate approximations of statistical characteristics of higher order than first one as to ε. The basis of our considerations is a system of linear equations of the form (5.20),

$$\big(A_0 + B(\omega)\big)\, X = b_0 + c(\omega). \tag{5.27}$$

The definitions of A_0, b_0, $B(\omega)$, and $c(\omega)$ are given with (5.20). It is

$$X(\omega) = X_0 - X_1(\omega) + X_2(\omega) - X_3(\omega) \pm \cdots$$

where X_r, $r = 0, 1, 2, 3$, can be written as

$$X_0 = A_0^{-1} b_0 ,$$

$$X_r(\omega) = (A_0^{-1}B)^r X_0 - (A_0^{-1}B)^{r-1} A_0^{-1} c \quad \text{for} \quad r = 1, 2, 3 .$$

Furthermore, we obtain

$$X_{0,i} = \sum_{p=1}^{m} \alpha_{ip} b_p ,$$

$$X_{1,i} = \sum_{p,q=1}^{m} \alpha_{ip} X_{0,q} b_{pq}(\omega) - \sum_{p=1}^{m} \alpha_{ip} c_p(\omega) ,$$

$$\begin{aligned} X_{2,i} &= \sum_{p,q=1}^{m} \alpha_{ip} b_{pq}(\omega)\, X_{1,q}(x) \\ &= \sum_{p,q,u,v=1}^{m} \alpha_{ip}\alpha_{qu} X_{0,v} b_{pq}(\omega)\, b_{uv}(\omega) - \sum_{p,q,u=1}^{m} \alpha_{ip}\alpha_{qu} b_{pq}(\omega)\, c_u(\omega) , \end{aligned} \tag{5.28}$$

$$\begin{aligned} X_{3,i} &= \sum_{p,q=1}^{m} \alpha_{ip} b_{pq}(\omega)\, X_{2,q}(\omega) \\ &= \sum_{p,q,u,v,k,h=1}^{m} \alpha_{ip}\alpha_{qu}\alpha_{vk} X_{0,h} b_{pq}(\omega)\, b_{uv}(\omega)\, b_{kh}(\omega) - \sum_{p,q,u,v,k=1}^{m} \alpha_{ip}\alpha_{qu}\alpha_{vk} b_{pq}(\omega)\, b_{uv}(\omega)\, c_k(\omega) \end{aligned}$$

where $b_{pq}(\omega)$ and $c_p(\omega)$ are given by

$$\begin{aligned} b_{pq}(\omega) &= \big(L_1(\omega)\, \varphi_p, \varphi_q\big) = \sum_{a=0}^{n-1} (\bar{f}_{a\varepsilon}, \varphi_p^{(a)} \varphi_q^{(a)}) \\ &= \sum_{a=0}^{n-1} \int_0^1 \varphi_p^{(a)}(x)\, \varphi_q^{(a)}(x)\, \bar{f}_{a\varepsilon}(x, \omega)\, \mathrm{d}x , \end{aligned} \tag{5.29}$$

$$c_p(\omega) = (\bar{g}_\varepsilon, \varphi_p) = \int_0^1 \varphi_p(x)\, \bar{g}_\varepsilon(x, \omega)\, \mathrm{d}x .$$

We put

$$\begin{aligned} r_{pq\varepsilon}(\omega) &\doteq \sum_{a=0}^{n-1} r_{pqa\varepsilon}(\omega) , \\ r_{pqa\varepsilon}(\omega) &\doteq \int_0^1 \varphi_p^{(a)}(x)\, \varphi_q^{(a)}(x)\, \bar{f}_{a\varepsilon}(x, \omega)\, \mathrm{d}x . \end{aligned} \tag{5.30}$$

The representation (1.142) as to the components of $X(\omega)$ follows with the help of (5.28) and using (5.30) we can apply all results of Section 1.7 to the solution of the considered system (5.27).

Deal with the mean of the solution $X(\omega)$. Using (1.169) it follows

$$\langle X_i - X_{i,0} \rangle = \left[\sum_{p,q,u,v=1}^{m} \alpha_{ip}\alpha_{qu} X_{0,v}\; {}^2\hat{A}_1'(pq, uv) - \sum_{p,q,u=1}^{m} \alpha_{ip}\alpha_{qu}\; {}^2\hat{A}_1'(pq, u) \right] \varepsilon + o(\varepsilon)$$

where the expressions of ${}^2\hat{A}_1$ are given by

$$ {}^2\hat{A}_1'(pq, uv) = \sum_{a,b=0}^{n-1} \int_0^1 \varphi_p^{(a)}\varphi_q^{(a)}\varphi_u^{(b)}\varphi_v^{(b)} a_{ab} \, \mathrm{d}x , $$

$$ {}^2\hat{A}_1'(pq, u) = \sum_{a=0}^{n-1} \int_0^1 \varphi_p^{(a)}\varphi_q^{(a)}\varphi_u a_{ag} \, \mathrm{d}x . $$

Now, the eigenfunctions $w_i(x)$ as to $\langle L \rangle$ and the boundary conditions from (5.1) are chosen as coordinate functions. Using

$$ \alpha_{ij} = \frac{1}{\lambda_j} \delta_{ij}, \qquad i, j = 1, 2, \ldots, m , $$

where λ_i, $i = 1, 2, \ldots, m$, denote the eigenvalues we can show that

$$ \langle X_i \rangle = \frac{1}{\lambda_i} \left[b_i + \left\{ \sum_{p,q=1}^{m} \frac{1}{\lambda_p \lambda_q} b_q \, {}^2\hat{A}_1'(ip, pq) - \sum_{p=1}^{m} \frac{1}{\lambda_p} \, {}^2\hat{A}_1'(ip, p) \right\} \varepsilon \right] + o(\varepsilon) . $$

Assuming $\bar{g}_\varepsilon(x, \omega)$ and $\left(\bar{f}_{0\varepsilon}(x, \omega), \ldots, \bar{f}_{n-1\varepsilon}(x, \omega)\right)$ are independent then

$$ {}^2\hat{A}_1'(pq, u) = 0 $$

and $\langle X_i \rangle$ can be reduced to the first summand.

Consider Example 5.3. The adequate eigenfunctions and eigenvalues are

$$ w_i(x) = \sqrt{2} \sin (i\pi x) ; \qquad \lambda_i = (i\pi)^2 + b . \tag{5.31} $$

Furthermore, we have

$$ b_i = (\langle g \rangle, w_i) = \frac{\sqrt{2}c}{i\pi} \left(1 - (-1)^i\right) = \begin{cases} \dfrac{2\sqrt{2}c}{i\pi} & \text{for} \quad i = 1, 3, 5, \ldots, \\ 0 & \text{for} \quad i = 2, 4, 6, \ldots \end{cases} $$

It is

$$ \begin{aligned} {}^2\hat{A}_1'(ip, pq) &= a_{00} \int_0^1 w_i w_p^2 w_q \, \mathrm{d}x \\ &= 4a_{00} \int_0^1 \sin (i\pi x) \sin^2 (p\pi x) \sin (q\pi x) \, \mathrm{d}x \\ &= a_{00} \int_0^1 \left[\cos \left((i - q) \pi x\right) - \frac{1}{2} \cos \left((2p - i + q) \pi x\right) - \frac{1}{2} \cos \left((2p + i - q) \pi x\right) \right. \\ &\quad \left. + \frac{1}{2} \cos \left((2p - i - q) \pi x\right) \right] \mathrm{d}x \end{aligned} $$

if the relation

$$ \int_0^1 \cos (k\pi x) \, \mathrm{d}x = 0 \quad \text{for} \quad k \neq 0 $$

is taken into consideration. Hence we obtain

$$ {}^2\hat{A}_1'(ip, pq) = a_{00} \left[\delta_{0\,i-q} - \frac{1}{2} \delta_{0\,2p-i+q} - \frac{1}{2} \delta_{0\,2p+i-q} + \frac{1}{2} \delta_{0\,2p-i-q} \right] $$

and assuming m, i odd then

$$\sum_{p,q=1}^{m} \frac{1}{\lambda_p\lambda_q} b_q\, {}^2\hat{A}_1'(ip, pq)$$

$$= a_{00}\left[\frac{1}{\lambda_i} b_i \sum_{\substack{p=1\\p\neq i}}^{m} \frac{1}{\lambda_p} + \frac{3}{2}\frac{1}{\lambda_i^2} b_i - \frac{1}{2}\sum_{p=1}^{\frac{i-1}{2}} \frac{1}{\lambda_p\lambda_{i-2p}} b_{i-2p} - \frac{1}{2}\sum_{p=1}^{\frac{m-i}{2}} \frac{1}{\lambda_p\lambda_{i+2p}} b_{i+2p}\right.$$

$$\left. + \frac{1}{2}\sum_{p=\frac{1}{2}(i+1)}^{\frac{1}{2}(i+m)} \frac{1}{\lambda_p\lambda_{2p-i}} b_{2p-i}\right],$$

and

$$\sum_{p,q=1}^{m} \frac{1}{\lambda_p\lambda_q} b_q\, {}^2\hat{A}_1'(ip, pq) = 0 \quad \text{for} \quad i \text{ even}$$

since $b_q = 0$ for q even.

Setting $b = -9$, $c = 1$ for $m = 3$ we have

$$\langle X_1\rangle = 1.03532 + 2.78973\varepsilon a_{00} + o(\varepsilon),$$

$$\langle X_2\rangle = o(\varepsilon),$$

$$\langle X_3\rangle = 0.00376 - 0.00719\varepsilon a_{00} + o(\varepsilon)$$

and for $m = 5$

$$\langle X_1\rangle = 1.03532 + 2.80272\varepsilon a_{00} + o(\varepsilon),$$

$$\langle X_i\rangle = o(\varepsilon) \quad \text{for} \quad i = 2, 4,$$

$$\langle X_3\rangle = 0.00376 - 0.00719\varepsilon a_{00} + o(\varepsilon),$$

$$\langle X_5\rangle = 0.00076 - 0.00005\varepsilon a_{00} + o(\varepsilon).$$

Using (5.18) it is

$$u_0^{(m)}(x) = \sum_{p=1}^{m} X_{0,p} w_p(x) = \begin{cases} 1.03532 w_1(x) + 0.00376 w_3(x) & \text{for} \quad m = 3, \\ 1.03532 w_1(x) + 0.00376 w_3(x) & \\ \qquad + 0.00076 w_5(x) & \text{for} \quad m = 5 \end{cases}$$

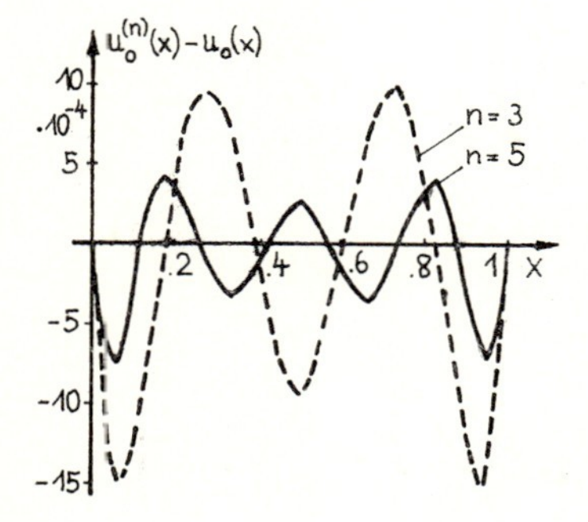

Fig. 5.5. Difference between the n-th Ritz-approximation $u_0^{(n)}(x)$ and $u_0(x)$

and the exact solution $u_0(x)$ is given by

$$u_0(x) = \frac{c}{\bar{\beta}^2 \sin(\bar{\beta})} \left(\sin(\bar{\beta}x) - \sin(\bar{\beta}) + \sin\left(\bar{\beta}(1-x)\right)\right)$$

where $\bar{\beta}^2 = -b$. Fig. 5.5 shows the difference between the m-th Ritz-approximation $u_0^{(m)}(x)$ of the solution of the averaged problem as to Example 5.3 and the exact solution of this averaged problem. There is a very good coincidence between $u_0(x)$ and $u_0^{(m)}(x)$ for $m = 3$ or $m = 5$.

Furthermore, the averaging problem can be solved approximately by

$$\langle u_m(x)\rangle - u_0^{(m)}(x) = \sum_{p=1}^{m} (\langle X_p\rangle - X_{p,0})\, w_p(x)$$

$$= \begin{cases} \left(2.78973 w_1(x) - 0.00719 w_3(x)\right) \varepsilon a_{00} + o(\varepsilon) & \text{for} \quad m = 3, \\ \left(2.80272 w_1(x) - 0.00719 w_3(x) - 0.00005 w_5(x)\right) \varepsilon a_{00} + o(\varepsilon) & \text{for} \quad m = 5. \end{cases}$$

The difference

$$\frac{1}{\varepsilon a_{00}} \left(\langle u_m(x)\rangle - u_0^{(m)}(x)\right)$$

is plotted in Fig. 5.6 where terms of the order $O(\varepsilon)$ are neglected. These results were determined with $b = -9$, $c = 1$.

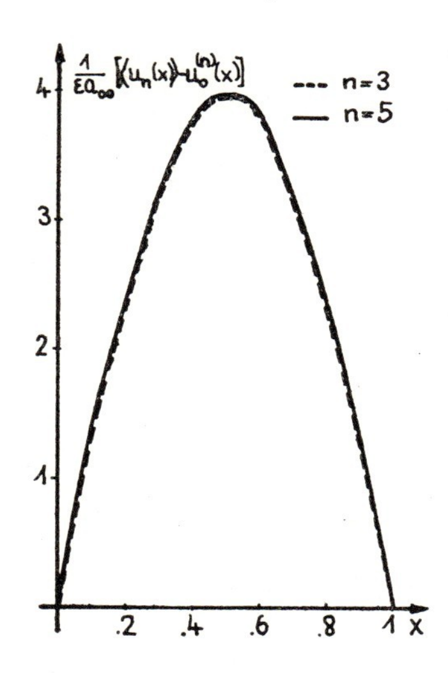

Fig. 5.6. Approximative solution of the averaging problem of a boundary value problem

We deal with second moments. Relation (1.174) leads to

$$\langle (X_i - \langle X_i\rangle)(X_j - \langle X_j\rangle)\rangle = C_{ij}^1 + C_{ij}^2 + o(\varepsilon^2)$$

where C_{ij}^1 and C_{ij}^2 are given by

$$C_{ij}^1 = \left[\sum_{p,q,u,v=1}^{m} \alpha_{ip}\alpha_{ju} X_{0,q} X_{0,v} B_{pq,uv} - \sum_{p,q,u=1}^{m} X_{0,q}\alpha_{ip}\alpha_{ju}(B_{pq,u} + B_{p,uq}) + \sum_{p,q=1}^{m} \alpha_{ip}\alpha_{jq} B_{p,q}\right] \varepsilon$$

with

$$B_{pq,uv} = {}^2\hat{A}_1'(pq, uv),$$

$$B_{pq,u} = {}^2\hat{A}_1'(pq, u),$$

$$B_{p,q} = {}^2\hat{A}_1'(p, q) = \int_0^1 \varphi_p \varphi_q a_{gg} \, \mathrm{d}x$$

and

$$C_{ij}^2 = [C_{ij}^{2,1} + C_{ij}^{2,2} + C_{ji}^{2,2} + C_{ij}^{2,3} + C_{ji}^{2,3} + C_{ij}^{2,4}] \, \varepsilon^2$$

where

$$C_{ij}^{2,1} \doteq \sum_{p,q,u,v=1}^{m} \alpha_{ip}\alpha_{ju} X_{0,q} X_{0,v} \, {}^2\hat{A}_2'(pq, uv)$$
$$- \sum_{p,q,u=1}^{m} \alpha_{ip}\alpha_{ju} X_{0,q} \big({}^2\hat{A}_2'(pq, u) + {}^2\hat{A}_2'(p, uq)\big) + \sum_{p,q=1}^{m} \alpha_{ip}\alpha_{jq} \, {}^2\hat{A}_2'(p, q),$$

$$C_{ij}^{2,2} \doteq \sum_{p,q,u,v,k,h=1}^{m} \alpha_{ip}\alpha_{jk}\alpha_{qu} X_{0,v} X_{0,h} \, {}^3\hat{A}_2'(pq, uv, kh)$$
$$- \sum_{p,q,u,v,k=1}^{m} \alpha_{ip}\alpha_{jk}\alpha_{qu} X_{0,v} \big({}^3\hat{A}_2'(pq, uv, k) + {}^3\hat{A}_2'(pq, kv, u)\big)$$
$$+ \sum_{p,q,u,v=1}^{m} \alpha_{ip}\alpha_{qu}\alpha_{jv} \, {}^3\hat{A}_2'(pq, u, v),$$

$$C_{ij}^{2,3} \doteq \sum_{p,q,u,v,k,h,l,s=1}^{m} \alpha_{ip}\alpha_{jl}\alpha_{qu}\alpha_{vk} X_{0,h} X_{0,s} \, [B_{pq,uv} B_{kh,ls} + B_{pq,kh} B_{uv,ls} + B_{pq,ls} B_{uv,kh}]$$
$$- \sum_{p,q,u,v,k,h,l=1}^{m} \alpha_{ip}\alpha_{jl}\alpha_{qu}\alpha_{vk} X_{0,h} \, [B_{pq,uv}(B_{kh,l} + B_{k,lh}) + B_{pq,kh} B_{uv,l}$$
$$+ B_{pq,lh} B_{uv,k} + B_{pq,k} B_{uv,lh} + B_{pq,l} B_{uv,kh}]$$
$$+ \sum_{p,q,u,v,k,h=1}^{m} \alpha_{ip}\alpha_{jh}\alpha_{qu}\alpha_{vk} \, [B_{pq,uv} B_{k,h} + B_{pq,k} B_{uv,h} + B_{pq,h} B_{uv,k}], \tag{5.32}$$

$$C_{ij}^{2,4} \doteq \sum_{p,q,u,v,k,h,l,s=1}^{m} \alpha_{ip}\alpha_{jk}\alpha_{qu}\alpha_{hl} X_{0,v} X_{0,s} \, [B_{pq,kh} B_{uv,ls} + B_{pq,ls} B_{uv,kh}]$$
$$- \sum_{p,q,u,v,k,h,l=1}^{m} \alpha_{ip}\alpha_{jk}\alpha_{qu}\alpha_{hl} X_{0,v} [B_{pq,kh}(B_{uv,l} + B_{u,lv}) + B_{pq,l} B_{uv,kh} + B_{pq,lv} B_{u,kh}]$$
$$+ \sum_{p,q,u,v,k,h=1}^{m} \alpha_{ip}\alpha_{jk}\alpha_{qu}\alpha_{hv} [B_{pq,kh} B_{u,v} + B_{pq,v} B_{u,kh}]. \tag{5.33}$$

${}^2\hat{A}_2'$ and ${}^3\hat{A}_2'$ are given by

$${}^2\hat{A}_2'(pq, uv) = \sum_{a,b=0}^{n-1} {}^2A_2(\varphi_p^{(a)} \varphi_q^{(a)}, \varphi_u^{(b)} \varphi_v^{(b)})_{\bar{f}_a \bar{f}_b},$$

$${}^2\hat{A}_2'(pq, u) = \sum_{a=0}^{n-1} {}^2A_2(\varphi_p^{(a)} \varphi_q^{(a)}, \varphi_u)_{\bar{f}_a \bar{g}},$$

$${}^2\hat{A}_2'(p, q) = {}^2A_2(\varphi_p, \varphi_q)_{\bar{g}\bar{g}},$$

$$^3\hat{A}_2'(pq, uv, kh) = \sum_{a,b,c=0}^{n-1} \int_0^1 \varphi_p^{(a)}\varphi_q^{(a)}\varphi_u^{(b)}\varphi_v^{(b)}\varphi_k^{(c)}\varphi_h^{(c)} a_{abc}\, dx\,,$$

$$^3\hat{A}_2'(pq, uv, k) = \sum_{a,b=0}^{n-1} \int_0^1 \varphi_p^{(a)}\varphi_q^{(a)}\varphi_u^{(b)}\varphi_v^{(b)}\varphi_k a_{abg}\, dx\,,$$

$$^3\hat{A}_2'(pq, u, v) = \sum_{a=0}^{n-1} \int_0^1 \varphi_p^{(a)}\varphi_q^{(a)}\varphi_u\varphi_v a_{agg}\, dx\,,$$

$$^3\hat{A}_2'(p, q, u) = \int_0^1 \varphi_p\varphi_q\varphi_u a_{ggg}\, dx\,.$$

Consider Example 5.3. We choose the eigenfunctions (5.31) as coordinate functions. The processes $\bar{f}_{0\varepsilon}(x, \omega)$, $\bar{g}_\varepsilon(x, \omega)$ are assumed to be symmetrically distributed so that $^3\hat{A}_2(.,.,.) = 0$ and then $C_{ij}^{2,2} = 0$. Because of the boundary conditions we have $w_p(0) = w_p(1) = 0$ and using Theorem 1.9 we have $^2\hat{A}_2(.,.) = 0$.

We want to evaluate the correlation function of $u_m(x, \omega)$,

$$C_m(x, y) \doteq \left\langle \left(u_m(x) - \langle u_m(x)\rangle\right) \left(u_m(y) - \langle u_m(y)\rangle\right)\right\rangle$$
$$= \sum_{i,j=1}^{m} \left\langle (X_i - \langle X_i\rangle)\,(X_j - \langle X_j\rangle)\right\rangle w_i(x)\, w_j(y)\,.$$

As to the first order we obtain

$$C_{ij}^1 = \frac{1}{\lambda_i\lambda_j}\left[\sum_{p,q=1}^{m} \frac{1}{\lambda_p\lambda_q}\, b_p b_q\, {}^2\hat{A}_1'(ip, jq) + {}^2\hat{A}_1'(i, j)\right]\varepsilon$$

because of the independence of $\bar{f}_{0\varepsilon}(x, \omega)$, $\bar{g}_\varepsilon(x, \omega)$ where $^2\hat{A}_1'(i, j)$ is given by

$$^2\hat{A}_1'(i, j) = \int_0^1 \varphi_i\varphi_j a_{gg}\, dx = \begin{cases} a_{gg} & \text{for } i = j\,, \\ 0 & \text{for } i \neq j \end{cases} \tag{5.34}$$

and $^2\hat{A}_1'(ip, jq)$ by

$$^2\hat{A}_1'(ip, jq) = a_{00}\int_0^1 w_i w_p w_j w_q\, dx$$
$$= 4a_{00}\int_0^1 \sin(i\pi x)\sin(p\pi x)\sin(j\pi x)\sin(q\pi x)\, dx$$
$$= \frac{1}{2}\, a_{00}[\delta_{0\,i-j-p+q} + \delta_{0\,i+j-p-q} - \delta_{0\,i-j-p-q} - \delta_{0\,i+j-p+q} - \delta_{0\,i-j+p+q}$$
$$- \delta_{0\,i+j+p-q} + \delta_{0\,i-j+p-q}]\,. \tag{5.35}$$

Then it is

$$C_{ij}^1 = \frac{1}{\lambda_i\lambda_j}\left[\frac{1}{2}\, a_{00}\left\{\sum_{\substack{p,q=1\\p-q=i-j}}^{m} \frac{b_p b_q}{\lambda_p\lambda_q} + \sum_{\substack{p,q=1\\p+q=j-i}}^{m} \frac{b_p b_q}{\lambda_p\lambda_q} + \sum_{\substack{p,q=1\\p-q=j-i}}^{m} \frac{b_p b_q}{\lambda_p\lambda_q} - \sum_{\substack{p,q=1\\p+q=i-j}}^{m} \frac{b_p b_q}{\lambda_p\lambda_q}\right.\right.$$
$$\left.\left. - \sum_{\substack{p,q=1\\p-q=i+j}}^{m} \frac{b_p b_q}{\lambda_p\lambda_q} - \sum_{\substack{p,q=1\\p+q=j-i}}^{m} \frac{b_p b_q}{\lambda_p\lambda_q} - \sum_{\substack{p,q=1\\p-q=-i-j}}^{m} \frac{b_p b_q}{\lambda_p\lambda_q}\right\} + a_{gg}\delta_{ij}\right]\varepsilon$$

and particularly

$$C_{ii}^1 = \frac{1}{\lambda_i^2}\left[\frac{1}{2}a_{00}\left\{2\sum_{p=1}^{m}\left(\frac{b_p}{\lambda_p}\right)^2 + \sum_{\substack{p,q=1\\p+q=2i}}^{m}\frac{b_pb_q}{\lambda_p\lambda_q} - \sum_{\substack{p,q=1\\p-q=2i}}^{m}\frac{b_pb_q}{\lambda_p\lambda_q} - \sum_{\substack{p,q=1\\q-p=2i}}^{m}\frac{b_pb_q}{\lambda_p\lambda_q}\right\} + a_{gg}\right]\varepsilon .$$

It is easy to see that

$$C_{ij}^1 = C_{ji}^1$$

and in order to determine

$$C_m^1(x, y) = \sum_{i,j=1}^{m} C_{ij}^1 w_i(x)\, w_j(y)$$

we have to evaluate $m(m+1)/2$ coefficients C_{ij}^1. The approximative variances $C_m^1(x, x)$ for $m = 1, \infty$ are plotted in Fig. 5.7 assuming $b = -9$, $c = 1$. We can observe a very

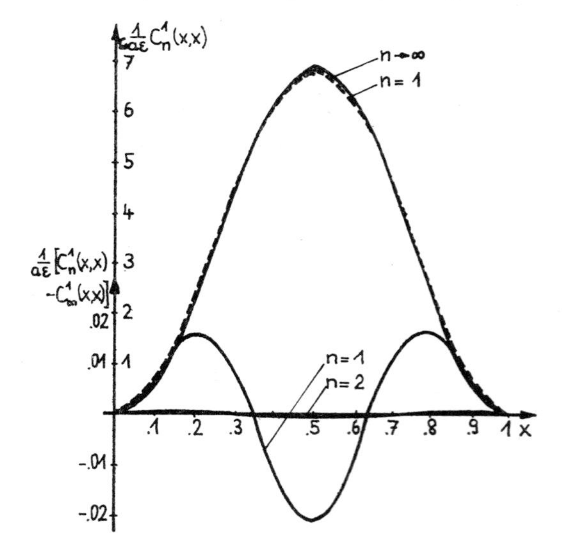

Fig. 5.7. Variances $C_n^1(x, x)$

good coincidence between $C_m^1(x, x)$ for $m = 1, 2$ and $C^1(x, x)$ (see also Fig. 5.3). These differences are also plotted in Fig. 5.7.

Now we deal with the second approximation as to ε. With the aid of the given conditions, (5.32) leads to

$$C_{ij}^{2,3} = \frac{1}{\lambda_i\lambda_j}\left\{\sum_{p,q,u,v=1}^{m}\frac{1}{\lambda_p\lambda_q}\frac{b_ub_v}{\lambda_u\lambda_v}\left[B_{ip,pq}B_{qu,jv} + B_{ip,qu}B_{pq,jv} + B_{ip,jv}B_{pq,qu}\right] + \sum_{p,q=1}^{m}\frac{1}{\lambda_p\lambda_q}B_{ip,pq}B_{q,j}\right\}$$

and (5.33) to

$$C_{ij}^{2,4} = \frac{1}{\lambda_i\lambda_j}\left\{\sum_{p,q,u,v=1}^{m}\frac{1}{\lambda_p\lambda_q}\frac{b_ub_v}{\lambda_u\lambda_v}\left[B_{ip,jq}B_{pu,qv} + B_{ip,qv}B_{pu,jq}\right] + \sum_{p,q=1}^{m}\frac{1}{\lambda_p\lambda_q}B_{ip,jq}B_{p,q}\right\}.$$

It is
$$B_{p,q} = {}^2\hat{A}_1'(p,q) \quad \text{and} \quad B_{pq,uv} = {}^2\hat{A}_1'(pq,uv)$$
and these terms are given by (5.34) and (5.35), respectively. With this we get
$$\begin{aligned}
C_{ij}^{2,3} = \frac{a^2}{\lambda_i\lambda_j}\Bigg\{&\frac{1}{4}\sum_{p,q,u,v=1}^{m}\frac{1}{\lambda_p\lambda_q}\frac{b_u b_v}{\lambda_u\lambda_v}\\
&\times [(\delta_{0i-2p+q}+\delta_{0i-q}-\delta_{0i+q}-\delta_{0i-2p-q}+\delta_{0i-q}-\delta_{0i+q}-\delta_{0i+2p-q})\\
&\times(\delta_{0q-j-u+v}+\delta_{0q-j+u-v}-\delta_{0q-j+u+v}-\delta_{0q-j-u-v}+\delta_{0q+j-u-v}\\
&-\delta_{0q+j-u+v}-\delta_{0q+j+u-v})\\
&+(\delta_{0i-q-p+u}+\delta_{0i-q+p-u}-\delta_{0i-q+p+u}-\delta_{0i-q-p-u}+\delta_{0i+q-p-u}\\
&-\delta_{0i+q-p+u}-\delta_{0i+q+p-u})\\
&\times(\delta_{0p-j-q+v}+\delta_{0p-j+q-v}-\delta_{0p-j+q+v}-\delta_{0p-j-q-v}+\delta_{0p+j-q-v}\\
&-\delta_{0p+j-q+v}-\delta_{0p+j+q-v})\\
&+(\delta_{0i-j-p+v}+\delta_{0i-j+p-v}-\delta_{0i-j+p+v}-\delta_{0i-j-p-v}+\delta_{0i+j-p-v}\\
&-\delta_{0i+j-p+v}-\delta_{0i+j+p-v})\\
&\times(\delta_{0p-2q+u}+\delta_{0p-u}-\delta_{0p+u}-\delta_{0p-2q-u}+\delta_{0p-u}-\delta_{0p+u}-\delta_{0p+2q-u})]\\
&+\frac{1}{2\lambda_j}\sum_{p=1}^{m}\frac{1}{\lambda_p}[\delta_{0i+j-2p}+2\delta_{0i-j}-\delta_{0i-j-2p}-\delta_{0i-j+2p}]\Bigg\},
\end{aligned}$$
$$\begin{aligned}
C_{ij}^{2,4} = \frac{a^2}{\lambda_i\lambda_j}\Bigg\{&\frac{1}{4}\sum_{p,q,u,v=1}^{m}\frac{1}{\lambda_p\lambda_q}\frac{b_u b_v}{\lambda_u\lambda_v}\\
&\times [(\delta_{0i-j-p+q}+\delta_{0i-j+p-q}-\delta_{0i-j+p+q}-\delta_{0i-j-p-q}+\delta_{0i+j-p-q}\\
&-\delta_{0i+j-p+q}-\delta_{0i+j+p-q})\\
&\times(\delta_{0p-q-u+v}+\delta_{0p-q+u-v}-\delta_{0p-q+u+v}-\delta_{0p-q-u-v}+\delta_{0p+q-u-v}\\
&-\delta_{0p+q-u+v}-\delta_{0p+q+u-v})\\
&+(\delta_{0i-q-p+v}+\delta_{0i-q+p-v}-\delta_{0i-q+p+v}-\delta_{0i-q-p-v}+\delta_{0i+q-p-v}\\
&-\delta_{0i+q-p+v}-\delta_{0i+q+p-v})\\
&\times(\delta_{0p-j-u+q}+\delta_{0p-j+u-q}-\delta_{0p-j+u+q}-\delta_{0p-j-u-q}+\delta_{0p+j-u-q}\\
&-\delta_{0p+j-u+q}-\delta_{0p+j+u-q})\\
&+\frac{1}{2}\sum_{p=1}^{m}\frac{1}{\lambda_p^2}[2\delta_{0i-j}+\delta_{0i+j-2p}-\delta_{0i-j-2p}-\delta_{0i-j+2p}]\Bigg\}.
\end{aligned}$$
The simplest case is investigated, $m = 1$. Thus
$$C_{11}^{2,3} = \frac{a^2}{\lambda_1^4}\left\{\frac{27}{4}\left(\frac{b_1}{\lambda_1}\right)^2+\frac{3}{2}\right\},$$
$$C_{11}^{2,4} = \frac{a^2}{\lambda_1^4}\left\{\frac{9}{2}\left(\frac{b_1}{\lambda_1}\right)^2+\frac{3}{2}\right\}$$
and furthermore
$$\begin{aligned}
C_{11} &= C_{11}^1 + C_{11}^2 + o(\varepsilon^2)\\
&= \frac{a\varepsilon}{\lambda_1^2}\left(\frac{3}{2}\left(\frac{b_1}{\lambda_1}\right)^2+1\right)+9\left(\frac{a\varepsilon}{\lambda_1^2}\right)^2\left(2\left(\frac{b_1}{\lambda_1}\right)^2+\frac{1}{2}\right)+o(\varepsilon^2).
\end{aligned}$$

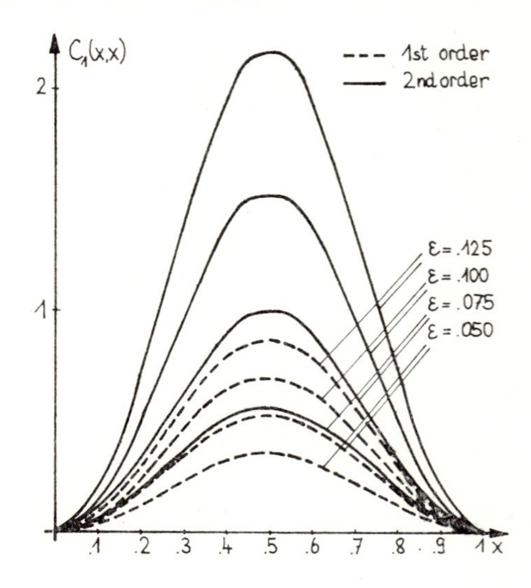

Fig. 5.8. Variances $C_n(x, x)$ for different values of ε, $n = 1$, $a = 1$

The relation

$$u_1(x, \omega) = X_1(\omega)\, w_1(x)$$

leads to

$$C_1(x, y) = C_{11}^1 w_1(x)\, w_1(y) + C_{11}^2 w_1(x)\, w_1(y) + o(\varepsilon^2).$$

We have plotted the approximative variances

$$C_{11}^1 w_1^2(x) \quad \text{and} \quad (C_{11}^1 + C_{11}^2)\, w_1^2(x)$$

for different values of ε setting $b = -9$, $c = 1$, $a = 1$ and assuming $\bar{f}_{0\varepsilon}(x, \omega)$, $\bar{g}_\varepsilon(x, \omega)$ are independent in Fig. 5.8. A remarkable difference between the results of first order and second order can be observed.

5.2. Partial differential operators

This section has the aim to obtain similar results as in the previous section for partial differential equations. The boundness of the Green's function as to an ordinary differential operator is used essentially. Investigating the Green's function of a partial differential equation we see that it is singular and the order of the singularity plays an important role.

First, we deal with the Dirichlet problem for the Laplace operator on $\mathbb{R}^n$ with $n \geqq 2$. The boundary conditions are assumed to be non-random and the inhomogeneous term of the differential equation is a random function.

We consider the boundary value problem of first kind

$$-\Delta u = g(x, \omega); \qquad u|_{\partial\mathcal{D}} = u_0; \qquad x \in \mathcal{D} \subset \mathbb{R}^n \tag{5.36}$$

on a bounded domain $\mathcal{D}$ the boundary $\partial\mathcal{D}$ of which is supposed to be smooth. The investigation of stationary transverse vibrations of a homogeneous membrane with constant density leads to such a problem in the case of $n = 2$. In this example, $u(x)$ denotes the vertical displacement and g the density of external loads. Stationary heat conduction and diffusion problems also lead to such inhomogeneous Dirichlet problems.

The solution of (5.36) has the form

$$u(x, \omega) = -\int_{\partial\mathcal{D}} \frac{\partial G}{\partial n} u_0 \, dS + \int_{\mathcal{D}} G(x, y) \, g(y, \omega) \, dy$$

where $G(x, y)$ denotes the Green's function of the Dirichlet problem (5.36) and $\partial G/\partial n$ the normal derivative of $G(x, y)$ as to y on $\partial\mathcal{D}$. The function $w(x)$ is the solution of the averaged problem

$$-\Delta w = \langle g(x) \rangle; \qquad w|_{\partial\mathcal{D}} = u_0$$

belonging to the boundary value problem (5.36). Now we deal with the difference between $u(x, \omega)$ and $w(x)$,

$$\bar{u}(x, \omega) \doteq u(x, \omega) - w(x) = \int_{\mathcal{D}} G(x, y) \, \bar{g}(y, \omega) \, dy \tag{5.37}$$

where $\bar{g}$ is assumed to be

$$\bar{g}(x, \omega) \doteq g(x, \omega) - \langle g(x) \rangle .$$

The Green's function including in (5.37) has the structure

$$G(x, y) = \begin{cases} -\dfrac{1}{2\pi} \ln |x - y| + H_2(x, y) & \text{for} \quad n = 2, \\[2ex] \dfrac{1}{(n-2) \, S_n\big(\mathcal{K}_1(0)\big)} \dfrac{1}{|x - y|^{n-2}} + H_n(x, y) & \text{for} \quad n \geqq 3 \end{cases}$$

and $H_n(x, y)$ denotes a harmonic function and $S_n\big(\mathcal{K}_1(0)\big)$ the area of the surface of $\mathcal{K}_1(0)$.

In order to deal with the solution $\bar{u}(x, \omega)$ functionals of the form

$$r_{i\varepsilon}(x, \omega) = \frac{1}{\sqrt{\varepsilon}^{\,n}} \int_{\mathcal{D}_i} F_i(x, y) \, g_\varepsilon(y, \omega) \, dy$$

with

$$F_i(x, y) = \begin{cases} J_i(x, y) \ln |x - y| & \text{for} \quad n = 2, \\[2ex] \dfrac{J_i(x, y)}{|x - y|^{\mu_i}} & \text{for} \quad n \geqq 3 \end{cases}$$

have to be investigated where $\mathcal{D}_i$ are bounded domains from $\mathbb{R}^n$ with continuous boundaries and $0 \leqq \mu_i < n$. $g_\varepsilon(x, \omega)$ denotes a weakly correlated function and $J_i(x, y)$ a continuous function. Some results as to $r_{i\varepsilon}(x, \omega)$ are contained in the following theorem.

Theorem 5.3. (1) *Assuming $x_1 \neq x_2$ second moments satisfy the limit relation*

$$\lim_{\varepsilon \downarrow 0} \langle r_{1\varepsilon}(x_1) \, r_{2\varepsilon}(x_2) \rangle = \int_{\mathcal{D}_1 \cap \mathcal{D}_2} F_1(x_1, y) \, F_2(x_2, y) \, a(y) \, dy \tag{5.38}$$

where $a(x)$ denotes the intensity as to the weakly corretated function $g_\varepsilon(x, \omega)$. In the case of $x_1 = x_2 \in \mathcal{D}_1 \cap \mathcal{D}_2$ under the additional condition

$$\mu_1 + \mu_2 < n$$

the limit relation (5.38) *is also fulfilled.*

(2) *Defining* $r_{i\varepsilon}(x_i, \omega) \doteq r_{i\varepsilon}$ *for* $i = 1, 2, \ldots, k$ *the k-th moment* $\left\langle \prod_{i=1}^{k} r_{i\varepsilon} \right\rangle$ *satisfies the limit relation*

$$\lim_{\varepsilon \downarrow 0} \left\langle \prod_{i=1}^{k} r_{i\varepsilon} \right\rangle = \begin{cases} \sum\limits_{\{i_1,j_1\},\ldots,\{i_{k/2},j_{k/2}\}} \prod\limits_{s=1}^{k/2} \lim\limits_{\varepsilon \downarrow 0} \langle r_{i_s\varepsilon} r_{j_s\varepsilon} \rangle & \text{for } k \text{ even}, \\ 0 & \text{for } k \text{ odd}. \end{cases} \tag{5.39}$$

If points x_i *agree then the inequality*

$$\max_s \left\{ \sum_{i=1}^{p_s} \mu_{l_{si}} \right\} < \frac{kn}{2}$$

has to be demanded. $\{l_{s1}, \ldots, l_{sp_s}\}$, $s = 1, 2, \ldots, m$, *denote the indices of the agreeing points* x_i. *It is*

$$\sum_{s=1}^{m} p_s = k.$$

For example, the relation (5.39) *used for* $\langle r_{1\varepsilon}(x_1)\, r_{2\varepsilon}(x_1)\, r_{3\varepsilon}(x_2) \rangle$ *demands the condition*

$$\max \{\mu_1 + \mu_2, \mu_3\} < \frac{3n}{2}.$$

(3) *The random function*

$$r_\varepsilon(x, \omega) = \frac{1}{\sqrt{\varepsilon^n}} \int_{\mathscr{D}} F(x, y)\, g_\varepsilon(y, \omega)\, \mathrm{d}y$$

with

$$F(x, y) = \frac{J(x, y)}{|x - y|^\mu}$$

converges in distribution as $\varepsilon \downarrow 0$ *to a Gaussian function* $\xi(x, \omega)$ *having the moments*

$$\langle \xi(x) \rangle = 0,$$
$$\langle \xi(x_1)\, \xi(x_2) \rangle = \int_{\mathscr{D}} F(x_1, y)\, F(x_2, y)\, a(y)\, \mathrm{d}y$$

if the condition $\mu < n/2$ *is fulfilled. In this,* $\mathscr{D}$ *denotes a bounded domain and* $J(x, y)$ *a continuous function on* $\mathscr{D}$ *as to* y.

This theorem is also valid for functions $F_{(i)}(x, y)$ with logarithmic singularities,

$$F_{(i)}(x, y) = J_{(i)}(x, y) \ln |x - y|.$$

Proof. (1) We deal with

$$\langle r_{1\varepsilon}(x_1)\, r_{2\varepsilon}(x_2) \rangle$$
$$= \frac{1}{\varepsilon^n} \int_{\mathscr{D}_1} \int_{\mathscr{D}_2} F_1(x_1, y_1)\, F_2(x_2, y_2)\, \langle g_\varepsilon(y_1)\, g_\varepsilon(y_2) \rangle\, \mathrm{d}y_1\, \mathrm{d}y_2, \quad x_1 \in \mathscr{D}_1,\, x_2 \in \mathscr{D}_2. \tag{5.40}$$

The set $\mathscr{C}_\varepsilon(\mathscr{B})$ is defined by

$$\mathscr{C}_\varepsilon(\mathscr{B}) = \{x \in \mathbb{R}^n : \operatorname{dist}\{x, \mathscr{B}\} \leqq \varepsilon\} \quad \text{for} \quad \mathscr{B} \subset \mathbb{R}^n.$$

If $y_1 \notin \mathscr{C}_\varepsilon(\mathscr{D}_1 \cap \mathscr{D}_2)$ or $y_2 \notin \mathscr{C}_\varepsilon(\mathscr{D}_1 \cap \mathscr{D}_2)$ then the correlation function

$$R_\varepsilon(y_1, y_2) \doteq \langle g_\varepsilon(y_1)\, g_\varepsilon(y_2)\rangle = 0$$

because of the property of the weak correlation of $g_\varepsilon(y, \omega)$.

(11) Let

$$\operatorname{dist}\{x_i, \mathscr{D}_1 \cap \mathscr{D}_2\} = \delta > 0 \quad \text{for} \quad i = 1, 2.$$

It follows that

$$F_1(x_1, y),\, F_2(x_2, y) \in \mathbf{L}_2\big(\mathscr{C}_{\delta/2}(\mathscr{D}_1 \cap \mathscr{D}_2)\big)$$

and using Theorem 1.6 we obtain the statement (1) in this case.

(12) Let

$$\operatorname{dist}\{x_2, \mathscr{D}_1 \cap \mathscr{D}_2\} = \delta > 0.$$

Then $F_2(x_2, y)$ is a continuous function for $y \in \mathscr{C}_{\delta/2}(\mathscr{D}_1 \cap \mathscr{D}_2) \cap \mathscr{D}_2$ and a step function $t_i(y)$ exists where

$$|t_i(y) - F_2(x_2, y)| < \eta \quad \text{for all } y \in \mathscr{C}_{\delta/2}(\mathscr{D}_1 \cap \mathscr{D}_2) \cap \mathscr{D}_2.$$

$t_i(y)$ is written as

$$t_i(y) = \sum_{p=1}^{p_i} c_{ip} 1_{\mathscr{C}_{ip}}(y)$$

with

$$\mathscr{C}_{ip} \subset \mathscr{C}_{\delta/2}(\mathscr{D}_1 \cap \mathscr{D}_2) \cap \mathscr{D}_2, \qquad \lambda(\partial \mathscr{C}_{ip}) = 0.$$

Supposing $\varepsilon < \delta/2$ we get

$$\begin{aligned} &\frac{1}{\varepsilon^n} \iint\limits_{\mathscr{D}_1 \mathscr{D}_2} F_1(x_1, y_1)\, t_i(y_2)\, R_\varepsilon(y_1, y_2)\, \mathrm{d}y_1\, \mathrm{d}y_2 \\ &= \frac{1}{\varepsilon^n} \sum_{p=1}^{p_i} c_{ip} \int\limits_{\mathscr{D}_1} F_1(x_1, y_1) \int\limits_{\mathscr{C}_{ip} \cap \mathscr{K}_\varepsilon(y_1)} R_\varepsilon(y_1, y_2)\, \mathrm{d}y_2\, \mathrm{d}y_1. \end{aligned} \tag{5.41}$$

With the aid of

$$\lim_{\varepsilon \downarrow 0} \frac{1}{\varepsilon^n} \int\limits_{\mathscr{C}_{ip} \cap \mathscr{K}_\varepsilon(y)} R_\varepsilon(y, z)\, \mathrm{d}z = a(y)\, 1_{\mathscr{D}_1 \cap \mathscr{C}_{ip}}(y)$$

almost surely for all $x \in \mathscr{D}_1$ we obtain

$$\begin{aligned} &\lim_{\varepsilon \downarrow 0} \frac{1}{\varepsilon^n} F_1(x_1, y_2) \int\limits_{\mathscr{C}_{ip} \cap \mathscr{K}_\varepsilon(y_1)} R_\varepsilon(y_1, y_2)\, \mathrm{d}y_2 \\ &= F_1(x_1, y_1)\, a(y_1)\, 1_{\mathscr{D}_1 \cap \mathscr{C}_{ip}}(y_1) \quad \text{a.s. as to all } y_1 \in \mathscr{D}_1 \end{aligned}$$

and

$$\left| \frac{1}{\varepsilon^n} F_1(x_1, y_1) \int\limits_{\mathscr{C}_{ip} \cap \mathscr{K}_\varepsilon(y_1)} R_\varepsilon(y_1, y_2)\, \mathrm{d}y \right| \leqq |F_1(x_1, y_1)|\, V_1 c_2$$

by means of similar considerations of the proof of Theorem 1.6. Since the majorant $|F_1(x_1, y_1)|\, V_1 c_2$ with $\mu_i < n$ is integrable Lebesgue's theorem leads to

$$\lim_{\varepsilon\downarrow 0} \frac{1}{\varepsilon^n} \int_{\mathcal{D}_1} F_1(x_1, y_1) \int_{\mathcal{C}_{ip} \cap \mathcal{K}_\varepsilon(y_1)} R_\varepsilon(y_1, y_2)\, dy_2\, dy_1 = \int_{\mathcal{D}_1 \cap \mathcal{C}_{ip}} F_1(x_1, y_1)\, a(y_1)\, dy_1$$

and furthermore by means of (5.41) to

$$\lim_{\varepsilon\downarrow 0} \frac{1}{\varepsilon^n} \iint_{\mathcal{D}_1 \mathcal{D}_2} F_1(x_1, y_1)\, t_i(y_2)\, R_\varepsilon(y_1, y_2)\, dy_1\, dy_2 = \int_{\mathcal{D}_1} F_1(x_1, y)\, t_i(y)\, a(y)\, dy\,.$$

Now we have proved the statement of (12) for step functions and go over to $F_2(x_2, y)$. We consider the inequality

$$\begin{aligned}
&\left| \frac{1}{\varepsilon^n} \iint_{\mathcal{D}_1 \mathcal{D}_2} F_1(x_1, y_1)\, F_2(x_2, y_2)\, R_\varepsilon(y_1, y_2)\, dy_1\, dy_2 - \int_{\mathcal{D}_1 \cap \mathcal{D}_2} F_1(x_1, y)\, F_2(x_2, y)\, a(y)\, dy \right| \\
&\leqq \left| \frac{1}{\varepsilon^n} \iint_{\mathcal{D}_1 \mathcal{D}_2} F_1(x_1, y_1) \big(F_2(x_2, y_2) - t_i(y_2)\big) R_\varepsilon(y_1, y_2)\, dy_1\, dy_2 \right| \\
&\quad + \left| \frac{1}{\varepsilon^n} \iint_{\mathcal{D}_1 \mathcal{D}_2} F_1(x_1, y_1)\, t_i(y_2)\, R_\varepsilon(y_1, y_2)\, dy_1\, dy_2 - \int_{\mathcal{D}_1 \cap \mathcal{D}_2} F_1(x_1, y)\, t_i(y)\, a(y)\, dy \right| \\
&\quad + \left| \int_{\mathcal{D}_1 \cap \mathcal{D}_2} F_1(x_1, y) \big(t_i(y) - F_2(x_2, y)\big)\, a(y)\, dy \right|.
\end{aligned}$$

The second summand on the right-hand side converges to zero as $\varepsilon \downarrow 0$. The first and third summand can be estimated by

$$\begin{aligned}
&\left| \frac{1}{\varepsilon^n} \iint_{\mathcal{D}_1 \mathcal{D}_2} F_1(x_1, y_1) \big(F_2(x_2, y_2) - t_i(y_2)\big) R_\varepsilon(y_1, y_2)\, dy_1\, dy_2 \right| \\
&\leqq V_1 \eta c_2 \int_{\mathcal{D}_1} |F_1(x_1, y)|\, dy\,, \\
&\left| \int_{\mathcal{D}_1 \cap \mathcal{D}_2} F_1(x_1, y) \big(t_i(y) - F_2(x_2, y)\big)\, a(y)\, dy \right| \leqq V_1 \eta c_2 \int_{\mathcal{D}_1 \cap \mathcal{D}_2} |F_1(x_1, y)|\, dy\,,
\end{aligned}$$

respectively. Since η can be an arbitrary positive real number the limit relation (5.38) is proved in the case of (12).

(13) Let

$$\operatorname{dist}\{x_1, \mathcal{D}_1 \cap \mathcal{D}_2\} = \delta > 0\,.$$

This case is analogous to (12).

(14) Let $x_1, x_2 \in \overline{\mathcal{D}}_1 \cap \overline{\mathcal{D}}_2$ and $x_1 \neq x_2$. Using $\varrho \doteq |x_1 - x_2|$ and

$$J(\mathcal{D}_1, \mathcal{D}_2; \varepsilon) \doteq \frac{1}{\varepsilon^n} \iint_{\mathcal{D}_1 \mathcal{D}_2} F_1(x_1, y_1)\, F_2(x_2, y_2)\, R_\varepsilon(y_1, y_2)\, dy_1\, dy_2$$

we have

$$\begin{aligned} J(\mathcal{D}_1, \mathcal{D}_2; \varepsilon) &= J\big(\mathcal{D}_1 \setminus [\mathcal{K}_{\varrho/4}(x_1) \cap \mathcal{D}_1], \mathcal{D}_2 \setminus [\mathcal{K}_{\varrho/4}(x_2) \cap \mathcal{D}_2]; \varepsilon\big) \\ &\quad + J\big(\mathcal{K}_{\varrho/4}(x_1) \cap \mathcal{D}_1, \mathcal{K}_{\varrho/4}(x_2) \cap \mathcal{D}_2; \varepsilon\big) \\ &\quad + J\big(\mathcal{D}_1 \setminus [\mathcal{K}_{\varrho/4}(x_1) \cap \mathcal{D}_1], \mathcal{K}_{\varrho/4}(x_2) \cap \mathcal{D}_2; \varepsilon\big) \\ &\quad + J\big(\mathcal{K}_{\varrho/4}(x_1) \cap \mathcal{D}_1, \mathcal{D}_2 \setminus [\mathcal{K}_{\varrho/4}(x_2) \cap \mathcal{D}_2]; \varepsilon\big). \end{aligned} \tag{5.42}$$

On the first summand of the right-hand side we apply the case (11) and on the third and fourth summand the cases (12) and (13), respectively. For $\varepsilon < \varrho/2$ we obtain

$$J\big(\mathcal{K}_{\varrho/4}(x_1) \cap \mathcal{D}_1, \mathcal{K}_{\varrho/4}(x_2) \cap \mathcal{D}_2; \varepsilon\big) = 0$$

because of

$$|y_1 - y_2| > \frac{\varrho}{2} > \varepsilon .$$

Then it is

$$\begin{aligned} &\lim_{\varepsilon \downarrow 0} J(\mathcal{D}_1, \mathcal{D}_2; \varepsilon) \\ &= \int\limits_{[\mathcal{D}_1 \setminus [\mathcal{K}_{\varrho/4}(x_1) \cap \mathcal{D}_1]] \cap [\mathcal{D}_2 \setminus [\mathcal{K}_{\varrho/4}(x_2) \cap \mathcal{D}_2]]} F_1(x_1, y)\, F_2(x_2, y)\, a(y)\, dy \\ &\quad + \int\limits_{[\mathcal{D}_1 \setminus [\mathcal{K}_{\varrho/4}(x_1) \cap \mathcal{D}_1]] \cap [\mathcal{K}_{\varrho/4}(x_2) \cap \mathcal{D}_2]} F_1(x_1, y)\, F_2(x_2, y)\, a(y)\, dy \\ &\quad + \int\limits_{[\mathcal{K}_{\varrho/4}(x_1) \cap \mathcal{D}_1] \cap [\mathcal{D}_2 \setminus [\mathcal{K}_{\varrho/4}(x_2) \cap \mathcal{D}_2]]} F_1(x_1, y)\, F_2(x_2, y)\, a(y)\, dy \\ &= \int\limits_{\mathcal{D}_1 \cap \mathcal{D}_2} F_1(x_1, y)\, F_2(x_2, y)\, a(y)\, dy \end{aligned}$$

so that we have also proved the statement of (1) for $x_1, x_2 \in \overline{\mathcal{D}}_1 \cap \overline{\mathcal{D}}_2$ and $x_1 \neq x_2$.

(15) Let $x \doteq x_1 = x_2 \in \overline{\mathcal{D}}_1 \cap \overline{\mathcal{D}}_2$. A relation as in (5.42) leads to

$$\begin{aligned} J(\mathcal{D}_1, \mathcal{D}_2; \varepsilon) &= J\big(\mathcal{D}_1 \setminus [\mathcal{K}_\varrho(x) \cap \mathcal{D}], \mathcal{D}_2 \setminus [\mathcal{K}_\varrho(x) \cap \mathcal{D}]; \varepsilon\big) \\ &\quad + J\big(\mathcal{D}_1 \setminus [\mathcal{K}_\varrho(x) \cap \mathcal{D}], \mathcal{K}_\varrho(x) \cap \mathcal{D}; \varepsilon\big) \\ &\quad + J\big(\mathcal{K}_\varrho(x) \cap \mathcal{D}, \mathcal{D}_2 \setminus [\mathcal{K}_\varrho(x) \cap \mathcal{D}]; \varepsilon\big) \\ &\quad + J\big(\mathcal{K}_\varrho(x) \cap \mathcal{D}, \mathcal{K}_\varrho(x) \cap \mathcal{D}; \varepsilon\big) \end{aligned}$$

where $\mathcal{D} \doteq \mathcal{D}_1 \cap \mathcal{D}_2$ and ϱ is an arbitrary positive real number. The first summand on the right-hand side converges to

$$\int\limits_{\mathcal{D} \setminus [\mathcal{K}_\varrho(x) \cap \mathcal{D}]} F_1(x, y)\, F_2(x, y)\, a(y)\, dy$$

and the second and third summand to zero. It remains to show that

$$\lim_{\varepsilon \downarrow 0} J\big(\mathcal{K}_\varrho(x) \cap \mathcal{D}, \mathcal{K}_\varrho(x) \cap \mathcal{D}; \varepsilon\big) = \int\limits_{\mathcal{K}_\varrho(x) \cap \mathcal{D}} F_1(x, y)\, F_2(x, y)\, a(y)\, dy . \tag{5.43}$$

This relation (5.43) shall be proved for

$$F_2(x, y) \doteq \frac{1_{\mathcal{B}}(y)}{|x - y|^{\mu_2}}$$

since from this the relation (5.43) is valid for a general function $F_2(x, y)$ by approximation by step functions. Now it follows

$$J(\mathcal{K}_\varrho(x) \cap \mathcal{D}, \mathcal{K}_\varrho(x) \cap \mathcal{D}; \varepsilon) = \frac{1}{\varepsilon^n} \int\limits_{\mathcal{K}_\varrho(x)\cap\mathcal{D}} F_1(x, y_1) \int\limits_{\mathcal{K}_\varrho(x)\cap\mathcal{D}\cap\mathcal{B}} \frac{1}{|x - y_2|^{\mu_2}} R_\varepsilon(y_1, y_2) \, dy_2 \, dy_1 .$$

If $x \notin \bar{\mathcal{B}}$ then $|x - y_2|^{-\mu_2}$ is bounded for $x \in \mathcal{K}_\varrho(x) \cap \mathcal{D} \cap \mathcal{B}$ and the desired limit relation can be obtained. Now we deal with the case of $x \in \bar{\mathcal{B}}$. It follows

$$J(\mathcal{K}_\varrho(x) \cap \mathcal{D} , \mathcal{K}_\varrho(x) \cap \mathcal{D}; \varepsilon) = J_1 + J_2$$

with

$$J_1 \doteq J([\mathcal{K}_\varrho(x) \setminus \mathcal{K}_{2\varepsilon}(x)] \cap \mathcal{D}, \mathcal{K}_\varepsilon(y_1) \cap \mathcal{K}_\varrho(x) \cap \mathcal{D} \cap \mathcal{B}; \varepsilon),$$

$$J_2 \doteq J(\mathcal{K}_{2\varepsilon}(x) \cap \mathcal{D}, \mathcal{K}_\varepsilon(y_1) \cap \mathcal{K}_\varrho(x) \cap \mathcal{D} \cap \mathcal{B}; \varepsilon).$$

J_1 satisfies the relation

$$\begin{aligned} J_1 &= \frac{1}{\varepsilon^n} \int\limits_{[\mathcal{K}_\varrho(x)\setminus\mathcal{K}_{2\varepsilon}(x)]\cap\mathcal{D}} F_1(x, y_1) \left(\int\limits_{\mathcal{K}_\varepsilon(y_1)\cap\mathcal{K}_\varrho(x)\cap\mathcal{D}\cap\mathcal{B}} \frac{R_\varepsilon(y_1, y_2)}{|x - y_2|^{\mu_2}} \, dy_2 \right) dy_1 \\ &= \frac{1}{\varepsilon^n} \int\limits_{[\mathcal{K}_\varrho(x)\setminus\mathcal{K}_{2\varepsilon}(x)]\cap\mathcal{D}} F_1(x, y_1) \frac{1}{|x - y_1|^{\mu_2}} \left(\int\limits_{\mathcal{K}_\varepsilon(y_1)\cap\mathcal{K}_\varrho(x)\cap\mathcal{D}\cap\mathcal{B}} R_\varepsilon(y_1, y_2) \, dy_2 \right) dy_1 \\ &\quad + \sum_{s=1}^{n} \frac{1}{\varepsilon^n} \int\limits_{[\mathcal{K}_\varrho(x)\setminus\mathcal{K}_{2\varepsilon}(x)]\cap\mathcal{D}} F_1(x, y_1) \\ &\quad \times \left(\int\limits_{\mathcal{K}_\varepsilon(y_1)\cap\mathcal{K}_\varrho(x)\cap\mathcal{D}\cap\mathcal{B}} \frac{\partial}{\partial y_{2s}} \frac{1}{|x - y_2|^{\mu_2}} \bigg|_{y_2=\tilde{y}_2} (y_{2s} - y_{1s}) \, R_\varepsilon(y_1, y_2) \, dy_2 \right) dy_1 \end{aligned}$$

where we have used the mean value theorem applied to $|x - y_2|^{-\mu_2}$,

$$\begin{aligned} |x - y_2|^{-\mu_2} &= |x - y_1|^{-\mu_2} + \sum_{s=1}^{n} \frac{\partial}{\partial y_{2s}} |x - y_2|^{-\mu_2}|_{y_2=\tilde{y}_2} (y_{2s} - y_{1s}) \\ &= |x - y_1|^{-\mu_2} + \mu_2 \sum_{s=1}^{n} (x_s - y_{2s}) \, |x - y_2|^{-\mu_2-2}|_{y_2=\tilde{y}_2} (y_{2s} - y_{1s}). \end{aligned}$$

The first summand J_{11} of J_1 can be written as

$$\begin{aligned} J_{11} &= \frac{1}{\varepsilon^n} \int\limits_{\mathcal{K}_\varrho(x)\cap\mathcal{D}} F_1(x, y_1) \frac{1}{|x - y_1|^{\mu_2}} \left(\int\limits_{\mathcal{K}_\varepsilon(y_1)\cap\mathcal{K}_\varrho(x)\cap\mathcal{D}\cap\mathcal{B}} R_\varepsilon(y_1, y_2) \, dy_2 \right) dy_1 \\ &\quad - \frac{1}{\varepsilon^n} \int\limits_{\mathcal{K}_{2\varepsilon}(x)\cap\mathcal{D}} F_1(x, y_1) \frac{1}{|x - y_1|^{\mu_2}} \left(\int\limits_{\mathcal{K}_\varepsilon(y_1)\cap\mathcal{K}_\varrho(x)\cap\mathcal{D}\cap\mathcal{B}} R_\varepsilon(y_1, y_2) \, dy_2 \right) dy_1 \end{aligned}$$

and we get

$$\lim_{\varepsilon \downarrow 0} J_{111} = \int\limits_{\mathcal{K}_\varrho(x) \cap \mathcal{D} \cap \mathcal{B}} F_1(x, y) \frac{1}{|x - y|^{\mu_2}} a(y) \, dy,$$

$$|J_{112}| \leqq \frac{1}{\varepsilon^n} d_2 V_1 \varepsilon^n A_1 \int\limits_{\mathcal{K}_{2\varepsilon}(x) \cap \mathcal{D}} |x - y|^{-\mu_1 - \mu_2} \, dy$$

$$\leqq d_2 V_1 A_1 S_n(\mathcal{K}_1(0)) \int\limits_0^{2\varepsilon} r^{-\mu_1 - \mu_2 + n - 1} \, dr$$

$$= d_2 V_1 A_1 S_n(\mathcal{K}_1(0)) \frac{1}{n - \mu_1 - \mu_2} (2\varepsilon)^{n - \mu_1 - \mu_2}$$

since $\mu_1 + \mu_2 < n$ and

$$|F_1(x, y)| \leqq A_1 |x - y|^{-\mu_1}.$$

Hence it is

$$\lim_{\varepsilon \downarrow 0} J_{11} = \int\limits_{\mathcal{K}_\varrho(x) \cap \mathcal{D} \cap \mathcal{B}} F_1(x, y) \frac{a(y)}{|x - y|^{\mu_2}} \, dy.$$

We turn to J_{12}. Under the given conditions the inequality

$$|(x_s - \tilde{y}_{2s}) |x - \tilde{y}_2|^{-\mu_2 - 2} (y_{2s} - y_{1s})| \leqq \varepsilon(|x - y_1| - \varepsilon)^{-\mu_2 - 1}$$

for $\tilde{y}_2 \in \mathcal{K}_\varepsilon(y_1)$ is valid because of

$$|x_s - \tilde{y}_{2s}| \leqq |x - \tilde{y}_2|; \qquad |y_{2s} - y_{1s}| \leqq \varepsilon; \qquad |x - \tilde{y}_2| \geqq |x - y_1| - \varepsilon.$$

Then we get

$$|J_{12}| \leqq n\mu_2 d_2 V_1 \varepsilon \int\limits_{[\mathcal{K}_\varrho(x) \setminus \mathcal{K}_{2\varepsilon}(x)] \cap \mathcal{D}} |F_1(x, y)| \, (|x - y| - \varepsilon)^{-\mu_2 - 1} \, dy$$

$$\leqq n\mu_2 d_2 V_1 A_1 \varepsilon S_n(\mathcal{K}_1(0)) \int\limits_{2\varepsilon}^{\varrho} \frac{r^{n-1}}{r^{\mu_1}(r - \varepsilon)^{\mu_2 + 1}} \, dr$$

$$\leqq n\mu_2 d_2 V_1 A_1 \varepsilon S_n(\mathcal{K}_1(0)) \, 2^{\mu_2 + 1} \int\limits_{2\varepsilon}^{\varrho} r^{n - 2 - \mu_1 - \mu_2} \, dr$$

$$= n\mu_2 d_2 V_1 A_1 \varepsilon S_n(\mathcal{K}_1(0)) \, 2^{\mu_2 + 1} \begin{cases} \left. \dfrac{r^{n - 1 - \mu_1 - \mu_2}}{n - 1 - \mu_1 - \mu_2} \right|_{2\varepsilon}^{\varrho} & \text{for} \quad \mu_1 + \mu_2 \neq n - 1, \\ \ln(r)|_{2\varepsilon}^{\varrho} & \text{for} \quad \mu_1 + \mu_2 = n - 1 \end{cases}$$

where we have used the relation $r - \varepsilon \geqq r/2$. From this we see that

$$\lim_{\varepsilon \downarrow 0} J_{12} = 0.$$

Finally, we consider J_2,

$$J_2 \doteq \frac{1}{\varepsilon^n} \int\limits_{\mathcal{K}_{2\varepsilon}(x) \cap \mathcal{D}} F_1(x, y_1) \left(\int\limits_{\mathcal{K}_\varepsilon(y_1) \cap \mathcal{K}_\varrho(x) \cap \mathcal{D} \cap \mathcal{B}} \frac{R_\varepsilon(y_1, y_2)}{|x - y_2|^{\mu_2}} \, dy_2 \right) dy_1$$

and obtain

$$|J_2| \leqq \frac{1}{\varepsilon^n} d_2 A_1 \int\limits_{\mathcal{K}_{2\varepsilon}(x)} |x - y_1|^{-\mu_1} \left(\int\limits_{\mathcal{K}_{\varepsilon}(y_1)} |x - y_2|^{-\mu_2} \, dy_2 \right) dy_1$$

$$\leqq \frac{1}{\varepsilon^n} d_2 A_1 \int\limits_{\mathcal{K}_{2\varepsilon}(x)} |x - y_1|^{-\mu_1} \, dy_1 \int\limits_{\mathcal{K}_{3\varepsilon}(x)} |x - y_2|^{-\mu_2} \, dy_2$$

$$\leqq d_2 A_1 \big(S_n(\mathcal{K}_1(0))\big)^2 \, 2^{n-\mu_1} 3^{n-\mu_2} \frac{1}{(n - \mu_1)(n - \mu_2)} \varepsilon^{n-\mu_1-\mu_2}.$$

Summarizing it is

$$\lim_{\varepsilon \downarrow 0} \frac{1}{\varepsilon^n} \int\limits_{\mathcal{K}_\varrho(x) \cap \mathcal{D}} F_1(x, y_1) \left(\int\limits_{\mathcal{K}_\varrho(x) \cap \mathcal{D} \cap \mathcal{B}} \frac{R_\varepsilon(y_1, y_2)}{|x - y_2|^{\mu_2}} \, dy_2 \right) dy_1$$

$$= \int\limits_{\mathcal{K}_\varrho(x) \cap \mathcal{D} \cap \mathcal{B}} F_1(x, y) \frac{a(y)}{|x - y|^{\mu_2}} \, dy$$

and the first statement of Theorem 5.3 is proved.

(2) In a first step we prove

$$\lim_{\varepsilon \downarrow 0} \frac{1}{\sqrt{\varepsilon}^{nk}} \int\limits_{\mathcal{E}(\{1,2,\ldots,k\})} \prod_{i=1}^{k} F_i(x_i, y_i) \left\langle \prod_{i=1}^{k} g_\varepsilon(y_i) \right\rangle dy_1 \ldots dy_k = 0$$

where the inequality

$$\max_s \left\{ \sum_{i=1}^{p_s} \mu_{l_{si}} \right\} < \frac{kn}{2}$$

is assumed to be fulfilled. It follows

$$\left| \frac{1}{\sqrt{\varepsilon}^{nk}} \int\limits_{\mathcal{E}(\{1,2,\ldots,k\})} \prod_{i=1}^{k} F_i(x_i, y_i) \left\langle \prod_{i=1}^{k} g_\varepsilon(y_i) \right\rangle dy_1 \ldots dy_k \right|$$

$$\leqq \frac{d_k}{\sqrt{\varepsilon}^{nk}} \prod_{i=1}^{k} A_i \int\limits_{\mathcal{D}_1} |x_1 - y_1|^{-\mu_1} \int\limits_{\mathcal{K}_{k\varepsilon}(y_1)} |x_2 - y_2|^{-\mu_2} \ldots \int\limits_{\mathcal{K}_{k\varepsilon}(y_1)} |x_k - y_k|^{-\mu_k} \, dy_k \ldots dy_1$$

$$\doteq J(\mathcal{D}_1; \varepsilon).$$

ϱ is given by

$$\varrho \doteq \begin{cases} \dfrac{1}{3} \min\limits_{x_i \neq x_j} \{|x_i - x_j|\} & \text{in the case that not all } x_i,\ i = 1, 2, \ldots, k, \text{ are equal,} \\ \text{arbitrary positive number} & \text{in the case that all } x_i,\ i = 1, 2, \ldots, k, \text{ are equal.} \end{cases}$$

Then we have

$$J(\mathcal{D}_1; \varepsilon) \leqq J\left(\mathcal{D}_1 \setminus \bigcup_{i=1}^{k} \mathcal{K}_\varrho(x_i); \varepsilon\right) + \sum_{i=1}^{k} J\big(\mathcal{K}_\varrho(x_i) \setminus \mathcal{K}_{(k+1)\varepsilon}(x_i)\big)$$

$$+ \sum_{i=1}^{k} J\big(\mathcal{K}_{(k+1)\varepsilon}(x_i)\big).$$

The first summand on the right-hand side of this inequality can be estimated by use of

$$|x_i - y_i| \geqq |x_i - y_i| - |y_i - y_1| \geqq \varrho - k\varepsilon \quad \text{for} \quad i = 1, 2, \ldots, k$$

by

$$J\left(\mathcal{D}_1 \setminus \bigcup_{i=1}^{k} \mathcal{K}_\varrho(x_i);\, \varepsilon\right)$$

$$\leqq \frac{1}{\sqrt{\varepsilon}^{nk}}\, d_k \prod_{i=1}^{k} A_i \prod_{i=1}^{k} (\varrho - k\varepsilon)^{-\mu_i}\, V(\mathcal{D}_1) \left((k\varepsilon)^n\, V_1\right)^{k-1}$$

$$= d_k \prod_{i=1}^{k} A_i V(\mathcal{D}_1)\, k^{n(k-1)}\, V_1^{k-1} \prod_{i=1}^{k} (\varrho - k\varepsilon)^{-\mu_i}\, \varepsilon^{\frac{n}{2}(k-2)}$$

and for $k \geqq 3$ it follows

$$\lim_{\varepsilon \downarrow 0} J\left(\mathcal{D}_1 \setminus \bigcup_{i=1}^{k} \mathcal{K}_\varrho(x_i);\, \varepsilon\right) = 0\,.$$

Consider $J\big(\mathcal{K}_\varrho(x_i) \setminus \mathcal{K}_{(k+1)\varepsilon}(x_i)\big)$ for an arbitrary $i \in \{1, 2, \ldots, k\}$. $j_1, \ldots, j_q$ denote the indices of the points x_j which agree with x_i; i.e. $x_{j_s} = x_i$ for $s = 1, 2, \ldots, q$. $j_{q+1}, \ldots, j_k$ denote the indices of the other points. Then we have

$$|x_{j_s} - y_{j_s}| \geqq |x_i - y_1| - |y_1 - y_{j_s}| \geqq |x_i - y_1| - k\varepsilon$$

$$\geqq |x_i - y_1| - \frac{k}{k+1}|x_i - y_1| = \frac{1}{k+1}|x_i - y_1| \quad \text{for } s = 1, 2, \ldots, q,$$

$$|x_{j_s} - y_{j_s}| \geqq |x_{j_s} - y_1| - |y_1 - y_{j_s}| \geqq |x_{j_s} - x_i| - |x_i - y_1| - |y_1 - y_{j_s}|$$

$$\geqq 3\varrho - \varrho - \frac{k}{k+1}\,\varrho \geqq \varrho \quad \text{for} \quad s = q+1, \ldots, k\,.$$

Using

$$Q \doteq \sum_{s=1}^{q} \mu_{j_s}, \qquad \bar{Q} \doteq \sum_{s=q+1}^{k} \mu_{j_s}$$

the considered term can be estimated by

$$J\big(\mathcal{K}_\varrho(x_i) \setminus \mathcal{K}_{(k+1)\varepsilon}(x_i)\big)$$

$$\leqq \frac{d_k}{\sqrt{\varepsilon}^{nk}} \prod_{j=1}^{k} A_j \prod_{s=q+1}^{k} [V_1 (k\varepsilon)^n\, \varrho^{-\mu_{j_s}}] \prod_{\substack{s=1\\ j_s \neq 1}}^{q} \left[\left(\frac{1}{k+1}\right)^{-\mu_{j_s}} (k\varepsilon)^n\, V_1\right]$$

$$\times \int_{\mathcal{K}_\varrho(x_1) \setminus \mathcal{K}_{(k+1)\varepsilon}(x_1)} |x_1 - y_1|^{-Q}\, dy_1$$

$$\leqq \bar{C}\varepsilon^{\frac{n}{2}(k-2)} \begin{cases} \dfrac{1}{n-Q}\left(\varrho^{n-Q} - \big((k+1)\,\varepsilon\big)^{n-Q}\right) & \text{for} \quad n \neq Q\,, \\ \ln(\varrho) - \ln\big((k+1)\,\varepsilon\big) & \text{for} \quad n = Q \end{cases}$$

for $x_1 = x_i$ and for $x_1 \neq x_i$ by

$$J\big(\mathcal{K}_\varrho(x_i) \setminus \mathcal{K}_{(k+1)\varepsilon}(x_i)\big)$$
$$\leqq \frac{d_k}{\sqrt{\varepsilon}^{nk}} \prod_{j=1}^{k} A_j \prod_{\substack{s=q+1\\ j_s \neq 1}}^{k} [V_1(k\varepsilon)^n \varrho^{-\mu_{j_s}}] \prod_{s=1}^{q} \left[\left(\frac{1}{k+1}\right)^{-\mu_{j_s}} (k\varepsilon)^n V_1\right] \varrho^{-\mu_1}$$
$$\times \int_{\mathcal{K}_\varrho(x_1)\setminus\mathcal{K}_{(k+1)\varepsilon}(x_1)} |x_1 - y_1|^{-Q}\, dy_1$$
$$\leqq \bar{C}\varepsilon^{\frac{n}{2}(k-2)} \begin{cases} \dfrac{1}{n-Q}\,(\varrho^{n-Q} - \big((k+1)\,\varepsilon\big)^{n-Q} & \text{for } n \neq Q, \\ \ln(\varrho) - \ln\big((k+1)\,\varepsilon\big) & \text{for } n = Q. \end{cases}$$

We get

$$\lim_{\varepsilon\downarrow 0} J\big(\mathcal{K}_\varrho(x_i) \setminus \mathcal{K}_{(k+1)\varepsilon}(x_i)\big) = 0.$$

Now we have to investigate the term $J\big(\mathcal{K}_{(k+1)\varepsilon}(x_i)\big)$ for an arbitrary $i \in \{1, 2, \ldots, k\}$. The same notations are chosen as for the previous term. It is

$$J\big(\mathcal{K}_{(k+1)\varepsilon}(x_i)\big)$$
$$\leqq \frac{d_k}{\sqrt{\varepsilon}^{nk}} \prod_{j=1}^{k} A_j \int_{\mathcal{K}_{2(k+1)\varepsilon}(x_i)} |x_1 - y_1|^{-\mu_1} \int_{\mathcal{K}_{2(k+1)\varepsilon}(x_i)} |x_2 - y_2|^{-\mu_2} \ldots$$
$$\times \int_{\mathcal{K}_{2(k+1)\varepsilon}(x_i)} |x_k - y_k|^{-\mu_k}\, dy_k \ldots dy_2\, dy_1.$$

For $s = q + 1, \ldots, k$ we obtain

$$|x_{j_s} - y_{j_s}| \geqq |x_{j_s} - x_i| - |x_i - y_{j_s}| \geqq 3\varrho - 2(k+1)\,\varepsilon \geqq \varrho$$

and for $s = 1, 2, \ldots, q$ the relation

$$\int_{\mathcal{K}_{2(k+1)\varepsilon}(x_i)} |x_{j_q} - y_{j_q}|^{-\mu_{j_q}}\, dy_{j_q} = \frac{S_n\big(\mathcal{K}_1(0)\big)}{n - \mu_{j_q}} \big(2(k+1)\,\varepsilon\big)^{n-\mu_{j_q}}.$$

From this it follows

$$J\big(\mathcal{K}_{(k+1)\varepsilon}(x_i)\big)$$
$$\leqq \frac{d_k}{\sqrt{\varepsilon}^{nk}} \prod_{j=1}^{k} A_j \prod_{s=1}^{q} \left[\frac{S_n\big(\mathcal{K}_1(0)\big)}{n - \mu_{j_s}} \big(2(k+1)\,\varepsilon\big)^{n-\mu_{j_s}}\right] \prod_{s=q+1}^{k} \left[\varrho^{-\mu_{j_s}} S_n\big(\mathcal{K}_1(0)\big) \big(2(k+1)\varepsilon\big)^n\right]$$
$$\leqq \tilde{C}\varepsilon^{\frac{nk}{2}-Q}$$

and furthermore

$$\lim_{\varepsilon\downarrow 0} J\big(\mathcal{K}_{(k+1)\varepsilon}(x_i)\big) = 0$$

if the condition

$$\max_s \left\{\sum_{i=1}^{p_s} \mu_{l_{si}}\right\} < \frac{kn}{2}$$

is taken into consideration.

The proof of (2) can be continued in a similar way to the proof of Theorem 1.7 and we do not need to repeat these considerations. The proof of (3) follows from the proof of Theorem 1.8. Theorem 5.3 is shown completely. ◀

An example is given for which the condition

$$\mu_1 + \mu_2 < n$$

of Theorem 5.3 is necessary in order to obtain limit relations. We choose a correlation function $R_\varepsilon(y_1, y_2)$ for which

$$\frac{1}{\varepsilon^n} \int\limits_{\mathcal{K}_\varepsilon(y_1)} R_\varepsilon(y_1, y_2)\, dy_2 \geqq a > 0 \quad \text{for all} \quad \varepsilon \leqq \varepsilon_0 .$$

Then the second moment $\langle \bar{r}_{1\varepsilon}(x)\, \bar{r}_{2\varepsilon}(x) \rangle$ is studied where

$$\bar{r}_{i\varepsilon}(x) = \frac{1}{\sqrt{\varepsilon^n}} \int\limits_{\mathcal{K}_\vartheta(x)} F_i(x, y)\, g_\varepsilon(y)\, dy; \qquad F_i(x, y) = \frac{1}{|x - y|^{\mu_i}} .$$

It follows

$$\begin{aligned} &\langle \bar{r}_{1\varepsilon}(x)\, \bar{r}_{2\varepsilon}(x) \rangle \\ &= \frac{1}{\varepsilon^n} \int\limits_{\mathcal{K}_\vartheta(x)} \frac{1}{|x - y_1|^{\mu_1}} \left(\int\limits_{\mathcal{K}_\vartheta(x) \cap \mathcal{K}_\varepsilon(y_1)} \frac{1}{|x - y_2|^{\mu_2}} R_\varepsilon(y_1, y_2)\, dy_2 \right) dy_1 \\ &\geqq \frac{1}{\varepsilon^n} \int\limits_{\mathcal{K}_\vartheta(x)} \frac{1}{|x - y_1|^{\mu_1}} \frac{1}{(|x - y_1| + \varepsilon)^{\mu_2}} \left(\int\limits_{\mathcal{K}_\vartheta(x) \cap \mathcal{K}_\varepsilon(y_1)} R_\varepsilon(y_1, y_2)\, dy_2 \right) dy_1 \end{aligned}$$

where the inequalities

$$|x - y_2| \leqq |x - y_1| + |y_1 - y_2| \leqq |x - y_1| + \varepsilon$$

were used. Furthermore, we have

$$\begin{aligned} \langle \bar{r}_{1\varepsilon}(x)\, \bar{r}_{2\varepsilon}(x) \rangle &\geqq a \int\limits_{\mathcal{K}_{\vartheta/2}(x)} \frac{1}{|x - y_1|^{\mu_1}} \frac{1}{(|x - y_1| + \varepsilon)^{\mu_2}}\, dy_1 \\ &= a S_n(\mathcal{K}_1(0)) \int\limits_0^{\vartheta/2} r^{n-1-\mu_1} (r + \varepsilon)^{-\mu_2}\, dr \\ &= a S_n(\mathcal{K}_1(0)) \int\limits_\varepsilon^{\vartheta/2+\varepsilon} (r - \varepsilon)^{n-1-\mu_1} r^{-\mu_2}\, dr . \end{aligned}$$

For $\mu_1 \geqq n - 1$; $\mu_1 + \mu_2 \geqq n$ it is

$$\begin{aligned} \langle \bar{r}_{1\varepsilon}(x)\, \bar{r}_{2\varepsilon}(x) \rangle &\geqq a S_n(\mathcal{K}_1(0)) \int\limits_\varepsilon^{\vartheta/2+\varepsilon} r^{n-1-\mu_1-\mu_2}\, dr \\ &\geqq a S_n(\mathcal{K}_1(0)) \int\limits_\varepsilon^{\vartheta/2+\varepsilon} \frac{1}{r}\, dr = a S_n(\mathcal{K}_1(0)) \ln \left(\frac{\varepsilon + \vartheta/2}{\varepsilon} \right) \end{aligned}$$

and for $\mu_1 < n - 1$; $\mu_1 + \mu_2 \geqq n$

$$\langle \bar{r}_{1\varepsilon}(x)\, \bar{r}_{2\varepsilon}(x) \rangle \geqq a S_n(\mathcal{K}_1(0)) \int\limits_{\varepsilon}^{\vartheta/2+\varepsilon} \frac{1}{r} \left(1 - \frac{\varepsilon}{r}\right)^n \mathrm{d}r$$

$$\geqq a S_n(\mathcal{K}_1(0)) \left[\ln \left(\frac{\varepsilon + \vartheta/2}{\varepsilon} \right) + \sum_{s=1}^{n} \binom{n}{s} \frac{1}{s} (-1)^{s+1} \left(\left(\frac{2\varepsilon}{\vartheta + 2\varepsilon} \right)^s - 1 \right) \right].$$

Hence, the relation

$$\lim_{\varepsilon \downarrow 0} \langle \bar{r}_{1\varepsilon}(x)\, \bar{r}_{2\varepsilon}(x) \rangle = \infty \quad \text{with} \quad \mu_1 + \mu_2 \geqq n$$

is shown.

The condition

$$\mu < \frac{n}{2}$$

follows from the demanded inequality in Theorem 5.3 (2) if all observation points x_i agree. In this case it is

$$\max_s \{\mu p_s\} = k\mu < \frac{kn}{2}, \quad \text{hence} \quad \mu < \frac{n}{2}.$$

From this considerations it is clear that limit theorems as to the solution of a Dirichlet problem (5.37) in $\mathbb{R}^n$ can be only obtained for such dimensions for which the inequality

$$n - 2 < \frac{n}{2} \quad \text{or} \quad n < 4$$

is valid. The cause is the singularity of the order $n - 2$ which the Green's function possesses.

Now, we formulate a result with respect to the boundary value problem (5.36).

Theorem 5.4. *For the solution $u_\varepsilon(x, \omega)$ of the Dirichlet problem* (5.36) *with the inhomogeneous term $g_\varepsilon(x, \omega)$ the limit relation*

$$\lim_{\varepsilon \downarrow 0} \frac{1}{\sqrt{\varepsilon^n}} \left(u_\varepsilon(x, \omega) - w(x) \right) = \xi(x, \omega) \quad \textit{in distribution}$$

is fulfilled for $n = 2, 3$ where

$$\bar{g}_\varepsilon(x, \omega) \doteq g_\varepsilon(x, \omega) - \langle g_\varepsilon(x) \rangle$$

is assumed to be a weakly correlated function. The random function $\xi(x, \omega)$ is Gaussian having zero-mean and the correlation function

$$\langle \xi(x_1)\, \xi(x_2) \rangle = \int\limits_{\mathcal{D}} G(x_1, y)\, G(x_2, y)\, a(y)\, \mathrm{d}y.$$

Furthermore, the correlation function can also be written as

$$\langle \xi(x_1)\, \xi(x_2) \rangle = \sum_{i,j=1}^{\infty} \frac{w_i(x_1)\, w_j(x_2)}{\lambda_i \lambda_j} (w_i, a w_j)$$

if $w_i(x)$, λ_i, $i = 1, 2, \ldots$, *denote the eigenfunctions and eigenvalues of the eigenvalue problem*

$$-\Delta w = \lambda w, \qquad w|_{\partial \mathcal{D}} = 0,$$

respectively. In particular, supposing a wide-sense homogeneous random function $\bar{g}_\varepsilon(x, \omega)$ *the intensity* a *is constant and the correlation function has the special form*

$$\langle \xi(x_1)\, \xi(x_2) \rangle = a \sum_{i=1}^{\infty} \frac{w_i(x_1)\, w_i(x_2)}{\lambda_i^2}.$$

Proof. The essential part of Theorem 5.4 follows immediately from Theorem 5.3. The eigenfunctions $\{w_i(x)\}_{i=1,2,\ldots}$ from

$$-\Delta w = \lambda w, \qquad w|_{\partial \mathcal{D}} = 0$$

are a complete orthonormal system as to $\mathbf{L}_2(\mathcal{D})$. Because of $n \leqq 3$ we have

$$|G(x, .)|_{\mathbf{L}_2(\mathcal{D})} < \infty$$

and the expansion

$$G(x, y) = \sum_{i=1}^{\infty} \big(G(x, .), w_i(.)\big)_{\mathbf{L}_2(\mathcal{D})}\, w_i(y) = \sum_{i=1}^{\infty} \frac{1}{\lambda_i}\, w_i(x)\, w_i(y)$$

where the convergence of the series means convergence as to $\mathbf{L}_2(\mathcal{D})$. Then, it follows

$$\lim_{k \to \infty} \left(\sum_{i=1}^{k} \frac{1}{\lambda_i}\, w_i(x_1)\, w_i(.), a(.) \sum_{j=1}^{k} \frac{1}{\lambda_j}\, w_j(x_2)\, w_j(.) \right)_{\mathbf{L}_2(\mathcal{D})}$$

$$= \sum_{i,j=1}^{\infty} \frac{1}{\lambda_i \lambda_j}\, w_i(x_1)\, w_j(x_2)\, (w_i, a w_j)_{\mathbf{L}_2(\mathcal{D})}$$

$$= \big(G(x_1, .), a G(x_2, .)\big)_{\mathbf{L}_2(\mathcal{D})}.$$

In the case of constant intensity a we notice that

$$(w_i, a w_j) = a(w_i, w_j) = a \delta_{ij}.$$

Now, Theorem 5.4 is proved completely. ◀

Example 5.4. Consider small stationary transverse vibrations of a fixed embedded circular membrane with constant density. The inhomogeneous term is assumed to be weakly correlated. For the displacement $u(x, y, \omega)$ of the membrane the boundary value problem (5.36) has to be solved where

$$\mathcal{D} = \{x \in \mathbb{R}^2 : |x| < R\}, \qquad u_0 = 0.$$

$g(x, \omega)$ denotes the density of the external force and $\bar{g}_\varepsilon(x, \omega)$ is supposed to be weakly correlated and wide-sense stationary. Then, the correlation function of the random limit function $\xi(x, \omega)$ can be determined by Theorem 5.4.

The eigenvalues of $-\Delta w = \lambda w$, $w|_{\partial \mathcal{D}} = 0$ are obtained by

$$\lambda_{ij} = \frac{1}{R^2}\, v_{ij}^2$$

and the adequate normalized eigenfunctions by

$$w_{ij,1}(r, \varphi) = f_{ij}(r) \sin(i\varphi),$$
$$w_{ij,2}(r, \varphi) = f_{ij}(r) \cos(i\varphi) \quad \text{for} \quad i = 0, 1, 2, \ldots; j = 1, 2, \ldots$$

with

$$f_{ij}(r) \doteq \sqrt{\frac{2}{\pi}} \frac{J_i\left(v_{ij} \frac{r}{R}\right)}{R\,|J_i'(v_{ij})|}$$

where

$$x_1 = r \cos(\varphi), \qquad x_2 = r \sin(\varphi)$$

and J_i denotes the i-th Bessel function and v_{ij} the j-th zero of J_i. From Theorem 5.4 it follows

$$\begin{aligned}
&\langle \xi(x_1)\, \xi(x_2) \rangle \\
&= a \sum_{i=0}^{\infty} \sum_{j=1}^{\infty} \frac{1}{\lambda_{ij}^2} \left[w_{ij,1}(x_1)\, w_{ij,1}(x_2) + w_{ij,2}(x_1)\, w_{ij,2}(x_2) \right] \\
&= \frac{2}{\pi} R^2 a \sum_{i=0}^{\infty} \sum_{j=1}^{\infty} \frac{J_i\left(\frac{1}{R} v_{ij} r_1\right) J_i\left(\frac{1}{R} v_{ij} r_2\right)}{v_{ij}^4 J_i'^2(v_{ij})} \cos\left(i(\varphi_1 - \varphi_2)\right) \qquad (5.44)
\end{aligned}$$

where

$$x_i = (x_{i1}, x_{i2}) = \left(r_i \cos(\varphi_i), r_i \sin(\varphi_i)\right) \quad \text{for} \quad i = 1, 2.$$

It is to see from (5.44) that the correlation function is invariant against a rotary motion as to the centre of $\mathscr{D}$. The variance is given by

$$\langle \xi^2(x) \rangle = \frac{2}{\pi} R^2 a \sum_{i=0}^{\infty} \sum_{j=1}^{\infty} \frac{J_i^2\left(v_{ij} \frac{r}{R}\right)}{v_{ij}^4 J_i'^2(v_{ij})} \doteq \sigma_R^2(r)$$

and this function is only dependent on r.

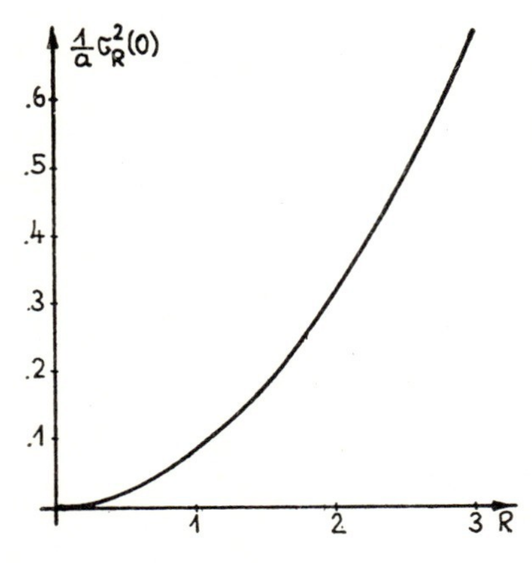

Fig. 5.9. $\sigma_R^2(0)$ in dependence on R

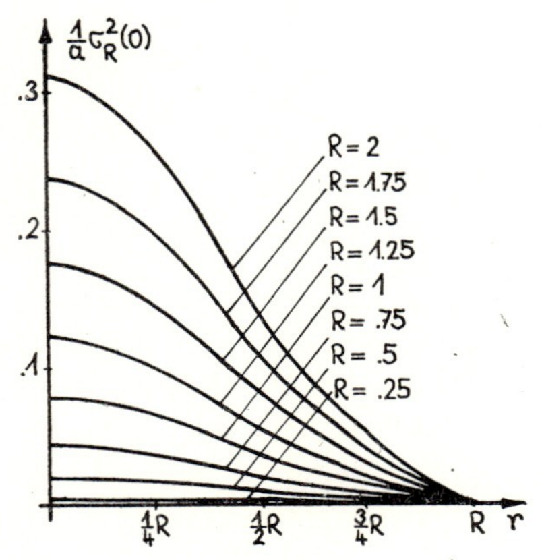

Fig. 5.10. $\sigma_R^2(r)$ as function of r for different values R

Fig. 5.9 shows $\sigma_R^2(r)$ for $r = 0$ as a function of R. $\sigma_R^2(r)$ in dependence on r is illustrated in Fig. 5.10 for different values R. The function $\sigma_R^2(r)$ possesses a maximum for $r = 0$ and $\sigma_R^2(0)$ increases if R goes to greater values. This property follows from the influence of the boundary condition $u|_{\partial\mathcal{D}} = 0$.

Consider the random boundary-value problem

$$Lu + f(x, \omega)\, u = g(x, \omega); \qquad \left.\frac{\partial^i u}{\partial n^i}\right|_{\partial\mathcal{D}} = 0, \quad i = 0, 1, \ldots, m-1 \tag{5.45}$$

on a bounded, sufficiently smooth domain $\mathcal{D} \subset \mathbb{R}^n$. The non-random operator L is defined by

$$Lu \doteq \sum_{k=1}^{m} \sum_{|\alpha|=|\beta|=k} (-1)^k D^\alpha\big(f_{\alpha\beta}(x)\, D^\beta u\big)$$

where α, β denote multi-indices

$$\alpha = (\alpha_1, \ldots, \alpha_n); \qquad \beta = (\beta_1, \ldots, \beta_n); \qquad |\alpha| = \sum_{p=1}^{n} \alpha_p .$$

The averaged problem belonging to (5.45) has the form

$$L_0 w \doteq Lw + \langle f(x)\rangle\, w = \langle g(x)\rangle; \qquad \left.\frac{\partial^i w}{\partial n^i}\right|_{\partial\mathcal{D}} = 0, \quad i = 0, 1, \ldots, m-1. \tag{5.46}$$

Let L_0 be a positive definite, symmetric, elliptic operator of the order $2m$ with symmetric coefficients

$$f_{\alpha\beta}(x) = f_{\beta\alpha}(x).$$

The Green's function $G(x, y)$ as to

$$L_0 w = 0; \qquad \left.\frac{\partial^i w}{\partial n^i}\right|_{\partial\mathcal{D}} = 0, \quad i = 0, 1, \ldots, m-1,$$

is supposed to be logarithmically singular, i.e.

$$G(x, y) = A(x, y) \ln |x - y|$$

or weakly singular, i.e.

$$G(x, y) = B(x, y)\, |x - y|^{-\mu} \quad \text{with} \quad \mu < \frac{n}{2},$$

where the functions $A(x, y)$, $B(x, y)$ are to be continuous. We suppose an expansion of the form

$$u(x, \omega) = \sum_{j=0}^{\infty} u_j(x, \omega) \tag{5.47}$$

where $u_j(x, \omega)$ denotes the terms of the solution $u(x, \omega)$ which are homogeneous of j-th order as to

$$\bar{f}(x, \omega) \doteq f(x, \omega) - \langle f(x)\rangle, \qquad \bar{g}(x, \omega) \doteq g(x, \omega) - \langle g(x)\rangle.$$

By means of (5.47) the boundary-value problems

$$L_0 u_0 = \langle g \rangle; \quad \left.\frac{\partial^i u_0}{\partial n^i}\right|_{\partial\mathcal{D}} = 0, \qquad i = 0, 1, \ldots, m-1,$$

$$L_0 u_1 = -\bar{f} u_0 + \bar{g}; \quad \left.\frac{\partial^i u_1}{\partial n^i}\right|_{\partial\mathcal{D}} = 0, \qquad i = 0, 1, \ldots, m-1,$$

$$L_0 u_j = -\bar{f} u_{j-1}; \quad \left.\frac{\partial^i u_j}{\partial n^i}\right|_{\partial\mathcal{D}} = 0, \quad i = 0, 1, \ldots, m-1 \quad \text{for} \quad j = 2, 3, \ldots$$

have to be solved as to $u_0, u_1, \ldots$. There u_0 denotes the solution of the averaged problem. It follows

$$\begin{aligned} u_0(x) &= \int_{\mathcal{D}} G(x, y) \langle g(y) \rangle \, dy, \\ u_1(x, \omega) &= -\int_{\mathcal{D}} G(x, y) [-\bar{g}(y, \omega) + \bar{f}(y, \omega) u_0(y)] \, dy, \\ u_j(x, \omega) &= (-1)^j \int_{\mathcal{D}} \ldots \int_{\mathcal{D}} G(x, y_1) G(y_1, y_2) \ldots G(y_{j-1}, y_j) \\ &\quad \times \bar{f}(y_1) \bar{f}(y_2) \ldots \bar{f}(y_{j-1}) [\bar{f}(y_j) u_0(y_j) - \bar{g}(y_j)] \, dy_1 \ldots dy_j . \end{aligned} \tag{5.48}$$

Now, we assume that the solution $u(x, \omega)$ of (5.45) can be written in the form (5.47) where the functions $u_j(x, \omega)$ are given by (5.48).

Theorem 5.5. *The random boundary-value problem* (5.45) *on $\mathcal{D}$ is considered with the conditions:*

(1) *$\mathcal{D}$ is a bounded, sufficiently smooth domain $\mathcal{D} \subset \mathbb{R}^n$ with* diam $(\mathcal{D}) = \vartheta$ *and* $n \leqq 3$.

(2) *The averaged operator $L_0 = L + \langle f \rangle$ is assumed to be a positive definite, symmetric operator of the order 2m with symmetric coefficients. Let the Green's function $G(x, y)$ belonging to the averaged problem* (5.46) *be logarithmically singular or weakly singular.*

(3) *The random functions $f(x, \omega)$, $g(x, \omega)$ are supposed to have smooth sample functions. The vector function $\left(\bar{f}_\varepsilon(x, \omega), \bar{g}_\varepsilon(x, \omega)\right)$ is assumed to be weakly correlated connected with correlation length ε. Let be*

$$|\bar{f}_\varepsilon(x, \omega)| \leqq \eta$$

where η is a small positive number.

(4) *The solution $u(x, \omega)$ of* (5.45) *is assumed to have the form* (5.47) *where the functions $u_j(x, \omega)$, $j = 0, 1, \ldots$, are given by* (5.48). *The random function*

$$\frac{1}{\sqrt{\varepsilon^n}} \left(u(x, \omega) - u_0(x)\right)$$

converges in distribution to a Gaussian random function $\xi(x, \omega)$ with mean $\langle \xi(x) \rangle = 0$ and correlation function

$$\langle \xi(x) \, \xi(y) \rangle = \int_{\mathcal{D}} G(x, z) G(y, z) \left\{a_{gg}(z) - u_0(z) \left(a_{gf}(z) + a_{fg}(z)\right) + u_0^2(z) a_{ff}(z)\right\} dz .$$

$u_0(z)$ denotes the solution of the averaged problem and can be determined by (5.48). *The intensities as to $\left(\bar{f}_\varepsilon(x, \omega), \bar{g}_\varepsilon(x, \omega)\right)$ are given by a_{gg}, a_{gf}, a_{fg}, a_{ff}.*

Proof. Let

$$^k u(x, \omega) = \sum_{j=0}^{k} u_j(x, \omega).$$

First, the proof is produced for

$$\frac{1}{\sqrt{\varepsilon^n}} \left({}^k u(x, \omega) - u_0(x) \right)$$

with $k \geqq 1$. We obtain

$$\frac{1}{\sqrt{\varepsilon^{pn}}} \left\langle \prod_{q=1}^{p} \left({}^k u(x_{i_q}) - u_0(x_{i_q}) \right) \right\rangle$$

$$= \frac{1}{\sqrt{\varepsilon^{pn}}} \left\langle \prod_{q=1}^{p} u_1(x_{i_q}) \right\rangle + \sum_{\substack{j_1, \ldots, j_p = 1 \\ \sum_{q=1}^{p} j_q > p}}^{p} \frac{1}{\sqrt{\varepsilon^{pn}}} \left\langle \prod_{q=1}^{p} u_{j_q}(x_{i_q}) \right\rangle. \tag{5.49}$$

Using the results of Section 1.3 the first summand on the right-hand side of (5.49) has the limit value

$$\lim_{\varepsilon \downarrow 0} \frac{1}{\sqrt{\varepsilon^{pn}}} \left\langle \prod_{q=1}^{p} u_1(x_{i_q}) \right\rangle = \begin{cases} \sum\limits_{\{u_1, v_1\}, \ldots, \{u_{p/2}, v_{p/2}\}} \prod\limits_{q=1}^{p/2} {}^2A_{1, u_q v_q} & \text{for } p \text{ even}, \\ 0 & \text{for } p \text{ odd} \end{cases}$$

where ${}^2A_{1, u_q v_q}$ is given by

$${}^2A_{1, u_q v_q} = \lim_{\varepsilon \downarrow 0} \frac{1}{\varepsilon^n} \left\langle u_1\left(x_{i_{u_q}}\right) u_1\left(x_{i_{v_q}}\right) \right\rangle$$

$$= \int_{\mathcal{D}} G\left(x_{i_{u_q}}, y\right) G\left(x_{i_{v_q}}, y\right) \left[a_{gg}(y) - u_0(y) \left(a_{fg}(y) + a_{gf}(y) \right) + u_0^2(y)\, a_{ff}(y) \right] \mathrm{d}y$$

if we use $u_1(x, \omega)$ from (5.48).

Now, we show that the second summand on the right-hand side of (5.49) converges to zero as $\varepsilon \downarrow 0$, i.e. we prove

$$\lim_{\varepsilon \downarrow 0} \frac{1}{\sqrt{\varepsilon^{pn}}} \left\langle \prod_{q=1}^{p} u_{j_q}(x_{i_q}) \right\rangle = 0 \quad \text{for} \quad \sum_{q=1}^{p} j_q > p. \tag{5.50}$$

Using (5.48),

$$u_j(x, \omega) = (-1)^j \int_{\mathcal{D}} \ldots \int_{\mathcal{D}} G(x, y_1) \prod_{u=1}^{j-1} G(y_u, y_{u+1}) \prod_{u=1}^{j-1} \bar{\bar{f}}_\varepsilon(y_u)\, \bar{\bar{f}}_\varepsilon(y_j)\, \mathrm{d}y_1 \ldots \mathrm{d}y_j$$

for $j = 1, 2, \ldots$ where $\bar{\bar{f}}_\varepsilon$ is defined by

$$\bar{\bar{f}}_\varepsilon(y, \omega) \doteq \bar{f}_\varepsilon(y, \omega)\, u_0(y) - \bar{g}_\varepsilon(y, \omega),$$

we get

$$\left\langle \prod_{q=1}^{p} u_{j_q}(x_{i_q}) \right\rangle = \underbrace{\int_{\mathcal{D}} \ldots \int_{\mathcal{D}}}_{\sum_{q=1}^{p} j_q = j\text{-times}} J({}^1y_1, \ldots, {}^1y_{j_1}, \ldots, {}^py_1, \ldots, {}^py_{j_p})\, \mathrm{d}^1y_1 \ldots \mathrm{d}^py_{j_p}$$

where the integrand J is given by

$$J = J({}^1y_1, \ldots, {}^1y_{j_1}, \ldots, {}^py_1, \ldots, {}^py_{j_p})$$
$$= \prod_{q=1}^{p} \left[(-1)^{j_q} G(x_{i_q}, {}^qy_1) \prod_{u=1}^{j_q-1} G({}^qy_u, {}^qy_{u+1}) \right] \left\langle \prod_{q=1}^{p} \left[\prod_{u=1}^{j_q-1} \bar{f}_\varepsilon({}^qy_u)\, \bar{\bar{f}}_\varepsilon({}^qy_{j_q}) \right] \right\rangle .$$

By means of (1.7) and (1.4) it follows

$$\Bigg| \underbrace{\int_{\mathcal{D}} \cdots \int_{\mathcal{D}}}_{j\text{-times}} J \, \mathrm{d}^1y_1 \ldots \mathrm{d}^py_{j_p} \Bigg| = \sum_{s=1}^{j/2} \sum_{\{\mathcal{J}_1, \ldots, \mathcal{J}_s\}} \int_{\mathcal{B}(\mathcal{J}_1, \ldots, \mathcal{J}_s)} |J| \, \mathrm{d}^1y_1 \ldots \mathrm{d}^py_{j_p}$$

$$\leqq \sum_{s=1}^{j/2} \sum_{\{\mathcal{J}_1, \ldots, \mathcal{J}_s\}} \int_{\mathcal{E}(\mathcal{J}_1)} \cdots \int_{\mathcal{E}(\mathcal{J}_s)} |J| \, \mathrm{d}^1y_1 \ldots \mathrm{d}^py_{j_p} \tag{5.51}$$

where the second sum is taken over all nonequivalent separations $\{\mathcal{J}_1, \ldots, \mathcal{J}_s\}$ of the indices of $\{{}^1y_1, \ldots, {}^1y_{j_1}, \ldots, {}^py_1, \ldots, {}^py_{j_p}\}$. First, we use the inequality

$$\left\langle \prod_{q=1}^{p} \left[\prod_{u=1}^{j_q-1} \bar{f}_\varepsilon({}^qy_u)\, \bar{\bar{f}}_\varepsilon({}^qy_{j_q}) \right] \right\rangle \leqq \left[\prod_{q=1}^{p} \prod_{u=1}^{j_q-1} \langle \bar{f}_\varepsilon^j({}^qy_u) \rangle \langle \bar{\bar{f}}_\varepsilon^j({}^qy_{j_q}) \rangle \right]^{\frac{1}{j}} \leqq d_j$$

where d_j denotes a bound of the absolute j-th moment of $\bar{f}_\varepsilon(x, \omega)$ and $\bar{\bar{f}}_\varepsilon(x, \omega)$, i.e.

$$\langle |\bar{f}_\varepsilon(x)|^j \rangle \leqq d_j \quad \text{for all } x \in \mathcal{D},$$

$$\langle |\bar{\bar{f}}_\varepsilon(x)|^j \rangle = \langle |\bar{f}_\varepsilon(x)\, u_0(x) - \bar{g}_\varepsilon(x)|^j \rangle \leqq d_j \quad \text{for all } x \in \mathcal{D}.$$

It is

$$\mathcal{E}(\mathcal{J}) = \left\{ (z_1, \ldots, z_j) \in \mathop{\times}_{q=1}^{j} \mathcal{D} : \{z_1, \ldots, z_j\} \ \varepsilon\text{-neighbouring} \right\}$$

and

$$\mathcal{E}(\mathcal{J}) \subset \mathcal{D} \times \mathcal{K}_{j\varepsilon}(z_1) \times \mathcal{K}_{j\varepsilon}(z_1) \times \ldots \times \mathcal{K}_{j\varepsilon}(z_1).$$

We obtain

$$\int_{\mathcal{E}(\mathcal{J}_1)} \cdots \int_{\mathcal{E}(\mathcal{J}_s)} |J| \, \mathrm{d}^1y_1 \ldots \mathrm{d}^py_{j_p}$$
$$\leqq \int_{\mathcal{D}} \mathrm{d}^1z_1 \int_{\mathcal{K}_{j\varepsilon}({}^1z_1)} \mathrm{d}^1z_2 \ldots \int_{\mathcal{K}_{j\varepsilon}({}^1z_1)} \mathrm{d}^1z_{k_1}$$
$$\times \int_{\mathcal{D}} \mathrm{d}^2z_1 \int_{\mathcal{K}_{j\varepsilon}({}^2z_1)} \mathrm{d}^2z_2 \ldots \int_{\mathcal{K}_{j\varepsilon}({}^2z_1)} \mathrm{d}^2z_{k_2} \ldots \int_{\mathcal{D}} \mathrm{d}^sz_1 \int_{\mathcal{K}_{j\varepsilon}({}^sz_1)} \mathrm{d}^sz_2 \ldots \int_{\mathcal{K}_{j\varepsilon}({}^sz_1)} \mathrm{d}^sz_{k_s} |J| \tag{5.52}$$

if

$$\mathcal{E}(\mathcal{J}_1) \text{ corresponds to } ({}^1z_1, \ldots, {}^1z_{k_1}),$$

$$\mathcal{E}(\mathcal{J}_2) \text{ to } ({}^2z_1, \ldots, {}^2z_{k_2}), \ldots, \mathcal{E}(\mathcal{J}_s) \text{ to } ({}^sz_1, \ldots, {}^sz_{k_s})$$

from

$$({}^1y_1, \ldots, {}^1y_{j_1}, \ldots, {}^py_1, \ldots, {}^py_{j_p}) \quad \text{with} \quad \sum_{q=1}^{s} k_q = j.$$

For further considerations we need the inequalities

$$\int_{\mathcal{D}} \frac{\mathrm{d}z}{|x - z|^\mu} \leqq \frac{1}{n - \mu} S_n\big(\mathcal{K}_1(0)\big)\, \vartheta^{n-\mu},$$

$$\int\limits_{\mathcal{K}_{l\varepsilon}(y)} \frac{\mathrm{d}z}{|x-z|^\mu} \leqq \begin{cases} \dfrac{1}{n-\mu} S_n(\mathcal{K}_1(0))\,(l\varepsilon)^{n-\mu} & \text{for } x \notin \mathcal{K}_{l\varepsilon}(y), \\ \displaystyle\int\limits_{\mathcal{K}_{2l\varepsilon}(x)} \frac{\mathrm{d}z}{|x-z|^\mu} \leqq \frac{1}{n-\mu} S_n(\mathcal{K}_1(0))\,(2l\varepsilon)^{n-\mu} & \text{for } x \in \mathcal{K}_{l\varepsilon}(y) \end{cases}$$

$$\leqq \frac{1}{n-\mu} S_n(\mathcal{K}_1(0))\,(2l\varepsilon)^{n-\mu}$$

where $S_n(\mathcal{K}_1(0))$ denotes the area of the surface of $\mathcal{K}_1(0)$. Then, it follows

$$\int\limits_{\mathcal{D}} |G(x,z)|\,\mathrm{d}z \leqq c_1, \tag{5.53}$$

$$\int\limits_{\mathcal{K}_{l\varepsilon}(y)} |G(x,z)|\,\mathrm{d}z \leqq \begin{cases} c_2\varepsilon^{n-\alpha} & \text{in the case of a logarithmic singularity of } G, \\ c_3\varepsilon^{n-\mu} & \text{in the case of a weak singularity of } G \end{cases}$$

and α is an arbitrary positive number. We only deal with the case of a weak singularity since the case of a logarithmic singularity is contained since $\alpha > 0$ is arbitrary.

Using (5.52) it is

$$\int\limits_{\mathcal{E}(\mathcal{J}_1)} \cdots \int\limits_{\mathcal{E}(\mathcal{J}_s)} |J|\,\mathrm{d}^1y_1 \ldots \mathrm{d}^py_{j_p}$$

$$\leqq \prod_{q=1}^{s} \left[\int\limits_{\mathcal{D}} \mathrm{d}z_q \int\limits_{\mathcal{K}_{j\varepsilon}(z_q)} \mathrm{d}y_q\, |G(\bar{z}_q, z_q)\, G(\bar{y}_q, y_q)|\,(c_3\varepsilon^{n-\mu})^{k_q-2}\right] \tag{5.54}$$

where

$$\bar{z}_p, \bar{y}_p \in \{{}^1y_1, \ldots, {}^1y_{j_1}, \ldots, {}^py_1, \ldots, {}^py_{j_p}\}.$$

We use the results of Section 1.6.1 and get

$$\frac{1}{\sqrt{\varepsilon}^{pn}} \int\limits_{\mathcal{E}(\mathcal{J}_1)} \cdots \int\limits_{\mathcal{E}(\mathcal{J}_s)} |J|\,\mathrm{d}^1y_1 \ldots \mathrm{d}^py_{j_p} \leqq c\varepsilon^{sn+(n-\mu)(j-2s)-np/2} = c\varepsilon^{(n/2-\mu)(j-2s)+n(j-p)/2}$$

and by means of

$$\frac{n}{2} - \mu > 0; \qquad j - 2s \geqq 0; \qquad j - p > 0$$

it follows

$$\lim_{\varepsilon\downarrow 0} \frac{1}{\sqrt{\varepsilon}^{pn}} \int\limits_{\mathcal{E}(\mathcal{J}_1)} \cdots \int\limits_{\mathcal{E}(\mathcal{J}_s)} |J|\,\mathrm{d}^1y_1 \ldots \mathrm{d}^py_{j_p} = 0$$

and then (5.50). The statement of Theorem 5.5 is proved for ${}^ku(x,\omega)$ with $k \geqq 1$.

Using the proof of Theorem 5.2 we see that Theorem 5.5 is proved if the relation

$$\lim_{\varepsilon\downarrow 0} \frac{1}{\varepsilon^n} \langle({}^ku(x) - u(x))^2\rangle = 0 \quad \text{uniformly as to } \varepsilon \tag{5.55}$$

is established. For this, we deal with

$$\frac{1}{\varepsilon^n} \langle u_i(x)\, u_j(x)\rangle \quad \text{where} \quad i + j \geqq N_0.$$

This term can be written as

$$\frac{1}{\varepsilon^n}\langle u_i(x)\, u_j(x)\rangle$$

$$= \frac{1}{\varepsilon^n}(-1)^{i+j}\underbrace{\int_{\mathcal{D}}\cdots\int_{\mathcal{D}}}_{(i+j)\text{-times}} G(x, y_1)\prod_{u=1}^{i-1} G(y_u, y_{u+1})\, G(x, z_1)\prod_{u=1}^{j-1} G(z_u, z_{u+1})$$

$$\times \left\langle \prod_{u=1}^{i-1} \bar{f}_\varepsilon(y_u)\, \bar{\bar{f}}_\varepsilon(y_i) \prod_{u=1}^{j-1} \bar{f}_\varepsilon(z_u)\, \bar{\bar{f}}_\varepsilon(z_j)\right\rangle \mathrm{d}y_1 \ldots \mathrm{d}y_i\, \mathrm{d}z_1 \ldots \mathrm{d}z_j. \tag{5.56}$$

The moment contained in (5.56) can be estimated by

$$\left|\left\langle \prod_{u=1}^{i-1} \bar{f}_\varepsilon(y_u)\, \bar{\bar{f}}_\varepsilon(y_i) \prod_{u=1}^{j-1} \bar{f}_\varepsilon(z_u)\, \bar{\bar{f}}_\varepsilon(z_j)\right\rangle\right| \leqq \eta^{i+j-2}[\langle \bar{\bar{f}}{}_\varepsilon^2(y_i)\rangle \langle \bar{\bar{f}}{}_\varepsilon^2(z_j)\rangle]^{1/2} \leqq \eta^{i+j-2} d_2$$

where the inequality

$$|u_0(x)| \leqq c_0$$

is applied. Assuming

$$(i+j)^{2n} \geqq \frac{1}{\varepsilon^n}$$

we obtain

$$\left|\frac{1}{\varepsilon^n}\langle u_i(x)\, u_j(x)\rangle\right| \leqq \frac{1}{\varepsilon^n}\,\eta^{i+j-2} c_1^{i+j} d_2 \leqq c_1^2 d_2 (c_1\eta)^{i+j-2}(i+j)^{2n} \doteq a_{i+j}$$

where we have used the first inequality of (5.53).

Now we suppose the relation

$$(i+j)^{2n} < \frac{1}{\varepsilon^n}$$

and then (5.54) leads to

$$\left|\frac{1}{\varepsilon^n}\langle u_i(x)\, u_j(x)\rangle\right|$$

$$\leqq \eta^{i+j-2}\, d_2 \frac{1}{\varepsilon^n} \sum_{s=1}^{(i+j)/2} \sum_{\{\mathcal{J}_1,\ldots,\mathcal{J}_s\}} \int_{\mathcal{E}(\mathcal{J}_1)}\cdots\int_{\mathcal{E}(\mathcal{J}_s)}$$

$$\times \left|G(x, y_1) \prod_{u=1}^{i-1} G(y_u, y_{u+1})\, G(x, z_1) \prod_{u=1}^{j-1} G(z_u, z_{u+1})\right| \mathrm{d}y_1 \ldots \mathrm{d}y_i\, \mathrm{d}z_1 \ldots \mathrm{d}z_j$$

$$\leqq d_2\eta^{i+j-2}\frac{1}{\varepsilon^n}\sum_{s=1}^{(i+j)/2}\sum_{\{\mathcal{J}_1,\ldots,\mathcal{J}_s\}} \left(\tilde{c}(2\varepsilon)^n\right)^s \left(\tilde{c}_3\left(2(i+j)\varepsilon\right)^{n-\mu}\right)^{i+j-2s} \tag{5.57}$$

if we use similar considerations which are contained in (5.54) and the inequality

$$\int_{\mathcal{D}} \mathrm{d}z \int_{\mathcal{K}_{2\varepsilon}(z)} \mathrm{d}y\, |G(\bar{z}, z)\, G(\bar{y}, y)| \leqq \tilde{c}(2\varepsilon)^n.$$

k_q, $q = 1, 2, \ldots, s$, denote the number of elements contained in $\mathcal{J}_q$. If $(k_1, \ldots, k_s)$ is fixed then the sum $\sum_{\{\mathcal{J}_1,\ldots,\mathcal{J}_s\}}$ has $\frac{(i+j)!}{k_1!\ldots k_s!}$ summands and we do not have more than $(i+j)^s$

possibilities to choose $(k_1, \dots, k_s)$. Hence

$$\sum_{s=1}^{(i+j)/2} \sum_{\{\mathcal{J}_1,\dots,\mathcal{J}_s\}} 1 \leqq \sum_{s=1}^{(i+j)/2} \frac{(i+j)!}{2^s} (i+j)^s = (i+j)! \left[\frac{\left(\frac{i+j}{2}\right)^{(i+j)/2+1} - 1}{\frac{i+j}{2} - 1} - 1 \right]$$

$$\leqq 2(i+j)!\,(i+j)^{(i+j)/2}$$

and every summand of (5.57) can be estimated by

$$S \doteq d_2 \eta^{i+j-2} \frac{1}{\varepsilon^n} \left(\tilde{c}(2\varepsilon)^n\right)^s \left(\tilde{c}_3\left(2(i+j)\,\varepsilon\right)^{n-\mu}\right)^{i+j-2s}$$

$$\leqq d_2 2^{n(i+j)/2} (\eta \tilde{c} \tilde{\tilde{c}}_3)^{i+j-2} \left((i+j)\,\varepsilon\right)^{(n-\mu)(i+j-2s)} \varepsilon^{n(s-1)}$$

if we note

$$s \leqq \frac{i+j}{2}; \qquad \tilde{\tilde{c}}_3 \doteq 2^{n-\mu} \tilde{c}_3 \geqq 1.$$

Furthermore, using $\varepsilon < (i+j)^{-2}$ it follows

$$S \leqq d_2 2^{n(i+j)/2} (\eta \tilde{c} \tilde{\tilde{c}}_3)^{i+j-2} (i+j)^{2n - n(i+j) + \mu(i+j-2s)}$$

$$\leqq 2^n d_2 (2^{n/2} \eta \tilde{c} \tilde{\tilde{c}}_3)^{i+j-2} (i+j)^{2n-(n-\mu)(i+j)}.$$

Now, (5.57) leads to

$$\left| \frac{1}{\varepsilon^n} \langle u_i(x)\, u_j(x) \rangle \right| \leqq 2(i+j)!\,(i+j)^{(i+j)/2}\, 2^n d_2 (2^{n/2} \eta \tilde{c} \tilde{\tilde{c}}_3)^{i+j-2} (i+j)^{2n-(n-\mu)(i+j)}$$

$$\leqq 2^{n+1} d_2 (2^{n/2} \eta \tilde{c} \tilde{\tilde{c}}_3)^{i+j-2} \frac{(i+j)!\,(i+j)^{2n}}{(i+j)^{(n-\mu-1/2)(i+j)}}$$

$$\leqq 2^{n+1} d_2 (2^{n/2} \eta \tilde{c} \tilde{\tilde{c}}_3)^{i+j-2} \frac{(i+j)!\,(i+j)^{2n}}{(i+j)^{(i+j)}} \doteq b_{i+j}$$

if we take into consideration the inequality

$$n - \mu - \frac{1}{2} \geqq 1 \quad \text{for} \quad n \geqq 2.$$

We see that

$$\sum_{i,j=1}^{\infty} b_{i+j} = \sum_{p=2}^{\infty} \sum_{q=1}^{p-1} b_p = \sum_{p=2}^{\infty} (p-1)\, b_p$$

and this series is convergent. This fact follows from the ratio test for small η,

$$\lim_{p\to\infty} \left| \frac{C^{p-1}(p+1)!\,(p+1)^{2n} p p^p}{(p+1)^{p+1} C^{p-2} p!\, p^{2n} (p-1)} \right| \leqq \lim_{p\to\infty} C \left(\frac{p+1}{p}\right)^{2n} \frac{p+1}{p-1} = C,$$

where we have put

$$C = 2^{n/2} \eta \tilde{c} \tilde{\tilde{c}}_3.$$

The series $\sum_{i,j=1}^{\infty} a_{i+j}$ is also convergent.

We have obtained

$$\left| \frac{1}{\varepsilon^n} \langle u_i(x)\, u_j(x) \rangle \right| \leqq a_{i+j} + b_{i+j}$$

and the series

$$\sum_{i,j=1}^{\infty} (a_{i+j} + b_{i+j})$$

is convergent. Hence, the uniform convergence (5.55) as to ε is shown and Theorem 5.5, too. ◀

The correlation function of the limit process $\xi(x, \omega)$ can be determined with the help of the Green's function $G(x, y)$ (see Theorem 5.5). This Green's function can be expanded as to the eigenfunctions $w_i(x)$ of the eigenvalue problem

$$L_0 w = \lambda w; \qquad \left.\frac{\partial^i w}{\partial n^i}\right|_{\partial \mathcal{D}} = 0, \qquad i = 0, 1, \ldots, m-1,$$

and we find

$$\langle \xi(x)\, \xi(y) \rangle = \sum_{i,j=1}^{\infty} \frac{1}{\lambda_i \lambda_j} w_i(x)\, w_j(y)$$
$$\times \int_{\mathcal{D}} \left(a_{gg}(z) - u_0(z) \left(a_{gf}(z) + a_{fg}(z)\right) + u_0^2(z)\, a_{ff}(z)\right) w_i(z)\, w_j(z)\, \mathrm{d}z .$$

We want to use this form of the limit correlation function in a following example.

Example 5.5. Consider the boundary value problem

$$-\Delta u + f(x, \omega)\, u = g(x, \omega); \qquad u|_{\partial \mathcal{D}} = 0$$

on a rectangular domain

$$\mathcal{D} = \{x \in \mathbb{R}^2 : 0 < x_1 < r, 0 < x_2 < s\}.$$

Hence, we deal with small transverse vibrations of a rectangular, fixedly, elastically embedded membrane. Furthermore, we assume that

$$\langle f(x) \rangle = f_0 > 0, \qquad \langle g(x) \rangle = g_0 \sin\left(\frac{1}{r}\, \pi x_1\right) \sin\left(\frac{1}{s}\, \pi x_2\right)$$

as to the averaged modulus of foundation and the averaged density of the external force. Now, the averaged problem (5.46) can be written as

$$-\Delta u_0 + f_0 u_0 = g_0 \sin\left(\frac{1}{r}\, \pi x_1\right) \sin\left(\frac{1}{s}\, \pi x_2\right); \qquad u_0|_{\partial \mathcal{D}} = 0$$

and this boundary value problem has the solution

$$u_0(x) = \frac{g_0}{\pi^2 \left(\frac{1}{r^2} + \frac{1}{s^2}\right) + f_0} \sin\left(\frac{1}{r}\, \pi x_1\right) \sin\left(\frac{1}{s}\, \pi x_2\right).$$

The eigenvalues and normalized eigenfunctions of the eigenvalue problem

$$-\Delta w + f_0 w = \lambda w; \qquad w|_{\partial \mathcal{D}} = 0$$

are calculated to have the form

$$\lambda_{ij} = \pi^2 \left(\left(\frac{i}{r}\right)^2 + \left(\frac{j}{s}\right)^2\right) + f_0; \qquad w_{ij}(x) = \frac{2}{\sqrt{rs}} \sin\left(\frac{i}{r}\, \pi x_1\right) \sin\left(\frac{j}{s}\, \pi x_2\right)$$

for $i, j = 1, 2, \ldots$, respectively.

Assuming that $\bar{f}_\varepsilon$, $\bar{g}_\varepsilon$ are wide-sense homogeneous independent random functions then by means of Theorem 5.5 the random function

$$\frac{1}{\varepsilon}\left(u(x,\omega) - u_0(x)\right)$$

converges in distribution to a zero-mean Gaussian function $\xi(x,\omega)$ as $\varepsilon \downarrow 0$. The correlation function of $\xi(x,\omega)$ can be determined as

$$\langle \xi(x)\,\xi(y)\rangle = a_{gg}\sum_{i,j=1}^{\infty}\frac{4\sin\left(\frac{i}{r}\,\pi x_1\right)\sin\left(\frac{j}{s}\,\pi x_2\right)\sin\left(\frac{i}{r}\,\pi y_1\right)\sin\left(\frac{j}{s}\,\pi y_2\right)}{rs\left[\pi^2\left(\left(\frac{i}{r}\right)^2+\left(\frac{j}{s}\right)^2\right)+f_0\right]^2}$$

$$+\,a_{ff}\sum_{i,j,p,q=1}^{\infty}\frac{4\sin\left(\frac{i}{r}\,\pi x_1\right)\sin\left(\frac{j}{r}\,\pi x_2\right)\sin\left(\frac{p}{s}\,\pi y_1\right)\sin\left(\frac{q}{s}\,\pi y_2\right)}{rs\left[\pi^2\left(\left(\frac{i}{r}\right)^2+\left(\frac{j}{s}\right)^2\right)+f_0\right]\left[\pi^2\left(\left(\frac{p}{r}\right)^2+\left(\frac{q}{s}\right)^2\right)+f_0\right]}\,c_{ijpq}$$

where c_{ijpq} is defined by

$$c_{ijpq} \doteq (u_0^2 w_{ij}, w_{qp})$$

$$= \frac{4g_0^2}{rs\left[\pi^2\left(\frac{1}{r^2}+\frac{1}{s^2}\right)+f_0\right]^2}\int_0^r\int_0^s \sin\left(\frac{i}{r}\,\pi z_1\right)\sin\left(\frac{j}{s}\,\pi z_2\right)\sin\left(\frac{p}{r}\,\pi z_1\right)$$

$$\times\,\sin\left(\frac{q}{s}\,\pi z_2\right)\left(\sin\left(\frac{1}{r}\,\pi z_1\right)\sin\left(\frac{1}{s}\,\pi z_2\right)\right)^2 \mathrm{d}z_1\,\mathrm{d}z_2\,.$$

In particular, the variance of $\xi(x)$ is given by

$$\langle \xi^2(x)\rangle = a_{gg}\sum_{i,j=1}^{\infty}\left[\frac{2\sin\left(\frac{i}{r}\,\pi x_1\right)\sin\left(\frac{j}{s}\,\pi x_2\right)}{\sqrt{rs}\left[\pi^2\left(\left(\frac{i}{r}\right)^2+\left(\frac{j}{s}\right)^2\right)+f_0\right]}\right]^2$$

$$+\,a_{ff}\sum_{i,j,p,q=1}^{\infty}\frac{2\sin\left(\frac{i}{r}\,\pi x_1\right)\sin\left(\frac{j}{s}\,\pi x_2\right)}{\sqrt{rs}\left[\pi^2\left(\left(\frac{i}{r}\right)^2+\left(\frac{j}{s}\right)^2\right)+f_0\right]}$$

$$\times\,\frac{2\sin\left(\frac{p}{r}\,\pi x_1\right)\sin\left(\frac{q}{s}\,\pi x_2\right)}{\sqrt{rs}\left[\pi^2\left(\left(\frac{p}{r}\right)^2+\left(\frac{q}{s}\right)^2\right)+f_0\right]}\,c_{ijpq}\,.$$

It is clear that the value of $\langle \xi(x)\,\xi(y)\rangle$ decreases if f_0 goes to greater values. For numerical calculations we deal with

$$c_{ijpq} = \frac{4g_0^2}{\left[\pi^2\left(\frac{1}{r^2}+\frac{1}{s^2}\right)+f_0\right]^2}\,c_{ip}(r)\,c_{jq}(s)$$

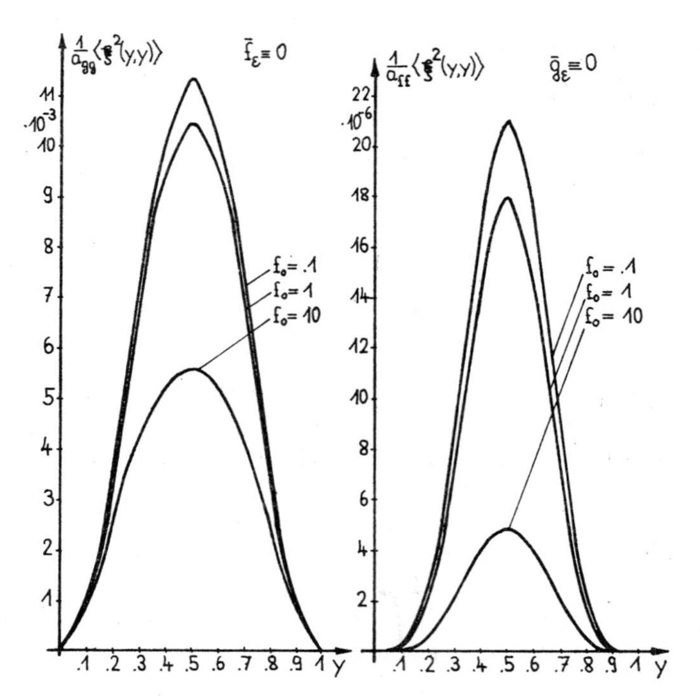

Fig. 5.11. Variances of $\xi(x, \omega)$ as function of y, $x = (y, y)$, for different values of f_0 and $r = s = 1$

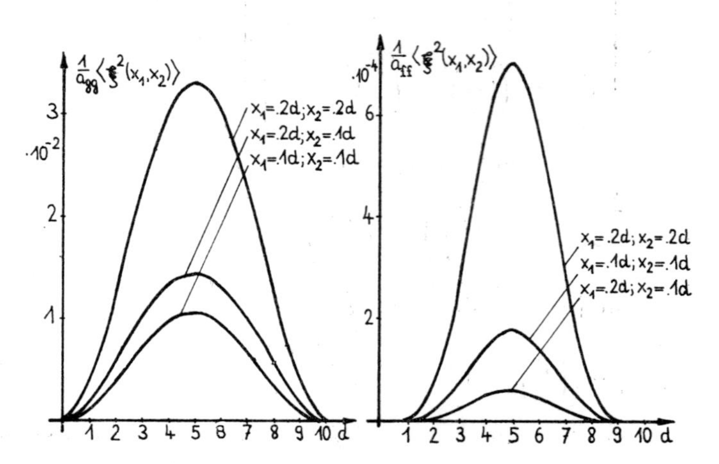

Fig. 5.12. Variances of $\xi(x, \omega)$ as function of d with $x_1 = \frac{rd}{10}$, $x_2 = \frac{sd}{10}$ for different r, s

where c_{ip} is defined by

$$c_{ip}(r) \doteq \frac{1}{r} \int_0^r \sin\left(\frac{i}{r}\pi z\right) \sin\left(\frac{p}{r}\pi z\right) \sin^2\left(\frac{z}{r}\pi\right) dz.$$

It is

$$c_{ip}(r) = \frac{1}{4}\,\delta_{0\,i-p} - \frac{1}{8}\,(\delta_{0\,i-p-2} + \delta_{0\,i-p+2} - \delta_{0\,i+p-2})$$

and using

$$v_{ij} \doteq \frac{1}{\lambda_{ij}}\, w_{ij}(x); \qquad v_{ij} \doteq 0 \quad \text{for} \quad i < 1 \text{ or/and } j < 1,$$

$$c_0 \doteq \frac{4g_0^2}{\left[\pi^2\left(\frac{1}{r^2} + \frac{1}{s^2}\right) + f_0\right]^2}$$

we obtain

$$\sum_{i,j,p,q=1}^{\infty} \frac{w_{ij}(x)\, w_{qp}(x)}{\lambda_{ij}\lambda_{pq}}\, c_{ijpq}$$

$$= \frac{c_0}{64}\left[\sum_{i,j=1}^{\infty} v_{ij}\,\{4v_{ij} - 2v_{ij-2} - 2v_{ij+2} - 2v_{i-2j} + v_{i-2j-2} + v_{i-2j+2} - 2v_{i+2j} + v_{i+2j-2} + v_{i+2j+2}\} + \sum_{i=1}^{\infty} v_{i1}\{2v_{i1} - v_{i-21} - v_{i+21}\} + \sum_{i=1}^{\infty} v_{1i}\{2v_{1i} - v_{1i-2} - v_{1i+2}\} + v_{11}^2\right].$$

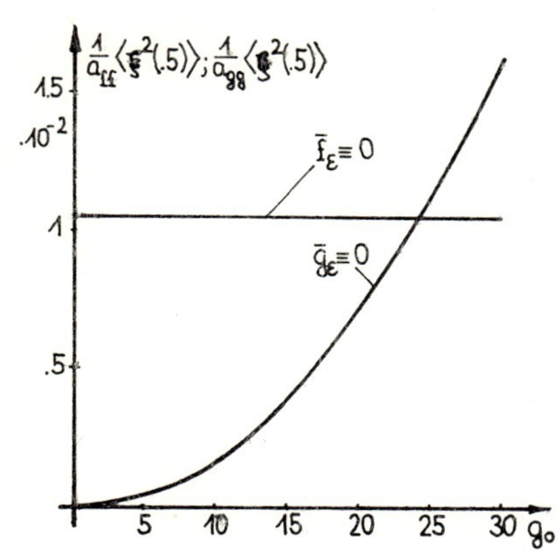

Fig. 5.13. Variances of $\xi(x, \omega)$ as function of g_0 in the cases of $\bar{f}_\varepsilon \equiv 0$, $\bar{g}_\varepsilon \not\equiv 0$ and $\bar{f}_\varepsilon \not\equiv 0$, $\bar{g}_\varepsilon \equiv 0$

Fig. 5.11 illustrates variances of $\xi(x, \omega)$ in dependence on y with $x = (y, y)$ where we have put $\bar{f}_\varepsilon(x, \omega) \equiv 0$ on the left-hand side of this figure and on the right-hand side $\bar{g}_\varepsilon(x, \omega) \equiv 0$. This values were calculated for $r = s = 1$. The dependence of the variances on the domain $\mathcal{D}$ is plotted in Fig. 5.12 where we have put $f_0 = 1$, $g_0 = 1$. Finally, Fig. 5.13 shows the variances of $\xi(x, \omega)$ as a function of g_0 with $f_0 = 1$; $r = s = 1$; $x_1 = x_2 = 0.5$. We see that this variances are independent of g_0 in the case of $\bar{f}_\varepsilon(x, \omega) \equiv 0$. Assuming $\bar{g}_\varepsilon(x, \omega) \equiv 0$ then $\langle \xi^2(x) \rangle$ is a quadratic function of g_0.

Bibliography

ADOMIAN, G.

[1] Stochastic Green's functions. Proc. Symp. Appl. Math. **16** (1964), 1–39.

[2] Nonlinear stochastic differential equations. J. Math. Anal. and Appl. **55** (1976), 441–452.

[3] Stochastic Systems. Academic Press, New York 1983.

[4] A new approach to the heat equation — An application to the decomposition method. J. Math. Anal. and Appl. **113** (1986), 202–209.

AHLBEHRENDT, N., and KEMPE, V.

[1] Analyse stochastischer Systeme. Akademie-Verlag, Berlin 1984.

ANAND, G. V., and RICHARD, K.

[1] Non-linear response of a string to random excitation. Internat. J. Non-Linear Mechs. **9** (1974), 251–260.

ARNOLD, L.

[1] Stochastische Differentialgleichungen. R. Oldenbourg Verlag, München/Wien 1973.

BARRY, M. R., and BOYCE, W. E.

[1] Numerical solution of a class of random boundary value problems. J. Math. Anal. and Appl. **67** (1978) 1, 96–119.

BÉCUS, G. A.

[1] Random generalized solutions to the heat equation. J. Math. Anal. and Appl. **60** (1977), 93–102.

[2] Solutions to the random heat equation by the method of successive approximations. J. Math. Anal. and Appl. **64** (1978), 277–296.

[3] Variational formulation of some problems for the random heat equation. In: ADOMIAN, G. (editor), Applied Stochastic Processes, Academic Press, New York 1980, 23–35.

BÉCUS, G. A., and COZZARELLI, F. A.

[1] The random steady-state diffusion problem, I. Random generalized solutions to Laplace's equation. SIAM J. Appl. Math. **31** (1976), 134–147.

[2] The random steady-state diffusion problem, II. Random solutions to nonlinear, inhomogeneous, steady-state diffusion problems. SIAM J. Appl. Math. **31** (1976), 148–158.

[3] The random steady-state diffusion problem, III. Solutions to random diffusion problems by the method of random successive approximations. SIAM J. Appl. Math. **31** (1976), 159–178.

BENASSI, A.

[1] Le problème de Dirichlet stochastique. C. r. Acad. Sci., Paris, Sér. A **286** (1978), 463–465.

BHARUCHA-REID, A. T.

[1] On the theory of random equations. Prob. Symp. Appl. Math. 16th, 1963, Amer. Math. Soc., Providence, R. I. 1964, 40–69.

[2] (editor) Probabilistic Methods in Applied Mathematics. Vol. 1. Academic Press, New York 1968.

[3] Random Integral Equations. Academic Press, New York 1972.

[4] (editor) Probabilistic Methods in Applied Mathematics. Vol. 2. Academic Press, New York 1970.

[5] (editor) Approximate Solution of Random Equations. North-Holland, New York/Oxford 1979.

BHARUCHA-REID, A. T., and SAMBANDHAM, M.

[1] Random Polynomials. Academic Press, Orlando/San Diego/New York/Austin/London/Montreal/Sydney/Tokyo/Toronto 1986. In: BIRNBAUM, Z. W., and LUKACS, E. (editors), Probability and Mathematical Statistics.

BILLINGSLEY, P.

[1] Convergence of Probability Measures. John Wiley & Sons, New York **1968**.

BLIVEN, D. O., and SOONG, T. T.

[1] On frequencies of elastic beams with random imperfections. J. Frankl. Institut **287** (1969), 297—304.

BOGDANOFF, J. L., and GOLDBERG, J. E.

[1] On the Bernoulli-Euler beam theory with random excitation. J. Aero/Space Sci. **5** (1960) 27, 371—376.

BOUC, R., and DEFILIPPI

[1] Stochastic linearization of truck Dynamics. Proc. X. Intern. Conf. on Nonlinear Oscillations, Varna 1984; Publishing House of the Bulgarian Academy of Sciences, Sofia 1985, 544—547.

BOYCE, W. E.

[1] Random vibrations of elastic strings and bars. Proc. U.S. Nat. Congr. Appl. Mech., 4th. Berkeley 1962, 77—85, Amer. Soc. Mech. Eng., New York 1962.

[2] Stochastic nonhomogeneous Sturm-Liouville problems. J. Frankl. Institut **282** (1966), 206—215.

[3] Approximate solution of random ordinary differential equations. Adv. Appl. Prob. **10** (1978), 172—184.

BOYCE, W. E., and DAY, W. B.

[1] On the relationship between the solution of a stochastic boundary value problem and parameters in the boundary conditions. SIAM J. Appl. Math. **19** (1970), 221—234.

BOYCE, W. E., and GOODWIN, B. E.

[1] Random transverse vibrations of elastic beams. SIAM J. **12** (1964), 613—629.

BOYCE, W. E., and NING-MAO XIA

[1] The approach to normality of the solution of random boundary and eigenvalue problems with weakly correlated coefficients. Quarterly of Appl. Math. **XL** (1983) 4, 419—445.

BRAUN, H.

[1] Untersuchungen von Fahrbahnunebenheiten und Anwendungen der Ergebnisse. Diss., Braunschweig (Techn. Univ.) 1969.

[2] Spektraldichten von Fahrbahnunebenheiten. Tagung Akustik und Schwingungstechnik, VDI-Verlag, Düsseldorf 1970.

BUNKE, H.

[1] Statistische Linearisierung. Z. Angew. Math. und Mech. **52** (1972), 79—84.

[2] Gewöhnliche Differentialgleichungen mit zufälligen Parametern. Akademie-Verlag, Berlin 1972.

CARSLOW, H. S., and JAEGER, J. C.

[1] Conduction of Heat in Solids. Oxford at the Clarendon Press 1973.

CAUGHEY, T. K.

[1] Response of a nonlinear string to random loading. J. Appl. Mech. **26** (1959), 341—344.

[2] Random excitation of a loaded nonlinear string. J. Appl. Mech. **27** (1960), 275—278.

[3] Equivalent linearization techniques. J. Acoust. Soc. Amer. **35** (1963), 1706—1711.

[4] Derivation and application of the Fokker-Planck equation to discrete nonlinear dynamic systems subjected to white random excitation. J. Acoust. Soc. Amer. **35** (1963), 1683—1692.

[5] On the response of a class of nonlinear oscillators to stochastic excitation. Proc. Collop. Inst. du Centre National de la Recherche Scientifique Marseille **148** (1964), 393–402.
[6] Nonlinear theory of random vibrations. Adv. in Appl. Mechs., Vol. 11, Academic Press, New York 1971, 209–253.

CHOW, P. L.
[1] Applications of function space integrals to problems in wave propagation in random media. J. Math. Phys. **13** (1972), 1224–1236.

COLLATZ, L.
[1] Eigenwertprobleme und ihre numerische Behandlung. Chelsea, New York 1948.
[2] Eigenwertaufgaben mit technischen Anwendungen. Akad. Verlagsgesellschaft, Geest & Portig, Leipzig 1963.

CRANDALL, S. H.
[1] (editor) Random Vibration. Vol. 1. M.I.T. Press, Cambridge/Mass. 1958.
[2] Random vibration. Appl. Mech. Reviews **12** (1959), 739–742.
[3] Perturbation techniques for random vibrations of nonlinear systems. J. Acoust. Soc. Amer. **35** (1963), 1700–1705.
[4] Zero Crossings, peaks, and other statistical measures of random responses. J. Acoust. Soc. Amer. **35** (1963), 1693–1699.
[5] Nonlinear problems in random vibration. In: SCHMIDT, G. (editor), Proc. VII. Internat. Konferenz über nichtlineare Schwingungen, Berlin 1975, Akademie-Verlag, Vol. II, Berlin 1977, 215–224.
[6] Heuristic and equivalent linearization techniques for random vibration of nonlinear oscillators. Proc. of the VII. Int. Conf. on Nonlinear Oscillations, Vol. I, Prag 1978, Academica, Prag 1979, 211–226.

CRANDALL, S. H., and MARK, W. D.
[1] Random vibration in mechanical systems. Academic Press, New York/London 1963.

CRANDALL, S. H., and YILDIZ, A.
[1] Random vibrations of beams. J. Appl. Mech. **29** (1962), 267–278.

DAY, W. B.
[1] A monotone property of the solution of a stochastic boundary value problem. Quart. Appl. Math. **28** (1970), 411–425.

DRECHSEL, E., HASE, W., NEUMANN, K.-H., VOM SCHEIDT, J., and WÖHRL, U.
[1] Dynamic systems with weakly correlated excitation. 10th IAVSD Symposium on Dynamics of Vehicles on Roads and Tracks, Extensive Summaries, Praha 1987, 45–46.

EINSTEIN, A.
[1] Über die von der molekularkinetischen Theorie der Wärme geforderte Bewegung von in ruhenden Flüssigkeiten suspendierten Teilchen. Ann. der Physik **17** (1905), 549–560.

EISENREICH, G.
[1] Lineare Algebra und analytische Geometrie. Akademie-Verlag, Berlin 1980.

EL MADANY, M. M., and DOKAINISH, M. A.
[1] Articulated vehicle dynamic analyses using equivalent linearization technique. SAE-Transactions 1980, Section 4, Varrandale **89** (1980) 4, 4506–4517.

ERINGEN, A. C.
[1] Transverse impact on beams and plates. J. Appl. Mech., Trans. ASME **75** (1953), 461–468.
[2] Response of beams and plates to random loads. J. Appl. Mech. **24** (1957) 1, 46–52.

ERMAKOV, S. M. (ЕРМАКОВ, С. М.)
[1] Метод Монте-Карло и смежные вопросы. Наука, Москва 1971. (German translation: Die Monte-Carlo-Methode und verwandte Fragen. Deutscher Verlag der Wissenschaften, Berlin 1975.)

FABIAN, L.
[1] Zufallsschwingungen und ihre Behandlung. Springer-Verlag, Berlin/Heidelberg/New York 1972.

FELLENBERG, B., and VOM SCHEIDT, J.

[1] Temperaturverteilung in technischen Bauteilen bei stochastischer Wärmezuführung. Lectures of the 2nd Conf. Mathematische Statistik in der Technik, Freiberg **1985**, 55–60.

[3] Stochastische Temperaturausbreitung. Proc. 3rd Conf. Stochastische Vorgänge und Zuverlässigkeit, Jena 1984, Heft 1, AdW der DDR, Inst. für Mechanik, S-Reihe, Nr. 2, Berlin 1985, 15–24.

[3] Linear integral equations of the 2nd kind with weakly correlated functions — Some principles and results. Proc. 2nd Conf. Stochastic Analysis, Zwickau 1986, 13–18.

[4] Probabilistic analysis of random temperature fields. FMC-Series No. 19, Karl-Marx-Stadt 1986, 15–24.

[5] Integralgleichungsmethoden für zufällige Modelle mit schwach korrelierten Funktionen. Proc. 11th Congress Anwendungen der Mathematik in den Ingenieurwissenschaften, Vol. 4, Weimar 1987, 50–53.

[6] On random boundary-initial value problems for parabolic differential equations. Bulletins for Appl. Math., Budapest (in print).

[7] Heat propagation with a random sliding input of heat. Bulletins for Appl. Math., Budapest (in print).

FISCHER, U., and STEPHAN, W.

[1] Mechanische Schwingungen. VEB Fachbuchverlag, Leipzig 1984.

FISZ, M.

[1] Wahrscheinlichkeitsrechnung und mathematische Statistik. VEB Deutscher Verlag der Wissenschaften, Berlin 1976.

FRIEDMAN, A.

[1] Partial Differential Equations of Parabolic Type. Prentice-Hall, Englewood Cliffs, N.J. 1964.

FRIEDRICH, H.

[1] Näherungsverfahren zur Berechnung von schwachnichtlinearen, stochastisch zwangserregten Schwingungssystemen mit einem Freiheitsgrad. In: SCHMIDT, G. (editor), Beiträge zur Schwingungstheorie, Akademie-Verlag, Berlin 1974, 47–70.

FRIEDRICH, H., HEIMANN, B., MARTINS, H., and RENGER, A.

[1] Diskrete Schwingungssysteme unter stochastischer Fremderregung. In: Tagung Dynamik und Getriebetechnik, Band C, VEB Fachbuchverlag, Leipzig 1973.

GASPARINI, D. A., and DEB CHAUDHURY, A.

[1] Dynamic response to nonstationary nonwhite excitation. J. of Eng. Mech. Div. **106**, EM6 (1980), 1233–1248.

GIHMAN, I. I., and SKOROCHOD, A. V. (Гихман, И. И., Скороход, А. В.)

[1] Случайные Процессы, Том 1, Наука, Москва 1971. (Englisch translation: The Theory of Stochastic Processes 1. Springer-Verlag, Berlin/Heidelberg/New York 1974.)

[2] Стохастические Дифференциальные Уравнения. Наукова Думка, Киев 1968 (German translation: Stochastische Differentialgleichungen, Akademie-Verlag, Berlin 1971.)

GOODWIN, B. E., and BOYCE, W. E.

[1] Vibrations of random elastic strings: method of integral equations. Quart. Appl. Math. **22** (1964), 261–266.

GOPALSAMY, K., and BHARUCHA-REID, A. T.

[1] On a class of parabolic differential equations driven by stochastic point processes. J. Appl. Prob. **12** (1975), 98–106.

GOSSMANN, E.

[1] Kovarianzanalyse mechanischer Zufallsschwingungen bei Darstellung der mehrfachkorrelierten Erregungen durch stochastische Differentialgleichungen. Ruhr-Universität Bochum, Mitteilungen aus dem Institut für Mechanik, Nr. 24, 1981.

GOSSMANN, E., and WALLER, H.

[1] Zur Behandlung von Zufallsschwingungen mechanischer Bauteile infolge mehrfachkorrelierter stochastischer Erregungen. Ing. Archiv **52** (1982), 131–141.

HAINES, C. W.
[1] Hierarchy methods for random vibrations of elastic strings and beams. J. Eng. Math. **1** (1967), 293—305.
HEIMANN, B.
[1] Übertragung von Zufallsfunktionen durch nichtlineare (trägheitslose) Systeme. In: SCHMIDT, G. (editor), Beiträge zur Schwingungstheorie, Akademie-Verlag, Berlin 1974, 1—18.
HEINRICH, W., and HENNIG, K.
[1] Zufallsschwingungen mechanischer Systeme. Akademie-Verlag, Berlin 1977.
HELMS, H.
[1] Grenzen der Verbesserungsfähigkeit von Schwingungskomfort und Fahrsicherheit an Fahrzeugen. Diss., Braunschweig (TU) 1974.
HENNIG, K. (editor)
[1] Random Vibrations and Reliability. Proc. of the IUTAM Symposium, Frankfurt (GDR) 1982, Akademie-Verlag, Berlin 1983.
HERBERT, R. E.
[1] Random vibrations of a nonlinear elastic beam. J. Acoust. Soc. Amer. **36** (1964) 11, 2090 to 2094.
IWAN, W. D., and YANG, I.-M.
[1] Application of statistical linearization techniques to nonlinear multidegree-of-freedom systems. J. Appl. Mech. **39** (1972), 545—550.
JETSCHKE, G.
[1] Different approaches to stochastic differential equations. Proc. 10th Winter School, Supplemento ai Rendiconti del Circolo Matematico di Palermo, Serie II-numero 2-1982, 161—169.
KOTULSKI, Z., and SOBCZYK, K.
[1] Linear systems and normality. J. Statist. Phys. **24** (1981) 2, 359—373.
KREUZER, E., and RILL, R.
[1] Vergleichende Untersuchungen von Fahrzeugschwingungen an räumlichen Ersatzmodellen. Ing. Archiv **52** (1982), 205—219
LENK, D., and ZSCHERPEL, W.
[1] Untersuchung des instationären Temperaturfeldes zweier Linienquellen. Wiss. Beitr. IH Zwickau **4** (1978) 1, 113—118.
LENK, D., NITTEL, J., and ZSCHERPEL, W.
[1] Berechnung und Messung von instationären Temperaturfeldern bei bewegten Reibungswärmequellen an Festkörperoberflächen. Schmierungstechnik **10** (1979) 9, 267—269.
LIESE, F.
[1] A limit theorem for sequences of weakly dependent random fields. In: VOM SCHEIDT, J. (editor), Problems of Stachastic Analysis in Applications, Wiss. Beitr. IH Zwickau, Special issue 1983, 290—305.
LIESE, F., and VOM SCHEIDT, J.
[1] A limit theorem for sequences of weakly dependent stochastic processes. Serdica Bulgaricae Mathematicae Publications **9** (1983), 18—30, Also in: VOM SCHEIDT, J. (editor), Problems of Stochastic Analysis in Applications, Wiss. Beitr. IH Zwickau, Special issue 1983, 267—289
LINGENER, A.
[1] Analysis of mechanical systems excited by random vibrations. In: HENNIG, K. (editor), Random Vibrations and Reliability, Proc. of the IUTAM Symposium, Frankfurt (GDR) 1982, Akademie-Verlag, Berlin 1983, 173—184.
LIU, S. C.
[1] Solutions of Fokker-Planck equations with applications in nonlinear random vibrations. Bel System Tech. J. **48** (1969), 2031—2051.
LYON, R. H.
Response of a nonlinear string to random excitation. J. Acoust. Soc. Amer. **32** (1960) 953—960.

MACVEAN, D. B.

[1] Response of vehicle accelerating over random profile. Ing. Archiv **49** (1980), 375–380.

MAKAROV, B. P.

[1] Eine Variationsmethode für die Lösung nichtlinearer stochastischer Aufgaben. Z. Angew. Math. und Mech. **55** (1975), 185–187.

MANTHEY, R.

[1] Schwache Konvergenz von Lösungen der stochastischen Wärmeleitungsgleichung mit Gaußschem Rauschen. Forschungsergebnisse der FSU Jena N/82/72, 1982.

[2] Weak convergence of solutions of the heat equation with Gaussian noise. Math. Nachr. **123** (1985), 157–168.

[3] On reaction-diffusion equation driven by white noise. Proc. 2nd Conf. on Stochastic Analysis, Zwickau 1986, 51–55.

[4] Existence and uniqueness of a solution of a reaction-diffusion equation with polynomial nonlinearity and white noise disturbance. Math. Nachr. **125** (1986), 121–133.

MEUSEL, B.

[1] Grenzverteilungsaussagen über die Lösungen stochastischer Rand- und Rand-Anfangswertprobleme partieller Differentialgleichungen mit schwach korrelierten Feldern. Diss. A, Leipzig 1981.

[2] Ebene und kugelsymmetrische Wellenausbreitung im Raum unter schwach korrelierter Erregung. Wiss. Zeitschr. TH Leuna-Merseburg **27** (1985) 5, 590–595.

[3] Ebene Wellen mit zufälliger Ausbreitungsgeschwindigkeit. Proc. 2nd Conf. on Stochastic Analysis, Zwickau 1986, 56–64.

MEUSEL, B., and VOM SCHEIDT, J.

[1] Transversalschwingungen eines dünnen Balkens unter Zufallsbelastung. Wiss. Beitr. IH Zwickau **6** (1980) 2, 64–67.

[2] Limit theorems for integral transformations of partially weakly correlated fields. In: VOM SCHEIDT, J. (editor), Problems of Stochastic Analysis in Applications, Wiss. Beitr. IH Zwickau, Special issue 1983, 94–117.

[3] Boundary-value problems of random differential equations with weakly correlated random excitation. In: VOM SCHEIDT, J. (editor), Problems of Stochastic Analysis in Applications, Wiss. Beitr. IH Zwickau, Special issue 1983, 118–133.

[4] Eindimensionale Wellenausbreitung unter zufälliger, schwachkorrelierter Erregung. Wiss. Beitr. IH Zwickau **10** (1984) 4, 81–88.

MIHLIN, S. G. (Михлин, С. Г.)

[1] Вариационные методы математической физики. Наука, Москва 1957. (German translation: Variationsmethoden der Mathematischen Physik. Akademie-Verlag, Berlin 1962.)

[2] Численная реализация вариационных методов. Наука, Москва 1966. (German translation: Numerische Realisierung von Variationsmethoden, Akademie-Verlag, Berlin 1969; English translation: The numerical performance of variational methods, Wolters-Noordhoff Publishing, Groningen 1971.)

MITSCHKE, M.

[1] Dynamik der Kraftfahrzeuge. Springer-Verlag, Berlin/Heidelberg/New York 1972.

[2] Einfluß der Radaufhängung auf Radlasten und Seitenkräfte. Institut für Fahrzeugtechnik, TU Braunschweig, Vortragsausdrucke der II. IfF-Tagung, Braunschweig 1978, 73–104.

[3] Beurteilung von Fahrzeugschwingungen. Institut für Fahrzeugtechnik, TU Braunschweig, Vortragsausdrucke der IV. IfF-Tagung, Braunschweig 1982, 93–117.

[4] Fahrverhalten von Personenkraftwagen auf unebener Straße. Automobiltechnische Zeitschrift **85** (1983) 11, 695–697.

[5] Beurteilungskriterien und Anforderungen an das Schwingungssystem Kraftfahrzeug. VDI Berichte 546 „Fahrdynamik und Federungskomfort“, Conf. Wolfsburg 1984, 1–24.

Mühe, P.

[1] Der Einfluß von Nichtlinearitäten in Feder- und Dämpferkennlinie auf die Schwingungseigenschaften von Kraftfahrzeugen. Diss., Braunschweig (TU) 1968.

Müller, K.-H.

[1] Anwendung der Theorie stochastischer Schwingungen zur Analyse seismisch erregter Bauwerke. Proc. 2nd Conf. on Stochastic Analysis, Zwickau 1986, 65–69.

Müller, P. C., Popp, K., and Schiehlen, W. O.

[1] Berechnungsverfahren für stochastische Fahrzeugschwingungen. Ing. Archiv **49** (1980), 235–254.

Parkus, H.

[1] Wärmespannungen bei zufallsabhängiger Oberflächentemperatur. Z. Angew. Math. und Mech. **42** (1962), 499–507.

[2] Random Processes in Mechanical Sciences. Intern. Centre for Mechanical Sciences, Springer-Verlag, Wien/New York 1969.

Pekala, W., and Szopa, J.

[1] The application of Green's multi-dimensional function to investigate the stochastic vibrations of dynamical systems. Ing. Archiv **54** (1984), 91–97.

Pevsner, J. M., Gridasov, G. G., Konev, A. D., and Pletnev, A. E. (Певзнер, Я.М., Гридасов, Г. Г., Конев, А. Д., Плетнев, А. Е.)

[1] Колебания автомобиля. Изгательство Машиностроение, Москва 1979 (Vehicle Vibrations).

Piszczek, K., and Niziol, J.

[1] Random vibration of mechanical systems. PWN-Polish Scientific Publishers, Warszawa; Ellis Horwood Limited Publishers, Chichester 1986.

Purkert, W., and vom Scheidt, J.

[1] Stochastische Eigenwertprobleme. Diss. B, Leipzig 1978.

[2] Randwertprobleme mit schwach korrelierten Prozessen als Koeffizienten. Transactions of the 8th Prague Conf. on Information Theory, Statistical Decision Functions, Random Processes, Prag 1978, 107–118.

[3] Ein Grenzverteilungssatz für stochastische Eigenwertprobleme. Z. Angew. Math. und Mech. **59** (1979), 611–623.

[4] Schwach korrelierte Prozesse und ihre Anwendungen. Sitzungsberichte AdW der DDR, Mathematik-Naturwissenschaften-Technik 1980, No. 23/N, Akademie-Verlag, Berlin 1980.

Remke, B.

[1] Untersuchungen zur Anwendung von Grenzverteilungssätzen auf die Lösungen einer Klasse von Randanfangswertproblemen mit schwach korreliertem inhomogenen Term. Diss. A, Zwickau 1986.

[2] Korrelationsfunktionen von Lösungen stochastischer Differentialgleichungen mit schwach korreliertem inhomogenen Term. Proc. 2nd Conf. on Stochastic Analysis, Zwickau 1986, 97–101.

Renger, A.

[1] Berechnung von zufallserregten Schwingungssystemen mittels Distributionenkalküls. In: Schmidt, G. (editor), Beiträge zur Schwingungstheorie, Akademie-Verlag, Berlin 1974, 19–46.

Renger, A., and Gupta, K. N.

[1] Ein programmiertes und allgemeingültiges Berechnungsverfahren zur Bestimmung des dynamischen Verhaltens von linearen Schwingungssystemen mit vielen Freiheitsgraden bei stationärer Zufallserregung. Wiss. Zeitschr. TH Magdeburg **15** (1971) 6, 589–601.

Samuels, J. C.

[1] Heat conduction in solids with random external temperatures and/or random internal heat generation. Int. J. Heat Mass Transfer **9** (1966), 301–314.

SAMUELS, J. C., and ERINGEN, A. C.

[1] Response of a simply supported Timoshenko beam to a purely random Gaussian process. J. Appl. Mech. **25** (1958) 4, 496–500.

[2] On stochastic linear systems. J. Math. Physics **38** (1959), 83–103.

VOM SCHEIDT, J.

[1] Lineare Schwingungsgleichungen mit schwach korreliertem inhomogenen Term. Wiss. Beitr. IH Zwickau **4** (1978) 2, 63–77.

[2] Grenzwertsätze bei Differentialgleichungen mit stochastischen Koeffizienten. Proc. 7th Conf. Probleme und Methoden der Mathematischen Physik, Vol. II, Karl-Marx-Stadt 1979, 121–126.

[3] A limit theorem of solutions of stochastic boundary-initial-value problems. In: ARATÓ, M., VERMES, D., and BALAKRISHNAN, A. V. (editors), Stochastic Differential Systems, Proc. 3rd IFIP-WG 7/1 Working Conf. Visegrád 1980, Springer-Verlag, Berlin/Heidelberg/New York 1981, 189–201.

[4] Stochastische Stabilität technischer Systeme. Wiss. Beitr. IH Zwickau **7** (1981) 1, 56–67.

[5] Some applications of limit theorems for weakly correlated fields to random eigenvalue problems. Trans. of the 9th Prague Conf. on Inf. Theory, Statistical Decision Functions, Random Processes, Prag 1983, 169–175.

[6] Simulation of random eigenvalue problems. In: VOM SCHEIDT, J. (editor), Problems of Stochastic Analysis in Applications, Wiss. Beitr. IH Zwickau, Special issue 1983, 134 to 172.

[7] Random vibrations generated by weakly correlated loads. In: VOM SCHEIDT, J. (editor), Problems of Stochastic Analysis in Applications, Wiss. Beitr. IH Zwickau, Special iussue 1983, 173–247.

[8] Zur Verteilung linearer Funktionale schwach korrelierter Felder. In: Proc. 3rd Conf. Stochastische Vorgänge und Zuverlässigkeit, Jena 1984, Vol. 1, AdW of GDR, Institute of Mechanics, S-series, No. 2, Berlin 1985, 145–155.

[9] Ein stochastisches Randanfangswertproblem für parabolische Differentialgleichungen. Wiss. Zeitschr. TH Leuna-Merseburg **27** (1985) 5, 596–604.

[10] Simulation results for linear functionals of weakly correlated random processes. IMACS-11 Congress System Simulation and Scientific Computation, Oslo 1985.

[11] Simulationsergebnisse für lineare Funktionale schwach korrelierter Prozesse. FMC-Series, Fracture Mechanics, Micromechanics, Coupled Fields, AdW of GDR, Institut for Mechanics, No. 19, Karl-Marx-Stadt 1986, 1–14.

[12] Random vibrations of supporting elements. Proc. 11th Congress Anwendungen der Mathematik in den Ingenieurwissenschaften, Vol 4, Weimar 1987, 76–80.

VOM SCHEIDT, J., and BHARUCHA-REID, A. T.

[1] On the distribution of the roots of random algebraic polynomials. In: VOM SCHEIDT, J. (editor), Problems of Stochastic Analysis in Applications, Wiss. Beitr. IH Zwickau, Special issue 1983, 44–82.

VOM SCHEIDT, J., FELLENBERG, B., and WÖHRL, U.

[1] To the distribution of linear functionals of weakly correlated fields — Application to random temperature fields. In: Differential Equations and Applications II, Proc. 3rd Conf., Rousse 1985, 923–926.

VOM SCHEIDT, J., and FELLENBERG, B.

[1] Some simulation results for functions of linear functionals of weakly correlated processes. Proc. 2nd Conf. Stochastic Analysis, Zwickau 1986, 112–119.

[2] Some simulation results as to weakly correlated processes. Mathematics and Computers in Simulation **29** (1987), 191–208.

[3] On the distribution of functionals of stochastic fields. In: ENGELBERT, H. J., and SCHMIDT, W. (editors), Stochastic Differential Systems, Proc. IFIP-WG 7/1 Working Conf., Eisenach 1986, Springer-Verlag, Berlin/Heidelberg/New Yok/London/Paris/Tokyo 1987, 99–108.

[4] On the distribution of solutions of random equations with weakly correlated fields. Memorial Volume honoring A. T. Bharucha-Reid, University of Georgia (in print).

vom Scheidt, J., and Purkert, W.

[1] Ein Eigenwertproblem mit weißem Rauschen als Koeffizienten. Wiss. Beitr. IH Zwickau **6** (1980) 4, 74–82.

[2] Limit theorems for solutions of stochastic differential equation problems. Intern. J. Math. & Math. Sci. **3** (1980) 1, 113–149.

[3] Random Eigenvalue Problems. Akademie-Verlag, Berlin 1983; also in: North Holland Series in "Probability and Applied Mathematics" (Bharucha-Reid, A. T., editor), New York/Amsterdam/Oxford 1983.

vom Scheidt, J., Seifert, H., and Baumgärtel, Ch.

[1] On the probabilistic distribution of parameters concerning a method of fuel-injection for a two-stroke combustion engine. In: vom Scheidt, J. (editor), Problems of Stochastic Analysis in Applications, Wiss. Beitr. IH Zwickau, Special issue 1983, 306–326.

vom Scheidt, J., and Wöhrl, U.

[1] Untersuchungen stochastischer Prozesse bezüglich Niveauüberschreitungen. Wiss. Beitr. IH Zwickau **6** (1980) 1, 72–77.

[2] Schwingungen an Fahrzeugen mit linearen Feder- und Dämpferkennlinien und stochastischer Erregung. Wiss. Beitr. IH Zwickau **9** (1983) 1, 51–58.

[3] Einmassenmodelle mit nichtlinearen Feder- und Dämpferkennlinien und stochastischer Erregung. Wiss. Beitr. IH Zwickau **9** (1983) 2, 91–96.

[4] A nonlinear vibration differential equation with weakly correlated excitation. In: vom Scheidt, J. (editor), Problems of Stochastic Analysis in Applications, Wiss. Beitr. IH Zwickau, Special issue 1983, 248–266.

[5] Grenzwertsätze für lineare Funktionale schwach korrelierter Vektorfelder — Anwendung auf dynamische Systeme. Proc. 2nd Conf. on Stochastic Analysis, Zwickau 1986, 120–135.

[6) Approximation der Verteilung linearer Funktionale schwach korreliert verbundener Vektorprozesse — Anwendung auf Schwingungsdifferentialgleichungen. Z. Angew. Math. und Mech. **67** (1987), 607–615.

Schiehlen, W. O.

[1] Nonstationary random vibrations. In: Hennig, K. (editor), Proc. of the IUTAM Symposium, Frankfurt (GDR) 1982, Akademie-Verlag, Berlin 1983, 295–305.

Schmidt, G.

[1] (editor) Beiträge zur Schwingungstheorie. Akademie-Verlag, Berlin 1974.

Sobczyk, K.

[1] Random vibrations of statistically inhomogeneous elastic systems. Proc. Vibr. Probl. **4** (1970) 11, 369–380.

[2] Stochastic Wave Propagation. PWN-Polish Scientific Publishers, Warszawa 1984; Elsevier, Amsterdam/Oxford/New York/Tokyo 1984.

Sobczyk, K., Macvean, D. B., and Robson, J. D.

[1] Response to profile-imposed excitation with randomly varying traversal velocity. Journal of Sound and Vibration **52** (1977), 37–49.

Socha, L.

[1] Application of moment equations to sensitivity analysis of the stochastic dynamical systems. In: Proc. 3rd Conf. Stochastische Vorgänge und Zuverlässigkeit, Jena 1984, Vol. 1, AdW of GDR, Institute of Mechanics, S-series, No. 2, Berlin 1985, 157–166.

Soong, T. T.

[1] Random Differential Equations in Science and Engineering. Academic Press, New York/London 1973; In: Bellman, R. (editor), Mathematics in Science and Engineering.

Stakgold, I.

[1] Boundary Value Problems of Mathematical Physics. Vol. II, Macmillan, New York 1967.

SZOPA, J.

[1] The comparison between Green's function method and integral equation method and their application to stochastic dynamical systems. In: Proc. 3rd Conf. Stochastische Vorgänge und Zuverlässigkeit, Jena 1984, Vol. 1, AdW of GDR, Institute of Mechanics, S-series, No. 2, Berlin 1985, 177–186.

THOMAS, J. H.

[1] Random vibration of thin elastic plates. Z. Angew. Math. Phys. **6** (1968) 19, 921–926.

UHLENBECK, G. E., and ORNSTEIN, L. S.

[1] On the theory of the Brownian motion. Physical Review **36** (1930), 823–841.

VAN LEAR, G. A., and UHLENBECK, G. E.

[1] The Brownian motion of strings and elastic rods. Physical Review **38** (1931), 1583–1598.

WALLENTOWITZ, H., BALASUBRAMANIAN, BIESINGER, H., and MEIER, G.

[1] Simulation von Fahrzeugschwingungen unter Berücksichtigung gemessener Fahrbahnunebenheiten. VDI-Berichte 537 „Berechnung im Automobilbau", Tagung Fellbach 1984, 169–190.

WALLRAPP, O., and SCHWARZ, W.

[1] Simulation des Störverhaltens von Fahrzeugen auf unebenen Straßen mit dem Mehrkörperprogramm MEDUSA. VDI-Berichte 537 „Berechnung im Automobilbau", Tagung Fellbach 1984, 151–168.

WEDIG, W.

[1] Zur Integration stochastischer Systeme mit stückweise linearen Kennlinien. Ing. Archiv **49** (1980), 201–215.

[2] Kovarianzanalyse stochastischer Schwingungssysteme. VDI-Bildungswerk BW 3198.

WÖHRL, U.

[1] Approximation von Fahrbahnunebenheiten. Proc. 2nd Conf. Stochastic Analysis, Zwickau 1986, 177–180.

[2] Zufällige Schwingungen. Diss. B, Zwickau 1988.

WÖHRL, U., and VOM SCHEIDT, J.

[1] Nonlinear vibration differential equations with weakly correlated excitation. Proc. ICNO-X Varna 1984, 806–809.

[2] Schwingungsmodelle mit schwach korrelierter Erregung. Proc. 10th Congress Anwendungen der Mathematik in den Ingenieurwissenschaften, Vol. 6, Weimar 1984, 91–94.

[3] Nichtlineare Schwingungsmodelle mit schwach korrelierter Erregung. In: Proc. 3rd Conf. Stochastische Vorgänge und Zuverlässigkeit, Jena 1984, Vol. 1, AdW of GDR, Institute of Mechanics, S-series, No. 2, Berlin 1985, 209–219.

[4] Limit theorems for linear functionals of weakly correlated processes and applications to vibration differential equations. Differential Equations and Applications II, Proc. of the 3rd Conf., Rousse 1985, 995–998.

[5] Einige Bemerkungen über das Einschwingverhalten linearer Schwingungsdifferentialgleichungen mit schwach korrelierter Erregung. Wiss. Beitr. IH Zwickau **12** (1986) 2, 70–76.

[6] Stochastisch erregte Schwingungen an Mehrmassenmodellen. Proc. 11th Congress Anwendungen der Mathematik in den Ingenieurwissenschaften, Vol. 4, Weimar 1987 (to appear).

ZEMAN, J. L.

[1] Approximate Analysis of Stochastic Processes in Mechanics. Intern. Centre for Mechanical Sciences, Courses and Lectures, No. 95, Undine 1971, Springer-Verlag, Wien/New York 1971.

ZURMÜHL, R.

[1] Matrizen und ihre technischen Anwendungen. Springer-Verlag, Berlin/Göttingen/Heidelberg 1964.

[2] Praktische Mathematik für Ingenieure und Physiker. Springer-Verlag, Berlin/Heidelberg/New York 1965.

Author Index

Subject Index